Lecture Notes in Computer Science 14895

The series Lecture Notes in Computer Science (LNCS), including its subseries Lecture Notes in Artificial Intelligence (LNAI) and Lecture Notes in Bioinformatics (LNBI), has established itself as a medium for the publication of new developments in computer science and information technology research, teaching, and education.

LNCS enjoys close cooperation with the computer science R & D community, the series counts many renowned academics among its volume editors and paper authors, and collaborates with prestigious societies. Its mission is to serve this international community by providing an invaluable service, mainly focused on the publication of conference and workshop proceedings and postproceedings. LNCS commenced publication in 1973.

Tianqing Zhu · Yannan Li
Editors

Information Security and Privacy

29th Australasian Conference, ACISP 2024
Sydney, NSW, Australia, July 15–17, 2024
Proceedings, Part I

Editors
Tianqing Zhu
City University of Macau
Macau, China

Yannan Li
University of Wollongong
Wollongong, NSW, Australia

ISSN 0302-9743 ISSN 1611-3349 (electronic)
Lecture Notes in Computer Science
ISBN 978-981-97-5024-5 ISBN 978-981-97-5025-2 (eBook)
https://doi.org/10.1007/978-981-97-5025-2

This Springer imprint is published by the registered company Springer Nature Singapore Pte Ltd.
The registered company address is: 152 Beach Road, #21-01/04 Gateway East, Singapore 189721, Singapore

Preface

It is our great pleasure to present the proceedings of ACISP 2024, the 29th Australasian Conference on Information Security and Privacy, held in Sydney, Australia, during 15–17 July, 2024.

ACISP was first held at the University of Wollongong in 1996 and since then has been organized in various cities in Australia and New Zealand – Wollongong (1996, 1999, 2003, 2008, 2012, 2014, 2018, 2022), Sydney (1997, 2001, 2004, 2010), Melbourne (2002, 2006, 2011, 2016), Brisbane (1998, 2000, 2005, 2009, 2013, 2015, 2021, 2023), Townville (2007), Auckland (2017), and Christchurch (2019). This annual event has gained its place among prestigious security and privacy conferences in the world.

ACISP 2024 had two rounds of submission, which attracted 91 papers in Round 1 and 141 papers in Round 2, respectively, resulting in 232 submissions in total. All papers were reviewed by at least three reviewers. The double-blind review phase was followed by a 10-day discussion that generated additional comments from the program committee members and the external reviewers. After the discussions, 23 papers were selected from Round 1 and 47 papers were selected from Round 2. Finally, 70 papers were accepted into the program, including one SOK paper, leading to an acceptance rate of 30.17%. These 70 papers were presented during the conference.

Among the accepted papers, the papers with the highest weighted review mark in each round received the Best Paper Award. The Best Papers were *DualRing-PRF: Post-Quantum (Linkable) Ring Signatures from Legendre and Power Residue PRFs* by Xinyu Zhang, Ron Steinfeld, Joseph Liu, Muhammed F. Esgin, Dongxi Liu and Sushmita Ruj and *Pairing-free ID-based Signatures as Secure as Discrete Logarithm in AGM* by Jia-Chng Loh, Fuchun Guo and Willy Susilo.

The conference would not have been possible without the hard work of the 58 program committee members from 14 different countries and the 130 external reviewers, who took part in the process of reviewing and subsequent discussions. We take this opportunity to thank the program committee members and the external reviewers for their tremendous job resulting in the current program. It has been an honour to work with them. We would also like to express our appreciation to Springer for their active cooperation and timely production of these conference proceedings. In addition to the technical talks, the program included four keynote talks. Finally, we would like to thank all the authors who submitted their work to ACISP 2024 and all the information security and privacy practitioners and enthusiasts who attended the event. Without your spirited participation, the conference would not be a success.

July 2024

Tianqing Zhu
Yannan Li

Organization

General Co-chairs

Jie Lu	University of Technology Sydney, Australia
Shui Yu	University of Technology Sydney, Australia

Program Co-chairs

Tianqing Zhu	City University of Macau, China
Yannan Li	University of Wollongong, Australia

Publication Co-chairs

Bo Liu	University of Technology Sydney, Australia
Angela Huo	University of Technology Sydney, Australia

Organisation Chairs

Dayong Ye	University of Technology Sydney, Australia
Sheng Shen	University of Sydney, Australia

Program Committee

Cristina Alcaraz	University of Málaga, Spain
Man Ho Au	Hong Kong Polytechnic University, China
Shi Bai	Florida Atlantic University, USA
Rishiraj Bhattacharyya	University of Birmingham, UK
Xiaofeng Chen	Xidian University, China
Josep Domingo-Ferrer	Universitat Rovira i Virgili, Spain
Keita Emura	Kanazawa University, Japan
Qiong Huang	South China Agricultural University, China
Ryoma Ito	National Institute of Information and Communications Technology, Japan
Dongseong Kim	University of Queensland, Australia

Jongkil Kim	Ewha Womans University, South Korea
Hyung Tae Lee	Chung-Ang University, South Korea
Qinyi Li	Queensland University of Technology, Australia
Yingjiu Li	University of Oregon, USA
Chi Liu	City University of Macau, China
Dongxi Liu	CSIRO, Data61, Australia
Shengli Liu	Shanghai Jiao Tong University, China
Wei Liu	University of Technology Sydney, Australia
Siqi Ma	University of New South Wales, Australia
Chris Mitchell	Royal Holloway, University of London, UK
Kirill Morozov	University of North Texas, USA
Ngoc Khanh Nguyen	King's College London, UK
Jianbing Ni	Queen's University, Canada
Udaya Parampalli	University of Melbourne, Australia
Tran Phuong	University of Arkansas at Little Rock, USA
Josef Pieprzyk	CSIRO/Data61, Australia
Youming Qiao	University of Technology Sydney, Australia
Elizabeth Quaglia	Royal Holloway, University of London, UK
Partha Sarathi Roy	University of Wollongong, Australia
Reihaneh Safavi-Naini	University of Calgary, Canada
Suranga Seneviratne	University of Sydney, Australia
Sheng Shen	University of Sydney, Australia
Leonie Simpson	Queensland University of Technology, Australia
Daniel Slamanig	Universität der Bundeswehr München, Germany
Nasrin Sohrabi	Deakin University, Australia
Yulei Sui	University of New South Wales, Australia
Atsushi Takayasu	University of Tokyo, Japan
Qiang Tang	University of Sydney, Australia
Viet Cuong Trinh	Hong Duc University, Vietnam
Damien Vergnaud	Sorbonne Université, France
Derek Wang	CSIRO, Data61, Australia
Huaxiong Wang	Nanyang Technological University, Singapore
Yu Wang	Guangzhou University, China
Yanhong Xu	University of Calgary, Canada
Jason Xue	CSIRO's Data61, Australia
Liang Xue	University of Guelph, Canada
Guomin Yang	Singapore Management University, Singapore
Rupeng Yang	University of Wollongong, Australia
Xun Yi	RMIT University, Australia
Zuobin Ying	City University of Macau, China
Xin Yu	University of Technology Sydney, Australia
Yong Yu	Shaanxi Normal University, China

Xingliang Yuan	University of Melbourne, Australia
Tsz Hon Yuen	Monash University, Australia
Lefeng Zhang	City University of Macau, China
Yudi Zhang	University of Wollongong, Australia

Additional Reviewers

Ahmed, Faisal
Alsaedi, Abdullah
Arora, Sanidhay
Asano, Kyoichi
Avizheh, Seoideh
Avizheh, Sepdieh
Bamiloshin, Michael
Biswas, Chinmoy
Blanco-Justicia, Alberto
Celi, Sofia
Chan, Kwan Yin
Chang, Lun-Ching
Chen, Chen
Chen, Jiaqiang
Chen, Shiyao
Chen, Xinjian
Chen, Zhili
Dang, Hai-Van
Daudén-Esmel, Cristofol
Dehghani Tezerjani, Mohammad
Deng, Jiaqi
Deng, Jiawei
Denipitiyage, Dishanika
Dey, Kunal
Ding, Xiaoyu
Duong, Dung Hoang
Dutta, Priyanka
Dutta, Sabyasachi
Fathi Rabooki, Saba
Fauzi, Prastudy
Feng, Ruijun
Florez-Gutierrez, Antonio
Gao, Yansong
Geren, Hasan
Glaeser, Noemi
Gong, Borui
Guiot, Miquel
Gunawardena, Ravin
Guo, Zhihao
Haffar, Rami
Haffey, Preston
Han, Shuai
Hu, David
Huang, Jianye
Jangir, Hansraj
Kanaoka, Akira
Karunanayake, Naveen
Kelarev, Andrei
Khan, Younas
Killeen, Brayden
Lai, Qiqi
Lai, Zhenzhi
Le, Huy Quoc
Li, Hongbo
Li, Jinhui
Li, Xinyu
Li, Yamin
Li, Yinan
Lin, Fuchun
Lin, Zhanren
Liu, Guozhen
Liu, Haotian
Liu, Yusen
Luo, Junwei
Lyu, Jiazhuo
Lyu, You
Martinez, Sergio
Ngo, Tran
Nguyen, Chau
Nguyen, Khoa
Niroshan, Akila
Pakana, Fitrio

Pan, Jing
Pan, Shimin
Panja, Somnath
Paulet, Russell
Qiu, Shaojian
Rashid, Fariza
Sakamoto, Kosei
Samadi, Yasaman
Shen, Jun
Shi, Junbin
Shi, Yiwen
Shiba, Rentaro
Sim, Jun Jie
Sugio, Nobuyuki
Sui, Yulei
Sun, Hui
Sun, Shifeng
Tairi, Erkan
Tang, Gang
Tang, Khai Hanh
Tian, Guohua
Tran, Minh Trung
Tsuchida, Hikaru
Tunc, Cihan
Wang, Dabao
Wang, Faxing
Wang, Jiabo
Wang, Jing
Wang, Liping
Wang, Weizhe
Wang, Xinqian
Wei, Jianghong
Win, Hsu Myat
Wu, Mingli
Xiao, Meiyan
Xie, Zhikang
Xiong, Qi
Xu, Feng
Xu, Lei
Xu, Pei
Xue, Haiyang
Yang, Haochen
Yang, Jia
Yao, Zihao
Yin, Yifan
Youmans, William
Yuan, Jiaming
Yuan, Liang
Yuchao, Yao
Zhang, Chengru
Zhang, Jiahao
Zhao, Yanqi
Zheng, Jinwei
Zhou, Yanwei
Zhou, Yunxiao
Zhu, Congcong
Zhu, Xiaogang
Ziaur, Rahman
Zong, Wei

Contents – Part I

Encryption and Its Applications

Digital Signatures

Cryptographic Primitives

Contents – Part II

Post-Quantum Cryptography

Cryptanalysis

Secure Protocols

Application Security

Contents – Part III

Blockchain Technology

Privacy Enhancing Technologies

AI Security

Symmetric-Key Cryptography

The Offline Quantum Attack Against Modular Addition Variant of Even-Mansour Cipher

Fangzhou Liu[1,2], Xueqi Zhu[1,2], Ruozhou Xu[1,2], Danping Shi[1,2], and Peng Wang[1,2,3](✉)

[1] Key Laboratory of Cyberspace Security Defense, Institute of Information Engineering, CAS, Beijing, China
{liufangzhou,zhuxueqi,xurouzhou,shidanping}@iie.ac.cn, w.rocking@gmail.com

[2] School of Cyber Security, University of Chinese Academy of Sciences, Beijing, China

[3] School of Cryptology, University of Chinese Academy of Sciences, Beijing, China

Abstract. At Eurocrypt 2017, the Even-Mansour (EM) cipher was modified to thwart the attack using Simon's algorithm: replace the XOR operation with modular addition. We call it Even-Mansour+ (EM+) cipher. Kuperberg's algorithm can recover the key of EM+ in sub-exponential time, but it requires quantum queries (Q2 mode), making it difficult to apply in practice. In this paper, we introduce a new attack against EM+, using only classical queries and offline quantum computations (Q1 mode). The key problem we solve is how to determine whether two functions have a shift, so that by combining Kuperberg's algorithm with Grover's algorithm, we can recover the key of EM+ in $O(n2^{n/3+\sqrt{n}})$ time.

Keywords: Even-Mansour cipher · modular addition · Kuperberg's algorithm · Q1 model · symmetric cryptography · hidden shift problem

1 Introduction

In recent years, with the development of quantum technology, it has been found that quantum computers have fundamental breakthroughs in computing and information processing compared to classical computers. Public-key cryptography is one of the most affected disciplines, because its security is based on hard problems, and some of these problems can be solved within polynomial time in the quantum setting. For example, RSA, the most commonly used public-key algorithm, is no longer secure, due to the fact that the hardness of factoring large integers has been broken by Shor's quantum algorithm [20]. In 2016, the National Institute of Standards and Technology (NIST) launched a worldwide effort to find post-quantum cryptographic algorithms to replace current public-key cryptography standards, and recently released three draft FIPS standards.

T. Zhu and Y. Li (Eds.): ACISP 2024, LNCS 14895, pp. 3–19, 2024.
https://doi.org/10.1007/978-981-97-5025-2_1

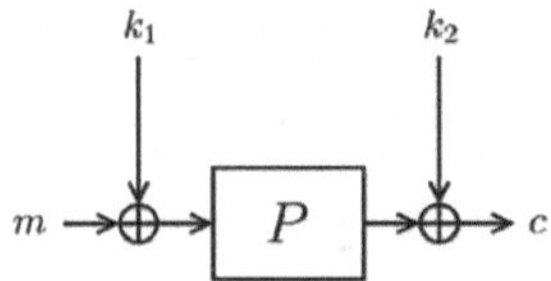

Fig. 1. Even-Mansour cipher

As the age of quantum computing approaches, it is urgent to investigate the quantum security of existing cryptographic algorithms. It was once believed that quantum computing had a limited impact on symmetric ciphers and Grover's algorithm [11] was the only threat, which offers quadratic acceleration for brute-force search. Therefore, simply doubling the key length would achieve quantum security similar to classical security.

However, some recent work has shown that Grover search is not the only way. If the attacker can make quantum queries, i.e. query the quantum oracle of the symmetric cipher by superposition states, then a large number of symmetric ciphers can be broken within polynomial time. For example, Even-Mansour cipher [16], three-round Feistel cipher [17], CBC-MAC [13], PMAC [4], OCB [4], GCM [13], etc., which have all been proven to be secure in the classical setting, are broken in polynomial time through quantum queries.

Simon's algorithm [21], which can recover the hidden period of a function in polynomial time, is one of the most widely used quantum algorithms in these attacks. The crucial step is to construct a periodic function and then use Simon's algorithm to recover the period. The period is often related to some secret information, based on which the security fails.

EM and EM+ Ciphers. The Even-Mansour cipher [9] (EM for short) is a block cipher based on public permutation:

$$EM_{k_1,k_2}(x) = P(x \oplus k_1) \oplus k_2,$$

where P is the permutation on $\{0,1\}^n$ and k_1, k_2 are keys. Due to its simplicity, the EM cipher has attracted significant attention in the crypto community, including security studies in settings of iteration [6], related-key [10], multi-key [18], known-key [7], post-quantum [1], etc. (Fig. 1).

When P is a public random permutation, EM is secure in the classical setting and its security bound is a typical birthday bound of $O(2^{n/2})$, which is the least time complexity required to break EM. However, a function with a hidden period can be constructed: $f(x) = EM_{k_1,k_2}(x) \oplus P(x) = P(x \oplus k_1) \oplus P(x) \oplus k_2$, its period is k_1, which means that $f(x) = f(x \oplus k_1)$. Then we can use Simon's quantum algorithm to recover the key.

At Eurocyrpt 2017, in order to thwart the attack using Simon's algorithm, Alagic et al. [2] proposed the Even-Manmour+ (EM+ for short) cipher: replace the XOR operation (bitwise mod 2 addition) in the EM with modulo 2^n addition, so that Simon's algorithm cannot be applied. At Asiacrypt 2018, Bonnetain et al. [5] gave an attack using Kuperberg's algorithm [15]. Although its time

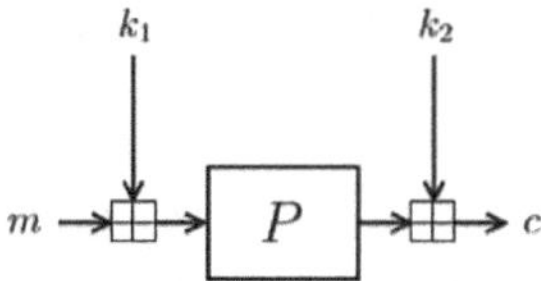

Fig. 2. Even-Mansour+ cipher

complexity is $O(n2^{1.8\sqrt{n}})$, it still has a significant advantage compared to the classical complexity of $O(2^{n/2})$ (Fig. 2).

Q2 vs. Q1. Quantum attacks can be classified into two categories according to the abilities of attackers: the Q1 model and the Q2 model. In the Q1 model, attackers can use a quantum computer to perform offline computation, but can only perform classical queries to the cryptographic algorithm. Offline computation refers to a computing process without interaction with quantum oracles of cryptographic algorithms. However, in the Q2 model, attackers can perform not only quantum computations, but also quantum queries.

The Q2 model has the ability to break a wide range of ciphers with a relatively low level of complexity. For example, in the Q2 model, Simon's algorithm can recover the hidden period of a function in polynomial time, while Kuperberg's algorithm can recover the hidden shift of two functions in sub-exponential time [15]. However, the Q2 model also has limitations: attaining a quantum oracle of the keyed cryptographic algorithm is difficult to achieve in practical applications.

In contrast, in the Q1 model, the attacker only has the quantum computation capability and can query the classical oracle. It can also construct quantum oracles for public algorithms without keys, such as the permutation in the EM cipher. Therefore, attacks in the Q1 model are more realistic, and many scholars [1] refer to the post-quantum model as the Q1 model.

Due to abundant attacking results in the Q2 model, one approach to get attacks in the Q1 model is to convert the algorithm in the Q2 model to the one in the Q1 model [3,12,16]. Converted algorithms are more realistic, but the corresponding time cost is also greater. An example is Simon's algorithm, when attacking special objects such as EM, can be converted from the Q2 model to the Q1 model [3]. The core idea is to generate a quantum superposition state through classical queries, and its time complexity increases from $O(n^3)$ in the Q2 model to $O(n^3 2^{n/3})$ in the Q1 model. Whether Kuperberg's algorithm can be converted into the Q1 model and what about its time complexity is the core issue to be discussed in this article (Table 1).

Our Contributions. We propose an attack against EM+ in the Q1 model. By combining Kuperberg's algorithm with Grover's algorithm, we can recover the key of EM+, which requires much less than $O(2^{n/2})$ queries. This attack is the first application of Kuperberg's algorithm in the Q1 model.

Table 1. Q1 and Q2 Quantum Attacks on EM and EM+.

Target	Model	Queries	Time	Reference
EM	Q2	$O(n)$	$O(n^3)$	[17]
EM	Q1	$O(2^{n/3})$	$O(n^3 2^{n/3})$	[3]
EM+	Q2	$O(n2^{1.8\sqrt{n}})$	$O(n2^{1.8\sqrt{n}})$	[5]
EM+	Q1	$O(n2^{n/3+\sqrt{n}})$	$O(n2^{n/3+\sqrt{n}})$	Sect. 3

The time complexity of this attack is $O(n2^{n/3+\sqrt{n}})$. If we take the permutation with block length of 1600 bits in SHA-3 [8,14] as the underlying permutation, the attack takes less than 2^{584} time complexity to complete, which is much faster than that in the classical setting, which is 2^{800}.

Organizations. Section 2 gives preliminaries. Section 3 shows our attack in the Q1 model. Section 4 gives conclusions.

2 Preliminaries

In this section, we will introduce some concepts of quantum computation [19] and review Grover's algorithm and Kuperberg's algorithm.

2.1 Quantum Computation

Qubits. An individual qubit is the fundamental objects in quantum computing. Unlike classical bits that can only be in the states of 0 or 1, qubits can be in linear combinations of these states, known as the superposition state: $|\psi\rangle = \alpha|0\rangle + \beta|1\rangle$. A qubit can be thought of as a vector in a two-dimensional complex vector space; $|0\rangle$ and $|1\rangle$ are computational basis states, which form an orthonormal basis for this vector space.

If we measure $|\psi\rangle = \alpha|0\rangle + \beta|1\rangle$, we may obtain the result 0 with probability $|\alpha|^2$ and the result 1 with probability $|\beta|^2$, such that the probabilities sum up to 1 ($|\alpha|^2 + |\beta|^2 = 1$), which is known as normalization. For simplicity, we sometimes omit the normalization factor.

More generally, we can consider a system consisting of n qubits that corresponds to a Hilbert space on the 2^n-dimensional complex field. The basis states of this system are of the form $|x_1x_2...x_n\rangle$, where there are 2^n possible quantum states in total.

Quantum Gates. The unitary matrix (also known as the quantum gate) U can convert one quantum $|\psi_1\rangle$ state into another state $|\psi_2\rangle$ ($|\psi_2\rangle = U|\psi_1\rangle$). Next, we will introduce two specific quantum gates.

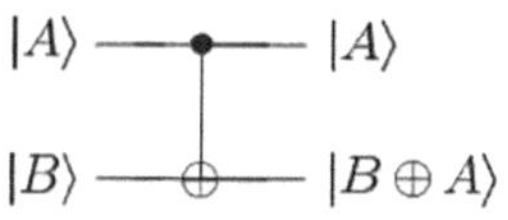

$|A\rangle$ —●— $|A\rangle$

$|B\rangle$ —⊕— $|B \oplus A\rangle$

Fig. 3. CNOT Gate

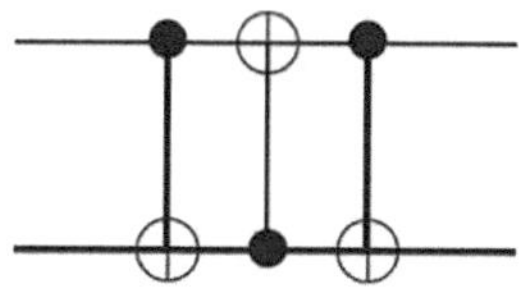

Fig. 4. Circuit swapping two qubits

- **Hadamard Gate H**

  The Hadamard gate acting on a single qubit is

$$H = \frac{1}{\sqrt{2}} \begin{bmatrix} 1 & 1 \\ 1 & -1 \end{bmatrix}.$$

  This gate turns a $|0\rangle$ into $(|0\rangle + |1\rangle)/\sqrt{2}$, and turns $|1\rangle$ into $(|0\rangle - |1\rangle)/\sqrt{2}$. Easy to verify, the inverse of H is itself, so $H^2|0\rangle = |0\rangle$, $H^2|1\rangle = |1\rangle$.

  We use $H^{\otimes n}$ to represent applying the Hadamard transform H to each qubit in the n qubits system. If the base state of n qubits is $|x\rangle$, then

$$H^{\otimes n}|x\rangle = \frac{1}{\sqrt{2^n}} \sum_{y \in \{0,1\}^n} (-1)^{x \cdot y} |y\rangle.$$

  For simplicity, we often write $H^{\otimes n}$ as H.

- **CNOT Gate**

  CNOT gate, or known as controlled-NOT gate, is the prototypical multi-qubit quantum logic gate. This gate has two input qubits, known as the control qubit and the target qubit. In Fig. 3, the top line represents the control qubit, while the bottom line represents the target qubit. If the control qubit is set to 0, then the target qubit is left alone. If the control qubit is set to 1, then the target qubit is flipped.

Quantum Circuits. Quantum circuits are composed of a series of quantum gates, through which we can evolve, measure, and manipulate qubits, thereby performing various quantum computing tasks.

The circuit is read from left to right, and each line in the circuit represents a wire in the quantum circuit. For example, the circuit in Fig. 4 accomplishes a

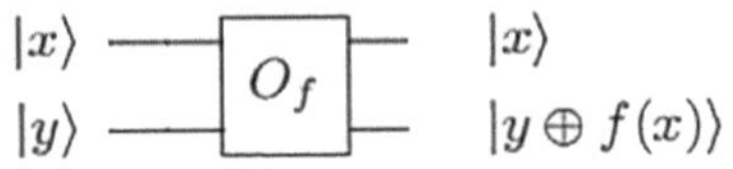

$$|x\rangle \; O_f \; |x\rangle \qquad |y\rangle \; \to \; |y \oplus f(x)\rangle$$

Fig. 5. Quantum oracle of f

simple but useful task - it swaps the states of the two qubits. Suppose that the state input to this quantum circuit is $|a, b\rangle$, then the sequence of the state is

$$|a, b\rangle \rightarrow |a, a \oplus b\rangle \rightarrow |b, a \oplus b\rangle \rightarrow |b, a\rangle .$$

The effect of the circuit, therefore, is to interchange the state of the two qubits.

Quantum Fourier Transform. The quantum Fourier transform on an orthonormal basis $|0\rangle, ..., |N-1\rangle$ is defined to be a linear operator with the following action on the basis states, which is the key ingredient for quantum factoring and many other interesting quantum algorithms.

$$|j\rangle \rightarrow \frac{1}{\sqrt{N}} \sum_{k=0}^{N-1} e^{2\pi ijk/N} |k\rangle .$$

Quantum Oracle. Many quantum algorithms require the attacker to have the ability to access some quantum oracles, which are also unitary transformations. For a classical function f, its corresponding quantum oracle is as follows: $O_f |x\rangle |y\rangle \rightarrow |x\rangle |y \oplus f(x)\rangle$. If we have access to O_f, we say that we can perform a quantum query to f (Fig. 5).

2.2 Grover's Algorithm

Grover's algorithm [11] gives a quadratic acceleration for exhaustive search problems:

Problem 1. Consider a boolean function $f : \{0,1\}^n \rightarrow \{0,1\}$, find a x that satisfies $f(x) = 1$.

In general, if f maps t inputs to 1 and all other inputs to 0, assuming that evaluating the value of the function f has a time complexity $O(1)$, Grover's algorithm can find a solution within the time complexity of $O(\sqrt{2^n/t})$. In the special case where there is only one x that satisfies $f(x) = 1$, then finding x requires $O(\sqrt{2^n})$ time complexity.

Sometimes, calculating the value of $f(x)$ also requires a certain amount of time complexity, assuming that the required time is T, then the time complexity of Grover's algorithm is $O(T \cdot \sqrt{2^n/t})$.

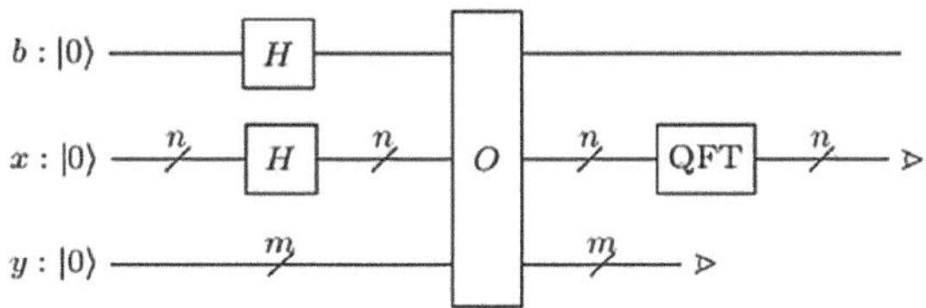

Fig. 6. Quantum circuit for Kuperberg's algorithm

2.3 Kuperberg's Algorithm

Kuperberg's algorithm [15] is used to solve the hidden shift problem, its modulo 2^n addition version is defined as follows (use + to represent modulo 2^n addition):

Problem 2. Let f, g two injective functions: $\{0,1\}^n \to \{0,1\}^m$, given the promise that there exists s such that, for all x, $f(x) = g(x+s)$, retrieve s.

Kuperberg's algorithm can be divided into two parts: the first part calls f and g to obtain some qubits with information related to the hidden shift, and the second part combines these qubits to extract the exact hidden shift. The expression of quantum oracle used in the first part is as follows [5]:

$$O : |b\rangle |x\rangle |y\rangle \to \begin{cases} |0\rangle |x\rangle |y \oplus f(x)\rangle & \text{if b=0,} \\ |1\rangle |x\rangle |y \oplus g(x)\rangle & \text{if b=1.} \end{cases}$$

This circuit (shown in Fig. 6) produces a uniform superposition in the registers b and x with Hadamard gates H, feeds them to the oracle O, and then measures the register y. This measurement gives a result y_0 and collapses the b and x registers in the state $\sum_{f(x)=y_0} |0\rangle |x\rangle + \sum_{g(x)=y_0} |1\rangle |x\rangle$. Since the functions f and g have a shift s, the state can be expressed as $|0\rangle |x_0\rangle + |1\rangle |x_0 + s\rangle$, but we do not know the exact value of x_0.

Then we apply a quantum Fourier transform (QFT) on the x register and measure the result. We get an l with a uniform probability, and the b register collapses to the state $|\psi_l\rangle = |0\rangle + e^{2\pi i l s/N} |1\rangle$ $(N = 2^n)$.

It can be seen that the property of $|\psi_l\rangle$ is related to s, but it is not intuitive and cannot be extracted directly. We can observe that this state becomes interesting when $l = N/2 = 2^{n-1}$, as $|\psi_{N/2}\rangle = |0\rangle + e^{\pi i s} |1\rangle$. If the lowest bit of s is 0, then $|\psi_{N/2}\rangle = |+\rangle$, otherwise, $|\psi_{N/2}\rangle = |-\rangle$. Therefore, if we measure it in the $\{|+\rangle, |-\rangle\}$ basis, we get one bit of s.

Recall that the probability of l is uniform, so we cannot directly obtain $|\psi_{N/2}\rangle$. We use a circuit to combine two states $|\psi_{l_1}\rangle$ and $|\psi_{l_2}\rangle$. Before the measurement, the system is in the state $\mathsf{CNOT} |\psi_{l_1}\rangle |\psi_{l_2}\rangle =$

$$|00\rangle + e^{2\pi i(l_1+l_2)s/N} |10\rangle + e^{2\pi i l_2 s/N} \left(|01\rangle + e^{2\pi i(l_1-l_2)s/N} |11\rangle\right).$$

If we measure a 0 we will get the qubit $|\psi_{l_1+l_2}\rangle$, and if we measure a 1 we will get $|\psi_{l_1-l_2}\rangle$. Our goal is to obtain $|\psi_{N/2}\rangle$, so we need to repeat the above process

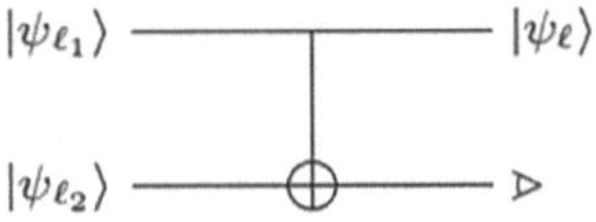

Fig. 7. Combination circuit

to obtain several random states $|\psi_l\rangle$ and combine them. If we only consider the variable l, it can be simplified as a mathematical problem: Given a set of random numbers ranging from 0 to 2^n, we can combine any two numbers by the operation $+$ or $-$, which we discover afterwards. All numbers can only be used once, and the goal is to obtain 2^{n-1} (Fig. 7).

The idea of combination is that if $a \neq \pm b$ are both multiples of 2^k and not multiples of 2^{k+1}, then $a+b$ and $a-b$ must be non-zero and also a multiple of 2^{k+1}. Therefore, we can start with odd numbers and continuously combine pairs to obtain numbers divisible by higher powers of 2 until we obtain 2^{n-1}.

After obtaining $|\psi_{2^{n-1}}\rangle$, we can get the lowest significant bit of the hidden shift, s_0. We then use a recursive procedure to recover the other bits of s. We can construct the functions $f'(x) = f(2x)$, $g'(x) = g(2x + s_0)$, which have the hidden shift $s' = (s - s_0)/2$. Observing that the second bit of s is the lowest bit of s', we can reapply the routine until we get all the bits.

The details of this algorithm are given in [15]. The paper also presents a sketch proof of its complexity $\widetilde{O}(2^{\sqrt{2\log_2(3)n}})$. The paper only focuses on the asymptotic exponent complexity, so the polynomial part is not well known. Bonnetain et al. [5] deduce that the complexity of recovering the whole hidden shift is $O(n2^{\sqrt{2\log_2(3)n}})$, as $\sqrt{2\log_2(3)} \approx 1.8$.

3 An Attack Against EM+ in the Q1 Model

3.1 Attack in the Q2 Model

At Eurocrypt 2017, Alagic et al. proposed EM+ to counteract attacks by Simon's algorithm. This cipher replaces the XOR operation in EM with the addition of modulo 2^n, expressed as: $EM{+}_{k_1,k_2}(x) = P(x + k_1) + k_2$, where P is a public random permutation.

In 2018, Bonnetain et al. [5] used Kuperberg's algorithm to attack EM+. Since Kuperberg's algorithm can find the hidden shift between two functions, we need to construct two shifted functions f, g, which hold that $f(x) = g(x+s)$ for all x, and the shift s is related to secret information such as keys.

The construction in [5] is $f(x) = EM{+}_{k_1,k_2}(x) + P(-x), g(x) = EM{+}_{k_1,k_2}(-x) + P(x)$, then $f(x) = g(x + k_1)$. Therefore, using Kuperberg's algorithm, the key can be recovered in sub-exponential time.

3.2 Attack in the Q1 Model

Kuperberg's algorithm makes quantum queries to two functions f and g, which contain the key k_1, k_2, so it is an attack in the Q2 model. To convert it to Q1, a common method [3] is to use classical queries to prepare the quantum state.

Assuming that the domain of f is $\{0,1\}^n$, it requires 2^n classical queries to construct $\sum_{x\in\{0,1\}^n} |x\rangle |f(x)\rangle$, with a time complexity of $O(2^n)$, which is comparable to the complexity required for brute-force attacks. To achieve a better time complexity, we try to find a way to reduce the number of classical queries. We achieve this using the following two steps:

The first step is to separate the parts containing keys from public parts. Although two functions given in [5] have a shift, they do not satisfy our requirement. Our new construction is

$$f(x) = EM{+}_{k_1,k_2}(x+1) - EM{+}_{k_1,k_2}(x), g(x) = P(x+1) - P(x).$$

We can see that $f(x) = g(x + k_1)$ and the shift is k_1. g only contains the public permutation P, so that the attacker can construct the corresponding quantum oracle.

The second step is to reduce the domain size of f in order to decrease the number of classical queries. The remaining domain is accelerated by Grover's algorithm to search, and the core idea is as follows: For convenience in notation, we write $s = k_1$. We split s into two parts $s_1, s_2 : s = s_1 \,||\, s_2$. Assuming that $s_1 \in \{0,1\}^u, s_2 \in \{0,1\}^{n-u}$, we have $s = s_1 \cdot 2^{n-u} + s_2$. From the original functions, we construct new functions $f', g'_i : \{0,1\}^u \to \{0,1\}^n$, which are defined as $f'(x) = f(2^{n-u} \cdot x), g'_i(x) = g(2^{n-u} \cdot x + i)$.

Observe that $f'(x) = f(2^{n-u} \cdot x) = g(2^{n-u} \cdot x + s) = g[2^{n-u} \cdot (x + s_1) + s_2]$. If $i = s_2$, there is $f'(x) = g'_{s_2}(x + s_1)$.

To obtain the value of s_2, we need to perform a Grover search on i. The condition for a successful search is to determine whether f' and g'_i have a shift. Assume that there is a shift detection algorithm S that takes input i and returns 1 when there is a shift between f' and g'_i, or 0 otherwise.

$$S(i) = \begin{cases} 1 & f' \text{ and } g'_i \text{ have a shift,} \\ 0 & \text{Otherwise.} \end{cases}$$

Then we can use S as an iterative part of Grover's algorithm. Roughly speaking, Grover's algorithm can measure the value $i_0 = s_2$, and we can then perform Kuperberg's algorithm on f', g'_{s_2} to obtain its shift s_1, and finally output $s = s_1 \,||\, s_2$. Algorithm 1 shows the pseudocode of the attack.

3.3 Shift Detection Algorithm

The goal of the shift detection algorithm S is to determine whether there is a shift between f' and g'_i. It is natural to think that even if we don't know whether

Algorithm 1. Recover the key of EM+ in the Q1 model

1: Let $f(x) = EM+_{k_1,k_2}(x+1) - EM+_{k_1,k_2}(x), g(x) = P(x+1) - P(x)$, construct $f'(x) = f(2^{n-u} \cdot x), g_i'(x) = g(2^{n-u} \cdot x + i)$.

2: Using 2^u classical queries to f', create the state:

$$|\psi_{f'}\rangle = \sum_{x \in \{0,1\}^u} |x\rangle |f'(x)\rangle$$

(The details of generating this state are shown in Algorithm 2.)

3: Create the uniform superposition of i:

$$\sum_{i \in \{0,1\}^{n-u}} |i\rangle$$

4: Apply Grover iterations. We can input $|\psi_{f'}\rangle$ and $|i\rangle$ into S, which can determine whether f' and g_i' have a shift or not. If there is a shift, it returns $|b \oplus 1\rangle$ on input $|b\rangle$. Otherwise it returns $|b\rangle$. (Algorithm 3 gives the details for S in the case that i is fixed.)

5: After $O(2^{\frac{n-u}{2}})$ Grover iterations, measure the index i and get the value i_0.

6: Apply Kuperberg's algorithm on f', g_{i_0}' to obtain its shift s_1, output $k_1 = s_1 \,||\, i_0$.

Algorithm 2. Generating the state $|\psi_{f'}\rangle$ [3]

1: Start with the zero state

$$|0\rangle |0\rangle$$

2: Apply Hadamard gates to the zero state:

$$\sum_{x \in \{0,1\}^u} |x\rangle$$

3: For each $x \in \{0,1\}^u$, query f', then apply a unitary that writes f' in the second register if the first contains the value x.

f' and g_i' have a shift, we can still perform the first iteration of Kuperberg's algorithm (written as **one-round-Kuperberg**) on them and obtain a result s_0.

If they have a shift s, the value of s_0 depends on whether s is even or odd. In other words, even if we repeat one-round-Kuperberg r times on the two functions, the value of s_0 will not change. If they do not have a shift, after r times of repetitions, the value of s_0 will change with a high probability (r is polynomial). Hence, we can use this property to determine whether the functions have a shift.

In the Q1 model, we first need to construct the quantum states of f' and g_i'. Observing that f' is invariant and its domain is $\{0,1\}^u$, we can construct its uniform superposition state using 2^u classical queries. $g_i'(x) = P(2^{n-u} \cdot x + i + 1) - P(2^{n-u} \cdot x + i)$ only contains public functions, so we can directly construct the corresponding quantum oracle.

Since P is a public random permutation, $g(x) = P(x+1) - P(x)$ can be considered as a random function. Set $a = 2^{n-u}$, we can partition the domain of

Algorithm 3. The procedure S that checks if functions f' and g'_i have a shift

1: Using the superposition query to g'_i, build the state:

$$\sum_{x\in\{0,1\}^u} |x\rangle |g'_i(x)\rangle$$

2: Apply the first iteration of Kuperberg's algorithm on f' and g'_i, compute the value of s'_0.

3: Repeat step 2 r times. If all the value of s'_0 are the same, output $|b \oplus 1\rangle$, otherwise output $|b\rangle$.

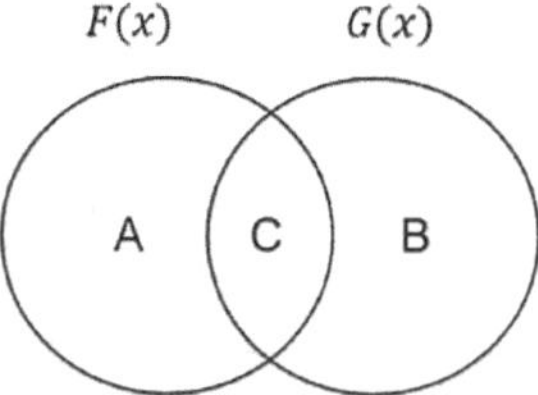

Fig. 8. Range of F and G

g and get a different functions: $g(ax), g(ax+1), ..., g(ax+a-1)$. They can be considered as independent random functions.

Recall that $f'(x) = f(ax) = g(ax+s) = g[a(x+s_1)+s_2], g'_i(x) = g(ax+i)$, $0 \le s_2, i \le a-1$. When $i \ne s_2$, $g(ax+i)$ is independent of $g(ax+s_2)$, so f' and g'_i can be considered as two independent random functions.

To study the output properties of the one-round-Kuperberg when the two functions do not have a shift, we can fix an input of $i \ne s_2$, then f' and g'_i can be regarded as two independent functions on $\{0,1\}^u \to \{0,1\}^n (u < n)$. For the convenience of symbol representation, we write these two functions as F and G. Now we apply one-round-Kuperberg to them:

We first produce the uniform superposition in the registers b and x with Hadamard gates H, feed them to the oracle O, and then measure register y. This measurement gives a result y_0 and collapses the b and x registers in the state $\sum_{F(x)=y_0} |0\rangle |x\rangle + \sum_{G(x)=y_0} |1\rangle |x\rangle$.

As shown in Fig. 8, the range of F and G can be partitioned into the part monopolized by F, denoted as A; the part monopolized by G, denoted as B; and the overlapping part, denoted as C. Then the state can be represented as:

$$|b\rangle |x\rangle = \begin{cases} \sum_{F(x)=y_0} |0\rangle |x\rangle & \text{if } y_0 \in A, \\ \sum_{G(x)=y_0} |1\rangle |x\rangle & \text{if } y_0 \in B, \\ \sum_{F(x)=y_0} |0\rangle |x\rangle + \sum_{G(x)=y_0} |1\rangle |x\rangle & \text{if } y_0 \in C. \end{cases}$$

Then we apply a quantum Fourier transform (QFT) on the x register. Let $N = 2^u$, then

$$|b\rangle|x\rangle \xrightarrow{QFT} \begin{cases} \frac{1}{\sqrt{N}}\sum_{j=0}^{N}\sum_{F(x)=y_0} e^{\frac{2\pi ijx}{N}}|0\rangle|j\rangle & \text{if } y_0 \in A, \\ \frac{1}{\sqrt{N}}\sum_{j=0}^{N}\sum_{G(x)=y_0} e^{\frac{2\pi ijx}{N}}|1\rangle|j\rangle & \text{if } y_0 \in B, \\ \frac{1}{\sqrt{N}}\sum_{j=0}^{N}\left(\sum_{F(x)=y_0} e^{\frac{2\pi ijx}{N}}|0\rangle + \sum_{G(x)=y_0} e^{\frac{2\pi ijx}{N}}|1\rangle\right)|j\rangle & \text{if } y_0 \in C. \end{cases}$$

When we measure the x register, we can get an $l \in \{0,1\}^u$, and the b register will collapse to the state:

$$|\psi_l\rangle = \begin{cases} |0\rangle & \text{if } y_0 \in A, \\ |1\rangle & \text{if } y_0 \in B, \\ \sum_{F(x)=y_0} e^{\frac{2\pi ilx}{N}}|0\rangle + \sum_{G(x)=y_0} e^{\frac{2\pi ilx}{N}}|1\rangle & \text{if } y_0 \in C. \end{cases}$$

Although the expression when $y_0 \in C$ is very complex, we can simply consider it as a superposition state: $|\psi_l\rangle = \alpha|0\rangle + \beta|1\rangle$ when $y_0 \in C$.

To obtain the state $|\psi_{N/2}\rangle$, we prepare multiple states $|\psi_{l_i}\rangle$ and use the CNOT circuit to combine them 2 by 2. Three situations occur.

The first situation: $|\psi_{l_1}\rangle = |0\rangle$ or $|1\rangle$, let $|\psi_{l_2}\rangle = \alpha|0\rangle + \beta|1\rangle$, then

$$\mathsf{CNOT}\,|\psi_{l_1}\rangle|\psi_{l_2}\rangle = \begin{cases} \alpha|00\rangle + \beta|01\rangle & |\psi_{l_1}\rangle = |0\rangle, \\ \beta|10\rangle + \alpha|11\rangle & |\psi_{l_1}\rangle = |1\rangle. \end{cases}$$

It can be seen that whether the measurement of the second register is 0 or 1, the final combined result is $|0\rangle$ or $|1\rangle$.

The second situation: $|\psi_{l_2}\rangle = |0\rangle$ or $|1\rangle$, let $|\psi_{l_1}\rangle = \alpha|0\rangle + \beta|1\rangle$, then

$$\mathsf{CNOT}\,|\psi_{l_1}\rangle|\psi_{l_2}\rangle = \begin{cases} \alpha|00\rangle + \beta|11\rangle & |\psi_{l_1}\rangle = |0\rangle, \\ \alpha|10\rangle + \beta|01\rangle & |\psi_{l_1}\rangle = |1\rangle. \end{cases}$$

Similarly, the final combined result is $|0\rangle$ or $|1\rangle$.

The third situation: $|\psi_{l_1}\rangle = \alpha_1|0\rangle + \beta_1|1\rangle$, $|\psi_{l_2}\rangle = \alpha_2|0\rangle + \beta_2|1\rangle$, then

$$\mathsf{CNOT}\,|\psi_{l_1}\rangle|\psi_{l_2}\rangle = \alpha_1|0\rangle(\alpha_2|0\rangle + \beta_2|1\rangle) + \beta_1|1\rangle(\alpha_2|1\rangle + \beta_2|0\rangle)$$

$$= (\alpha_1\alpha_2|0\rangle + \beta_1\beta_2|1\rangle)|0\rangle + (\alpha_1\beta_2|0\rangle + \beta_1\alpha_2|1\rangle)|1\rangle.$$

When we measure 0, we obtain the state $|\psi_l\rangle = \alpha_1\alpha_2|0\rangle + \beta_1\beta_2|1\rangle$, otherwise we obtain the state $|\psi_l\rangle = \alpha_1\beta_2|0\rangle + \beta_1\alpha_2|1\rangle$. If both $|\psi_{l_1}\rangle$ and $|\psi_{l_2}\rangle$ are superposition states (all α and β are not zero), then $|\psi_l\rangle$ is also a superposition state.

We can conclude that the result of the combination of two states is a superposition state only when they are both superposition states. Then we can analyze the probability if $|\psi_{N/2}\rangle$ is a superposition state:

The case where $|\psi_{N/2}\rangle$ is a superposition state must satisfy the condition that the states participating in the combination are all superposition states. Suppose that there are m states participating in the combination that ultimately forms

$|\psi_l\rangle$, and the probability that any $|\psi_{l_i}\rangle$ is a superposition state is p, then the final probability is p^m. Although we cannot determine the value of m, we know that $m \geq 1$. $m = 1$ represents the case where we measure $l = 2^{n-1}$ directly without the need for combinations.

Next, we will find an upper bound for p, $p = \Pr[y_0 \in C]$. Recall that F and G can be considered as two independent random functions on $\{0,1\}^u \to \{0,1\}^n$, and the ranges of both functions can be seen as sets that contain 2^u random values from $\{0,1\}^n$.

Let the range of F be D, then the number of elements in D is $\#D \leq 2^u$, the probability that any value in G coincides with a value in D is $\#D/2^n \leq 2^u/2^n$. And knowing that G has 2^u values in total, we can infer that $\#\{x|G(x) \in C\} \leq 2^u \cdot 2^u/2^n$. Similarly, $\#\{x|F(x) \in C\} \leq 2^u \cdot 2^u/2^n$. Knowing that

$$p = \Pr[y_0 \in C] = \frac{\#\{x|F(x) \in C\} + \#\{x|G(x) \in C\}}{2^u + 2^u},$$

so

$$p \leq \left(\frac{2^{2u}}{2^n} + \frac{2^{2u}}{2^n}\right) / (2^u + 2^u) = 2^{u-n}.$$

The probability that $|\psi_{N/2}\rangle$ is a superposition state is

$$\Pr[|\psi_{N/2}\rangle \text{ is a superposition state}] = p^m \leq p \leq 2^{u-n},$$

so

$$\Pr[|\psi_{N/2}\rangle \text{ is } |0\rangle \text{ or } |1\rangle] \geq 1 - p \geq 1 - 2^{u-n}.$$

In the last step, we measure $|\psi_{N/2}\rangle$ in the $\{|+\rangle, |-\rangle\}$ basis, and obtain a result s_0: if we measure $|+\rangle$, then $s_0 = 0$, otherwise $s_0 = 1$. We will repeat this procedure r times to obtain $s_{01}, s_{02}, ..., s_{0r}$. If they are all 0 or all 1, it is considered that F and G have a shift. Recall that we assume F and G don't have a shift, we call the algorithm fails at this case.

Let the offset is $\epsilon = \left|\Pr[s_0 = 0] - \frac{1}{2}\right|$, then the probability of failure is $\Pr_{\mathsf{bad}} = \left(\frac{1}{2} + \epsilon\right)^r + \left(\frac{1}{2} - \epsilon\right)^r$. It can be seen that $0 < \epsilon < 1/2$, and $\Pr_{\mathsf{bad}}$ increases monotonically as ϵ increases. So to get the upper bound of $\Pr_{\mathsf{bad}}$, we need to find the upper bound of ϵ.

When $|\psi_{N/2}\rangle$ is $|0\rangle$ or $|1\rangle$, the probability of measuring $|+\rangle$ is always $1/2$. Therefore, $\Pr[s_0 = 0]$ depends on the probability of measuring $|+\rangle$ $(0 < \Pr_+ < 1)$ when $|\psi_{N/2}\rangle$ is a superposition state. Then

$$\Pr[s_0 = 0]_{\mathsf{max}} \leq p \cdot 1 + (1-p) \cdot \frac{1}{2} = \frac{1}{2}(1+p),$$

$$\Pr[s_0 = 0]_{\mathsf{min}} \geq p \cdot 0 + (1-p) \cdot \frac{1}{2} = \frac{1}{2}(1-p).$$

From this, it can be inferred that $\epsilon = \left|\Pr[s_0 = 0] - \frac{1}{2}\right| \leq \frac{1}{2}p$, then

$$\Pr_{\mathsf{bad}} = \left(\frac{1}{2} + \epsilon\right)^r + \left(\frac{1}{2} - \epsilon\right)^r \leq \left(\frac{1}{2}\right)^r \cdot [(1+p)^r + (1-p)^r].$$

Since $p \leq 2^{u-n}$ is a small quantity, $\Pr_{\mathsf{bad}}$ can be Taylor expanded: Let $a(x) = (1+x)^r$, then

$$a(p) = 1 + rp + \frac{r(r-1)p^2}{2} + o(p^2),$$
$$a(-p) = 1 - rp + \frac{r(r-1)p^2}{2} + o(p^2),$$
$$a(p) + a(-p) = 2 + r(r-1)p^2 + o(p^2) < 2 + r^2p^2 \leq 2 + r^2 2^{2u-2n}.$$

We can obtain that

$$\Pr_{\mathsf{bad}} < (\frac{1}{2})^{r-1} + (\frac{1}{2})^r \cdot r^2 \cdot 2^{2u-2n}.$$

$\Pr_{\mathsf{bad}}$ will be smaller when r is larger. In the next section, we will discuss the value of r that can meet the requirements of practical applications.

3.4 Time Complexity and Failure Probability

Reference to the Algorithm 1, the whole attack can be separated into three parts: generating the state $|\psi_{f'}\rangle$, applying Grover iterations, and applying Kuperberg's algorithm.

Generating the state $|\psi_{f'}\rangle$: Since the domain of f' is $\{0,1\}^u$, generating the superposition states requires 2^u classical queries, with a time complexity of $O(2^u)$.

Applying Grover iterations: assuming that the search space of i is E and the running time of algorithm S is T, the total time complexity of Grover's algorithm is $T \cdot (\#E)^{1/2}$. Since $i \in \{0,1\}^{n-u}$, then $\#E = 2^{n-u}$. We also know that algorithm S means that repeating one-round-Kuperberg for r times, so $T = O(r \cdot 2^{1.8\sqrt{u}})$, and it can be inferred that the time complexity of Grover search is $O(2^{\frac{n-u}{2}}) \cdot O(r2^{1.8\sqrt{u}})$.

Finally, we apply Kuperberg's algorithm on f', g'_{i_0}, which has a time complexity of $O(u2^{1.8\sqrt{u}})$. Therefore, the time complexity of the attack is

$$O(2^u) + O(2^{\frac{n-u}{2}}) \cdot O(r2^{1.8\sqrt{u}}) + O(u2^{1.8\sqrt{u}})$$
$$< O(r) \cdot [O(2^u) + O(2^{\frac{n-u}{2}+1.8\sqrt{u}})].$$

Since r is polynomial, its impact on time complexity is much smaller than exponential parts. Therefore, according to the analysis of derivatives, the time complexity reaches a minimum when $u = \frac{n-u}{2} + 1.8\sqrt{u}$. Solving the equation, we get that

$$u = \frac{n}{3} + \frac{13 + \sqrt{169 + 156n}}{18}.$$

It can be seen that $u > n/3$, in order to make the time complexity of this algorithm better than Grover's algorithm, it also needs to satisfy $u < n/2$, which has a solution of $n > 26$. However, in practical use, n is generally 128-bit or larger, so this algorithm is feasible.

To be more intuitive, we also get that when $n > 7.76$, $\frac{13+\sqrt{169+156n}}{18} < \sqrt{n}$. So the time complexity is

$$O(r2^{\frac{n}{3}+\sqrt{n}}), \text{ if } n > 26.$$

We call the attack fails when the recovered $s = k_1$ does not match the true key of the EM+. The reason for the failure of is that the Grover search measured $i_0 \neq s_2$, and naturally, it was unable to recover the correct shift s.

The algorithm used by Grover search is S:

$$S(i) = \begin{cases} 1 & f' \text{ and } g_i' \text{ have a shift,} \\ 0 & \text{Otherwise.} \end{cases}$$

Recall that the failure probability of this algorithm is $\Pr_{\mathsf{bad}}$, $i \in \{0,1\}^{n-u}$, it is expected that there are a total of $2^{n-u} \cdot \Pr_{\mathsf{bad}}$ wrong solutions. Then the success of the attack means that the unique correct solution is measured among all solutions, otherwise the attack fails, so the probability of attack failure can be expressed as

$$\Pr_{\mathsf{fail}} = \Pr[i_0 \neq s_2] = \frac{2^{n-u} \cdot \Pr_{\mathsf{bad}}}{2^{n-u} \cdot \Pr_{\mathsf{bad}} + 1} < 2^{n-u} \cdot \Pr_{\mathsf{bad}}.$$

We also know that $\Pr_{\mathsf{bad}} < (\frac{1}{2})^{r-1} + (\frac{1}{2})^r \cdot r^2 \cdot 2^{2u-2n}$, and $n/3 < u < n/2$, so when there is $r < 2^{n/2}$, we can infer that $r^2 \cdot 2^{2u-2n} < 2^{2u-n} < 1$, then

$$\Pr_{\mathsf{bad}} < (\frac{1}{2})^{r-1} + (\frac{1}{2})^r = 3 \cdot (\frac{1}{2})^r, \text{ when } r < 2^{n/2}.$$

So

$$\Pr_{\mathsf{fail}} < 3 \cdot 2^{n-u-r} < 3 \cdot 2^{\frac{2n-3r}{3}}, \text{ when } r < 2^{n/2}.$$

It can be seen that the probability of attack failure decreases as r increases, but it will also affect the time required for the algorithm. Therefore, we should balance the value of r from the time complexity and probability of failure. For example, we can let $r = \frac{5}{3}n$ (polynomial), then $\Pr_{\mathsf{fail}} < 3 \cdot 2^{-n}$, and this probability is already comparable to the probability of guessing the key k_1 randomly. Finally, according to the derivative analysis, when $n > 8$, $5n/3 < 2^{n/2}$ always holds.

Therefore, when $n > 26$, let $r = \frac{5}{3}n$, the time complexity is $O(n2^{\frac{n}{3}+\sqrt{n}})$, and the probability of failure $\Pr_{\mathsf{fail}} < 3 \cdot 2^{-n}$.

4 Conclusions

We propose an attack against EM+ in the Q1 model. In the case where only classical queries can be made to the key-containing part, we recover the key of

EM+. The time complexity is $O(n2^{\frac{n}{3}+\sqrt{n}})$, and the upper bound of the probability of failure can be reduced to $3 \cdot 2^{-n}$. The key problem we solve is how to determine whether two functions have a shift. Based on the two constructed functions with hidden shift, we turn the attack using Kuperberg's algorithm in the Q2 model into the one in the Q1 model.

The security bound of EM is given in [1]: $2^{n/3}$ queries are necessary to attack EM in the Q1 model. In more detail, let q_P denote the number of quantum queries to P, P^{-1}, q_E denote the number of classical queries to EM, EM^{-1}, then any attack with a constant success probability requires either $q_P^2 \cdot q_E = \Omega(2^n)$ or $q_E^2 \cdot q_P = \Omega(2^n)$.

Observing that the security proof did not use the specificity of XOR operations, we can replace it with operations of other groups, such as modular addition, and this would not change the bound. Therefore, the security bound of EM+ in the Q1 model is also $2^{n/3}$. However, our attack requires $O(n2^{\frac{n}{3}+\sqrt{n}})$ queries, which does not meet the bound. Whether the bound is tight for EM+ is still an open problem.

Acknowledgments. The authors thank the anonymous reviewers for many helpful comments. The work of this paper is supported by the National Key Research and Development Program of China (No. 2018YFA0704704), the National Natural Science Foundation of China (No. 62172410), and the Youth Innovation Promotion Association of Chinese Academy of Sciences.

References

1. Alagic, G., Bai, C., Katz, J., Majenz, C.: Post-quantum security of the Even-Mansour cipher. In: Dunkelman, O., Dziembowski, S. (eds.) EUROCRYPT 2022, Part III. LNCS, vol. 13277, pp. 458–487. Springer, Cham (2022). https://doi.org/10.1007/978-3-031-07082-2_17
2. Alagic, G., Russell, A.: Quantum-secure symmetric-key cryptography based on hidden shifts. In: Coron, J.-S., Nielsen, J.B. (eds.) EUROCRYPT 2017. LNCS, vol. 10212, pp. 65–93. Springer, Cham (2017). https://doi.org/10.1007/978-3-319-56617-7_3
3. Bonnetain, X., Hosoyamada, A., Naya-Plasencia, M., Sasaki, Y., Schrottenloher, A.: Quantum attacks without superposition queries: the offline Simon's algorithm. CoRR abs/2002.12439 (2020). https://arxiv.org/abs/2002.12439
4. Bonnetain, X., Leurent, G., Naya-Plasencia, M., Schrottenloher, A.: Quantum linearization attacks. In: Tibouchi, M., Wang, H. (eds.) ASIACRYPT 2021. LNCS, vol. 13090, pp. 422–452. Springer, Cham (2021). https://doi.org/10.1007/978-3-030-92062-3_15
5. Bonnetain, X., Naya-Plasencia, M.: Hidden shift quantum cryptanalysis and implications. In: Peyrin, T., Galbraith, S. (eds.) ASIACRYPT 2018. LNCS, vol. 11272, pp. 560–592. Springer, Cham (2018). https://doi.org/10.1007/978-3-030-03326-2_19
6. Chen, S., Steinberger, J.: Tight security bounds for key-alternating ciphers. In: Nguyen, P.Q., Oswald, E. (eds.) EUROCRYPT 2014. LNCS, vol. 8441, pp. 327–350. Springer, Heidelberg (2014). https://doi.org/10.1007/978-3-642-55220-5_19

7. Cogliati, B., Seurin, Y.: Strengthening the known-key security notion for block ciphers. In: Peyrin, T. (ed.) FSE 2016. LNCS, vol. 9783, pp. 494–513. Springer, Heidelberg (2016). https://doi.org/10.1007/978-3-662-52993-5_25
8. Dworkin, M.J.: SHA-3 standard: permutation-based hash and extendable-output functions. In: NIST Special Publication (2015)
9. Even, S., Mansour, Y.: A construction of a cipher from a single pseudorandom permutation. J. Cryptol. **10**(3), 151–162 (1997). https://doi.org/10.1007/s001459900025
10. Farshim, P., Procter, G.: The related-key security of iterated Even–Mansour ciphers. In: Leander, G. (ed.) FSE 2015. LNCS, vol. 9054, pp. 342–363. Springer, Heidelberg (2015). https://doi.org/10.1007/978-3-662-48116-5_17
11. Grover, L.K.: A fast quantum mechanical algorithm for database search. In: Miller, G.L. (ed.) Proceedings of the Twenty-Eighth Annual ACM Symposium on the Theory of Computing, pp. 212–219. ACM (1996). https://doi.org/10.1145/237814.237866
12. Hosoyamada, A., Sasaki, Y.: Cryptanalysis against symmetric-key schemes with online classical queries and offline quantum computations. IACR Cryptology ePrint Archive, p. 977 (2017). http://eprint.iacr.org/2017/977
13. Kaplan, M., Leurent, G., Leverrier, A., Naya-Plasencia, M.: Breaking symmetric cryptosystems using quantum period finding. In: Robshaw, M., Katz, J. (eds.) CRYPTO 2016. LNCS, vol. 9815, pp. 207–237. Springer, Heidelberg (2016). https://doi.org/10.1007/978-3-662-53008-5_8
14. Kelsey, J., Chang, S.J., Perlner, R.: SHA-3 derived functions: cSHAKE, KMAC, TupleHash, and ParallelHash. NIST Special Publication, **800**, 185 (2016)
15. Kuperberg, G.: A subexponential-time quantum algorithm for the dihedral hidden subgroup problem. SIAM J. Comput. **35**(1), 170–188 (2005). https://doi.org/10.1137/S0097539703436345
16. Kuwakado, H., Morii, M.: Quantum distinguisher between the 3-round Feistel cipher and the random permutation. In: IEEE International Symposium on Information Theory, ISIT 2010, pp. 2682–2685. IEEE (2010). https://doi.org/10.1109/ISIT.2010.5513654
17. Kuwakado, H., Morii, M.: Security on the quantum-type Even-Mansour cipher. In: Proceedings of the International Symposium on Information Theory and its Applications, ISITA 2012, pp. 312–316. IEEE (2012). https://ieeexplore.ieee.org/document/6400943/
18. Mouha, N., Luykx, A.: Multi-key security: the Even-Mansour construction revisited. In: Gennaro, R., Robshaw, M. (eds.) CRYPTO 2015. LNCS, vol. 9215, pp. 209–223. Springer, Heidelberg (2015). https://doi.org/10.1007/978-3-662-47989-6_10
19. Nielsen, M.A., Chuang, I.L.: Quantum Computation and Quantum Information (10th Anniversary Edition). Cambridge University Press (2016)
20. Shor, P.W.: Algorithms for quantum computation: discrete logarithms and factoring. In: 35th Annual Symposium on Foundations of Computer Science, pp. 124–134. IEEE Computer Society (1994). https://doi.org/10.1109/SFCS.1994.365700
21. Simon, D.R.: On the power of quantum computation. In: 35th Annual Symposium on Foundations of Computer Science, pp. 116–123. IEEE Computer Society (1994). https://doi.org/10.1109/SFCS.1994.365701

Known-Key Attack on GIFT-64 and GIFT-64[g_0^c] Based on Correlation Matrices

Xiaomeng Sun[1], Wenying Zhang[1(✉)], René Rodríguez[2], and Huimin Liu[1]

[1] School of Information Science and Engineering, Shandong Normal University, Jinan, China
zhangwenying@sdnu.edu.cn
[2] Famnit & IAM, University of Primorska, Koper, Slovenia

Abstract. Block ciphers are often used as building blocks for one-way compression functions, which in turn, can be employed to construct hash functions. Two well-known important methods in the design of one-way compression function from block ciphers are the Davies-Meyer compression and the Myagushi-Preneel compression. To verify the security of such a construction, it is necessary to evaluate the robustness of the underlying block cipher against, e.g., the secret key, which is the so-called known-key model. In this paper, we evaluate the security of the lightweight block cipher GIFT-64 in the known-key setting, when used as a building block of hash functions. We significantly improve the known-key distinguisher to full GIFT-64. The distinguisher is composed of truncated differentials over 13 rounds and a meet-in-the-middle distinguisher over 15 rounds. We leverage a relationship between truncated differentials and multiple linear approximations cryptanalysis. It allows us to transfer searching for truncated differentials to constructing multiple linear approximations, resulting in the improved probability of truncated differentials.

To collect the related linear approximations as many as possible, we use a relatively low-dimensional binary correlation matrix where the hamming weight of the linear mask for the internal state can reach 8. Employing this correlation matrix, we precisely calculate all linear approximations that satisfy the specific conditions. To obtain a full-round distinguisher, these approximations are pre-filtered by a meet-in-the-middle distinguisher via a new matching method called *rotational recombination.* We would like to highlight that our distinguisher can apply successfully to the full-round GIFT-64 with a time complexity of 2^{60}. Furthermore, we implement the attack successfully on a variant of GIFT-64, GIFT-64[g_0^c], proposed at Eurocrypt 2022.

Keywords: Correlation Matrices · Known-Key Distinguisher · Multiple Linear Cryptanalysis · Meet-in-the-Middle Attack · GIFT-64

1 Introduction

The emergence of lightweight cryptography has led to new avenues of research. Particularly, the development of block ciphers are efficient in both software and

T. Zhu and Y. Li (Eds.): ACISP 2024, LNCS 14895, pp. 20–40, 2024.
https://doi.org/10.1007/978-981-97-5025-2_2

hardware. Among these, PRESENT block cipher [6] has certainly gained prominence. However, Banik et al. [3] have revisited its design approach and introduced GIFT, which outperforms PRESENT in terms of efficiency. The comprehensive design of the nonlinear layer and linear layer of GIFT have made it one of the most energy-efficient ciphers available. Its exceptional performance in both software and hardware has led to its adoption as a fundamental cryptographic primitive for various lightweight authenticated encryption with associated data schemes, such as GIFT-COFB [2], and GIFT-64-like [17]. Remarkably, GIFT-COFB has been selected as one of the top ten finalists in the NIST Lightweight Cryptography standardization project[1].

Hash functions (one-way compression functions, in general) are often built from block ciphers. The Davies-Meyer (DM) construction and the Matyas-Meyer-Oseas (MMO) construction are two the most important modes to build hash function from block cipher [15]. In these hashing modes, the message or the initial values play the role of the key. For example, Bogdanov et al. constructed hash functions using PRESENT [7]. Over the past years, the evaluation of the security of block ciphers in hashing modes has been an important topic in cryptanalysis [8].

A typical assumption to analyse the security of these modes of operation is that the attacker knows the initial values or the messages since the security of the protocol relies on a flawless block cipher. Hence, giving the attacker secret information should not substantially leak more information of the hash values than not knowing these values. When the key is known to the attacker, this paradigm is the so-called Known-Key (KK) scenario or model. Thus, a KK attack can be any type of attack that aims to effectively capture a specific property under the assumption that the attacker can obtain any random secret key. While, the same property can not hold for a random block cipher. KK attack is first used to (essentially) distinguish 7-round AES from a random permutation based on integral cryptanalysis [13]. This attack takes about 2^{56} encryptions, which is a notable complexity compared to other attacks. At Asiacrypt 2014, Gilbert et al. proposed a more detailed description of the KK model and extended the integral-based KK attack to 10-round AES-128 with a complexity of 2^{64} encryption operations [10]. We note that the KK model is assumed in preimage attacks on block-cipher-based hash functions [15]. Furthermore, the hash function derived from the lightweight block cipher using the MMO or DM, is applied to our KK model.

At Eurocrypt 2022, Sun et al. proposed a variant of GIFT-64, GIFT-64[g_0^c], which provide a better protection against differential cryptanalysis and low-energy consumption, even when compared to GIFT-64 [17]. The design of new variants is a priority when it comes to the security evaluation of block ciphers, since variants usually meet better resistance against known attacks or they may be more efficient while keeping the same level of security as the original version.

Our Contributions. We present a novel method to compute multiple linear (ML) approximations of GIFT-64, based on a correlation

[1] https://csrc.nist.gov/projects/lightweight-cryptography.

matrix. The proposed matrix is a 2080×2080 binary matrix designed to efficiently estimate a large number of linear approximations for which the hamming weight of masks can reach 8. Comparing to our matrix, the 2080×2080 correlation matrix used in linear approximations of PRESENT could only make use of masks with hamming weights 1 or 2 [9]. Furthermore, the proposed matrix also solves the problem of exponential growth of the size of the matrix with respect to the weight of the involved mask, a problem encountered previously. The source codes can be found at https://anonymous.4open.science/r/The-improved-matrix-with-2080x2080-5773/README.md.

For the First Time, We Significantly Apply a KK Attack to the Full GIFT-64 by Using a Hybrid Approach. In the Truncated Differential (TD) attack, we leverage a relationship between TD and ML cryptanalysis, which allows us to transfer searching for TD to constructing ML approximations, resulting in the improved probability of TD. Then, we pre-add a Meet-in-the-Middle (MitM) attack to filter out the inputs for the TD layer, covering 15 rounds. The matching for both forward and backward directions in the MitM attack is achieved using a new regrouping method termed *rotational recombination.*

Our Distinguisher on the Full GIFT-64 with Success Probability is 51.79% and is Equal to 99.59% When Considering a Version Reduced to 27 Rounds in 2^{60} Encryptions. Note that the distinguisher on full PRESENT in [5] is considered as successful with success probability 50.5%. To further support the suitability of our distinguisher, we implement the distinguishing attack on the lighter and more efficient variant GIFT-64$[g_0^c]$, where the success probability of full rounds increases to 56.80%. We compare our results with the previous analysis in Table 1, where SK, RK, and KK stand for Single-Key, Related-Key, and Known-Key, respectively.

Organization. In Sect. 2, we recall the MitM preimage attack against reduced AES used in a hashing mode. Then, we outline the general idea of our attack on GIFT-64. In Sect. 3, the approach to improving the TD distinguisher using correlation matrices is given. Details about the new regrouping technique worked on the MitM layer are provided in Sect. 4. Estimates on time complexity and success probability of our distinguisher on GIFT-64 are explained in Sect. 5. In Sect. 6, we present the distinguishing attack on the GIFT variant GIFT-64$[g_0^c]$. Finally, we make some conclusions in Sect. 7.

Notations. The following notations will be used throughout the paper:

- δ: the input difference of the MitM layer;
- Δ: the output difference of the MitM layer;
- Γ: the output difference of the TD layer;
- X_i/Y_i: the input state of the i-th round/the output state of the i-th round;
- $\Omega^{(r)}(\alpha \rightsquigarrow \beta)$: the set of all r-round linear approximations with the input mask α and output mask β;
- $L^{(r)}(\alpha \rightarrow \beta)$: one of r-round linear approximation contained in $\Omega^{(r)}(\alpha \rightsquigarrow \beta)$;
- $c(\alpha \rightarrow \beta)$: the correlation of $L^{(1)}(\alpha \rightarrow \beta)$;
- hw_α: the Hamming weight of α;

Table 1. Summary of known attacks on GIFT-64.

# Rounds	Attack	Setting	Time	Data	Memory	Ref.
19	Linear	SK	$2^{127.11}$	$2^{62.96}$	$2^{60.00}$	[19]
20	Differential	SK	$2^{125.50}$	$2^{62.58}$	$2^{62.58}$	[19]
15	MitM	SK	$2^{120.00}$	$2^{64.00}$	-	[3]
23	Boomerang	RK	$2^{126.60}$	$2^{63.30}$	-	[14]
24	Rectangle	RK	$2^{106.00}$	$2^{63.78}$	$2^{64.10}$	[12]
25	Differential	RK	$2^{123.23}$	$2^{60.96}$	$2^{102.86}$	[18]
28	**Truncated Differential**	**KK**	-	$2^{60.00}$	-	**[this paper]**

2 Preliminaries

2.1 Meet-in-the-Middle Attack on Hashing Modes of Reduced AES

Sasaki et al. proposed meet-in-the-middle attack on hashing modes, the hash function built using the Davies-Meyer construction can be expressed as

$$F(x, k) = E_k(x) \oplus x,$$

where $E_k(x) : \{0,1\}^n \rightarrow \{0,1\}^n$ is the permutation obtained from the encryption function of an n-bit block cipher using the key k. For any integer L with $L \gg n$, a (iterated) hash function $H : \{0,1\}^L \rightarrow \{0,1\}^n$ can be recursively defined using the DM structure as follows. The input $M \in \{0,1\}^L$ to be hashed is split in $s = \frac{L}{n}$ blocks, say, $M = (M_0, \ldots, M_{s-1})$ (padding may be needed). Then set an initial value for H_0, define

$$H_i = F(H_{i-1}, M_{i-1}) = E_{M_{i-1}}(H_{i-1}) \oplus H_{i-1}$$

and $H(M) = H_s$. This recursive definition is referred as the Merkle-Damgård construction.

In a prototypical MitM attack on hash function, the hash function is divided into two parts, such that a subset of input bits of the messages (state #0 computed by state #9) only affects the forward values (encryption) while another subset of input bits of the messages (state #9) affect the backward values (decryption). Based on this approach, Sasaki applied a MitM preimage attack to reduced AES [15].

The general idea of Sasaki's attack is to find the chaining value H_{i-1} while fixing the key value M_{i-1} to a randomly chosen constant (at a given stage i), until the given target hash value H_i is matched in state #5. The details of this attack are depicted in Fig. 1, where the state #16 represents the ciphertext $E_{M_{i-1}}(H_{i-1})$ encrypted by state #9, and the state #0 represents the variable value H_{i-1} XORed by state #16 and the target value. When the values of state #5 encrypted by #0 are matched with state #5 decrypted by #9, the attack is successful.

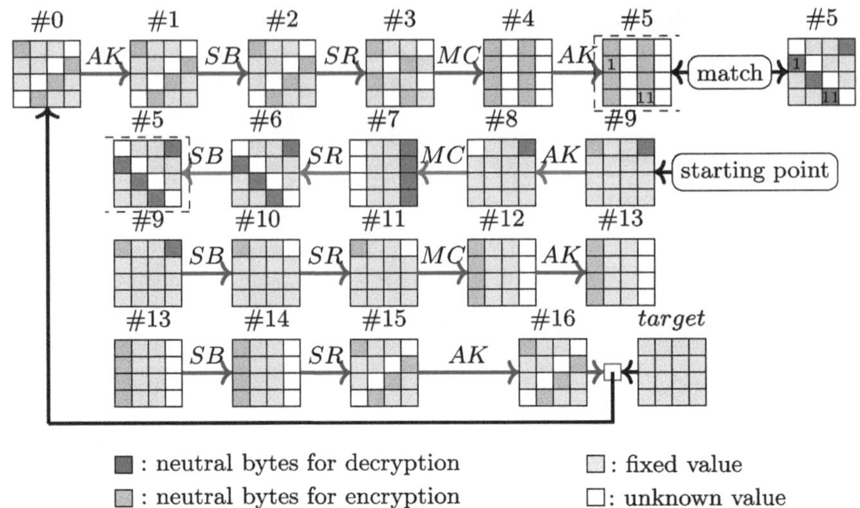

Fig. 1. Basic MitM attack on reduced 4-round AES in hashing mode. (Color figure online)

Specially, in Fig. 1, the attacker starts at state #9. The byte #9[0] is active for encryption, whereas the byte #9[12] is active for decryption only. These bytes are called *neutral* bytes. Other bytes in state #9 are (randomly) fixed for encryption and decryption. The attack goes as follows: encryption covers states #9 to #16 and #0 to #5 (marked in blue) and decryption works backward covering states #9 to #5 (marked in red). The 8 known bytes (blue bytes of #5) are obtained after 3-round encryption and 12 known bytes (4 red bytes and 8 grey bytes of #5) are known after 1-round decryption. The forward and backward procedure meet at state #5, where the two bytes #5[1, 11] are overlapped between the results from both directions. This allows the attacker to efficiently check this match to mount the attack.

Inspired by the described attack on hash function, we propose a new MitM attack on the hashing modes of GIFT-64.

2.2 The General Framework of Our Attack

The proposed attack utilizes a well-established structure that has previously been used to successfully attack other block ciphers [5,11]. Specifically, we leverage a strong relation between the TD ($\Delta \rightarrow \Gamma$) and the ML ($\Omega^{(r)}(\alpha \rightsquigarrow \beta)$) over 13 rounds, as demonstrated in Sect. 3.2. To extend the number of rounds, we pre-add a MitM layer ($\delta \rightarrow \Delta$) into the TD distinguisher. The MitM layer is designed to meet the requirement that the output difference of the MitM layer equals the input difference of the TD distinguisher, where these differences are predetermined. Building upon the approaches outlined in [11,15], we divide the MitM layer into two parts, each with two distinct start points. The first start

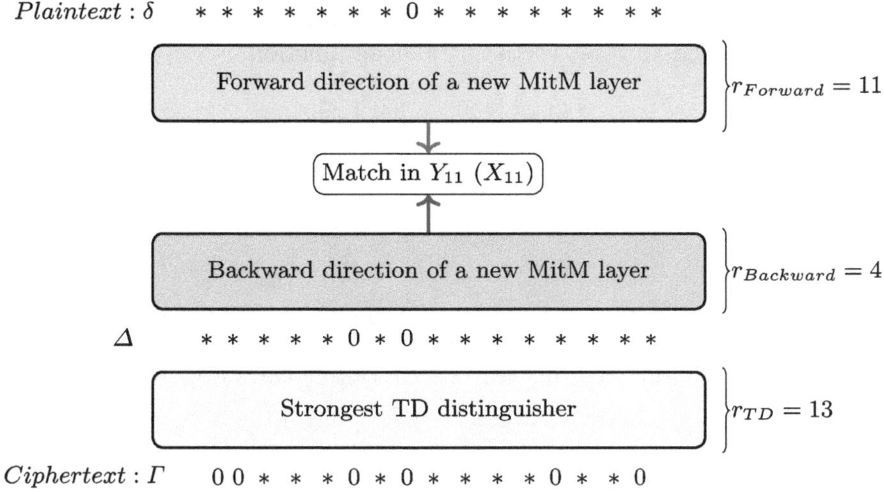

Fig. 2. The construction of full round distinguisher.

point is the plaintext with the 8-th inactive S-box, which serves as the input for the first round of the 11-round encryption. The second start point is the internal state of encrypting plaintext over 15-round, which is also the input for the last 4-round decryption. The state has the 8-th and 10-th inactive S-boxes. These two parts are matched in the middle state (X_{11} and Y_{11}) in opposite directions. It significantly enhances the efficiency of the attack (Fig. 2).

3 Transfer Searching for Truncated Differentials to Constructing Multiple Linear Trails

In this section, the TD part of our attack is shown in detail. Since the probability of the TD distinguisher depends on the capacity of ML approximations (in Sect. 3.2), we aim to maximize this capacity by constructing a suitable 2080×2080 correlation matrix (shown in Sect. 3.3). In Sect. 3.4, we will introduce how to use the correct rows and columns weighted no more than 2 to count the number of approximations accurately within input and output mask weighted no more than 8.

3.1 Evaluation of Truncated Differentials

Blondeau et al. [5] have noted that differential properties are much easier to handle than linear properties in a KK setting. As a result, recent research in [4] has focused on establishing connections between ML and TD cryptanalysis. For instance, by transforming TD into ML approximations.

Given two n-bit vectors $x = (x_0, \ldots, x_{n-1})$ and $y = (y_0, \ldots, y_{n-1})$, the inner product of x and y is defined as $x \cdot y = x_0y_0 \oplus \cdots \oplus x_{n-1}y_{n-1}$. Let r, s, t, q, n be positive integers such that $n = s + t = q + r$. In the following, we use the same notions as [4]. Given an n-bit vectorial Boolean function

$$G : \mathbb{F}_2^s \times \mathbb{F}_2^t \to \mathbb{F}_2^q \times \mathbb{F}_2^r.$$

We denote by (x_s, x_t) and (y_q, y_r) its input and output respectively. A truncated differential $(\Delta \rightsquigarrow \Gamma) \in \mathbb{F}_2^n \times \mathbb{F}_2^n$ is a set of differentials given by

$$(\Delta \rightsquigarrow \Gamma) = \{(0, \Delta_t) \rightsquigarrow (0, \Gamma_r), \Delta_t \in \mathbb{F}_2^t, \Gamma_r \in \mathbb{F}_2^r\}.$$

It is composed of 2^t input differences $(0, \Delta_t)$ and 2^r output differences $(0, \Gamma_r)$, having a total of 2^{t+r} truncated differentials.
The probability of truncated differentials Pr can be calculated as follows:

$$Pr = 2^{-t} \sum_{\Delta_t \in \mathbb{F}_2^t, \Gamma_r \in \mathbb{F}_2^r} P\left[(0, \Delta_t) \overset{G}{\rightsquigarrow} (0, \Gamma_r)\right]. \tag{1}$$

3.2 The Relationship Between Truncated Differentials and Multiple Linear Cryptanalysis

Multiple Linear (ML) approximations $\Omega^{(r)}(\alpha \rightsquigarrow \beta) \in \mathbb{F}_2^n \times \mathbb{F}_2^n$ are defined as

$$\Omega^{(r)}(\alpha \rightsquigarrow \beta) = \{(a_s, 0) \rightsquigarrow (b_q, 0), a_s \in \mathbb{F}_2^s, b_q \in \mathbb{F}_2^q\},$$

so that it is composed of 2^s input masks $(a_s, 0)$ and 2^q output masks $(a_q, 0)$, where s (respectively q) denotes the number of active bits of input mask (output mask). In other words, $\Omega^{(r)}(\alpha \rightsquigarrow \beta)$ includes 2^{s+q} linear approximations over $\mathbb{F}_2^n$.

Given a Boolean function f on $\mathbb{F}_2^n$, its correlation $\mathrm{cor}(f(x))$ is defined as

$$2^{-n}(\#\{x \in \mathbb{F}_2^n : f(x) = 0\} - \#\{x \in \mathbb{F}_2^n : f(x) = 1\}).$$

The capacity C of ML approximations are defined as the quadratic sum of the expected correlations of all the linear approximations included in the n-dimensional approximations [4]. It is also determined by the number of linear approximations contained in the input mask $(a_s, 0)$ and output mask $(b_q, 0)$. Moreover, we get

$$C = \sum_{(a_s, b_q) \neq (0,0)} \mathrm{cor}^2\left(a_s \cdot x_s \oplus b_q \cdot y_q\right). \tag{2}$$

It can be shown that the probability of truncated differentials $Pr(TD)$ can be written as follows,

Lemma 1. *([4]) For all $\delta_s \in \mathbb{F}_2^s$ and $\Delta_q \in \mathbb{F}_2^q$,*

$$Pr(TD) = 2^{-q} \sum_{(a_s, b_q)} (-1)^{a_s \cdot \delta_s \oplus b_q \cdot \Delta_q} \mathrm{cor}^2\left((a_s, 0) \cdot x \oplus (b_q, 0) \cdot G(x)\right). \tag{3}$$

When $\delta_s = 0$ and $\Delta_q = 0$ (i.e. the position where the inactive S-boxes in difference equals the position of the active S-boxes in the mask), combining Eq. (1) and Eq. (2), we get

$$Pr(TD) = 2^{-q}(C + 1). \tag{4}$$

The powerful mathematical relation between the probability of the TD and ML capacity has been exploited to derive strong distinguishers. In other words, $Pr(TD)$ is no longer determined by the input difference δ_s and output difference Δ_q, but determined by the proximity of output mask β to input mask α. Thus, the focus is shifted from finding high probability differentials to finding n-dimension linear approximations with strong correlation, i.e., to find as many linear approximations as possible. Throughout the strong relation between $Pr(TD)$ and C, we give the definition of the bias of a TD (shown in Theorem 1), the degree of proximity to the theoretical probability of TD.

We use the SAT solver to get approximations with correlations larger than 2^{-50} based on fixed input mask 0x0000 0x0c0c 0x0000 0x0000 and output mask 0x4000 0x2000 0x0000 0x0080. The candidate linear approximation of GIFT-64 is provided by Sun et al. [19]. There are 61729 approximations with fixed masks improving the capacity C reach to $2^{-59.839}$.

In the following, we propose Theorem 1 based on the above candidate linear approximation, which shows the bias of TD.

Theorem 1. *Following the above notation, we have* $q = 3$ *and* $C = 2^{-59.839}$. *Then,* $Pr(TD)$ *is* $2^{-3}(2^{-59.839} + 1)$. *Thereby, its bias is* $2^{-(3+59.839)}$.

Proof. Let the i-th (respectively j-th) S-box in the input (respectively output) mask be denoted by I_i (respectively J_j). These ML approximations whose output mask have three active bits at S-boxes J_1, J_{11} and J_{15}, i.e. 7-th, $(4 \times 11 + 2)$-th, and $(4 \times 15 + 2)$-th active bits. Moreover, once the output mask is fixed, 2^{-q} becomes a constant value. Similarly, if the input mask is fixed further, the value of C is completely determined by the number of linear approximations. Therefore $Pr(TD)$ can be computed by Eq. (4). Consequently, we can easily get the degree of proximity to theoretical TD probability $2^{-q}C$, which is the bias $2^{-(3+59.839)}$.

In view of Eq. (4), the larger C and the smaller q, the higher probability of the TD. However, whether it is the value $C = 2^{-61.607}$ obtained by Sun et al. or the further improved value $C = 2^{-59.839}$ obtained by us, they are actually too small to mount an effective KK distinguisher, since the success probability of the distinguisher is less than 50% (shown in Sect. 5 for this computation).

3.3 Correlation Matrix of Multiple Linear Approximations

In order to enhance the capacity of ML approximations, it is essential to incorporate a well-designed correlation matrix to capture linear approximations as effectively as possible.

We now describe the idea to construct the 2080×2080 binary correlation matrix M to search for 1-round linear approximations, where 2080 is the number of 64-dimensional binary vectors with Hamming weight 1 or 2, i.e., $\binom{64}{1} + \binom{64}{2} = 2080$. The general form of the matrix M is given as follows.

$$\begin{array}{c} \\ \{0\} \\ \{1\} \\ \vdots \\ \{63\} \\ \{0,1\} \\ \vdots \\ \{0,63\} \\ \vdots \\ \{62,63\} \end{array} \begin{array}{c} \begin{array}{cccccccccc} \{0\} & \{1\} & \cdots & \{63\} & \{0,1\} & \cdots & \{0,63\} & \cdots & \{62,63\} \end{array} \\ \left(\begin{array}{ccccccccc} * & * & \cdots & * & * & \cdots & * & \cdots & * \\ * & * & \cdots & * & * & \cdots & * & \cdots & * \\ \vdots & \vdots & & \vdots & \vdots & & \vdots & & \vdots \\ * & * & \cdots & * & * & \cdots & * & \cdots & * \\ * & * & \cdots & * & * & \cdots & * & \cdots & * \\ \vdots & \vdots & & \vdots & \vdots & & \vdots & & \vdots \\ * & * & \cdots & * & * & \cdots & * & \cdots & * \\ \vdots & \vdots & & \vdots & \vdots & & \vdots & & \vdots \\ * & * & \cdots & * & * & \cdots & * & \cdots & * \end{array} \right) \end{array}.$$

We mark the entry * at the intersection of a row and a column, which corresponds to input mask α and output mask β of 1-round linear approximation $L^{(1)}(\alpha \rightarrow \beta)$. The rows and columns of the matrix are both indexed by the set

$$I = \{\{i\}, \{i, j | i < j\} : i, j \in \{0, \ldots 63\}\}.$$

The indices $\{i\}$ of the first 64 rows or columns represent the i-th (single) active bit in the input and output mask. Then, the indices $\{i, j\}$ of the remaining rows (columns) represent both the i-th and the j-th bits are active in the input (output) mask.

We assume that the μ-th row (respectively the ν-th column) indexed by set I, ordered according to the integers

$$\mu \text{ (or } \nu) = |i - j| + \sum_{a=0}^{\min(\{i,j\})} (64 - a). \tag{5}$$

M is obtained by multiplying two 2080×2080 matrices. Namely, $M = M_P \cdot M_S$. The first matrix is used to describe the propagation of the linear bias through S-boxes, denoted as M_S. The second matrix is used to describe the propagation of the linear bias of the P-layer, denoted as M_P. To construct M_S, we refer to the linear approximation table (LAT). The Lemma 2 is concluded by LAT. Let $\alpha \xrightarrow{S} \beta$ denote a linear approximation of the GIFT S-box with input mask α and output mask β and the correlation of $\alpha \xrightarrow{S} \beta$ is denoted by $\rho(\alpha \rightarrow \beta)$.

Lemma 2. *The GIFT S-box has the following properties:*

S1. *For* $hw_\alpha = hw_\beta = 1$, *i.e.*, $\alpha, \beta \in \{1, 2, 4, 8\} : \rho(\alpha \to \beta) = 0$ *except that* $\rho(1 \to 8) = \pm 2^{-1}$ *and* $\rho(1 \to 4) = \rho(2 \to 8) = \pm 2^{-2}$.

S2. *For* $hw_\alpha = hw_\beta = 2$, *i.e.*, $\alpha, \beta \in \{3, 5, 6, 9, a, c\} : \rho(\alpha \to \beta) = \pm 2^{-1}$ *or* $\pm 2^{-2}$.

S3. *For* $hw_\alpha = 1$, $hw_\beta = 2$, *i.e.*, $\alpha \in \{1, 2, 4, 8\}, \beta \in \{3, 5, 6, 9, a, c\} : \rho(\alpha \to \beta) = \pm 2^{-1}$ *or* $\pm 2^{-2}$.

S4. *For* $hw_\alpha = 2$, $hw_\beta = 1$, *i.e.*, $\alpha \in \{3, 5, 6, 9, a, c\}, \beta \in \{1, 2, 4, 8\} : \rho(\alpha \to \beta) = \pm 2^{-1}$ *or* $\pm 2^{-2}$.

The entry $m^S_{\mu,\nu}$ in M_S is equal to 1 if and only if $\alpha \xrightarrow{S} \beta$ has a nonzero bias. The support of input mask α denoted by *supp*(α), equals to the μ-th row, which is computed by Eq. (5). A similar approach is used in computing the support of output mask β denoted by *supp*(β). More precisely,

$$m^S_{\mu,\nu} = \begin{cases} 1, & \text{if } \rho\left(\alpha \to \beta\right) \neq 0 \text{ and } supp(\alpha) = \mu, supp(\beta) = \nu \\ 0, & \text{otherwise} \end{cases}.$$

Similarly, the entry $m^P_{\mu,\nu}$ in M_P is equal to 1 when the μ-th row indexed by active bits' set $\{i_1, i_2\}$ is mapped to the ν-th column indexed by active bits' set $\{j_1, j_2\}$ under the P-layer, namely:

$$m^P_{\mu,\nu} = \begin{cases} 1, & \text{if } \{P(i_1), P(i_2)\} = \{j_1, j_2\} \text{ and } i_1, i_2, j_1, j_2 \in I \\ 0, & \text{otherwise} \end{cases}.$$

Eventually, the value of the entry $m^M_{\mu,\nu}$ is equal to 1, when the 1-round approximation propagates from the input mask α of the S-box to the output mask β of the P-layer. We have the expression of $m^M_{\mu,\nu}$,

$$m^{M^r}_{\mu,\nu} = \begin{cases} 1, & \text{if } c\left(\alpha \to \beta\right) \neq 0, supp(\alpha) = \mu, supp(\beta) = \nu \\ 0, & \text{otherwise} \end{cases}.$$

In simple terms, if the correlation of a 1-round approximation is not equal to 0, the corresponding entries are set to 1. For example, we consider a 1-round approximation $L^{(1)}$(0x00000x00000x00000x0a00 $\to$ 0x00020x00000x00000x0000), where the input mask is indexed by $\{9, 11\}$, corresponding to $\mu = 597$-th row. Similarly, the $\nu = 50$-th column corresponds to the output mask indexed by $\{50\}$. It is observed that $c(\alpha \to \beta) \neq 0$, when $m^M_{597,50}$ is set to 1 in M.

Further, we propose two observations on the M.

Observation 1. *Essentially, the key of M searching accurately for the number of linear approximations over 1-round, is to ensure the entries set to 1 at correct rows and columns matched with corresponding input mask and output mask.*

Observation 2. *When the input mask has two active S-boxes following S_2 of Lemma 2, the columns of entries set to 1 on M_S are indexed by active bits' sets* *determined by output mask. After P-layer, each two active bits from two active S-boxes (active bits' sets) recombine into an active S-box (combination of active bits' set). This combination of active bits' set indexes correct output mask and correct position where entries are set to 1 (see case 3 in Sect. 3.4).*

Based on the two observations, we draw a conclusion about the matrix M. When active bits recombine after P-layer, the column of entries set to 1 on M should be indexed by the combination active bits' set matched to the output mask of P-layer. Only in this way can the number of approximations of a ML with fixed input mask and fixed output mask be computed accurately by M.

However, the matrix proposed in [9] can not stress Observation 2. Thus the corresponding column of entries set to 1 on M are not correct. It is the reason why the matrix only make use of three different sets of approximations with masks weight 1 or 2 of 1-round. In contrast, our improved matrix (proposed in Sect. 3.4), with smaller size, motivated by above observations addresses above problem and shows the linear bias of masks for four active S-boxes with hamming weight reach 8.

3.4 Strengthen the Capacity by Combining the Improved Matrix and SAT Solver

We use the optimal 13-round linear approximations with correlation 2^{-34} obtained by [19] to optimize the improved bias of our attack. Then, we filter these approximations satisfying following conditions. Out of these, we have 47 approximations remaining. We choose one of them with input mask 0x0000 0x0c0c 0x0000 0x0000 and output mask 0x2200 0x0104 0x0000 0x4001 as the candidate.

Condition 1. *There are at most two active S-boxes in the input of an oracle.*
Condition 2. *There are at most four active S-boxes in the 5-th to 7-th round.*
Condition 3. *There are at most two active bits in each active S-box.*

Some Improvements on M. We now illustrate how entries in M are reset to 1 by using a new rule. We break down the candidate into three cases. These differ in the number of active S-boxes for the input mask and output mask in a 1-round internal approximation by using different methods setting entries to 1.

Case 1. *The input and output mask of the internal state have 4 active S-boxes and each of them has 2 active bits.*

Example 1. Table 2 shows a 1-round linear approximation with input mask 0x0000 0x0505 0x0000 0x0505 of ΓX^5 and output mask 0x0a0a 0x0000 0x0a0a 0x0000 of ΓX^6. Both masks have four active S-boxes, with Hamming weight 8.

In Table 2, the third column of the table shows the indices of rows and columns—active bits' sets. The fourth column of the table shows the row and column corresponding to indices. We use four colors (red, blue, orange, and black) to denote the propagation of four active S-boxes from ΓX^5 to ΓX^6. The active bits in the same S-box of the input mask and corresponding rows and columns

Table 2. The propagation from ΓX^5 to ΓX^6 in the improved matrix M.

State	Linear Mask	Active Bits' Set	Row or Column	Entries Set to 1 at (row, column)	Value
ΓX^5	0x0000 0x0505 0x0000 0x0505	{0, 2}, {24, 26}, {48, 50}, {56, 58}	row: 65, 1301, 1961, 2053		
output of S-box (input of P-layer)	0x0000 0x0a0a 0x0000 0x0a0a	{1, 3}, {9, 11}, {33, 35}, {41, 43}	column of M_S: 128, 596, 1616, 1828 (row of M_P: 128, 596, 1616, 1828)		
ΓX^6	0x0a0a 0x0000 0x0a0a 0x0000	<17, 51>, <19, 49>, <25, 59>, < 27,57 > → {17, 19}, {25, 27}, {49, 51},{57, 59}	column: 1000, 1340, 1961, 2060	(65, 1000), (65, 1961), (1301, 1000), (1301, 1961) (1961, 1340), (1961, 2060), (2053, 1340), (2053, 2060)	1

1 (μ, ν): The μ-th row and ν-th column.
2 $\{i, j\}$: A subset of set I, where the i-th and j-th active bits are in the same S-box.
3 $< i, j >$: A set of two active bits, where the i-th and j-th active bits are in different S-box.
4 $\rightarrow$: The active bits' set $< i, j >$ transform into the active bits' set $\{i, j\}$.
5 1: Entries are reset to 1 at a new position (μ, ν).

are marked with the same color. The color remains the same after the S-box and P-layer, regardless of where the bits go to. The indices of rows and columns on the matrix are also marked with the same color as the corresponding indices of active bits. In this case, the linear propagation of an S-box satisfies S_2 of Lemma 2. After the P-layer, two active bits of an active S-box propagate to two different S-boxes through the P-layer. In ΓX^5, the four active bits' sets $\{0,2\}$, $\{24,26\}$, $\{48,50\}$, $\{56,58\}$ are propagate to $\{1,3\}$, $\{25,27\}$, $\{49,51\}$, $\{57,59\}$ one by one. When going through the P-layer, the active bits' sets are propagated into $< 17,51 >, < 19,49 >, < 25,59 >, < 27,57 >$ respectively. However, the mask of ΓX^6 does not indexed by these sets. Thus, our improved matrix transforms these sets into new recombination sets $\{17,19\}$, $\{25,27\}$, $\{49,51\}$, $\{57,59\}$. It indicates that the first element of set $\{0,2\}$ go to the first element of set $\{17,19\}$ and the second element of set $\{0,2\}$ go to the second element of $\{49,51\}$ after P-layer. The set $\{0,2\}$ corresponds to the row number 65, and the two sets of ΓX^6 correspond to column number 1000 and 1976 computed by Eq. (5). Thereby, we set the entries $m^M_{65,1000}$ and $m^M_{65,1976}$ to 1. Meanwhile, the same rule of setting entries to 1 is also applied to remaining masks of the approximation.

Case 2. *The input mask of the internal state has 2 active S-boxes and each of them has 2 active bits. While the output mask of the internal state has 4 active S-boxes and each of them has 1 active bit.*

Example 2. Table 3 shows a 1-round linear approximation with input mask 0x0000 0x0050 0x0000 0x0050 and output mask 0x0000 0x0808 0x0000 0x0202. The input mask has two active S-boxes and output mask has four active S-boxes, with Hamming weight 4. In this case, the linear propagation of an S-box satisfies S_2 of Lemma 2. However, it is ruled out the recombination of active bits after the P-layer. Thus, it uses the generic method to set the entries to 1. For example, the entries $m^M_{311,1}$ and $m^M_{311,35}$ are set to 1, indicating the propagation from mask 0xa indexed by active bits' set {4, 6} of ΓX^3 to two masks 0x8 indexed by active bits 1 and 35 of ΓX^4 respectively.

Case 3. *The input mask of the internal state has 2 active S-boxes and each of them has 2 active bits. While the output mask of the internal state has 1 active S-boxes and each of them has 1 active bit.*

Table 3. The propagation from ΓX^3 to ΓX^4 on the improved M.

State	Linear Mask	Active Bits' Sets	Row or Column	Entries Set to 1 at (row, column)	Value
ΓX^3	0x0000 0x0050 0x0000 0x0050	{4, 6}, {36, 38}	row: 311 column: 1730		
output of S-box (input of P-layer)	0x0000 0x00a0 0x0000 0x00a0	{5, 7}, {37, 39}	column of M_S: 370, 1730 row of M_P: 370, 1730		
ΓX^4	0x0000 0x0808 0x0000 0x0202	1, 9, 35, 43	column: 1, 9, 35, 43	(311, 1), (311, 35), (1703, 9), (1703, 43)	1

Table 4 shows a 1-round linear approximation with the input mask 0x0000 0x0000 0xa0a0 0x0000 and the output mask 0x0000 0x0000 0x0000 0x00a0, where the input mask has two active S-boxes and output mask has one active S-box. In this case, the propagation of S-box satisfies S_4. After the P-layer, the two active bits go to S_1 of ΓX^8, forming a new active bits' set. Note that only when both values of $m^M_{1178,370}$ and $m^M_{1486,370}$ are both set to 1, it will be check that $\pi(\Gamma X^7 \to \Gamma X^8) \neq 0$.

Table 4. The propagation from ΓX^7 to ΓX^8 on the improved M.

State	Linear Mask	Active Bits' Sets	Row or Column	Entries Set to 1 at (row, column)	Value
ΓX^7 (input of S-box)	0x0000 0x0000 0xa0a0 0x0000	{21, 23}, {29, 31}	row: 1178, 1486		
output of S-box (input of P-layer)	0x0000 0x0000 0x8020 0x0000	21, 31	column of M_S: 21, 31 (row of M_P: 21, 31)		
ΓX^8 (output of P-layer)	0x0000 0x0000 0x0000 0x00a0	⟦5, 7⟧	column: 370	(1178, 370), (1486, 370)	1

Getting More Approximations by Using M^r. By using the powers of improved M, M^r, we can describe r-round ML approximations denoted by

$$L^{(\mathrm{r})}(\alpha \to \beta) = \underbrace{L^{(1)} \circ L^{(1)} \circ \ldots \circ L^{(1)}}_{r \text{ times}}.$$

The value of $m^{M^r}_{\mu,\nu}$ at the intersection of the μ-th row and the ν-th column, represents the number of linear approximations over r-round with fixed input mask and output mask. It is essential to choose the minimum value $m^{M^r}_{\mu,\nu}$ to ensure that all active bits' sets of input mask in the first round can reach the output mask in the final round. The details of how M^r evaluates the largest number of approximations within the candidate for 13-round GIFT-64 are shown in Table 5. Finally, we derive the improved TD probability in the following Theorem 2 based on the above linear approximations. Note that this bias is greater than the one obtained in Theorem 1.

Theorem 2. *The improved capacity of the proposed ML approximations is at least* $2^{-53.16}$. *Even though the number of output active bits is 6, the bias of the given TD over 13 rounds is at least* $2^{-(53.16+6)}$.

Table 5. Approximations over 13 rounds by using M^{13}.

State	Linear Mask	Active Bits' Sets	Row or Column	The number of approximations
ΓX^0	0x0000 0x0c0c 0x0000 0x0000	{34, 35} {42, 43}	1645, 1849	-
ΓX^1	0x0000 0x0100 0x0000 0x0100	8, 40	(1645, 8), (1849, 40)	7
ΓX^2	0x0000 0x0000 0x0808 0x0000	19, 27	(8, 19), (40, 27)	21
ΓX^3	0x0000 0x0050 0x0000 0x0050	$<4,38>$, $<6,36>$ $\rightarrow$ {4, 6}, {36, 38}	(19, 311), (19, 1703), (27, 311), (27, 1703)	8
ΓX^4	0x0000 0x0808 0x0000 0x0202	1, 9, 35, 43	(311, 1),(311, 35), (1703, 9), (1703, 43)	612
ΓX^5	0x0000 0x0505 0x0000 0x0505	$<0,34>$, $<2,32>$,$<8,42>$, $<10,40>$ $\rightarrow$ {0, 2}, {24, 26}, {48, 50}, {56, 58}	(1, 65), (1, 1301), (9, 65), (9, 1301) (35, 1961), (35, 2053), (43, 1961), (45, 2053)	420
ΓX^6	0x0a0a 0x0000 0x0a0a 0x0000	$<17,51>$, $<19,49>$, $<25,59>$, $<27,57>$ $\rightarrow$ {17, 19}, {25, 27}, {49, 51}, {57,59}	(65, 1000), (65, 1961), (1301, 1000), (1301, 1961) (1961, 1340), (1961, 2060), (2053, 1340), (2053, 2060)	3072
ΓX^7	0x0000 0x0000 0xa0a0 0x0000	{21, 23}, {29, 31}	(1000, 1178), (1340, 1178), (1961, 1486), (2060, 1486)	5290
ΓX^8	0x0000 0x0000 0x0000 0x00a0	{5, 7}	(1178, 370), (1486, 370)	11408
ΓX^9	0x0000 0x0000 0x0000 0x0002	1	(370, 1)	16812
ΓX^{10}	0x0008 0x0000 0x0000 0x0000	51	(1, 51)	23388
ΓX^{11}	0x0000 0x4000 0x0000 0x1000	12, 46	(51, 12), (51, 46)	179131
ΓX^{12}	0x0400 0x0200 0x0000 0x0008	3, 41, 58	(12, 3), (46, 41), (46, 58)	134874
ΓX^{13}	0x2200 0x0104 0x0000 0x4001	0, 14, 34, 40, 57, 61	(3, 0), (3, 34), (41, 40), (41, 57) (58, 14), (58, 61)	731976

4 The Meet-in-the-Middle Layer

In this layer, we generate N pairs of input plaintexts (PT) that satisfy the input difference δ of the first round. These PT are then encrypted forward for 10 rounds to state X_{11}. Then, N^{TD} pairs messages are obtained by encrypting PT that satisfy the output difference Δ of the last round covered by this layer. While these satisfied messages are then decrypted backward for 4 rounds to state Y_{11}. The comprehensive mechanism of the MitM approach is illustrated below:

$$PT \xrightarrow[10\ rounds]{Encryption} X_{11} \underset{11th-round}{\overset{matching}{\longleftrightarrow}} Y_{11} \xleftarrow[3\ rounds]{Decryption} N^{TD} messages.$$

The new matching method we proposed as following can achieve full GIFT-64.

4.1 A New Matching Method: Rotational Recombination

The basic idea of the proposed matching method originates from the fixslicing representation of the permutation P-layer presented in [1]. This fixslicing representation divides the state of the block cipher into 4 slices: for each nibble $b_3b_2b_1b_0$, the bit b_0 at position i is placed in slice i for $0 \leq i \leq 3$. Then, the fixslicing technique represents P-layer as the composition of four rotational operations on rows and columns per slice. This ensures that all bits within a slice remain in the same slice after the P-layer. Moreover, the process is fully synchronized after 4 consecutive rounds.

For our recombination method, we call the 4 bits of an S-box as a cell. The i-th cell is denoted as S_i. For any selected cell $S_i = b_3b_2b_1b_0$, the bit b_0 is placed in the group G_0 (colored by blue), bit b_1 in group G_1 (colored by purple), bit b_2 in group G_2 (colored by pink) and bit b_3 in group G_3 (colored by yellow). This recombination partition is called O_1, whose formal description is

$$\cup_{O_1} = (G_3, G_2, G_1, G_0),$$

where representing the i-th bit with b_i, $G_I = \{b_i | i \in \{0, \dots, 63\}, i \bmod 4 = I\}$ for $0 \leq I \leq 3$.

Additionally, a second partition of bits, called O_2, will be described as

$$\cup_{O_2} = (G'_3, G'_2, G'_1, G'_0),$$

where $G'_i = \{b_{16i}, \ldots, b_{16i+15}\}$ for $i \in \{0, 1, 2, 3\}$.

This procedure is illustrated in Fig. 3. For instance, when selecting cells S_2, S_6, S_{10}, S_{14} involved in G'_2 in the input state X_3, the first bit of each selected cells are recombined into G_0 in X_4. Then these bits go to 4 consecutive cells S_8, S_9, S_{10}, S_{11} in X_5 also belonged to G'_2. In general, the bits of G'_0 (respectively G'_1, G'_2, G'_3) are colored in pink (respectively yellow, blue, purple) for this selected cell. For our purposes, it will suffice to focus on the groups highlighted in blue only, while the other groups will be ignored. Note that the 16 bits of G'_2 in X_3 map to the same cells in X_5, X_7, X_9 and X_{11}. In other words, the selected bits of X_5 are obtained from selected bits of X_3 by following the two operations O_1 in X_3 and O_2 in X_4. The operations O_1 and O_2 cover two rounds as a rotational recombination operation. It can be rotated n times with $\lfloor n \leqslant \frac{r_0}{2} \rfloor$, where r_0 denotes the number of rounds for encryption. That is, the recombination method makes the selected bits from the initial round go to the same cells after two rounds.

4.2 Rotational Recombination Method for the MitM

In the subsection, we show how to apply these rotational operations to the MitM layer.

Encryption for the First Two Rounds. We fix the 4 input bits of the cell S_8 (colored in blue) in state X_0 to random values, which satisfy a given input difference δ. Then, the values of bits $X_1[8, 25, 42, 59]$, which denote the 8-th, 25-th, 42-th and 59-th bits of X_1 are obtained. Furthermore, we guess the other 12 bits of S_2, S_6, S_{10}, S_{14} of X_1. We use the operation O_1 to partition these 16 bits into four groups and only G_0 is selected. Continue encrypting the bits of G_0 to get bits $\{4i + 0 | 8 \leqslant i \leqslant 11\}$, where i denotes i-th S-box (or cell). These 16 bits are called "determined bits". To summarize, we get 2^{16} values of X_2 and each value has 16 determined bits.

Decryption for the Last Two Rounds. We fix 8 output bits of S_8 and S_{10} (colored in pink) in Y_{14} to random values. The bits $Y_{14}[34, 35, 42, 43]$ in the input of the 14-th round are selected for decryption. Guess the other 12 bits of S_0, S_1, S_8, S_9 to get 16 determined bits

$$Y_{12}[0, 2, 5, 7, 8, 10, 13, 15, 16, 18, 21, 23, 24, 26, 29, 31].$$

Similarly, as for decryption, we get a collection of 2^{16} values of Y_{12}, and each value has 16 determined bits.

Matching Stage Using Rotational Recombination. After the forward and backward calculations, we obtain some matching pairs for X_2 and Y_{12}. To encrypt

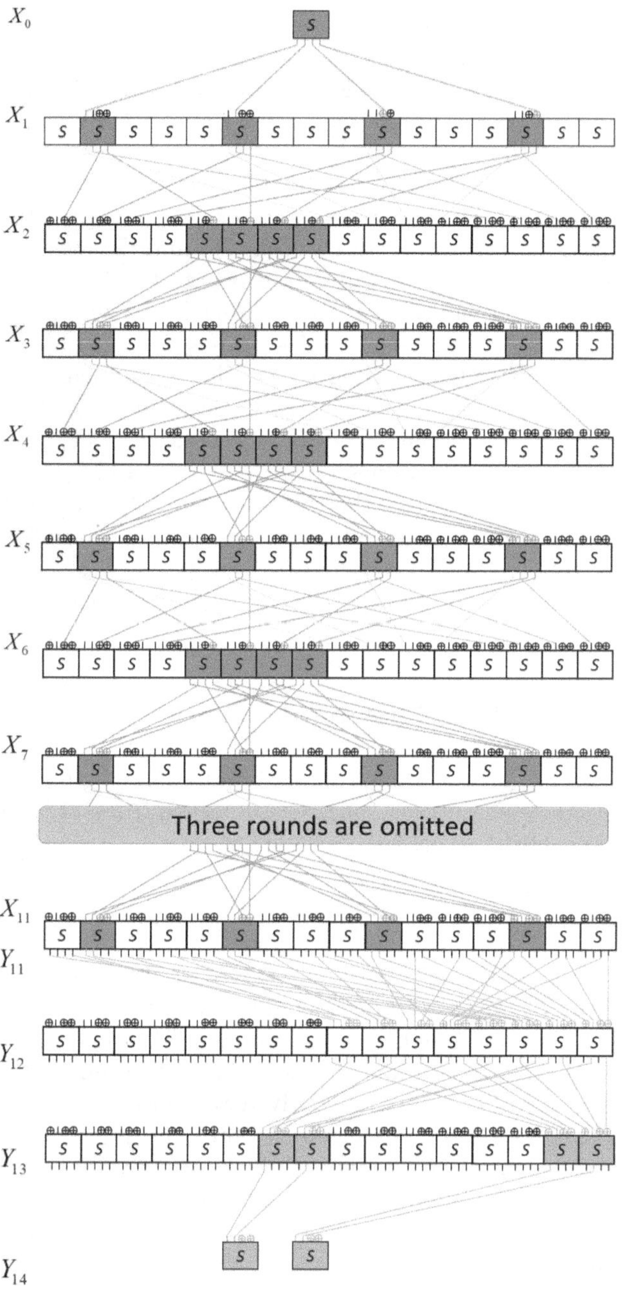

Fig. 3. The model of the MitM layer.

the values of X_2 into the next round, we exhaustively determine remaining 12 bits in each group of X_2. To reach the 11-th round, we rotate 5 recombination operations in total. In this case, we select the bits in G_0 and G'_2 for each rotational operation. Similarly, we also exhaustively determine 16 bits in Y_{12} and decrypt these to Y_{11}.

Initialize a table T_{E_i} for $0 \leqslant i \leqslant 3$ to store the partially determined values of X_{11} computed from the determined bits of X_2. That is, T_{E_0} will contain the selected bits in G_0 (respectively G_1, G_2, G_3) and G'_2 (respectively G'_3, G'_0, G'_1) for each rotational operation, which propagate to X_{11}. Similarly, we store the partially determined values of Y_{11} in a table T_D. After that, we merge the corresponding T_{E_i} and T_D into new tables T_i, $0 \leqslant i \leqslant 3$, obtaining a collection of fully determined bits when looking at Y_{11}. Each T_{E_i} shares 8 common bits with T_D, as seen in T_i. We proceed in the same fashion by merging tables T_0 and T_1, T_2 and T_3, to store their values in new tables $T_{0,1}$, $T_{2,3}$. On average, we get 2^{16} values in $T_{0,1}$ and 2^{16} values in $T_{2,3}$, since T_0 and T_1 as well as T_2 and T_3 share 16 common bits. At last, the tables $T_{0,1}$ and $T_{2,3}$ are merged to obtain 2^{32} values in X_{11} with fully determined bits, since they share 32 common bits. It is important to note that by running over all pairs of X_2 and Y_{12}, we filter out all the plaintexts that cannot satisfy the difference in input/output of the MitM layer.

5 Evaluation of Our Distinguisher

The data complexity and the success probability of the distinguisher are evaluated in this section based on the capacity C obtained from TD layer and filtered messages from the MitM layer.

5.1 The Data Complexity of the Distinguisher

The data complexity of the whole attack is attributed to the MitM layer, which is pre-added to the TD layer. The data complexity is 2^{60} since we guess 12 bits of X_1 and 48 bits of X_2. Additionally, we go through 2^{16+16} determined values of X_2 and Y_{12} to get partially determined values in the 11-th round in both directions. During the matching phase, each partially determined value is encrypted and decrypted, resulting in 2^{36} fully determined values in Y_{11}. Finally, we obtain 2^{56} values by reverse calculation from the fully determined values of X_{11}, which also satisfy the restrictions imposed on the input/output of an oracle for the MitM layer. On average, 2^{56} values can be input to the TD layer, resulting in 2^{111} pairs to observe a bias.

5.2 The Success Probability of the Distinguisher

Inspired by the analysis done in [4], we propose a generic algorithm to distinguish a cipher from random permutations. In Algorithm 1, the superscript i in CT^i denotes the i-th obtained value in the last round, where $i = 1, 2, \ldots, N^{TD}$.

Algorithm 1. Known-Key Distinguisher.

Input: N plaintext pairs with difference δ for an oracle to fulfil the TD attack.
Output: The boolean value 0 or 1.

1: Set an all-zero table T of size N^{TD} and a counter $D = 0$.
2: Iterate over the N plaintext pairs $(P, P + \delta)$ to generate N^{TD} pairs of values in X_{15} with difference Δ.
3: Iterate the 2^s truncated values (x_s, x_t) in X_{15} to generate ciphertexts $CT^i = (y_q, y_r)$ with difference Γ on q bits and set $T[x_t] = y_q$.
4: Iterate over $(x_s, x_t), (x'_s, x'_t)$ for $1 \leqslant i, j \leqslant N^{TD}$, if $T[x_t] = T[x'_t]$ then D += 1.
5: If $D > \mathcal{T}$, where $\mathcal{T}$ is the threshold defined in [5], then we can distinguish the cipher from a random permutation.
   return 1

In [16], the authors proposed a practical evaluation of the success probability P_S. When the total number of pairs N^{TD} derived from N pairs plaintexts—the inputs of the TD layer—is large enough, the success probability P_S is approximately determined by N^{TD} and the capacity C, according to the central normal distribution. It is then computed as follows:

$$P_S \approx \Phi(\sqrt{\mu}) = \Phi\left(\frac{\sqrt{2^{-q}N^{TD}} \cdot C}{2}\right), \tag{6}$$

where Φ denotes the cumulative distribution function of the standard normal distribution. When $P_S > 0.5$, the KK attack is regarded as "successful". Note that the distinguisher on full PRESENT in [5] is considered as successful with success probability 50.5%. We get the success probability of full GIFT-64 is 51.79% according to $q = 6, C = 2^{-53.16}$ and $N^{TD} = 2^{111}$. For 27-round GIFT-64 (15-round MitiM and 12-round TD), the success probability is up to 99.59% with capacity $2^{-51.36}$.

6 Experiments on GIFT-64[g_0^c]

In this section, we apply our distinguisher to the new proposed variant of GIFT-64, called GIFT-64[g_0^c], which is based on "group mappings" [17]. A graphical depiction of GIFT-64[g_0^c] is shown in [17]. To mount our distinguisher on GIFT-64[g_0^c], we first search for all the optimal linear approximations with correlation 2^{-34} over 13 rounds using the SAT solver. For TD layer, the approximation with the input mask 0x100a 0x0000 0x0000 0x00000 and output mask 0x1008 0x0000 0x4000 0x2000 has the maximum number of linear approximations among candidates—calculated using our well-designed matrix. There are 6342431 linear approximations over 13 rounds, so that the capacity C equals $2^{-55.43}$. The value q of the output oracle drops from 6 to 4 compared with GIFT-64. Define the output set $\tilde{\Gamma} = \{\gamma_{63}\gamma_{62}\cdots\gamma_0 \mid \gamma_b = 0$ for $b \in \{$ 13, 30, 51, 60$\}$ and the input set $\tilde{\Delta} = \{\delta_{63}\delta_{62}\cdots\delta_0 | \delta_a = 0$ for $a \in \{53, 55, 60\}\}$ of the TD-layer. The active S-box are then restricted, so the bias (as in Theorem 1) equals $2^{-(4+55.43)}$. Therefore, the TD distinguisher $(\tilde{\Delta} \to \tilde{\Gamma})$ has probability $Pr(TD) - 2^{-4}(2^{-55.43} + 1)$. In

MitM layer, we carry out 2^{12+9} guesses for values in both directions to get 16 determined bits in X_2 and 12 determined bits in X_{12}. The complexity of getting each determined value is dominated by merging the tables $T_{0,1}$ and $T_{2,3}$, which requires a complexity of 2^{36}. Moreover, there are 2^{113} pairs of values that satisfy the constraints on the MitM layer and the input of the TD layer. Compared with GIFT-64, the capacity C and the numbers of input values into MitM are both better so that the success probability increases to 56.80%.

7 Conclusions

In this paper, we have evaluated the security of GIFT-64 block cipher and its variant GIFT-64[g_0^c], under the Known-Key scenario—an important measure security for hashing modes of block ciphers. This is the very first non-random property exhibited the full GIFT-64 and GIFT-64-like ciphers. The distinguisher includes two layers: a TD layer with a pre-added MitM layer. To increase the probability of the TD, all the differentials are turned into linear approximations using well-known techniques from ML cryptanalysis, resulting in a good capacity for ML approximations. The idea is to find more approximations within the ML approximations elaborates on correlation matrices. The proposed matrix takes into account linear masks of hamming weight up to 8. Moreover, in the MitM layer, we have introduced a new method to match the forward and backward direction, called the rotational recombination method. This rotational technique helps to filter out the inputs of the truncated differentials so we can cover full round GIFT-64 and full round GIFT-64[g_0^c]. As a matter of fact, the designers of GIFT [3] phrase that GIFT satisfies "bad output must go to good input". However, our MitM attack suggests otherwise.

Acknowledgements. The authors would like to thank anonymous reviewers for their helpful comments and suggestions. This paper is supported by the National Natural Science Foundation of China (Grants No. 62272282, 62071280) and the Natural Science Foundation of Shandong Province (Grants No. ZR2020KF011, ZR2020MF056). The third author is partially supported by the Slovenian Research Agency (Projects J1-4084, J1-2451 and N1-0159).

References

1. Adomnicai, A., Najm, Z., Peyrin, T.: Fixslicing: a new GIFT representation fast constant-time implementations of GIFT and GIFT-COFB on ARM Cortex-M. IACR Trans. Cryptogr. Hardw. Embed. Syst. **2020**, 402–427 (2020). https://doi.org/10.13154/tches.v2020.i3.402-427
2. Banik, S., et al.: GIFT-COFB. IACR Cryptology ePrint Archive, p. 738 (2020). https://eprint.iacr.org/2020/738
3. Banik, S., Pandey, S.K., Peyrin, T., Sasaki, Yu., Sim, S.M., Todo, Y.: GIFT: a small present. In: Fischer, W., Homma, N. (eds.) CHES 2017. LNCS, vol. 10529, pp. 321–345. Springer, Cham (2017). https://doi.org/10.1007/978-3-319-66787-4_16

4. Blondeau, C., Nyberg, K.: New links between differential and linear cryptanalysis. In: Johansson, T., Nguyen, P.Q. (eds.) EUROCRYPT 2013. LNCS, vol. 7881, pp. 388–404. Springer, Heidelberg (2013). https://doi.org/10.1007/978-3-642-38348-9_24
5. Blondeau, C., Peyrin, T., Wang, L.: Known-key distinguisher on full PRESENT. In: Gennaro, R., Robshaw, M. (eds.) CRYPTO 2015. LNCS, vol. 9215, pp. 455–474. Springer, Heidelberg (2015). https://doi.org/10.1007/978-3-662-47989-6_22
6. Bogdanov, A., et al.: PRESENT: an ultra-lightweight block cipher. In: Paillier, P., Verbauwhede, I. (eds.) CHES 2007. LNCS, vol. 4727, pp. 450–466. Springer, Heidelberg (2007). https://doi.org/10.1007/978-3-540-74735-2_31
7. Bogdanov, A., Leander, G., Paar, C., Poschmann, A., Robshaw, M.J.B., Seurin, Y.: Hash functions and RFID tags: mind the gap. In: Oswald, E., Rohatgi, P. (eds.) CHES 2008. LNCS, vol. 5154, pp. 283–299. Springer, Heidelberg (2008). https://doi.org/10.1007/978-3-540-85053-3_18
8. Dong, X., Guo, J., Li, S., Pham, P.: Triangulating rebound attack on AES-like hashing. In: Dodis, Y., Shrimpton, T. (eds.) CRYPTO 2022. LNCS, vol. 13507, pp. 94–124. Springer, Santa Barbara (2022). https://doi.org/10.1007/978-3-031-15802-5_4
9. Flórez-Gutiérrez, A., Naya-Plasencia, M.: Improving key-recovery in linear attacks: application to 28-round PRESENT. In: Canteaut, A., Ishai, Y. (eds.) EUROCRYPT 2020. LNCS, vol. 12105, pp. 221–249. Springer, Cham (2020). https://doi.org/10.1007/978-3-030-45721-1_9
10. Gilbert, H.: A simplified representation of AES. In: Sarkar, P., Iwata, T. (eds.) ASIACRYPT 2014. LNCS, vol. 8873, pp. 200–222. Springer, Heidelberg (2014). https://doi.org/10.1007/978-3-662-45611-8_11
11. Hao, Y., Meier, W.: Truncated differential based known-key attacks on round-reduced SIMON. Des. Codes Cryptogr. **2017**(83), 467–492 (2017). https://doi.org/10.1007/s10623-016-0242-3
12. Ji, F., Zhang, W., Zhou, C., Ding, T.: Improved (related-key) differential cryptanalysis on GIFT. In: Dunkelman, O., Jacobson, Jr., M.J., O'Flynn, C. (eds.) SAC 2020. LNCS, vol. 12804, pp. 198–228. Springer, Cham (2021). https://doi.org/10.1007/978-3-030-81652-0_8
13. Knudsen, L.R., Rijmen, V.: Known-key distinguishers for some block ciphers. In: Kurosawa, K. (ed.) ASIACRYPT 2007. LNCS, vol. 4833, pp. 315–324. Springer, Heidelberg (2007). https://doi.org/10.1007/978-3-540-76900-2_19
14. Liu, Y., Sasaki, Yu.: Related-key boomerang attacks on GIFT with automated trail search including BCT effect. In: Jang-Jaccard, J., Guo, F. (eds.) ACISP 2019. LNCS, vol. 11547, pp. 555–572. Springer, Cham (2019). https://doi.org/10.1007/978-3-030-21548-4_30
15. Sasaki, Yu.: Meet-in-the-middle preimage attacks on AES hashing modes and an application to whirlpool. In: Joux, A. (ed.) FSE 2011. LNCS, vol. 6733, pp. 378–396. Springer, Heidelberg (2011). https://doi.org/10.1007/978-3-642-21702-9_22
16. Selçuk, A.A.: On probability of success in linear and differential cryptanalysis. J. Cryptol. **21**(1), 131–147 (2008). https://doi.org/10.1007/s00145-007-9013-7
17. Sun, L., Preneel, B., Wang, W., Wang, M.: A greater GIFT: strengthening GIFT against statistical cryptanalysis. Cryptology Accepted by Eurocrypt, p. 243 (2022). https://eprint.iacr.org/2022/243

18. Sun, L., Wang, W., Wang, M.: Accelerating the search of differential and linear characteristics with the SAT method. IACR Trans. Symmetric Cryptol. **2021**(1), 269–315 (2021). https://doi.org/10.46586/tosc.v2021.i1.269-315
19. Sun, L., Wang, W., Wang, M.: Improved attacks on GIFT-64. In: AlTawy, R., Hülsing, A. (eds.) SAC 2021. LNCS, vol. 13203, pp. 246–265. Springer, Cham (2022). https://doi.org/10.1007/978-3-030-99277-4_12

On the Security Bounds for Block Ciphers Without Whitening Key Addition Against Integral Distinguishers

Fanyang Zeng and Tian Tian(✉)

Information Engineering University, Zhengzhou 450001, China
tiantian_d@126.com

Abstract. At ASIACRYPT 2021, Phil Hebborn *et al.*, developed a powerful theory for block ciphers against integral distinguishers under the assumption of independent round keys and a whitening key XORed with the full state. Nevertheless, for certain block ciphers, say SIMON and Simeck, a whitening key is not part of their design. In this paper, we provide an improved theory on the resistance of integral distinguishers, which can be applied to block ciphers without a whitening key addition. As an illustration, we apply our arguments to SIMON32 and Simeck32 and obtain tight security bounds for them against integral distinguishers. To the best of our knowledge, this is the first time that the truly tight security bounds against integral distinguishers are derived for both ciphers.

Keywords: Block ciphers · Integral distinguishers · SIMON · Simeck

1 Introduction

The integral attack is an important cryptanalytic approach against block ciphers, which emerging from the square attack [1] and first proposed by Knudsen and Wagner at FSE 2002 [2]. Generally speaking, a critical step of integral attacks is searching integral distinguishers, i.e., search a subset of plaintexts such that the sum of corresponding ciphertexts is a constant.

To date, the most effective strategy for searching integral distinguishers is the division property, along with its variants. The division property was first introduced by Todo at EUROCRYPT 2015 [3], and has been widely used in searching integral distinguishers against many block ciphers. Employing this novel methodology, Todo revisited the 6-round integral characteristic on MISTY1 previously proposed in [4], achieving the first theoretical cryptanalysis of the full MISTY1 [3]. Subsequently, to further refine the use of S-box properties, Todo and Morii introduced the bit-based division property and a variant employing three subsets [5]. Utilizing the latter, the authors discovered three balanced bits in 14-round SIMON32 [5]. However, there is a limitation in the bit-based division property,

This work was supported by the National Natural Science Foundation of China under Grant No. 62372464.

T. Zhu and Y. Li (Eds.): ACISP 2024, LNCS 14895, pp. 41–56, 2024.
https://doi.org/10.1007/978-981-97-5025-2_3

that is, it only works on small block ciphers such as SIMON32 [6] and Simeck32 [7], due to its substantial memory complexity. To address this limitation, Xiang et al. combined the MILP (Mixed Integer Linear Programming) modeling technique with the bit-based division property [8]. They introduced the concept of the division trail and fundamental propagation rules for logical operations including AND, COPY, and XOR within the MILP framework. This approach enabled them to explore zero-sum distinguishers by solving MILP models covering all potential division trails. Their MILP-aided division property method yielded best results for a variety of block ciphers, namely SIMON, SIMECK, PRESENT [9], RECTANGLE [10], LBlock [11], and TWINE [12].

On the other hand, it is highly desirable to evaluate the security bounds for a block cipher against integral distinguishers. A breakthrough came with [13], where the authors, for the first time, demonstrated how to compute influential lower bounds on the degree of round-reduced block ciphers, based on the division property established in [14]. With these lower bounds, they proved that any monomial of degree $n-1$ presents in the Algebraic Normal Form (ANF) of any linear combination of output bits, for at least one key on 13-round SKINNY-64, 11-round GIFT and 11-round PRESENT, under the assumption of independent round keys. Notably, this proof also illustrates that there can be no integral distinguisher if the input set is a subspace of the plaintext space. Later, in [15], Hebborn et al. further advanced their theoretical framework for evaluating block ciphers security bounds against integral distinguishers, introducing a vital tool known as the integral-resistance matrix. Based on their result, the security bounds of block ciphers against integral distinguishers, under the assumption of independent round keys and a whitening key XORed to the full state, could be evaluated by providing a full rank integral-resistance matrix. In particular, the security bounds against integral distinguishers given by their theory are valid both for any possible input set and any possible output mask. They provide security bounds for integral distinguishers against SKINNY-64, CRAFT, GIFT-64, PRESENT, SIMON, and Simeck.

However, there are block ciphers not incorporating a whitening key addition, such as SKINNY-64, GIFT-64, SIMON, and Simeck. As a result, the security bounds offered in [15] may not be the true security bounds for these block ciphers. Hence, it is crucial to develop a more comprehensive theory on the integral-resistance property, which not only can be applied to block ciphers with a whitening key addition but also can be applied to block ciphers without a whitening key XORed to the full state.

1.1 Our Contribution

In this paper, we aim to propose a comprehensive theory on the security bounds against integral distinguishers, which is also known as the integral-resistance property. The main focus is the block ciphers without a whitening key addition.

To begin with, we present a general description of block ciphers, which incorporates block ciphers with and without whitening key addition as well as block ciphers with full or non-full round key addition. This includes the introduction of two parameters: a binary variable K_0, which indicates the presence of a whitening key, and an index set I, which represents the bits of the internal state that are XORed with the round key bits. Based on this general description of block ciphers, we redefine the integral-resistance property and establish a necessary and sufficient condition for it.

Next, under different values of K_0 and I, we further simplify the sufficient and necessary conditions for the integral-resistance property of block ciphers. As a result, all these sufficient and necessary conditions are given by the linear independence of polynomials in the set defined by K_0 and I.

Furthermore, in practice, the necessary and sufficient conditions is equivalent to the construction of a full-rank matrix. We analyze the computational complexity required to construct the full-rank matrix for different values of K_0 and I.

Finally, we apply our new theory to SIMON32 and Simeck32 and show that no integral distinguishers for 16 rounds and more for both ciphers. The resource codes, key patterns and integral-resistance matrices can be found at https://github.com/BLOCKCIPHERAES/SIMONandSimeck.

2 Preliminaries

2.1 Notations

We use the following notation. Let $\mathbb{F}_2$ denote the finite field of two elements. For a positive integer n, let $\mathbb{F}_2^n$ denote the n-dimensional vector space over $\mathbb{F}_2$. A vector $x \in \mathbb{F}_2^n$ is further denoted by $x = (x_1, x_2, \ldots, x_n)$, and the number of 1's in $(x_1, x_2, \ldots, x_n)$ is called the Hamming weight of x which is denoted by $\mathrm{wt}(x)$. The set of non-zero bit positions of the vector x is denoted by $\mathrm{Sup}(x)$, i.e.,

$$\mathrm{Sup}(x) = \{i | x_i = 1\}.$$

For two vectors $x, y \in \mathbb{F}_2^n$, we denote by

$$\langle x, y \rangle = \sum_i x_i y_i$$

the canonical inner product.

2.2 Block Ciphers and Its Algebraic Norm Form

A block cipher E can be considered as a function from $\mathbb{F}_2^n \times \mathbb{F}_2^m$ to $\mathbb{F}_2^n$, i.e.,

$$\begin{aligned} E : \mathbb{F}_2^n \times \mathbb{F}_2^m &\rightarrow \mathbb{F}_2^n \\ (x, k) &\mapsto E(x, k), \end{aligned}$$

where x represents plaintext variables, k represents key variables, and n, m are two positive integers. Under the assumption of independent round keys, key variables k also can be seen as all round key variables. *We assume that all round keys are independent in the following paper.* The block cipher E is also usually treated as a family of functions indexed by the values of k, i.e., $\{E_k(x) \mid k \in \mathbb{F}_2^m\}$ where

$$\begin{aligned} E_k : \mathbb{F}_2^n &\to \mathbb{F}_2^n \\ x &\mapsto E(x, k). \end{aligned}$$

For each fixed $k \in \mathbb{F}_2^m$, the function E_k is a bijection from $\mathbb{F}_2^n$ to $\mathbb{F}_2^n$. We could describe the algebraic normal form (ANF) of E by

$$E(x, k) = \sum_{u \in \mathbb{F}_2^n \setminus \{1\}} \sum_{v \in \mathbb{F}_2^m} \lambda_{u,v} k^v x^u \tag{1}$$

where $x^u = \prod_i x_i^{u_i}$, $k^v = \prod_i k_i^{v_i}$, $\lambda_{u,v} = (\lambda_{u,v}^{(1)}, \lambda_{u,v}^{(2)}, \ldots, \lambda_{u,v}^{(n)}) \in \mathbb{F}_2^n$. Let us denote $\sum_{v \in \mathbb{F}_2^m} \lambda_{u,v} k^v$ in (1) by $p_u(k)$, i.e.,

$$\begin{aligned} p_u(k) &= (p_u^{(1)}(k), p_u^{(2)}(k), \ldots, p_u^{(n)}(k)) \\ &= (\sum_{v \in \mathbb{F}_2^m} \lambda_{u,v}^{(1)} k^v, \sum_{v \in \mathbb{F}_2^m} \lambda_{u,v}^{(2)} k^v, \ldots, \sum_{v \in \mathbb{F}_2^m} \lambda_{u,v}^{(n)} k^v). \end{aligned} \tag{2}$$

Then the ANF of E can be rewritten

$$E(x, k) = \sum_{u \in \mathbb{F}_2^n \setminus \{1\}} p_u(k) x^u. \tag{3}$$

2.3 The Integral-Resistance Property

The notion of integral-resistance property was first proposed by Hebborn et al., in [15], which implies non-existence of any integral distinguishers. More specifically, for a block cipher E, the sum $\sum_{x \in M} \langle \beta, E_k(x) \rangle$ is key-dependent for any possible set M (excluding only the whole input space and the empty set) and any possible non-zero mask β.

2.4 The Division Property

Here, we only briefly introduce the division property with its propagation rules and the relationship between division trails and the coefficients $\lambda_{u,v}^{(i)}$, for further details please refer to [15].

Definition 1. *Let $\mathbb{X} \subseteq \mathbb{F}_2^n$ be a set. We define the division property of $\mathbb{X}$ as*

$$\mathcal{U}(\mathbb{X}) := \left\{ u \in \mathbb{F}_2^n \middle| \sum_{x \in \mathbb{X}} x^u = 1 \right\}.$$

Remark 1. $\mathcal{U}$ is a linear mapping.

Definition 2 (Propagation). *Let* $F : \mathbb{F}_2^n \to \mathbb{F}_2^n$ *be defined as*

$$F(x) = F(x_1, x_2, \ldots, x_n) = (y_1, y_2, \ldots, y_n) = y$$

where y_i *are multivariate polynomials over* $\mathbb{F}_2$ *in the variables* x_i. *For division properties* $a = (a_1, a_2, \ldots, a_n), b = (b_1, b_2, \ldots, b_n) \in \mathbb{F}_2^n$, *we say that* a *propagates to* b, *denoted by* $a \xrightarrow{F} b$, *if and only if* $b \in \mathcal{U}(F(\mathcal{U}(\{a\})))$, *which is equivalent to that* y^b *contains the monomial* x^a.

In general, F is actually given as the composition of many functions as $F = F_R \circ \cdots \circ F_2 \circ F_1$. A more general definition above is given as follows.

Definition 3 (Trail). *Given* $F : \mathbb{F}_2^n \to \mathbb{F}_2^n$ *as* $F = F_R \circ \cdots \circ F_2 \circ F_1$ *and division properties* $a_0, a_1, \ldots, a_R \in \mathbb{F}_2^n$, *we call* $(a_0, \ldots, a_R)$ *a division trail for the compositions of* F *into the* F_i *if and only if*

$$\forall i \in \{1, \ldots, R\}, a_{i-1} \xrightarrow{F_i} a_i.$$

We denote such a trail by $a_0 \xrightarrow{F_1} a_1 \xrightarrow{F_2} \ldots \xrightarrow{F_R} a_R$.

The relationship between division trails and the coefficients $\lambda_{u,v}^{(i)}$ are given by the following corollary.

Corollary 1. *Let* $F : \mathbb{F}_2^n \to \mathbb{F}_2^n$ *be a function with the algebraic normal form given by*

$$F(x) = \sum_{u \in \mathbb{F}_2^n} \lambda_u x^u$$

where $\lambda_u = (\lambda_u^{(1)}, \lambda_u^{(2)}, \ldots, \lambda_u^{(n)}) \in \mathbb{F}_2^n$. *Then*

$$\lambda_l^{(i)} = c \Leftrightarrow \#\{a_1, \ldots, a_{R-1} | l \xrightarrow{F_1} a_1 \xrightarrow{F_2} \ldots \xrightarrow{F_R} e_j\} \equiv c \pmod 2,$$

where $c \in \{0, 1\}$.

3 The Integral-Resistance Property for More Generalized Block Ciphers

The integral-resistance proposed in [15] primarily relies on two assumptions: the presence of independent round keys and the utilization of a whitening key XORed to the full state. However, it should be noted that not all block ciphers have a whitening key. In this section, we will expand the concept of integral-resistance to encompass a broader range of generalized block ciphers. Before delving into that, it is necessary to provide a comprehensive description of the structure of block ciphers.

3.1 Description of the Structure of Block Ciphers

Let E^r be an r-round iterative block cipher. Let $s_t = (s_t^{(1)}, s_t^{(2)}, \ldots, s_t^{(n)})$ be the n internal state bits after t rounds and let $s_0 = (s_0^{(1)}, s_0^{(2)}, \ldots, s_0^{(n)})$ be the initial state. The keys k can be regarded as a composition of r independent round keys denoted by $k_t = (k_t^{(1)}, k_t^{(2)}, \ldots, k_t^{(n)})$, where $1 \le t \le r$. Then the round function of E can be expressed as

$$s_t^{(i)} = F(s_{t-1}^{(1)}, s_{t-1}^{(2)}, \ldots, s_{t-1}^{(n)}) \oplus k_t^{(i)} \tag{4}$$

where F is a bijective function from $\mathbb{F}_2^n$ to $\mathbb{F}_2^n$, $1 \le t \le r$ and $1 \le i \le n$.

It is worth noting that before the first round of encryption, there may be a whitening key $k_0 = (k_1^{(0)}, k_2^{(0)}, \ldots, k_n^{(0)})$ XORed to the initial state, i.e.,

$$s_1^{(i)} = F(s_0^{(1)} \oplus k_0^{(1)}, s_0^{(2)} \oplus k_2^{(2)}, \ldots, s_0^{(n)} \oplus k_0^{(n)}) \oplus k_1^{(i)},$$

where $1 \le i \le n$. If there is no whitening key, then we will supplement the declaration $(k_0^{(1)}, k_0^{(2)}, \ldots, k_0^{(n)}) = (0, 0, \ldots, 0)$. Therefore, we introduce a new parameter K_0, where $K_0 = 0$ denotes the absence of a whitening key XORed on the full state, while $K_0 = 1$ denotes the presence of a whitening key XORed on the full state.

Furthermore, while updating the internal state s_t, it is possible that not every state bit is XORed with a corresponding round key bit $k_t^{(i)}$ for certain block ciphers. To handle this, we introduce an index set I, which is a subset of $\{1, 2, \ldots, n\}$, indicating the positions where a XOR operation with a round key bit is performed. Then, for each element $i \notin I$, $k_t^{(i)} = 0$ for $1 \le t \le r$. It is evident that when $I = \{1, 2, \ldots, n\}$, each round key is XORed with the full state, and we assume that I cannot be an empty set.

Having established a generalized description for block ciphers using Eq. (4), as well as incorporating a parameter K_0 and an index set I, we will now proceed to introduce our theory of integral-resistance, which could be used to a large range of block ciphers.

3.2 Integral-Resistance Property Aimed at Generalized Block Ciphers

Firstly, let us redefine the concept of integral-resistance property in order to generalize it for block ciphers described in the last section.

Definition 4. *Let E^r be an r-round iterated block cipher described as Eq. (4) with $K_0 \in \{0, 1\}$ and an index set $I \subseteq \{1, 2, \ldots, n\}$. If for any possible set M (excluding only the whole input space and the empty set) and any possible non-zero mask β, the sum*

$$\sum_{x \in M} \langle \beta, E_{k,k_0}^r(x) \rangle$$

is key-dependent, then we say that E^r satisfies the integral-resistance property.

Next, we will explore the sufficient and necessary conditions for E^r to satisfy the integral-resistance property by starting with its ANF. We define the ANF of E^r as follows,

$$E^r_{k,k_0}(x) = \sum_{v\in\mathbb{F}_2^n\setminus\{1\}} q_v(k,k_0)x^v, \tag{5}$$

where $q_v(k,k_0) = (q_v^{(1)}(k,k_0), q_v^{(2)}(k,k_0), \ldots, q_v^{(n)}(k,k_0))$, and $q_v^{(i)}(k,k_0)$ are some polynomials on variables k and k_0 for $1 \le i \le n$. The sufficient and necessary conditions are given by Theorem 1.

Theorem 1. *Let E^r be defined as above. Then E^r has the integral-resistance property if and only if all polynomials in*

$$\{q_v^{(i)}(k,k_0)|v \in \mathbb{F}_2^n\setminus\{1\}; 1 \le i \le n\}$$

are linearly independent.

Proof. It can be seen that

$$\begin{aligned}
\sum_{x\in M}\langle\beta, E_{k,k_0}(x)\rangle &= \sum_{x\in M}\sum_{i\in\mathrm{Sup}(\beta)}\sum_{v\in\mathbb{F}_2^n\setminus\{1\}} q_v^{(i)}(k,k_0)x^v \\
&= \sum_{i\in\mathrm{Sup}(\beta)}\sum_{v\in\mathbb{F}_2^n\setminus\{1\}}\left(\sum_{x\in M} x^v\right) q_v^{(i)}(k,k_0) \\
&= \sum_{i\in\mathrm{Sup}(\beta)}\sum_{l\in\mathcal{U}(M)} q_l^{(i)}(k,k_0).
\end{aligned} \tag{6}$$

Since β is any possible non-zero mask, $\mathrm{Sup}(\beta)$ can be any subset of $\{1,2,\ldots,n\}$ excluding the empty set. Similarly, as M is non-empty and not the full space, $\mathcal{U}(M)$ can be any subset of $\mathbb{F}_2^n$ excluding the empty set and the set $\{(1,1,\ldots,1)\}$. Therefore, $\sum_{i\in Sup(\beta)}\sum_{l\in\mathcal{U}(M)} q_l^{(i)}(k,k_0)$ can be any linear combination of these polynomials in $\{q_v^{(i)}(k,k_0)|v \in \mathbb{F}_2^n\setminus\{1\}, 1 \le i \le n\}$. According to Eq. (6), the theorem is proved.

However, determining whether these $2^n - 1$ polynomials are linearly independent is a highly challenging task. Therefore, we will discuss different cases based on the different values of parameters K_0 and I, hoping to obtain more practical and simpler necessary and sufficient conditions.

The First Case. The first case is that $K_0 = 1$ and I can be any subset of $\{1,2,\ldots,n\}$. Then, utilizing Eqs. (3), the ANF of $E^r_{k,k_0}(x)$ can also be written in the following form,

$$\begin{aligned}
E^r_{k,k_0}(x) &= E^r_k(x \oplus k_0) \\
&= \sum_{u\in\mathbb{F}_2^n\setminus\{1\}} p_u(k)(x\oplus k_0)^u \\
&= \sum_{v\in\mathbb{F}_2^n\setminus\{1\}}\left(\sum_{u\succeq v} p_u(k)k_0^{u\oplus v}\right)x^v,
\end{aligned} \tag{7}$$

where $u \succeq v$ holds true if and only if $u^{(i)} \geq v^{(i)}$ for $1 \leq i \leq n$. Clearly, there exists the following relationship between $q_v(k, k_0)$ and $p_u(k)$,

$$q_v^{(i)}(k, k_0) = \sum_{u \succeq v} p_u^{(i)}(k) k_0^{u \oplus v}, \tag{8}$$

where $1 \leq i \leq n$.

According to Eq. (8), with the presence of the independent whitening key k_0, we can easily obtain the following theorem,

Theorem 2. *Let E^r be a block cipher with $K_0 = 1$ and $I \subseteq \{1, 2, \ldots, n\}$. Then all polynomials in*

$$\{q_v^{(i)}(k, k_0) | v \in \mathbb{F}_2^n \backslash \{1\}; 1 \leq i \leq n\}$$

are linearly independent if and only if all polynomials in

$$T(K_0, I) = \{p_u^{(i)}(k) | wt(u) = n - 1; 1 \leq i \leq n\}$$

are linearly independent. Furthermore, E^r has the integral-resistance property if and only if all polynomials in $T(K_0, I)$ are linearly independent.

Proof. The necessity is apparent, and so we omit it here. Our main concern is sufficiency. Assume that there are t polynomials $q_{v_1}^{(i_1)}(k, k_0), q_{v_2}^{(i_2)}(k, k_0), \ldots,$ $q_{v_t}^{(i_t)}(k, k_0)$ that are linearly dependent, where $1 \leq t \leq n$ and $1 \leq i_l \leq n$ for $1 \leq l \leq t$. Then we have that

$$\begin{aligned}
\sum_{l=1}^{t} q_{v_l}^{(i_l)}(k, k_0) &= \sum_{l=1}^{t} \sum_{u \succeq v_l} p_u^{(i_l)}(k) k_0^{u \oplus v_l} \\
&= \sum_{w \in \mathbb{F}_2^n \backslash \{1\}} \left(\sum_{\substack{l=1 \\ \mathrm{Sup}(v_l) \cap \mathrm{Sup}(w) = \emptyset \\ \mathrm{wt}(v_l \oplus w) \leq n-1}}^{t} p_{v_l \oplus w}^{(i_l)}(k) \right) k_0^w \\
&= 0.
\end{aligned} \tag{9}$$

According to Eq. (9), we know that for each fixed w, it holds that

$$\sum_{\substack{l=1 \\ \mathrm{Sup}(v_l) \cap \mathrm{Sup}(w) = \emptyset \\ \mathrm{wt}(v_l \oplus w) \leq n-1}}^{t} p_{v_l \oplus w}^{(i_l)}(k) = 0.$$

Here, we assume that v_m is one of $\{v_1, v_2, \ldots, v_t\}$ with the smallest weight and we consider a vector w such that $\mathrm{Sup}(v_m) \cap \mathrm{Sup}(w) = \emptyset$ and $\mathrm{wt}(v_m \oplus w) = n - 1$. Then, the polynomial $p_{v_m \oplus w}^{(i_m)}(k)$ actually is in the set $T(K_0, I)$. Since the other vectors v_l satisfy $\mathrm{wt}(v_l) \geq \mathrm{wt}(v_m)$ for $1 \leq l \leq t$ and $l \neq m$, it follows that if

$\mathrm{Sup}(v_l) \cap \mathrm{Sup}(w) = \emptyset$, then $\mathrm{wt}(v_l \oplus w) \geq n-1$; otherwise $\mathrm{Sup}(v_l) \cap \mathrm{Sup}(w) \neq \emptyset$. Therefore, we can view

$$\sum_{\substack{l=1 \\ \mathrm{Sup}(v_l) \cap \mathrm{Sup}(w)=\emptyset \\ \mathrm{wt}(v_l \oplus w) \leq n-1}}^{t} p_{v_l \oplus w}^{(i_l)}(k)$$

as a summation of at least one polynomials from set $T(K_0, I)$ under this fixed w. Since all polynomials in $T(K_0, I)$ are linearly independent, we have

$$\sum_{\substack{l=1 \\ \mathrm{Sup}(v_l) \cap \mathrm{Sup}(w)=\emptyset \\ \mathrm{wt}(v_l \oplus w) \leq n-1}}^{t} p_{v_l \oplus w}^{(i_l)}(k) \neq 0,$$

under the fixed w, and thus

$$\sum_{l=1}^{t} q_{v_l}^{(i_l)}(k, k_0) \neq 0,$$

which means that $q_{v_1}^{(i_1)}(k, k_0), q_{v_2}^{(i_2)}(k, k_0), \ldots, q_{v_t}^{(i_t)}(k, k_0)$ are linearly independent. Therefore, all polynomials in set $\{q_v^{(i)}(k, k_0) | v \in \mathbb{F}_2^n \backslash \{1\}; 1 \leq i \leq n\}$ are linearly independent. Then, according to Theorem 1, the theorem is true.

The Second Case. Next, let us consider the second case where $K_0 = 0$ and $I = \{1, 2, \ldots, n\}$. For the purpose of clarity, we need to add an extra round, denoted as E^{r+1}. Moreover, we define $k = (k_1, k_2, \ldots, k_r, k_{r+1})$ and $k' = k \backslash k_1$. Then, we have that

$$E_{k,k_0}^{r+1}(x) = E_k^{r+1}(x) = E_{k'}^r(F(x) \oplus k_1).$$

Using Eq. (7), we can express the ANF of $E_{k'}^r(F(x) \oplus k_1)$ as follows,

$$\begin{aligned} E_{k'}^r(F(x) \oplus k_1) &= \sum_{u \in \mathbb{F}_2^n \backslash \{1\}} p_u(k')(F(x) \oplus k_1)^u \\ &= \sum_{v \in \mathbb{F}_2^n \backslash \{1\}} \left(\sum_{u \succeq v} p_u(k') k_1^{u \oplus v} \right) (F(x))^v, \end{aligned}$$

and we can easily obtain the Theorem 3.

Theorem 3. *Let E^{r+1} be a block cipher with $K_0 = 0$ and $I = \{1, 2, \ldots, n\}$. Then E^{r+1} has the integral-resistance property if and only if all polynomials*

$$T(K_0, I) = \{p_u^{(i)}(k') | wt(u) = n-1, 1 \leq i \leq n\}$$

are linearly independent.

Proof. According to Theorem 2, we know that for any possible set M and any possible non-zero mask β, the sum $\sum_{F(x)\in M}\langle\beta, E^r_{k'}(F(x)\oplus k_1)\rangle$ is key-dependent if and only if the polynomials $p^{(j)}_{u_i}(k')$ are linearly independent, where $1\le i,j,\le n$ and $wt(u_i)=n-1$. Consider the following equation

$$\sum_{F(x)\in M}\langle\beta, E^r_{k'}(F(x)\oplus k_1)\rangle = \sum_{x\in F^{-1}(M)}\left\langle\beta, E^{r+1}_{k,k_0}(x)\right\rangle.$$

Since M is an arbitrary set and F is a bijective function, $F^{-1}(M)$ can be any possible set. Therefore, the statement of the theorem is true.

The Third Case. Finally, there is one more case remaining, where $K_0=0$ and $I\subsetneq\{1,2,\ldots,n\}$. Similar to the second case, we need to consider an additional round, i.e., E^{r+1}, and obtain the integral-resistance property from $E^r_{k'}(F(x)\oplus k_1)$. Here, since I is a subset of $\{1,2,\ldots,n\}$, the ANF of $E^r_{k'}(F(x)\oplus k_1)$ can be rewritten,

$$\begin{aligned} E^r_{k'}(F(x)\oplus k_1) &= \sum_{v\in\mathbb{F}_2^n\backslash\{1\}} q_v(k',k_1)(F(x))^v \\ &= \sum_{v\in\mathbb{F}_2^n\backslash\{1\}}\left(\sum_{\substack{u\succeq v\\ \mathsf{Sup}(u\oplus v)\subseteq I}} p_u(k')k_1^{u\oplus v}\right)(F(x))^v. \end{aligned} \tag{10}$$

According to Theorem 1, we know that the sum $\sum_{x\in M}\langle\beta, E^r_{k'}(F(x)\oplus k_1)\rangle$ is key-dependent if and only if the polynomials $q^{(i)}_v(k',k_1)$ are linearly independent for $1\le i\le n$ and $v\in\mathbb{F}_2^n\backslash\{1\}$, where M is an arbitrary set (excluding only the whole input space and the empty set) and β is an arbitrary non-zero mask. Thus, we will have the following theorem

Theorem 4. *Let E^{r+1} be a block cipher with $K_0=0$ and $I\subsetneq\{1,2,\ldots,n\}$. Then the polynomials in*

$$\{q^{(i)}_v(k',k_1)|v\in\mathbb{F}_2^n\backslash\{1\};1\le i\le n\}$$

are linearly independent if and only if all polynomials in

$$T(K_0,I)=\{p^{(i)}_u(k')|I\subseteq Sup(u)\ or\ wt(u)=n-1;1\le i\le n\}$$

are linearly independent. Moreover, E^{r+1} has the integral-resistance property if and only if all polynomials in $T(K_0,I)$ are linearly independent.

Proof. Similar to Theorem 2, we only need to prove the sufficiency. We assume that there exist t polynomials $q^{(i_1)}_{v_1}(k',k_1), q^{(i_2)}_{v_2}(k',k_1),\ldots,q^{(i_t)}_{v_t}(k',k_1)$ that are linearly dependent. Then, combining Eq. (10), we have that

$$\begin{aligned}\sum_{l=1}^{t} q_{v_l}^{(i_l)}(k', k_1) &= \sum_{l=1}^{t} \sum_{\substack{u \succeq v_l \\ \text{Sup}(u \oplus v_l) \subseteq I}} p_u^{(i_l)}(k') k_1^{u \oplus v_l} \\ &= \sum_{w \subseteq I} \left(\sum_{\substack{l=1 \\ \text{Sup}(v_l) \cap \text{Sup}(w) = \emptyset \\ \text{wt}(v_l \oplus w) \leq n-1}}^{t} p_{v_l \oplus w}^{(i_l)}(k') \right) k_1^w \\ &= 0.\end{aligned}$$

Thus, for each fixed $w \subseteq I$, it holds that

$$\sum_{\substack{l=1 \\ \text{Sup}(v_l) \cap \text{Sup}(w) = \emptyset \\ \text{wt}(v_l \oplus w) \leq n-1}}^{t} p_{v_l \oplus w}^{(i_l)}(k') = 0.$$

We denote $u_{\overline{I}}$ as a vector such that $\text{Sup}(u_{\overline{I}}) = \{1, 2, \ldots, n\} \backslash I$. Next, for each vector v_l, we calculate the parameter $\text{wt}(v_l|I)$ as follows,

$$\text{wt}(v_l|I) = \begin{cases} |\text{Sup}(v_l) \cap I| + 1 & \text{if } \text{Sup}(u_{\overline{I}}) \subseteq \text{Sup}(v_l), \\ |\text{Sup}(v_l) \cap I| & \text{others,} \end{cases}$$

where $1 \leq l \leq n$. Here, we assume that v_m is one of $\{v_1, v_2, \ldots, v_t\}$ with the smallest value of $\text{wt}(v_m|I)$. Then we can select a vector w such that $\text{Sup}(v_m) \cap \text{Sup}(w) = \emptyset$ and $\text{wt}(v_m \oplus w) = n - 1$, if $\text{Sup}(u_{\overline{I}}) \subseteq \text{Sup}(v_m)$; otherwise, we can select a vector w satisfying that $\text{Sup}(v_m) \cap \text{Sup}(w) = \emptyset$ and $I \subseteq \text{Sup}(v_m \oplus w)$. At this point, we know that $p_{v_m \oplus w}^{(i_m)}(k')$ is a polynomial from the set $T(K_0, I)$. Since the other vectors satisfy $wt(v_l|I) \geq wt(v_m|I)$ for $1 \leq l \leq t$ and $l \neq m$, it follows that if $\text{Sup}(v_l) \cap \text{Sup}(w) = \emptyset$, then $\text{wt}(v_l \oplus w) \geq n - 1$ or $I \subseteq (v_l \oplus w)$; otherwise $\text{Sup}(v_l) \cap \text{Sup}(w) \neq \emptyset$. Therefore,

$$\sum_{\substack{l=1 \\ \text{Sup}(v_l) \cap \text{Sup}(w) = \emptyset \\ \text{wt}(v_l \oplus w) \leq n-1}}^{t} p_{v_l \oplus w}^{(i_l)}(k')$$

is a summation of at least one polynomials in $T(K_0, I)$ under this fixed w that we selected. The remaining proof is exactly the same as that of Theorem 2, which we will omit here.

3.3 The Complexity of Verifying the Integral-Resistance Property

In this section, we adopted the approach described in [15] to verify the integral-resistance property, which involves constructing a full-rank matrix. Here we will briefly introduce this approach, focusing primarily on complexity analysis.

Initially, we label the vectors in $T(K_0, I)$ that satisfy the conditions with $u_1, u_2, \dots, u_d$, where $d = |T(K_0, I)|/n$. Then, according to Eq. (2), we know that

$$p_u^{(i)}(k) = \sum_{v \in \mathbb{F}_2^m} \lambda_{u,v}^{(i)} k^v.$$

Subsequently, we need to find a set of key patterns $v_1, v_2, \dots, v_t$ with $t \geq |T(K_0, I)|$ and evaluate each value of $\lambda_{u_i,v_l}^{(j)}$ for $1 \leq i \leq d, 1 \leq j \leq n$ and $1 \leq l \leq t$. These values can be compiled to form a matrix of size $t \times |T(K_0, I)|$, represented as

$$M(T(K_0,I)) = \begin{pmatrix} \lambda_{u_1,v_1}^{(1)} & \lambda_{u_2,v_1}^{(1)} & \cdots & \lambda_{u_d,v_1}^{(1)} & \lambda_{u_1,v_1}^{(2)} & \lambda_{u_2,v_1}^{(2)} & \cdots & \lambda_{u_i,v_1}^{(j)} & \cdots & \lambda_{u_d,v_1}^{(n)} \\ \lambda_{u_1,v_2}^{(1)} & \lambda_{u_2,v_2}^{(1)} & \cdots & \lambda_{u_d,v_2}^{(1)} & \lambda_{u_1,v_2}^{(2)} & \lambda_{u_2,v_2}^{(2)} & \cdots & \lambda_{u_i,v_2}^{(j)} & \cdots & \lambda_{u_d,v_2}^{(n)} \\ \vdots & \vdots & & \vdots & \vdots & \vdots & & \vdots & & \vdots \\ \lambda_{u_1,v_t}^{(1)} & \lambda_{u_2,v_t}^{(1)} & \cdots & \lambda_{u_d,v_t}^{(1)} & \lambda_{u_1,v_t}^{(2)} & \lambda_{u_2,v_t}^{(2)} & \cdots & \lambda_{u_i,v_t}^{(j)} & \cdots & \lambda_{u_d,v_t}^{(n)} \end{pmatrix},$$

which is called *an integral-resistance matrix*. It is evident that the polynomials in set $T(K_0, I)$ are linearly independent if and only if there exists a set of key patterns $v_1, v_2, \dots, v_t$ such that the matrix $M(T(K_0, I))$ is full rank.

To construct an integral-resistance matrix $M(T(K_0, I))$ with full rank, for each row of $M(T(K_0, I))$, we need to search a critical pattern v_l that guarantees at least one coefficient $\lambda_{u_i,v_l}^{(j)}$ to be equal to 1, where $1 \leq l \leq t$. The method of searching these key patterns, which is referred to as "trail extension", is detailedly given in [13]; this has been omitted from our discussion for brevity. To calculate the remaining coefficients when a particular key pattern v_l is chosen, we thoroughly enumerate all possible division trails, checking the parity of the number of these trails.

Therefore, the computational complexity comes from two aspects. On the one hand, it takes time to search the key patterns. On the other hand, it involves the computation of the remaining coefficients under a fixed key pattern. Since the time required for these two processes is different, we denote the time for searching one key pattern as t_1 and the time for computing one coefficient under a fixed key as t_2. The total time required for the entire process is at least

$$|T(K_0, I)| * t_1 + |T(K_0, I) - 1|^2 * t_2.$$

More specifically, in the first and second case, the complexity is at least

$$n^2 * t_1 + (n^2 - 1)^2 * t_2,$$

only related to the block length n. In the third case, the complexity is at least

$$(n * (2^{n-|I|} + |I| - 1)) * t_1 + (n * (2^{n-|I|} + |I| - 1) - 1)^2 * t_2$$

related to the size of I and the block length n.

To provide a more intuitive understanding of this formula, let us consider SKINNY-64 and GIFT-64 as examples. In SKINNY-64 and GIFT-64, we have

$K_0 = 0, |I| = 32$ and $n = 64$. Thus, the time required to verify the security bound against integral distinguishers for SKINNY-64 and GIFT-64 is

$$64 * (2^{64-32} + 32 - 1) * t_1 + (64 * (2^{64-32} + 32 - 1) - 1)^2 * t_2 \approx 2^{40} t_1 + 2^{80} t_2,$$

which is challenging to accomplish. Therefore, we will apply our theory to SIMON32 and Simeck32 that the computational complexity will be lower.

Note that the validity of Theorems 2, 3, and 4 is based on the assumption that round keys are independent, whereas Theorem 1 does not rely on the independence of round keys. If the assumption of independent round keys is invalid for a block cipher, then only Theorem 1 could be used to evaluate the integral-resistance property, which is an extremely challenging task.

4 Applications

4.1 SIMON32 and Simeck32

SIMON is a family of block ciphers based on a two-branch balanced Feistel network with simple round functions consisting of three operations: AND, XOR, and rotation. We denote the SIMON using n-bit words by SIMON2n, where $2n \in \{32, 48, 64, 96, 128\}$. The ith internal state and the ith round key are denoted by

$$s_i = (sl_i, sr_i) = (s_i^{(1)}, \dots, s_i^{(n)}, s_i^{(n+1)}, \dots, s_i^{(2n)}),$$

and

$$k_i - (kl_i, kr_i) = (k_i^{(1)}, \dots, k_i^{(n)}, k_i^{(n+1)}, \dots, k_i^{(2n)}),$$

respectively. The structure of one round SIMON2n encryption is given by,

$$\begin{aligned} s_{i+1} &= (sl_{i+1}, sr_{i+1}) \\ &= ((sl_i \lll a) \cdot (sl_i \lll b) \oplus (sl_i \lll c) \oplus sr_i, sl_i) \oplus (kl_i, kr_i) \end{aligned}$$

with $K_0 = 0$ and $I = \{n+1, n+2, \cdots, 2n\}$, where $a = 1$, $b = 8$ and $c = 2$.

Simeck is also a family of lightweight block ciphers, which is very similar to SIMON, just replacing the rotation constants by $a = 0$, $b = 5$ and $c = 1$. Let us denote the Simeck with $2n$-bit block length by Simeck2n, where $2n \in \{32, 48, 64\}$.

4.2 Constructing Full-Rank Integral-Resistance Matrices

Here, we aim to demonstrate the integral-resistance property of 16-round SIMON32 and 16-round Simeck32. Based on the structure of SIMON32/Simeck32, we need to use Theorem 4 to verify the integral-resistance property. Namely, we need to search $32 \times (2^{32-16} + 16 - 1) = 32 \times 65551$ key patterns at least, aimed at 15-round SIMON32/Simeck32, and compute the remaining $(32 \times 65551 - 1)^2$ coefficients to construct a full-rank integral-resistance matrix.

We attempt to construct the following mode matrix to reduce computational complexity,

$$M(T(K_0, I)) = \begin{pmatrix} A_1 & 0 & \dots & 0 \\ * & A_2 & \dots & 0 \\ \vdots & \vdots & & \vdots \\ * & * & \dots & A_{32} \end{pmatrix}, \quad (11)$$

where $*$ represents an undetermined matrix. For each A_i, it is a 65551×65551 matrices given as follows,

$$A_i = \begin{pmatrix} \lambda^{(i)}_{u_1, v_{d(i-1)+1}} & \lambda^{(i)}_{u_2, v_{d(i-1)+1}} & \dots & \lambda^{(i)}_{u_d, v_{d(i-1)+1}} \\ \lambda^{(i)}_{u_1, v_{d(i-1)+2}} & \lambda^{(i)}_{u_2, v_{d(i-1)+2}} & \dots & \lambda^{(i)}_{u_d, v_{d(i-1)+2}} \\ \vdots & \vdots & & \vdots \\ \lambda^{(i)}_{u_1, v_{d(i-1)+d}} & \lambda^{(i)}_{u_2, v_{d(i-1)+d}} & \dots & \lambda^{(i)}_{u_d, v_{d(i-1)+d}} \end{pmatrix},$$

where $d = 65551$ and $1 \le i \le 32$. Thanks to their simple structures, we only need to choose the different 14th and 13th round key patterns aimed at the different A_i to obtain such a matrix $M(T(K_0, I))$ in Eq. (11).

Next, we only need to ensure that each matrix A_i is full rank. Here, we might as well stipulate that $wt(u_j) \le wt(u_l)$, when $j \le l$. Then, it becomes apparent that the round keys pattern v we find aimed at u_j, i.e., $\lambda^{(i)}_{u_j, v} = 1$, always results in other coefficients $\lambda^{(i)}_{u_l, v} = 1$, where $j \le l$. These operations allow us to quickly obtain full-rank matrices $A_i (1 \le i \le 32)$, with following mode

$$A_i = \begin{pmatrix} 1 & 0 & \dots & 0 \\ * & 1 & \dots & 0 \\ \vdots & \vdots & & \vdots \\ * & * & \dots & 1 \end{pmatrix}_{65551 \times 65551},$$

where $*$ is an unknown value.

Finally, we constructed the full-rank integral-resistance matrices $M(T(K_0, I))$ for 15-round SIMON32 and 15-round Simeck32. According to our theory, the security bounds of SIMON32 and Simeck32 against integral distinguishers are 16 rounds. The codes for finding the key patterns along with the key patterns and full-rank matrices can be found at https://github.com/BLOCKCIPHERAES/SIMONandSimeck.

5 Conclusion

In this paper, we introduce a comprehensive theoretical framework for analyzing the integral-resistance properties of block ciphers, specifically those ciphers without the whitening key addition. Utilizing this framework, we provide the true security bounds for SIMON32 and Simeck32 against integral distinguishers. Besides, it is found that for a block cipher without the whitening key layer, the computation complexity involved in verifying the integral-resistance property tends to increase as the block size increases or the number of round key additions decreases. A rough estimation on the computational complexity indicates that constructing the full-rank integral-resistance matrices for GIFT-64 and SKINNY-64 would cost at least 2^{40} for searching key patterns and 2^{80} for computing coefficients respectively, setting a significant challenge. Hence, it would be interesting to explore methods for constructing a special type of full-rank integral-resistance matrices that do not need to compute all coefficients by exploiting certain structural features of these block ciphers.

References

1. Daemen, J., Knudsen, L., Rijmen, V.: The block cipher Square. In: Biham, E. (ed.) FSE 1997. LNCS, vol. 1267, pp. 149–165. Springer, Heidelberg (1997). https://doi.org/10.1007/BFb0052343
2. Knudsen, L., Wagner, D.: Integral cryptanalysis. In: Daemen, J., Rijmen, V. (eds.) FSE 2002. LNCS, vol. 2365, pp. 112–127. Springer, Heidelberg (2002). https://doi.org/10.1007/3-540-45661-9_9
3. Todo, Y.: Integral cryptanalysis on full MISTY1. In: Gennaro, R., Robshaw, M. (eds.) CRYPTO 2015. LNCS, vol. 9215, pp. 413–432. Springer, Heidelberg (2015). https://doi.org/10.1007/978-3-662-47989-6_20
4. Matsui, M.: New block encryption algorithm MISTY. In: Biham, E. (ed.) FSE 1997. LNCS, vol. 1267, pp. 54–68. Springer, Heidelberg (1997). https://doi.org/10.1007/BFb0052334
5. Todo, Y., Morii, M.: Bit-based division property and application to SIMON family. In: Peyrin, T. (ed.) FSE 2016. LNCS, vol. 9783, pp. 357–377. Springer, Heidelberg (2016). https://doi.org/10.1007/978-3-662-52993-5_18
6. Beaulieu, R., Shors, D., Smith, J., Treatman-Clark, S., Weeks, B., Wingers, L.: The SIMON and SPECK families of lightweight block ciphers. IACR Cryptology ePrint Archive, p. 404 (2013)
7. Yang, G., Zhu, B., Suder, V., Aagaard, M.D., Gong, G.: The Simeck family of lightweight block ciphers. In: Güneysu, T., Handschuh, H. (eds.) CHES 2015. LNCS, vol. 9293, pp. 307–329. Springer, Heidelberg (2015). https://doi.org/10.1007/978-3-662-48324-4_16
8. Xiang, Z., Zhang, W., Bao, Z., Lin, D.: Applying MILP method to searching integral distinguishers based on division property for 6 lightweight block ciphers. In: Cheon, J.H., Takagi, T. (eds.) ASIACRYPT 2016. LNCS, vol. 10031, pp. 648–678. Springer, Heidelberg (2016). https://doi.org/10.1007/978-3-662-53887-6_24
9. Bogdanov, A., et al.: PRESENT: an ultra-lightweight block cipher. In: Paillier, P., Verbauwhede, I. (eds.) CHES 2007. LNCS, vol. 4727, pp. 450–466. Springer, Heidelberg (2007). https://doi.org/10.1007/978-3-540-74735-2_31
10. Zhang, W., Bao, Z., Lin, D., Rijmen, V., Yang, B., Verbauwhede, I.: RECTANGLE: a bit-slice lightweight block cipher suitable for multiple platforms. Sci. China Inf. Sci. **58**(12), 1–15 (2015)
11. Wu, W., Zhang, L.: LBlock: a lightweight block cipher. In: Lopez, J., Tsudik, G. (eds.) ACNS 2011. LNCS, vol. 6715, pp. 327–344. Springer, Heidelberg (2011). https://doi.org/10.1007/978-3-642-21554-4_19

12. Suzaki, T., Minematsu, K., Morioka, S., Kobayashi, E.: *TWINE*: a lightweight block cipher for multiple platforms. In: Knudsen, L.R., Wu, H. (eds.) SAC 2012. LNCS, vol. 7707, pp. 339–354. Springer, Heidelberg (2013). https://doi.org/10.1007/978-3-642-35999-6_22
13. Hebborn, P., Lambin, B., Leander, G., Todo, Y.: Lower bounds on the degree of block ciphers. In: Moriai, S., Wang, H. (eds.) ASIACRYPT 2020. LNCS, vol. 12491, pp. 537–566. Springer, Cham (2020). https://doi.org/10.1007/978-3-030-64837-4_18
14. Hao, Y., Leander, G., Meier, W., Todo, Y., Wang, Q.: Modeling for three-subset division property without unknown subset. In: Canteaut, A., Ishai, Y. (eds.) EUROCRYPT 2020. LNCS, vol. 12105, pp. 466–495. Springer, Cham (2020). https://doi.org/10.1007/978-3-030-45721-1_17
15. Hebborn, P., Lambin, B., Leander, G., Todo, Y.: Strong and tight security guarantees against integral distinguishers. In: Tibouchi, M., Wang, H. (eds.) ASIACRYPT 2021. LNCS, vol. 13090, pp. 362–391. Springer, Cham (2021). https://doi.org/10.1007/978-3-030-92062-3_13

Tight Multi-user Security of Ascon and Its Large Key Extension

Bishwajit Chakraborty[1], Chandranan Dhar[2](✉), and Mridul Nandi[2]

[1] Nanyang Technological University, Singapore, Singapore
bishwajit.chakrabort@ntu.edu.sg
[2] Indian Statistical Institute, Kolkata, India
chandranandhar@gmail.com

Abstract. The ASCON cipher suite has recently become the preferred standard in the NIST Lightweight Cryptography standardization process. Despite its prominence, the initial dedicated security analysis for the ASCON mode was conducted quite recently. This analysis demonstrated that the ASCON AEAD mode offers superior security compared to the generic Duplex mode, but it was limited to a specific scenario: single-user nonce-respecting, with a capacity strictly larger than the key size. In this paper, we eliminate these constraints and provide a comprehensive security analysis of the ASCON AEAD mode in the multi-user setting, where the capacity need not be larger than the key size. Regarding data complexity D and time complexity T, our analysis reveals that ASCON achieves AEAD security when T is bounded by $\min\{2^{\kappa}/\mu, 2^{c}\}$ (where κ is the key size, and μ is the number of users), and DT is limited to 2^{b} (with b denoting the size of the underlying permutation, set at 320 for ASCON). Our results align with NIST requirements, showing that ASCON allows for a tag size as small as 64 bits while supporting a higher rate of 192 bits, provided the number of users remains within recommended limits. However, this security becomes compromised as the number of users increases significantly. To address this issue, we propose a variant of the ASCON mode called LK-ASCON, which enables doubling the key size. This adjustment allows for a greater number of users without sacrificing security, while possibly offering additional resilience against quantum key recovery attacks. We establish tight bounds for LK-ASCON, and furthermore show that both ASCON and LK-ASCON maintain authenticity security even when facing nonce-misuse adversaries.

Keywords: ASCON · Multi-user Security · 256-bit Key · AEAD · tight security · lightweight cryptography

1 Introduction

Authenticated Encryption (AE) serves as a fundamental component of symmetric cryptography, enabling the simultaneous encryption and authentication of a

The full version of this paper can be found in [CDN24].

T. Zhu and Y. Li (Eds.): ACISP 2024, LNCS 14895, pp. 57–76, 2024.
https://doi.org/10.1007/978-981-97-5025-2_4

plaintext. Often, AE provides the capability to authenticate supplementary data, which, unlike the plaintext, is transmitted without encryption. In this context, AE is referred to as Authenticated Encryption with Associated Data (AEAD). Extensive research has been dedicated in recent years to developing and scrutinizing efficient and secure AEAD algorithms. Notably, widely utilized AEAD mechanisms across the Internet include AES-GCM [MV04], which is used in TLS [SCM08], and Chacha20-Poly1305, XSalsa20-Poly1305 [Ber05,Ber08b,Ber08a].

A specific area of focus involves the application of AEAD schemes in lightweight cryptography, which has garnered significant attention in research over the past decade. This exploration has been primarily inspired by the CAESAR competition [Com14], emphasizing authenticated encryption design, and subsequently by the lightweight cryptography (LWC) competition [NIS18] hosted by the US National Institute of Standards and Technology (NIST).

Of particular interest in the realm of lightweight cryptography is the Ascon cipher suite, which has emerged as the winner of the CAESAR competition (in the lightweight applications category), and more recently in the NIST LWC competition. Initially proposed as a candidate in Round 1 of the CAESAR competition [Com14], subsequent iterations of Ascon (v1.1 and v1.2) incorporated minor modifications to the original design (version 1 [DEMS14]). The latest version, v1.2 [DEMS19], acclaimed as the victor of the NIST LWC competition, encompasses the Ascon-128 and Ascon-128a authenticated ciphers, alongside the Ascon-Hash hash function and the Ascon-Xof extendable output function. All the components within the suite ensure 128-bit security and utilize a shared 320-bit permutation internally, facilitating the implementation of both duplex-based AEAD and sponge-based extendable-output hashing using a single lightweight primitive.

1.1 Existing Security Analysis

The authenticated encryption mode of Ascon is based on the duplex construction [BDPA11], specifically the MonkeyDuplex construction [BDPA12]. However, in contrast to MonkeyDuplex, Ascon's mode integrates double-keyed initialization and double-keyed finalization to reinforce its resilience. For an elaborate depiction of the Ascon AEAD mode, please refer to Sect. 3.

Until recently, security analyses of Ascon predominantly regarded it as a variant of Duplex construction (as indicated in [DEMS19]). A common constraint in the existing analyses of Duplex constructions, is the condition $DT \ll 2^c$ (or comparable variations where D might be substituted by q_d), where D is the data complexity and T is the time complexity, c is the capacity of the underlying sponge and q_d is the number of decryption queries.

At Asiacrypt 2023, Chakraborty et al. [CDN23a] conducted the first dedicated security analysis of Ascon. They leveraged the double-keyed initialization and finalization of Ascon, demonstrating the removal of the term $DT/2^c$ for the Ascon AEAD. They achieved a bound of the order

$$\mathcal{O}\left(\frac{T}{2^{\kappa}} + \frac{D}{2^{\tau}} + \frac{DT}{2^{b}}\right).$$

The authors also demonstrated that their bound is tight. However, this bound was only attainable in the single-user nonce-respecting setting, where nonces cannot be reused across encryption queries. Additionally, their analysis assumed that $\kappa < c$, i.e. the key size is strictly lesser than the capacity.[1]

In a concurrent work [ML23], Lefevre and Mennink also presented a dedicated security analysis of Ascon. While they focus on various settings (nonce-based confidentiality and authenticity, authenticity under nonce misuse and state recovery), they could only show the impact of strengthened initialization and finalization of Ascon in the case of authenticity under state recovery. However, in the case of conventional multi-user nonce-based authenticity, their bounds reduce to $\frac{q_d T}{2^c}$.

1.2 Our Contribution

In this work, we present a comprehensive analysis of the Ascon AEAD mode. Our first result establishes a tight AEAD security bound for Ascon in the multi-user nonce-respecting setting. Considering the number of users μ, tag size τ bits, key size κ bits, capacity c bits, and state size b bits, the derived bound is of the order

$$\mathcal{O}\left(\frac{\mu T}{2^\kappa} + \frac{D}{2^\tau} + \frac{DT}{2^b}\right).$$

Comparing this with the results of [CDN23a], we can see that although there is some multi-user security degradation, the term $DT/2^c$ can be overcome in this setting as well, thus improving over [ML23]. We also show that the achieved bound is tight. As a bonus, we establish the aforementioned result for the $\kappa = c$ scenario as well, thus broadening the assumption to $\kappa \leq c$. This extension presents an added benefit: considering the NIST LWC requirements ($D \leq 2^{53}, T \leq 2^{112}$, $\kappa \geq 128$, $\tau \geq 64$), as long as the number of users does not become too large, our findings suggest that a capacity size of $c = 128$ (given $b = 320$) and $\tau = 64$ are adequate to ensure sufficient security for Ascon. This selection allows for a higher rate of 192 bits (constrained to 184 when $\kappa < c$), significantly enhancing efficiency while maintaining security within the parameters of the random permutation model.

In the nonce-misuse setting, where nonces can be reused for encryption queries, confidentiality cannot be guaranteed. However, our second result shows that as far as authenticity security is concerned, the bounds are of the order

$$\mathcal{O}\left(\frac{\mu T}{2^\kappa} + \frac{D}{2^\tau} + \frac{DT}{2^b} + \frac{D^2}{2^c}\right).$$

This is also an improvement over [ML23], where the authors could not overcome the hurdle of $DT/2^c$ under any attack setting.

[1] In the original work [CDN23a], they initially assume $\kappa \leq c$ but later revise their assertions to $\kappa < c$ in the modified e-print version [CDN23b]. We delve into the intricacies of the case when $\kappa = c$ in Sect. 3.4.

A significant drawback of the ASCON AEAD mode is its compromised security as the number of users increases, with the term $\mu T/2^\kappa$ becoming the dominant factor (due to the 128-bit size of the key). One simple solution to this limitation would be to increase the key size of ASCON. Moreover, a larger key size has the capability to enhance resilience against key recovery attacks that utilize Grover's algorithm. However, key-size cannot be directly increased in the ASCON mode. For instance, if we opt for a nonce size of 128 bits, along with an extra 64-bit IV, the key-size becomes confined to 128 bits. The ASCON-80pq scheme was introduced as a component of the ASCON cipher suite, aiming to tackle this challenge. However, it should be noted that ASCON-80pq is only capable of accommodating a 160-bit key. As our final result, we introduce a novel AEAD mode, akin to ASCON, labeled as LK-ASCON (representing Large Key ASCON). This mode facilitates the doubling of the key size from 128 bits to 256 bits without requiring an increase in capacity, thus maintaining both security and efficiency. The resulting bound is of the order

$$\mathcal{O}\left(\frac{\mu T}{2^\kappa} + \frac{T}{2^c} + \frac{D}{2^{\min\{\tau,c\}}} + \frac{DT}{2^b}\right).$$

This bound is also tight. When nonces can be misused, the authenticity bound is again of the order

$$\mathcal{O}\left(\frac{\mu T}{2^\kappa} + \frac{D^2+T}{2^c} + \frac{D}{2^\tau} + \frac{DT}{2^b}\right).$$

A comparison among our results and the results of [CDN23a] and [ML23] can be found in Fig. 1.

Setting	**Security**	[CDN23a]	[ML23]	**This work**
su nr ASCON	AEAD	$\frac{T}{2^\kappa} + \frac{DT}{2^b}$	$\frac{T}{2^\kappa} + \frac{\sigma_d T}{2^c}$	$\frac{T}{2^\kappa} + \frac{DT}{2^b}$
mu nr ASCON	AEAD	-	$\frac{\mu T}{2^\kappa} + \frac{\sigma_d T}{2^c}$	$\frac{\mu T}{2^\kappa} + \frac{DT}{2^b}$
mu nm ASCON	Authenticity	-	$\frac{\mu T}{2^\kappa} + \frac{DT}{2^c}$	$\frac{\mu T}{2^\kappa} + \frac{DT}{2^b} + \frac{D^2}{2^c}$
mu nr LK-ASCON	AEAD	-	-	$\frac{\mu T}{2^\kappa} + \frac{DT}{2^b} + \frac{T}{2^c}$
mu nm LK-ASCON	Authenticity	-	-	$\frac{\mu T}{2^\kappa} + \frac{DT}{2^b} + \frac{D^2+T}{2^c}$

Fig. 1. Security Analysis Comparison. The expression $D/2^\tau$ is a common factor in all entries. "su" and "mu" represent single-user and multi-user, respectively. "nr" and "nm" denote nonce-respecting and nonce-misuse, respectively. The term σ_d refers to the data complexity of decryption queries.

1.3 Organization of the Paper

In Sect. 2, we define the basic notations used in the paper. We give a brief description of the AEAD security in the random permutation model, and also briefly describe the H-coefficient technique. Moving forward, in Sect. 3, we present a detailed examination of the Ascon AEAD scheme. We present one of our two primary results, the security bound of Ascon, and establish its significance in relation to the NIST LWC criteria. To support our claims, we provide an interpretation of our findings within the context of the NIST guidelines, and discuss the tightness. Then, in Sect. 4, we present the authenticity security of Ascon in the nonce misuse setting. In Sect. 5, we define LK-Ascon, a variant of Ascon and state the security of LK-Ascon, along with a proof outline. Finally, in Sect. 6, we conclude the paper.

2 Preliminaries

2.1 Notations

Let $\{0,1\}^n$ represent the set of bit strings of length n, and $\{0,1\}^+$ denote the set of bit strings of arbitrary length. The empty string is denoted by λ, and we define $\{0,1\}^* = \{\lambda\} \cup \{0,1\}^+$. For any integers $a \leq b \in \mathbb{N}$, $[b]$ and $[a,b]$ denote the sets $\{1,2,\ldots,b\}$ and $\{a,a+1,\ldots,b\}$, respectively. For $n,k \in \mathbb{N}$ with $n \geq k$, the falling factorial is defined as $(n)_k := n(n-1)\cdots(n-k+1)$. It's worth noting that $(n)_k \leq n^k$.

For any bit string $x = x_1x_2\cdots x_k \in \{0,1\}^k$ of length k, and for $n \leq k$, we use $\lceil x \rceil_n := x_1 \cdots x_n$ (and $\lfloor x \rfloor_n := x_{k-n+1}\cdots x_k$) to denote the most (and least) significant n bits of x. The bit concatenation operation is denoted by $\|$. The notation $(x_1,\ldots,x_r)$ is also used to represent the bit concatenation operation $x_1\|\cdots\|x_r$, where $x_i \in \{0,1\}^*$. For instance, if $V := x\|z := (x,z) \in \{0,1\}^r \times \{0,1\}^c$, then $\lceil V \rceil_r = x$ and $\lfloor V \rfloor_c = z$. The bitwise XOR operation is denoted by $\oplus$.

For a finite set $\mathcal{X}$, $\mathsf{X} \xleftarrow{\$} \mathcal{X}$ denotes the uniform and random sampling of X from $\mathcal{X}$, and $\mathsf{X} \xleftarrow{\mathsf{wor}} \mathcal{X}$ denotes sampling without replacement of X from $\mathcal{X}$.

Padding and Parsing a Bit String. Let $r > 0$ be an integer and $X \in \{0,1\}^*$. Let $d = |X| \bmod r$ (the remainder while dividing $|X|$ by r).

$$\mathsf{pad}_1(X) = \begin{cases} \lambda & \text{if } |X| = 0 \\ X\|1\|0^{r-1-d} & \text{otherwise} \end{cases}$$

and

$$\mathsf{pad}_2(X) = X\|1\|0^{r-1-d}.$$

Given $X \in \{0,1\}^*$, let $x = \lceil \frac{|X|+1}{r} \rceil$. We define $(X_1,\ldots,X_x) \xleftarrow{r}_* X$ as $X_1\|\cdots\|X_x = X$, $|X_1| = \cdots = |X_{x-1}| = r$ and

$$X_x = \begin{cases} \lambda & \text{if } |X| = r(x-1) \\ \lfloor X \rfloor_{|X|-r(x-1)} & \text{otherwise} \end{cases}.$$

For $N \geq 4,\ n = \log_2 N$, we define

$$\mathsf{mcoll}(q, N) = \begin{cases} 3 & \textit{if } 4 \leq q \leq \sqrt{N} \\ \frac{4 \log_2 q}{\log_2 \log_2 q} & \textit{if } \sqrt{N} < q \leq N \\ 5n \left\lceil \frac{q}{nN} \right\rceil & \textit{if } N < q. \end{cases}$$

2.2 Authenticated Encryption with Associated Data: Definition and Security Model

An authenticated encryption scheme with associated data functionality, abbreviated as AEAD, is characterized by a tuple of algorithms $\mathsf{AE} = (\mathsf{E}, \mathsf{D})$. These algorithms, referred to as the encryption and decryption algorithms, operate over the *key space* $\mathcal{K}$, *nonce space* $\mathcal{N}$, *associated data space* $\mathcal{A}$, *message space* $\mathcal{M}$, *ciphertext space* $\mathcal{C}$, and *tag space* $\mathcal{T}$. The functionalities are defined as follows:

$$\mathsf{E} : \mathcal{K} \times \mathcal{N} \times \mathcal{A} \times \mathcal{M} \rightarrow \mathcal{C} \times \mathcal{T} \quad \text{and} \quad \mathsf{D} : \mathcal{K} \times \mathcal{N} \times \mathcal{A} \times \mathcal{C} \times \mathcal{T} \rightarrow \mathcal{M} \cup \{\mathsf{rej}\}.$$

Here, rej signifies that the tag-ciphertext pair is invalid and consequently rejected. Additionally, the correctness condition is imposed:

$$\mathsf{D}(K, N, A, \mathsf{E}(K, N, A, M)) = M \text{ for any } (K, N, A, M) \in \mathcal{K} \times \mathcal{N} \times \mathcal{A} \times \mathcal{M}.$$

For a key $K \in \mathcal{K}$, we use $\mathsf{E}_K(\cdot)$ and $\mathsf{D}_K(\cdot)$ to denote $\mathsf{E}(K, \cdot)$ and $\mathsf{D}(K, \cdot)$, respectively. In this paper, we consider $\mathcal{K} = \{0,1\}^\kappa, \mathcal{N} = \{0,1\}^\nu, \mathcal{T} = \{0,1\}^\tau$, and $\mathcal{A}, \mathcal{M} = \mathcal{C} \subseteq \{0,1\}^*$.

AEAD Security in the Random Permutation Model

Let $\mathsf{Perm}(b)$ denote the set of all permutations over $\{0,1\}^b$ and $\mathsf{Func}(\mathcal{N} \times \mathcal{A} \times \mathcal{M}, \mathcal{M} \times \mathcal{T})$ denote the set of all functions from (N, A, M) to (C, T) such that $|C| = |M|$. We consider the AEAD security in the multi-user (mu) setting, parameterized by the number of users μ. Let:

- $\Pi \xleftarrow{\$} \mathsf{Perm}(b)$ (we use the superscript $\pm$ to denote bidirectional access to Π),
- $\Gamma_1, \ldots, \Gamma_\mu \xleftarrow{\$} \mathsf{Func}(\mathcal{N} \times \mathcal{A} \times \mathcal{M}, \mathcal{M} \times \mathcal{T})$,
- rej denotes the degenerate function from $(\mathcal{N}, \mathcal{A}, \mathcal{M}, \mathcal{T})$ to $\{\mathsf{rej}\}$, and
- $K_1, \ldots, K_\mu \xleftarrow{\$} \mathcal{K}$.

We have the following definition:

Definition 1. *Let* AE_Π *be an AEAD scheme based on the random permutation* Π, *defined over* $(\mathcal{K}, \mathcal{N}, \mathcal{A}, \mathcal{M}, \mathcal{T})$. *The mu-AEAD advantage of an adversary* $\mathscr{A}$ *against* AE_Π *is defined as*

$$\mathbf{Adv}_{\mathsf{AE}_\Pi}^{\mathsf{mu-aead}}(\mathscr{A}) := \left| \Pr_{\substack{(\mathsf{K_i})_{i=1}^{\mu} \xleftarrow{\$} \mathcal{K} \\ \Pi^\pm}} \left[\mathscr{A}^{(E_{\mathsf{K_i}}, D_{\mathsf{K_i}})_{i=1}^{\mu}, \Pi^\pm} = 1 \right] - \Pr_{\substack{(\Gamma_i)_{i=1}^{\mu} \\ \Pi^\pm}} \left[\mathscr{A}^{(\Gamma_i)_{i=1}^{\mu}, \mathsf{rej}, \Pi^\pm} = 1 \right] \right|.$$

Here $\mathscr{A}^{\mathsf{E}_{\mathsf{K}_i},\mathsf{D}_{\mathsf{K}_i},\Pi^{\pm}}$ denotes $\mathscr{A}$'s response after its interaction with $\mathsf{E}_{\mathsf{K}_i}$, $\mathsf{D}_{\mathsf{K}_i}$, and $\Pi^{\pm}$ (i.e., both forward and backward queries to Π) respectively. Similarly, $\mathscr{A}^{\Gamma_i,\mathsf{rej},\Pi^{\pm}}$ denotes $\mathscr{A}$'s response after its interaction with Γ_i, rej, and $\Pi^{\pm}$ respectively.

In this paper, we assume that the adversary is adaptive. This means that the adversary neither issues duplicate queries nor requests information for which the response is already known due to some previous query. Let q_e, q_d, and q_p represent the number of queries made across all $\mathsf{E}_{\mathsf{K}_i}$, all $\mathsf{D}_{\mathsf{K}_i}$, and $\Pi^{\pm}$, respectively. Furthermore, let σ_e and σ_d denote the sum of input lengths (including associated data and message) across all encryption and decryption queries, respectively. Additionally, let $\sigma := \sigma_e + \sigma_d$ represent the combined resources for construction queries.

Remark 1. Here σ corresponds to the online or data complexity, and q_p corresponds to the offline or time complexity of the adversary. An adversary adhering to the specified resource constraints is referred to as an $(q_p, \sigma_e, \sigma_d)$-adversary.

Separation into Confidentiality and Authenticity. For any AEAD scheme, the security can be separated into confidentiality and authenticity. In the mu setting, we have the following definitions:

Definition 2. *Let* AE_Π *be an AEAD scheme based on the random permutation* Π, *defined over* $(\mathcal{K}, \mathcal{N}, \mathcal{A}, \mathcal{M}, \mathcal{T})$. *The mu-confidentiality advantage of an adversary* $\mathscr{A}$ *against* AE_Π *is defined as*

$$\mathbf{Adv}_{\mathsf{AE}_\Pi}^{\mathsf{mu-conf}}(\mathscr{A}) := \left| \Pr_{\substack{(\mathsf{K}_i)_{i=1}^{\mu} \xleftarrow{\$} \mathcal{K} \\ \Pi^{\pm}}} \left[\mathscr{A}^{(E_{\mathsf{K}_i})_{i=1}^{\mu},\Pi^{\pm}} = 1 \right] - \Pr_{\substack{(\Gamma_i)_{i=1}^{\mu} \\ \Pi^{\pm}}} \left[\mathscr{A}^{(\Gamma_i)_{i=1}^{\mu},\Pi^{\pm}} = 1 \right] \right|,$$

and the mu-authenticity advantage of an adversary $\mathscr{A}$ *against* AE_Π *is defined as*

$$\mathbf{Adv}_{\mathsf{AE}_\Pi}^{\mathsf{mu-auth}}(\mathscr{A}) := \Pr_{\substack{(\mathsf{K}_i)_{i=1}^{\mu} \xleftarrow{\$} \mathcal{K} \\ \Pi^{\pm}}} \left[\mathscr{A}^{(E_{\mathsf{K}_i}, D_{\mathsf{K}_i})_{i=1}^{\mu},\Pi^{\pm}} \textit{ forges} \right].$$

In the context of authenticity, we use the term "$\mathscr{A}$ forges" to describe a situation where $\mathscr{A}$ successfully makes a query to one of its decryption oracles, and this query is not the result of a previous encryption query. By an easy reduction [DNT19], it can be shown that

$$\mathbf{Adv}_{\mathsf{AE}_\Pi}^{\mathsf{mu-auth}}(\mathscr{A}) \leq \left| \Pr_{\substack{(\mathsf{K}_i)_{i=1}^{\mu} \xleftarrow{\$} \mathcal{K} \\ \Pi^{\pm}}} \left[\mathscr{A}^{(\mathsf{E}_{\mathsf{K}_i}, \mathsf{D}_{\mathsf{K}_i})_{i=1}^{\mu},\Pi^{\pm}} = 1 \right] - \Pr_{\substack{(\Gamma_i)_{i=1}^{\mu} \\ \Pi^{\pm}}} \left[\mathscr{A}^{(\Gamma_i)_{i=1}^{\mu},\mathsf{rej},\Pi^{\pm}} = 1 \right] \right|.$$

Proposition 1 ([BN08]). *There exist adversaries* $\mathscr{A}$, $\mathscr{A}'$ *and* $\mathscr{A}''$ *having same query complexities such that*

$$\mathbf{Adv}_{\mathsf{AE}_\Pi}^{\mathsf{mu-aead}}(\mathscr{A}) \leq \mathbf{Adv}_{\mathsf{AE}_\Pi}^{\mathsf{mu-conf}}(\mathscr{A}') + \mathbf{Adv}_{\mathsf{AE}_\Pi}^{\mathsf{mu-auth}}(\mathscr{A}'').$$

2.3 H-Coefficient Technique

Consider an adversary $\mathscr{A}$, which is deterministic and computationally unbounded, attempting to distinguish between the real oracle, denoted as $\mathcal{O}_{\text{re}}$, and the ideal oracle, denoted as $\mathcal{O}_{\text{id}}$. The interaction of $\mathscr{A}$ with its oracle is captured by the query-response tuple denoted as ω. In certain scenarios, after the query-response phase of the game, the oracle may choose to reveal additional information to the distinguisher. In such cases, the extended definition of the transcript may include that additional information. Let Θ_{re} (respectively, Θ_{id}) represent the random transcript variable when $\mathscr{A}$ interacts with $\mathcal{O}_{\text{re}}$ (respectively, $\mathcal{O}_{\text{id}}$). The probability of realizing a specific transcript ω in the security game with an oracle $\mathcal{O}$ is referred to as the *interpolation probability* of ω with respect to $\mathcal{O}$. Given the determinism of $\mathscr{A}$, this probability depends solely on the oracle $\mathcal{O}$ and the transcript ω. A transcript ω is considered *realizable* if $\Pr[\Theta_{\text{id}} = \omega] > 0$. In this paper, $\mathcal{O}_{\text{re}} = (\mathsf{E_K}, \mathsf{D_K}, \Pi^{\pm})$, $\mathcal{O}_{\text{id}} = (\Gamma, \mathsf{rej}, \Pi^{\pm})$, and the adversary aims to distinguish $\mathcal{O}_{\text{re}}$ from $\mathcal{O}_{\text{id}}$ in an AEAD sense.

Proposition 2 (H-coefficient technique [Pat91,Pat08]**).** *Let Ω be the set of all realizable transcripts. For some $\epsilon_{\mathsf{bad}}, \epsilon_{\mathsf{ratio}} > 0$, suppose there is a set $\Omega_{\mathsf{bad}} \subseteq \Omega$ satisfying the following:*

- $\Pr[\Theta_{\text{id}} \in \Omega_{\mathsf{bad}}] \leq \epsilon_{\mathsf{bad}}$;
- *For any $\omega \notin \Omega_{\mathsf{bad}}$,*

$$\frac{\Pr[\Theta_{\text{re}} = \omega]}{\Pr[\Theta_{\text{id}} = \omega]} \geq 1 - \epsilon_{\mathsf{ratio}}.$$

Then for any adversary $\mathscr{A}$, we have the following bound on its AEAD distinguishing advantage:

$$\mathbf{Adv}^{\mathsf{aead}}_{\mathcal{O}_{\text{re}}}(\mathscr{A}) \leq \epsilon_{\mathsf{bad}} + \epsilon_{\mathsf{ratio}}.$$

A proof of Proposition 2 can be found in multiple papers including [Pat08,CS14, MN17].

2.4 Partial XOR-Function Graph

A *partial function* $\mathcal{L} : \{0,1\}^b \dashrightarrow \{0,1\}^c$ is a subset $\mathcal{L} = \{(p_1, q_1), \ldots, (p_t, q_t)\} \subseteq \{0,1\}^b \times \{0,1\}^c$ with distinct p_i values. An *injective partial function* has distinct q_i values. Define

$$\mathsf{domain}(\mathcal{L}) = \{p_i : i \in [t]\}, \quad \mathsf{range}(\mathcal{L}) = \{q_i : i \in [t]\}.$$

We write $\mathcal{L}(p_i) = q_i$ and for all $p \notin \mathsf{domain}(\mathcal{L})$, $\mathcal{L}(p) = \bot$.

Consider a partial function $\mathcal{P} : \{0,1\}^b \dashrightarrow \{0,1\}^b$, $r \in [b-1]$. Define $\mathcal{P}^{\oplus} : \{0,1\}^b \times \{0,1\}^r \dashrightarrow \{0,1\}^b$ as

$$\mathcal{P}^{\oplus}(u, x) = \mathcal{P}(u' \oplus x) \| u''),$$

where $u = u' \| u''$ and $u' \in \{0,1\}^r$. Define $G^{\oplus} := G^{\mathcal{P}^{\oplus}}$ with labeled edges denoted as $u \xrightarrow{x}_{\oplus} v$. A more detailed discussion on partial function graphs appears in the full version of this paper.

3 The ASCON AEAD Mode

In this section, we introduce the ASCON AEAD [DEMS19] mode of operation, which is essentially a modified version of the Duplex construction. Let b represent the state size of the underlying permutation π, and consider $0 < r < b$ as the number of bits of associated data/message processed per permutation call. The term r is referred to as the rate of the ASCON construction, while $c = b - r$ is known as the capacity.

Let κ, ν, τ denote the key size, nonce size, and tag size, respectively, with the constraints: (i) $\tau \leq \kappa \leq c$, and (ii) $\kappa + \nu \leq b$. Note that, unlike [CDN23a], we assume $\kappa \leq c$, thus encompassing the $\kappa = c$ case.

We fix an $IV \in \{0,1\}^{b-\kappa-\nu}$. The AEAD utilizes a permutation π (the ASCON permutation), modeled as a random permutation for security analysis. Below, we describe the encryption and decryption algorithms of the ASCON AEAD mode. For a visual representation of the encryption algorithm, refer to Fig. 2.

Encryption Algorithm. It receives an input of the form $(N, A, M) \in \{0,1\}^\nu \times \{0,1\}^* \times \{0,1\}^*$ and a key $K \in \{0,1\}^\kappa$. We divide the encryption algorithm into three phases: (i) initialization, (ii) associated data and message processing, and (iii) tag generation, run sequentially.

INITIALIZATION. In this phase, we first apply the following function

$$\textsc{Init}^\pi(K, N) = \pi(IV\|K\|N) \oplus (0^{b-\kappa}\|K) := V_0.$$

ASSOCIATED DATA AND MESSAGE PROCESSING. We first parse the associated data and message:

$$(A_1, \ldots, A_a) \xleftarrow{r} \mathsf{pad}_1(A), \quad (M_1, \ldots, M_m) \xleftarrow{r} \mathsf{pad}_2(M).$$

Then, using the XOR-function graph corresponding to the function $\pi^\oplus$, we obtain a walk

$$V_0 \xrightarrow{A_1}_\oplus V_1 \xrightarrow{A_2}_\oplus \cdots \xrightarrow{A_a}_\oplus V_a, \quad V_a \oplus 0^*1 \xrightarrow{M_1}_\oplus V_{a+1} \cdots \xrightarrow{M_{m-1}}_\oplus V_{a+m-1}.$$

We define the ciphertext as follows:

$$C_i = \lceil V_{a+i-1} \rceil_r \oplus M_i, \quad \forall i \in [m], \quad C = \lceil C_1 \| \cdots \| C_m \rceil_{|M|}.$$

We denote the above process as

$$\mathsf{AM\_Proc}^\pi(V_0, A, M) \to \big(C, F := V_{t-1} \oplus (M_m\|0^c)\big).$$

TAG GENERATION. Finally, we compute

$$T := \textsc{Tag}^\pi(K, F) = \lfloor \pi\big(F \oplus (0^r\|K\|0^{c-\kappa})\big) \rfloor_\tau \oplus \lfloor K \rfloor_\tau.$$

The ASCON AEAD returns (C, T).

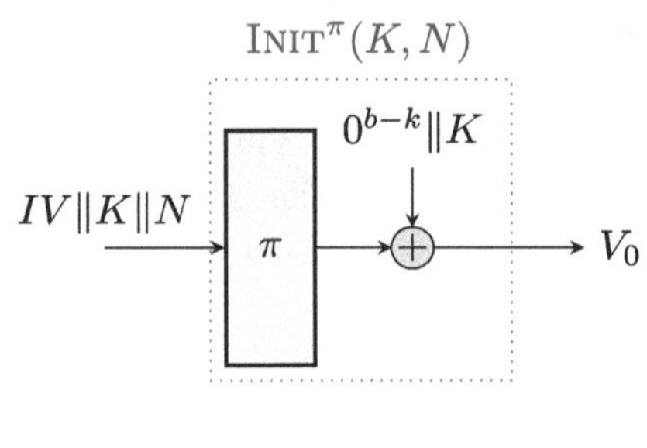

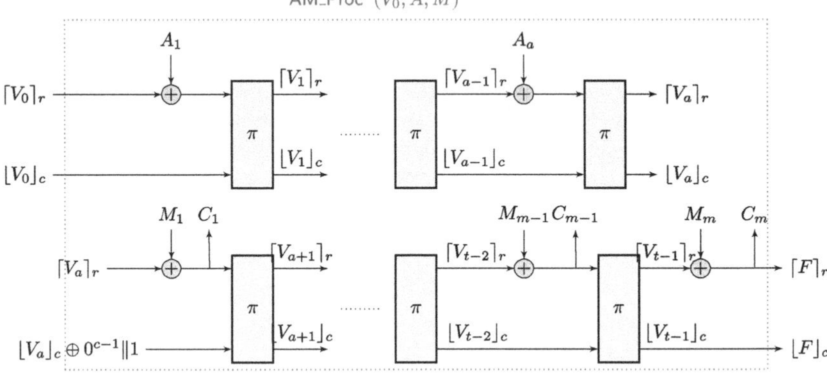

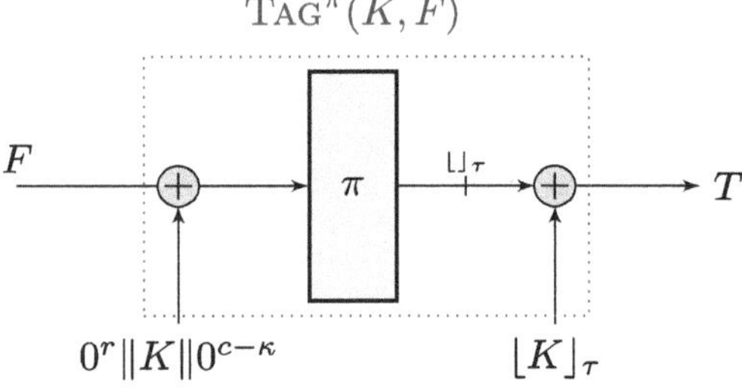

Fig. 2. Encryption in Ascon AEAD. The final ciphertext is $C = \lceil C_1 \| \cdots \| C_m \rceil_{|M|}$, and tag is T. Here $t := a + m$.

Decryption Algorithm. The decryption algorithm consists of two steps. It first performs a verification algorithm to ensure the correctness of the ciphertext and tag pair. Upon successful verification, the algorithm advances to generate the corresponding message. Our emphasis is primarily on the verification process itself, rather than the specific steps involved in message computation.

On receiving an input of the form $(N, A, C, T) \in \{0,1\}^\nu \times \{0,1\}^* \times \{0,1\}^* \times \{0,1\}^\tau$ and a key $K \in \{0,1\}^\kappa$, the steps of the verification process is outlined below:

1. $(A_1, \ldots, A_a) \xleftarrow{r} \mathsf{pad}_1(A)$ and $(C_1, \ldots, C_l) \xleftarrow{r} \mathsf{pad}_2(C)$.
2. Compute $V_0 := \textsc{Init}^\pi(K, N)$.

3. Compute the walk for the permutation π

$$V_0 \xrightarrow{A_1}_{\oplus} V_1 \xrightarrow{A_2}_{\oplus} \cdots \xrightarrow{A_a}_{\oplus} V_a.$$

4. Let $C_l = C_l' \| 10^*$ for some C_l' (may be the empty string) and $|C_l'| = d$. Let $z_a = \lfloor V_a \oplus 0^*1 \rfloor_c$.
   - Case $l = 1$: Define $F = C_l' \parallel (\lfloor V_a \rfloor_{b-d} \oplus 10^*1)$.
   - Case $l \geq 2$: Compute

$$z_a \xrightarrow{C_1} z_{a+1} \xrightarrow{C_2} \cdots \xrightarrow{C_{l-2}} z_{a+l-2}.$$

   Define $F = C_l' \parallel \lfloor \pi(C_{l-1} \| z_{a+l-2}) \oplus 10^* \rfloor_{b-d}$.
5. Reject if $T \neq \textsc{Tag}^\pi(K, F)$, otherwise, accept.

3.1 Security Bound of Ascon

Theorem 1. *Consider a nonce-respecting AEAD adversary $\mathscr{A}$ making q_p permutation queries, q_e encryption queries with a total number of σ_e data blocks, and q_d decryption queries with a total number of σ_d data block. Define $\sigma := \sigma_e + \sigma_d$. Then, we can upper bound the mu-AEAD advantage of $\mathscr{A}$ against* Ascon *as follows:*

$$\begin{aligned}\mathbf{Adv}^{\text{mu-AEAD}}_{\text{ASCON}}(\mathscr{A}) \leq\ & \frac{\mu^2}{2^\kappa} + \frac{2q_d}{2^\tau} + \frac{\sigma_e^2}{2^b} + \frac{\sigma_d(q_p+\sigma_d)}{2^b} + \frac{\mathsf{mcoll}(\sigma_e, 2^r)(\sigma_d + q_p)}{2^c} \\ & + \frac{\mu(q_p+\sigma)}{2^\kappa} + \frac{\mathsf{mcoll}(q_e, 2^\tau)q_d}{2^c} + \frac{\mathsf{mcoll}(\sigma + q_p, 2^\tau)q_d}{2^\kappa} \\ & + \frac{q_e^2 + q_d^2 + q_e q_d + (q_e+q_d)(\sigma+q_p)}{2^b} + \frac{q_e(\sigma+q_p)}{2^b} \\ & + \frac{\mathsf{mcoll}(q_e, 2^{r+c-\kappa})(\sigma+q_p)}{2^\kappa} + \frac{q_d(\sigma+q_p)}{2^{c+\tau}}.\end{aligned}$$

3.2 Interpretation of Theorem 1

First, note that when $\mu = 1$ (corresponding to the su setting), the aforementioned bound aligns with the security bound established in [CDN23a]. In that work, the authors interpreted their bound in the context of the NIST LWC requirements, demonstrating the security of the Ascon mode even when $c = 136$ and $\tau = 64$, as opposed to the original description's requirement of $c \geq 192$ and $\tau = 128$.

As highlighted by the authors in [CDN23b], they were unable to reduce the capacity to 128 because their result was proven under the additional assumption that $\kappa < c$. Given that Ascon has a key size of 128 bits, the capacity needed to exceed 128 bits.

In this current work, we enhance the proof technique, and our proof encompasses the $\kappa = c$ scenario as well. Consequently, the Ascon mode remains secure even when $c = 128$. It is crucial to note that the design of ASCON prohibits $\kappa > c$, as the keys are XOR-ed in the capacity part.

Next, coming to the mu setting, note that the only extra terms in the security bound as compared to the su setting are $\frac{\mu^2}{2^\kappa}$ and $\frac{\mu(q_p+\sigma)}{2^\kappa}$. While the term $\frac{\mu^2}{2^\kappa}$ does not pose any threat, the term $\frac{\mu(q_p+\sigma)}{2^\kappa}$ reduces the security significantly as the number of users becomes large. For ASCON, the key size is 128, and according to NIST LWC specifications, q_p can be of the order 2^{112}. This does not leave room for a very large μ. For example, if the number of users is around 2^{15}, the advantage is less than half.

Hence, it is evident that in the mu setting, the security of the ASCON mode persists even when $c = 128$ and $\tau = 64$, provided the number of users does not reach excessively large values.

3.3 Tightness of the Bounds

We derive a bound of the following order:

$$\frac{\mu q_p}{2^\kappa} + \frac{\mu^2}{2^\kappa} + \frac{q_p}{2^c} + \frac{q_d}{2^c} + \frac{\sigma_e^2}{2^b} + \frac{\sigma_d^2}{2^b} + \frac{q_d}{2^\tau} + \frac{q_p \sigma_d}{2^b} + \frac{q_d}{2^\kappa}$$

As observed above, the only additional terms compared to the su setting, are $\mu q_p/2^\kappa$ and $\mu^2/2^\kappa$.

- The term $\frac{\mu q_p}{2^\kappa}$ corresponds to generic attacks which guess one of the keys in primitive calls.
- The term $\frac{\mu^2}{2^\kappa}$ corresponds to generic attacks which guess key collisions.

As attacks for the remaining terms were already shown in [CDN23a], our bounds are tight.

3.4 A Special Case: $\kappa = c$

Before delving into the proof of Theorem 1, let's first examine why the authors of [CDN23a] operated under the assumption $\kappa < c$. In the scenario where $\kappa = c$, in the special case of having only one message block and no associated data, the keys in the output of the initialization phase and the input of the finalization phase nullify each other. This interaction is illustrated in Fig. 3.

We acknowledge this as a special case before initiating our proof to explain why the proof in [CDN23a] cannot be directly adapted to the multi-user setting with minor adjustments. Nevertheless, it is worth mentioning that our subsequent proof adheres to the general structure of [CDN23a], and some overlap is inevitable.

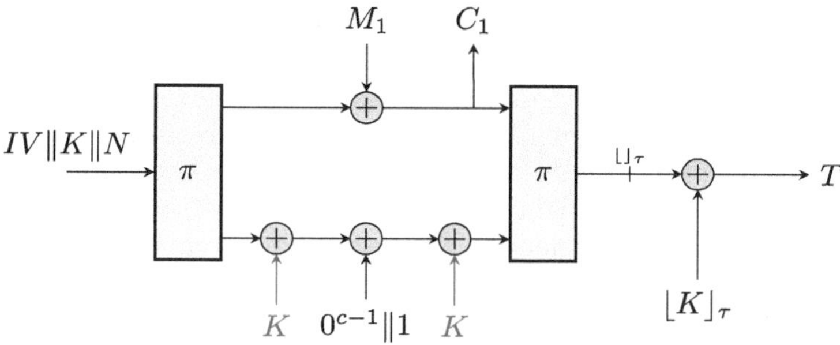

Fig. 3. Encryption in ASCON AEAD when $\kappa = c$, and we have only a single block message without any associated data. The two keys additions in red cancel each other out. The final ciphertext is $C = \lceil C_1\rceil_{|M|}$, and tag is T. (Color figure online)

3.5 Proof Overview of Theorem 1

We employ the H-coefficient technique for our proof. In the real world, μ keys $K_1, \ldots, K_\mu$ and a random permutation Π are sampled independently. All queries are then responded to honestly. The extended transcript consists of:

- all inputs and outputs corresponding to encryption, decryption, and primitive queries,
- all inputs and outputs of the permutation calls corresponding to encryption and decryption queries.

The ideal world consists of an online phase and an offline phase. In the online phase:

- encryption queries responded to randomly,
- all decryption queries are rejected, and
- permutation queries are responded to faithfully.

The offline phase samples intermediate variables, and generates an extended transcript. It proceeds in stages:

1. Start with permutation query transcript P.
2. Sample intermediate variables for encryption queries to obtain permutation input-output pairs $\mathsf{P_E}$.
3. Randomly extend P to P_1 by setting input-outputs for decryption queries. Set $\mathsf{P}_2 := \mathsf{P}_1 \cup \mathsf{P_E}$.
4. Finally, sample keys $K_1, \ldots, K_\mu$. Set input-output pairs for initialization first, and then move on to the finalization phase. Update P_2 twice to obtain $\mathsf{P}_{\mathsf{fin}}$.

In the offline phase of the ideal world, bad events occur when

- Variables sampled are not permutation-compatible.
- We have a correct forging.
- Decryption queries are not rejected.

Bounding the bad events concludes the proof. A detailed proof of the theorem appears in the full version of this paper.

4 Authenticity in the Nonce Misuse Setting

Up until now, we only considered the nonce-respecting setting, where no two encryption queries to the same user had the same nonce, although repetition of nonce across decryption queries was allowed. If the adversary reuses nonces for encryption queries, confidentiality cannot be guaranteed anymore, but we can still aim for authenticity. Considering the authenticity security of ASCON against adversaries that can possibly misuse nonces. We have the following result:

Theorem 2. *Consider a possibly nonce-misusing authentication adversary* $\mathscr{A}$ *making* q_p *permutation queries,* q_e *encryption queries with a total number of* σ_e *data blocks, and* q_d *decryption queries with a total number of* σ_d *data blocks. Define* $\sigma := \sigma_e + \sigma_d$. *Then, we can upper bound the mu-auth advantage of* $\mathscr{A}$ *against* ASCON *as follows:*

$$\begin{aligned}\mathbf{Adv}^{\mathrm{mu-auth}}_{\mathrm{ASCON}}(\mathscr{A}) \leq\; & \frac{\mu^2}{2^\kappa} + \frac{2q_d}{2^\tau} + \frac{\sigma_e^2}{2^b} + \frac{\sigma_d(q_p+\sigma_d)}{2^b} + \frac{\mathsf{mcoll}(\sigma_e, 2^r)(\sigma_d + q_p)}{2^c} \\ & + \frac{\mu(q_p+\sigma)}{2^\kappa} + \frac{\mathsf{mcoll}(q_e, 2^\tau)q_d}{2^c} + \frac{\mathsf{mcoll}(\sigma + q_p, 2^\tau)q_d}{2^\kappa} \\ & + \frac{q_e^2 + q_d^2 + q_e q_d + (q_e+q_d)(\sigma+q_p)}{2^b} + \frac{q_e(\sigma+q_p)}{2^b} \\ & + \frac{\mathsf{mcoll}(q_e, 2^{r+c-\kappa})(\sigma+q_p)}{2^\kappa} + \frac{q_d(\sigma+q_p)}{2^{c+\tau}}.\end{aligned}$$

This bound is similar to that of Theorem 1, only the term $\sigma_e^2/2^b$ is replaced by $\sigma_e^2/2^c$. Note that in the lightweight setting, $\sigma_e^2 \ll 2^{128}$. This means ASCON maintains the authenticity security even under nonce misuse. The proof is very similar to that of the nonce-respecting setting and appears in the full version of this paper. To show that the bound is tight, in the full version, we demonstrate a forgery that establishes the $\sigma_e^2/2^c$ bound.

5 Large Key ASCON

One of the major limitations of ASCON is its compromised security as the user count scales up. This issue is highlighted in Sect. 3.2, where the term $\frac{\mu q_p}{2^\kappa}$ emerges as the dominant factor in Theorem 1, thereby limiting the number of users. The most straightforward remedy for this constraint would involve augmenting the key size of ASCON. Additionally, a larger key size has the potential to fortify the resistance against key recovery attacks that leverage Grover's algorithm. It is essential to note that increasing the key size does not necessarily bolster security against quantum attacks in a generalized context, and the quantum security of any ASCON or related scheme must be assessed independently.

The key size cannot be directly increased in the original ASCON construction because of the constraints $\kappa + \nu \leq b$ and $\kappa \leq c$, where ν denotes the nonce size. For instance, if we opt for a nonce size of 128 bits, along with an extra 64-bit IV, the key size becomes confined to 128 bits. This limitation remains unchanged if we aim to permit a rate of 192 bits. Consequently, we introduce a novel AEAD mode akin to ASCON, known as the LK-ASCON mode, allowing for an arbitrary key size denoted as $\tau \leq \kappa < b$, where τ signifies the tag size. To maintain consistency, we select an $IV \in \{0,1\}^{b-\kappa}$. The AEAD uses a permutation π (can be the same ASCON permutation), modeled to be the random permutation while we analyze its security.

Remark 2. While we define the mode for any $\tau \leq \kappa < b$, we call this LK-ASCON (for Large Key ASCON) as this enables us to increase the key size from 128 bits to up to 320 bits. Of particular interest is the variant with a key size 256 bits (allowing a 64-bit IV). We call this variant ASCON-256. Note that $\tau \leq \kappa$ is necessary for masking the full tag.

Encryption Algorithm. It receives an input of the form $(N, A, M) \in \{0,1\}^{\nu} \times \{0,1\}^* \times \{0,1\}^*$ and a key $K \in \{0,1\}^{\kappa}$. We divide the encryption algorithm into the same three phases: (i) initialization, (ii) nonce, associated data and message processing, and (iii) tag generation, run sequentially.

INITIALIZATION. In this phase, we first apply the following function

$$\textsc{Init}^{\pi}(K) = \pi(IV\|K) \oplus (0^{b-\kappa}\|K) := V_0.$$

Note that the initialization process no longer takes the nonce N as an input. Here, it is processed with the associated data in the next step.

NONCE, ASSOCIATED DATA AND MESSAGE PROCESSING. We first parse them:

$$(A_1, \ldots, A_a) \xleftarrow{r} \mathsf{pad}_1(N, A), \quad (M_1, \ldots, M_m) \xleftarrow{r} \mathsf{pad}_2(M).$$

Note that a cannot be zero here, as even if there is no associated data, the nonce is parsed. As before, $m \geq 1$. Using the XOR-function graph corresponding to the function $\pi^{\oplus}$, we obtain a walk

$$V_0 \xrightarrow{A_1}_{\oplus} V_1 \xrightarrow{A_2}_{\oplus} \cdots \xrightarrow{A_a}_{\oplus} V_a, \quad V_a \oplus 0^*1 \xrightarrow{M_1}_{\oplus} V_{a+1} \cdots \xrightarrow{M_{m-1}}_{\oplus} V_{a+m-1}.$$

We define the ciphertext as follows:

$$C_i = \lceil V_{a+i-1} \rceil_r \oplus M_i, \quad \forall i \in [m], \quad C = \lceil C_1 \| \cdots \| C_m \rceil_{|M|}.$$

We denote the above process as

$$\mathsf{AM\_Proc}^{\pi}(V_0, N, A, M) \rightarrow (C, F := V_{t-1} \oplus (M_m \| 0^c)).$$

TAG GENERATION. Finally, we compute

$$T := \textsc{Tag}^{\pi}(K, F) = \lfloor \pi(F \oplus (K\|0^{b-\kappa})) \rfloor_{\tau} \oplus \lfloor K \rfloor_{\tau}$$

The LK-ASCON AEAD returns (C, T). A pictorial description of the encryption algorithm of LK-ASCON can be found in Fig. 4.

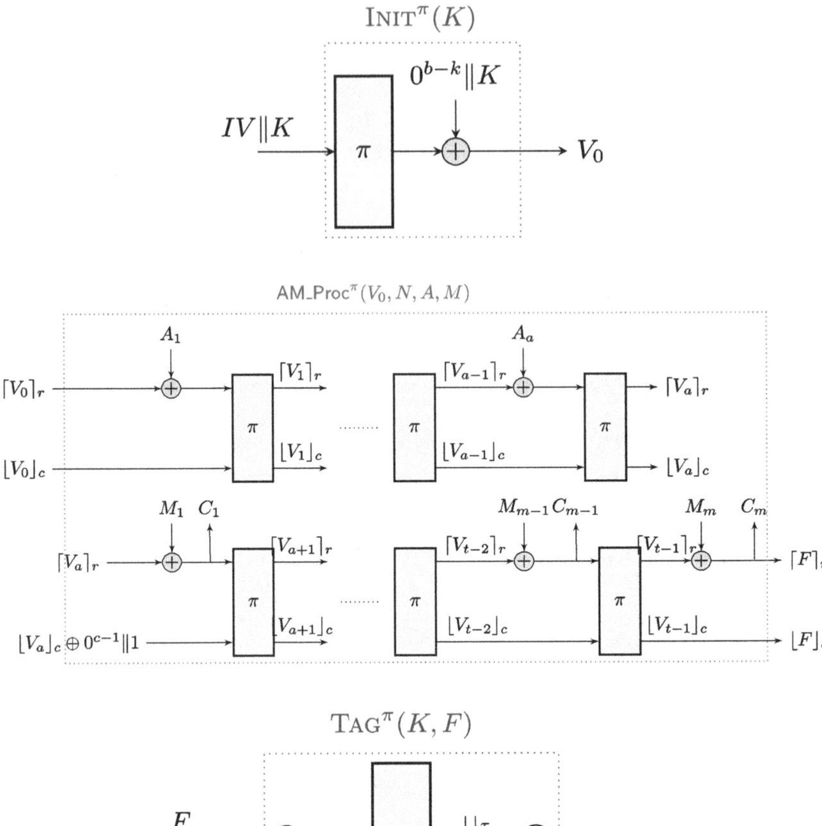

Fig. 4. Encryption in LK-ASCON AEAD. The difference with conventional ASCON is that there is no nonce at the input of INIT. The nonce is parsed with the associated data, and fed at the AM_Proc step. Also, the key addition at the input of TAG is $K\|0^{b-\kappa}$, meaning even if $\kappa \leq c$, it is xored at the rate part.

Decryption Algorithm. As before, our focus lies primarily on the verification process itself, rather than the specific steps involved in message computation. On receiving an input of the form $(N, A, C, T) \in \{0,1\}^\nu \times \{0,1\}^* \times \{0,1\}^* \times \{0,1\}^\tau$ and a key $K \in \{0,1\}^\kappa$, the steps of the verification process is outlined below:

1. $(A_1, \ldots, A_a) \xleftarrow{r} \mathsf{pad}_1(N, A)$ and $(C_1, \ldots, C_l) \xleftarrow{r} \mathsf{pad}_2(C)$.
2. Compute $V_0 := \textsc{Init}^\pi(K)$.

3. Compute the walk for the permutation π

$$V_0 \xrightarrow{A_1}_{\oplus} V_1 \xrightarrow{A_2}_{\oplus} \cdots \xrightarrow{A_a}_{\oplus} V_a$$

4. Let $C_l = C_l' \| 10^*$ for some C_l' (may be the empty string) and $|C_l'| = d$. Let $z_a = \lfloor V_a \oplus 0^*1 \rfloor_c$.
   - Case $l = 1$: Define $F = C_l' \parallel (\lfloor V_a \rfloor_{b-d} \oplus 10^*1)$.
   - Case $l \geq 2$: Compute

$$z_a \xrightarrow{C_1} z_{a+1} \xrightarrow{C_2} \cdots \xrightarrow{C_{l-2}} z_{a+l-2}$$

   We define $F = C_l' \parallel \lfloor \pi(C_{l-1} \| z_{a+l-2}) \oplus 10^* \rfloor_{b-d}$.
5. Reject if $T \neq \text{TAG}^\pi(K, F)$, otherwise, accept.

Remark 3. The functions INIT^π, $\mathsf{AM\_Proc}^\pi$ and TAG^π are different for ASCON and LK-ASCON. In fact, for the first two, even the domains are different. We have intentionally reused the notations to emphasize the similarity in the processes of the two AEAD modes.

5.1 Security Bounds on LK-ASCON

We give two security bounds on ASCON: AEAD advantage for nonce-respecting multi-user LK-ASCON, and authenticity advantage for nonce-misuse multi-user LK-ASCON, along with their interpretations.

Theorem 3. *Consider a nonce-respecting AEAD adversary $\mathscr{A}$ making q_p permutation queries, q_e encryption queries with a total number of σ_e data blocks, and q_d decryption queries with a total number of σ_d data block. Define $\sigma := \sigma_e + \sigma_d$. Then, we can upper bound the mu-AEAD advantage of $\mathscr{A}$ against LK-ASCON as follows:*

$$\begin{aligned}\mathbf{Adv}^{\text{mu-AEAD}}_{LK\text{-ASCON}}(\mathscr{A}) \leq\ & \frac{\mu^2}{2^\kappa} + \frac{2q_d}{2^\tau} + \frac{\sigma_e^2}{2^b} + \frac{\sigma_d(q_p + \sigma_d)}{2^b} + \frac{\mathsf{mcoll}(\sigma_e, 2^r)(\sigma_d + q_p)}{2^c} \\ & + \frac{\mu(q_p + \sigma)}{2^\kappa} + \frac{\mathsf{mcoll}(q_e, 2^\tau) q_d}{2^c} + \frac{\mathsf{mcoll}(\sigma + q_p, 2^\tau) q_d}{2^\kappa} \\ & + \frac{q_e^2 + q_d^2 + q_e q_d + (q_e + q_d)(\sigma + q_p)}{2^b} + \frac{\omega_{r,\kappa}(\sigma + q_p)}{2^b},\end{aligned}$$

where

$$\omega_{r,\kappa} = \begin{cases} 2q_e, & \text{if } r \leq \kappa \\ q_e + \mathsf{mcoll}(q_e, 2^{r-\kappa}) \cdot 2^{r-\kappa} & \text{otherwise} \end{cases}.$$

Interpretation of the Theorem. The first thing we would like our modified construction to have is to achieve the same security as ASCON when $\kappa \leq c$ (though this necessitates a larger IV). Upon interpreting our bound within the context of the NIST LWC requirements, it becomes evident that we maintain the same level of security as previously, encompassing both a 128-bit rate and 192-bit rate, as well as both a 64-bit tag and 128-bit tag.

Next, note that for any arbitrary κ, the above bound is of the order

$$\mathcal{O}\left(\frac{\mu q_p}{2^\kappa} + \frac{q_d}{2^\tau} + \frac{q_p}{2^c} + \frac{\sigma q_p}{2^b}\right),$$

when interpreted with NIST parameters, which is exactly the same as that of multi-user AEAD security of conventional ASCON. Thus, LK-ASCON achieves the same security as ASCON, while enabling an increase in key size of up to 320 bits. Particularly, we would like to note that ASCON-256 achieves the same security as ASCON by just doubling the key size and keeping everything else the same (note that for other key sizes, we need to change the IV as well).

Theorem 4. *Consider a possibly nonce-misusing authentication adversary* $\mathscr{A}$ *making* q_p *permutation queries,* q_e *encryption queries with a total number of* σ_e *data blocks, and* q_d *decryption queries with a total number of* σ_d *data blocks. Define* $\sigma := \sigma_e + \sigma_d$*. Then, we can upper bound the mu-auth advantage of* $\mathscr{A}$ *against LK-*ASCON *as follows:*

$$\begin{aligned}\mathbf{Adv}^{\text{mu-auth}}_{LK\text{-ASCON}}(\mathscr{A}) \leq \frac{\mu^2}{2^\kappa} + \frac{2q_d}{2^\tau} + \frac{\sigma_e^2}{2^c} + \frac{\sigma_d(q_p+\sigma_d)}{2^b} + \frac{\mathsf{mcoll}(\sigma_e, 2^r)(\sigma_d+q_p)}{2^c} \\ + \frac{\mu(q_p+\sigma)}{2^\kappa} + \frac{\mathsf{mcoll}(q_e, 2^\tau)q_d}{2^c} + \frac{\mathsf{mcoll}(\sigma+q_p, 2^\tau)q_d}{2^\kappa} \\ + \frac{q_e^2+q_d^2+q_eq_d+(q_e+q_d)(\sigma+q_p)}{2^b} + \frac{\omega_{r,\kappa}(\sigma+q_p)}{2^b}.\end{aligned}$$

This shows that like ASCON, LK-ASCON too maintains authenticity under nonce-misuse. In the next section, we give a proof overview of Theorem 3. The proof of Theorem 4 is a straightforward extension, and we outline the sketch in the full version of this paper.

5.2 Proof Overview of Theorem 3

The proof follows the structure outlined in Sect. 3.5 for ASCON. However, there are notable differences, summarized as follows:

- The case $\kappa = c$ does not result in key nullification because the positions where keys are XOR-ed are distinct. This aspect simplifies the proof slightly.
- New challenges arise due to alterations in constraints. The nonce is now processed with associated data, leading to the identical output of the initialization phase, V_0, for each query to the same user. Additionally, if the nonce spans more than one block, up to a certain point i, all V_i values can be the same for two queries. This challenge is addressed by defining the longest common prefix, especially for encryption queries.

– The analysis diverges depending on whether $\kappa \geq r$. If $\kappa \geq r$, the final block of the ciphertext is fully masked by the key XOR in the input of the finalization phase. Otherwise, if $\kappa < r$, the final ciphertext block is only partially masked.

A detailed proof is provided in the full version of this paper.

6 Conclusion

In this paper, we derived a multi-user security bound for the Ascon AEAD mode, the winner of the recently concluded NIST LWC competition. This mode follows a Sponge type of construction. Notably, the inclusion of a key XOR operation during the tag generation phase allows us to derive a bound of the following order:

$$\frac{\mu T}{2^{\kappa}} + \frac{T}{2^{c}} + \frac{D}{2^{\tau}} + \frac{DT}{2^{b}}$$

where T is the time complexity and D is the data complexity of the adversary. We also show that Ascon maintains this authenticity security even in the nonce misuse setting, although confidentiality is not guaranteed.

Finally, we introduce a variant of Ascon, called LK-Ascon, which allows an increase in key size, up to 320 bits. Notably, increasing the key size enhances the security of Ascon in the multi-user setting in addition to providing better security against quantum key recovery attacks utilizing Grover's algorithm. Like Ascon, LK-Ascon also maintains its authenticity security in the nonce misuse setting.

Acknowledgements. Bishwajit Chakraborty is supported by the NRF-ANR project SELECT ("NRF-2020-NRF-ANR072"). Mridul Nandi is partially supported by SERB Matrics project (MTR/2023/000302) titled "Indistinguishability Security in Cryptography".

References

[BDPA11] Bertoni, G., Daemen, J., Peeters, M., Van Assche, G.: Duplexing the sponge: single-pass authenticated encryption and other applications. In: Miri, A., Vaudenay, S. (eds.) SAC 2011. LNCS, vol. 7118, pp. 320–337. Springer, Heidelberg (2012). https://doi.org/10.1007/978-3-642-28496-0_19

[BDPA12] Bertoni, G., Daemen, J., Peeters, M., Van Assche, G.: Permutation-based encryption, authentication and authenticated encryption. In: DIAC 2012 (2012)

[Ber05] Bernstein, D.J.: The poly1305-AES message-authentication code. In: Gilbert, H., Handschuh, H. (eds.) FSE 2005. LNCS, vol. 3557, pp. 32–49. Springer, Heidelberg (2005). https://doi.org/10.1007/11502760_3

[Ber08a] Bernstein, D.J.: Chacha, a variant of salsa20. Workshop Record of SASC **8**, 3–5 (2008)

[Ber08b] Bernstein, D.J.: The Salsa20 family of stream ciphers. In: Robshaw, M., Billet, O. (eds.) New Stream Cipher Designs. LNCS, vol. 4986, pp. 84–97. Springer, Heidelberg (2008). https://doi.org/10.1007/978-3-540-68351-3_8

[BN08] Bellare, M., Namprempre, C.: Authenticated encryption: Relations among notions and analysis of the generic composition paradigm. J. Cryptol. **21**(4), 469–491 (2008)

[CDN23a] Chakraborty, B., Dhar, C., Nandi, M.: Exact security analysis of ASCON. In: Guo, J., Steinfeld, R. (eds.) ASIACRYPT 2023, Part III. LNCS, vol. 14440, pp. 346–369. Springer, Cham (2023). https://doi.org/10.1007/978-981-99-8727-6_12

[CDN23b] Chakraborty, B., Dhar, C., Nandi, M.: Exact security analysis of ASCON. IACR Cryptol. ePrint Arch., p. 775 (2023)

[CDN24] Chakraborty, B., Dhar, C., Nandi, M.: Tight multi-user security of ascon and its large key extension. Cryptology ePrint Archive, Paper 2024/579 (2024). https://eprint.iacr.org/2024/579

[Com14] The CAESAR Committee. Caesar: competition for authenticated encryption: security, applicability, and robustness (2014)

[CS14] Chen, S., Steinberger, J.: Tight security bounds for key-alternating ciphers. In: Nguyen, P.Q., Oswald, E. (eds.) EUROCRYPT 2014. LNCS, vol. 8441, pp. 327–350. Springer, Heidelberg (2014). https://doi.org/10.1007/978-3-642-55220-5_19

[DEMS14] Dobraunig, C., Eichlseder, M., Mendel, F., Schläffer, M.: ASCON v1. Submission to the CAESAR Competition (2014)

[DEMS19] Dobraunig, C., Eichlseder, M., Mendel, F., Schläffer, M.: ASCON. Submission to NIST LwC Standardization Process (FINALIST) (2019)

[DNT19] Dutta, A., Nandi, M., Talnikar, S.: Beyond Birthday Bound Secure MAC in Faulty Nonce Model. In: Ishai, Y., Rijmen, V. (eds.) EUROCRYPT 2019, Part I. LNCS, vol. 11476, pp. 437–466. Springer, Cham (2019). https://doi.org/10.1007/978-3-030-17653-2_15

[ML23] Mennink, B., Lefevre, C.: Generic security of the ascon mode: on the power of key blinding. IACR Cryptol. ePrint Arch., p. 796 (2023)

[MN17] Mennink, B., Neves, S.: Encrypted Davies-Meyer and its dual: towards optimal security using mirror theory. In: Katz, J., Shacham, H. (eds.) CRYPTO 2017. LNCS, vol. 10403, pp. 556–583. Springer, Cham (2017). https://doi.org/10.1007/978-3-319-63697-9_19

[MV04] McGrew, D.A., Viega, J.: The security and performance of the Galois/counter mode (GCM) of operation. In: Canteaut, A., Viswanathan, K. (eds.) INDOCRYPT 2004. LNCS, vol. 3348, pp. 343–355. Springer, Heidelberg (2004). https://doi.org/10.1007/978-3-540-30556-9_27

[NIS18] NIST. Submission requirements and evaluation criteria for the Lightweight Cryptography Standardization Process (2018). https://csrc.nist.gov/CSRC/media/Projects/Lightweight-Cryptography/documents/final-lwc-submission-requirements-august2018.pdf

[Pat91] Patarin, J.: Etude des Générateurs de Permutations Pseudo-aléatoires Basés sur le Schéma du DES. Ph.D thesis, Université de Paris (1991)

[Pat08] Patarin, J.: The "coefficients H" technique. In: Avanzi, R.M., Keliher, L., Sica, F. (eds.) SAC 2008. LNCS, vol. 5381, pp. 328–345. Springer, Cham (2008). https://doi.org/10.1007/978-3-642-04159-4_21

[SCM08] Salowey, J., Choudhury, A., McGrew, D.A.: AES Galois counter mode (GCM) cipher suites for TLS. RFC **5288**, 1–8 (2008)

Differential Distinguishing Attacks on SNOW-V, SNOW-Vi and KCipher-2

Rikuto Kurahara[1], Kosei Sakamoto[2], Yuto Nakano[3], and Takanori Isobe[1](✉)

[1] University of Hyogo, Kobe, Japan
takanori.isobe@ai.u-hyogo.ac.jp
[2] Mitsubishi Electric Corporation, Kamakura, Japan
Sakamoto.Kosei@dc.MitsubishiElectric.co.jp
[3] KDDI Research, Inc., Fujimino, Japan

Abstract. In this paper, we evaluate the security against differential attacks for three important stream ciphers: SNOW-V, SNOW-Vi and KCipher-2. SNOW-V and SNOW-Vi are proposed as standard encryption schemes for the 5G mobile communication system, while KCipher-2 is a standard algorithm by ISO/IEC 18033-4 and CRYPTREC. We leverage state-of-the-art SAT-aided search tools. This enables us to evaluate bit-level differential characteristics during the initialization phase. Besides, to enhance search efficiency, we properly choose input differences, and eliminate ineffective differential transitions by carefully analyzing the differential behaviors in the nonlinear transformation. As a result, in the related-IV setting, we demonstrate that distinguishing attacks on 5/5/8 rounds for SNOW-V/SNOW-Vi/KCipher-2, respectively. In the related-key setting, we can extend to 7/8/8 rounds for SNOW-V/SNOW-Vi/KCipher-2, respectively. Our results are best attacks on the initialization phase of these ciphers, and first result in the related-key setting.

Keywords: Stream cipher · KCipher-2 · SNOW-V · SNOW-Vi · Differential attack · SAT Solver

1 Introduction

SNOW-V, which is a new variant of a family of SNOW stream ciphers [4], was proposed for a standard encryption scheme for the 5G mobile communication system in 2019 by Ekdahl et al [5]. To achieve the strong security requirements by the 3GPP standardization organization for the 5G system, SNOW-V provides a 256-bit security level against key recovery attacks with a 256-bit key and 128- bit IV (Initialization Vectors). SNOW-V consists of a Linear Feedback Shift Register (LFSR) and Finite State Machine (FSM). The overall structure of SNOW-V follows the design strategy of SNOW 2.0 and SNOW-3G. It takes advantage of AES-NI and some SIMD operations for efficient implementation in high-end software environments. Each round has two AES-round operations to update the states of the FSM.

A slightly modified version of SNOW-V stream cipher, called SNOW-Vi, was proposed by the same designers in 2021 [6]. The structural differences between

T. Zhu and Y. Li (Eds.): ACISP 2024, LNCS 14895, pp. 77–97, 2024.
https://doi.org/10.1007/978-981-97-5025-2_5

SNOW-V and SNOW-Vi are the LFSR update function and the location of the tap T2. The purpose of this change is to better accommodate a fast software implementation on lower-grade CPUs that only supports 128-bit wide SIMD registers. KCipher-2 [9] is a stream cipher proposed by Kiyomoto et al, in 2007, and supports a 128-bit key and a 128-bit IV. It was adopted as an ISO/IEC standard (ISO/IEC 18033-4) in 2012 as well as selected in the CRYPTREC Cipher List [1] in 2013. KCipher-2 follows the design of the SNOW-type structure but differs from others by adopting a dynamic feedback shift register (DFSR) that dynamically changes the feedback function.

1.1 Existing Work

For initialization phase, Hoki et al. demonstrated distinguishing attacks, which distinguish keystream form a random sequence, on SNOW-V and SNOW-Vi by Mixed Integer Linear Programming (MILP) [7,8]. Specifically, they presented differential attacks on 4/4 rounds of SNOW-V/SNOW-Vi, respectively, and integral attacks on 5/5/7 rounds of SNOW-V/SNOW-Vi/KCipher-2, respectively. However, their differential attacks face challenges in searching entire search space in a significant number of rounds due to computational limitations. Thus, all existing results are limited to a small number of rounds or limited search space. For the key stream generation phase, the security evaluation in the correlation attack was conducted by Shi et al [14], which shows that a correlation attack can recover the full state on SNOW-V and SNOW-Vi with time complexity of $2^{237.5}$.

1.2 Our Contribution

In this paper, we evaluate the security of differential attacks against SNOW-V/SNOW-Vi/KCipher-2 by employing the SAT evaluation method proposed by Sun et al [15]. This approach enables us to investigate bit-level differential characteristics for larger search space than existing results. In order to minimize the search cost for a large number of rounds while keeping our capability, we analyze the differential behaviors in the nonlinear transformation of the initialization algorithm, and properly choose input differences to be inserted into key and IV space, ensuring an optimized balance between comprehensiveness and efficiency.

As another technical contribution, we provide an efficient modeling method of the Dynamic Feedback Counter (DFC), which is part of the function of KCipher-2. To the best of our knowledge, there has been no research on developing a model for the DFC, which has posed challenges for conducting security evaluations of KCipher against differential cryptanalysis. Our new modeling method allows us to ignore the differential characteristics that are predicted to be ineffective for attacks in advance. This method enables to perform differential analysis of KCipher-2 for the first time. However, it needs to be careful that the evaluation result is not necessarily optimal because it is an evaluation that ignores some differential characteristics.

As a result, in the related-IV setting, we demonstrate that distinguishing attacks on 5/5/8 rounds for SNOW-V/SNOW-Vi/KCipher-2, respectively, as

Table 1. Summary of distinguishing attacks in the initialization phase.

Cipher	Attack Method	Round	Time/Data	ref
SNOW-V	Integral	5	2^{48}	[7]
	Differential (Related-IV)	3	2^{17}	
		4	2^{97}	
		3	2^{17}	Ours
		4	2^{91}	
		5	2^{169}	
	Differential (Related-Key)	**5**	2^{28}	
		6	2^{113}	
		7	2^{218}	
SNOW-Vi	Integral	5	2^{16}	[7]
	Differential (Related-IV)	3	2^{12}	
		4	2^{39}	
		4	2^{37}	Ours
		5	2^{117}	
	Differential (Related-Key)	**6**	2^{60}	
		7	2^{96}	
KCipher-2	Integral	7	2^{3}	[8]
	Differential (Related-IV)	**7**	2^{96}	Ours
		8	2^{123}	
	Differential (Related-Key)	**7**	2^{96}	
		8	2^{123}	

shown in Table 1. In the related-key setting, we can extend to 7/8/8 rounds for SNOW-V/SNOW-Vi/KCipher-2, respectively. Our results are best attacks on initialization phase of these ciphers, and first result in the related-key setting.

2 Preliminaries

In this section, we first describe differential attacks. Then, we explain the security evaluation using the SAT solver.

2.1 Differential Cryptanalysis

The differential attack [3] proposed by Biham et al. is one of the most powerful attack methods against symmetric-key block ciphers, which can also be applied to stream cipher as shown by Biham and Dunkelman [2]. When evaluating the security of differential attacks, we estimate the probability that the input differences reach the output differences. To estimate that probability, called *differential probability*, we often use a *differential characteristic*, a sequence of the internal differences in a primitive. Let $E(\cdot) = f_r(\cdot) \circ \cdots \circ f_1(\cdot)$ be the r-round iterated primitive, then differential characteristics are defined as follows:

Definition 1. (Differential Characteristics)
Differential characteristics are a sequence of differences over E defined as follows:

$$C = (c_0 \xrightarrow{f_1} c_1 \xrightarrow{f_2} \dots \xrightarrow{f_r} c_r),$$

where $(c_0, c_1, \dots, c_r)$ denotes the differences in the output of each round, and c_0 denotes the differences in the input of a primitive.

We can estimate the probability of differential characteristics as follows:

$$\Pr(C) = \prod_{i=1}^{r} \Pr(c_{i-1} \xrightarrow{f_i} c_i).$$

We often use *weight* to express the probability or differential characteristics for simplification. A weight is defined as follows:

Definition 2. (Weight)
A weight w is a negated value of the binary logarithm of the probability Pr defined as $w = -\log_2 P_r$

When we evaluate weight in stream ciphers, we consider weight related only to the first output key stream.

2.2 Evaluation Using SAT Solver

Security evaluation using mathematical solvers is known as a promising method to evaluate the security of symmetric-key primitives. In Inscrypt 2011, Mouha et al. proposed an automated method to evaluate the security against differential cryptanalysis using the mixed integer linear programming (MILP) [13]. In FSE 2022, Sun et al. proposed a SAT-based automated method, making it possible to evaluate more efficiently than evaluation by MILP [15].

SAT is a problem of determining whether there is a true variable assignment in the boolean formula given by a binary variable. It is known that SAT solvers can efficiently solve such a problem. The recently proposed SAT solver accepts the boolean formula written in Conjunctive normal form (CNF), which is the conjunction ($\wedge$) of the disjunction ($\vee$) on boolean variables.

The SAT Model to Evaluate Optimal Differential Characteristics. To explain the SAT models for each operation, we follow the explanation of Sun et al.'s work [15]. Since the ShiftRows operation, which is a part of the AES round function, involves byte-level swapping, it does not require SAT modeling. Additionally, MixColumns can be represented by 92 XOR operations as proposed by Lin et al. [10]. Therefore, we explain the models of an XOR, S-box, and modular addition.

XOR. Let α, β, and γ be boolean variables such that $\alpha \oplus \beta = \gamma$. The differential propagation over an XOR is expressed as follows: , then the differential holds if and only if the values of α, β, and γ validate all the assertions in the following:

$$\alpha \vee \beta \vee \overline{\gamma} = 1, \quad \alpha \vee \overline{\beta} \vee \gamma = 1, \qquad \overline{\alpha} \vee \beta \vee \gamma = 1, \quad \overline{\alpha} \vee \overline{\beta} \vee \overline{\gamma} = 1.$$

S-Box. Let $\boldsymbol{a} = (a_0, a_1, \ldots, a_{i-1})$, $\boldsymbol{b} = (b_0, b_1, \ldots, b_{i-1})$, and $\boldsymbol{p} = (p_0, p_1, \ldots, p_{j-1})$ be the input and output differences of an i-bit S-box and boolean variables to express the weight in an S-box where j is the maximum weight of the differential propagation, respectively.

To express the differential propagation and its weight in an S-box, we construct the following Boolean formula:

$$f(\boldsymbol{a}, \boldsymbol{b}, \boldsymbol{p}) = \begin{cases} 1 & \textit{if } \Pr(\boldsymbol{a} \rightarrow \boldsymbol{b}) = 2^{-\sum_{q=0}^{j-1} p_q}, \\ 0 & \textit{otherwise.} \end{cases}$$

A set A, which contains all vectors satisfying $f(\boldsymbol{x}, \boldsymbol{y}, \boldsymbol{z}) = 0$, is expressed as follows:

$$A = \{(\boldsymbol{x}, \boldsymbol{y}, \boldsymbol{z}) \in \mathbb{F}_2^{2i+j} \mid f(\boldsymbol{x}, \boldsymbol{y}, \boldsymbol{z}) = 0\},$$

where $\boldsymbol{x} = (x_0, x_1, \ldots, x_{i-1})$, $\boldsymbol{y} = (y_0, y_1, \ldots, y_{i-1})$, and $\boldsymbol{z} = (z_0, z_1, \ldots, z_{j-1})$. We must prohibit the propagation expressed A because it is equivalent to a set of invalid propagation pattern. It can be realized as follows:

$$\bigvee_{c=0}^{i-1} (a_c \oplus x_c) \vee \bigvee_{d=0}^{i-1} (b_d \oplus y_d) \vee \bigvee_{e=0}^{j-1} (p_e \oplus z_e) = 1, \; (\boldsymbol{x}, \boldsymbol{y}, \boldsymbol{z}) \in A. \tag{1}$$

Therefore, these clauses are enough to extract the differential propagation with corresponding weight in an i-bit S-box. We can convert Eq. (1) to

$$g(\boldsymbol{a}, \boldsymbol{b}, \boldsymbol{p}) = \bigwedge_{(\boldsymbol{x}, \boldsymbol{y}, \boldsymbol{z}) \in \mathbb{F}_2^{2i+j}} \left(g(\boldsymbol{x}, \boldsymbol{y}, \boldsymbol{z}) \vee \bigvee_{c=0}^{i-1} (a_c \oplus x_c^{\eta}) \vee \bigvee_{d=0}^{i-1} (b_d \oplus y_d^{\eta}) \vee \bigvee_{e=0}^{j-1} (p_e \oplus z_e^{\eta}) \right).$$

This equation is called the *product-of-sum* of g. We can reduce the number of clauses in g by several tools, such as Espresso logic minimizer[1].

Modular Addition. Let $\boldsymbol{\alpha} = (\alpha_0, \alpha_1, \ldots, \alpha_{31})$, $\boldsymbol{\beta} = (\beta_0, \beta_1, \ldots, \beta_{31})$ and $\gamma = (\gamma_0, \gamma_1, \ldots, \gamma_{31})$ be boolean variables such that $\alpha + \beta = \gamma$. A 32-bit modular addition is expressed as follows:

[1] https://ptolemy.berkeley.edu/projects/embedded/pubs/downloads/espresso/index.htm.

$$\alpha_i \vee \beta_i \vee \overline{\gamma_i} \vee \alpha_{i+1} \vee \beta_{i+1} \vee \gamma_{i+1} = 1,\ \alpha_i \vee \overline{\beta_i} \vee \gamma_i \vee \alpha_{i+1} \vee \beta_{i+1} \vee \gamma_{i+1} = 1,$$
$$\overline{\alpha_i} \vee \beta_i \vee \gamma_i \vee \alpha_{i+1} \vee \beta_{i+1} \vee \gamma_{i+1} = 1,\ \overline{\alpha_i} \vee \overline{\beta_i} \vee \overline{\gamma_i} \vee \alpha_{i+1} \vee \beta_{i+1} \vee \gamma_{i+1} = 1,$$
$$\alpha_i \vee \beta_i \vee \gamma_i \vee \overline{\alpha_{i+1}} \vee \overline{\beta_{i+1}} \vee \overline{\gamma_{i+1}} = 1,\ \alpha_i \vee \overline{\beta_i} \vee \overline{\gamma_i} \vee \overline{\alpha_{i+1}} \vee \overline{\beta_{i+1}} \vee \overline{\gamma_{i+1}} = 1,$$
$$\overline{\alpha_i} \vee \beta_i \vee \overline{\gamma_i} \vee \overline{\alpha_{i+1}} \vee \overline{\beta_{i+1}} \vee \overline{\gamma_{i+1}} = 1,\ \overline{\alpha_i} \vee \overline{\beta_i} \vee \gamma_i \vee \overline{\alpha_{i+1}} \vee \overline{\beta_{i+1}} \vee \overline{\gamma_{i+1}} = 1,$$
$$\alpha_{31} \oplus \beta_{31} \oplus \gamma_{31} = 0.$$

where $0 \leq i \leq 30$. In addition, corresponding weight $\boldsymbol{w} = (w_0, w_1, \ldots, w_{31})$ over 32-bit modular addition is determined for $0 \leq i \leq 30$ as follows:

$$\overline{\alpha_{i+1}} \vee \gamma_{i+1} \vee w_i = 1, \quad \beta_{i+1} \vee \overline{\gamma_{i+1}} \vee w_i = 1,$$
$$\alpha_{i+1} \vee \overline{\beta_{i+1}} \vee w_i = 1, \quad \alpha_{i+1} \vee \beta_{i+1} \vee \gamma_{i+1} \vee \overline{w_i} = 1,$$
$$\overline{\alpha_{i+1}} \vee \overline{\beta_{i+1}} \vee \overline{\gamma_{i+1}} \vee \overline{w_i} = 1.$$

Objective Function. We set the objective function to identify the weigh of differential characteristics. Let $\boldsymbol{w} = (w_0, w_1, \ldots, w_{n-1})$ be a set of weights in a primitive. It is enough to give the following objective function to identify the total weigh in a primitive:

$$\sum_{i=0}^{n-1} w_i \leq k.$$

We call a objective function *Boolean cardinality constraints* and can efficiently implement it by `kmtotalizer` [11].

3 Specifications of KCipher-2 and SNOW-V/SNOW-Vi

In this section, we introduce the specifications of SNOW-V, SNOW-Vi and KCipher-2. Let $\oplus$ be a bit-wise exclusive OR, and $\boxplus_{32}$ be an integer addition modulo 2^{32}, respectively. Stream ciphers first load the key and an initialization vector and initialize them in the initialization phase. The number of an initialization rounds are 16, 16 and 24 for KCipher-2, SNOW-V, and SNOW-Vi, respectively. Then, the key stream is generated by updating the internal states.

3.1 Specification of SNOW-V

SNOW-V takes the 128-bit initialization vector IV, where $IV = (iv_7, ..., iv_0) \in (\mathbb{F}_2^{16})^8$, and the 256-bit secret key K, $K = (k_{15}, ..., k_0) \in (\mathbb{F}_2^{16})^{16}$ as inputs. The internal state consists of the Linear Feedback Shift Registers (LFSR) part and Finite State Machine (FSM) part.

Fig 1 shows the LFSR structure in SNOW-V. The LFSR part is a circular structure consisting of two shift registers LFSR-A and LFSR-B. LFSR-A consists of 16 registers denoted $(a_{15}^r, ..., a_0^r)$, LFSR-B consists of 16 registers denoted

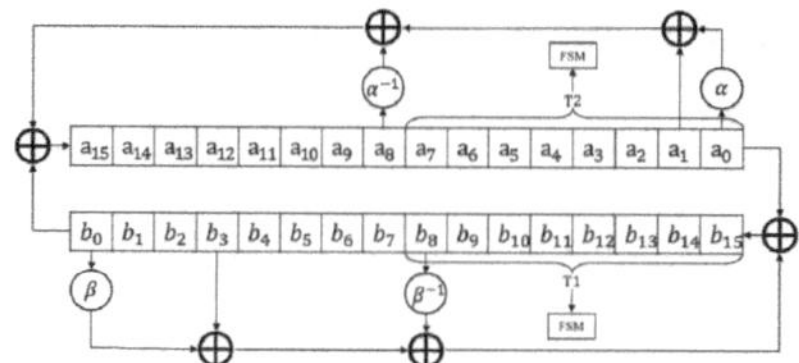

Fig. 1. LFSR structure in SNOW-V.

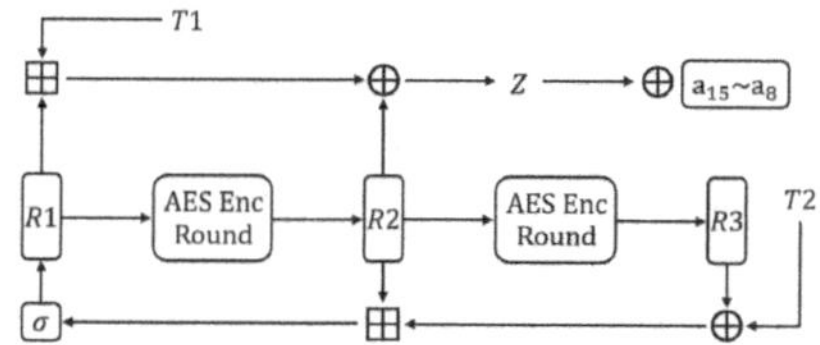

Fig. 2. FSR initialization algorithm

$(b^r_{15}, ..., b^r_0)$, where r for $0 \leq r \leq 15$ is the number of the rounds in the initialization algorithm. Each cell represents an element in $\mathbb{F}_2^{16}$, and LFSR-A and LFSR-B determine the initial values as follows:

$$(a^0_{15}, ..., a^0_0) = (k_7, ..., k_0, iv_7, ..., iv_0), \quad (b^0_{15}, ..., b^0_0) = (k_{15}, ..., k_8, 0, ..., 0).$$

The elements of LFSR-A and LFSR-B are respectively generated by the following polynomials:

$$\begin{aligned} g^A(x) =& x^{16} + x^{15} + x^{12} + x^{11} + x^8 + x^3 + x^2 + x + 1 \in \mathbb{F}_2[x], \\ g^B(x) =& x^{16} + x^{15} + x^{14} + x^{11} + x^8 + x^6 + x^5 + x + 1 \in \mathbb{F}_2[x]. \end{aligned}$$

Let $\alpha \in \mathbb{F}_2^{16}$ be a root of $g^A(x)$ and $\beta \in \mathbb{F}_2^{16}$ be a root of $g^B(x)$. At time $t \geq 0$, the states of the LFSRs are denoted as $(a^{(t)}_{15}, ..., a^{(t)}_0)$ and $(b^{(t)}_{15}, ..., b^{(t)}_0)$ respectively for LFSR-A and LFSR-B. These sequences are updated as follows:

$$\begin{aligned} a^{(t+1)}_{15} &= b^{(t)}_0 + \alpha a^{(t)}_0 + a^{(t)}_1 + \alpha^{(-1)} a^{(t)}_8 \mod g^A(\alpha), \quad & a^{(t+1)}_i &= a^{(t)}_{i+1}, \\ b^{(t+1)}_{15} &= a^{(t)}_0 + \beta b^{(t)}_0 + a^{(t)}_3 + \beta^{(-1)} b^{(t)}_8 \mod g^B(\beta), \quad & b^{(t+1)}_i &= b^{(t)}_{i+1}, \end{aligned}$$

for $i = 0, ..., 14$. LFSRs update the internal state eight times in a single step. i.e., 16 cells of the total 32 cells in the LFSR part can be updated in a single step, and the two taps $T1$ and $T2$ have the following values:

$$T1^{(t)} = (b^{(8t)}_{15}, ..., b^{(8t)}_8), \qquad T2^{(t)} = (a^{(8t)}_7, ..., a^{(8t)}_0).$$

The two taps obtained in the LFSR part, $T1$ and $T2$, are taken as inputs values in the FSM, and the FSM produces a 128-bits keystream block $z^{(t)}$ at time $t \geq 0$. The FSM structure is shown in Fig. 2.

The FSM part consists of three 128-bits registers, $(R1^r, R2^r, R3^r) \in (\mathbb{F}_2^{128})^3$, and these initial values are determined as $R1 = R2 = R3 = $ 0x00. The key stream, $z^{(t)}$ at time $t \geq 0$, is given by the following expression:

$$z^{(t)} = (R1^{(t)} \boxplus_{32} T1^{(t)}) \oplus R2^{(t)}.$$

The four 32-bit parts of the 128-bit words are added with carry, but the carry does not propagate from a lower 32-bit word to the higher. In the initialization algorithm, the keystream $z^{(t)}$ is not output, but is contained in the internal state as follows.

$$(a_{15}, ..., a_8) \leftarrow (a_{15}, ..., a_8) \oplus z^{(t)}$$

Registers $R2$ and $R3$ are updated through a full AES encryption round function as **SubBytes**, **ShiftRows**, **MixColumns**, and **AddRoundKey**, which are denoted by $AES^R(IN, KEY)$. Each register is updated by the following expressions.

$$\begin{aligned} R1^{(t+1)} &= \sigma(R2^{(t)} \boxplus_{32} (R3^{(t)} \oplus T2^{(t)})), \\ R2^{(t+1)} &= AES^R(R1^{(t)}, 0), \quad R3^{(t+1)} = AES^R(R2^{(t)}, 0), \end{aligned}$$

where σ is a byte-oriented permutation given by

$$\sigma = [0, 4, 8, 12, 1, 5, 9, 13, 2, 6, 10, 14, 3, 7, 11, 15].$$

When the initialization algorithm is r-round, before the last $(r-1)$th round and $(r-2)$th round, the divided keys are xored to the register R1 as follows:

$$R1^{r-2} = R1^{r-2} \oplus (k_7, ..., k_0), \quad R1^{r-1} = R1^{r-1} \oplus (k_{15}, ..., k_8).$$

3.2 Structure of SNOW-Vi

The basic structure of SNOW-Vi is almost the same as SNOW-V, and only the LFSR update and the scope of $T2$ are different. The LFSR structure SNOW-Vi is shown in Fig. 3. In SNOW-Vi, the elements of LFSR-A and LFSR-B are respectively generated by the following polynomials:

$$\begin{aligned} g^A(x) =& x^{16} + x^{14} + x^{11} + x^9 + x^6 + x^5 + x^3 + x^2 + 1 \in \mathbb{F}_2[x], \\ g^B(x) =& x^{16} + x^{15} + x^{14} + x^{11} + x^{10} + x^7 + x^2 + x + 1 \in \mathbb{F}_2[x]. \end{aligned}$$

At time $t \geq 0$, the states of the LFSRs are denoted as $(a_{15}^{(t)}, ..., a_0^{(t)})$, and $(b_{15}^{(t)}, ..., b_0^{(t)})$ respectively for LFSR-A and LFSR-B. These sequences are updated as follows:

$$\begin{aligned} a_{15}^{(t+1)} &= b_0^{(t)} + \alpha a_0^{(t)} + a_7^{(t)} \mod g^A(\alpha), \ a_i^{(t+1)} = a_{i+1}^{(t)}, \\ b_{15}^{(t+1)} &= a_0^{(t)} + \beta b_0^{(t)} + b_8^{(t)} \mod g^B(\beta), \ b_i^{(t+1)} = b_{i+1}^{(t)}, \end{aligned}$$

The two taps $T1$ and $T2$ have the following values:

$$T1^{(t)} = (b_{15}^{(8t)}, ..., b_8^{(8t)}), \quad T2^{(t)} = (a_{15}^{(8t)}, ..., a_8^{(8t)}).$$

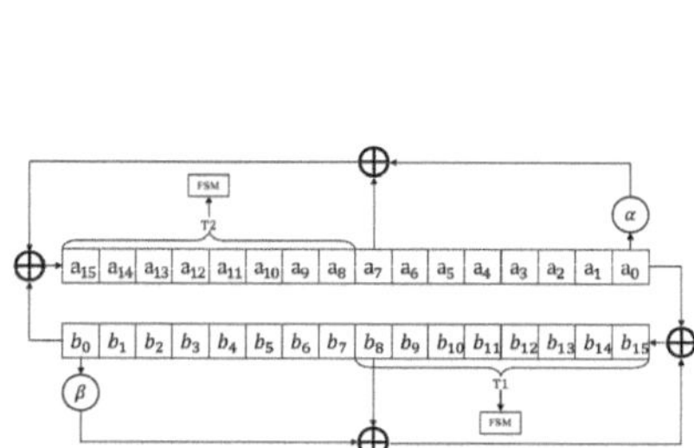

Fig. 3. LFSR structure in SNOW-Vi.

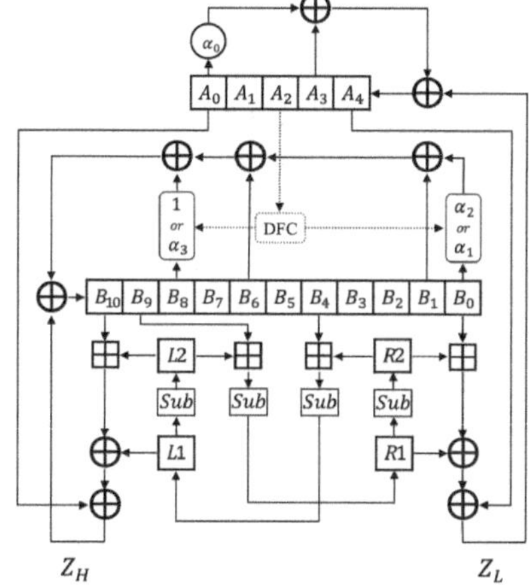

Fig. 4. Structure of KCipher-2.

3.3 Structure of KCipher-2

The overall structure of KCipher-2 is shown in Fig. 4. KCipher-2 takes the 128-bit initialization vector IV, where $IV = (IV_0, IV_1, IV_2, IV_3) \in (\mathbb{F}_2^{32})^4$, and the 128-bit secret key IK, $IK = (IK_0, IK_1, IK_2, IK_3) \in (\mathbb{F}_2^{32})^4$ as inputs. The internal state consists of Linear Feedback Shift Registers (LFSR) part, Dynamic Feedback Shift Register (DFSR) part and Finite State Machine (FSM) part.

Before loading the initial value, the secret key IK is extended to a 384-bit internal key $K = (K_0, ..., K_{12}) \in (\mathbb{F}_2^{32})^{12}$ by the key scheduling. The key scheduling is shown as follows:

$$\begin{aligned}
&K_0 = IK_0, K_1 = IK_1, K_2 = IK_2, K_3 = IK_3,\\
&K_4 = K_0 \oplus Sub((K_3 \lll 8) \oplus (K_3 \ggg 24)) \oplus RC0,\\
&K_5 = K_1 \oplus K_4, K_6 = K_2 \oplus K_5, K_7 = K_3 \oplus K_6,\\
&K_8 = K_4 \oplus Sub((K_7 \lll 8) \oplus (K_7 \ggg 24)) \oplus RC1,\\
&K_9 = K_5 \oplus K_8, K_{10} = K_6 \oplus K_9, K_{11} = K_7 \oplus K_{10},
\end{aligned}$$

where $RC0$ and $RC1$ are constant values, $RC0 = $ 0x01000000, $RC1 =$ 0x02000000, and Sub function is defined as a combination of SubBytes and MixColumns.

LFSR consists of 5 registers denoted $(A_0^r, ..., A_4^r)$, DFSR consists of 11 registers denoted $(B_0^r, ..., B_{10}^r)$, where r for $0 \leq r \leq 23$ is the number of the rounds in the initialization phase. Each cell represents an element in $\mathbb{F}_2^{32}$, and LFSR and DFSR determine the initial values as follows:

$$A_m = K_{4-m}(m = 0, ..., 4),$$
$$B_0 = K_{10}, B_1 = K_{11}, B_2 = IV_0, B_3 = IV_1,$$
$$B_4 = K_8, B_5 = K_9, B_6 = IV_2, B_7 = IV_3,$$
$$B_8 = K_7, B_9 = K_5, B_{10} = K_6.$$

At time $t \geq 0$, update the internal states of LFSR and DFSR respectively are shown as follows:

$$A_i^{t+1} = \begin{cases} A_{i+1}^t & \text{(i=0,...,3)} \\ A_3^t \oplus \alpha_0 A_0^t & \text{(i=4)}, \end{cases}$$
$$B_i^{t+1} = \begin{cases} B_{i+1}^t & \text{(i=0,...,9)} \\ (\alpha_1^{cl1} + \alpha_2^{1-cl1} - 1)B_0^t + B_1^t + B_6^t + \alpha_3^{cl2} B_8^t & \text{(i=10)}, \end{cases}$$

where, $cl1$ and $cl2$ are determined by DFC (dynamic feedback counter) and defined by $cl1 = A_2[1]$ and $cl2 = A_2[0]$. Let $A_x[y]$ be the yth bit of A_x where $A_x[0]$ is the most significant bit of A_x. α_0 , α_1 , α_2 , and α_3 are the roots of the following irreducible polynomials:

$$\alpha_0 : x^4 + \beta^{24}x^3 + \beta^3x^2 + \beta^{12}x + \beta^{71} \in (\mathbb{F}_2^8)[x],$$
$$\alpha_1 : x^4 + \gamma^{230}x^3 + \gamma^{156}x^2 + \gamma^{93}x + \gamma^{29} \in (\mathbb{F}_2^8)[x],$$
$$\alpha_2 : x^4 + \delta^{34}x^3 + \delta^{16}x^2 + \delta^{199}x + \delta^{248} \in (\mathbb{F}_2^8)[x],$$
$$\alpha_3 : x^4 + \zeta^{157}x^3 + \zeta^{253}x^2 + \zeta^{56}x + \zeta^{16} \in (\mathbb{F}_2^8)[x],$$

where β, γ, δ, and ζ are the roots of the following irreducible polynomials:

$$\beta : x^8 + x^7 + x^6 + x + 1 \in (\mathbb{F}_2)[x], \quad \gamma : x^8 + x^5 + x^3 + x^2 + 1 \in (\mathbb{F}_2)[x],$$
$$\delta : x^8 + x^6 + x^3 + x^2 + 1 \in (\mathbb{F}_2)[x], \quad \zeta : x^8 + x^6 + x^5 + x^2 + 1 \in (\mathbb{F}_2)[x].$$

The FSM part consists of three 128-bits registers, $(L1^r, L2^r, R1^r, R2^r) \in (\mathbb{F}_2^{32})^4$, and these initial values are determined as $L1 = L2 = R1 + R2 = 0x00$. Each register is updated by the following expressions.

$$L1^{r+1} = Sub(R2^r \boxplus_{32} B_4^r), L2^{r+1} = Sub(L1^r),$$
$$R1^{r+1} = Sub(L2^r \boxplus_{32} B_9^r), R2^{r+1} = Sub(R1^r),$$

where the Sub function is the same as one used in key schedule. The key stream, $Z_H^{(t)}$ and $Z_L^{(t)}$ at time $t \geq 0$, is given by the following expression:

$$Z_H^{(t)} = B_{10}^{(t)} \boxplus_{32} L2^{(t)} \oplus L1^{(t)} \oplus A_0^{(t)}, \quad Z_L^{(t)} = B_0^{(t)} \boxplus_{32} R2^t \oplus R1^t \oplus A_4^{(t)}.$$

In the initialization algorithm, the keystream $Z_H^{(t)}$ and $Z_L^{(t)}$ are not output, but are contained in the internal state as follows.

$$B_{10} \leftarrow B_{10} \oplus Z_H^{(t)}, \quad A_4 \leftarrow A_4 \oplus Z_L^{(t)}.$$

3.4 Limitations of Existing Results

The security evaluation for differential attacks on SNOW-V and SNOW-Vi has already been carried out by Hoki et al. using MILP as an evaluation method for differential characteristics [7]. Existing studies of SNOW-V and SNOW-Vi show that distinguishing attacks are feasible with up to 4 rounds in related-IV setting attacks. However, we identify that there are following two limitation in their evaluations.

Limitation 1: Restricted Search Space. In search for differential characteristics by MILP, it is difficult to evaluate for whole search space in a significant number of rounds because it requires huge computer resources.

To address this issue, they conducted an evaluation within a highly restricted search area. More precisely, the search was confined to instances where the Hamming weight for the input difference is limited to 1 out of 128, allowing them to explore only a fraction, $128/2^{128}$, of the entire space. In addition, the scope of input difference is only initialization vector IV, and related-key attacks are not considered.

Limitation 2: Modeling of DFC. As the modeling of DFC in KCipher-2 is complex, it has not been considered so far. In particular, depending on the differential characteristics, it is necessary to consider the differential transitions of 32-bit input and output. Thus considering all patterns is time-consuming problem. This is one of the reasons why the detailed differential analysis of KCipher-2 could not be carried out so far.

4 Efficient Search for Differential Characteristics

To address limitations in existing results, we take two approaches described in Sect. 4.1. Moreover, we choose the input differences by exploiting the internal structures for SNOW-V and SNOW-Vi.

4.1 Addressing the Limitations in the Existing Results

Solution for Limitation 1. It is known that a MILP-based automatic search method contributes to reducing the evaluation time to identify the weight of differential characteristics [12]. The previous studies on the differential distinguisher on SNOW-V and SNOW-Vi utilize an MILP-based method to identify optimal differential characteristics on as long rounds as possible [7], but they can evaluate only up to 4 rounds for both SNOW-V and SNOW-Vi even through they restrict their search space.

Recent studies show that a SAT-based approach is much more efficient than a MILP-based one regarding the evaluation of optimal differential characteristics. In this paper, we evaluate optimal differential characteristics not only in the single-key setting but also in the related-key setting. In the related-key setting, the attacker can control the differences not only in the initialization vector but also in the key, leading to a more powerful attack. To evaluate the security in this setting, we need to consider the space of key differences, which increases the evaluation cost. Considering the increase in the evaluation cost and the capability of a MILP-based method for SNOW-V and SNOW-Vi shown in [7], it seems better to use a SAT-based one rather than an MILP-based one. Therefore, we employ an SAT-based one in this paper to overcome this limitation.

In addition, to minimize the search cost for a large number of rounds while keeping our capability for finding best characteristics, we analyze the differential behaviors in the nonlinear transformation of the initialization algorithm and properties of internal structures, and properly choose input differences to be inserted into key and IV space, ensuring an optimized balance between comprehensiveness and efficiency. Details are described in 4.2.

Solution for Limitation 2. As described in Sect. 3.4, the operation of the DFC in KCipher-2 is determined by differential transitions, which makes the security evaluation against differential cryptanalysis harder. To address this problem, we only consider the differential characteristics where the difference values of $A_2[1]$ and $A_2[0]$, which determine the operation of the DFC, are both inactive.

If the difference of $A_2[0]$ (resp. $A_2[1]$) is active, which means the values of $A_2[0]$ (resp. $A_2[1]$) are different such that $(A_2[0], A_2^*[0]) = (0,1)$ or $(1,0)$ where $A_2[0] \oplus A_2[0]^* = \Delta A_2[0]$ (resp. $(A_2[1], A_2^*[1]) = (0,1)$ or $(1,0)$ where $A_2[1] \oplus A_2[1]^* = \Delta A_2[1]$), the operations of the DFC in each value behaves different operations.

In that case, it is hard to grasp the transition of the DFC because we cannot know the values in the DFC, and how the DFC behaves depends on the values.

However, in the case of that the DFC behaves α_2 or α_1, the output differences can be calculated independently by 8 bits. It allows us to find a byte that can have any output difference for all input differences, bringing at least 8 weights

In contrast, if the differences of $A_2[0]$ and $A_2[1]$ are inactive, the operation of the DFC in both values behave the same operation as the multiplication of α_2 or α_1 and α_3 or 1 for $A_2[1]$ and $A_2[0]$, respectively. In that case, we can calculate the weight that occurred in the DFC because the operations of the DFC are consistent in both values. Moreover, if the most significant byte of ΔB_8 is inactive, the DFC of $A_2[1]$ behaves as a linear operation that does not bring additional weight. This is because that the operations of α_1, α_2, and α_3 can be viewed as a linear transformation, and both α_2 and α_1 have 0 output difference if the difference of the most significant byte of ΔB_8 is inactive.

If the DFC behaves α_3 or 1, the weight brought from the DFC will be zero only if all differences in B_8 are inactive.

Fig 5 shows the difference transition of α_1 or α_2.

In the second case, the differences of $A_2[0]$ and $A_2[1]$ are inactive, occurs with the probability 2^{-1}, which can be realized by posing one weight in the

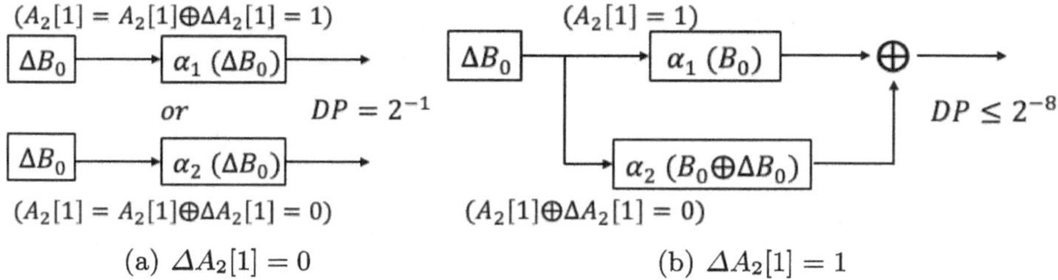

Fig. 5. Difference transitions of α_1 or α_2

DFC. Therefore, we always pose one weight in the DFC to only consider the differential transition in the second case.

4.2 Choosing Input Differences by Exploiting Internal Structures

We employ a two-step search strategy to identify the best differential characteristics for SNOW-V, SNOW-Vi and KCipher-2.

1. We first conduct a comprehensive search across the entire space of IV and key up to the maximum number of rounds that can be evaluated within a feasible time. If we cannot reach the number of rounds where the differential probability is less than 2^{-128} and 2^{-256} for KCipher-2 and SNOW-V/SNOW-Vi, respectively, we proceed to the next step to evaluate more rounds.
2. To reduce the search cost, we constrain the search space for IV and key. To properly restrict the search space, we examine the properties of optimal differential characteristics across the entire search space obtained in step 1, and impose restrictions in such a way that better differential characteristics can be found even in the limited search space. The details of our search space and limitations are shown in Table 2. In the following, we will explain reason of these restrictions for each target.

SNOW-V. In the related-IV setting, it is possible to evaluate the search space comprehensively up to four rounds. Analyzing best differential characteristics for 1–4 rounds, we observe that all input differences exhibit a Hamming Weight HW of two or less. The reason why $HW \leq 2$ is more appropriate than $HW = 1$ is explained as follows. In the case where there are two active bytes in the output of the first round (R1) entering the round function of the second round of AES, the differences can cancel each other out during the MixColumns operation. Thus, with $HW = 2$, it is possible to limit this to three active bytes, while with $HW = 1$, it results in at least four active bytes in R2 of the second rounds as illustrated in Fig. 6. These significantly affect the differential propagation in the subsequent rounds, thereby contributing to a reducing in the total number of active S-boxes. Thus, we restrict our input difference to HW of two or less.

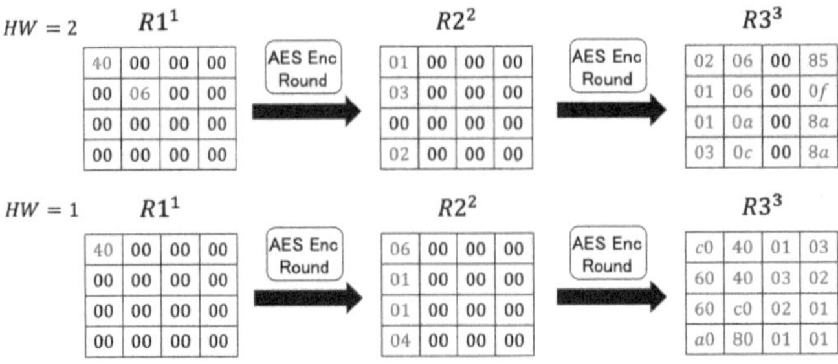

Fig. 6. Differential Propagation when $HW = 1$ and $HW = 2$

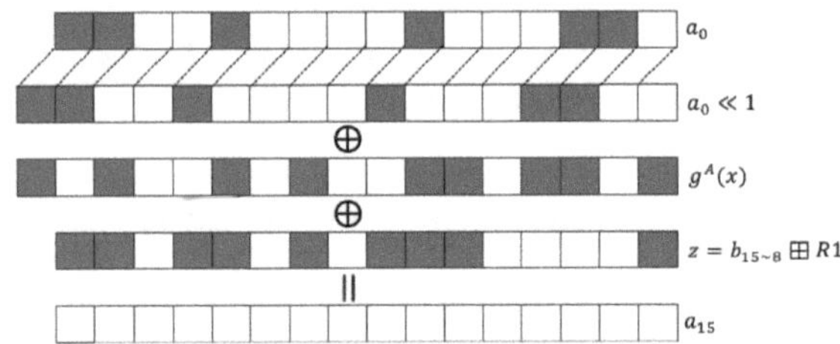

Fig. 7. Cancellation in the primitive polynomials

In the related-key setting, the position of all input differences is mandated to be between k_8 and k_{15}, with a HW of one. It is because that k_8 and k_{15} are loaded into b_8 and b_{15}, respectively, and any difference between b_8 and b_{15} is input into the S-box layer, at the latest, three steps later. In this case, search space is estimated as $\binom{128}{2} + \binom{128}{1} = 8256$.

SNOW-Vi. In the related-IV setting, it is possible to evaluate the search space comprehensively up to five rounds. Analyzing best differential characteristics for 1–5 rounds,we observe that all input differences exhibit a Hamming Weight $HW = 1$. So, we utilize the limit of $HW = 1$.

In the related-key setting, the position of all input differences is mandated to be in $k_8 - k_{15}$ and IV, and active bytes should be less than 4 bytes. In this case, search space is estimated as $\binom{32}{4} \times 2^{32} = 2^{47.13}$. The reasons of this limitation are given as follows.

- In contrast to SNOW-V, the introducing differences into IV is deemed more suitable. This difference comes from to the different ranges of T_2, which is inputs of FSM, as shown in Fig. 1 and 3. In SNOW-V, IV is loaded into a_0 to a_7, and these are directly inputted to the FSM, leading to expansions of differences in the FSM at an early stage. On the other hand, in SNOW-Vi, since a_8 to a_{15} are inputted into the FSM, the introduction of differences in the IV does not have a significant impact to increase the active Sboxes.

- More importantly, generating cancellations to avoid introducing differences into a_8 to a_{15} during the LFSR updates can significantly reduce the number of active S-boxes. This is realized by the cancellation of differences between a_0 to a_{15} (corresponding to IV) and b_8 to b_{15} (corresponding to k_8 and k_{15}).
- In addition, the reason why a restriction based on HW is not appropriate is due to the inclusion of Galois Field operations in the LFSR update equation. In order to cancel out in the primitive polynomials, it requires at least four or fewer active bytes as shown in Fig. 7.

KCipher-2. In the case of KCipher-2, we were able to perform a comprehensive search across the entire space of IV and key up to the round where the probability is less than 2^{128}.

5 Applications

This section shows the best differential characteristic probability of SNOW-V, SNOW-Vi and KCipher-2 in related-IV and related-key settings, which are derived by our search shown in Sec. 4. To simplify the analysis, each evaluation focuses on differential characteristics from inputs of IV and/or key to the first output byte of key stream, by which we perform distinguishing attacks on keystream. In the following section, we will explain results of each algorithm.

Table 2. The search space.

Cipher	Attack Type	HW	Positions	Search Space	ref
SNOW-V	Related-IV	No Limit	No Limit	2^{128}	[7] and Ours
		$HW = 1$	No Limit	128	[7]
		$HW \leq 2$	No Limit	8256	Ours
	Related-Key	No Limit	No Limit	2^{384}	Ours
		$HW = 1$	k_8-k_{15}	128	Ours
SNOW-Vi	Related-IV	No Limit	No Limit	2^{128}	[7] and Ours
		$HW = 1$	No Limit	128	Ours
	Related-Key	No Limit	No Limit	2^{384}	Ours
		Active-Byte ≤ 4	k_8-k_{15} and IV	$2^{47.13}$	Ours
KCipher-2	Related-IV	No Limit	No Limit	2^{128}	Ours
	Related-Key	No Limit	NoLimit	2^{256}	Ours

5.1 Results for SNOW-V

Related-IV Attacks. Table 3 shows our search results with existing search results [7] of related-IV and related-key attacks against SNOW-V. The existing results estimate the best differential characteristic probability up to 3 rounds across entire IV space, and derive best differential probability over 4 rounds in

Table 3. Differential characteristic probability on related-IV attack and related-Key attack for SNOW-V where each result is given in weight.

Related-IV											
HW	Positions	Search Space	1R	2R	3R	4R	5R	6R	7R	8R	Ref
No Limit	No Limit	2^{128}	0	8	17	-	-	-	-	-	[7]
$HW = 1$	No Limit	128	0	8	17	97	-	-	-	-	[7]
No Limit	No Limit	2^{128}	0	**7**	17	**91**	$\leq$ **170**	-	-	-	Ours
$HW \leq 2$	No Limit	8256	0	**7**	17	**91**	**169**	**256<**	-	-	Ours
Related-Key											
HW	Positions	Search Space	1R	2R	3R	4R	5R	6R	7R	8R	Ref
No Limit	No Limit	2^{384}	0	1	1	14	28	113	-	-	Ours
$HW = 1$	k_8-k_{15}	128	1	2	1	14	28	113	217	256<	Ours

the restricted IV search space, where HW of IV difference is fixed to 1. Our method enables us to identify the best differential characteristic probability up to 4 rounds in the entire IV space, and reveal that optimal probability is 2^{91}, which is improved from previous best one of 2^{97}. For 5 rounds, although it is infeasible to search across entire IV, we find the differential characteristics with a probability of 2^{170}.

To extend the search to more rounds, we restricted HW of the IV difference to 2 or less. Consequently, we discover a 5-round differential characteristics with a probability of 2^{-169} as depicted in Table 7. In addition, during this evaluation, we cannot find the differential characteristic with probability of more than 2^{-256} in 6 rounds.

Distinguishing Attack. According to these results, we can mount distinguishing attacks against SNOW-V up to 5 rounds with time complexity of 2^{169} in the related-IV setting. Under the limitation of 2^{64}, a 3-round attack is feasible with time complexity of 2^{17}.

Related-Key Attacks. According to Table 3, we obtain best differential characteristics up to 6 rounds in the entire IV and Key space, and reveal that optimal probability of 6 rounds is 2^{-113}. To extend the search to more rounds, we constrain the input difference to positions k_8-k_{15} and further limit HW to 1. k_8-k_{15} has HW of 1. Consequently, we discover a 7-round differential characteristics with a probability of 2^{-217} as depicted in Table 8. In addition, during this evaluation, we cannot find the differential characteristic with probability of more than 2^{-256} in 8 rounds.

Distinguishing Attack. According to these results, we can mount a distinguishing attack against SNOW-V up to 7 rounds with time complexity of 2^{217} in the related-key setting. Under the limitation of 2^{64}, a 5-round attack is feasible with time complexity of 2^{17}.

5.2 Results for SNOW-Vi

Related-IV Attacks. Table 4 compare our search results with existing search results [7] of related-IV attacks against SNOW-Vi. The existing results estimate the best differential characteristic probability up to 4 rounds across entire IV space.

Our method enables us to identify the best differential probability up to 5 rounds in the entire IV space, and reveal that optimal probability is 2^{117} as shown in Table 9. In addition, we improve the probability of one of 4 rounds from 2^{39} to 2^{37}. To extend the search to more rounds, we restricted the HW of the IV difference to one. However, we cannot find the differential characteristic with probability of more than 2^{-256} in 6 rounds.

Distinguishing Attack. According to these results, we can mount a distinguishing attack against SNOW-Vi up to 5 rounds in the related-IV setting as the differential probability in 5 rounds is more than 2^{256}.

Related-Key Attacks. According to Table 4, we find the best differential probability up to 7 rounds in the whole IV and Key space, and reveal that optimal probability is 2^{-96}. To extend the search to more rounds, we constrain the input difference to positions k_8-k_{15} and further limit the the number of active-bytes to 4 or less. This restriction is based on our observation that the optimal characteristics over 3–7 rounds start from the k_8-k_{15} and IV difference where active-bytes to 4 or less. Consequently, we discover a 8-round differential characteristics with a probability of 2^{-179} as depicted in Table 10.

Distinguishing Attack. According to these results, we can mount a distinguishing attack against SNOW-Vi up to 8 rounds with time complexity of 2^{179} in the related-key setting. Under the limitation of 2^{64}, a 6-round attack is feasible with time complexity of 2^{37}.

5.3 Results for KCipher-2

Table 4. Differential characteristic probability on related-IV attack and related-Key attack for SNOW-Vi where each result is given in weight.

Related-IV											
HW	Positions	Search Space	1R	2R	3R	4R	5R	6R	7R	8R	ref
No Limit	No Limit	2^{128}	0	2	12	39	-	-	-	-	[7]
No Limit	No Limit	2^{128}	0	2	**11**	**37**	117	-	-	-	Ours
$HW = 1$	No Limit	128	0	2	**11**	**37**	117	$\mathbf{256<}$	-	-	Ours
Related-Key											
HW	Positions	Search Space	1R	2R	3R	4R	5R	6R	7R	8R	ref
No Limit	No Limit	2^{384}	0	0	4	17	35	60	96	≤ 256	Ours
Active-Byte ≤ 4	k_8-k_{15} and IV	$35960 * 2^{32}$	0	2	4	17	35	60	96	179	Ours

Table 5. Differential characteristic probability on related-IV attack and related-Key attack for KCipher-2 where each result is given in weight.

Related-IV												
HW	Position	Search Space	1R	2R	3R	4R	5R	6R	7R	8R	9R	ref
No Limit	No Limit	2^{128}	0	0	0	3	11	49	81	123	159	Ours
Related-Key												
HW	Position	Search Space	1R	2R	3R	4R	5R	6R	7R	8R	9R	ref
No Limit	No Limit	2^{256}	0	0	0	3	11	49	81	123	159	Ours

Related-IV Attacks. Table 5 shows our search results in the related-IV setting against KCipher-2. We estimate the best differential characteristic probability up to 9 rounds in the whole IV space and reveal that optimal probability in 8-round is 2^{-123} as depicted in Table 6, and in 9-round is 2^{-159}.

Distinguishing Attack. According to this results, we can mount a distinguishing attack against KCipher-2 up to 8 rounds in the related-IV setting with time complexity of 2^{123}.

Table 6. Differential characteristic for 8-rounds of KCipher-2 on related-IV attack.

	$r = 0$	$r = 1$	$r = 2$	$r = 3$	$r = 4$	$r = 5$	$r = 6$	$r = 7$	$r = 8$
A_0	0x00000000	0x00000000	0x00000000	0x00000000	0x00000000	0x00000000	0x00000000	0x00000002	0x00000202
A_1	0x00000000	0x00000000	0x00000000	0x00000000	0x00000000	0x00000000	0x00000002	0x00000202	0x00000200
A_2	0x00000000	0x00000000	0x00000000	0x00000000	0x00000000	0x00000002	0x00000202	0x00000200	0x00000002
A_3	0x00000000	0x00000000	0x00000000	0x00000000	0x00000002	0x00000202	0x00000200	0x00000002	0x00000202
A_4	0x00000000	0x00000000	0x00000000	0x00000002	0x00000202	0x00000200	0x00000002	0x00000202	0x00008f02
B_0	0x00000000	0x00000000	0x00000002	0x00000200	0x00000000	0x0c020408	0x04000408	0x0402040a	0x00000200
B_1	0x00000000	0x00000002	0x00000200	0x00000000	0x00000000	0x00020000	0x0c020408	0x04000408	0x0402040a
B_2	0x00000002	0x00000200	0x00000000	0x00000000	0x00000000	0x00000000	0x00020000	0x00020000	0x04000408
B_3	0x00000200	0x00000000	0x00000000	0x00000000	0x00000002	0x00000000	0x00000000	0x00000000	0x0c020408
B_4	0x00000000	0x00000000	0x00000000	0x00000002	0x00000000	0x00000000	0x00000000	0x00000000	0x00020000
B_5	0x00000000	0x00000000	0x00000002	0x00000000	0x00000000	0x00000000	0x00000000	0x00000000	0x00000000
B_6	0x00000000	0x00000002	0x00000000	0x00000000	0x00000000	0x00000000	0x00000000	0x00000000	0x00000000
B_7	0x00000002	0x00000000	0x00000000	0x00000000	0x00000000	0x00000000	0x00000000	0x00000000	0x00000000
B_8	0x00000000	0x00000000	0x00000000	0x00000000	0x00000000	0x00000002	0x00000000	0x00000000	0x00000000
B_9	0x00000000	0x00000000	0x00000000	0x00000000	0x00000000	0x00000000	0x00000002	0x00000000	0x00000000
B_{10}	0x00000000	0x00000000	0x00000000	0x00000000	0x00020000	0x00000000	0x00000000	0x00000002	0x00000000
$L2$	0x00000000	0x00000000	0x00000000	0x00000000	0x00000000	0x00020000	0x00000000	0x00000000	0x00000000
$L1$	0x00000000	0x00000000	0x00000000	0x00000000	0x0c040408	0x00000000	0x00000000	0x00000000	0x00000000
$R2$	0x00000000	0x00000000	0x00000000	0x00000000	0x00000000	0x00000000	0x00000000	0x00000000	0x10102030
$R1$	0x00000000	0x00000000	0x00000000	0x00000000	0x00000000	0x00000000	0x00000000	0x00008f00	0x5d6afb9f
Z_H	0x00000000	0x00000000	0x00000000	0x00000000	0x0c020408	0x04000408	0x04000408	0x0c020408	0x00000002
Z_L	0x00000000	0x00000000	0x00000002	0x00000202	0x00000202	0x00000200	0x00000002	0x00008d00	0x4d7a54ad

Table 7. Differential characteristic for 5-rounds of SNOW-V on related-IV attack.

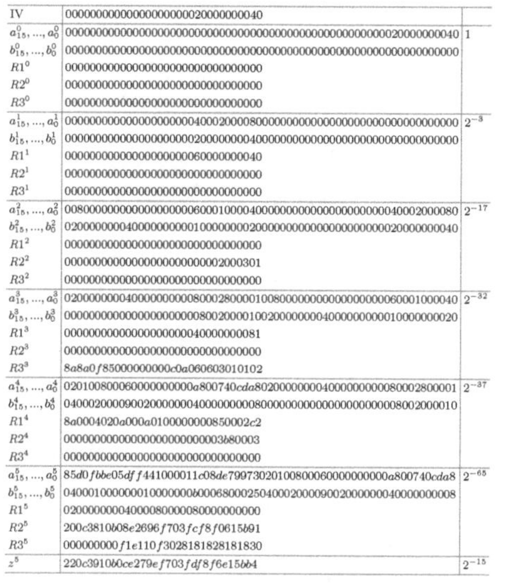

IV	00000000000000000000002000000040	
$a^0_{15},\dots,a^0_0$	0000000000000000000000000000000000000000000000000000002000000040	1
$b^0_{15},\dots,b^0_0$	0000000000000000000000000000000000000000000000000000000000000000	
$R1^0$	00000000000000000000000000000000	
$R2^0$	00000000000000000000000000000000	
$R3^0$	00000000000000000000000000000000	
$a^1_{15},\dots,a^1_0$	0000000000000000000004000200008000000000000000000000000000000000	2^{-3}
$b^1_{15},\dots,b^1_0$	0000000000000000000000200000004000000000000000000000000000000000	
$R1^1$	00000000000000000000006000000040	
$R2^1$	00000000000000000000000000000000	
$R3^1$	00000000000000000000000000000000	
$a^2_{15},\dots,a^2_0$	0080000000000000000000600010000400000000000000000000040002000080	2^{-17}
$b^2_{15},\dots,b^2_0$	0200000000400000000001000000002000000000000000000000002000000040	
$R1^2$	00000000000000000000000000000000	
$R2^2$	00000000000000000000000002000301	
$R3^2$	00000000000000000000000000000000	
$a^3_{15},\dots,a^3_0$	0200000000400000000008000280000100800000000000000000060001000040	2^{-32}
$b^3_{15},\dots,b^3_0$	0000000000000000000000800200001002000000004000000000010000000020	
$R1^3$	00000000000000000000004000000081	
$R2^3$	00000000000000000000000000000000	
$R3^3$	8a8a0f85000000000c0a060603010102	
$a^4_{15},\dots,a^4_0$	020100800060000000000a800740cda802000000004000000000080002800001	2^{-37}
$b^4_{15},\dots,b^4_0$	0400020000900200000004000000000800000000000000000000008002000010	
$R1^4$	8a0004020a000a0100000000850002c2	
$R2^4$	00000000000000000000000003b80003	
$R3^4$	00000000000000000000000000000000	
$a^5_{15},\dots,a^5_0$	85d0fbbe05dff441000011c08de79973020100800060000000000a800740cda8	2^{-65}
$b^5_{15},\dots,b^5_0$	040001000000010000000b000680002504000200009002000000040000000008	
$R1^5$	02000000004000080000080000000000	
$R2^5$	200c3810b08e2696f703fcf8f0615b91	
$R3^5$	000000000f1e110f3028181828181830	
z^5	220c3910b0ce279ef703fdf8f6e15bb4	2^{-15}

Table 8. Differential characteristic for 7-rounds of SNOW-V on related-key attack.

IV	00000000000000000000000000000000	
Key	0000000000000000004000000000000000000000000000000000000000000000	
$a^0_{15},\dots,a^0_0$	0000000000000000000000000000000000000000000000000000000000000000	1
$b^0_{15},\dots,b^0_0$	0000000000000000004000000000000000000000000000000000000000000000	
$R1^0$	00000000000000000000000000000000	
$R2^0$	00000000000000000000000000000000	
$R3^0$	00000000000000000000000000000000	
$a^1_{15},\dots,a^1_0$	0000000000000000004000000000000000000000000000000000000000000000	2^{-1}
$b^1_{15},\dots,b^1_0$	0000000000000000002000000000000000000000000000000040000000000000	
$R1^1$	00000000000000000000000000000000	
$R2^1$	00000000000000000000000000000000	
$R3^1$	00000000000000000000000000000000	
$a^2_{15},\dots,a^2_0$	0000000000000000000000000000000000000000000000000040000000000000	2^{-2}
$b^2_{15},\dots,b^2_0$	0000000000000000009000000000004000000000000000000020000000000000	
$R1^2$	00000000000000000000000000000000	
$R2^2$	00000000000000000000000000000000	
$R3^2$	00000000000000000000000000000000	
$a^3_{15},\dots,a^3_0$	0000000000000000003000400000004000000000000000000000000000000000	2^{-4}
$b^3_{15},\dots,b^3_0$	0000000000400000000480000000000000000000000000000090000000000040	
$R1^3$	00000000000040000000000000000000	
$R2^3$	00000000000000000000000000000000	
$R3^3$	00000000000000000000000000000000	
$a^4_{15},\dots,a^4_0$	0040000000c04000034000200000006000000000000000000030004000000040	2^{-15}
$b^4_{15},\dots,b^4_0$	0000000000200000010400000000001000000000040000000048000000000000	
$R1^4$	00000000000000000000000000000000	
$R2^4$	00000000000000000101020300000000	
$R3^4$	00000000000000000000000000000000	
$a^5_{15},\dots,a^5_0$	0040000000002000109c2090004000a00040000000c040000340002000000060	2^{-40}
$b^5_{15},\dots,b^5_0$	0000000000800000000220000000000000000000002000000104000000000010	
$R1^5$	00001000000000000000020000001040	
$R2^5$	00000000000000000000000000000000	
$R3^5$	0102030180c040400301010220204060	
$a^6_{15},\dots,a^6_0$	00001040012090c04fe83348000010c00040000000002000109c2090004000a0	2^{-46}
$b^6_{15},\dots,b^6_0$	0040000000c04000015900000000014400000000008000000002200000000000	
$R1^6$	00800020c2004120010000400140e200	
$R2^6$	404080c0404080c000000000804020a0	
$R3^6$	00000000000000000000000000000000	
$a^7_{15},\dots,a^7_0$	504088800110898026a74858a110088400001040012090c04fe83348000010c0	2^{-83}
$b^7_{15},\dots,b^7_0$	0060000000240000f4c52010004000200040000000c040000159000000000144	
$R1^7$	4040108000408400 80e0202040409000	
$R2^7$	10e57fe3332175c6ceec475e00000000	
$R3^7$	14d7c8089a8d7c588393572d20e9571d	
z^7	50056f633305f1c6db09474e40009020	2^{-26}

Related-Key Attacks. According to Table 5, we find the best differential probability up to 9 rounds in the whole IV and key space and reveal that optimal probability in 8-round is 2^{-123}, and in 9-round is 2^{-159}. These are the same as those of related-IV setting. Therefore, it is a better strategy to enter the difference only in IV for distinguishing attacks against KCipher-2.

Table 9. Differential characteristic for 5-rounds of SNOW-Vi on related-IV attack.

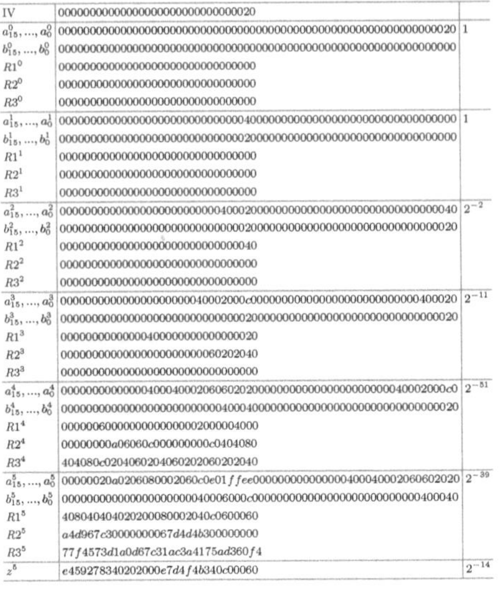

IV	00000000000000000000000000000020	
$a_{15}^{0}, \ldots, a_{0}^{0}$	0000000000000000000000000000000000000000000000000000000000000020	1
$b_{15}^{0}, \ldots, b_{0}^{0}$	0000000000000000000000000000000000000000000000000000000000000000	
$R1^{0}$	00000000000000000000000000000000	
$R2^{0}$	00000000000000000000000000000000	
$R3^{0}$	00000000000000000000000000000000	
$a_{15}^{1}, \ldots, a_{0}^{1}$	0000000000000000000000000000004000000000000000000000000000000000	1
$b_{15}^{1}, \ldots, b_{0}^{1}$	0000000000000000000000000000002000000000000000000000000000000000	
$R1^{1}$	00000000000000000000000000000000	
$R2^{1}$	00000000000000000000000000000000	
$R3^{1}$	00000000000000000000000000000000	
$a_{15}^{2}, \ldots, a_{0}^{2}$	0000000000000000000000000040002000000000000000000000000000000040	2^{-2}
$b_{15}^{2}, \ldots, b_{0}^{2}$	0000000000000000000000000000002000000000000000000000000000000020	
$R1^{2}$	00000000000000000000000000000040	
$R2^{2}$	00000000000000000000000000000000	
$R3^{2}$	00000000000000000000000000000000	
$a_{15}^{3}, \ldots, a_{0}^{3}$	000000000000000000000040002000c000000000000000000000000000400020	2^{-11}
$b_{15}^{3}, \ldots, b_{0}^{3}$	0000000000000000000000000000002000000000000000000000000000000020	
$R1^{3}$	00000000000000400000000000000020	
$R2^{3}$	00000000000000000000000060202040	
$R3^{3}$	00000000000000000000000000000000	
$a_{15}^{4}, \ldots, a_{0}^{4}$	00000000000000400040002060602020000000000000000000000040002000c0	2^{-51}
$b_{15}^{4}, \ldots, b_{0}^{4}$	0000000000000000000000000040004000000000000000000000000000000020	
$R1^{4}$	00000060000000000000002000004000	
$R2^{4}$	00000000a06060c000000000c0404080	
$R3^{4}$	404080c0204060204060202060202040	
$a_{15}^{5}, \ldots, a_{0}^{5}$	00000020a0206080002060c0e01ffee000000000000000400040002060602020	2^{-39}
$b_{15}^{5}, \ldots, b_{0}^{5}$	000000000000000000000040006000c000000000000000000000000000400040	
$R1^{5}$	408040404020200080002040c0600060	
$R2^{5}$	a4d967c30000000067d4d4b300000000	
$R3^{5}$	77f4573d1a0d67c31ac3a4175ad360f4	
z^{5}	e459278340202000e7d4f4b340c00060	2^{-14}

Table 10. Differential characteristic for 8-rounds of SNOW-Vi on related-key attack.

IV	0000000000000000000000000000c846	
Key	00000000000000000000000000006e2100000000000000000000000000000000	
$a_{15}^{0}, \ldots, a_{0}^{0}$	000000000000000000000000000000000000000000000000000000000000c846	1
$b_{15}^{0}, \ldots, b_{0}^{0}$	00000000000000000000000000006e2100000000000000000000000000000000	
$R1^{0}$	00000000000000000000000000000000	
$R2^{0}$	00000000000000000000000000000000	
$R3^{0}$	00000000000000000000000000000000	
$a_{15}^{1}, \ldots, a_{0}^{1}$	0000000000000000000000000000000000000000000000000000000000000000	2^{-11}
$b_{15}^{1}, \ldots, b_{0}^{1}$	0000000000000000000000000000a66700000000000000000000000000006e21	
$R1^{1}$	00000000000000000000000000000000	
$R2^{1}$	00000000000000000000000000000000	
$R3^{1}$	00000000000000000000000000000000	
$a_{15}^{2}, \ldots, a_{0}^{2}$	0000000000000000000000000000000000000000000000000000000000000000	2^{-11}
$b_{15}^{2}, \ldots, b_{0}^{2}$	00000000000000000000000000007a250000000000000000000000000000a667	
$R1^{2}$	00000000000000000000000000000000	
$R2^{2}$	00000000000000000000000000000000	
$R3^{2}$	00000000000000000000000000000000	
$a_{15}^{3}, \ldots, a_{0}^{3}$	0000000000000000000000000000000000000000000000000000000000000000	2^{-12}
$b_{15}^{3}, \ldots, b_{0}^{3}$	0000000000000000000000000000fa6c00000000000000000000000000007a25	
$R1^{3}$	00000000000000000000000000000000	
$R2^{3}$	00000000000000000000000000000000	
$R3^{3}$	00000000000000000000000000000000	
$a_{15}^{4}, \ldots, a_{0}^{4}$	0000000000000000000000000000000100000000000000000000000000000000	2^{-10}
$b_{15}^{4}, \ldots, b_{0}^{4}$	00000000000000000000000000000e260000000000000000000000000000fa6c	
$R1^{4}$	00000000000000000000000000000000	
$R2^{4}$	00000000000000000000000000000000	
$R3^{4}$	00000000000000000000000000000000	
$a_{15}^{5}, \ldots, a_{0}^{5}$	0000000000000000000000000000000200000000000000000000000000000001	2^{-14}
$b_{15}^{5}, \ldots, b_{0}^{5}$	000000000000000000000000000036790000000000000000000000000000 0e26	
$R1^{5}$	00000000000000000000000000000001	
$R2^{5}$	00000000000000000000000000000000	
$R3^{5}$	00000000000000000000000000000000	
$a_{15}^{6}, \ldots, a_{0}^{6}$	00000000000000000000000000021c0c00000000000000000000000000000002	2^{-17}
$b_{15}^{6}, \ldots, b_{0}^{6}$	00000000000000000000000000002a3400000000000000000000000000003679	
$R1^{6}$	00000000000000000000000000000002	
$R2^{6}$	000000000000000000000000120e0e1c	
$R3^{6}$	00000000000000000000000000000000	
$a_{15}^{7}, \ldots, a_{0}^{7}$	0000000000000000000000020e02021700000000000000000000000000021c0c	2^{-54}
$b_{15}^{7}, \ldots, b_{0}^{7}$	000000000000000000000000000046c400000000000000000000000000002a34	
$R1^{7}$	00000012000000000000000e00000000	
$R2^{7}$	0000000000000000000000000c040408	
$R3^{7}$	404080c002040602040602020a06060c	
$a_{15}^{8}, \ldots, a_{0}^{8}$	0000000e0000000000020e000e172f180000000000000000000000020e020217	2^{-34}
$b_{15}^{8}, \ldots, b_{0}^{8}$	000000000000000000000000000020ea000000000000000000000000000046c4	
$R1^{8}$	40020400400402008002020040026e20	
$R2^{8}$	e5aaaa4f0000000043c8c88b00000000	
$R3^{8}$	e8e8cb236ad4be6a21bc9d9df55353a6	
z^{8}	a5a8ae4f40040200c3caca8b40001c80	2^{-16}

6 Conclusion

In this paper, we evaluated the security against differential attacks for three important stream ciphers:SNOW-V, SNOW-Vi and KCipher-2. As a result, in the related-IV setting, we demonstrated that distinguishing attacks on 5/5/8 rounds for SNOW-V/SNOW-Vi/KCipher-2, respectively. In the related-key setting, we could extend to 7/8/8 rounds for SNOW-V/SNOW-Vi/KCipher-2, respectively. Our results are best attacks on initialization phase of these ciphers, and first result in the related-key setting.

Acknowledgments. This research was in part conducted under a contract of "Research and development on new generation cryptography for secure wireless communication services" among "Research and Development for Expansion of Radio Wave Resources (JPJ000254)", which was supported by the Ministry of Internal Affairs and Communications, Japan.

References

1. CRYPTREC Cipher List. https://www.cryptrec.go.jp/list.html/
2. Biham, E., Dunkelman, O.: Differential cryptanalysis in stream ciphers. IACR Cryptol. ePrint Arch. 218 (2007). http://eprint.iacr.org/2007/218
3. Biham, E., Shamir, A.: Differential cryptanalysis of DES-like cryptosystems. J. Cryptol. **4**(1), 3–72 (1991)
4. Ekdahl, P., Johansson, T.: A new version of the stream cipher SNOW. In: Nyberg, K., Heys, H. (eds.) SAC 2002. LNCS, vol. 2595, pp. 47–61. Springer, Heidelberg (2003). https://doi.org/10.1007/3-540-36492-7_5
5. Ekdahl, P., Johansson, T., Maximov, A., Yang, J.: A new SNOW stream cipher called SNOW-V. IACR Trans. Symmetric Cryptology **2019**(3), 1–42 (2019).https://doi.org/10.13154/tosc.v2019.i3.1-42
6. Ekdahl, P., Maximov, A., Johansson, T., Yang, J.: SNOW-Vi: an extreme performance variant of SNOW-V for lower grade CPUs. In: Proceedings of the 14th ACM Conference on Security and Privacy in Wireless and Mobile Networks, pp. 261–272. WiSec 2021, Association for Computing Machinery, New York, NY, USA (2021).https://doi.org/10.1145/3448300.3467829
7. Hoki, J., Isobe, T., Ito, R., Liu, F., Sakamoto, K.: Distinguishing and key recovery attacks on the reduced-round SNOW-V and SNOW-Vi. J. Inf. Secur. Appl. **65**, 103100 (2022). https://doi.org/10.1016/j.jisa.2021.103100, https://www.sciencedirect.com/science/article/pii/S2214212621002763
8. Hoki, J., Sakamoto, K., Minematsu, K., Isobe, T.: Practical integral distinguishers on SNOW 3G and KCipher-2. IEICE Trans. Fundam. Electron. Commun. Comput. Sci. **E104.A**(11), 1603–1611 (2021).https://doi.org/10.1587/transfun.2020EAP1102
9. Kiyomoto, S., Tanaka, T., Sakurai, K.: K2: A stream cipher algorithm using dynamic feedback control. In: SECRYPT 2007, pp. 204–213. INSTICC Press (2007)
10. Lin, D., Xiang, Z., Zeng, X., Zhang, S.: A framework to optimize implementations of matrices. In: Paterson, K.G. (ed.) CT-RSA 2021. LNCS, vol. 12704, pp. 609–632. Springer, Cham (2021). https://doi.org/10.1007/978-3-030-75539-3_25
11. Martins, R., Joshi, S., Manquinho, V., Lynce, I.: Incremental cardinality constraints for MaxSAT. In: O'Sullivan, B. (ed.) CP 2014. LNCS, vol. 8656, pp. 531–548. Springer, Cham (2014). https://doi.org/10.1007/978-3-319-10428-7_39
12. Mouha, N., Wang, Q., Gu, D., Preneel, B.: Differential and linear cryptanalysis using mixed-integer linear programming. In: Wu, C.-K., Yung, M., Lin, D. (eds.) Inscrypt 2011. LNCS, vol. 7537, pp. 57–76. Springer, Heidelberg (2012). https://doi.org/10.1007/978-3-642-34704-7_5
13. Mouha, N., Wang, Q., Gu, D., Preneel, B.: Differential and linear cryptanalysis using mixed-integer linear programming. In: Wu, C.K., Yung, M., Lin, D. (eds.) Inf. Secur. Cryptology, pp. 57–76. Springer, Berlin Heidelberg, Berlin, Heidelberg (2012)
14. Shi, Z., Jin, C., Zhang, J., Cui, T., Ding, L., Jin, Y.: A correlation attack on full snow-v and snow-vi. In: Dunkelman, O., Dziembowski, S. (eds.) Advances in Cryptology - EUROCRYPT 2022, pp. 34–56. Springer International Publishing, Cham (2022)
15. Sun, L., Wang, W., Wang, M.: Accelerating the search of differential and linear characteristics with the SAT method. IACR Trans. Symmetric Cryptology **2021**(1), 269–315 (Mar 2021).https://doi.org/10.46586/tosc.v2021.i1.269-315, https://tosc.iacr.org/index.php/ToSC/article/view/8840

Efficient Search for Optimal Permutations of Refined Type-II Generalized Feistel Structures

Xiaodan Li[1,2,3](✉), Wenling Wu[1,3](✉), Yuhan Zhang[1,3], and Ee Duan[1,3]

[1] Trusted Computing and Information Assurance Laboratory, Institute of Software Chinese Academy of Sciences, Beijing 100190, China
{xiaodan2018,wenling}@iscas.ac.cn

[2] China Satellite Network System Institute Co., Ltd., Beijing 100083, China

[3] University of Chinese Academy of Sciences, Beijing 100049, China

Abstract. Type-II Generalized Feistel Structures are widely used to design block ciphers benefit from their simplicity and high parallelism. However, there is a trade-off between efficiency (i.e. the number of rounds) and compactness (i.e. the partition number). Hence, Suzaki et al. (in FSE2010), Cauchois et al. (in FSE2019) and Derbez et al. (in FSE2019) studied how to find optimal permutations for Type-II Generalized Feistel Structures to improve the diffusion property.

In this paper, we further investigate how to find the optimal permutations for Refined Type-II Generalized Feistel Structures (RGFS) using k blocks and a permutation of size sk. First, we propose pair-equivalent relations and permutational equivalent relations and combine these two strategies to exhaustively search permutations that achieve optimal diffusion rounds. Then, to reduce the search space, we focus on the even-odd permutations. Based on even-odd permutations, we reveal the relations between the block size k, the permutation size s and the optimal full diffusion rounds and show Type-II GFS needs at least 4 rounds to achieve full diffusion. Besides, we also give the conditions that the block size k and the sub-block size s need to satisfy to achieve full diffusion after 4 rounds. Moreover, using pair-equivalent relations on even-odd permutations, we give the upper bound on the number of such equivalent classes. And then, using the search strategies we give some optimal permutations in different sizes of Type-II RGFS. Finally, we conduct a security analysis of our results with respect to impossible differentials and integral attacks.

Keywords: Feistel · Generalized Feistel Structure · Permutation · Diffusion Round

1 Introduction

The Feistel structure is one of the most popular categories of block ciphers, which was invented by Feistel when designing Lucifer and made popular with

Supported by the National Natural Science Foundation of China (No. 62072445).

T. Zhu and Y. Li (Eds.): ACISP 2024, LNCS 14895, pp. 98–117, 2024.
https://doi.org/10.1007/978-981-97-5025-2_6

the use of data encryption standard DES [1]. Feistel ciphers divide a message into two blocks and is also used in Camellia [2] and SIMON [3]. The Generalized Feistel Structure(GFS) which was used to design CAST-256 [4], CLEFIA [5] and Piccolo [6] divide a message into k blocks, where $k > 2$. These structures have the advantage that the encryption and decryption algorithms are similar and thus allow small implementations on hardware. Type-II GFS [7] is one popular form of GFS, where the output of a single round of Type-II GFS for input $(m_0, m_1, \cdots, m_{k-1})$ is

$$(c_0, c_1, \cdots, c_{k-1}) = (F_0(m_0) \oplus m_1, m_2, F_1(m_2) \oplus m_3, m_4, \\ \cdots, F_{k-2/2}(m_{k-2}) \oplus m_{k-1}, m_0),$$

where F_i are round functions, $0 \leq i \leq (k-2)/2$. It is equivalent to applying Feistel transformation for every two blocks and then following by a cyclic shift of the blocks. Then several block ciphers are based on Type-II GFS for its simplicity and high parallelism, such as CLEFIA and HIGHT [8].

The Type-II GFS with a large k is suitable for small-scale implementations but leading to low diffusion which can be exploited by some attacks, such as impossible differential attacks [9]. Hence, Type-II GFS generally needs a large number of rounds. To reduce the number of rounds to attain a sufficient level of security, Nyberg [10] used a different permutation instead of the cyclic shift. In FSE'10, Suzaki and Minematsu [11] showed the diffusion property of Type-II GFS can be improved by only changing the internal block shuffle from the cyclic shift. And they used *diffusion round* to compare permutations which is the minimum round to ensure full diffusion. Moreover, they searched all optimum shuffles up to $k \leq 16$ and confirmed that such block shuffles can be used to improve the resistance of GFS against some cryptanalysis, such as impossible differential attack and saturation attack [12], and improve the efficiency with respect to pseudorandomness. Further, Cauchois et al. [13] investigated the equivalence classes of even-odd permutations, i,e., the images of even numbers are odd, which preserve cryptographic properties and gave optimal permutations for $k \leq 24$. Besides, they also used tree representations to find permutations with good intermediate diffusion properties and led to some optimal even-odd permutations for $k = 26, 32, 64$ and 128. Later, Derbez et al. [14] proposed an efficient algorithm based on a new characterization for the permutation to have full diffusion after a given number of rounds and went further to construct all optimal even-odd permutations for $k = 28, 30, 32, 36$ and proved better lower bounds for $k = 34, 38, 40$ and 42. Such constructions were used to design TWINE [15]. To improve the diffusion of GFS, Piccolo [6] used Type-II GFS with 4 blocks and utilizes an 8-bit word-based permutation between rounds instead of a 16-bit word-based cyclic shift used in the standard Type-II GFS. It breaks the 16-bit word structure and thus improves the security against cryptanalysis exploiting strong word-based structures such as saturation attacks. This structure makes Piccolo achieve both high security and notably compact implementation in hardware and be one of the competitive ultra-lightweight block ciphers which are suitable for extremely constrained environments such as RFID tags and sensor nodes. Besides, Simon's algorithm is a well-known quantum algorithm which can achieve an exponential acceleration over a classical algorithm. It has been widely used in quantum cryptanalysis of Type-II GFS [16]. This motivates us to investigate the Type-II GFS with a fine-grained permutation called Refined Type-II GFS (Type-II RGFS) to improve the diffusion further.

Our Contribution. In this paper, we aim to find the optimal permutations for Type-II RGFS when achieving optimal full diffusion rounds. To reduce the search complexity, we present two equivalence relations on permutations, pair-equivalent and permutational equivalent, drawn from equivalence classes of Type-II RGFS that preserve the same cryptographic properties. First, we propose the pair-equivalent relations and permutational equivalent relations and combine these two strategies to exhaustively search the permutations that achieve optimal diffusion rounds. Then, to reduce the search space, we focus on the even-odd permutations. Based on even-odd permutations, we reveal the relations between the block sizes k, the permutation sizes s and the optimal full diffusion rounds and show Type-II RGFS needs at least 4 rounds to achieve full diffusion. Besides, we also give the conditions that the block sizes k and the permutation sizes s need to satisfy to achieve full diffusion after 4 rounds. Moreover, using pair-equivalent relations on even-odd permutations, we give the upper bound on the number of such equivalent classes. And then, using the search strategies we give some optimal permutations in different sizes of Type-II RGFS. The comparison of optimal diffusion round with the best-known permutations before is given in Table 1. Finally, we give some evaluation for impossible differentials and integral cryptanalysis and compared with classical Type-II RGFS, the permutations we found perform well in this case.

Table 1. The comparison of optimal diffusion round with the best known permutations

k	s	*Diffusion Round*	*Lower Bound of Diffusion Round*	*Ref*
4	1	4	4	[11]
	2	4	4	Sec.3 and Sec.4
	3	4	4	Sec.4
	4	4	4	Sec.4
6	1	5	5	[11]
	2	5	5	Sec.4
	3	4	4	Sec.4
	4	4	4	Sec.4
8	1	6	6	[11]
	2	5	5	Sec.4
	4	4	4	Sec.4
10	1	7	6	[11]
	2	5	5	Sec.4
12	1	8	7	[11]
	2	6	5	Sec.4

Structure of the Paper. The rest of paper is organised as follows. In Sect. 2, we give some notation and definitions used throughout the paper. Section 3 introduces two methods to divide permutations into equivalence classes. Then we focus on even-odd permutations to find the optimal permutations in Sect. 4. In Sect. 5, we evaluated the resistance against impossible differential and integral cryptanalysis of the optimal permutations we found. Finally, we conclude the paper in Sect. 6.

2 Preliminaries

In this section, we give some notation and definitions that will be used in this paper.

2.1 Type-II Generalized Feistel Structure

From the definition of Type-II GFS, we give the definition of Refined Type-II GFS next.

Definition 1. *Let s, r be positive integers. A plaintext is divided to k blocks, and represented by $(X_0, X_1, \cdots, X_{k-1})$, where $k \geq 4$ is an even integer. And each block is divided into s sub-blocks, i.e., $X_i (0 \leq i \leq k-1)$ can be represented by $(X_{i,0}, X_{i,1}, \cdots, X_{i,s-1})$. Let π be a permutation of $\{0, 1, \cdots, sk-1\}$, $\{F_{i,j}\}_{i \in \{1,\cdots,r\}, j \in \{0,\cdots,\frac{k}{2}-1\}}$ be a sufficiently good non-linear function. Then the i-th round function of Refined Type-II GFS Γ is given by $M_i = \Pi \circ R_i (1 \leq i \leq r)$, where*

$$R_i : (X_0, \cdots, X_{k-1}) \rightarrow \left(X_0, X_1 \oplus F_{i,0}(X_0), \cdots, X_{k-2}, X_{k-1} \oplus F_{i,\frac{k}{2}-1}(X_{k-2})\right),$$
$$\Pi = (X_{0,0}, \cdots, X_{0,s-1}, \cdots, X_{k-1,0}, \cdots, X_{k-1,s-1}) \rightarrow \left(X_{\pi(0)}, X_{\pi(1)}, \cdots, X_{\pi(sk-1)}\right).$$

We assume that the underlying F-function $F_{i,j}$ as an arbitrary S-box S, then the i-th round function of Refined Type-II GFS is illustrated in Fig. 1. Note that the overall structure of Piccolo [6] is Type-II RGFS with $k = 4$ and $s = 2$.

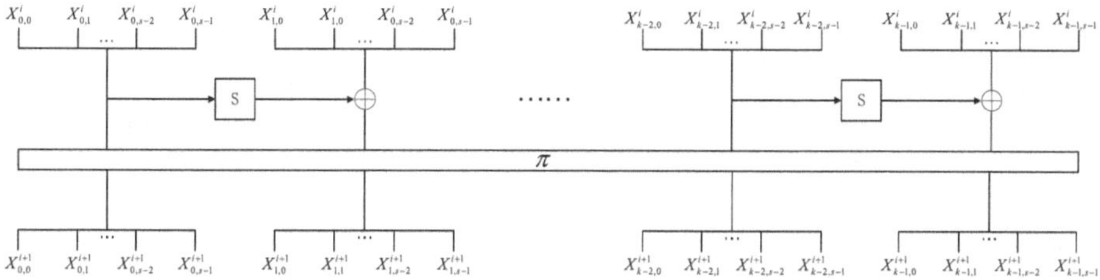

Fig. 1. Refined Type-II Generalized Feistel Structure

2.2 Diffusion Round

Diffusion was first defined by Shannon [17], and the Diffusion Round of GFS was defined in [11] and represented as DR. This is defined as the minimal number of rounds such that every sub-block of the ciphertext depends on every sub-block of the plaintext. And the authors in [11] also showed that the number of rounds required to achieve resistance against both saturation and impossible differential attacks is closely related to the DR. And they proved that a given scheme needs at least $2DR+1$ to resist the above attacks. Thus, Dffusion Round is an important indicator when designing a block cipher and a smaller DR would imply a faster, better diffusion. Next, we give the following definition.

Definition 2. *For a Refined Type-II Generalized Feistel Structure built from a permutation π, we use $DR_{i,j}(\pi)$ and $DR_{i,j}(\pi^{-1})$ to denote the minimal number of encryption rounds and decryption rounds such that the $(i,j)-th$ input sub-block of the first round $X^0_{(i,j)}$ is fully diffused to all output sub-blocks, respectively. Then the full diffusion round of a permutation π is*

$$DR_{max}(\pi) = \max_{0\leq i\leq k-1, 0\leq j\leq s-1}\{DR_{(i,j)}(\pi), DR_{(i,j)}(\pi^{-1})\}.$$

Thanks to Definition 2, a natural fact is that different permutations lead to different diffusion round. As we will see, if full diffusion has not been attained a certain attack is possible. Hence, a Refined Type-II Generalized Feistel Structure built from a permutation with a small DR_{max} is desirable. And we call a permutation of sk elements is optimal if its full diffusion round is minimal among all permutations of sk elements.

3 Equivalence Classes of Permutations

Let S_{sk} denote the set of all permutations of $\{0, 1, \cdots, sk-1\}$. To find the optimal permutations, a simple way is to go through all of the permutations in S_{sk} and then check which one is optimal. However, there are $(sk)!$ permutations that should be checked quickly beyond acceptable search complexity. To reduce the search complexity, we consider dividing permutations into equivalence classes. First, we give the following natural definition.

Definition 3. *Let $\Gamma_1 = M_{1,i}, \Gamma_2 = M_{2,i}$ be any two Refined Type-II Generalized Feistel Structure. We say that Γ_1 and Γ_2 are equivalent if there exists a permutation φ in S_{sk} such that $\forall i \in \{1, 2, \cdots, n\}$:*

$$M_{1,i} = \varphi \circ M_{2,i} \circ \varphi^{-1},$$

where n is the number of rounds of Γ.

The two structures with equivalence relation defined in Definition 3 have the same cryptographic properties. Next, we give two equivalence relations on Refined Type-II Generalized Feistel Structures.

3.1 Pair-Equivalent

First, we define a special permutation of pairs to lead to the equivalence relations on Refined Type-II Generalized Feistel Structures.

Definition 4. *The set of permutations of pairs S_{sk}^{p} is the subset of permutations in S_{sk}, and it is defined as*

$$S_{sk}^{p} = \{\varphi \in S_{sk} | \varphi(2si + t) = 2sj + t,$$

and

$$\varphi(2si + s + t) = \varphi(2si + t) + s\},$$

where $i, j = 0, 1, \cdots, \frac{k}{2} - 1$, $t = 0, 1, \cdots, s - 1$.

Then we can get following theorem.

Theorem 1. *Let $\Gamma_1 = RGFS((R_i)_{i\in\{1,\cdots,r\},\pi_1,r})$ and $\Gamma_2 = RGFS((R_i)_{i\in\{1,\cdots,r\},\pi_2,r})$ be two Refined Type-II Generalized Feistel structures. They are said to be equivalent if and only if there exists a permutation of pairs $\varphi \in S_{sk}^{p}$ such that*

$$\pi_1 = \varphi^{-1} \circ \pi_2 \circ \varphi.$$

Proof. For $M_{i,0}$, if $\Gamma_1 \sim \Gamma_2$, then

$$\pi_1 \circ R_0 = \varphi \circ \pi_2 \circ R_0 \circ \varphi^{-1}.$$

This equation holds if and only if $\forall x$ there holds

$$\pi_1(R_0(x)) = \varphi(\pi_2(R_0(\varphi^{-1}(x)))).$$

Let $y = R_0(x)$, then $x = R_0^{-1}(y)$, so

$$\pi_1(R_0(x)) = \varphi(\pi_2(R_0(\varphi^{-1}(x))))$$

is equivalent to $\pi_1(y) = \varphi(\pi_2(R_0(\varphi^{-1}(R_0^{-1}(y)))))$. Then we can get

$$\pi_1 = \varphi \circ \pi_2 \circ R_0 \circ \varphi^{-1} \circ R_0^{-1}.$$

Since $R_0 \circ \varphi^{-1} \circ R_0^{-1} = \varphi^{-1}$ means $R_0 = \varphi^{-1} \circ R_0 \circ \varphi$. This equality holds if and only if $\varphi \in S_{sk}^{p}$. Thus, $\Gamma_1 \sim \Gamma_2$ if and only if there exists a permutation of pairs $\varphi \in S_{sk}^{p}$ such that $\pi_1 = \varphi^{-1} \circ \pi_2 \circ \varphi$.

Benefit from Theorem 1, we can get following definition of equivalence relation on Refined Type-II Generalized Feistel Structures.

Definition 5. *Let $\Gamma_1 = M_{1,i}, \Gamma_2 = M_{2,i}$ be any two Refined Type-II Generalized Feistel Structures. We say that Γ_1 and Γ_2 are equivalent if there exist a permutation φ in S_{sk}^{p} such that $\forall i \in \{1, 2, \cdots, n\}$:*

$$M_{1,i} = \varphi \circ M_{2,i} \circ \varphi^{-1}.$$

Meanwhile, based on the equivalence relation on Refined Type-II Generalized Feistel Structures, we also can define pair-equivalent on permutations.

Definition 6. *Let π_1 and π_2 be two permutations in S_{sk}. They are said to be pair-equivalent if there exists a permutation of pairs φ in S^p_{sk} such that*

$$\pi_1 = \varphi \circ \pi_2 \circ \varphi^{-1}.$$

Note that two Refined Type-II Generalized Feistel Structures which are pair-equivalent have exactly the same cryptography properties.

3.2 Permutation-Equivalence Classes

Before introducing our equivalence class partitioning scheme, we first need to use matrices to redescribe the definition of the Refined Type-II Generalized Feistel Structures in Section II.

Let I be an $s \times s$ identity matrix, O be an $s \times s$ zero matrix and S be an $s \times s$ matrix whose all elements are 1. Then the i-th round function of Refined Type-II GFS(Fig. 1) Γ is given by $M_i = \Pi \circ R_i (1 \leq i \leq r)$, where Π is an $sk \times sk$ permutation matrix, R_i is an $sk \times sk$ matrix and

$$R_i = \begin{pmatrix} \mathrm{I} & \mathrm{O} & \mathrm{O} & \mathrm{O} & \cdots & \mathrm{O} & \mathrm{O} \\ \mathrm{S} & \mathrm{I} & \mathrm{O} & \mathrm{O} & \cdots & \mathrm{O} & \mathrm{O} \\ \mathrm{O} & \mathrm{O} & \mathrm{I} & \mathrm{O} & \cdots & \mathrm{O} & \mathrm{O} \\ \mathrm{O} & \mathrm{O} & \mathrm{S} & \mathrm{I} & \cdots & \mathrm{O} & \mathrm{O} \\ \vdots & \vdots & \vdots & \vdots & \ddots & \vdots & \vdots \\ \mathrm{O} & \mathrm{O} & \mathrm{O} & \mathrm{O} & \cdots & \mathrm{I} & \mathrm{O} \\ \mathrm{O} & \mathrm{O} & \mathrm{O} & \mathrm{O} & \cdots & \mathrm{S} & \mathrm{I} \end{pmatrix}.$$

In other words, given the values of s and k, R_i is uniquely determined. Then we have following conclusion.

Theorem 2. *Let $M_{1,i} = \Pi_1 \circ R_i, M_{2,i} = \Pi_2 \circ R_i$ be any two one round Refined Type-II Generalized Feistel Structures. If $M_{1,i}$ and $M_{2,i}$ is equivalent, then $M^r_{1,i} \sim M^r_{2,i}$.*

Proof. Let $M_{1,i} \sim M_{2,i}$, then their exists a permutation ϕ such that

$$M_{1,i} = \phi \circ M_{2,i} \circ \phi^{-1}.$$

Then

$$\begin{aligned} M^r_{1,i} &= \phi \circ M_{2,i} \circ \phi^{-1} \circ \phi \circ M_{2,i} \circ \phi^{-1} \circ \cdots \circ M_{2,i} \circ \phi^{-1} \\ &= \phi \circ M^r_{2,i} \circ \phi^{-1}, \end{aligned}$$

thus $M^r_{1,i} \sim M^r_{2,i}$.

Theorem 3. *Let* $\Gamma_1 = M_{1,i} = \Pi_1 \circ R_i, \Gamma_2 = M_{2,i} = \Pi_2 \circ R_i$ *be any two Refined Type-II Generalized Feistel Structures. We say that* Γ_1 *and* Γ_2 *is equivalent if there exists a permutation* ϕ *in* S_{sk} *such that*

$$\phi \circ R_i = R_i \circ \phi$$

and

$$\Pi_1 = \phi \circ \Pi_2 \circ \phi^{-1}.$$

Proof. Since

$$\phi \circ R_i = R_i \circ \phi \text{ and } \Pi_1 = \phi \circ \Pi_2 \circ \phi^{-1},$$

then

$$\begin{aligned}\Gamma_1 = M_{1,i} = \Pi_1 \circ R_i &= \phi \circ \Pi_2 \circ \phi^{-1} \circ R_i \\ &= \phi \circ \Pi_2 \circ R_i \circ \phi^{-1} = \phi \circ \Gamma_2 \circ \phi^{-1},\end{aligned}$$

thus Γ_1 and Γ_2 are equivalent.

It is important to note that two permutation-equivalent Refined Type-II Generalized Feistel structures share the same cryptographic properties. Next, we aim to classify all permutations in S_{sk} up to permutation equivalent, i.e. to find the permutation set S_{sk}^{eq} so that each permutation ϕ in S_{sk}^{eq} satisfy $\phi \circ R_i = R_i \circ \phi$. However, the search space is of size $(sk)!$. When sk grows, it is still too big to try an exhaustive search. In order to reach our goal, we first give an observation that R_i can be viewed as a block matrix of the following form:

$$R_i = \begin{pmatrix} \mathrm{V} & \mathrm{O} & \cdots & \mathrm{O} \\ \mathrm{O} & \mathrm{V} & \cdots & \mathrm{O} \\ \vdots & \vdots & \ddots & \vdots \\ \mathrm{O} & \mathrm{O} & \cdots & \mathrm{V} \end{pmatrix},$$

where $V = \begin{pmatrix} \mathrm{I} & \mathrm{O} \\ \mathrm{S} & \mathrm{I} \end{pmatrix}$.

If a permutation W satisfies $W \circ V = V \circ W$, then there exists a permutation

$$M = \begin{pmatrix} \mathrm{W} & \mathrm{O} & \cdots & \mathrm{O} \\ \mathrm{O} & \mathrm{W} & \cdots & \mathrm{O} \\ \vdots & \vdots & \ddots & \vdots \\ \mathrm{O} & \mathrm{O} & \cdots & \mathrm{W} \end{pmatrix}, \tag{1}$$

that satisfies $M \circ R_i = R_i \circ M$. That is to say, we only need to find the permutation set S_W such that each permutation W in S_W satisfies $W \circ V = V \circ W$. This leads to **Strategy 1**.

Strategy 1

1. For fixed s, k, finding the permutation set S_W that each permutation W in S_W satisfies $W \circ V = V \circ W$.

2. According to (1), using permutation set S_W to expand to permutation set S_M.
3. For any permutation M in S_M, if $\Pi_1 = M \circ \Pi_2 \circ M^{-1}$, then Π_1 and Π_1 are in the same equivalent class.

The search complexity is reduced to $((2s)!)^{k/2}$. When $k \geq 4$, it is a lot smaller than the total number of (sk)! Although the search complexity of Strategy 1 is bigger than that when only using pair-equivalent relations, we can use Strategy 1 to reduce the number of equivalent classes further. Therefore, in the actual search, we first use the pair-equivalent relations to narrow the optimal permutation to a certain range and then use Strategy 1 to further reduce the number of optimal permutation equivalence classes.

Example 1. Let $s = 2, k = 4$, the optimal full diffusion rounds is 4. And the number of equivalent classes with full diffusion rounds 4 is 24 when using pair-equivalent relations and Strategy 1. However, using pair-equivalent relations, the corresponding number of the equivalent class is 48.

4 The Even-Odd Case

In this section, we will focus on even-odd permutations. The even-odd permutation means the even blocks diffuse to odd blocks and odd blocks diffuse to even blocks, but every sub-block in one block can diffuse to different block. That is to say, X_{i_1,j_1} diffuses to X_{i_2,j_2}, X_{i_2,j_1} diffuses to X_{i_1,j_1}, where i_1 is even, i_2 is odd, $j_1, j_2 = 0, 1, \cdots, s-1$. And the even-odd permutation set is denoted by S_{sk}^{eo}. Then an even-odd permutation π of size sk will be denoted by the pair of permutations (p, q) of size $sk/2$ verifying $\forall i \in \{0, 1, \cdots, \frac{k}{2}-1\}, t \in \{0, 1, \cdots, s-1\}$,

$$\pi(2si + t) = p(si + t) + (\lfloor \frac{p(si + t)}{s} \rfloor + 1) \cdot s,$$

$$\pi((2i + 1)s + t) = q(si + t) + \lfloor \frac{q(si + t)}{s} \rfloor \cdot s.$$

Example 2. Let $k = 4, s = 3, \pi = p, q$ and $p = (1, 3, 0, 5, 2, 4), q = (0, 3, 1, 2, 5, 4)$, then

$$\pi(0) = p(0) + 1 \times 3 = 4,\ \pi(1) = p(1) + 2 \times 3 = 9,$$
$$\pi(2) = p(2) + 1 \times 3 = 3,\ \pi(3) = q(0) + 0 \times 3 = 0,$$
$$\pi(4) = q(1) + 1 \times 3 = 6,\ \pi(5) = q(2) + 0 \times 3 = 1,$$
$$\pi(6) = p(3) + 2 \times 3 = 11,\ \pi(7) = p(4) + 1 \times 3 = 5,$$
$$\pi(8) = p(5) + 2 \times 3 = 10,\ \pi(9) = q(3) + 0 \times 3 = 2,$$
$$\pi(10) = q(4) + 1 \times 3 = 8,\ \pi(11) = q(5) + 1 \times 3 = 7,$$

then $\pi = (4, 9, 3, 0, 6, 1, 11, 5, 10, 2, 8, 7)$. Thus, the Refined Type-II Generalized Feistel structure based on s, k, π is depicted in Fig. 2.

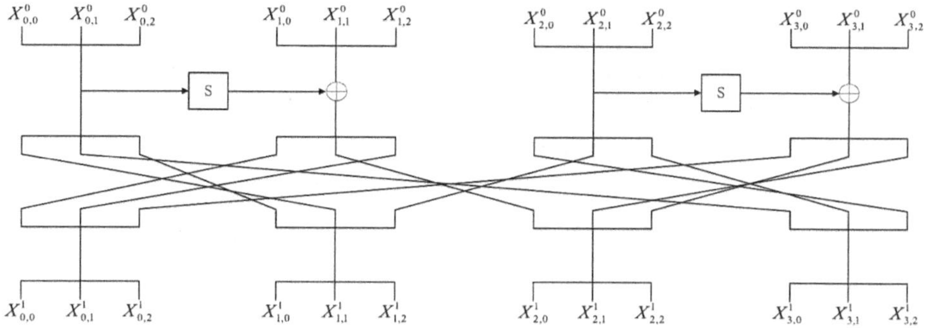

Fig. 2. RGFS base on $k = 4, s = 3, \pi = (4, 9, 3, 0, 6, 1, 11, 5, 10, 2, 8, 7)$

4.1 The Properties of Full Diffusion

First, we give some properties of full diffusion for even-odd permutation.

Theorem 4. *Let $\pi = (p, q)$ be an even-odd permutation over sk elements. Then π achieves full diffusion after r rounds if and only if each sub-block of each even block achieves full diffusion after $r - 1$ rounds.*

Proof. If each sub-block of each even block achieves full diffusion after $r - 1$ rounds, then each sub-block of each odd block achieves full diffusion after r rounds, i.e. π achieves full diffusion after r rounds.

Theorem 5. *Let $\pi = (p, q)$ be an even-odd permutation over sk elements. Then π achieves full diffusion after r rounds if and only if any sub-block of each even block diffuse to all even block after $r - 2$ rounds.*

Proof. If any sub-block of each even block diffuse to all even block after $r - 2$ rounds, then any sub-block of each even block achieve full diffusion after $r - 1$ rounds. From Theorem 4, π achieves full diffusion after r rounds.

Next, we investigate the relationship between sk and the full diffusion rounds and compute a lower bound on the minimal diffusion round for the RGFS structures with optimal even-odd permutations.

Theorem 6. *Let π be an even-odd permutation over sk elements, n be the full diffusion rounds. Then*

$$\frac{1}{\sqrt{s^2+4}}((\frac{\sqrt{s^2+4}+s}{2})^{n-1} - (\frac{s-\sqrt{s^2+4}}{2})^{n-1}) \geq \frac{sk}{2}. \tag{2}$$

Proof. After one round, one even position has diffused to one even position and s odd positions. One odd position has diffused to one even position after one round. Now, assume the input has only one even nonzero position and it diffuses to a_r odd positions and b_r even positions after r rounds. Then

$$a_r = b_{r-1},$$

$$b_r = s \cdot b_{r-1} + a_{r-1},$$

where $a_0 = 0,\ a_1 = 1$ and $b_0 = 1,\ b_1 = s$. Then we have $b_r = s \cdot b_{r-1} + b_{r-2}$, $b_{r+1} = s \cdot b_r + b_{r-1}$, then

$$\begin{aligned} & b_{r+1} + \frac{\sqrt{s^2+4}-s}{2} \cdot b_r \\ &= \frac{\sqrt{s^2+4}+s}{2}(b_r + \frac{\sqrt{s^2+4}-s}{2} \cdot b_{r-1}), \end{aligned}$$

then

$$b_{r+1} + \frac{\sqrt{s^2+4}-s}{2} \cdot b_r = (\frac{\sqrt{s^2+4}+s}{2})^{r+1}.$$

Further, we have

$$\begin{aligned} & b_{r+1} - \frac{\sqrt{s^2+4}+s}{2\sqrt{s^2+4}}(\frac{\sqrt{s^2+4}+s}{2})^{r+1} \\ &= \frac{s-\sqrt{s^2+4}}{2}(b_r - \frac{\sqrt{s^2+4}+s}{2\sqrt{s^2+4}}(\frac{\sqrt{s^2+4}+s}{2})^r), \end{aligned}$$

then we can get

$$b_r = \frac{1}{\sqrt{s^2+4}}((\frac{\sqrt{s^2+4}+s}{2})^{r+1} - (\frac{s-\sqrt{s^2+4}}{2})^{r+1}).$$

If one even position diffuses to all even positions, then $b_n \geq \frac{sk}{2}$. From Theorem 4 and 5, if the RGFS builded by π achieves full diffusion after n rounds, then

$$\frac{1}{\sqrt{s^2+4}}((\frac{\sqrt{s^2+4}+s}{2})^{n-1} - (\frac{s-\sqrt{s^2+4}}{2})^{n-1}) \geq \frac{sk}{2}.$$

Theorem 7. *Let π be an even-odd permutation over sk elements, then for any s, k, the RGFS requires at least 4 rounds to achieve full diffusion. Further, if the RGFS achieves full diffusion after 4 rounds, then*

$$k \leq 2(s + \frac{1}{s}).$$

Specially, when $k = 4$, $s \geq 1$; when $k > 4$, $s \geq \frac{k}{2}$.

Proof. Let n be full diffusion rounds. Then according to Theorem 6:

If $n = 1$, (2) is $0 \geq sk$. Obviously impossible.

If $n = 2$, (2) is $sk \leq 2$. Since $k \geq 4$, thus it is impossible.

If $n = 3$, (2) is $s \geq \frac{sk}{2}$. Since $s > 0$, (2) means $k \leq 2$. Thus $n = 3$ is impossible.

If $n = 4$, (2) is $k \leq 2(s + \frac{1}{s})$, i.e. $s + \frac{1}{s} \geq \frac{k}{2}$.

Specially, if $k = 4$, (2) is $s + \frac{1}{s} \geq 2$. It is always true when $s \geq 1$. If $k \geq 4$, $0 < \frac{1}{s} \leq 1$ since $s \geq 1$. Thus, if $s + \frac{1}{s} \geq \frac{k}{2}$ holds, we only need $s \geq \frac{k}{2}$.

According to Theorem 6, we summarize the lower bounds of the number of full diffusion rounds for a series of RGFS structures in Table 2. These theorems reveal the diffusion rules and significantly decrease the complexity of the exhaustive search. And for RGFS, if we want to achieve the lower bounds of full diffusion rounds 4, we need at least $s = k/2$.

Table 2. The lower bounds of full diffusion rounds for RGFS

RGFS size		lower bounds of full diffusion rounds	RGFS size		lower bounds of full diffusion rounds
k	s		k	s	
4	1	4	10	1	6
4	2	4	10	2	5
4	3	4	10	3	5
4	4	4	10	4	5
6	1	5	10	5	4
6	2	5	12	1	7
6	3	4	12	2	5
6	4	4	12	3	5
8	1	6	12	4	5
8	2	5	12	5	5
8	3	5	12	6	4
8	4	4			

4.2 Pair-Equivalence Classes of Even-Odd Permutations

In this section, we give a strategy to build a set of even-odd permutations such that for any equivalence class, there exists at least one permutation that belongs to this class. First, we recall that any permutation can be decomposed into a composition of cycles.

Definition 7. *Let π be a permutation in S_k. A cycle decomposition of π is a decomposition of π as a product of cycles with disjoint supports. If π has w_1 cycles of size $v_1, \cdots,$ and w_w cycles of size v_w, we say that π has decomposition type V_π, written as follows:*

$$V_\pi = ((v_1, w_1), \cdots, (v_w, w_w)).$$

Focus on the even-odd permutations, our search space is $(p!)^2$. However, because of the exponential growth of $(p!)^2$, the search quickly becomes intractable. Next, we further reduce the upper bound of the number of pair-equivalence classes for the even-odd permutation.

Theorem 8. *Let $sk = 2p$, and N_p be the number of distinct cycle decompositions of permutations in S_p. The number of pair-equivalence classes of even-odd permutations in S_{sk} is upper bounded by*

$$U_{sk}^{eo} = N_p \cdot p!.$$

Proof. There are two natural bijections $\Psi_1(S_{sk}^p \to S_p)$ and $\Psi_2(S_p \times S_p \to S_{sk}^{eo})$. For any $\varphi \in S_{sk}^p, \overline{\varphi} \in S_p$, $\Psi_1 : \varphi \mapsto \overline{\varphi}$, s.t.

$$\overline{\varphi}(si + t) = \varphi(2si + t) \ mod \ (s) + \lfloor \varphi(2si + t)/2\,s \rfloor \cdot s.$$

Since $\varphi \ mod \ (s) = \varphi - \lfloor \varphi/s \rfloor \cdot s$, then

$$\overline{\varphi}(si + t) = \varphi(2si + t) + (\lfloor \varphi(2si + t)/2\,s \rfloor - \lfloor \varphi(2si + t)/s \rfloor) \cdot s.$$

For $\varphi_1, \varphi_2 \in S_p, \varphi \in S_{sk}^{eo}$, $\Psi_2 : (\varphi_1, \varphi_2) \mapsto \varphi$, s.t.

$$\begin{cases} \varphi(2si + t) = \varphi_1(si + t) + (\lfloor \varphi_1(si + t)/s \rfloor + 1) \cdot s \\ \varphi((2i + 1) \cdot s + t) = \varphi_2(si + t) + \lfloor \varphi_2(si + t)/s \rfloor \cdot s \end{cases}$$

with $\Psi_2^{-1} : \varphi \mapsto (\varphi_1, \varphi_2)$, s.t.

$$\begin{cases} \varphi_1(si + t) = \varphi(2si + t) \ mod \ (s) + \lfloor \varphi(2si + t)/2\,s \rfloor \cdot s \\ \varphi_2(si + t) = \varphi((2i + 1) \cdot s + t) \ mod \ (s) + \lfloor \varphi((2i + 1) \cdot s + t)/2\,s \rfloor \cdot s \end{cases}$$

i.e.

$$\begin{cases} \varphi_1(si + t) = \varphi(2si + t) + (\lfloor \varphi(2si + t)/2s \rfloor - \lfloor \varphi(2si + t)/s \rfloor) \cdot s \\ \varphi_2(si + t) = \varphi((2i + 1) \cdot s + t) + (\lfloor \varphi((2i + 1) \cdot s + t)/2s \rfloor - \lfloor \varphi((2i + 1) \cdot s + t)/s \rfloor) \cdot s \end{cases}$$

Next, we prove any pair-equivalence class of even-odd permutations always belong to following set:

$$\{\Psi_2(\varphi_j, \theta), j \in \{1, \cdots, N_p\}, \theta \in S_p\}.$$

Let π be an even-odd permutation in S_{sk}^{eo}, $(\pi_1, \pi_2) = \Psi_2^{-1}(\pi)$. Let $j \in \{1, \cdots, N_p\}$ such that $V_{\varphi_j} = V_{\pi_1}$. Then there exists $\phi \in S_p$ such that

$$\phi \circ \pi_1 \circ \phi = \varphi_j.$$

Then, there exists $\psi \in S_p$ such that

$$\Psi_2^{-1}(\Psi_1^{-1}(\phi^{-1}) \circ \pi \circ \Psi_1^{-1}(\phi)) = (\varphi_j, \psi).$$

Theorem 8 leads to following Strategy to search the representation of each even-odd equivalent class.

Strategy 2

1. For all decomposition types v of size p. Fix an permutation p_v that satisfies this decomposition type.

2. For all permutations $q \in S_p$, constructing the permutation $\pi_{p_v,q}$:

$$\begin{cases} \pi_{p_v,q}(2si+t) = p_v(si+t) + (\lfloor p_v(si+t)/s \rfloor + 1) \cdot s \\ \pi_{p_v,q}(2(i+1) \cdot s + t) = q(si+t) + (\lfloor q(si+t)/s \rfloor) \cdot s \end{cases}$$

   where $s \geq 1, k \geq 4$ be even, $i \in \{0, 1, \cdots, \frac{k}{2} - 1\}, t \in \{0, 1, \cdots, s-1\}$.
3. For arbitrary permutation $\varphi \in S_{sk}$, there always exists $\pi_{p_v,q} \in \xi_{sk} = \{\pi_{p_v,q}\}_{v,q}$ such that $\pi_{p_v,q}$ and φ are in the same pair-equivalent class.

To reduce the number of equivalent classes further, we can also combine Strategy 1 and 2 that is called Strategy 3.

Applying Strategy 3, we find the optimal even-odd permutations for several sizes of RGFS, and the results are shown in Table 1. And in the appendix, we give some representatives that reach the optimal even-odd diffusion rounds. From the results, when $k \leq 10$, we have found at least one even-odd permutation that can reach its theoretical lower bound of full diffusion round. In particular, for $k = 10, s = 2$, there are only two optimal even-odd pair-equivalence classes. And we don't find even-odd permutations with diffusion round 5 when $k = 12, s = 2$.

5 Security Analysis

In order to estimate the cryptographic resistance of the schemes, designers usually ensure that they can resist known attacks. In [11], the authors showed a given scheme needs at least $2DR + 1$ to resist both saturation and impossible differential attacks. DR measures the quality of the diffusion of a round function of a given scheme and reveals it to be fundamental in the design of such primitives. In Table 1, we give the comparison of the optimal diffusion round of RGFS with the best-known permutations of GFS. Compared to the common Type-II GFS, the Refined Type-II Generalized Feistel Structures lead to greater diffusion property and improve security.

Besides, for the optimal permutations we found, we apply the automatic search algorithm to evaluate the resistance against impossible differential attack and integral attack.

Impossible differential cryptanalysis is one of the most powerful cryptanalytic techniques. Recently, the automatic search algorithm based on the MILP method [18] is an effective approach for searching impossible differential characteristics. We apply the automatic impossible differential characteristics search algorithm based on the MILP method to every Type-II Generalized Feistel Structures with k block and sk sizes permutations that we found to find the upper bound of the round of distinguisher, and the results are given in Appendix.

Division property is a generalized integral property, which is a new method for integral distinguisher search. Xiang [19] uses MILP method to search integral distinguishers based on bit-based division property. Sun [20] also applies SAT method in word-based division property, which can find integral characteristics for large size ciphers. Combined Xiang's modeling method and Sun's modeling method, we use the MILP method to search integral distinguishers based on the word-based division property for every Type-II Generalized Feistel Structures with k block and sk sizes permutations that we found and find the upper bound of the round of distinguisher. The results are given in Appendix.

6 Conclusion and Perspectives

In this paper, we present two equivalence relations on permutations, pair-equivalent and permutational equivalent, drawn from equivalence classes of Type-II GFS that preserve the same cryptographic properties. First, we propose the pair-equivalent relations and permutational equivalent relations and combine these two strategies to exhaustively search the permutations that achieve optimal diffusion rounds. Then, to reduce the search space, we focus on the even-odd permutations. Based on even-odd permutations, we reveal the relations between the block sizes k, the permutation sizes s and the optimal full diffusion rounds and show Type-II GFS needs at least 4 rounds to achieve full diffusion. Besides, we also give the conditions that the block sizes k and the sub-block sizes s need to satisfy to achieve full diffusion after 4 rounds. Moreover, using pair-equivalent relations on even-odd permutations, we give the upper bound on the number of such equivalent classes. And then, using the search strategies we give some optimal permutations in different sizes of Type-II RGFS. The comparison of optimal diffusion round with the best-known permutations before is given in Table 1.

Finally, we use MILP method to search impossible differential characteristics and integral distinguishers for every Type-II Generalized Feistel Structures that we found and find the upper bound of the round of distinguisher. The results are given in Appendix. If we fix the property of $F-$function, we can evaluate the resistance against differential and linear cryptanalysis, and we leave them for future work. Overall, compared to classical Type-II GFS, Type-II RGFS structures with optimal permutations lead to better diffusion and stronger security and we think the results presented in this work will be particularly useful for future construction of lightweight ciphers based on Type-II GFS.

Appendix A

In the following Tables, we exhibit a representative for each extended class of optimal permutations and their security evaluation, where Imp. and Intg. stand for impossible differential characteristic and integral characteristic, respectively (Tables 3, 4, 5, 6, 7 and 8).

Table 3. Optimal permutations for $k = 4, s = 2$ up to extended pair-equivalence

Optimal Permutations k = 4, s = 2, $DR(\pi) = 4$	Intg.	Imp.
{2,7,0,5,3,6,1,4}	4	5
{2,7,0,5,6,3,1,4}	4	5
{7,2,0,5,3,6,1,4}	4	5
{7,2,0,5,6,3,1,4}	4	5
{3,6,0,5,2,7,1,4}	4	5
{6,3,0,5,2,7,1,4}	4	5
{3,6,0,5,7,2,1,4}	4	5
{6,3,0,5,7,2,1,4}	4	5
{2,6,0,5,3,7,1,4}	5	5
{2,6,0,5,7,3,1,4}	5	5
{6,2,0,5,3,7,1,4}	5	5
{6,2,0,5,7,3,1,4}	5	5
{3,7,0,5,2,6,1,4}	5	5
{7,3,0,5,2,6,1,4}	5	5
{3,7,0,5,6,2,1,4}	5	5
{7,3,0,5,6,2,1,4}	5	5
{3,2,0,5,6,7,1,4}	6	6
{6,7,0,5,3,2,1,4}	6	6
{2,3,0,5,6,7,1,4}	6	7
{2,3,0,5,7,6,1,4}	6	7

Table 4. Some optimal permutations for $k = 6, s = 2$ up to extended pair-equivalence

Optimal Permutations k = 6, s = 2, $DR(\pi) = 5$	Imp.	Intg.
{10,2,0,9,3,6,1,4,7,11,5,8}	6	6
{10,2,0,9,3,7,1,4,6,11,5,8}	6	6
{11,2,0,9,3,7,1,4,6,10,5,8}	6	6
{10,3,0,9,2,6,1,4,7,11,5,8}	6	6
{10,3,0,9,2,7,1,4,6,11,5,8}	6	6
{2,7,0,4,3,11,1,9,10,6,5,8}	6	6
{3,6,0,4,2,10,1,9,11,7,5,8}	6	6
{3,6,0,4,2,11,1,9,10,7,5,8}	6	6
{3,7,0,4,2,10,1,9,11,6,5,8}	6	6
{3,7,0,4,2,11,1,9,10,6,5,8}	6	6
{11,2,0,9,3,6,1,4,7,10,5,8}	6	7
{11,3,0,9,2,6,1,4,7,10,5,8}	6	7
{11,3,0,9,2,7,1,4,6,10,5,8}	6	7
{2,6,0,4,3,10,1,9,11,7,5,8}	6	7
{2,6,0,4,3,11,1,9,10,7,5,8}	6	7
{2,7,0,4,3,10,1,9,11,6,5,8}	6	7
{2,6,0,9,3,10,1,4,7,11,5,8}	7	6
{2,10,0,9,3,6,1,4,7,11,5,8}	7	6
{2,10,0,9,3,6,1,4,11,7,5,8}	7	6
{2,7,0,9,3,10,1,4,6,11,5,8}	7	6

Table 5. Some optimal permutations for $k = 8, s = 2$ up to extended pair-equivalence

Optimal Permutations k = 8, s = 2, $DR(\pi) = 5$	Intg.	Imp.
{11,7,0,4,10,6,1,13,15,2,5,8,3,14,9,12}	6	8
{7,10,0,13,11,14,1,4,3,15,5,8,2,6,9,12}	7	7
{10,7,0,13,11,14,1,4,3,15,5,8,2,6,9,12}	7	7
{7,10,0,13,11,14,1,4,15,3,5,8,2,6,9,12}	7	7
{7,10,0,13,14,11,1,4,15,3,5,8,2,6,9,12}	7	7
{10,7,0,13,11,14,1,4,15,3,5,8,2,6,9,12}	7	7
{10,7,0,13,14,11,1,4,3,15,5,8,6,2,9,12}	7	7
{7,10,0,13,11,14,1,4,15,3,5,8,6,2,9,12}	7	7
{7,10,0,13,14,11,1,4,15,3,5,8,6,2,9,12}	7	7
{15,11,0,4,7,14,1,8,3,10,5,13,2,6,9,12}	7	7
{15,11,0,4,14,7,1,8,3,10,5,13,2,6,9,12}	7	7
{15,11,0,4,7,14,1,8,10,3,5,13,2,6,9,12}	7	7
{11,15,0,4,14,7,1,8,10,3,5,13,2,6,9,12}	7	7
{15,11,0,4,14,7,1,8,10,3,5,13,2,6,9,12}	7	7
{11,15,0,4,7,14,1,8,3,10,5,13,6,2,9,12}	7	7
{11,15,0,4,14,7,1,8,10,3,5,13,6,2,9,12}	7	7
{15,11,0,4,14,7,1,8,10,3,5,13,6,2,9,12}	7	7
{11,14,0,4,3,7,1,13,10,15,5,8,2,6,9,12}	7	7
{11,14,0,4,3,7,1,13,15,10,5,8,2,6,9,12}	7	7
{11,14,0,4,10,7,1,13,15,3,5,8,2,6,9,12}	7	7

Table 6. Some optimal permutations for $k = 8, s = 4$ up to extended pair-equivalence

Optimal Permutations k = 8, s = 4, $DR(\pi) = 4$	Intg.	Imp.
{4, 12, 20, 28, 0, 8, 16, 24, 5, 13, 21, 29, 1, 9, 17, 25, 6, 14, 22, 30, 2, 10, 18, 26, 7, 15, 23, 31, 3, 11, 19, 27}	4	5
{4, 12, 20, 28, 0, 8, 16, 24, 5, 13, 21, 29, 1, 9, 17, 25, 6, 14, 22, 30, 2, 10, 18, 26, 7, 15, 23, 31, 3, 11, 27, 19}	4	5
{4, 12, 20, 28, 0, 8, 16, 24, 5, 13, 21, 29, 1, 9, 17, 25, 6, 14, 22, 30, 2, 10, 18, 26, 7, 15, 23, 31, 3, 19, 11, 27}	4	5
{4, 12, 20, 28, 0, 8, 16, 24, 5, 13, 21, 29, 1, 9, 17, 25, 6, 14, 22, 30, 2, 10, 18, 26, 7, 15, 23, 31, 3, 27, 11, 19}	4	5
{4, 12, 20, 28, 0, 8, 16, 24, 5, 13, 21, 29, 1, 9, 17, 25, 6, 14, 22, 30, 2, 10, 18, 26, 7, 15, 23, 31, 3, 19, 27, 11}	4	5
{4, 12, 20, 28, 0, 8, 16, 24, 5, 13, 21, 29, 1, 9, 17, 25, 6, 14, 22, 30, 2, 10, 18, 26, 7, 15, 23, 31, 3, 27, 19, 11}	4	5
{4, 12, 20, 28, 0, 8, 16, 24, 5, 13, 21, 29, 1, 9, 17, 25, 6, 14, 22, 30, 2, 10, 18, 26, 7, 15, 23, 31, 11, 3, 19, 27}	4	5
{4, 12, 20, 28, 0, 8, 16, 24, 5, 13, 21, 29, 1, 9, 17, 25, 6, 14, 22, 30, 2, 10, 18, 26, 7, 15, 23, 31, 11, 3, 27, 19}	4	5
{4, 12, 20, 28, 0, 8, 16, 24, 5, 13, 21, 29, 1, 9, 17, 25, 6, 14, 22, 30, 2, 10, 18, 26, 7, 15, 23, 31, 19, 3, 11, 27}	4	5
{4, 12, 20, 28, 0, 8, 16, 24, 5, 13, 21, 29, 1, 9, 17, 25, 6, 14, 22, 30, 2, 10, 18, 26, 7, 15, 23, 31, 27, 3, 11, 19}	4	5
{4, 12, 20, 28, 0, 8, 16, 24, 5, 13, 21, 29, 1, 9, 17, 25, 6, 14, 22, 30, 2, 10, 18, 26, 7, 15, 23, 31, 19, 3, 27, 11}	4	5
{4, 12, 20, 28, 0, 8, 16, 24, 5, 13, 21, 29, 1, 9, 17, 25, 6, 14, 22, 30, 2, 10, 18, 26, 7, 15, 23, 31, 27, 3, 19, 11}	4	5
{4, 12, 20, 28, 0, 8, 16, 24, 5, 13, 21, 29, 1, 9, 17, 25, 6, 14, 22, 30, 2, 10, 18, 26, 7, 15, 23, 31, 11, 19, 3, 27}	4	5
{4, 12, 20, 28, 0, 8, 16, 24, 5, 13, 21, 29, 1, 9, 17, 25, 6, 14, 22, 30, 2, 10, 18, 26, 7, 15, 23, 31, 11, 27, 3, 19}	4	5
{4, 12, 20, 28, 0, 8, 16, 24, 5, 13, 21, 29, 1, 9, 17, 25, 6, 14, 22, 30, 2, 10, 18, 26, 7, 15, 23, 31, 19, 11, 3, 27}	4	5
{4, 12, 20, 28, 0, 8, 16, 24, 5, 13, 21, 29, 1, 9, 17, 25, 6, 14, 22, 30, 2, 10, 18, 26, 7, 15, 23, 31, 27, 11, 3, 19}	4	5
{4, 12, 20, 28, 0, 8, 16, 24, 5, 13, 21, 29, 1, 9, 17, 25, 6, 14, 22, 30, 2, 10, 18, 26, 7, 15, 23, 31, 19, 27, 3, 11}	4	5
{4, 12, 20, 28, 0, 8, 16, 24, 5, 13, 21, 29, 1, 9, 17, 25, 6, 14, 22, 30, 2, 10, 18, 26, 7, 15, 23, 31, 27, 19, 3, 11}	4	5
{4, 12, 20, 28, 0, 8, 16, 24, 5, 13, 21, 29, 1, 9, 17, 25, 6, 14, 22, 30, 2, 10, 18, 26, 7, 15, 23, 31, 11, 19, 27, 3}	4	5
{4, 12, 20, 28, 0, 8, 16, 24, 5, 13, 21, 29, 1, 9, 17, 25, 6, 14, 22, 30, 2, 10, 18, 26, 7, 15, 23, 31, 11, 27, 19, 3}	4	5

Table 7. Optimal permutations for $k = 10, s = 2$ up to extended pair-equivalence

Optimal Permutations k = 10, s = 2, $DR(\pi) = 5$	Intg.	Imp.
{15,19,0,4,14,7,1,8,18,11,5,12,6,2,9,17,3,10,13,16}	9	10
{18,11,0,4,19,15,1,8,10,3,5,12,14,7,9,17,6,2,13,16}	9	10

Table 8. Some optimal permutations for $k = 12, s = 2$ up to extended pair-equivalence

Optimal Permutations k = 12, s = 2, $DR(\pi) = 6$	Intg.	Imp.
{3,6,0,8,7,10,1,16,11,2,4,21,15,18,5,9,19,22,12,20,23,14,13,17}	10	11
{3,6,0,8,7,10,1,20,11,2,4,16,15,18,5,9,19,22,12,17,23,14,13,21}	10	11
{3,6,0,8,7,10,1,21,11,2,4,16,15,18,5,9,19,22,12,17,23,14,13,20}	10	11
{3,6,0,8,7,10,1,17,11,2,4,16,15,18,5,9,19,22,12,20,23,14,13,21}	10	11
{3,6,0,8,7,10,1,21,11,2,4,17,15,18,5,9,19,22,12,20,23,14,13,16}	10	11
{3,6,0,8,7,10,1,16,11,2,4,21,15,18,5,9,19,22,12,20,23,14,17,13}	10	11
{3,6,0,8,7,10,1,21,11,2,4,17,15,18,5,9,19,22,12,20,23,14,16,13}	10	11
{3,6,0,8,7,10,1,21,11,2,4,16,15,18,5,9,19,22,12,17,23,14,20,13}	10	11
{3,6,0,8,7,10,1,20,11,2,4,16,15,18,5,9,19,22,12,17,23,14,21,13}	10	11
{3,6,0,8,7,10,1,20,11,2,4,16,15,18,5,9,19,22,12,21,23,14,17,13}	10	11
{3,6,0,8,7,10,1,17,11,2,4,16,15,18,5,9,19,22,12,20,23,14,21,13}	10	11
{3,6,0,8,7,10,1,21,11,2,4,17,15,18,5,9,19,22,16,12,23,14,13,20}	10	11
{3,6,0,8,7,10,1,21,11,2,4,20,15,18,5,9,19,22,16,12,23,14,13,17}	10	11
{3,6,0,8,7,10,1,16,11,2,4,21,15,18,5,9,19,22,17,12,23,14,13,20}	10	11
{3,6,0,8,7,10,1,16,11,2,4,17,15,18,5,9,19,22,20,12,23,14,13,21}	10	11
{3,6,0,8,7,10,1,16,11,2,4,17,15,18,5,9,19,22,21,12,23,14,13,20}	10	11
{3,6,0,8,7,10,1,16,11,2,4,20,15,18,5,9,19,22,21,12,23,14,13,17}	10	11
{3,6,0,8,7,10,1,16,11,2,4,21,15,18,5,9,19,22,20,12,23,14,13,17}	10	11
{3,6,0,8,7,10,1,17,11,2,4,16,15,18,5,9,19,22,20,12,23,14,13,21}	10	11
{3,6,0,8,7,10,1,17,11,2,4,16,15,18,5,9,19,22,21,12,23,14,13,20}	10	11
{3,6,0,8,7,10,1,20,11,2,4,16,15,18,5,9,19,22,17,12,23,14,13,21}	10	11
{3,6,0,8,7,10,1,21,11,2,4,16,15,18,5,9,19,22,17,12,23,14,13,20}	10	11
{3,6,0,8,7,10,1,20,11,2,4,16,15,18,5,9,19,22,21,12,23,14,13,17}	10	11
{3,6,0,8,7,10,1,17,11,2,4,21,15,18,5,9,19,22,20,12,23,14,13,16}	10	11
{3,6,0,8,7,10,1,20,11,2,4,17,15,18,5,9,19,22,21,12,23,14,13,16}	10	11
{3,6,0,8,7,10,1,21,11,2,4,17,15,18,5,9,19,22,20,12,23,14,13,16}	10	11
{3,6,0,8,7,10,1,21,11,2,4,20,15,18,5,9,19,22,17,12,23,14,13,16}	10	11
{3,6,0,8,7,10,1,20,11,2,4,21,15,18,5,9,19,22,17,12,23,14,13,16}	10	11
{3,6,0,8,7,10,1,16,11,2,4,20,15,18,5,9,19,22,12,17,23,14,13,21}	11	11
{3,6,0,8,7,10,1,16,11,2,4,21,15,18,5,9,19,22,12,17,23,14,13,20}	11	11

References

1. National Bureau of Standards.: Data encryption standard. NBS FIPS PUB 46, U.S. Department of Commerce (1977)
2. Aoki, K., et al.: *Camellia*: a 128-bit block cipher suitable for multiple platforms — design and analysis. In: Stinson, D.R., Tavares, S. (eds.) SAC 2000. LNCS, vol. 2012, pp. 39–56. Springer, Heidelberg (2001). https://doi.org/10.1007/3-540-44983-3_4
3. Beaulieu, R., Shors, D., Smith, J., Treatman-Clarket, S., Weeks, B., Wingers, L.: The SIMON and SPECK lightweight block ciphers. In: Anne, C. (eds.) The 52nd Annual Design Automation Conference 2015, LNCS, pp. 1–6. ACM, New York (2015). https://doi.org/10.1145/2744769.2747946
4. Adams, C., Gilchrist, J.: The cast-256 encryption algorithm. Request Comments **2612**, 99–110 (2016)
5. Daemen, J., Van Assche, G.: Producing collisions for PANAMA, instantaneously. In: Biryukov, A. (ed.) FSE 2007. LNCS, vol. 4593, pp. 1–18. Springer, Heidelberg (2007). https://doi.org/10.1007/978-3-540-74619-5_1
6. Shibutani, K., Isobe, T., Hiwatari, H., Mitsuda, A., Akishita, T., Shirai, T.: *Piccolo*: an ultra-lightweight Blockcipher. In: Preneel, B., Takagi, T. (eds.) CHES 2011. LNCS, vol. 6917, pp. 342–357. Springer, Heidelberg (2011). https://doi.org/10.1007/978-3-642-23951-9_23
7. Zheng, Y., Matsumoto, T., Imai, H.: On the construction of block ciphers provably secure and not relying on any unproved hypotheses. In: Brassard, G. (ed.) CRYPTO 1989. LNCS, vol. 435, pp. 461–480. Springer, New York (1990). https://doi.org/10.1007/0-387-34805-0_42
8. Hong, D., et al.: HIGHT: a new block cipher suitable for low-resource device. In: Goubin, L., Matsui, M. (eds.) CHES 2006. LNCS, vol. 4249, pp. 46–59. Springer, Heidelberg (2006). https://doi.org/10.1007/11894063_4
9. Biham, E., Biryukov, A., Shamir, A.: Cryptanalysis of skipjack reduced to 31 rounds using impossible differentials. In: Stern, J. (ed.) EUROCRYPT 1999. LNCS, vol. 1592, pp. 12–23. Springer, Heidelberg (1999). https://doi.org/10.1007/3-540-48910-X_2
10. Nyberg, K.: Generalized Feistel networks. In: Kim, K., Matsumoto, T. (eds.) Advances in Cryptology ASIACRYPT 1996, LNCS, vol. 1163, pp. 91–104. Springer, Heidelberg (1996). https://doi.org/10.1007/BFb0034838
11. Suzaki, T., Minematsu, K.: Improving the generalized Feistel. In: Hong, S., Iwata, T. (eds.) FSE 2010. LNCS, vol. 6147, pp. 19–39. Springer, Heidelberg (2010). https://doi.org/10.1007/978-3-642-13858-4_2
12. Daemen, J., Knudsen, L., Rijmen, V.: The block cipher Square. In: Biham, E. (eds.) Fast Software Encryption 1997, LNCS, vol. 1267, pp. 149–165. Springer, Heidelberg (1997). https://doi.org/10.1007/BFb0052343
13. Cauchois, V., Gomez, C., ThomasV, G.: General diffusion analysis: How to find optimal permutations for generalized type-II Feistel schemes. IACR Trans. Symmetric Cryptol **2019**(1), 264–301 (2019)
14. Derbez, P., Fouque, P., Lambin, B., Mollimard, V.: Efficient search for optimal diffusion layers of generalized Feistel networks. IACR Trans. Symmetric Cryptol **2019**(2), 218–240 (2019)
15. Suzaki, T., Minematsu, K., Morioka, S., Kobayashi, E.: *TWINE*: a lightweight block cipher for multiple platforms. In: Knudsen, L.R., Wu, H. (eds.) SAC 2012. LNCS, vol. 7707, pp. 339–354. Springer, Heidelberg (2013). https://doi.org/10.1007/978-3-642-35999-6_22

16. Dong, X., Li, Z., Wang, X.: Quantum cryptanalysis on some generalized Feistel schemes. Sci. Chin. Inf. Sci. **62**(22501), 180–191 (2019)
17. Shannon, C.E.: Communication theory of secrecy systems. Bell Syst. Tech. J. **28**(4), 656–715 (1949)
18. Sasaki, Yu., Todo, Y.: New impossible differential search tool from design and cryptanalysis aspects. In: Coron, J.-S., Nielsen, J.B. (eds.) EUROCRYPT 2017. LNCS, vol. 10212, pp. 185–215. Springer, Cham (2017). https://doi.org/10.1007/978-3-319-56617-7_7
19. Xiang, Z., Zhang, W., Bao, Z., Lin, D.: New impossible differential search tool from design and cryptanalysis aspects. In: Cheon, J.H., Takagi, T. (eds.) ASIACRYPT 2016. LNCS, vol. 10031, pp. 648–678. Springer, Heidelberg (2016). https://doi.org/10.1007/978-3-662-53887-6_24
20. Sun, L., Wang, W., Wang, M.: Automatic search of bit-based division property for ARX ciphers and word-based division property. In: Takagi, T., Peyrin, T. (eds.) ASIACRYPT 2017. LNCS, vol. 10624, pp. 128–157. Springer, Cham (2017). https://doi.org/10.1007/978-3-319-70694-8_5

Homomorphic Encryption

F-FHEW: High-Precision Approximate Homomorphic Encryption with Batch Bootstrapping

Man Chen[1], YuYue Chen[2], Rui Zong[1(✉)], ZengPeng Li[3], and Zoe L. Jiang[2,4(✉)]

[1] International Digital Economy Academy, Shenzhen, China
zongrui@idea.edu.cn
[2] School of Computer Science and Technology, Harbin Institute of Technology, Shenzhen, China
[3] School of Cyber Science and Technology, Shandong University, Jinan, China
[4] Guangdong Provincial Key Laboratory of Novel Security Intelligence Technologies, Shenzhen, China
zoeljiang@hit.edu.cn

Abstract. Floating-point fully homomorphic encryption (FPFHE) supports arbitrary computation on ciphertexts and yields approximate results. On one hand, for the state-of-the-art, the CKKS-like scheme (Jutla et al., EUROCRYPT 2022) achieves a precision of 100-bit decimal as the messages originally include some encoding noise, as well as some additional errors will be generated by bootstrapping operation. Nonetheless, operations with higher precision are also appreciated by various kinds of applications, such as Semidefinite Programming requiring 128-bit precision. On the other hand, the CKKS-like scheme is very computationally intensive. Compared to processing the same data in clear, it is slower, less efficient, and more energy-consuming.

In this paper, we propose a high-precision approximate homomorphic encryption with batch bootstrapping based on the Gentry-Sahai-Waters scheme over rings (or RingGSW). Firstly, to support a precision of 128-bit decimal, mapping floating-point numbers to B-based cyclotomic polynomials with 128−fraction coefficients. Furthermore, we use polynomial truncation in homomorphic multiplication to support deep-level circuit with upper-bound depth $O(\log q/(B_g\sigma))$. Next, we utilize trace function computation to achieve batch multiplication, achieving an amortized multiplication complexity of $O(n^{1.75}\log q)$. Overall, the proposed scheme has half amortized multiplication time and supports deeper-level circuit $> O(\log q_0/(B_g\sigma))$, when compared to the CKKS-like scheme.

Keywords: FPFHE · Laurent polynomial · Truncation · Bootstrapping

1 Introduction

Floating-point fully homomorphic encryption (FPFHE) enables arbitrary operations for real number data encrypted without decrypting them. Various floating-point FHE applications have recently been explored to safeguard private data,

T. Zhu and Y. Li (Eds.): ACISP 2024, LNCS 14895, pp. 121–140, 2024.
https://doi.org/10.1007/978-981-97-5025-2_7

including secure cloud computing (SCC) [1], private set intersection (PSI) [2] and privacy-preserving machine learning (PPML) [3].

Currently, the CKKS-like schemes [4,5] are one of the most promising solutions for implementing a PPML system. However, in the CKKS-like scheme, noises as a means to ensure security, approximate errors are introduced into the encrypted messages. Without resolving such encoding error issues, the CKKS-like scheme may fail to deliver satisfactory results for applications that require complex and precise outputs in PPML. Moreover, in the CKKS-like scheme, the bootstrapping operation does not decrease noise but instead amplifies the errors. The CKKS-like scheme has achieved a maximum precision of 100 bits [5]. This level of precision is deemed insufficient for comprehensive application within deep learning systems. In certain applications such as Semidefinite Programming (SDP) [6], high precision operations with 128-bit precision are required to obtain accurate results. The mentioned limitation encourage us to ask the following question:

Is it feasible to encode floating-point numbers in homomorphic encryption while supporting high-precision operations?

The answer to this question is affirmative, as we propose an FHEW-like scheme abbreviated as F-FHEW that supports high-precision floating numbers computation. The encoding scheme is similar to that of IEEE754 [7], with a provision for a 128-bit fractional coefficient. Specifically, the floating-point number $p \in \mathbb{Q}$ is converted into a B-based representation, yielding a number y_B, which preserves a decimal fraction of 128 digits. Then B-based number y_B representation needs to be expressed as a Laurent polynomial $p(x) \in \mathbb{Z}_B[x, x^{-1}]$. Unfortunately, this mapping lacks the multiplicative homomorphism property. We have to establish an isomorphic mapping of the Laurent polynomial $\mathbb{Z}_B[x, x^{-1}]$ to the cyclotomic polynomial $\mathbb{Z}_B[x, -x]/\Phi_m(x)$ where $\Phi_m(x) = x^n + 1$. To support deeper-level circuit depth, we adopt polynomial truncation in the FHEW [10] scheme, as shown in Fig. 1:

Then, we use the batch F-FHEW [8,9] technique to reduce the amortized multiplication cost. The approach involves decomposing cyclotomic ring into three isomorphic cyclotomic rings $\mathbb{Z}[\xi_m] \cong \mathbb{Z}[\xi_q] \otimes \mathbb{Z}[\xi_p] \otimes \mathbb{Z}[\xi_t]$. We can map $x \in R_Q = \sum x_i\ u_i$ and $y \in R_Q = \sum y_i\ u_i^\vee v_i$, where u_i, $u_i^\vee$ are dual base in $\mathbb{Z}[\xi_p]$ and base $v_i \in \mathbb{Z}[\xi_t]$. Then utilizing dual bases and trace calculations to construct homomorphic computations over cyclotomic field.

1.1 Our Contributions

Below, we illustrate contributions based on the approaches mentioned above.

- **High-precision floating-number computation**. To improve the precision of FPFHE, we establish an isomorphic mapping of the Laurent polynomial to the cyclotomic polynomial. By encoding floating-point numbers as cyclotomic polynomials, it achieves 128−bit precision that exceeds the precision of the CKKS-like scheme [5].

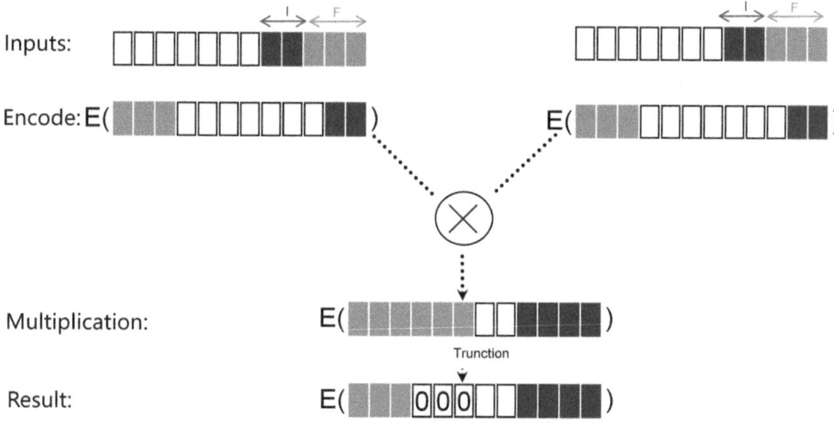

Fig. 1. Truncation process

- **Effective deployable performance**. Compared to the CKKS-like scheme [5], there are three main advantages guaranteeing the effectively deployable performance of F-FHEW. In more detail, (1) **Faster encoding process**: The proposed encoding approach doesnot require the process of solving linear equations to determine the encoding coefficients, which guarantees F-FHEW with 5–10 times faster than the CKKS-like scheme; (2) **Lower computational complexity**: F-FHEW doesnot involve pre-generating the evaluation key and executive key-switching during encoding. So the F-FHEW scheme reduces the computational complexity of homomorphic multiplication. Below, Table 1 shows the performance comparison between our proposed and the CKKS-like scheme. (3) **Deeper circuit depth**: In homomorphic multiplication, the low-order drop is performed when the coefficient terms in the fractional part exceed 128. When compared to the CKKS scheme, it obtains deeper-level circuit depth of $O(\log q_0/(B_g\sigma))$.

Table 1. Performance comparison between CKKS [4] and F-FHEW.

Scheme	Public key	Private key	Evaluation key
CKKS [4]	$\mathcal{O}(N\lceil\log q_l\rceil)$	$\mathcal{O}(N\lceil\log q_l\rceil)$	$\mathcal{O}(N\lceil\log q_l\rceil)$
F-FHEW	$\mathcal{O}(N\lceil\log q\rceil)$	$\mathcal{O}(N\lceil\log q\rceil)$	**0**
Scheme	Key switch complexity	Ciphertext size	Mult complexity
CKKS [4]	$\mathcal{O}(N^2\lceil\log q_l\rceil^2)$	$\mathcal{O}(N\lceil\log q_l\rceil)$	$\mathcal{O}(N^3\lceil\log q_l\rceil^3)$
F-FHEW	**0**	$\mathcal{O}(N\lceil\log q\rceil)$	$\mathcal{O}(N^2\lceil\log q\rceil^2)$

In the Microsoft SEAL library [13] of CKKS, the value of N is typically set to $2,048$, and the values of q_0 and $q_l = q_0 q^{l-1}$ are 2^{40} and $2^{40+20(l-1)}$. In our proposed scheme F-FHEW, the value of N is typically set to $2,048$, the value of q is 2^{256}.

- **Batch bootstraping.** To reduce the complexity of bootstrapping, we generalize F-FHEW with batch bootstrapping. The SIMD architecture enables parallel bootstrapping for $\tilde{\mathcal{O}}(\lambda^{0.25})$ ciphertexts. As a result, the amortized bootstrapping complexity is reduced from $\tilde{\mathcal{O}}(\lambda)$ to $\tilde{\mathcal{O}}(\lambda^{0.75})$ in our scheme. Remarkably, during bootstrapping, F-FHEW supports noise reduction while the CKKS-like schemes amplify the noise [5].

1.2 Related Work

Numerous efforts have been dedicated to researching the processing of FPFHE [14,15]. The CKKS scheme [4] was proposed that maps numbers in complex space $\mathbb{C}^{N/2}$ to a cyclotomic ring $\mathbb{Z}[X]/(\Phi_M(X))$. It designed a method to construct a homomorphic scheme for approximate arithmetic in that the precision range was 2^{-40}. Jutla et al. [5] used a sine series to approximate the mod function, resulting in 100-bit bootstrapping precision of CKKS. However, in the CKKS encoding process, noise is considered as part of the errors that impact the precision of the scheme.

On the other hand, Moon et al. [16] represented 32-bit floating-point numbers with sign, exponent, and mantissa, which similar to the $(-1)^{(\text{sign})} \cdot 1.(\text{FRACTION}) \cdot 2^{(\text{EXPONENT}(2)-127)}$ in IEEE 754 representation. And they integrate it into the TFHE and HEAAN algorithms. Lee et al. [17] represented 64-bit floating-point numbers in FHEW with bootstrapping. Since these methods require high amortized homomorphic multiplication complexity, they suffer from sluggish operational efficiency.

2 Preliminaries

We denote the real numbers as $\mathbb{R}$, the rational numbers as $\mathbb{Q}$, and the integer numbers as $\mathbb{Z}$, the complex numbers as $\mathbb{C}$. Vectors are typically represented by lower-case bold letters (e.g., $\mathbf{x}$), matrix symbols are upper-case bold letters (e.g., $\mathbf{X}$). The notation $\langle \mathbf{x}, \mathbf{y} \rangle$ stands for the dot product of two vectors. We define the l_p norm as $\|\mathbf{x}\|_p = (\sum_{i \in [n]} |x_i|^p)^{1/p}$, where $\|\mathbf{x}\|_\infty = \max_{i \in [n]} |x_i|$. Laurent polynomials form a ring denoted $\mathbb{Z}[x, x^{-1}]$, which differ from ordinary polynomials in that they have terms of negative degree.

2.1 Algebraic Number Theory Background

We provide essential foundational knowledge in algebraic number theory. This work extensively employs the concepts of cyclotomic ring, field, trace function and dual base, more additional details can be found in the work [19].

Number Fields. For $m > 0$, let ζ_m denote an element of multiplicative order m, i.e., a primitive m-th root of unity. The minimal polynomial of ζ_m is m-th cyclotomic polynomial

$$\Phi_m(x) = \prod_{i \in \mathbb{Z}_m^*} (x - \omega_m^i) \in \mathbb{Z}[x],$$

where $\omega_m = \exp(2\pi\sqrt{-1}/m)$. Therefore, there exists a natural isomorphism between the field K and $\mathbb{Q}(\zeta_m)$, given by $\zeta_m \to x$. The m-th cyclotomic field $\mathbb{Q}(\zeta_m)$ is the field obtained by adjoining ζ_m to the field of rationals

$$K = \mathbb{Q}[x]/(\Phi_m(x)) \cong \mathbb{Q}(\zeta_m).$$

Since $\Phi_m(x)$ has degree $n = |\mathbb{Z}_m^*| = \varphi(m)$, totient of m, we can consider K as a vector space of dimension n over $\mathbb{Q}$. It has a power basis $(\zeta_m^j)_{j\in[n]} = (1, \zeta_m, \ldots, \zeta_m^{n-1}) \in K^{[n]}$ of K. In particular, $n = m/2$ is a power of two, $\mathcal{R} = \mathbb{Z}[X]/(\Phi_m(X)) = \mathbb{Z}[X]/(x^n+1)$ is the cyclotomic ring, and $\mathcal{R}_q = \mathcal{R}/q\mathcal{R} \cong \mathbb{Z}_q[X]/(x^n+1)$ is the residue ring.

The Powerful Basis. The powerful basis $\mathbf{p}$, which is used to characterize the fields $K = \mathbb{Q}(\zeta_m)$ and the rings $R = \mathbb{Z}(\zeta_m)$, is defined as follows:

$$\mathbf{p} = (\zeta_m^j)_{j\in[\varphi(m)]}.$$

Let m have a prime-power factorization $m = \prod_\ell m_\ell$, then $K = \mathbb{Q}(\zeta_m)$ is isomorphic to the tensor product $\otimes_\ell \mathbb{Q}(\zeta_{m_\ell})$, and $\mathbf{p}$ as the tensor product $\mathbf{p} = \otimes_\ell \mathbf{p}_\ell$. For each $K' = \mathbb{Q}(\zeta'_m)$, there exists power bases $\mathbf{p}_\ell$.

Lemma 1 (The norm of power basis [19]). *Each element p_j in the vector $\mathbf{p}$ satisfies the condition that $||p_j||_\infty = 1$ in ℓ_∞ norm. Similarly, in ℓ_2 norm, denoted as $||p_j||_2 = \sqrt{\varphi(m)} = \sqrt{n}$.*

Trace. The trace of $\mathbf{Tr} = \mathrm{Tr}_{K/\mathbb{Q}}\colon K \to \mathbb{Q}$ is the sum of all the embeddings, $\mathrm{Tr}(a) = \sum_i \sigma_i(a)$. More particularly, if L/K is a Galois extension and a is in L, then the trace of a is the sum of all the Galois conjugates of a,

$$\mathrm{Tr}_{K/\mathbb{Q}}(\alpha) = \sum_{\sigma\in\mathrm{Gal}(K/\mathbb{Q})} \sigma(\alpha),$$

where $\mathrm{Gal}(K/\mathbb{Q})$ denotes the Galois group of $K/\mathbb{Q}$. The $\mathrm{Tr}_{K/\mathbb{Q}}$ is a linear function from K to $\mathbb{Q}$: $\mathrm{Tr}(a+b) = \mathrm{Tr}(a) + \mathrm{Tr}(b))$ and $\mathrm{Tr}(c \cdot a) = c \cdot \mathrm{Tr}(a)$ for all $a, b \in K$ and $c \in \mathbb{Q}$. It is also worth noting that

$$\mathrm{Tr}(a \cdot b) = \sum_i \sigma_i(a)\sigma_i(b) = \langle \sigma(a), \overline{\sigma(b)} \rangle,$$

which represents a symmetric bilinear form similar to the inner product computed from the embeddings of a and b. For any $\mathbb{Q}$-basis $B = \{b_j\}$ of K, we denote its dual basis $B^\vee = \{b_j^\vee\}$, for $1 \le i, j \le n$,

$$\mathrm{Tr}_{F/K}(b_i b_j^\vee) = \begin{cases} 0 & i \neq j \\ 1 & i = j. \end{cases}$$

2.2 Subgaussian

Let $\chi_{\mu,\sigma}$ is a Gaussian probability distribution with a mean value of μ and a standard deviation of σ. When $\mu = 0$, we simply state χ_σ.

Theorem 1 (Subgaussian [18]**).** *Let a random variable X is σ-subgaussian if there exists $\sigma > 0$ for any $t \in R$ such as:*

$$\mathbb{E}\left[\exp(tX)\right] \leq \exp(\frac{\sigma^2 t^2}{2}).$$

- *σ-subgaussian: Assume that X is σ-subgaussian, then the following statements are true: $\mathbb{E}(X) = 0$ and $\mathbb{E}(X^2) = Var(X) \leq \sigma^2$.*
- *Homogeneity: If X is σ-subgaussian, then αX is $\alpha\sigma$-subgaussian for $\alpha \in R$.*
- *Pythagorean additivity: Assume that X_1 and X_2 are independent and are σ_1 and σ_2 subgaussian, respectively, then $(X_1 + X_2)$ is $\sqrt{\sigma_1^2 + \sigma_2^2}$-subgaussian.*

For independent random variables, the subgaussian cloud be homogeneous. In the case of finite variables X_i is σ_i-subgaussian, $i \in [n]$, and this conclusion can be extended $\sqrt{\sum_{i=1}^{n} \sigma_i^2}$-subgaussian. In a following lemma for the preimage sampling, the distribution of $\mathbf{x}$ follows a subgaussian distribution.

Lemma 2 (Preimage sampling [10]**).** *For any integers $q > 0$, $B_g > 0$, $\ell = \lceil \log_{B_g} q \rceil$ and $\mathbf{g} = (1, \cdots, B_g^{\ell-1})$. Then there is a preimage sampling algorithm denoted as $f_{\mathbf{g}} : \mathbb{Z}_q \to \mathbb{Z}^\ell$ such that $f_{\mathbf{g}}(a) = \mathbf{x}$ with probability proportional to χ_σ, for the Gaussian parameter $\sigma \geq B_g \cdot \omega(\sqrt{\log n})$, satisfying $<\mathbf{g}, \mathbf{x}> = a \mod q$.*

Notably, the gadget vector $\mathbf{g}$ can be used to the gadget matrix case $\mathbf{G} = \mathbf{g} \otimes \mathbf{I}_n$ by the tensor product of $\mathbf{g}$ with the identity matrix $\mathbf{I}_n$.

2.3 Encryption Schemes

We first briefly review the state-of-the-art RGSW cryptosystem [10]. It supports homomorphic addition and multiplication on arbitrary integer vectors.

RLWE. Consider a public parameter $a \in R_q$, a secret key $s \in R$ and a message $m \in R_q$. The resulting RLWE ciphertext will be as follows:

$$\text{RLWE}_s(m) = (a, b) \in R_q^{1\times 2},$$

where $b = -a \cdot s + e + m$ and $e \leftarrow \chi^N$ is the noise error. The RLWE problem serves as a fundamental security foundation in the development of lattice-based homomorphic encryption schemes. Lyubashevsky et al. [20] provides compelling evidence of their computational hardness. However, RLWE encryption exhibits linear homomorphism, where noise also grows linearly. There is a risk of the error growing beyond bounds.

Gadget RLWE or RLWE′. A novel RLWE encryption scheme is introduced to improve the scalar multiplication operation, enabling controlled noise growth within acceptable bound:

$$\mathrm{RLWE}'_s(m) = (\mathrm{RLWE}_s(m), \mathrm{RLWE}_s(B_g m), \ldots, \mathrm{RLWE}_s(B_g^{\ell-1} m)) \in R_q^{k\times 2},$$

where the dimension of $\mathrm{RLWE}'_{s(m)}$is is $\ell = \log_{B_g} q$ and the modulus is power for some base B_g.

RGSW. The $\mathrm{RGSW}_s(m)$ ciphertexts can be equivalently written as

$$\mathrm{RGSW}_s(m) = (\mathrm{RLWE}^{'}_s(-s\cdot m), \mathrm{RLWE}^{'}_s(m)) \in R_q^{2k\times 2}.$$

We now summarize the operations that can be performed within our algorithm. In order to perform multiplication, one employs $\mathrm{RLWE}^{'}$ with a gadget vector $(d_0, d_1, \cdots, d_{\ell-1})$, with a constant number $d = \sum_i B_g^i d_i$. Let $\mathbf{c}_i = \mathrm{RLWE}(B_g^i m)$, and the new scalar multiplication operation $(\odot)$: $R \odot \mathsf{RLWE}' \rightarrow$ RLWE:

$$d \odot (\mathbf{c}_0, \ldots, \mathbf{c}_{k-1}) = \sum_{i\in[\ell]} d_i \cdot \mathbf{c}_i.$$

The Gadget Decompose is specifically denoted as follows:

$$(a_0, b_0, \cdots, a_{k-1}, b_{k-1}) \leftarrow \mathrm{Decompose}(a, b).$$

Given $(a, b) = \mathrm{RLWE}_s(m_0; e_0)$ and $(\mathbf{c}, \mathbf{c}') = \mathrm{RGSW}_s(m_1)$, we can compute external product $(\circ)$: RLWE $\times$ RGSW $\rightarrow$ RLWE:

$$\begin{aligned}\mathrm{RLWE}_s(m_0) \circ \mathrm{RGSW}_s(m_1) &:= a \odot \mathrm{RLWE}'_s(s\cdot m_1) + b \odot \mathrm{RLWE}'_s(m_1)\\ &= \mathrm{RLWE}'_s(a\cdot s\cdot m_1 + b\cdot m_1)\\ &= \mathrm{RLWE}_s(m_0\cdot m_1 + e_0\cdot m_1)\end{aligned}$$

It represents an encryption of the product $m_0 \cdot m_1$, augmented by an additional error term $e_0 \cdot m_1$. For any two ciphertexts $\mathrm{RGSW}_s(m_0)$ and $\mathrm{RGSW}_s(m_1)$, the definition of internal product

$$(\diamond)\colon \mathrm{RGSW} \times \mathrm{RGSW} \rightarrow \mathrm{RGSW}$$

is as follows:

$$\begin{aligned}&\mathrm{RGSW}_s(m_0) \diamond \mathrm{RGSW}_s(m_1)\\ &= (\mathrm{RLWE}^{'}_s(-sm_0) \diamond \mathrm{RGSW}_s(m_1), \mathrm{RLWE}^{'}_s(m_0) \diamond \mathrm{RGSW}_s(m_1))\\ &= (\mathrm{RLWE}^{'}_s(-sm_0m_1), \mathrm{RLWE}^{'}_s(m_0m_1))\\ &= \mathrm{RGSW}_s(m_0m_1).\end{aligned}$$

3 F-FHEW Scheme

In this section, the F-FHEW scheme with polynomial truncation is proposed. Due to the exponential growth of decimal in the size of the plaintext modulus, we use polynomial truncation method to support deeper circuit depth. Inspired by the work [21], plaintexts are encoded as Laurent polynomials into a ciphertext space $R_t = \mathbb{Z}_t[X]/(f(x))$. Moreover, we enable the FHEW-like schemes [10] to facilitate floating-point arithmetic operations.

3.1 Plaintexts Encoding with Polynomial Truncation

Let y be a floating-point number and denote by $y = y^+.y^-$ B-based representation. Let y^+ denotes the integer digits $b_{I^+}b_{I^+-1}\cdots b_1b_0$, and y^- denotes the fractional digits $b_{-1}b_{-2}\cdots b_{-I}$; thus we can write

$$y^+ = b_{I^+}\cdot B^{I^+} + b_{I^+-1}\cdot B^{I^+-1} + \cdots + b_1\cdot B + b_0,$$

$$y^- = b_{-I^-}\cdot B^{-I^-} + b_{-I^-+1}\cdot B^{-I^-+1} + \cdots + b_{-2}\cdot B^{-2} + b_{-1}\cdot B^{-1},$$

where $b_i \in (\frac{-(B-1)}{2}, \frac{(B-1)}{2})$. Then encode the floating-point number y in a Laurent polynomial

$$p = \sum_{0\le i\le I^+} X^i b_i + \sum_{0<i\le I^-} X^{-i} b_{-i} \in \mathbb{Z}[X, X^{-1}]. \tag{1}$$

Utilizing the ring equation $X^n + 1 = 0$, with n is a power of 2, the Laurent polynomial could be encoded in cyclotomic fields as follows

$$p = \sum_{0\le i\le I^+} X^i b_i - \sum_{0<i\le I^-} X^{n-i} b_{-i} \in \mathbb{Z}[X]/(X^n + 1). \tag{2}$$

A theorem below demonstrates that the mapping from floating-point numbers to polynomials preserves isomorphism.

Theorem 2. *ϕ is a ring isomorphism if there is a map $\phi : \mathbb{Z}[X, X^{-1}] \to \mathbb{Z}[X]/(X^n + 1)$, with $n - I^- > I^+$.*

Proof. Let $f(X), g(X) \in \mathbb{Z}[X, X^{-1}]$, and $r(X) = f(X) \bmod \Phi_m(X)$. We define $\phi(f(X)) = r(X)$, where ϕ is a ring homomorphism from $\mathbb{Z}[X, X^{-1}]$ to $\mathbb{Z}[X]/(X^n+1)$ satisfying $\phi(f(X)+g(X)) = \phi(f(X))+\phi(g(X))$, $\phi(f(X)\cdot g(X)) = \phi(f(X))\cdot\phi(g(X))$. For $\Phi_m(x) = \prod_{i\in\mathbb{Z}_m^*}(x - \omega_m^i) \in \mathbb{Z}[x]$, we assume that

$$h_1(X) = q_1(X)\Phi_m(X) + r_1(X),\ h_2(X) = q_2(X)\Phi_m(X) + r_2(X),$$

compute the polynomial multiplication and obtain

$$\begin{aligned} h_1(X)\cdot h_2(X) = {} & q_1(X)q_2(X)\Phi_m^2(X) \\ & + [q_1(X)r_2(X) + q_2(X)r_1(X)]\Phi_m(X) + r_1(X)r_2(X). \end{aligned}$$

Notably, both $\Phi_m(X)$ and $\Phi_m^2(X)$ are zero elements in R. Therefore, we have

$$\phi(h_1(X) \cdot h_2(X)) = \phi(h_1(X))\phi(h_2(X)).$$

To prove that the mapping ϕ is a bijection, we show that it is both injective and surjective. Assume we have two Laurent polynomials $p(X)$ and $q(X)$ such that $\phi(p(X)) = \phi(q(X)) = z(X)$. A floating-point number y can be uniquely represented as a B-based Laurent polynomial and thus $p(X) = q(X)$. For any $z(X) \in \mathbb{Z}[X]/(X^n + 1)$, we have $p(X) \in \mathbb{Z}[X, X^{-1}]$ so that the mapping is surjective.

Polynomial Truncation. The method of polynomial truncation (PT) can be defined as follows. Let $m \in R_B$ be plaintext, Δ represent least significant bits to be truncation. Polynomial Truncation

$$m_{\mathrm{PT}} = \mathrm{PT}(m, \Delta),$$

means plaintext m discarding the least significant Δ bits. Based on whether the most significant bit in the discarding ratio vector Δ is 0 or 1, the plaintext m_{PT} selectively adds a carry. In this paper, we could reserve 128 coefficients for the fractional part. If the most significant bit of $\mathbf{c}_{discard} = \frac{B-1}{2}$, then round up by 1, otherwise round down to 0. The following Fig. 2 is an example polynomial truncation.

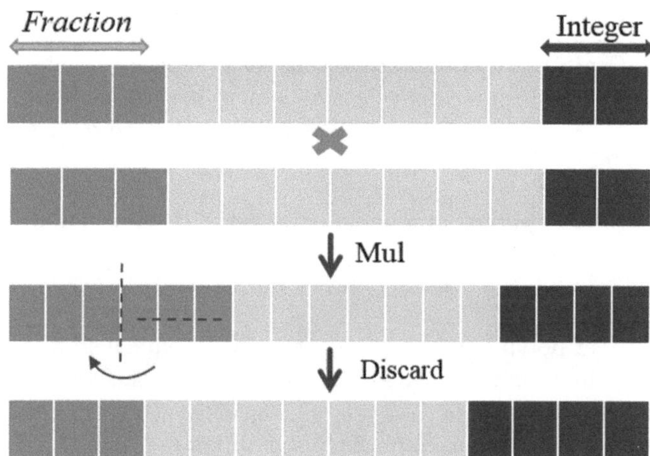

Fig. 2. Polynomial Truncation method.

Here's an example that explains how floating-point numbers are encoded.

Example 1. Let 3 will be encoded as X, while 1/3 will be encoded as $1/X = -X^{N-1}$. Consider $y = 6.333333 = 1\bar{1}0.1_3$ and $y' = 3.3478 = 1.1001_3$, with

$\bar{1} = -1$. For $p = X^2 - X + X^{-1}$ and $p^{'} = X + X^{-1} + X^{-4}$ be the polynomial expressions of y and $y^{'}$ respectively.

$$\begin{aligned}\phi(p) \cdot \phi(p^{'}) &= (X^2 - X - X^{N-1}) \cdot (X - X^{N-1} - X^{N-4}) \\ &= X^3 - X^2 - X^N - X^{N+1} + X^N + X^{2N-2} - X^{N-2} + X^{N-3} + X^{2N-5} \\ &= X^3 - X^2 + X - X^{N-1} + X^{N-2} + X^{N-3} - X^{N-5} \bmod (X^N + 1)\end{aligned}$$

Decoding yields the result $3^3 - 3^2 + 3 + 3^{-1} - 3^{-2} - 3^{-3} + 3^{-5} = 21.19$ that we except.

3.2 Floating-Point Homomorphic Encryption

This section details the proposed F-FHEW supporting floating-point homomorphic computation. Following the method introduced in Sect. 3.1, floating-point numbers are encoded onto B-based balanced polynomials to be used as plaintext in the F-FHEW encryption scheme.

F-FHEW Scheme:

- F-FHEW.Setup $(1^\lambda, 1^L)$: Give the security parameter λ and the circuit depth L. Choose a modulus $q \geq 2$, a polynomial degree n which is a power of 2, the mth cyclotomic ring $R = \mathbb{Z}[x] / (x^n + 1)$, the residue ring $R_q = R/qR = \mathbb{Z}_q[x] / (x^n + 1)$, and the error distribution χ_σ as an error probability distribution over R, where σ is a standard deviation. Let public parameters are (n, q, χ_σ).
- F-FHEW.SkGen (1^λ): The secret key is $SK := s$ from a ternary set in $\{-1, 0, 1\}^N$ with Hamming weight h.
- F-FHEW.PkGen(SK, Δ): Let Δ represent least significant bits to be truncation. Randomly and uniformly select a element $a \in R_q$, a secret key s, and a noise error $e \in \chi_\sigma$, calculating $b = -as + e$. Let $\mathbf{A} = (b, a) \in R_q^2$ and output public key $PK := \mathbf{A} \in R_q^2$.
- F-FHEW.Ecd(p, n):
  - Given a floating-point number $p \in \mathbb{Q}$, get a B-based number $p^{'}$ through base conversion;
  - Convert a number $p^{'}$ into a Laurent polynomial $p(x) \in \mathbb{Z}_B[x, -x]$ by the Eq. 1;
  - The Laurent polynomial $p(x)$ maps to a cyclotomic field $m \in \mathbb{Z}_B[x]/(x^n + 1)$ by the ring isomorphism ϕ. The coefficients of the resulting polynomial m in a particular range $[\frac{-(B-1)}{2}, \ldots, \frac{B-1}{2}]$.
- F-FHEW.Dcd(m, n): Evaluating the polynomial at $x = B$, $x^n = -1$ and calculating the resulting value $p \in \mathbb{Q}$. The decoding process is as follows.

$$\mathbb{Z}_B[x]/\Phi_m(x) \xrightarrow{\phi^{-1}} \mathbb{Z}_B[x, x^{-1}] \xrightarrow{x=B} p \in \mathbb{Q}.$$

- F-FHEW.Enc(PK, m):We encode $m \in R_B$ as $\tilde{m} = ((q/B)m) \in R_q$. Given a message $\tilde{m} \in R_q$, output ciphertext

$$\boldsymbol{c} = (\mathrm{RLWE}_s^{'}(-s \cdot \tilde{m}), \mathrm{RLWE}_s^{'}(\tilde{m})).$$

- F-FHEW.Dec($SK, \boldsymbol{c}$): The ciphertexts allow to recover the message m by first decrypting RLWE($B_g^{k-1}\tilde{m}$) to recover (m mod B_g). Then subtracting $B_g^{k-2} \cdot$ (m mod B_g) from RLWE($B_g^{k-2}m$) to recover (m mod B_g^2), and so on.
- F-FHEW.Add($\boldsymbol{C}_1, \boldsymbol{C}_2$): For $\boldsymbol{C}_1, \boldsymbol{C}_2 \in \mathcal{R}_q^{2\ell \times 2}$, output $\boldsymbol{C}_{\text{add}} \leftarrow \boldsymbol{C}_1 + \boldsymbol{C}_2 \pmod{q}$.
- F-FHEW.Mul($\boldsymbol{C}_1, \boldsymbol{C}_2$): Multiplication between RLWE ciphertexts is supported by computing the RLWE decryption function homomorphically.

$$(\diamond)\colon \mathsf{RGSW} \times \mathsf{RGSW} \to \mathsf{RGSW} \quad (\mathbf{C}_1, \mathbf{C}_2) \diamond \mathbf{C} = (\mathbf{C}_1 \diamond \mathbf{C}, \mathbf{C}_2 \diamond \mathbf{C}).$$

Given a random matrix $\mathbf{Z} = [\text{RLWE}(0), \text{RLWE}(0), \cdots, \text{RLWE}(0)]^T$ and the gadget matrix $\mathbf{G}$, Gadget_Decompose$(a, b) \cdot \mathbf{G} = (a, b)$. The operation of RGSW external product is as follows

$$\begin{aligned}
&\text{Gadget_Decompose}(\text{RLWE}(m_0)) \cdot \text{RGSW}(m_1) \\
&= (a_0, \cdots, a_{d_{g-1}}, b_0, \cdots, b_{d_{g-1}}) \cdot (\mathbf{Z} + m_1 \cdot \mathbf{G}) \\
&= \text{RLWE}(0) + m_1 \cdot \text{RLWE}(m_0) \\
&= \text{RLWE}(0) + m_1 \cdot (\text{RLWE}(0) + (0, m_0)) \\
&= \text{RLWE}(0) + (0, \text{PT}(m_0 \cdot m_1, \Delta)) \\
&= \text{RLWE}(\text{PT}(m_0 \cdot m_1, \Delta)).
\end{aligned}$$

What needs to be noticed is that $\text{RLWE}(0) + m_1 \cdot (\text{RLWE}(0))$ can be seen as RLWE(0). So the operation of RGSW internal product is

$$\begin{aligned}
&\text{RGSW}(m_0) \diamond \text{RGSW}(m_1) \\
&= (\text{RLWE}^{'}(-sm_0), \text{RLWE}^{'}(m_0)) \diamond \text{RGSW}(m_1) \\
&= (\text{RLWE}^{'}(-sm_0) \diamond \text{RGSW}(m_1), \text{RLWE}^{'}(m_0) \diamond \text{RGSW}(m_1)) \\
&= (\text{RLWE}^{'}(\text{PT}(-sm_0m_1, \Delta)), \text{RLWE}^{'}(\text{PT}(m_0m_1, \Delta))) \\
&= \text{RGSW}(\text{PT}(m_0m_1, \Delta)).
\end{aligned}$$

The encryption and decryption noise of the scheme is analyzed below. According to the heuristic method [23,24], when $e \leftarrow \chi_\sigma$, we use 6σ as our bound. For $e_1 \leftarrow \sigma_1$, $e_2 \leftarrow \sigma_2$, we use $16\sigma_1\sigma_2$ as bound of e_1e_2. When $a \leftarrow U_q$ then each coefficient has standard deviation $\sqrt{q^2/12}$.

Lemma 3 (Homomorphic Addition/Multiplication). *Let n, χ_σ, R, Δ be the parameter of the encryption scheme as described above $\boldsymbol{c}_{add} \leftarrow$* F-FHEW. Add$(\boldsymbol{c_1}, \boldsymbol{c_2})$ *and $\boldsymbol{c}_{mult} \leftarrow$* FHEW. Mult$(\boldsymbol{c_1}, \boldsymbol{c_2})$. *The upper bound of $\|E_{add}\| \leq 12\sigma$ and $\|E_{mult}\| \leq B/2 \cdot 6\sigma$ respectively. For the add noise $\|E_{add}\| \leq q/2$ and the multiplication noise $\|E_{mult}\| \leq q/2$, we can correctly obtain that: $m_1 \leftarrow$* F-FHEW. Dec$(\boldsymbol{c}_{add}, s)$, *$m_2 \leftarrow$* F-FHEW. Dec$(\boldsymbol{c}_{mul}, s)$.

Proof. Let the ciphertext $\boldsymbol{c_1}$, $\boldsymbol{c_2}$ as inputs, then compute

$$\begin{aligned}
\boldsymbol{c}_0 + \boldsymbol{c}_1 &= \text{RGSW}_s(m_0) + \text{RGSW}_s(m_1) \\
&= (\text{RLWE}_s^{'}(-s \cdot m_0), \text{RLWE}_s^{'}(m_0)) + (\text{RLWE}_s^{'}(-s \cdot m_1), \text{RLWE}_s^{'}(m_1)),
\end{aligned}$$

where $\mathbf{e}_0 = \{e_{0,i}\}_{i\in[\ell]}, \mathbf{e}_0' = \{e_{0,i}'\}_{i\in[\ell]} \leftarrow \text{Error}(\text{RGSW}_s(m_0)), \mathbf{e}_1 = \{e_{1,i}\}_{i\in[\ell]}, \mathbf{e}_1' = \{e_{1,i}'\}_{i\in[\ell]} \leftarrow \text{Error}(\text{RGSW}_s(m_1))$, we get

$$\|E_{add}\| = \text{Error}(\mathbf{c}_0 + \mathbf{c}_1) = \|\mathbf{e}_0 + \mathbf{e}_1, \mathbf{e}_0' + \mathbf{e}_1'\|_\infty \leq 12\sigma.$$

Now analyze the noise in homomorphic multiplication.

Note that Gadget_Decompose($\text{RLWE}_s(m_0)$) $= (a_0, b_0 \cdots, a_{d_{g-1}}, b_{d_{g-1}})$. The variable a_i, b_i follows a (discrete) uniform distribution centered around 0 within the range $[-B_g/2, B_g/2]$, we can obtain that

$$\begin{aligned}
\text{Error}\{\text{RLWE}(m_0) \circ \text{RGSW}(m_1)\} &= \text{Error}\{(a_0, \cdots, a_{d_g-1}, b_0, \cdots, b_{d_g-1}) \cdot (Z + m_1 \cdot G)\} \\
&= \text{Error}\{\text{RLWE}(0) + (0, m_0 m_i)\} \\
&\approx \text{Error}\{(\text{RLWE}(0) + (0, \text{PT}(m_0 m_i, \Delta))\} \\
&= \|\sum_{i=0}^{\ell-1} \text{PT}(a_i e_i, \Delta) + \sum_{i=0}^{\ell-1} \text{PT}(b_i e_i', \Delta) + \text{PT}(m_1 e, \Delta)\|_\infty \\
&\leq 2 \cdot n \cdot \ell \cdot 16 \cdot (\sqrt{\frac{B_g^2}{12}}) \cdot \sigma + n \cdot \ell \cdot B/2 \cdot 6\sigma,
\end{aligned}$$

where $\{e_1\}, \{e_1'\} \leftarrow \text{Error}(\text{RGSW}(m_1)), e = \text{Error}(\text{RGSW}(m_0))$. The noise in a matrix can be simply measured by the noise in the vector,

$$\text{Error}(\text{RCSW}(m_0) \diamond \text{RGSW}(m_1)) \leq 2 \cdot n \cdot \ell \cdot 16 \cdot (\sqrt{\frac{B_g^2}{12}}) \cdot \sigma + n \cdot \ell \cdot B/2 \cdot 6\sigma.$$

After obtaining the noise E_{mult}, we now proceed to calculate the circuit depth L. If the parameters ensure correct decryption results, we compute

$$(n \log q)^L (2 \cdot 16 \cdot (\sqrt{\frac{B_g^2}{12}}) \cdot \sigma + B/2 \cdot 6\sigma) < q/2,$$

and estimate the circuit depth to be $L = O(\log q/(B_g \sigma))$.

4 Batch Scheme with Bootstrapping

4.1 Batch Scheme

This section describes batch fully homomorphic encryption scheme with bootstrapping for floating-point numbers. It introduces a batch [11,12] framework in new RLWE ciphertexts. Particularly, let the m-th root of unity for $m = q\rho\tau$ for co-prime factors q, p, t. Then the cyclotomic ring $\mathbb{Z}[\xi_m]$ is isomorphic to the tensor of the three smaller and linearly disjoint cyclotomic rings, i.e.,

$$\mathbb{Z}[\xi_m] \cong \mathbb{Z}[\xi_q] \otimes \mathbb{Z}[\xi_p] \otimes \mathbb{Z}[\xi_t].$$

In this case, define sub-rings $\mathcal{R}_1$, $\mathcal{R}_2$, $\mathcal{R}_3$ as $\mathbb{Z}[\xi_q]$, $\mathbb{Z}[\xi_p]$, $\mathbb{Z}[\xi_t]$, respectively. $\mathbb{Z}[\xi_p]$ has dimension $\rho = \varphi(p)$ and $\mathbb{Z}[\xi_t]$ has dimension $\tau = \varphi(t)$, where $\varphi(\cdot)$

is the Euler's function. Moreover, the total dimension $N = \varphi(M) = \varphi(q)\varphi(\tau)$. For $r = \min\{\rho, \tau\}$, $\mathcal{R}_1 = \mathbb{Z}[\xi_q]$, $\mathcal{R}_2$ has basis $\{\mathsf{v}_1, \mathsf{v}_2, \ldots, \mathsf{v}_\rho\}$, $\mathcal{R}_3$ has basis $\{\mathsf{w}_1, \mathsf{w}_2, \ldots, \mathsf{w}_\tau\}$. Let $(\mathsf{v}_1^\vee, \mathsf{v}_2^\vee, \ldots, \mathsf{v}_\rho^\vee)$ and $(\mathsf{w}_1^\vee, \mathsf{w}_2^\vee, \ldots, \mathsf{w}_\tau^\vee)$ denote the $\mathbb{Z}$-bases of the dual spaces $\mathcal{R}_2^\vee$ and $\mathcal{R}_3^\vee$. Given the plaintexts $(x_1, x_2, \cdots, x_r) \in \mathbb{Z}[\xi_q]^r$ and $(y_1, y_2, \cdots, y_r) \in \mathbb{Z}[\xi_q]^r$, then pack

$$\mathrm{Pack}_1(\mathbf{x}) = \sum_{i=1}^{r} x_i \cdot \mathsf{v}_i \in \mathcal{R}_1 \otimes \mathcal{R}_2,$$

and

$$\mathrm{Pack}_2(\mathbf{y}) = \sum_{i=1}^{r} y_i \cdot \mathsf{v}_i^\vee \mathsf{w}_i \in \mathcal{R}_1 \otimes \mathcal{R}_2^\vee \otimes \mathcal{R}_3.$$

For simplicity, the ring $\mathcal{R} \equiv \mathcal{R}_1 \otimes \mathcal{R}_2 \otimes \mathcal{R}_3$, $\mathcal{R}_{12} \equiv \mathcal{R}_1 \otimes \mathcal{R}_2$, and $\mathcal{R}_{13} \equiv \mathcal{R}_1 \otimes \mathcal{R}_3$. Similarly, the field $K \equiv K_1 \otimes K_2 \otimes K_3$, $K_{12} \equiv K_1 \otimes K_2$, and $K_{13} \equiv K_1 \otimes K_3$. Denote the trace functions as

$$\mathrm{Tr}_{K/K_{12}} : K \to K_{12} \quad \text{and} \quad \mathrm{Tr}_{K/K_{13}} : K \to K_{13}.$$

Compute

$$\begin{aligned}\mathrm{Trace}_{\mathcal{R}/\mathcal{R}_{13}}(\mathbf{x} \cdot \mathbf{y}) &= \mathrm{Trace}_{\mathcal{R}/\mathcal{R}_{13}}\Big(\sum \mathrm{PT}(x_i y_i, \Delta) \cdot \mathsf{v}_i \mathsf{v}_i^\vee \mathsf{w}_i + \sum_{i \neq j} \mathrm{PT}(x_i y_i, \Delta) \cdot \mathsf{v}_i \mathsf{v}_i^\vee \mathsf{w}_j\Big) \\ &= \sum \mathrm{PT}(x_i y_i, \Delta) \cdot \mathsf{w}_i \in \mathcal{R}_1 \otimes \mathcal{R}_3\end{aligned}.$$

4.2 Batch F-FHEW Scheme

Batch F-FHEW Encryption:

- Let $\mathrm{Pack}_1(\mathbf{x}) = \sum_{i=1}^{r} x_i \cdot \mathsf{v}_i \in \mathcal{R}_1 \otimes \mathcal{R}_2$, and $\mathrm{Pack}_2(\mathbf{y}) = \sum_{i=1}^{r} y_i \cdot \mathsf{v}_i^\vee \mathsf{w}_i \in \mathcal{R}_1 \otimes \mathcal{R}_2^\vee \otimes \mathcal{R}_3$.
- $\mathrm{CT} = \text{F-RLWE.Enc}(\mathrm{Pack}_1(\mathbf{x})) \in \mathcal{R}_1 \otimes \mathcal{R}_2$, F-RLWE is the second-row element in the F-FHEW matrix.
- $\mathrm{CT}' = \text{F-FHEW.Enc}(\mathrm{Pack}_2(\mathbf{y})) \in (\mathcal{R}_1 \otimes \mathcal{R}_2^\vee \otimes \mathcal{R}_3)^{2 \times 2\ell}$;
- $\mathrm{CT}^* = \mathsf{EvalTr}_{K/K_{13}}(\mathrm{CT} \circ \mathrm{CT}') \in \mathcal{R}_1 \otimes \mathcal{R}_3$, for an algorithm $\mathsf{EvalTr}_{K/K_{13}}$ that input a F-RLWE ciphertext $\mathbf{c} \in \text{F-RLWE}(\mu)$ and outputs a F-RLWE ciphertext $\mathbf{c}' \in \text{F-RLWE}(\mathsf{Tr}_{K/K_{13}}(\mu))$.

$$\begin{aligned}\mathrm{CT}^* &= \mathsf{EvalTr}_{K/K_{13}}(\mathrm{CT} \circ \mathrm{CT}') \\ &= \mathsf{EvalTr}_{K/K_{13}}(\text{F-RLWE}(\mathrm{Pack}_1(\mathbf{x}) \cdot \mathrm{Pack}_2(\mathbf{y}))) \\ &= \text{F-RLWE}(\mathsf{Tr}_{K/K_{13}}(\mathrm{Pack}_1(\mathbf{x}) \cdot \mathrm{Pack}_2(\mathbf{y}))) \\ &= \text{F-RLWE}(\mathrm{PT}(x_i y_i, \Delta)\mathsf{w}_i) \in \mathcal{R}_1 \otimes \mathcal{R}_3.\end{aligned}$$

- Computations between mode $\mathcal{R}_1 \otimes \mathcal{R}_3$ and mode $\mathcal{R}_1 \otimes \mathcal{R}_2$ can be switched. Let $\mathbf{z} = \sum z_i \cdot \mathsf{v}_i \mathsf{w}_i^\vee \in \mathcal{R}_1 \otimes \mathcal{R}_2 \otimes \mathcal{R}_3^\vee$, we can perform the conversion from mode $R_1 \otimes R_3$ to mode $R_1 \otimes R_2$:

$$\mathrm{Trace}_{R/R_{12}}(\mathbf{x}\mathbf{y} \cdot \mathbf{z}) = \sum \mathrm{PT}(\mathrm{PT}(x_i \cdot y_i, \Delta) \cdot z_i, \Delta) \cdot \mathsf{v}_i \in \mathcal{R}_1 \otimes \mathcal{R}_2.$$

The following theorem provides a rigorous proof of the correctness and homomorphic noise computation of the F-FHEW batch processing scheme.

Theorem 3 (Noise Errors in the External Product). *Consider $\mathbf{c}_1, \ldots, \mathbf{c}_r$ as F-RLWE ciphertexts with corresponding error terms $\mathrm{e}_1, \ldots, \mathrm{e}_r$ and messages $x_1, \ldots, x_r \in \mathcal{R}_1$. Additionally, there are F−RGSW ciphertexts denoted as $\boldsymbol{C}_1, \ldots, \boldsymbol{C}_r$, accompanied by error terms $\mathbf{e}_1', \ldots, \mathbf{e}_r'$ and messages $y_1, \ldots, y_r \in \mathcal{R}_1$. Let $\mathrm{CT}^* = \mathsf{EvalTr}_{\mathrm{K}/\mathrm{K}_{13}}(\mathsf{F\text{-}RLWE}(\mathsf{Pack}(\mathbf{x}) \cdot \mathsf{Pack}(\mathbf{y})))$, the error noise is upper bounded by*

$$\|\mathrm{Err}(\mathrm{CT}^*)\| \le \frac{2\rho(p_1-1)}{p} r\{2\ell N 16(\sqrt{\frac{B_g^2}{12}})\sigma\} + \frac{2\rho(p_2-1)}{t} r\{\ell NB/2 \cdot 6\sigma\} + 3pE_{mult}.$$

Proof. The homomorphic noise computation process of the F-FHEW scheme is similar in [11]. Suppose choosing prime integer $p = p_1^{d_1}$ and $t = p_2^{d_2}$, we have $\|\mathrm{v}_i^{\vee}\|_\infty \le 2(p_1-1)/p$ and $\|\mathrm{v}_i\|_\infty = 1$, as well as $\|\mathrm{w}_i^{\vee}\|_\infty \le 2(p_2-1)/t$ and $\|\mathrm{w}_i\|_\infty = 1$. According to the theorem 4.6 in [11], $\mathsf{Err}(\mathrm{CT}^*)) = \mathsf{Tr}_{K/K_{13}}(\mathrm{CT}^*) + e''$, and $\|\mathrm{e}''\|_\infty \le 3\rho E_{mult}$,

$$\mathrm{Err}(\mathrm{EvalTr}_{K/K_{13}}(\mathrm{CT}^*)) = \mathrm{Tr}_{K/K_{13}}(\mathrm{CT}^*) + 3\rho E_{mult}.$$

And for trace function $\mathcal{R}_1 \otimes \mathcal{R}_2 \otimes \mathcal{R}_3$ to $\mathcal{R}_1 \otimes \mathcal{R}_3$,

$$\begin{aligned} \mathrm{Tr}_{K/K_{13}}(\mathrm{CT}^*) &= \mathsf{Tr}_{K/K_{13}}\left(\sum_i \mathrm{Decompose}^{-1}(\mathrm{CT})\mathrm{e}_i'\mathrm{v}_i^{\vee}\mathrm{w}_i + \mathrm{Pack}_2(\mathbf{y}) \cdot (e_i \mathrm{v}_i)\right) + 3pE_{mult} \\ &\le \frac{2\rho(p_1-1)}{p} \sum \|\mathrm{Decompose}^{-1}(\mathrm{CT})\mathrm{e}_i'\|_\infty + \rho\|\mathrm{Pack}_2(\mathbf{y})\| \sum \|e_i\|_\infty + 3pE_{mult} \\ &\le \frac{2\rho(p_1-1)}{p} r\{2\ell N16(\sqrt{\frac{B_g^2}{12}})\sigma\} + \frac{2\rho(p_2-1)}{t} r\{\ell NB/2 \cdot 6\sigma\} + 3pE_{mult}. \end{aligned}$$

4.3 Bootstrapping

In order to reduce the complexity of bootstrapping, we propose a batch bootstrapping technique to make F-FHEW encryption a fully homomorphic encryption scheme, see Algorithm 1.

Definition 1 ([11]). *For the packed messages $m_1, \cdots, m_k$ in the Batch RGSW Encryption, and let $e \in \chi_\sigma$ be some input. Define*

$$T^{2j}(e) = \begin{cases} \mathsf{Tr}_{K/K_{13}}\left(m_k \mathsf{Tr}_{K/K_{12}}\left(m_{k-1} \cdot e\right)\right) & for\ j = 1 \\ T^{2j-2}\left(\mathsf{Tr}_{K/K_{13}}\left(m_{k-2j+1}\mathsf{Tr}_{K/K_{12}}\left(m_{k-2j} \cdot e\right)\right)\right) & for\ j \in [2, k/2] \end{cases}$$

To unfold the recursive formula, we can obtain the following equation $e_k = \widetilde{e}_k + T^2(\widetilde{e}_{k-2}) + T^4(\widetilde{e}_{k-4}) + \cdots + T^k(e_0)$. For $j \le k/2$, any $e \in \chi_\sigma$, prime integer $p = p_1^{d_1}$ and $t = p_2^{d_2}$, then $\|T^{2j}(e)\|_\infty \le 4p_1p_2r^2 \cdot \|e\|_\infty$.

Algorithm 1. Batch Bootstrapping

Input: r RLWE ciphertexts $(b_i, \boldsymbol{a}_i) = (b_i, a_{i1}, ..., a_{in}) \in \mathsf{RLWE}_s(\mu_i)$, given $\text{Pack}(\text{BK}_i) = \text{F-FHEW.Enc}(\text{Pack}(\mathbf{s}))$, $\text{ACC}_0 = \text{F-RLWE.Enc}\left(\text{Pack}(\xi_q^{b_i})\right)$
Output: A F-RLWE ciphertext $\boldsymbol{c} \in \text{F-RLWE}_s(\text{Tr}_{K/K_{13}}(\mu))$.
1: $\text{Pack}(\text{BK}_i) = \text{F-FHEW.Enc}(\text{Pack}(\xi_q^{\mathbf{a}_i \mathbf{s}}))$
2: $k = \log_B q$
3: **for do**$i = 1, \ldots, k$
4: $\quad \text{ACC}_k = \mathsf{EvalTr}_{K/K_{13}}(\text{Pack}(\text{BK}_i) \circ \text{ACC}_{k-1})$
5: **end for**
6: Output ACC_k and then Extract

Firstly, analyze the correctness of the bootstrapping algorithm 1. For y being a power of ξ_q, the equation is $\sum_{0 \le i < q} y^i = 1 + \mathsf{Tr}_{K/K_{23}}(y)$, $\mathsf{Tr}_{K/K_{23}} : K \to K_{23}$. Given $\xi_q^z = \xi_q^{b_i - \langle \boldsymbol{a}_i \mathbf{s} \rangle}$, $\mathsf{test} = (\sum_{y \in \mathbb{Z}_q \& \lfloor y \rceil_2 = 1} \xi_q^{-y})$, $z \in \mathbb{Z}_q$, let $x = \mathsf{test} \cdot (q^{-1} \xi_q^z)$, so we can get

$$q^{-1} + \text{Tr}_{K/K_{23}}(x) = \begin{cases} 1 \text{ if } \lfloor z \rceil_2 = 1 \\ 0 \text{ otherwise} \end{cases}.$$

According to the aforementioned equation and batch computation, the algorithm finally output $\sum_{i \in [r]} \lfloor b_i - \langle \boldsymbol{a}_i, \mathbf{s} \rangle \rceil_B \cdot \mathsf{v}_i$. For $\sum_{i \in [r]} q^{-1} \xi_q^{b_i - \langle \boldsymbol{a}_i, \mathbf{s} \rangle / (\frac{B}{2^i})} \mathsf{v}_i, i = 1, \cdots, \log_2 B$. By batch computation framework, we have

$$\begin{aligned}
\mathsf{Err}(\mathsf{ACC}_k) &= \mathsf{Err}(\mathsf{Eval\text{-}Tr}(\mathsf{BK}_k \circ \mathsf{ACC}_{k-1})) \\
&= e_k' + \mathsf{Tr}_{K/K_{13}}(\mathsf{Err}(\mathsf{BK}_k)\mathbf{G}^{-1}(\mathsf{ACC}_{k-1})) + \mathsf{Tr}_{K/K_{13}}(m_k \cdot e_{k-1}) \\
&= e_k' + \mathsf{Tr}_{K/K_{13}}(\mathsf{Err}(\mathsf{BK}_k)\mathbf{G}^{-1}(\mathsf{ACC}_{k-1})) \\
&+ \mathsf{Tr}_{K/K_{13}}\left(m_k \cdot \left(e_{k-1}'' + \mathsf{Tr}_{K/K_{12}}(m_{k-1} \cdot e_{k-2})\right)\right) \\
&= e_k' + \mathsf{Tr}_{K/K_{13}}(\mathsf{Err}(\mathsf{BK}_k)\mathbf{G}^{-1}(\mathsf{ACC}_{k-1})) + \text{Tr}_{K/K_{13}}\left(m_k \cdot e_{k-1}''\right) \\
&+ \text{Tr}_{K/K_{13}}\left(m_k \cdot \text{Tr}_{K/K_{12}}(m_{k-1} \cdot e_{k-2})\right) \\
&= \widetilde{e}_k + \text{Tr}_{K/K_{13}}\left(m_k \cdot \text{Tr}_{K/K_{12}}(m_{k-1} \cdot e_{k-2})\right).
\end{aligned}$$

We can get $\widetilde{e}_k = e_k' + \mathsf{Tr}_{K/K_{13}}(\mathsf{Err}(\mathsf{BK}_k)\mathbf{G}^{-1}(\mathsf{ACC}_{k-1})) + \text{Tr}_{K/K_{13}}\left(m_k \cdot e_{k-1}''\right)$, where the additive error e_k' is obtained from the key-switching algorithm. According to the Definition 1,

$$\text{Tr}_{K/K_{13}}(m_k \cdot \text{Tr}_{K/K_{12}}(m_{k-1} \cdot e_{k-2}) \le 24 p_1 p_2 r^2 \sigma.$$

Assuming that the errors of the key-switch keys are upper bounded by E, then the error term $\widetilde{e}_k$ is constrained by a subgaussian distribution with a parameter less than $O(\ell\sqrt{N \log Q} \cdot E)$.

5 Efficiency

In this chapter, we conduct an HE security assessment to obtain specific parameters. The prefix "STD" refers to HE security level [22] resistance to three types

Table 2. Specific Parameters

Parameter Set	Dim $N = \phi(m)$ $= \phi(q) \times \rho \times \tau$	Batch param	q	h	$\log_2 Q$	$\log_2 Q_{ks}$	B_{ks}	B_g	B_r
STD128	$256 \times 4 \times 6 = 6144$	4	257	131	25	13	128	2^7	32
STD192	$256 \times 6 \times 6 = 9216$	6	257	131	35	17	128	2^{13}	32
STD256	$256 \times 6 \times 6 = 9216$	6	257	160	54	27	28	2^8	46
STD128Q	$256 \times 4 \times 6 = 6144$	4	257	131	27	13	32	2^8	32
STD192Q	$256 \times 6 \times 6 = 9216$	6	257	131	37	19	64	2^{25}	32
STD256Q	$256 \times 6 \times 6 = 9216$	6	257	160	58	29	16	2^7	32

of attack: uSVP (uniform Shortest Vector Problem), dec (decoding attack), and dual(dual attack). In contrast, the suffix "Q" refers to the resistance to BKZ attacks, whereas BKZ can resist quantum attacks. For block size β, the BKZ algorithm with SVP oracle has the sieving cost $2^{0.292\beta+16.4}$ [22]. To establish the required block size β, we can get the root-Hermite factor [25]

$$\log \delta_0 = log\left(\frac{\beta}{2\pi e}(\pi\beta)^{\frac{1}{\beta}}\right) \cdot \frac{1}{2(\beta-1)} = \frac{\log^2\left(\epsilon' \tau \, \alpha \sqrt{2e}\right)}{4n \log q},$$

for some $\epsilon' > 1$ and some fixed $\tau \leq 1$. The following Table 2 provides the estimated parameters for security levels STD128, STD192, STD256, STD128Q, STD192Q, and STD256Q using the LWE estimator[1]. We applied the parameter from [10]: for the RLWE parameters n, small (RLWE) modulus q, RLWE/RGSW modulus Q, RLWE modulus used for key switching Q_{ks}, gadget base for the Gadget matrix B_g, bootstrapping key parameters B_r, Key switching parameters B_{ks}. Every experimentation was performed on a machine with Intel(R) Core(TM) i5-10400F CPU @ 2.90 GHz processor. PALISADE compiled with the following CMake flags: WITH-NATIVEOPT = ON (machine-specific optimizations were applied by the compiler) and WITH-INTEL-HEXL = ON (AVX-512 acceleration was used) by MinGw64. The performance of the CKKS and F-FHEW used to implement the encoding process is compared in Figs. 3 and 4. The batch size of CKKS is 8. It shows that the encoding efficiency of F-FGSW remains unaffected by the security level and faster than CKKS. Figure 5 describes the performance comparison of logic and arithmetic operations between F-TFHE, F-FHEW, CKKS, with the base $B = 2$. Figure 6 describes the performance comparison of F-FHEW when base $B = 2, 3$. As a result, choosing a larger base $B = 3$ improves the runtime efficiency.

We arbitrary choose double and single precision messages x without encoding error i.e., Decode(Encode(x)) = x as follows:

$$x_B^1 = 6.6666623789 * 10^{-9}, x_B^2 = 2.566659875123 * 10^{-13}$$

$$x_B^3 = 8.656454564545645 * 10^{-14}, x_B^4 = 4.5456487845457845 * 10^{-17}$$

[1] https://bitbucket.org/malb/lwe-estimator/src/master/.

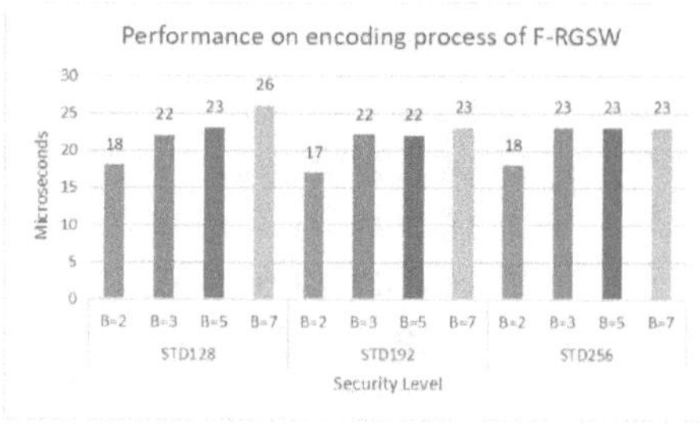

Fig. 3. F-FHEW encoding process

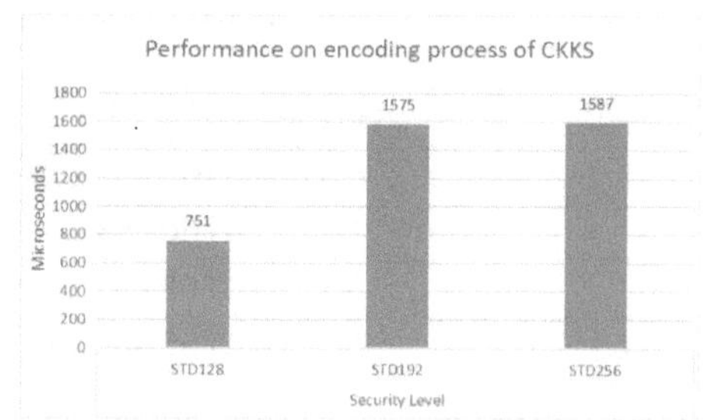

Fig. 4. CKKS encoding process

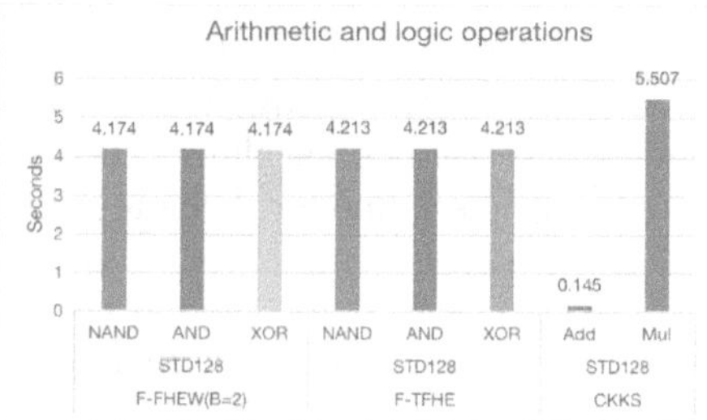

Fig. 5. CKKS,F-TFHE and F-FHEW

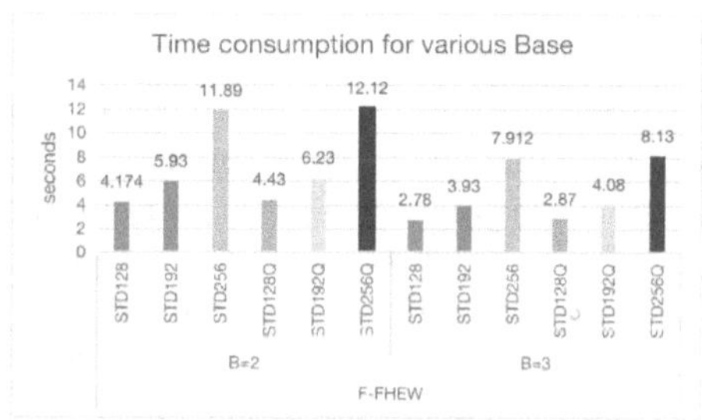

Fig. 6. The NAND operation of F-FHEW

Table 3. Precision operation results

Add	$x_B^1 + x_B^2$	$x_B^3 - x_B^4$
$B = 2$	**6.666919044887538 ∗ 10^{-9}**	**8.651908915761106 ∗ 10^{-14}**
$B = 3$	**6.666919044887511 ∗ 10^{-9}**	**8.651908915761105 ∗ 10^{-14}**
$B = 5$	**6.666919044887513 ∗ 10^{-9}**	**8.651908915761111 ∗ 10^{-14}**
$B = 7$	**6.666919044887513 ∗ 10^{-9}**	**8.651908915761105 ∗ 10^{-14}**
Correct value	**6.666919044887513 ∗ 10^{-9}**	**8.651908915761102 ∗ 10^{-14}**

Then we evaluate $z_1 = (x_B^1 + x_B^2)$, $z_2 = (x_B^3 - x_B^4)$, $z_3 = z_1 \cdot z_2$, and $z_4 = z_2^3$ on the ciphertext domain, and results are listed in Tables 3 and 4. The correct calculation values are listed which are rounded down.

Table 4. Precision operation results

Mul	$(x_B^1 + x_d^2)(x_B^3 - x_B^4)$	$[(x_B^1 + x_d^2)(x_B^3 - x_B^4)]^2$
$B = 2$	$\mathbf{5.768157632511932 * 10^{-22}}$	$\mathbf{3.327164247350524 * 10^{-43}}$
$B = 3$	$\mathbf{5.768157632511956 * 10^{-22}}$	$\mathbf{3.327164247350601 * 10^{-43}}$
$B = 5$	$\mathbf{5.768157632511979 * 10^{-22}}$	$\mathbf{3.327164247350601 * 10^{-43}}$
$B = 7$	$\mathbf{5.768157632511991 * 10^{-22}}$	$\mathbf{3.327164247300603 * 10^{-43}}$
Correct value	$\mathbf{5.768157632511974 * 10^{-22}}$	$\mathbf{3.327164247350614 * 10^{-43}}$

6 Conclusion

In conclusion, this paper establishes a ring isomorphism between the Laurent and cyclotomic representations of polynomials, effectively linking them in the context of floats. Based on B-based cyclotomic polynomials as plaintexts, the paper introduces F-FHEW, a homomorphic encryption scheme for floating-point numbers that provides 128-bit precision. In addition, the complexity of homomorphic multiplication is analyzed to be $O(n^{1.75} \log q)$ by utilizing the method of polynomial truncation. With the use of field packing, a more efficient batch bootstrapping scheme for multi bits is proposed. For the future work, it is expected to apply F-FHEW to privacy-preserving machine learning.

Acknowledgement. This work is supported by the National Natural Science Foundation of China under Grant No. 62272131 and the Guangdong Provincial Key Laboratory of Novel Security Intelligence Technologies under Grant No. 2022B1212010005.

References

1. Kara, M., Laouid, A., Hammoudeh, M., Bounceur, A.: One Digit checksum for data integrity verification of cloud-executed homomorphic encryption operations. Cryptology ePrint Archive (2023)
2. Hu, J., Chen, J., Dai, W., Wang, H.: Fully homomorphic encryption-based protocols for enhanced private set intersection functionalities. Cryptology ePrint Archive (2023)
3. Dai, T., Duan, L., Jiang, Y., Li, Y., Mei, F., Sun, Y.: Force: making 4PC > 4× PC in privacy preserving machine learning on GPU. Cryptology ePrint Archive (2023)
4. Cheon, J.H., Kim, A., Kim, M., Song, Y.: Homomorphic encryption for arithmetic of approximate numbers. In: Takagi, T., Peyrin, T. (eds.) ASIACRYPT 2017, Part I 23. LNCS, vol. 10624, pp. 409–437. Springer, Cham (2017). https://doi.org/10.1007/978-3-319-70694-8_15
5. Jutla, C.S., Manohar, N.: Sine series approximation of the mod function for bootstrapping of approximate HE. In: Advances in Cryptology-EUROCRYPT 2022: Proceedings of the 41st Annual International Conference on the Theory and Applications of Cryptographic Techniques, Trondheim, Norway, 30 May–3 June 2022, Part I, pp. 491–520. Springer, Cham (2022). https://doi.org/10.1007/978-3-031-06944-4_17

6. Kono, F., Nakasato, N., Nakata, M.: Accelerating 128-bit floating-point matrix multiplication on FPGAs. arXiv preprint arXiv:2306.04087 (2023)
7. Markstein, P.: The new IEEE-754 standard for floating point arithmetic. In: Dagstuhl Seminar Proceedings. Schloss Dagstuhl-Leibniz-Zentrum für Informatik (2008)
8. Chen, H., Chillotti, I., Song, Y.: Improved bootstrapping for approximate homomorphic encryption. In: Ishai, Y., Rijmen, V. (eds.) EUROCRYPT 2019. LNCS, vol. 11477, pp. 34–54. Springer, Cham (2019). https://doi.org/10.1007/978-3-030-17656-3_2
9. Chillotti, I., Gama, N., Georgieva, M., Izabachène, M.: Faster fully homomorphic encryption: bootstrapping in less than 0.1 seconds. In: Cheon, J.H., Takagi, T. (eds.) ASIACRYPT 2016. LNCS, vol. 10031, pp. 3–33. Springer, Heidelberg (2016). https://doi.org/10.1007/978-3-662-53887-6_1
10. Micciancio, D., Polyakov, Y.: Bootstrapping in FHEW-like cryptosystems. In: Proceedings of the 9th on Workshop on Encrypted Computing and Applied Homomorphic Cryptography, pp. 17–28 (2021)
11. Liu, F.H., Wang, H.: Batch bootstrapping I: a new framework for SIMD bootstrapping in polynomial modulus. In: Proceedings of the 42nd Annual International Conference on the Theory and Applications of Cryptographic Techniques. Advances in Cryptology, EUROCRYPT 2023, Part III, Lyon, France, 23–27 April 2023, pp. 321–352. Springer, Cham (2023). https://doi.org/10.1007/978-3-031-30620-4_11
12. Liu, F.-H., Wang, H.: Batch bootstrapping II: bootstrapping in polynomial modulus only requires O (1) FHE multiplications in amortization. In: Proceedings of the 42nd Annual International Conference on the Theory and Applications of Cryptographic Techniques, EUROCRYPT 2023, Part III, Lyon, France, 23–27 April 2023, pp. 353–384. Springer, Cham (2023). https://doi.org/10.1007/978-3-031-30620-4_12
13. Chen, H., Laine, K., Player, R.: Simple encrypted arithmetic library - SEAL v2.1. In: Brenner, M., et al. (eds.) FC 2017. LNCS, vol. 10323, pp. 3–18. Springer, Cham (2017). https://doi.org/10.1007/978-3-319-70278-0_1
14. Dowlin, N., Gilad-Bachrach, R., Laine, K., Lauter, K., Naehrig, M., Wernsing, J.: Manual for using homomorphic encryption for bioinformatics. Proc. IEEE **105**(3), 552–567 (2017)
15. Costache, A., Smart, N.P., Vivek, S., Waller, A.: Fixed-point arithmetic in SHE schemes. In: 23rd International Conference on Selected Areas in Cryptography, SAC 2016, St. John's, NL, Canada, 10–12 August (2017)
16. Moon, S., Lee, Y.: An efficient encrypted floating-point representation using HEAAN and TFHE. Secur. Commun. Netw. **2020**, 1–18 (2020)
17. Lee, S., Shin, D.J.: Overflow-detectable Floating-point Fully Homomorphic Encryption. Cryptology ePrint Archive (2022)
18. Rudelson, M., Vershynin, R.: Hanson-Wright inequality and sub-gaussian concentration (2013)
19. Lyubashevsky, V., Peikert, C., Regev, O.: A toolkit for ring-LWE cryptography. In: Johansson, T., Nguyen, P.Q. (eds.) EUROCRYPT 2013. LNCS, vol. 7881, pp. 35–54. Springer, Heidelberg (2013). https://doi.org/10.1007/978-3-642-38348-9_3
20. Lyubashevsky, V., Peikert, C., Regev, O.: On ideal lattices and learning with errors over rings. J. ACM (JACM) **60**(6), 1–35 (2013)
21. Castryck, W., Iliashenko, I., Vercauteren, F.: Homomorphic SIM^{2}D operations: single instruction much more data. In: Nielsen, J.B., Rijmen, V. (eds.) EUROCRYPT 2018. LNCS, vol. 10820, pp. 338–359. Springer, Cham (2018). https://doi.org/10.1007/978-3-319-78381-9_13

22. Albrecht, M., et al.: Homomorphic encryption standard. In: Lauter, K., Dai, W., Laine, K. (eds.) Protecting Privacy through Homomorphic Encryption, pp. 31–62. Springer, Cham (2021). https://doi.org/10.1007/978-3-030-77287-1_2
23. Costache, A., Smart, N.P.: Which ring based somewhat homomorphic encryption scheme is best? In: Sako, K. (ed.) CT-RSA 2016. LNCS, vol. 9610, pp. 325–340. Springer, Cham (2016). https://doi.org/10.1007/978-3-319-29485-8_19
24. Gentry, C., Halevi, S., Smart, N.P.: Homomorphic evaluation of the AES circuit. In: Safavi-Naini, R., Canetti, R. (eds.) CRYPTO 2012. LNCS, vol. 7417, pp. 850–867. Springer, Heidelberg (2012). https://doi.org/10.1007/978-3-642-32009-5_49
25. Albrecht, M.R., Player, R., Scott, S.: On the concrete hardness of learning with errors. J. Math. Cryptol. **9**(3), 169–203 (2015)

NTRU-Based FHE for Larger Key and Message Space

Robin Jadoul[1], Axel Mertens[1(✉)], Jeongeun Park[2], and Hilder V. L. Pereira[3]

[1] COSIC, KU Leuven, Leuven, Belgium
{robin.jadoul,axel.mertens}@esat.kuleuven.be
[2] Norwegian University of Science and Technology (NTNU), Trondheim, Norway
jeongeun.park@ntnu.no
[3] Universidade Estadual de Campinas, Campinas, Brazil
hilder@unicamp.br

Abstract. The NTRU problem has proven a useful building block for efficient bootstrapping in Fully Homomorphic Encryption (FHE) schemes, and different such schemes have been proposed. FINAL (ASIACRYPT 2022) first constructed FHE using homomorphic multiplexer (CMux) gates for the blind rotation operation. Later, XZD+23 (CRYPTO 2023) gave an asymptotic optimization by changing the ciphertext format to enable ring automorphism evaluations. In this work, we examine an adaptation to FINAL to evaluate CMux gates of higher arity and the resulting tradeoff to running times and bootstrapping key sizes. In this setting, we can compare the time and space efficiency of both bootstrapping protocols with larger key space against each other and the state of the art.

Keywords: FHE · FINAL · NTRU · Bootstrapping

1 Introduction

All current constructions for fully homomorphic encryption (FHE) schemes rely on noisy encryptions for which the noise accumulates and grows as homomorphic operations are applied. When the noise grows too big, however, correct decrypted value is no longer guaranteed. Therefore, a form of homomorphic decryption that resets the noise level back to a point where more operations can be applied is necessary, the so called *bootstrapping* operation, which is the most computationally expensive part of the scheme. Since the first introduction of FHE, there have been continuous attempts [5,6,8,10,13] to make bootstrapping algorithms more practical and overcome this bottleneck. Initially, FHE schemes had large message spaces that could pack many messages into a single ciphertext, offering good amortized costs, since each homomorphic operation acts in parallel

J. Park This work was done when the author was at COSIC, KU Leuven.

T. Zhu and Y. Li (Eds.): ACISP 2024, LNCS 14895, pp. 141–160, 2024.
https://doi.org/10.1007/978-981-97-5025-2_8

on all packed messages. However, the bootstrapping for these is very expensive in terms of both memory and time. One well-known scheme in this family, BGV [3], performs bootstrapping with a typical running time of a few minutes on a common commercial personal computer, even for modern and optimized implementations [12].

A very different approach was taken by [1], which proposed an efficient bootstrapping for ciphertexts encrypting a single bit instead of many larger messages. That method was improved in [8] and finally implemented, yielding, for the first time, a bootstrapping algorithm that runs in less than one second on a common laptop. [6] improved further upon this method, with the introduction of the TFHE scheme, for which bootstrapping runs in merely a few miliseconds.

After a hiatus of a few years, another scheme, called FINAL [2], improved upon TFHE and achieved single-bit bootstrapping that runs in about 70% of TFHE's running time, using bootstrapping keys of only about half the size. The core of this improvement lies in the introduction of NTRU-based ciphertexts for the bootstrapping algorithm. As a concurrent work, Kluczniak [18] introduced similar scheme, which was broken due to their choice of ciphertext space [16] called cyclic NTRU ring. With the negacyclic version, still their construction has survived against existing attacks, their ciphertext form has additional larger noise than the ciphertext of FINAL, which is less efficient than FINAL in terms of noise management.

Using a slightly different ciphertext form, still based on the NTRU problem, Xiang et al. [20] introduced a new scheme, which we call by XZD+23 from now on, that allows for the usage of ring automorphisms in the bootstrapping phase, leading to asymptotically lower memory usage and bigger key sizes that could allow for smaller base ciphertexts. Their improvement to the running time when compared to TFHE is, however, overall not as pronounced as FINAL's and while the memory usage is asymptotically better, for most practical parameter sets, it can be seen to be higher. As an alternative to enable larger key sizes and smaller lattice dimensions, Joye and Paillier [17] suggested a version of the CMux gate with higher arity for the TFHE scheme. As FINAL's bootstrapping makes use of the same CMux construction, it could be amenable to a similar change, but no descriptions of this change have been given yet, nor has the comparison between both methods to increase the key space been made for NTRU-based FHE schemes.

Although those schemes achieve the fastest bootstrapping known to date, using them to evaluate arbitrary circuits can end up being very expensive, because they only encrypt bits. Even simple operations, such as integer addition, have to be represented by binary circuits and each gate of the circuit requires one bootstrapping. Thus, to overcome this limitation, TFHE was extended to support messages in $\mathbb{Z}_{2^k}$, that is, k-bit messages. Simultaneously, it gained the ability to perform *programmable bootstrapping* (PBS) (also sometimes referred to as *functional bootstrapping*). In this paradigm, every bootstrapping operation can not only perform a noise reduction; it can also apply an arbitrary function f, represented as a lookup table, to the underlying message of a ciphertext, at no additional cost [7].

Our Contributions

In this work, we study the n-ary CMUX from [17] and how it can be employed in FINAL. We also adapt other techniques and improvements from the literature to FINAL. For example, we notice that by carefully changing the definition of scalar NGS ciphertexts from FINAL, it is possible to use an approximate gadget decomposition (introduced first in TFHE) without increasing the final noise in a noticeable way, thereby reducing the cost of the bootstrappring almost for free. We apply the same optimizations to the XZD+23 scheme, in order to compare the two main NTRU-based FHEs on equal footing.

While the bigger potential key size in XZD+23 brings a slight reduction in its memory usage and running time, we observe that it will theoretically still need approximately double the amount of significant operations (as measured by counting the number of required FFTs and FFT-domain multiplications). We also find that, depending on the employed FFT-library and the subsequent timings for an FFT-operation and a pointwise multiplication, FINAL comes out on top in efficiency-comparison with XZD+23. We further give theoretical bounds for the key space that result in higher efficiency than the ternary key setting.

We note that NTRU based PBS for binary LWE secret key was briefly introduced in [18], however, as we mentioned above, their ciphertext has larger noise than FINAL, therefore, FINAL can naturally achieve the same functionality without increasing parameters. We stress that we examine PBS of FINAL, further, with larger secret key in this paper.

Roadmap

After the preliminaries in Sect. 2, we discuss our improvements and PBS for FINAL in Sect. 3. We also give evidence that bootstrapping by n-ary CMUX is not a guaranteed improvement for FINAL, and needs to be carefully evaluated per implementation or even supporting hardware. Then, in Sect. 4, we apply similar improvements and PBS to the scheme from [20]. Finally, in Sect. 5, the two schemes are compared.

2 Preliminaries

2.1 Notation and Mathematical Background

Vectors, Polynomials, and Norms. We use lower-case bold letters for vectors and upper-case bold letters for matrices, e.g. $\mathbf{a}$ and $\mathbf{A}$. For any vector $\mathbf{u}$, $\|\mathbf{u}\|$ denotes the infinity norm. Throughout the paper, N is always a power of two and $\mathcal{R} := \mathbb{Z}[X]/\langle X^N + 1\rangle$ is the $(2N)$-th cyclotomic ring. Any element f of $\mathcal{R}$ can be always represented by a unique polynomial of degree smaller than N, hence, writing $f = \sum_{i=0}^{N-1} f_i \cdot X^i$ is unambiguous and we can define the coefficient vector of f as $\phi(f) := (f_0, \ldots, f_{N-1}) \in \mathbb{Z}^N$. We define the infinity norm of f as $\|f\| := \|\phi(f)\|$. For any $Q \in \mathbb{Z}^+$, let $\mathcal{R}_Q := \mathcal{R}/Q\mathcal{R} = \mathbb{Z}_Q[X]/\langle X^N + 1\rangle$.

Finally, we define $\mathbb{M} := \{\pm m \cdot X^k \mid m \in \mathbb{Z}_p, k \in \mathbb{N}\}$, which will be used as the plaintext space of the vector ciphertexts defined in Sect. 2.4. Here, $\mathbb{Z}_p$ is the overall plaintext space, which for the original FINAL scheme is $\mathbb{Z}_2$.

Discrete Gaussian Distribution. We first describe the discrete Gaussian distribution where our secret elements are sampled from. Typically, a discrete Gaussian distribution is defined as a distribution over $\mathbb{Z}$, where every element in $\mathbb{Z}$ is sampled with probability proportional to its probability mass function value under a Gaussian distribution over $\mathbb{R}$. Let the Gaussian function $\rho_{\sigma,c}$ be defined as $\rho_{\sigma,c}(x) = \exp(-\frac{|x-c|^2}{2\cdot\sigma^2})$ for $\sigma, c \in \mathbb{R}_{>0}$. Then, $\rho_{\sigma,c}(\mathbb{Z}) = \sum_{i=-\infty}^{\infty} \rho_{\sigma,c}(i)$. The discrete Gaussian distribution with standard deviation σ and mean c is a distribution on $\mathbb{Z}$ with the probability of $x \in \mathbb{Z}$ given by $\rho_{\sigma,c}(x)/\rho_{\sigma,c}(\mathbb{Z})$. For $c = 0$, we denote this distribution by χ_σ.

Subgaussian Distribution. For the analysis of encryption parameters, we need subgaussian random variables over $\mathbb{R}$.

Definition 2.1. *A random variable V over $\mathbb{R}$ is α-subgaussian if its moment generating function satisfies*

$$\mathbb{E}[\exp(t \cdot V)] \leq \frac{1}{2}\exp(\alpha^2 \cdot t^2)$$

for all $t \in \mathbb{R}$. The tails of such a subgaussian are dominated by the tails of a Gaussian.

We define the variance $\mathsf{Var}(a)$ (resp. $\mathsf{Var}(\mathbf{x})$) of $a \in \mathcal{R}$ (resp. $\mathbf{x} \in \mathbb{Z}^n$) to be the maximum variance over the coefficients of a (resp. the components of $\mathbf{x}$). The product of two polynomials $a, b \in \mathcal{R}$ has a variance $\mathsf{Var}(a \cdot b) = N \cdot \mathsf{Var}(a) \cdot \mathsf{Var}(b)$. Similarly, the maximum variance of each column of a matrix $\mathbf{X}$ is $\mathsf{Var}(\mathbf{X})$. From the definition of a subgaussian distribution, it can be proven that the variance of an α-subgaussian random variable V is bounded by $\mathsf{Var}(V) \leq \alpha^2$. Informally, this says that the tails of V are dominated by a Gaussian function with standard deviation α. The following lemma is from [15].

Lemma 2.1. *If $\mathbf{x}$ is a discrete random vector over $\mathbb{R}^n$ such that each component x_i of $\mathbf{x}$ is α_i-subgaussian, then the vector $\mathbf{x}$ is a β-subgaussian vector where $\beta = \max_{i\in[n]} \alpha_i$.*

Subgaussian random variables possess a significant property known as Pythagorean additivity. When considering two random variables, α-subgaussian X and β-subgaussian Y, alongside integers $a, b \in \mathbb{Z}$, the random variable $a \cdot X + b \cdot Y$ is known to be $\sqrt{a^2 \cdot \alpha^2 + b^2 \cdot \beta^2}$-subgaussian. This in turn implies that

$$\mathsf{Var}(a \cdot X) + \mathsf{Var}(b \cdot Y) \leq a^2 \cdot \mathsf{Var}(X) + b^2 \cdot \mathsf{Var}(Y) \leq a^2 \cdot \alpha^2 + b^2 \cdot \beta^2.$$

Hardness Assumptions. Most FHE schemes base their security on the well-known LWE and RLWE problems. FINAL also uses the NTRU problem. We use the standard definitions and recall them in the full version for completeness.

2.2 Decompositions

For fixed integers q and B, we set $\ell := \lceil \log_B q \rceil$ and define the gadget vector $\mathfrak{g}_{q,B} := (B^0, \ldots, B^{\ell-1})$. When q and B are clear from the context, we write $\mathfrak{g}$. Then, for any $k \in \mathbb{Z}_q$, we represent k by an integer in $[-q/2, q/2)$ and define its signed decomposition in base B as $\mathfrak{g}^{-1}(k) = (k_0, \ldots, k_{\ell-1})$ for each integer $|k_i| \leq B/2$ for $i \in [\ell]$. It is easy to see that $\mathfrak{g}^{-1}(k) \cdot \mathfrak{g} = k$. For any $f \in R_Q$, we define $\mathfrak{g}^{-1}(f) := \sum_{i=0}^{N-1} \mathfrak{g}^{-1}(f_i) X^i$. It is clear that

$$\mathfrak{g}^{-1}(f) \cdot \mathfrak{g} = \sum_{i=0}^{N-1} \mathfrak{g}^{-1}(f_i) \cdot \mathfrak{g} \cdot X^i = \sum_{i=0}^{N-1} f_i \cdot X^i = f.$$

Alternatively, a randomized gadget decomposition [11,14,15] can be used not to rely on Heuristic assumption for independent behaviours of each element of the output noise vector.

2.3 LWE Encryption Scheme

FHE schemes like FHEW [8], TFHE [6], and FINAL [2] are actually composed of two homomorphic schemes: a simple one, that encrypts the messages and performs the homomorphic operations, which we call the base scheme, and a second scheme, called the accumulator. The former is used to encrypt the actual data and is commonly based on the LWE problem, while the latter is used in the bootstrapping of the base scheme. In this section, we present the LWE scheme with message space $\mathbb{Z}_p$.

As first discussed in [4], an encryption scheme can be constructed from the LWE problem as follows:

Definition 2.2. (LWE encryption). *Let $n, q, p \in \mathbb{N}$, $\sigma \in \mathbb{R}_{>0}$, and χ_{key} a distribution over $\mathbb{Z}^n$.*

We define the scaling factor as $\Delta := q/p$. If q and p, respectively the ciphertext modulus and the plaintext modulus, are not both powers of two, a rounding must be applied at the moment of encryption. We call the following an LWE encryption of $m \in \mathbb{Z}_p$ under the secret key $s \leftarrow \chi_{\mathsf{key}}$: $(\mathbf{a}, \mathbf{b} = \sum_{i=0}^{n-1} \langle a_i, s_i \rangle + \Delta \cdot m + e)$ where elements of $\mathbf{a}$ are sampled uniformly random from $\mathbb{Z}_q$ and e is sampled from χ_σ. Other than binary.

A ciphertext of this format can be decrypted by computing $m = \left\lfloor \frac{\mathbf{b} - \sum_{i=0}^{n-1} a_i \cdot s_i}{\Delta} \right\rceil$. Decryption will be successful on condition that $|e| \leq \Delta/2$.

Intuitively, during encryption, we lift the message to a higher range by multiplying it with Δ, and add noise in the lower range. This means that m is mapped to a bucket around the "clean" value. Unless the noise becomes so big that the ciphertext enters another bucket, the message can be decrypted correctly.

Note that, originally, this encryption scheme used χ_{key} as a distribution over $\{0,1\}^n$, but we shall use other key distributions later on too.

2.4 Ciphertexts for FINAL

In this section we recall the NTRU-based, GSW-like Scheme (NGS) that was presented in FINAL [2]. It is used in the accumulator in the bootstrapping algorithm. The scheme includes two encryption functions with different ciphertext forms, the first of which (NGS.EncS) encrypts a plaintext m as an element of $\mathcal{R}_Q$, and the second (NGS.EncVec) encrypts m as a vector of elements in $\mathcal{R}_Q$. The message m can be any polynomial in $\mathcal{R}_p$ for NGS.EncS, but for NGS.EncVec it is assumed that $m \in \mathbb{M} = \{\pm b \cdot X^k \mid b \in \mathbb{Z}_p, k \in \mathbb{N}\}$. As the NGS ciphertexts are never decrypted in the bootstrapping mechanism, we omit the corresponding description here.

NGS can be described by the following four procedures:

- NGS.ParamGen(1^λ): From the security parameter, output Param $:= (N, p, Q, \zeta, B, l)$, with N the dimension of $\mathcal{R}$, p and Q respectively the plaintext and ciphertext modulus, ζ the standard deviation for the key, B the decomposition base and $l = \lceil \log_B(Q) \rceil$ the dimension over $\mathcal{R}_Q$ of a vector ciphertext.
- NGS.KeyGen(Param): Sample $f' \leftarrow \chi_\zeta^N$ and set $f := 1 + p \cdot f'$ until f^{-1} exists in $\mathcal{R}_Q$. Output $sk := f$.
- NGS.EncS(sk, m): Sample $g \leftarrow \chi_\zeta^N$, set $\Delta := Q/p$, and return $c = g/f + \Delta \cdot m \in \mathcal{R}_Q$. This c is called a *scalar encryption of* m.
- NGS.EncVec(sk, m): Sample $g_i \leftarrow \chi_\zeta^N$ for $0 \leq i < l$. Set $\mathbf{g} := (g_0, \ldots, g_{l-1})$ and $\mathfrak{g} = (B^0, \ldots, B^{l-1})$. Return $\mathbf{c} = \mathbf{g}/f + \mathfrak{g} \cdot m \in \mathcal{R}_Q^l$. This $\mathbf{c}$ is called a *vector encryption of* m.

An *external product* can be defined as an operation that takes an NGS scalar ciphertext encrypting a message u and an NGS ciphertext encrypting a message v, returning an NGS scalar ciphertext encrypting the message $u \cdot v$.

Definition 2.3. (External product *[2]***).** *From a scalar encryption $c := g/f + \Delta \cdot u \in \mathcal{R}_Q$ of the ternary polynomial u and a vector encryption $\mathbf{c} := \mathbf{g}/f + \mathfrak{g} \cdot v \in \mathcal{R}_Q^l$ of the message v, we define the vector product to be*

$$c \boxdot \mathbf{c} := \mathfrak{g}^{-1}(c) \cdot \mathbf{c} \in \mathcal{R}_Q \tag{1}$$

As $\mathfrak{g}^{-1}(c) \cdot \mathfrak{g} = c$, note that $c_{mult} = c \boxdot \mathbf{c}$ is equal to

$$c_{mult} := (\mathfrak{g}^{-1}(c) \cdot \mathbf{g})/f + (\mathfrak{g}^{-1}(c) \cdot \mathfrak{g} \cdot v) = \underbrace{(\mathfrak{g}^{-1}(c) \cdot \mathbf{g} + g \cdot v)}_{g_{mult}}/f + \Delta \cdot u \cdot v. \tag{2}$$

c_{mult} is then a valid scalar encryption of $u \cdot v$ if the noise term g_{mult} is small enough.

The noise growth in the external product can be bounded by $\mathsf{Var}(\mathsf{Err}(c_{mult})) \leq N\ell\gamma^2 \cdot \mathsf{Var}(\mathbf{g}) + ||v||_2^2 \cdot \mathsf{Var}(g) + 4\zeta^2$, as proven in lemma 4 of FINAL [2]. Here, $\mathbf{g}^{-1}(a)$ is assumed γ-subgaussian for $a \in \mathcal{R}_Q$ and some $\gamma = O(B)$. For k consecutive external products with fresh vector ciphertexts, the noise is bound by $\mathsf{Var}(\mathsf{Err}(c')) \leq N\ell\gamma^2 \cdot \sum_{i=1}^{k} \mathsf{Var}(\mathbf{g}_i) + \mathsf{Var}(g_0) + 4\zeta^2$.

In Sect. 3.3, we introduce a modification to this definition, to reduce the cost (both computational and memory) of bootstrapping, at the cost of only insignificant noise growth.

3 Improvements for FINAL

In this section, we suggest three modifications and additions to the original FINAL scheme: we suggest how to use a larger key space, we discuss approximate decomposition in external products, and finally we explain how to enable PBS after growing the plaintext space.

3.1 Bootstrapping FINAL

Before describing our modifications, we first recall a high-level overview of FHE bootstrapping based on a blind rotation procedure, as applied in TFHE [6], FINAL [2] and XZD+23 [20]. Starting from an LWE ciphertext $c = (\langle \mathbf{a}, \mathbf{s} \rangle, b) \bmod 2N$ in the base scheme, let the *test polynomial* $T(X) = \sum_{0 \leq i < N} \lfloor i/\Delta \rfloor \cdot X^i$. Observe that the negacyclic rotation of T,

$$X^{2N/p - q/2N(b - \langle \mathbf{a}, \mathbf{s} \rangle)} \cdot T(X)$$

has a decryption of c in the constant coefficient.

By computing this rotation *blindly* – that is, without revealing $\mathbf{s}$, usually by having some encryption of it as extra information: the *bootstrapping* or *evaluation key* – and performing a final key switching step that extracts this coefficient as a new LWE ciphertext under $\mathbf{s}$, we have performed a bootstrap. For a properly tuned blind rotation procedure, the noise will be sufficiently reduced to allow further homomorphic operation.

In practice, the blind rotation will accumulate rotations of X^{a_i}, rotated by a secret amount s_i, with an additive homomorphism in the exponent. These individual blind rotations are usually achieved through a multiplexing operation based on s_i, or by applying an appropriate ring automorphism to (an encryption of) X^{s_i}, potentially followed by some key-switching step to ensure all individual blind rotations are encryptions under the same accumulator key, so that the additive homomorphism applies.

3.2 Larger Key Space

Let us first recall the (binary) CMUX, as it is used in the blind rotation procedure of FINAL. It can be seen as the homomorphic version of the Multiplexer (MUX) gate, that has three inputs and one output. Based on the value of the first input (the *selector bit* b) one of the other two input values is returned as the output value. In the homomorphic variant, all the inputs and outputs are ciphertexts. The gate computes $a_b = (b-1) \cdot a_0 + b \cdot a_1 = a_0 + b \cdot (a_1 - a_0)$.

In FINAL, the values a_0, a_1 and a_b are scalar NGS ciphertexts, and b is a vector NGS ciphertext encrypting one bit. A general extension of the CMUX gate for ν values a_i is called an ν-ary CMUX. The selector for this is no longer just a bit, but rather an index with a value between zero and $\nu - 1$. This index can be encoded as ν indicator values b_i of which exactly one has a value of one, and every other one has value zero. Then the generalized CMUX computes

$$\sum_{i=0}^{\nu-1} b_i \cdot a_i$$

which can be rewritten to perform one multiplication fewer and have one indicator (b_0) fewer by remembering that $\sum_i b_i = 1$, namely:

$$a_0 + \sum_{i=1}^{\nu-1} b_i \cdot (a_i - a_0).$$

FINAL's base encryption scheme uses a binary LWE key. Given fixed level of security (say 128 bits of security), we can consider the tradeoff between the LWE dimension n and the size of the key space, or the key distribution χ_{key}. When we increase the key space (or increase the standard deviation for χ_{key}), the dimension n can be lowered without losing the wanted security guarantees. The advantage is that the number of required "individual" blind rotations during bootstrapping lowers, as n of those are performed, but this comes at a higher cost for each such rotation. Thus, by introducing more freedom in the choice of χ_{key} for FINAL, it could be possible to find a better tradeoff.

There are two different approaches in the literature that have introduced a similar extension for TFHE. One uses the aforementioned extension from binary to ν-ary CMux gates [17]. The other makes use of ring automorphisms [19]. The first approach works naturally when it is applied to FINAL. However, we have found the second approach, based on ring automorphisms, to be of limited use for FINAL, due to a seeming incompatibility with the NGS ciphertext format used, that is mostly incompatible with the required key-switching procedure to the accumulator key. [20] manages to bypass this limitation with a similar, but sufficiently different ciphertext format. We discuss this scheme later on in Sect. 4, alongside similar enhancements as discussed in this section.

General Extension of Ternary CMux Gate. In order to deal with a larger secret key space in FINAL, we employ an ν-ary CMux gate in the bootstrapping. We use the idea proposed by Joye and Paillier in [17]. Other literature such as [19] combines existing similar bootstrapping methods, resulting in a scheme that has a smaller noise growth than either. However, this method is slower than the one of [17]. The number of polynomial multiplications required in the bootstrapping algorithm is $2nl$ for [17], and $3nl$ for [19]. It must be noted, however, that the latter algorithm requires less space for key storage.

We now provide a concrete definition for the ν-ary CMux gate, as applied to FINAL's NGS ciphertexts. Let the LWE key $\mathbf{s} \leftarrow [-K, K]$, and let its components s_i be the selector values. We can then define $2K$ keys $\mathtt{bsk}_{i,j}$ for $j = -K, \dots -1, 1, \dots, K$ such that

$$\mathtt{bsk}_{i,j} := \begin{cases} \mathsf{NGS.EncVec}(1) & \text{when } s_i = j \\ \mathsf{NGS.EncVec}(0) & \text{otherwise} \end{cases}$$

The $(2 \cdot K + 1)$-ary CMux gate that computes the blind rotation $X^{c_i \cdot s_i}$ then becomes

$$\mathtt{CMux}_i(c_i) := 1 + \sum_{j=-K, j\neq 0}^{K} (X^{c_i \cdot j} - 1) \cdot \mathtt{bsk}_{i,j} \tag{3}$$

It is clear that for a given c_i, there will never be more than one term in the summation that is nonzero, and that correctness for the blind rotation is achieved.

With this definition, we can now determine a bound on the noise output of a single CMux gate, in function of the noise $\mathsf{Err}(\mathtt{bsk})$ of a fresh vector encryption. Let $c_{\mathsf{Mux}} := \mathtt{CMux}_i(c_i)$ for any $0 \leq i \leq n-1$. Then, the following holds:

$$\begin{aligned} \mathsf{Var}(\mathsf{Err}(c_{\mathsf{Mux}})) &\leq \sum_{k=-K, k\neq 0}^{K} ||X^{c_i} - 1||_2^2 \cdot \mathsf{Var}(\mathsf{Err}(\mathtt{bsk}_{i,k})) \\ &\leq 4K \cdot \mathsf{Var}(\mathsf{Err}(\mathtt{bsk})). \end{aligned}$$

In Algorithm 1, we present the bootstrapping from [2], as applied to our new context, with an updated definition of $\mathtt{CMux}$ and a generalized plaintext space. It can be verified that this algorithm corresponds to the high-level overview at the start of this section, and that therefore the constant term of $\mathsf{msg}(\mathsf{ACC})$ in line 6 is $\lfloor Q/p \rceil \cdot m$[1].

Noise Analysis. Now we consider the noise of the output of this bootstrapping algorithm. In the first line, we switch the modulus of the input ciphertext from q to $2N$. We denote the resulting ciphertext by (a_{2N}, b_{2N}). We then have:

$$\left| \frac{b - \langle \mathbf{a}, \mathbf{s} \rangle}{\Delta} - \frac{q}{2N} \cdot \frac{b_{2N} - \langle \mathbf{a}_{2N}, \mathbf{s} \rangle}{\Delta} \right| \leq \frac{q}{4 \cdot N} \cdot \left| \frac{b - \langle \mathbf{a}, \mathbf{s} \rangle}{\Delta} \right|$$

From line 3 to 5 of the bootstrapping algorithm, the output ACC is obtained by performing n external products with c_{Mux} whose noise variance is $\mathsf{Var}(\mathsf{Err}(c_{\mathsf{Mux}}))$. The variance of the final $\mathsf{Err}(\mathsf{ACC})$ can therefore be bounded by:

$$\begin{aligned} \mathsf{Var}(\mathsf{Err}(\mathsf{ACC})) &\leq n \cdot N \cdot l \cdot \gamma^2 \cdot \mathsf{Var}(\mathsf{Err}(c_{\mathsf{Mux}})) + 4 \cdot ||\mathsf{msg}(\mathsf{ACC})||_2^2 \cdot \zeta^2 \\ &\leq n \cdot N \cdot l \cdot \gamma^2 \cdot \mathsf{Var}(\mathsf{Err}(c_{\mathsf{Mux}})) + 4\, N p^2 \cdot \zeta^2 \end{aligned}$$

[1] For more details on the key switching procedure, we refer the reader to [2]

Algorithm 1: Bootstrapping algorithm for $m \in \mathbb{Z}_p$

Input: $\mathtt{ct} \in \mathbb{Z}_q^{n \times n}$, a base scheme ciphertext encrypting $m \in \mathbb{Z}_p$
$\{\mathtt{bsk}_{i,j}\}_{0 \leq i \leq n-1, j \in [-K,K] \setminus \{0\}}$, bootstrapping keys, where each $\mathtt{bsk}_{i,j} \in \mathcal{R}_{Q,N}^{l}$
$\mathtt{ksk}$, a key-switching key from the NGS secret key $f \in \mathcal{R}$ to the base scheme secret key $\mathbf{s} \in \mathbb{Z}^n$
Output: $\mathtt{ct}' \in \mathbb{Z}_q^n$, a base scheme ciphertext encrypting the same m.

1 $((a_0, \ldots, a_{n-1}), b) \leftarrow \lfloor \frac{2 \cdot N \cdot \mathtt{ct}}{q} \rceil$
2 $\mathsf{ACC} \leftarrow \lfloor \frac{Q}{p} \rceil \cdot X^{-b+N/p} \cdot \sum_{i=0}^{N-1} X^i$;
3 **for** $i = 0 \ldots n-1$ **do**
4 $\quad c_{\mathsf{Mux}} \leftarrow \mathtt{CMux}_i(a_i)$;
5 $\quad \mathsf{ACC} \leftarrow \mathsf{ACC} \boxdot c_{\mathsf{Mux}}$;
6 $\mathsf{ACC} \leftarrow \mathsf{ModSwitch}_{Q \to q}(\mathsf{ACC})$
7 $\mathtt{ct}' \leftarrow \mathsf{KeySwitch}(\mathsf{ACC}, \mathtt{ksk})11^2$
8 **return** $\mathtt{ct}'$

After this step, another modulus switch is applied. Following lemma 6 of [2], this leads to a bound of

$$\begin{aligned}\mathsf{Var}(\mathsf{Err}(\mathsf{ACC})) \leq & (q/Q)^2 \cdot n \cdot N \cdot l \cdot \gamma^2 \cdot \mathsf{Var}(\mathsf{Err}(c_{\mathsf{Mux}})) \\ & + (q/Q)^2 \cdot 4 \cdot Np^2 \cdot \zeta + 1 + 16 \cdot N \cdot \zeta^2\end{aligned}$$

After the external product with a key switching key $\mathtt{ksk}$ in line 7, the noise in the resulting $\mathtt{ct}'$ has a variance

$$\begin{aligned}\mathsf{Var}(\mathsf{Err}(\mathtt{ct})) & \leq N \cdot L \cdot \gamma^2 \cdot \mathsf{Var}(\mathsf{Err}(\mathtt{ksk})) + \mathsf{Var}(\mathsf{Err}(\mathsf{ACC})) \\ & \leq N \cdot L \cdot \gamma^2 \cdot \mathsf{Var}(\mathsf{Err}(\mathtt{ksk})) \\ & + (q/Q)^2 \cdot n \cdot N \cdot l \cdot \gamma^2 \cdot \mathsf{Var}(\mathsf{Err}(c_{\mathsf{Mux}})) \\ & + (q/Q)^2 \cdot 4 \cdot N \cdot \zeta^2 + 1 + 16 \cdot N \cdot \zeta^2\end{aligned} \tag{4}$$

where L is the dimension of the key switching key.

Under the central limit heuristic, we can bound the noise of the output $\mathtt{ct}'$ of this algorithm. Given an LWE ciphertext $\mathtt{ct}$ encrypting a message m, algorithm 1 outputs an LWE ciphertext $\mathtt{ct}'$ encrypting the same message. In addition, under the central limit heuristic, the noise contained in the output behaves as a Gaussian distribution, hence, with overwhelming probability, it satisfies the following bound

$$\begin{aligned}||\mathsf{Err}(\mathtt{ct}')||_2^2 \leq 6^2 \cdot (& N \cdot L \cdot \gamma^2 \cdot \varepsilon_{\mathtt{ksk}} + 4 \cdot K \cdot (q/Q)^2 \cdot n \cdot N \cdot l \cdot \gamma^2 \cdot \varepsilon_{\mathtt{bsk}} \\ & + (q/Q)^2 \cdot 4 \cdot N \cdot \zeta^2 + 1 + 16 \cdot N \cdot \zeta^2)\end{aligned}$$

where $\varepsilon_{\mathtt{ksk}} = O(\mathsf{Var}(\mathsf{Err}(\mathtt{ksk})))$ and $\varepsilon_{\mathtt{bsk}} = O(\mathsf{Var}(\mathsf{Err}(\mathtt{bsk})))$.

Parameters and Efficiency. This analysis of the ν-ary CMux was prompted by the hope that a larger key space could allow for a lower LWE dimension, and hence better efficiency without affecting the security level of the scheme. However, the efficiency does not solely depend on the LWE dimension n, but also on the computational cost of the CMUX evaluation, which grows linearly in K (when we let the secret key be sampled from $[-K, K]$). In particular, we only consider the computational cost of an FFT transformation (either forward or inverse) and that of a pointwise multiplication (in the FFT domain) as significant, and ignore all other costs. Denote the time cost for these as respectively T_{FFT} and T_{PM}. For a bootstrapping operation with LWE dimension n and secret key sampled in $[-K, K]$, we see the following two expensive parts of Algorithm 1: $2Kn\ell \cdot T_{\mathsf{PM}}$ for the CMux evaluations and $n(\ell+1) \cdot T_{\mathsf{FFT}} + n\ell \cdot T_{\mathsf{PM}}$ for the external products, leading to a total cost of

$$\mathsf{cost}_K := n\left((\ell+1) \cdot T_{\mathsf{FFT}} + (2K+1)\ell \cdot T_{\mathsf{PM}}\right).$$

When we compare this to the cost of a FINAL bootstrap with binary keys of LWE (resp. NGS vector ciphertext) dimension n' (resp. ℓ'):

$$\mathsf{cost}_{\mathsf{bin}} := n'\left((\ell'+1) \cdot T_{\mathsf{FFT}} + 2\ell' \cdot T_{\mathsf{PM}}\right).$$

It seems reasonable to assume that an FFT takes longer to compute than a pointwise multiplication of the same length, and as such that $T_{\mathsf{FFT}} > T_{\mathsf{PM}}$, but when increasing K, the corresponding value of n does not decrease as rapidly. Therefore, we can see that any decision as to which variant is preferable will depend on exactly how much faster the pointwise multiplication is. We can quantify this with the value $\alpha := \frac{T_{\mathsf{FFT}}}{T_{\mathsf{PM}}}$, which will depend on the machine used to run the bootstrapping and the chosen FFT implementation.

Here, we provide both a bound on the optimal choice of K, depending on α and the respective dimensions for the ciphertexts, and, in Table 1, an overview of some values α for different FFT libraries. We use degree 2^{14} polynomials for these tests, as that is a reasonable upper bound for practical FHE parameters, as well as resulting in an expectation of a more pronounced asymptotic difference. The actual value of α will hence depend not only on implementation and hardware efficiency, but also on the choice of parameters. These measurements were all made on a `AMD Ryzen 5 PRO 3400G` CPU, in a single threaded context. Where possible, a choice for a real-to-complex variant of the FFT was made, to better match the usage in the scheme. Theoretically, the optimal choice for K would correspond to

$$K < \frac{\alpha(n'(\ell'+1) - n(\ell+1)) + 2n'\ell' - n\ell}{2nl}.$$

For the sake of intuition, if we assume the extremely generous situation where we could achieve $(n, \ell) = (n'/2, \ell')$ with $\ell' = 1$, the obtained bound is $K < \alpha + \frac{3}{2}$, implying that T_{FFT} should be roughly K times as slow as T_{PM}. As can be seen in Table 1, even under this generous assumption, the gain when choosing an efficient FFT implementation is very limited.

Table 1. Observed running times for FFT and pointwise multiplication under different programming languages and for different FFT libraries, running on degree 2^{14} polynomials.

Language	FFT Library	T_{FFT} (ms)	T_{PM} (ms)	$\alpha := \frac{T_{\mathsf{FFT}}}{T_{\mathsf{PM}}}$
C++	spqlios	59	21	2.81
C++	ffnt	245		11.67
C++	fftw	46		2.19
C++	nayuki	165		7.86
Rust	concrete_fft	86	26	3.31
Rust	concrete_fftw	42		1.62

The Issue with Automorphisms. The other approach to introduce larger keyspaces, as considered in [19], makes use of ring automorphisms on $\mathcal{R}$: $\mathtt{ct}(X) \mapsto \mathtt{ct}(X^k)$. These are cheap to compute, as they only involve a reordering and some sign changes of the coefficients of a polynomial. However it also changes the encryption key from $f(X)$ to $f(X^k)$. To account for this, the automorphism should be followed by a key switching procedure back to $f(X)$. That is, we want some (efficient) procedure $\mathsf{NGS.EncS}(f(X), m) = \mathsf{KeySwitch}(\mathsf{NGS.EncS}(f(X^k), m), \mathtt{ksk})$ that does not change the underlying message. Additionally, the additional noise from the key switching should be limited to fit within the bootstrapping noise budget.

In other words, we start with a scalar NGS ciphertext which is a form of $c := g/f + \Delta m$, and end up with another scalar NGS ciphertext $g'/f' + \Delta m$. An insightful approach is to do an external product with an NGS vector ciphertext encrypting f under f'. If we limit ourselves to choices of $\mathtt{ksk}$ such that $\mathsf{KeySwitch}(c, \mathtt{ksk}) = c \boxdot \mathtt{ksk}$, as an arbitrary limit to mean "efficient" that simultaneously giving an easy evaluation of the noise growth, it can be seen that the result generally falls into either of two cases. It does not have the correct form to be a scalar NGS ciphertext, or the noise growth is too high.

It should be noted that [20] manages to perform a blind rotation with ring automorphisms, by choosing an alternative scalar ciphertext form that allows for key switching with an external product: $Enc_f(m) = \frac{\tau \cdot g + \Delta \cdot u}{f} \in \mathcal{R}_Q$. We extend our results for FINAL to this alternative form in Sect. 4 and afterwards compare the resulting FHE schemes in Sect. 5.

3.3 Approximate Decomposition

Recall the definition of a scalar NGS ciphertext encrypting a message m_0. We now define a *noisy scalar NGS ciphertext*, which has an extra noise term $e \in \mathcal{R}_Q$:

$$\hat{c} = \mathsf{NGS.EncS}_e(f, m_0) = \frac{g}{f} + \Delta \cdot m_0 + e.$$

The decryption error for a noisy ciphertext then becomes $g + e \cdot f$, and in order to have the additional error due to e be comparable to the error also present in a regular ciphertext, we introduce the following condition, to enforce that the variance of the newly introduced error term does not exceed the noise already present normally:

$$\mathsf{Var}(e) \leq \frac{\mathsf{Var}(g)}{N\mathsf{Var}(f)} \tag{5}$$

The external product can be extended to cover multiplication between a noisy scalar ciphertext and a (regular) vector ciphertext encrypting m_1 without any further changes. It can be verified that the result is again a noisy scalar NGS ciphertext.

$$\begin{aligned}\hat{c} \boxdot \mathbf{c} = \langle \mathbf{g}_B^{-1}(\hat{c}), \mathbf{c} \rangle &= \sum_{i=0}^{\ell-1} \frac{c_i g_i}{f} + \hat{c} m_1 \\ &= \frac{\sum_{i=0}^{\ell-1} c_i g_i + m_1 g}{f} + \Delta m_0 m_1 + m_1 e \\ &= \frac{\bar{g}}{f} + \Delta m_0 m_1 + \bar{e} \end{aligned} \tag{6}$$

We can now also define the *approximate decomposition* of a (either noisy or regular) scalar ciphertext c as the decomposition $\mathbf{g}_B^{-1}(c)$, with the k least significant digits set to 0. Alternatively, this can be expressed as the decomposition $\mathbf{g}_B^{-1}(\hat{c})$ of the noisy scalar ciphertext $\hat{c} = c - [c]_{B^k}$. When we consider the external product that uses an approximate decomposition, or equivalently, the external product $\hat{c} \boxdot \mathbf{c}$,[3] we see from Eq. 6 that the resulting noise terms become $\bar{g} = \sum_{i=k}^{l-1} c_i g_i + m_1 g$ and $\bar{e} = m_1(e - [c]_{B^k})$. We can hence compute a bound for the variance of the resulting errors, where we assume $\mathbf{c}$ to be a fresh vector encryption.

$$\begin{aligned}\mathsf{Var}(\bar{g}) &\leq \mathsf{Var}(m_1 g) + \mathsf{Var}\left(\sum_{i=k}^{\ell-1} c_i g_i\right) \\ &\leq \mathsf{Var}(g) + NB^2(\ell - k)\mathsf{Var}(g_i) \\ &\leq p^2\mathsf{Var}(g) + NB^2(\ell - k)\mathsf{Var}(f)\end{aligned}$$

$$\begin{aligned}\mathsf{Var}(\bar{e}) &\leq \mathsf{Var}(m_1 e) + \mathsf{Var}(m_1 [c]_{B^k}) \\ &\leq \frac{p^2\mathsf{Var}(g)}{N\mathsf{Var}(f)} + p^2 B^k\end{aligned}$$

[3] We allow the ciphertext being decomposed approximately to already be noisy, so the extra noise term of $\hat{c}$ becomes $e - [c]_{B^k}$.

Entering this into inequality 5, we get a bound on k, the number of digits that can be ignored by the approximate decomposition:

$$\frac{p^2\mathsf{Var}(g)}{N\mathsf{Var}(f)} + p^2B^{2k} \le \frac{p^2\mathsf{Var}(g) + NB^2(\ell - k)\mathsf{Var}(f)}{N\mathsf{Var}(f)}$$
$$B^{2k-2} \le \ell - k$$
$$k \le \frac{1}{2}\log_B(\ell - k) - log_B(p) + 1$$

The advantage gained from this external product with approximate decomposition is that we can omit the k least significant components of any vector ciphertext involved as well, without further effect. In practice, this amounts to a saving of k forward FFT computations and k pointwise multiplications. The computation and memory costs are now identical to having only vector dimension $d - k$, rather than d. This does come at the cost of a limited noise growth, which has no practical impact since a bootstrapping operation is performed after every homomorphic operation.

3.4 Programmable Bootstrapping

After combining the bootstrapping algorithm with plaintext space $\mathbb{Z}_p$ as in Algorithm 1 and enhancing it with a free choice of ν-ary CMux gates and approximate decompositions, we can introduce programmable bootstrapping (PBS) into the mix. Recall that the bootstrapping procedure makes use of a blind rotation followed by a keyswitch back to the base scheme. The goal of the blind rotation is to rotate the coefficients of a test polynomial $T(X) \in \mathcal{R}_Q$ by a decryption of an LWE ciphertext, in such a way that the constant coefficient of the underlying plaintext polynomial ends up containing the decryption of the LWE ciphertext. In the description of Algorithm 1, the initial value of ACC represents (a noiseless encryption of) the polynomial $T(X)$.

When considering PBS, this notion is enhanced such that the constant coefficient no longer just contains the message m, but the evaluation of an arbitrary function $F(m)$, usually represented by a lookup table of size p.[4]

To encode the function or lookup table F into a polynomial $T(X)$, we observe that

$$(T \cdot X^{\langle \mathbf{a}, \mathbf{s} \rangle - b})(0) = F(m) = F\left(\left\lfloor \frac{\langle \mathbf{a}, \mathbf{s}, \rangle - b}{\Delta/2} \right\rceil\right)$$

and so, when we let $T(X) = t_0 + t_1X + \ldots + t_{N-1}X^{N-1}$, choosing

$$t_i = F\left(\left\lfloor \frac{2i}{\Delta} + \frac{1}{2} \right\rceil\right)$$

[4] Since the memory and computational cost for FHE tend to grow rapidly as the plaintext modulus p grows, the memory requirements to represent and store this lookup table are easily met. In practice, $\mathbb{Z}_p$ will seldom be much larger than a single byte at once.

satisfies this requirement. We note here that we consider the decryption after a modulus switch to $2N$, and that as such $\Delta = 2N/p$ here. Applying this construction to the identity function $F(x) = x$, we indeed recover the previous non-programmable bootstrapping algorithm. When the blind rotation and keyswitching procedures work correctly and with appropriate noise growth, it then follows that the addition of PBS preserves correctness and adds no further noise growth to the final bootstrapped ciphertext.

Note that for LWE, here we use $\frac{\Delta}{2}$ rather than Δ. This extra bit is needed to allow programmable bootstrapping with non-negacyclic functions. Alternatively, when F is a negacyclic function, its compatibility with the negacyclic nature of the cyclotomic ring $\mathcal{R}$ permits some extra efficiency in the plaintext space within ciphertexts. It is then possible to use the regular choice of $\Delta = q/p$ (or $\Delta_{2N} = 2N/p$ after modulo switching) for the LWE ciphertexts.

4 Improvements for XZD+23

Since [20] follows a comparable blueprint to [2], with a modification to the scalar NTRU ciphertext and the introduction of ring automorphisms in the blind rotation, the natural question becomes whether the modification made for FINAL can also be applied to the XZD+23 scheme. By construction, the scheme already supports arbitrary distributions for the LWE key, so we need only consider the approximate decomposition and the programmable bootstrapping.

4.1 Approximate Decomposition

Parallel to the approximate decomposition for FINAL, we first recall that a scalar ciphertext encrypting a message m_0 has the form

$$c = \frac{\tau g + \Delta m_0}{f}$$

and then define a *noisy scalar ciphertext* to have an additional noise term, such that

$$\hat{c} = \frac{\tau g + \Delta m_0}{f} + e.$$

The external product with a vector ciphertext $\mathbf{c} = \tau \frac{\mathbf{g}}{f} + \mathbf{g} m_1$ then becomes

$$\begin{aligned}
\hat{c} \boxdot \mathbf{c} = \langle \mathbf{g}_B^{-1}(\hat{c}), \mathbf{c} \rangle &= \sum_{i=0}^{\ell-1} \tau \frac{\hat{c}_i g_i}{f} + \hat{c} m_1 \\
&= \frac{\tau \left(\sum_{i=0}^{\ell-1} \hat{c}_i g_i + m_1 g \right) + \Delta m_0 m_1}{f} + m_1 e \\
&= \frac{\tau \bar{g} + \Delta m_0 m_1}{f} + \bar{e} \qquad (7)
\end{aligned}$$

To ensure the extra noise e does not grow too big or break correctness, we seek to bound its variance by that of g, and as such require that

$$\mathsf{Var}(e) \leq \frac{\tau^2 \mathsf{Var}(g)}{N\mathsf{Var}(f)} \tag{8}$$

Now let $\hat{c}$ represent an approximation of some (potentially noisy) ciphertext c, such that $\hat{c} = c - [c]_{B^k}$. That is, $\mathbf{g}_B^{-1}(\hat{c})$ corresponds to $\mathbf{g}^{-1}(c)$ with the k least significant digits set to zeros. Then we can additionally assume that the k corresponding elements of $\mathbf{c}$ are arbitrary, or even absent, which means we need less memory to store $\mathbf{c}$ and less time to compute $\hat{c} \boxdot \mathbf{c}$. From 7, we can determine bounds on the variance of the corresponding $\bar{g}$ and $\bar{e}$, under the assumption that $\mathbf{c}$ is a fresh vector ciphertext and that hence $\mathsf{Var}(g_i) = \mathsf{Var}(f)$.

$$\mathsf{Var}(\bar{g}) \leq p^2 \mathsf{Var}(g) + NB^2(\ell - k)\mathsf{Var}(f)\mathsf{Var}(\bar{e}) \leq \frac{\tau^2 p^2 \mathsf{Var}(g)}{N\mathsf{Var}(f)} + p^2 B^k \tag{9}$$

If we then further constrain $\mathsf{Var}(\bar{g})$ and $\mathsf{Var}(\bar{e})$ to follow 8, we derive

$$k \leq \frac{1}{2}\log_B(\ell - k) - log_B(p) + 1 \tag{10}$$

as a limit on the number of components we can omit from any stored vector ciphertexts and from the external product calculations. It can be observed that the gain in efficiency is equivalent to that which we achieve with the optimization for FINAL. It should be noted that XZD+23 can also be easily adapted to support programmable bootstrapping following the same blueprint as described in Sect. 3.4.

5 Comparison

Comparing performance of different FHE schemes without implicitly comparing implementations along with their respective amount of processor-specific optimizations and implementation-defined peculiarities is all but impossible. We eschew this approach in favor of a more theoretical comparison. In particular, we examine the external products performed during a single bootstrapping operation, as that is generally the most expensive operation to be performed. We quantify the number of FFT operations (either forward or inverse) and the number of polynomial pointwise products in the FFT domain, incurred by the external product and otherwise. As discussed for the evaluation of our ν-ary CMux in Sect. 3.2, it is possible to express the difference in timing between these two operations by some factor α, but since this factor again highly depends on choice of implementation and hardware, we do not further combine these two terms in our comparison. We present a side-by-side comparison of these metrics for the FINAL and XZD+23 schemes in Table 2. Note that in XZD+23, the authors report a clear improvement when compared to TFHE based on implementations of both schemes. The difference of this comparison is more pronounced than we would expect from our theoretical results counting the number of required FFT and pointwise multiplications.

For the memory requirements of a bootstrapping operation, we refer solely to the size of the bootstrapping or evaluation keys needed, as the keyswitching procedure from the accumulator back to the base ciphertext is similar enough between all examined schemes. We express all memory sizes in amount of complex numbers needed,[5] as everything can be precomputed and stored in the FFT domain. These numbers are also presented in Table 2.

As indicated, parameter choices will differ between schemes. In Table 3 we give the LWE and NTRU parameters for our adapted FINAL scheme for 2, 4 and 8 message bits respectively. The LWE parameters were chosen equal to those of TFHE-rs (for the same message spaces, respectively). FINAL will achieve (at least) the same message accuracy p, as FINAL has less noise for the same parameters. Next, we ensure a security level of 128 bits, by defining the NTRU parameters using the NTRU Fatigue estimator [9]. We furthermore guarantee a maximal decryption error probability of 2^{-40}, with respect to the noise bound on a bootstrapped ciphertext (Eq. 4).

Finally, in Table 4, we show how increasing the key space influences the LWE dimension, and how this in turn influences the run-time of the bootstrapping algorithm. We find that α values under 10, it is never beneficial for the performance to have $K > 1$. Also, as given in Table 2, larger K also increases the required memory.

It is difficulty to fairly compare our improved version of FINAL with that of XDZ+23 due to effects of the increased key space on the LWE dimension. If we use a uniform distribution over $[-3, 3]$ (that is, we set $K = 3$) for our version of FINAL, as an approximation of the Gaussian distribution with variance $4/3$ (as is given as a parameter set in [20]), we can compare our schemes both with our theoretical results and with their experimental outcome. We take the parameter set described by XZD+23 as *P128G*, which is more lenient having allowing a decryption error probability up to 2^{-34}. We then find, from the formulas in Table 2, that for this setting, our algorithm requires one third of the FFTs, but 2.3 times the number of pointwise multiplications, while consuming 16.5% less memory. For α-values larger than 0.96. The instantiation of FINAL with our modifications is hence more performant than our instantiation of XZD+23.

The effects of our modifications on the two schemes are not just found in performance, as they expand the functionality as well. Still, in both schemes, using the approximate decomposition will grant a decrease of ℓ with at least 1, depending on the other parameters. Furthermore, we repeat that the ν-ary CMux provides a trade-off between pointwise multiplications and FFTs, which could be beneficial depending on the FFT library used. The performance of unaltered FINAL can be found by fixing K in Table 2.

By examining the noise growth for each individual scheme, it can be observed that for binary LWE keys, the parameter choice be close to identical. When growing the key space either by the application of ν-ary CMux gates or by taking

[5] This would usually be equivalent to 16 bytes each.

Table 2. Overview of the number of FFTs, pointwise multiplications (PM) and memory for the bootstrapping and evaluation keys (expressed in amount of complex numbers) per FHE scheme. A + superscript indicates a scheme incorporating our improvements.

Scheme	#FFT	#PM	Memory
FINAL$^+$	$n_F(\ell_F + 1)$	$(2K + 1)n_F\ell_F$	$(2K)n_F N_F \ell_F$
XZD+23$^+$	$2n_X(\ell_X + 1)$	$2n_X\ell_X$	$N_X^2\ell_X$

Table 3. The parameters for FINAL with our modifications, for different plaintext size p, using ternary secret keys.

p	n	q	B	ℓ	N	Q
2^2	656	2^{64}	2^8	2	2^{10}	2^{20}
2^4	742	2^{64}	2^{23}	1	2^{10}	2^{20}
2^8	1017	2^{64}	2^{15}	2	2^{11}	2^{30}

advantage of efficient ring automorphism computation being possible, the LWE dimension can be reduced without dropping below a wanted security threshold. It is however the case that this improvement does not generally weigh up against the constant factors seen in the different counts of operations. The XZD+23 scheme has the advantage that its memory usage is entirely independent from the choice of key space, and its running time only decreases as the LWE dimension decreases. For the tradeoffs involved in the adaptation of FINAL to larger key sizes, we refer the reader back to Sect. 3.2.

To get an idea about the concrete memory requirements presented here, we need to be able to compare some typical values for the NTRU dimension N and the LWE dimension n, as they have a strong impact on the interpretation of Table 2. For a security target level of 128 bits, a typical choice of n would be some value $2^9 \leq n \leq 2^{10}$. The NTRU dimension N needs to grow more as the plaintext space $\mathbb{Z}_p$ grows, though it must remain a power of two to preserve correctness. For $p = 2$,[6] the choice $N = 2^{10}$ is made. It then grows along a growing p, and for moderately sized p, already values such as $N = 2^{14}$ are required. We thus observe that generally $N > n$, and a term quadratic in N in the memory requirements can quickly become prohibitive for larger plaintext sizes.

Table 4. The influence of key space K on n, and the minimum value of α where run-time is better than $K = 1$.

p	2^2				2^4				2^8			
K	2	3	5	10	2	3	5	10	2	3	5	10
n	568	550	530	500	656	635	615	585	935	915	890	860
α	10.9	18.8	31.7	55.7	13.3	21.8	36.7	65.1	20.8	33.9	54.1	96.6

[6] That is, the case where no PBS is possible or needed, as described in [2,20].

6 Conclusion

In this paper, we have suggested improvements for the two leading FHE schemes with NTRU-based bootstrapping, FINAL [2] and XZD+23 [20]. For both schemes, we described an approximate decomposition technique that allows improvements in both memory requirements and the performance of external products, by decreasing the dimensions of vector ciphertexts. Additionally, we established the feasibility of increases in the size of the plaintext space, further enabling programmable bootstrapping for arbitrary functions $\mathbb{Z}_p \to \mathbb{Z}_p$. In FINAL, we established the potential for an increase in the key space of the base scheme, based up techniques by Joye and Paillier [17], finding its application to provide only limited benefits.

Finally, we provided a comparison of the two main approaches to an increased key space for NTRU-based FHE. In this, we found FINAL to be the better choice in most choices, due to a constant factor difference in amount of required external products and the difference in memory required for the bootstrapping keys, despite the increase in key size coming at a higher cost than XZD+23 in general.

Acknowledgements. This work has been supported in part by the FWO under an Odysseus project GOH9718N, and CyberSecurity Research Flanders with reference number VR20192203.

References

1. Alperin-Sheriff, J., Peikert, C.: Faster bootstrapping with polynomial error. In: Garay, J.A., Gennaro, R. (eds.) CRYPTO 2014. LNCS, vol. 8616, pp. 297–314. Springer, Heidelberg (2014). https://doi.org/10.1007/978-3-662-44371-2_17
2. Bonte, C., Iliashenko, I., Park, J., Pereira, H.V.L., Smart, N.P.: FINAL: faster FHE instantiated with NTRU and LWE. In: Agrawal, S., Lin, D. (eds.) ASIACRYPT 2022, Part II. LNCS, vol. 13792, pp. 188–215. Springer, Heidelberg (2022). https://doi.org/10.1007/978-3-031-22966-4_7
3. Brakerski, Z., Gentry, C., Vaikuntanathan, V.: (Leveled) fully homomorphic encryption without bootstrapping. In: Goldwasser, S. (ed.) ITCS 2012, pp. 309–325. ACM, January 2012. https://doi.org/10.1145/2090236.2090262
4. Brakerski, Z., Vaikuntanathan, V.: Efficient fully homomorphic encryption from (standard) LWE. In: Ostrovsky, R. (ed.) 52nd FOCS, pp. 97–106. IEEE Computer Society Press, October 2011. https://doi.org/10.1109/FOCS.2011.12
5. Chen, H., Han, K.: Homomorphic lower digits removal and improved FHE bootstrapping. In: Nielsen, J.B., Rijmen, V. (eds.) EUROCRYPT 2018. LNCS, vol. 10820, pp. 315–337. Springer, Cham (2018). https://doi.org/10.1007/978-3-319-78381-9_12
6. Chillotti, I., Gama, N., Georgieva, M., Izabachène, M.: Faster fully homomorphic encryption: bootstrapping in less than 0.1 seconds. In: Cheon, J.H., Takagi, T. (eds.) ASIACRYPT 2016. LNCS, vol. 10031, pp. 3–33. Springer, Heidelberg (2016). https://doi.org/10.1007/978-3-662-53887-6_1

7. Chillotti, I., Joye, M., Paillier, P.: Programmable bootstrapping enables efficient homomorphic inference of deep neural networks. In: Dolev, S., Margalit, O., Pinkas, B., Schwarzmann, A. (eds.) CSCML 2021. LNCS, vol. 12716, pp. 1–19. Springer, Cham (2021). https://doi.org/10.1007/978-3-030-78086-9_1
8. Ducas, L., Micciancio, D.: FHEW: bootstrapping homomorphic encryption in less than a second. In: Oswald, E., Fischlin, M. (eds.) EUROCRYPT 2015. LNCS, vol. 9056, pp. 617–640. Springer, Heidelberg (2015). https://doi.org/10.1007/978-3-662-46800-5_24
9. Ducas, L., van Woerden, W.: NTRU fatigue: how stretched is overstretched? In: Tibouchi, M., Wang, H. (eds.) ASIACRYPT 2021. LNCS, vol. 13093, pp. 3–32. Springer, Cham (2021). https://doi.org/10.1007/978-3-030-92068-5_1
10. Geelen, R., Iliashenko, I., Kang, J., Vercauteren, F.: On polynomial functions modulo p^e and faster bootstrapping for homomorphic encryption. In: Hazay, C., Stam, M. (eds.) EUROCRYPT 2023, Part III. LNCS, vol. 14006, pp. 257–286. Springer, Heidelberg (2023). https://doi.org/10.1007/978-3-031-30620-4_9
11. Genise, N., Micciancio, D., Polyakov, Y.: Building an efficient lattice gadget toolkit: subgaussian sampling and more. In: Ishai, Y., Rijmen, V. (eds.) EUROCRYPT 2019. LNCS, vol. 11477, pp. 655–684. Springer, Cham (2019). https://doi.org/10.1007/978-3-030-17656-3_23
12. Halevi, S., Shoup, V.: Bootstrapping for `HElib`. In: Oswald, E., Fischlin, M. (eds.) EUROCRYPT 2015. LNCS, vol. 9056, pp. 641–670. Springer, Heidelberg (2015). https://doi.org/10.1007/978-3-662-46800-5_25
13. Halevi, S., Shoup, V.: Bootstrapping for HElib. J. Cryptol. **34**(1), 7 (2021). https://doi.org/10.1007/s00145-020-09368-7
14. Jeon, S., Lee, H.S., Park, J.: Efficient lattice gadget decomposition algorithm with bounded uniform distribution. IEEE Access **9**, 17429–17437 (2021). https://doi.org/10.1109/ACCESS.2021.3053288
15. Jeon, S., Lee, H.S., Park, J.: Practical randomized lattice gadget decomposition with application to FHE. In: Tsudik, G., Conti, M., Liang, K., Smaragdakis, G. (eds.) Computer Security - ESORICS 2023, pp. 353–371. Springer, Cham (2023). https://doi.org/10.1007/978-3-031-50594-2_18
16. Joye, M.: On NTRU-ν-um modulo $X^N - 1$. Cryptology ePrint Archive, Report 2022/1092 (2022). https://eprint.iacr.org/2022/1092
17. Joye, M., Paillier, P.: Blind rotation in fully homomorphic encryption with extended keys. In: Dolev, S., Katz, J., Meisels, A. (eds.) CSCML 2022, vol. 13301, pp. 1–18. Springer, Cham (2022). https://doi.org/10.1007/978-3-031-07689-3_1
18. Kluczniak, K.: NTRU-v-um: secure fully homomorphic encryption from NTRU with small modulus. In: Yin, H., Stavrou, A., Cremers, C., Shi, E. (eds.) ACM CCS 2022, pp. 1783–1797. ACM Press, November 2022. https://doi.org/10.1145/3548606.3560700
19. Lee, Y., Micciancio, D., Kim, A., Choi, R., Deryabin, M., Eom, J., Yoo, D.: Efficient FHEW bootstrapping with small evaluation keys, and applications to threshold homomorphic encryption. In: Hazay, C., Stam, M. (eds.) EUROCRYPT 2023, Part III. LNCS, vol. 14006, pp. 227–256. Springer, Heidelberg (2023). https://doi.org/10.1007/978-3-031-30620-4_8
20. Xiang, B., Zhang, J., Deng, Y., Dai, Y., Feng, D.: Fast blind rotation for bootstrapping FHEs. In: Handschuh, H., Lysyanskaya, A. (eds.) CRYPTO 2023, Part IV. LNCS, vol. 14084, pp. 3–36. Springer, Heidelberg (2023). https://doi.org/10.1007/978-3-031-38551-3_1

An Efficient Integer-Wise ReLU on TFHE

Yi Huang[1], Junping Wan[1], Zoe L. Jiang[1,2](✉), Jun Zhou[3], Junbin Fang[4], and Zhenfu Cao[3]

[1] Harbin Institute of Technology, ShenZhen, China
{huangyi,wanjunping97}@stu.hit.edu.cn, zoeljiang@hit.edu.cn

[2] Guangdong Provincial Key Laboratory of Novel Security Intelligence Technologies, Shenzhen, China

[3] East China Normal University, Shanghai, China
{jzhou,zfcao}@sei.ecnu.edu.cn

[4] Jinan University, Guangzhou, China
tjunbinfang@jnu.edu.cn

Abstract. Fully homomorphic encryption (FHE) enables users to process encrypted data, while preserving data privacy throughout the data computation process. It develops ways to privately execute neural networks. Although bit-wise FHE over the torus (TFHE) was originally proposed to support non-linear functions, such as ReLU operation which is often used in neural networks, the computational complexity of the homomorphic ReLU operation is linearly to data precision. Integer-wise TFHE enables integer bootstrapping with homomorphic addition. However, it leaves an open problem to support homomorphic multiplication and ReLU due to negacyclicity limitation. In this paper, we first propose the $ExMultbyBin(x)$ algorithm for integer-wise multiplication by extending the data range from $\{0, \cdots, B/2-1\}$ to $\{-B, \cdots, B-1\}$. Then, we propose the idea of function transformation by equivalently transform the $ReLU(x)$ function to a new function $ExMultbyBin(x, f_{id}(x), sign(x) - B/2)$. Finally, we achieve a privacy-preserving ReLU function *IntReLU* with *integer-wise* TFHE, resulting in computational complexity *independent* of data precision. That is, when the data precision is n-bit, *IntReLU* has a computational complexity of $\mathcal{O}(1)$. Experimental results in the TFHE library indicate that, the operation time of our *intReLU* is reduced by 17% when the data precision is 6-bit compared to the bit-wise TFHE scheme.

Keywords: Fully homomorphic encryption (FHE) · FHE over the torus (TFHE) · ReLU

1 Introduction

Fully homomorphic encryption (FHE) allows performing homomorphic mathematical operations on encrypted data. Upon decrypting, the results match those obtained by operating on the corresponding plaintext data. Applying FHE in

T. Zhu and Y. Li (Eds.): ACISP 2024, LNCS 14895, pp. 161–179, 2024.
https://doi.org/10.1007/978-981-97-5025-2_9

cloud computing enables the separation of data processing rights and data ownership. Client retains data ownership, while cloud server holds data processing rights, thereby achieving outsourced privacy computation in cloud computing [1,2]. Specifically, the client encrypts local private data before uploading it to the cloud server. The cloud server can perform large-scale homomorphic operations on the ciphertext and then send the encrypted computation results back to the client. After decryption, the client can obtain the computed results.

In 2009, Gentry creatively proposed the first FHE scheme [3] based on the hardness problem on ideal lattices. Since then, numerous FHE schemes have been introduced, categorized into the first [3–6], second [7–9], and third generations [10–12]. Currently, mainstream FHE schemes include the Cheon-Kim-Kim-Song (CKKS) scheme [9] and Fully Homomorphic Encryption over the Torus (TFHE) scheme [12]. The CKKS scheme supports homomorphic arithmetic operations on real or complex numbers, suitable for linear operations such as matrix computations, yielding approximate results. Although CKKS can support homomorphic addition, multiplication, and polynomial function operations with relatively high precision, it does not directly support non-linear function homomorphic operations like ReLU. To handle non-linear functions, CKKS needs to approximate them using complex polynomial functions [13–15], which significantly increases computational costs. On the other hand, TFHE supports homomorphic Boolean operations on bit-wise data, making it suitable for non-linear functions. TFHE extends integers to the torus based on the Learning With Errors (LWE) problem [16], replacing internal product operations with external ones, significantly improving bootstrapping.

In TFHE scheme, each boolean operation is accompanied by a bootstrapping operation. Hence its bootstrapping operation is also referred to as gate bootstrapping or bit bootstrapping. Gate bootstrapping allows TFHE to directly support non-linear functions. However, the complexity of circuits also increases the computation time of the TFHE scheme. For example, when Lou et al. [17] implemented the ReLU function using TFHE, they needed to encrypt and operate on data bit by bit, i.e., taking the complement of the sign bit ciphertext and performing AND gate operations with each of the remaining bits. When the precision of plaintext data is n, gate bootstrapping needs to be performed $n-1$ times, resulting in a linear relationship between computational complexity and data precision. Thus, Bourse et al. [18] modified TFHE to achieve integer encryption. However, the modified scheme only supports integer addition and sign function. Subsequently, Okada et al. [19] proposed integer-wise TFHE based on TFHE, eliminating the need to encrypt and operate on integer data bit by bit, and directly supporting simple operations like MultbyBin and division on ciphertexts of positive integer data in the range $\{0, \cdots, B-1\}$.

However, integer-wise TFHE does not directly support ReLU operation because the TFHE scheme has the negacyclicity property, i.e., $f(x) = -f(x+B)$ when $x \in \{-B, \cdots, -1\}$. Recently, many works have extended TFHE to integers and proposed Full Domain Functional Bootstrap (FDFB). Kluczniak et al. [20] decomposed function $F(x)$ on the interval $\{-B, \cdots, B-1\}$ into two negacyclic functions $F_0(x)$ and $F_1(x)$. By this, a multiplication between a GSW ciphertext of one-bit plaintext b and a LWE ciphertext needs $log(Q)$ bootstrappings (Q represents the plaintext modulus), such that $F(x)$ is obtained by $GSW(1-b) \cdot LWE(F_0(x)) + GSW(b) \cdot LWE(F_1(x)) = F_b(x) = F(x)$. Yang et al. [21] introduced a function bootstrapping algorithm that can perform arbitrary function computations on $\mathbb{Z}_N$ with 2 bootstrappings and 1 addition. However, this work can only address bootstrapping on the domain $\{0, \cdots, B-1\}$. Pierre et al. [22] achieved homomorphic operations on $\{-B, \cdots, B-1\}$ through bootstrapping for both pseudo-odd and pseudo-even functions. They implemented homomorphic computations for arbitrary functions based on pseudo-odd and pseudo-even functions. While the work itself indicates the need for 4 bootstrappings, we find that it overlooks 2 more bootstrappings for the division by 2 operation.

1.1 Our Contributions

This paper proposes an efficient integer-wise homomorphic ReLU operation *IntReLU* on TFHE. It mainly comprises two contributions:

On one hand, the existing integer-wise TFHE scheme [19] employs the homomorphic multiplication algorithm *MultbyBin* based on integer bootstrapping, supporting homomorphic operations only for plaintext integers in the range $\{0, \cdots, B/2-1\}$. Consequently, errors occur in homomorphic operations when plaintext integers are in the range $\{-B, \cdots, -1\} \cup \{B/2, \cdots, B-1\}$. This paper introduces the *ExMultbyBin* algorithm, which addresses this issue by supporting homomorphic multiplication operations for negative integer plaintexts in the range $\{-B, \cdots, -1\}$ through *linear transformation* and for positive integer plaintexts in the range $\{B/2, \cdots, B-1\}$ through *odd slot mapping*. This enables homomorphic multiplication operations for all integer plaintexts in the range $\{-B, \cdots, B-1\}$.

Linear Transformation: In TFHE, due to the negacyclicity limitation ($f(x) = -f(x+B)$), when $x \in \{-B, \cdots, -1\}$, $f(x)$ is assigned as $-f(x+B)$, inconsistent with the expected $f(x)$. That is why TFHE cannot support negative integer. For instance, let $f(x) = x$, $B = 8$, then $f(-2) = -f(6) = -6 \neq f(-2) = -2$. Intuitively to solve it for $x \in \{-B, \cdots, -1\}$, the obtained $-f(x+B)$ can be simply nested with 1 additional f operation to get $f(-f(x+B)) = f(x)$. However, the nested f operation is equivalent to execute 1 extra bootstrapping operation. We find that $-f(x+B) - B = -f(x)$ when $f(x) = x$. Therefore, this paper utilizes $-(-f(x+B) - B)$, i.e., one subtraction and one negation to replace 1 extra bootstrapping, which we called the linear transformation.

Odd Slot Mapping: In TFHE, the *MultbyBin* algorithm produces intermediate results that can only be even, i.e., it can only take values in even slots. Consequently, all B numbers in the set $\{0, \cdots, B-1\}$ are mapped to $B/2$ even slots in the range $\{0, \cdots, B-1\}$. This inevitably results in 2 numbers being mapped to the same slot, causing errors in operations. Therefore, this paper transforms the $B/2$ numbers in the range $\{B/2, \cdots B-1\}$ by mapping them to odd slots, ensuring a one-to-one mapping and guaranteeing the correctness of operations.

On the other hand, ReLU is a piecewise function. Specifically, when $x < 0$, $ReLU(x) = 0$, and when $x \geq 0$, $ReLU(x) = f_{id}(x) = x$. For binary plaintext, the ReLU function based on TFHE is implemented by performing bit-wise AND operations on each bit after flipping the sign bit. Each gate operation requires a bootstrapping process, and for an input data precision of n, it requires $n-1$ bootstrappings. This paper introduces the concept of equivalent function transformation and proposes a new function, $ExMultbyBin(x, f_{id}(x), sign(x)-B/2)$, which is equivalent to the $ReLU(x)$ function. This approach ensures that the complexity of the homomorphic ReLU operation is independent of the data precision n. Further efficiency is achieved by introducing the PBSManyLUT technique [23], reducing one homomorphic operation. Specifically, for integer plaintext, this paper first introduces a new function, $ExMultbyBin(x, f_{id}(x), sign(x) - B/2)$, which is equivalent to the ReLU function. This transformation converts the ReLU function into the multiplication of integer ciphertext and binary ciphertext (0 or $-B$). By performing f_{id} and $sign$ operations on the integer x ciphertext separately, applying a simple subtraction to the sign operation result, mapping it to binary ciphertext of 0 or $-B$, and then invoking the $ExMultbyBin$ algorithm for the multiplication of integer ciphertext and binary ciphertext (0 or $-B$), the ReLU function is effectively implemented. The efficient $intReLU$ algorithm proposed in this paper requires a total of 5 bootstrappings, and with the introduction of the PBSManyLUT technique, it reduces to 4 bootstrappings.

1.2 Organization

After introducing the background knowledge of TFHE and the basic concepts of integer-wise TFHE in Sects. 2 and 3 outlines the $intReLU$ algorithm for implementing the ReLU function on integer ciphertexts. Specifically, Sect. 3.1 introduces the $ExMultbyBin$ algorithm, and Sect. 3.2 details the $intReLU$ algorithm based on the $ExMultbyBin$ algorithm. Section 4 presents the experimental results and analysis of this paper. Finally, in Sect. 5, we summarize the content of this paper.

2 Preliminaries

2.1 Background of TFHE

Notations. Denote the security parameter as λ. The set $\{0, 1\}$ will be written as $\mathbb{B}$, real numbers and integers are denoted as $\mathbb{R}$ and $\mathbb{Z}$, respectively. The real torus $\mathbb{R}/\mathbb{Z}$ is denoted by $\mathbb{T}$, which is a set of real numbers modulo 1. $\mathfrak{R}$ denotes the ring of polynomials $\mathbb{Z}[X]/(X^N + 1)$. $\mathbb{T}_N[X]$ denotes $\mathbb{R}[X]/(X^N + 1)$ mod 1. Write vectors in bold. Use $\mathbf{s} \xleftarrow{U} \mathcal{S}$ to denote the process of sampling $\mathbf{s}$ uniformly at random over $\mathcal{S}$, and $e \leftarrow \chi$ denotes the process of sampling error e according to the Gaussian probability distribution χ.

Learning with Errors. The Learning with Errors (LWE) problem was proposed by Regev in [16]. Let $n \geq 1$ be an integer, $\chi \in \mathbb{R}^+$ be a probability distribution for noise and a secret vector $\mathbf{s} \in \mathbb{B}^n$. Define the LWE distribution $\text{LWE}_{n,\chi,\mathbf{s}}$ as a sample $(\mathbf{a}, b)$, where $\mathbf{a} \xleftarrow{U} \mathbb{T}^n$, $b = \mathbf{a} \cdot \mathbf{s} + e \in \mathbb{T}$, and the error $e \leftarrow \chi$. The LWE assumption states that if $\mathbf{s} \xleftarrow{U} \mathcal{S}$, and $\mathcal{S} \in \mathbb{B}^n$, it is hard to distinguish between an LWE sample and a uniformly random sample from $\mathbb{T}^n \times \mathbb{T}$.

Sub-gaussian. A distribution χ_σ is σ-Sub-Gaussian iff it satisfies $\forall t \in \mathbb{R}$, $E(exp(t\chi)) \leq exp(\sigma^2 t^2/2)$. If χ and χ' are two independent σ and σ'-Sub-Gaussian variables, then for all $\alpha, \beta \in \mathbb{R}$, $\alpha\chi + \beta\chi'$ is $\sqrt{\alpha^2\chi^2 + \beta^2\chi'^2}$-Sub-Gaussian.

LWE-Based Encryption. Introduce the Regev's bit-wise encryption from [16]. Let $\mu \in \mathbb{B}$ be a plaintext and λ be the security parameter. The encryption and decryption algorithms are as follows:

LWE.Setup(λ): fix $n = n(\lambda)$, $\chi = \chi(\lambda)$, output $\mathbf{s} \xleftarrow{U} \mathbb{B}^n$

LWE.Enc($\mathbf{s}, \mu$): output sample $(\mathbf{a}, b)$, $\mathbf{a} \xleftarrow{U} \mathbb{T}^n$, $b = \mathbf{a} \cdot \mathbf{s} + e + \mu/2$, where $e \leftarrow \chi$

LWE.Dec($\mathbf{s}$,($\mathbf{a}, b$)): output $\lceil 2(b - \mathbf{a} \cdot \mathbf{s}) \rfloor$

When noise e is bounded and

$$|e| < 1/4 \tag{1}$$

the decryption algorithms is correct. Since $2(b - \mathbf{a} \cdot \mathbf{s}) = 2e + m$, $|2e| < \frac{1}{2}$, $\lceil 2(b - \mathbf{a} \cdot \mathbf{s}) \rfloor = m$. The bit-wise encryption scheme can be extended to integer-wise encryption which we will introduce in next section.

TLWE. TLWE is a generalization of LWE and Ring-LWE [24]. Let $k \geq 1$ be an integer, N be a power of 2 and χ be a probability distribution for error over $\mathbb{R}_N[X]$. A TLWE secret key $\bar{\mathbf{s}} \in \mathbb{B}_N[X]^K$ is a vector of k polynomials $\in \mathfrak{R} = \mathbb{Z}[X]/(X^N + 1)$ with binary coefficients. Given a polynomial plaintext $\mu \in \mathbb{T}_N[X]$, a TLWE encryption of μ under secret key $\mathbf{s}$ is a sample $(\bar{\mathbf{a}}, \bar{b}) \in \mathbb{T}_N[X]^k \times \mathbb{T}_N[X]$, $\bar{\mathbf{a}} \xleftarrow{U} \mathbb{T}_N[X]^k$, $\bar{b} = \bar{\mathbf{a}} \cdot \bar{\mathbf{s}} + e + \mu$, where $e \leftarrow \chi$. From a TLWE sample $(\bar{\mathbf{a}}, \bar{b})$ of a polynomial plaintext μ under secret key $\bar{\mathbf{s}}$, we can extract a LWE sample $(\mathbf{a}, b) = \mathit{Extract}((\bar{\mathbf{a}}, \bar{b}))$ of the constant term of μ under an extracted key $\mathbf{s} = \text{ExtractKey}(\bar{\mathbf{s}})$. The detail of *Extract* and ExtractKey can be found in [25]. For simplicity, in the rest of this paper, the dimension of TLWE sample k is set to $k = 1$, thus a TLWE sample ($\mathbf{a}$,b)$\in \mathbb{T}_N[X] \times \mathbb{T}_N[X]$.

TGSW. TGSW is a generalized version of GSW [10]. TGSW can be seen as the matrix equivalent of TLWE. More detail can be found in [25].

2.2 Integer-Wise TFHE

Integer-Wise Encryption. This idea was already used in previous works such as [18,19,26]. Let $B \in \mathbb{N}$, plaintext $\mu \in \{-B, \cdots, B-1\}$. The integer-wise encryption and decryption algorithms are as follow:

TFHE.Setup(λ): fix $n = n(\lambda)$, $\chi = \chi(\lambda)$, output $\mathbf{s} \overset{U}{\longleftarrow} \mathbb{B}^n$

TFHE.Enc($\mathbf{s}, \mu$): output sample $(\mathbf{a}, b)$, $\mathbf{a} \overset{U}{\longleftarrow} \mathbb{T}^n$, $b = \mathbf{a} \cdot \mathbf{s} + e + \frac{\mu}{2B}$, where $e \leftarrow \chi$

TFHE.Dec($\mathbf{s}$,($\mathbf{a}, b$)): output $\lceil 2B \cdot (b - \mathbf{a} \cdot \mathbf{s}) \rfloor$

When noise e is bounded and

$$|e| < \frac{1}{4B} \tag{2}$$

the decryption algorithm is correct. Since $2B \cdot (b - \mathbf{a} \cdot \mathbf{s}) = 2Be + m$, $|2Be| < \frac{1}{2}$, $\lceil 2B \cdot (b - \mathbf{a} \cdot \mathbf{s}) \rfloor = m$.

Integer-Wise Bootstrapping. We revisit the integer-wise bootstrapping in TFHE proposed in [19]. Algorithm 1 shows the integer-wise bootstrapping algorithm.

Algorithm 1: Integer-wise bootstrapping [19] in TFHE

Input: A ciphertext $C_{m_{in}}$:= an LWE sample ($\mathbf{a}$,b)$\in$ LWE$_\mathbf{s}(m_{in})$, where its plaintext $m_{in} \in \{-B, \cdots, B-1\}$, a bootstrapping key $\mathrm{BK}_{\mathbf{s} \to \mathbf{s}'', \alpha}$, a keyswitching key $\mathrm{KS}_{\mathbf{s}' \to \mathbf{s}, \gamma}$, where $\mathbf{s}' = KeyExtract(\mathbf{s}'')$, a constant function $f : \{-B, \cdots, B-1\} \to \{-B, B-1\}$, and a set of coefficients $\{\mu_0, \cdots, \mu_{N-1}\}$ of the testvector which correspond to the function f.

Output: An LWE sample LWE$_\mathbf{s}(m_{out})$:= $Bootstrap(C_{m_{in}}, f)$, where

$$m_{out} = f(m_{in}) := \begin{cases} f(m_{in}), m_{in} \in \{0, \cdots, B-1\} \\ -f(B + m_{in}), m_{in} \in \{-B, \cdots, -1\} \end{cases} \tag{3}$$

1: $\bar{b} := \lceil 2Nb \rfloor$, $\bar{a}_i := \lceil 2Na_i \rfloor$ for each $i \in [1, n]$
2: testv := $\mu_0 + \mu_1 X^{-1} + \cdots + \mu_{N-1} X^{-N-1} \in \mathbb{T}_N[X]$
3: ACC$\leftarrow X^{\bar{b}} \cdot$(0,testv)$\in \mathbb{T}_N[X] \times \mathbb{T}_N[X]$
4: **for** i=1 to n **do** ACC$\leftarrow$ $[\mathbf{h} + (X^{-\bar{a}_i} - 1) \cdot BK_i] \boxdot ACC$
5: $\mathbf{u}$:= $SampleExtract$(ACC)
6: **return** $KeySwitch_{KS}(\mathbf{u})$

The integer-wise bootstrapping algorithm consists of 5 following functions:

Rounding (line 1). After line 1 of rounding, obtain an LWE sample $(\bar{a}, \bar{b}) \in \mathbb{Z}_{2N}^n \times \mathbb{Z}_{2N}$, and

$$\bar{b} - \bar{\mathbf{a}} \cdot \mathbf{s} = \lceil 2Nb \rfloor - \sum_{i=1}^{n} \lceil 2Na_i \rfloor s_i = 2N(e + m_{in}/2B) + e_{ACC} \tag{4}$$

where e_{ACC} is the sum of rounding errors, and rounding errors are uniformly distributed in $(-\frac{1}{2}, \frac{1}{2})$. Then $e_{ACC} = 0$ and can be ignored.

Setup (line 2). Initialize the polynomial testv, the coefficients of which satisfy

$$\mu_{i\times N/B-N/2B} = \cdots = \mu_{i\times N/B+N/2B} = \frac{f(i)}{2B}, i \in \{0, \cdots, B-1\} \tag{5}$$

so that the N coefficients of testv are uniformly filled with B values from $\frac{f(0)}{2B}$ to $\frac{f(B-1)}{2B}$.

BlindRotate (line 3 and line 4). In line 3, (0, testv) is a trivial TLWE sample, the multiplication between $X^{\bar{b}}$ and (0, testv) is equivalent to the shift of the coefficients in polynomial. So, in line 3 ACC is a TLWE sample of a polynomial plaintext μ with constant term $\frac{f(\bar{b})}{2B}$.

In line 4, the external product $\boxdot$ performs the following mapping:

$$\boxdot : TGSW \times TLWE \rightarrow TLWE.$$

Roughly speaking, the external product $\boxdot$ multiples a TGSW sample of plaintext m_1 and a TLWE sample of plaintext m_2, outputs a TLWE sample of plaintext $m_1 \times m_2$. So after iteration, ACC is a TLWE sample of a polynomial plaintext with constant term $\mu_{\bar{b}-\bar{\mathbf{a}}\cdot\mathbf{s}}$. From (4), $\bar{b}-\bar{\mathbf{a}}\cdot\mathbf{s} = 2Ne+m_{in}\times N/B$. And according to (2), we can get $m_{in}\times N/B-N/2B < 2Ne+m_{in}\times N/B < m_{in}\times N/B+N/2B$. So $\mu_{\bar{b}-\bar{\mathbf{a}}\cdot\mathbf{s}}$ is between $m_{in}\times N/B-N/2B$ and $m_{in}\times N/B+N/2B$. From (5), we can obtain $\mu_{\bar{b}-\bar{\mathbf{a}}\cdot\mathbf{s}} = \frac{f(m_{in})}{2B}$, so ACC is a TLWE sample of a polynomial plaintext with constant term $\frac{f(m_{in})}{2B}$.

From Theorem 4.6 in [25] we obtain

$$||Err(ACC)||_\infty \leq 2n(k+1)lN\beta\alpha_{BK} + n(1+kN)\epsilon, \tag{6}$$

where $\beta = B_g/2$, $\epsilon = 1/2B_g^l$ are precision parameters of the gadget decomposition, $l \in \mathbb{N}$ and $B_g \in \mathbb{N}$. α_{BK} is a noise parameter of the bootstrapping key BK. More detail can be found in [25].

Extract(line 5). SampleExtract extracts the a LWE sample $\mathbf{u}$ from TLWE sample ACC. In detail, ACC is a TLWE sample of a polynomial plaintext μ under secret key $\mathbf{s}''$, the constant term of μ is $\frac{f(\bar{b}-\bar{\mathbf{a}}\cdot\mathbf{s})}{2B}$. $\mathbf{u}$ is a LWE sample of plaintext $\frac{f(\bar{b}-\bar{\mathbf{a}}\cdot\mathbf{s})}{2B}$ under secret key $\mathbf{s}''$. *SampleExtract* doesnot add extra noise, $||Err(\mathbf{u})||_\infty \leq ||Err(ACC)||_\infty$.

KeySwitch(line 6). After *KeySwitch*, we obtain a LWE sample $(\mathbf{a}', b')$ of plaintext $\frac{m_{out}}{2B} = \frac{f(m_{in})}{2B} \in \mathbb{T}$. We use the same *KeySwitch* algorithm as in [25]. From Lemma 4.3 in [25], we obtain the bound of the noise as $||Err((\mathbf{a}', b'))||_\infty \leq ||Err(ACC)||_\infty + kNt\gamma + kN2^{-(t+1)}$. According to (6), $||Err((\mathbf{a}', b'))||_\infty \leq 2n(k+1)lN\beta\alpha_{BK} + n(1+kN)\epsilon + kNt\gamma + kN2^{-(t+1)}$. We can ensure that the output sample $(\mathbf{a}', b')$ is a "fresh" LWE sample by selecting the parameters that satisfy the upper bound from (2) as follows:

$$2n(k+1)lN\beta\alpha_{BK} + n(1+kN)\epsilon + kNt\gamma + kN2^{-(t+1)} < 1/4B. \tag{7}$$

Homomorphic Integer Multiplication by Binary Number. We revisit the *MultbyBin* algorithm proposed in [19] which multiplies integer ciphertext $C_{m_{int}}$ with binary ciphertext $C_{m_{bin}}$, achieving a result of C_0 when $m_{bin} = 0$ and $C_{m_{int}}$ when $m_{bin} = B$. Algorithm 2 provides the detail.

Algorithm 2: Homomorphic multiplication by binary number [19]

Input: Ciphertext $C_{m_{int}}$, $C_{m_{bin}}$, where $m_{int} \in \{0, cdots, B-1\}$, $m_{bin} \in \{B, 0\}$
Output: $C_{m_{out}}$:=$MultbyBin(C_{m_{int}}, C_{m_{bin}})$, $m_{out} = m_{int}$ if $m_{bin} = B$, 0 else.
1: C_{tmp}=$Bootstrap(C_{m_{int}} + C_{m_{bin}} + (\mathbf{0}, \frac{B}{2B}), f_{id})$
2: **return** $Bootstrap(C_{m_{int}} + C_{tmp}, f_{half})$

In Algorithm 2, when $m_{bin} = B$, in line 1, $C_{m_{int}} + C_{m_{bin}} + (\mathbf{0}, \frac{B}{2B})$ is equal to $C_{m_{int}} + (\mathbf{0}, \frac{B}{2B}) + (\mathbf{0}, \frac{B}{2B})$. According to *Bootstrap* described in Algorithm 1, C_{tmp} is a "fresh" LWE sample and $tmp = m_{int}$. So, $C_{m_{int}} + C_{tmp}$ is equal to $C_{2\times m_{int}}$, after *Bootstrap* in line 2, we can get a LWE sample $C_{m_{out}}$ and $m_{ont} = m_{int}$. Similarly, when $m_{bin} = 0$, $C_{m_{int}} + C_{m_{bin}} + (\mathbf{0}, \frac{B}{2B})$ is equal to $C_{m_{int}} + (\mathbf{0}, \frac{B}{2B})$. After *Bootstrap*, C_{tmp} is a "fresh" LWE sample and $tmp = -m_{int}$. In line 2, $C_{m_{int}} + C_{tmp}$ is equal to C_0, and $m_{ont} = m_0$. At this point, multiplication of integer ciphertext and binary ciphertext has been achieved. It is worth noting that [19] states that B in Algorithm 2 must be an odd number, otherwise when $m_{int} \in \{B/2, \cdots, B-1\}$ and $m_{bin} = B$, arithmetic errors will occur in line 2. However, this paper finds that when $m_{int} \in \{B/2, \cdots, B-1\}$ and $m_{bin} = B$ in Algorithm 2, regardless of whether B is odd or even, there will be arithmetic errors in line 2. More details will be explained in Sect. 3.1.

3 Integer-Wise ReLU on TFHE

ReLU is a piecewise function which is widely used in neural networks. Specifically, when $x < 0$, $ReLU(x) = 0$, and when $x \geq 0$, $ReLU(x) = f_{id}(x) = x$. This function can be equivalently transformed to multiplication with binary number, that is multiplying 0 when $x < 0$ and multiplying 1 when $x \geq 0$. Homomorphic ReLU means the ReLU operation on encrypted data. Therefore, the *MultbyBin* algorithm [19] can be used to achieve homomorphic ReLU. However, *MultbyBin* can only support input data in the range of $\{B/2, \cdots, B-1\}$. Section 3.1 extends it by proposing the *ExMultbyBin* algorithm, which supports a larger integer interval $\{-B, \cdots, B-1\}$, followed by Sect. 3.2 to introduce the *intReLU* function which implement homomorphic ReLU for integers using *ExMultbyBin*.

3.1 Extend Homomorphic Integer Multiplication with a Binary

In Sect. 2, we introduce Algorithm 2, the *MultbyBin* algorithm for integer multiplication by a binary number. However, three drawbacks exist. Firstly, m_{bin}

belongs to $\{B, 0\}$ which is inconsistent with the interval supported by the integer bootstrapping scheme $\{-B, \cdots, B-1\}$. This implies that binary ciphertext $C_{m_{bin}}$ cannot be directly represented when $m_{bin} = B$. Secondly, the *MultbyBin* algorithm only supports ciphertext of positive integers and 0, while cannot be compatible with ciphertext of negative integers in $\{-B, \cdots, -1\}$. Finally, there will be arithmetic errors for ciphertext of integers in $\{B/2, \cdots, B-1\}$, regardless of whether B is odd or even. Therefore we can summarize that *MultbyBin* only supports ciphertext of $m_{int} \in \{0, \cdots, B/2-1\}$. To address these drawbacks, this paper proposed the *ExMultbyBin* algorithm as shown in Algorithm 3.

Algorithm 3: Extend integer multiplication with binary number

---

Input: Ciphertext $C_{m_{int}}, C_{m_{id}}, C_{m_{bin}}$, where $m_{int} \in \{-B, \cdots, B-1\}$, $m_{bin} \in \{-B, 0\}, C_{m_{id}} = Bootstrap(C_{m_{int}}, f_{id})$

Output: $C_{m_{out}} := ExMultbyBin(C_{m_{int}}, C_{m_{out}}), m_{out} = m_{int}$ if $m_{bin} = -B$, 0 else.

1: $C_{tmp1} = C_{m_{id}} + C_{m_{bin}} + (\mathbf{0}, -\frac{B}{2B})$

2: $C_{tmp2} = Bootstrap(C_{m_{int}} + C_{tmp1}, f_{sign1})$

3: $C_{tmp3} = Bootstrap(C_{tmp2} + (\mathbf{0}, \frac{1}{2B}), f_{half})$

4: **return** $Bootstrap(C_{m_{int}} + C_{tmp1} + C_{tmp3}, f_{half})$

As we mentioned in integer-wise bootstrapping in Sect. 2, the set of coefficients $\{\mu_0, \cdots, \mu_{N-1}\}$ of the testvector which correspond to the specific function f. In Algorithm 3, the function $f_{id} : \{0, \cdots, B-1\} \to \{0, \cdots, B-1\}$ is defined as $f_{id}(x) = x$. For f_{id} the coefficients $\mu_0, \cdots, \mu_{N-1} \in \mathbb{T}$ of testv are defined as follows:

$$\begin{cases} \mu_0, \cdots, \mu_{\lfloor \frac{N}{2B} \rfloor - 1} := \dfrac{f_{id}(0)}{2B} = 0, \\ \mu_{\lfloor \frac{(2i-1)N}{2B} \rfloor}, \cdots, \mu_{\lfloor \frac{(2i+1)N}{2B} \rfloor - 1} := \dfrac{f_{id}(i)}{2B} = \dfrac{i}{2B}, i = \{1, \cdots, B-1\}, \\ \mu_{N - \lfloor \frac{N}{2B} \rfloor}, \cdots, \mu_{N-1} := \dfrac{-f_{id}(0)}{2B} = 0. \end{cases}$$

f_{sign1} is defined as $f_{sign1}(x) = -1$ if $x \geq 0$, 1 otherwise. For f_{sign1} the coefficients $\mu_0, \cdots, \mu_{N-1} \in \mathbb{T}$ of testv are defined as follows:

$$\begin{cases} \mu_0, \cdots, \mu_{\lfloor \frac{N}{2B} \rfloor - 1} := \dfrac{f_{sign1}(0)}{2B} = -\dfrac{1}{2B}, \\ \mu_{\lfloor \frac{(2i-1)N}{2B} \rfloor}, \cdots, \mu_{\lfloor \frac{(2i+1)N}{2B} \rfloor - 1} := \dfrac{f_{sign1}(i)}{2B} = -\dfrac{1}{2B}, i = \{1, \cdots, B-1\}, \\ \mu_{N - \lfloor \frac{N}{2B} \rfloor}, \cdots, \mu_{N-1} := \dfrac{-f_{sign1}(0)}{2B} = \dfrac{1}{2B}. \end{cases}$$

The function f_{half} is defined as $f_{half}(x) = \frac{x}{2}$ if x is even, $-\frac{x+1}{2} - \frac{B-1}{2}$ otherwise. For f_{half} the coefficients $\mu_0, \cdots, \mu_{N-1} \in \mathbb{T}$ of testv are defined as follows:

$$\begin{cases} \mu_0, \cdots, \mu_{\lfloor \frac{N}{2B} \rfloor - 1} := \dfrac{f_{half}(0)}{2B} = 0, \\ \mu_{\lfloor \frac{(2i-1)N}{2B} \rfloor}, \cdots, \mu_{\lfloor \frac{(2i+1)N}{2B} \rfloor - 1} := \dfrac{f_{half}(i)}{2B}, i = \{1, \cdots, B-1\}, \\ \mu_{N - \lfloor \frac{N}{2B} \rfloor}, \cdots, \mu_{N-1} := \dfrac{-f_{half}(0)}{2B} = 0. \end{cases}$$

To address the first drawback of *MultbyBin*, we modify the binary interval of m_{bin} from $\{B, 0\}$ to $\{-B, 0\}$. Additionally, we adaptively adjust the LWE sample $(\mathbf{0}, \frac{B}{2B})$ in the line 1 of Algorithm 3 to LWE sample $(\mathbf{0}, -\frac{B}{2B})$.

To address the second drawback of *MultbyBin*, considering the negacyclicity property($f(x) = -f(x+B)$) in TFHE, when $x \in \{-B, \cdots, -1\}$, $f(x)$ will be assigned $-f(x+B)$, which is inconsistent with the expected $f(x)$. For instance, when $f(x) = x, B = 8, f(-2) = -f(6) = -6 \neq f(-2) = -2$. This inconsistency is mainly reflected in f_{id} in line 1 of Algorithm 2. Intuitively, we can achieve $f(-f(x+B)) = f(x)$ by nested f function, but this would introduce an expensive Bootstrapping Algorithm. Instead of bootstrapping, this paper employs a linear transformation. For $f_{id}(x) = x$ in Algorithm 2, $-f(x+B) - B = -f(x)$ as shown in line 1 of Algorithm 3. Specifically, when $m_{int} \in \{-B, \cdots, -1\}$ and $m_{bin} = 0$, $C_{tmp1} = C_{m_{id}-B} = C_{-m_{int}}$. Otherwise, when $m_{int} \in \{0, \cdots, B-1\}$ and $m_{bin} = -B$, $C_{tmp1} = C_{m_{id}-B-B} = C_{m_{id}} = C_{int}$. This is consistent with the result of Algorithm 2 when $m_{int} \in \{0, \cdots, B-1\}$. Therefore, we extend $m_{int} \in \{0, \cdots, B/2-1\}$ of *MultbyBin* to $m_{int} \in \{-B, \cdots, B/2-1\}$ in *ExMultbyBin*.

To address the third drawback of *MultbyBin*, observe that $C_{m_{int}} + C_{tmp} = C_{2m_{int}}$ in Algorithm 2, indicating that the intermediate result can only be an even numbers, restricting them mapped to even-slot values. When $m_{int} \in \{B/2, \cdots, B-1\}, 2m_{int} \in \{\lceil B \rfloor, \cdots, 2B-2\}$. After modulo $2B$, $2m_{int} \in \{-\lceil B \rfloor, \cdots, -2\}$. When $m_{int} \in \{0, \cdots, B/2-1\}$, $2m_{int} \in \{0, 2, \cdots, \lceil B \rfloor\}$. In other words, the B numbers $\{0, \cdots, B-1\}$ will be mapped to the B even numbers in $\{-B, \cdots, B-1\}$. However, due to the negacyclicity property, among the B even numbers, there are only $B/2$ truly valid numbers. In general, regardless of whether B is odd or even, when $m_{int} \in \{0, \cdots, B-1\}, 2m_{int}$ always maps to even slot, meaning that two numbers will map to the same slot, resulting arithmetic errors when $m_{int} \in \{B/2, \cdots, B-1\}$. For example, when $B = 4, m_{in} \in \{-B, \cdots, B-1\}$, the mapping relationship of f_{half} is as shown in the Fig. 1.

When $m_{in} = 3$, $2m_{int} = 6$ will be mapped to -2. After f_{half} operation, $f_{half}(-2) = -1 \neq 3$. Therefore this paper addresses this issue in Algorithm 3 by mapping $\{B/2, \cdots, B-1\}$ to the $B/2$ odd numbers in $\{-B, \cdots, B-1\}$ through odd slot mapping. Specifically, the core idea of the odd slot mapping is to change the modulus $2B$ to $2B-1$. The modulo $2B-1$ is achieved by subtracting $\frac{sign(x)-1}{2}$. In Algorithm 3, when $m_{int} \in \{B/2, .., B-1\}, C_{tmp2}$ is a LWE sample of plaintext -1, C_{tmp3} is a LWE sample of plaintext 1. When $m_{int} \in \{0, .., B/2-1\}, C_{tmp2}$ is a LWE sample of plaintext 1, C_{tmp3} is a LWE sample of

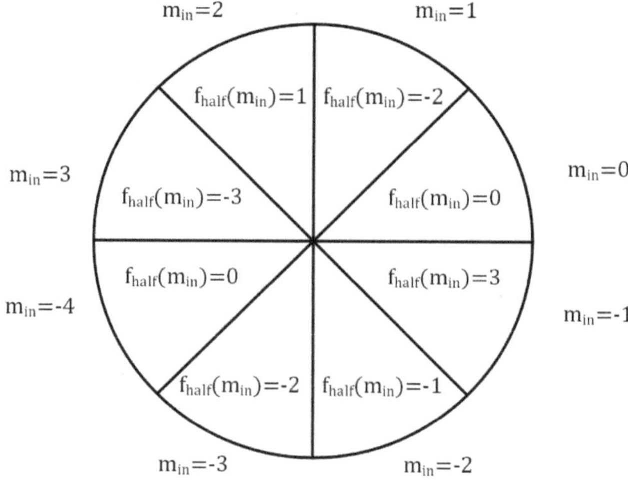

Fig. 1. the mapping relationship of f_{half} when $B = 4$

plaintext 0. In line 4 by using C_{tmp3}, when $m_{int} \in \{0, .., B/2-1\}$, map the result to even slot; when $m_{int} \in \{B/2, .., B-1\}$ map it to odd slot. This achieves a one-to-one mapping to ensure correct calculation.When $m_{in} = 3$, $2m_{int} = 6$ will be mapped to -1 by module $2B-1$. After f_{half} operation, $f_{half}(-1) = 3$. Therefore, we extend the $m_{int} \in \{0, \cdots, B/2-1\}$ of $MultbyBin$ to $m_{int} \in \{0, \cdots, B-1\}$ in $ExMultbyBin$. In summary, combining with linear transformation, we design the $ExMultbyBin$ algorithm which supports the ciphertext of integers in $\{-B, \cdots, B-1\}$.

In the rest of this subsection, we analyze the correctness of Algorithm 3.

Lemma 1. *Let $\beta \leq 1/16B$, input ciphertexts of Algorithm 3 are $C_{m_{int}}$, $C_{m_{bin}}$, $C_{m_{id}}$ such that $|Dec(C_{m_{int}})-m_{int}| < \beta$, $|Dec(C_{m_{bin}})-m_{bin}| < \beta$, $|Dec(C_{m_{id}})-m_{id}| < \beta$ respectively, output a ciphertext $C_{m_{out}}$ such that $|Dec(C_{m_{out}})-m_{out}| < \beta$, where $m_{out} = m_{int}$ if $m_{bin} = -B$, 0 else.*

Proof. We discuss three cases. When $m_{bin} = 0$ and $m_{int} \in \{-B, \cdots, 0\}$, C_{tmp1} in line 1 is a ciphertext of $m_{id} - B + 2\beta = -m_{int} + 2\beta$. In line 2, the input of Bootstrap is a ciphertext of $0 + 3\beta$, after Bootstrap C_{tmp2} is a ciphertext of $-1 + e_\beta$, where $e_\beta < \beta$. In line 3, the input of Bootstrap is a ciphertext of $0 + e_\beta < 0 + \beta$, after Bootstrap C_{tmp3} is a ciphertext of $0 + e_\beta < 0 + \beta$. In line 4, the input of Bootstrap is a ciphertext of $0 + 3\beta + e_\beta < 0 + 4\beta < 0 + 1/4B$, so the error will not exceed $1/4B$, which will not lead to computational errors. After Bootstrap, $C_{m_{out}}$ is a ciphertext of $0 + e_\beta$, hence $|Dec(C_{m_{out}}) - m_{out}| = |0 + e_\beta - 0| < \beta$ as claimed in the lemma.

When $m_{bin} = -B$ and $m_{int} \in \{1, \cdots, B/2-1\}$, C_{tmp1} in line 1 is a ciphertext of $m_{id} - 2B + 2\beta = m_{id} + 2\beta = m_{int} + 2\beta$. In line 2, the input of Bootstrap is a ciphertext of $2m_{int} + 3\beta$. Since $2m_{int} < B$, after Bootstrap C_{tmp2} is a ciphertext of $-1 + e_\beta$, where $e_\beta < \beta$. In line 3, the input of Bootstrap is a ciphertext of

$0+e_\beta < 0+\beta$, after Bootstrap C_{tmp3} is a ciphertext of $0+e_\beta < 0+\beta$. In line 4, the input of Bootstrap is a ciphertext of $2m_{int}+3\beta+e_\beta < 2m_{int}+4\beta < 2m_{int}+1/4B$, so the error will not exceed $1/4B$, which will not lead to computational errors. After Bootstrap, $C_{m_{out}}$ is a ciphertext of $m_{int}+e_\beta$, hence $|Dec(C_{m_{out}})-m_{out}| = |m_{int}+e_\beta - m_{int}| < \beta$ as claimed in the lemma.

When $m_{bin} = -B$ and $m_{int} \in \{B/2, \cdots, B-1\}$, in line 2, the input of Bootstrap is a ciphertext of $2m_{int}+3\beta$. Since $2m_{int} > B$, after Bootstrap C_{tmp2} is a ciphertext of $1+e_\beta$, where $e_\beta < \beta$. In line 3, the input of Bootstrap is a ciphertext of $2+e_\beta < 2+\beta$, after Bootstrap C_{tmp3} is a ciphertext of $1+e_\beta < 1+\beta$. In line 4, the input of Bootstrap is a ciphertext of $2m_{int}+1+3\beta+e_\beta < 2m_{int}+1+4\beta < 2m_{int}+1+1/4B$, so the error will not exceed $1/4B$, which will not lead to computational errors. After Bootstrap, $C_{m_{out}}$ is a ciphertext of $m_{int}+e_\beta$, hence $|Dec(C_{m_{out}}) - m_{out}| = |m_{int}+e_\beta - m_{int}| < \beta$ as claimed in the lemma. □

3.2 Integer-Wise ReLU

In this subsection, we introduce $ExMultbyBin$ from Sect. 3.1 to implement the ReLU function for integers $intReLU$ based on the idea of equivalent functions. The definition of $ReLU(x)$ is as follows:

$$ReLU(x) = \begin{cases} x, x \geq 0 \\ 0, x < 0 \end{cases} \tag{8}$$

To enable the ReLU operations on encrypted data, i.e., the homomorphic ReLU, plaintext m_{in} is encrypted by **TFHE.Enc** described in Sect. 2.2. The proposed algorithm $IntReLU$ takes the ciphertext $C_{m_{in}}$ as input, performs homomorphic ReLU operations and outputs $C_{m_{out}}$.

For the ciphertext of bit-wise plaintext, when implementing the ReLU function using TFHE, most works, such as [17], generally achieve this by inverting the sign bit and performing AND gates on the remaining bites one by one, as illustrated in Fig. 2. For convenient, refer it to bit-wise ReLU. Since each gate operation needs one bootstrapping, $n-1$ bootstrappings are required when the precision of the input date is n-bit.

For the ciphertext of integer-wise plaintext, this paper first proposes a new function $ExMultbyBin(x, f_{id}(x), sign(x)-B/2)$ which is equivalent to the ReLU function. It converts the ReLU function to multiplication $ExMultbyBin$ of integer ciphertext and binary ciphertext, as shown in Fig. 3 and Algorithm 4.

Specifically, integer plaintext m_{in} is encrypted by **TFHE.Enc** described in Sect. 2.2. The proposed algorithm $IntReLU$ takes the ciphertext $C_{m_{in}}$ as input. Firstly, a homomorphic $sign(x)$ operation needs to be applied to m_{in} to determine its sign. And due to the $ExMultbyBin$ algorithm requiring $f_{id}(x)$ as input, thus the homomorphic f_{id} operation is necessary. It's worth noting that implementing different homomorphic operations on the same data can be accomplished using the $PBSManyLUT$ technique.

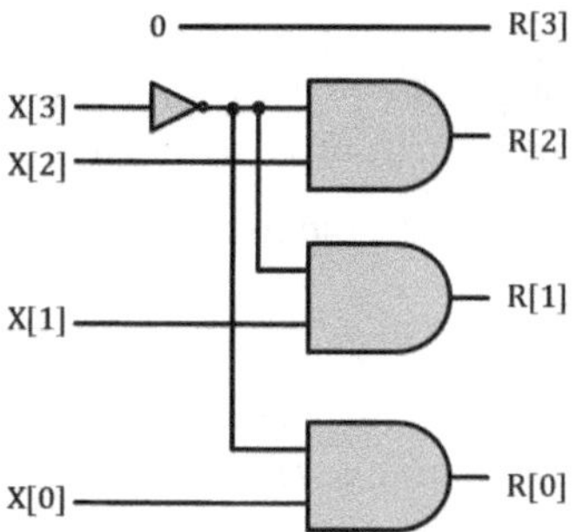

Fig. 2. A 4-bit ReLU circuit in [17] (X:input; R:output)

Algorithm 4: Integer ReLU on TFHE

Input: Ciphertext $C_{m_{in}}$, where $m_{in} \in \{-B, \cdots, B-1\}$
Output: $C_{m_{out}} := IntReLU(C_{m_{in}})$, $m_{out} = m_{in}$ if $m_{in} \in \{0, \cdots, B-1\}$, 0 else.
1: $PBSManyLUT(C_{id}, C_{sign}, C_{m_{in}}, f_{id}, f_{sign2})$
2: $C_{tmp} = C_{sign} - (\mathbf{0}, -\frac{B}{2} \times \frac{1}{2B})$
3: **return** $ExMultbyBin(C_{m_{in}}, C_{id}, C_{tmp})$

PBSManyLUT [23] can perform multiple f functions on a ciphertext in one bootstrapping, as shown in line 1 of Algorithm 4, simultaneously computing f_{id} and f_{sign2} in one bootstrapping. The function f_{sign2} is defined as $f_{sign2}(x) = -\frac{B}{2}$ if $x \geq 0$, $\frac{B}{2}$ otherwise. In *Bootstrap* of *PBSManyLUT*, assuming B is even, the coefficients $\mu_0, \cdots, \mu_{N-1} \in \mathbb{T}$ of testv are defined as follows:

$$
\begin{cases}
\mu_0, \mu_2 \cdots, \mu_{\frac{N}{2B}-2} := \dfrac{f_{id}(0)}{2B} = 0, \\
\mu_1, \mu_3 \cdots, \mu_{\frac{N}{2B}-1} := \dfrac{f_{sign2}(0)}{2B} = -\dfrac{B}{2} \times \dfrac{1}{2B}, \\
\mu_{\frac{(2i-1)N}{2B}}, \cdots, \mu_{\frac{(2i+1)N}{2B}-2} := \dfrac{f_{id}(i)}{2B} = \dfrac{i}{2B}, i = \{1, \cdots, B-1\}, \\
\mu_{\frac{(2i-1)N}{2B}+1}, \cdots, mu_{\frac{(2i+1)N}{2B}-1} := \dfrac{f_{sign2}(i)}{2B} = -\dfrac{B}{2} \times \dfrac{1}{2B}, i = \{1, \cdots, B-1\}, \\
\mu_{N-\frac{N}{2B}}, \cdots, \mu_{N-2} := \dfrac{-f_{id}(0)}{2B} = 0, \\
\mu_{N-\frac{N}{2B}+1}, \cdots, \mu_{N-1} := \dfrac{-f_{sign2}(0)}{2B} = \dfrac{B}{2} \times \dfrac{1}{2B}.
\end{cases}
$$

After *PBSManyLUT*, C_{id}, C_{sign} are LWE samples of plaintext $f_{id}(m_{in})$ and $f_{sign2}(m_{in})$, respectively. When $m_{in} \geq 0$, $f_{sign2}(m_{in}) = -\frac{B}{2}$, otherwise $f_{sign2}(m_{in}) = \frac{B}{2}$. In order to apply *ExMultbyBin* algorithm, it is necessary to convert $f_{sign2}(m_{in})$ into a binary data 0 or $-B$. Thus, a subtraction is performed in line 2 to accomplish this. So when $m_{in} \geq 0$, C_{tmp} is a LWE sample of plaintext

$-B$. According to $ExMultbyBin$, $m_{out} = m_{in}$. When $m_{in} < 0$, $f_{sign2}(m_{in}) = \frac{B}{2}$, so C_{tmp} is a LWE sample of plaintext 0. After $ExMultbyBin$, $m_{out} = 0$.

Then we analyze the correctness of Algorithm 4.

Lemma 2. *Let $\beta \leq 1/16B$, input ciphertexts of Algorithm 4 is $C_{m_{in}}$ such that $|Dec(C_{m_{in}}) - m_{in}| < \beta$, output a ciphertext $C_{m_{out}}$ such that $|Dec(C_{m_{out}}) - m_{out}| < \beta$, where $m_{out} = ReLU(m_{in})$.*

Proof. We discuss two cases. When $m_{in} \in \{-B, \cdots, -1\}$, in line 1 C_{id} and C_{sign} are ciphertexts of $m_{id} + e_\beta = -(m_{in} + B) + e_\beta$ and $B/2 + e_\beta$ respectively, where $e_\beta < \beta$. In line 2, C_{tmp} is a ciphertext of $0 + e_\beta$. By Lemma 1, ExMultbyBin outputs a ciphertext $C_{m_{out}}$ of $0+e_\beta$. Hence, $|Dec(C_{m_{out}}) - m_{out}| = |0+e_\beta - 0| < \beta$ as claimed in the lemma.

When $m_{in} \in \{0, \cdots, B-1\}$, in line 1 C_{id} and C_{sign} are ciphertexts of $m_{id} + e_\beta = m_{in} + e_\beta$ and $-B/2 + e_\beta$ respectively, where $e_\beta < \beta$. In line 2, C_{tmp} is a ciphertext of $-B + e_\beta$. By Lemma 1, ExMultbyBin outputs a ciphertext $C_{m_{out}}$ of $m_{in} + e_\beta$. Hence, $|Dec(C_{m_{out}}) - m_{out}| = |m_{in} + e_\beta - m_{in}| < \beta$ as claimed in the lemma. □

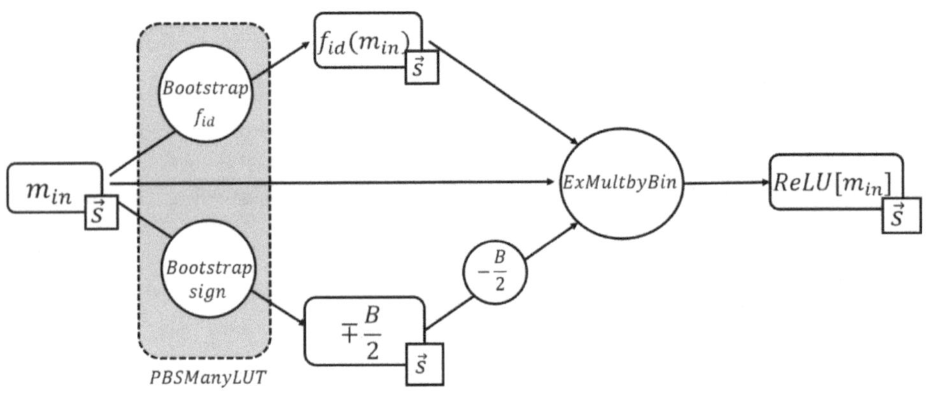

Fig. 3. Illustration of integer-wise ReLU on TFHE

From Algorithm 3, we observe that $ExMultbyBin$ requires three bootstrappings. After introducing the $PBSManyLUT$ algorithm which requires one bootstrapping, the integer-wise ReLU implementation algorithm requires four bootstrappings.

3.3 Security Analysis

In this section, we present the analyses on security.

Theorem 1. *For a LWE sample ($\mathbf{a}$,b) of message $m \in \{-B, \cdots, B-1\}$, and parameters n, k, l, N, β, α_{BK}, ϵ, γ, if the same parameters are chosen, then the security of IntReLU relies on the security of TFHE [25].*

Proof. Equation (7) provides the noise bound for *IntReLU*, the larger the integer range B, the smaller the noise upper bound. Given the specified parameters, when the upper bound of noise satisfies (7), the decryption result of *IntReLU* is correct. If the upper bound of noise does not satisfy (7), it is necessary to select a smaller parameter and conduct a security analysis again, otherwise excessive noise can lead to decryption errors. In other words, Eq. (7) can represent the correctness of *IntReLU*, and the parameters in Eq. (7) signify the security of *IntReLU*. This paper uses the same security and noise parameters as the original TFHE scheme, so the security of *IntReLU* relies on the original TFHE scheme, which is explained in detail in [25].

4 Experiment and Application

4.1 Experiment

This paper implements *IntReLU* in the TFHE library [25]. Our parameters are set according to the default values of the TFHE library, where the dimension of polynomials, LWE sample and TLWE sample are $N = 1,024$, $n = 500$ and $k = 1$ respectively. The decomposition basis and length for TGSW ciphertexts and KeySwitch are $B_g = 2^{10}, l = 2$ and $2^l = 4, t = 8$ respectively. The security of these parameters is estimated to be 128-bit security.

We ran all experiments on an Intel(R) Xeon(R) Platinum 8369HB CPU @ 3.30 GHz. We implemented *IntReLU* and bit-wise ReLU in [17] for integers with different precision, as shown in Table 1. As mentioned in Sect. 3.2, for data precision of n bits, bit-wise TFHE requires $n-1$ bootstrappings, while IntReLU requires four bootstrappings. Therefore, we commence our experiments with 5-bit precision.

Table 1. Comparison of bit-wise ReLU and IntReLU at different precision

Method	Library	Type	Time (ms)				
			5-bit	6-bit	7-bit	8-bit	9-bit
[17]	TFHE	Bit-wise	223.467	281.013	335.536	389.702	444.364
Ours (*IntReLU*)	TFHE	Integer-wise	234.438	**234.360**	237.011	234.686	234.574

In Table 1, we observe that the running time of bit-wise ReLU increases linearly with the increase of precision, while the running time of *IntReLU* is independent of precision. When the accuracy is 5-bit, the time of bit-wise ReLU and *IntReLU* is similar. This is consistent with our analysis that when precision is n-bit, *IntReLU* needs 4 bootstrappings, while bit-wise ReLU requires $n-1$ bootstrappings. When the data precision is 6-bit, our proposed *IntReLU* reduces time by 17% compared to the bit-wise ReLU.

Furthermore, we provide a comparison of algorithmic complexity between *IntReLU* and other works that have surpassed the TFHE negacyclicity constraint. Table 2 illustrates the theoretical comparison results. We can observe that, although [21] requires only 2 Bootstrap, it supports integer in $\{0, \cdots, B-1\}$ and cannot handle $\{-B, \cdots, -1\}$, failing to completely address the negacyclicity constraint of TFHE. For the integer range $\{-B, \cdots, B-1\}$, compared to [20] and [22], the proposed algorithm *IntReLU* in this paper has the lowest algorithmic complexity.

Table 2. Comparison of algorithmic complexity between *IntReLU* and existing works

Method	Supported integer	Algorithmic complexity
[20]	$\{-B, \cdots, B-1\}$	$1 + log(Q)$ Bootstrap
[21]	$\{0, \cdots, B-1\}$	2 Bootstrap + 1 Add
[22]	$\{-B, \cdots, B-1\}$	6 Bootstrap
Ours (*IntReLU*)	$\{-B, \cdots, B-1\}$	4 Bootstrap

Correctness. As the data precision increases, the upper bound of noise will decrease according to (7). In Sect. 3.3, we pointed out that if the security parameters remain unchanged, the upper bound of the noise will not satisfy (7), resulting in decryption errors. Overall, our proposed *IntReLU* balances data precision and decryption accuracy. Related works [20–22] also has the limitation. Through experimental verification, when precision is 5-bit and 6-bit, the decryption accuracy can reach 100%. But for 7-bit precision, the decryption accuracy decreases to 53%. However, this does not affect the security of *IntReLU*, as changes in precision will not affect security parameters. It is worth noting that when the data precision exceeds 6-bit, large integers can be decomposed into multiple 6-bit integers using either the radix or the CRT (Chinese Remainder Theorem) representations, thereby supporting ReLU operations on large integers.

4.2 Application

IntReLU can be widely utilized in privacy-preserving deep learning. For instance, when employing TFHE for privacy protection in neural networks, input data often come in integer type. Therefore, when performing ReLU activation, *IntReLU* is highly suitable. Compared to bit-wise TFHE, *IntReLU* can improve the time for homomorphic ReLU operations by 17%. Additionally, by adjusting parameters, it can be further adapted to specific privacy protection scenarios such as healthcare and transportation.

5 Conclusion

This paper proposes an efficient integer-wise homomorphic ReLU operation *IntReLU* on TFHE, capable of directly encrypting integer data and perform ReLU operations. We first extend integer multiplication on TFHE through linear transformation and odd slot mapping, proposing the algorithm *ExMultbyBin*. By employing the idea of equivalent functions, *ExMultbyBin* is used to achieve *IntReLU*, and the running time of *IntMultbyBin* is independent of data precision. When data precision is 6-bit, *IntReLU* reduces time by 17% compared to the bit-wise ReLU. For the future work, we will use the radix or the CRT representations to apply integer-wise ReLU on large data which precision exceeds 6-bit.

Acknowledgement. This work is supported by National Natural Science Foundation of China (No. 62272131) and Guangdong Provincial Key Laboratory of Novel Security Intelligence Technologies (No. 2022B1212010005).

References

1. Gilad-Bachrach, R., Dowlin, N., Laine, K., Lauter, K., Naehrig, M., Wernsing, J.: CryptoNets: applying neural networks to encrypted data with high throughput and accuracy. In: International Conference on Machine Learning, pp. 201–210. PMLR (2016)
2. Sanyal, A., Kusner, M., Gascon, A., Kanade, V.: TAPAS: tricks to accelerate (encrypted) prediction as a service. In: International Conference on Machine Learning, pp. 4490–4499. PMLR (2018)
3. Gentry, C.: Fully homomorphic encryption using ideal lattices. In: Proceedings of the Forty-First Annual ACM Symposium on Theory of Computing, pp. 169–178 (2009)
4. Van Dijk, M., Gentry, C., Halevi, S., Vaikuntanathan, V.: Fully homomorphic encryption over the integers. In: Gilbert, H. (ed.) EUROCRYPT 2010. LNCS, vol. 6110, pp. 24–43. Springer, Heidelberg (2010). https://doi.org/10.1007/978-3-642-13190-5_2
5. Smart, N.P., Vercauteren, F.: Fully homomorphic encryption with relatively small key and ciphertext sizes. In: Nguyen, P.Q., Pointcheval, D. (eds.) PKC 2010. LNCS, vol. 6056, pp. 420–443. Springer, Heidelberg (2010). https://doi.org/10.1007/978-3-642-13013-7_25
6. Stehlé, D., Steinfeld, R.: Faster Fully Homomorphic Encryption. In: Abe, M. (ed.) ASIACRYPT 2010. LNCS, vol. 6477, pp. 377–394. Springer, Heidelberg (2010). https://doi.org/10.1007/978-3-642-17373-8_22
7. Brakerski, Z., Gentry, C., Vaikuntanathan, V.: (leveled) fully homomorphic encryption without bootstrapping. ACM Trans. Comput. Theor. (TOCT) **6**(3), 1–36 (2014)
8. Fan, J., Vercauteren, F.: Somewhat practical fully homomorphic encryption. Cryptology ePrint Archive (2012)
9. Cheon, J.H., Kim, A., Kim, M., Song, Y.: Homomorphic encryption for arithmetic of approximate numbers. In: Takagi, T., Peyrin, T. (eds.) ASIACRYPT 2017, Part I 23. LNCS, vol. 10624, pp. 409–437. Springer, Cham (2017). https://doi.org/10.1007/978-3-319-70694-8_15

10. Gentry, C., Sahai, A., Waters, B.: Homomorphic encryption from learning with errors: conceptually-simpler, asymptotically-faster, attribute-based. In: Canetti, R., Garay, J.A. (eds.) CRYPTO 2013, Part I. LNCS, vol. 8042, pp. 75–92. Springer, Heidelberg (2013). https://doi.org/10.1007/978-3-642-40041-4_5
11. Ducas, L., Micciancio, D.: FHEW: bootstrapping homomorphic encryption in less than a second. In: Oswald, E., Fischlin, M. (eds.) EUROCRYPT 2015. LNCS, vol. 9056, pp. 617–640. Springer, Heidelberg (2015). https://doi.org/10.1007/978-3-662-46800-5_24
12. Chillotti, I., Gama, N., Georgieva, M., Izabachène, M.: TFHE: fast fully homomorphic encryption over the torus. J. Cryptol. **33**(1), 34–91 (2019). https://doi.org/10.1007/s00145-019-09319-x
13. Ao, W., Boddeti, V.N.: AutoFHE: automated adaption of CNNs for efficient evaluation over FHE. arXiv preprint arXiv:2310.08012 (2023)
14. Manino, E., Magri, B., Mustafa, M.A., Cordeiro, L.C.: Certified private inference on neural networks via Lipschitz-guided abstraction refinement⋆. In: 6th Workshop on Formal Methods for ML-Enabled Autonomous Systems, FoMLAS 2023, 17–18 July 2023, Paris, France (2023)
15. Park, J., Kim, M.J., Jung, W., Ahn, J.H.: AESPA: accuracy preserving low-degree polynomial activation for fast private inference. arXiv preprint arXiv:2201.06699 (2022)
16. Regev, O.: On lattices, learning with errors, random linear codes, and cryptography. J. ACM (JACM) **56**(6), 1–40 (2009)
17. Lou, Q., Jiang, L.: SHE: a fast and accurate deep neural network for encrypted data. In: Advances in Neural Information Processing Systems, vol. 32 (2019)
18. Bourse, F., Minelli, M., Minihold, M., Paillier, P.: Fast homomorphic evaluation of deep discretized neural networks. In: Shacham, H., Boldyreva, A. (eds.) CRYPTO 2018, Part III 38. LNCS, vol. 10993, pp. 483–512. Springer, Cham (2018). https://doi.org/10.1007/978-3-319-96878-0_17
19. Okada, H., Kiyomoto, S., Cid, C.: Integerwise functional bootstrapping on TFHE. In: Susilo, W., Deng, R.H., Guo, F., Li, Y., Intan, R. (eds.) ISC 2020. LNCS, vol. 12472, pp. 107–125. Springer, Cham (2020). https://doi.org/10.1007/978-3-030-62974-8_7
20. Kluczniak, K., Schild, L.: FDFB: full domain functional bootstrapping towards practical fully homomorphic encryption. arXiv preprint arXiv:2109.02731 (2021)
21. Yang, Z., Xie, X., Shen, H., Chen, S., Zhou, J.: TOTA: fully homomorphic encryption with smaller parameters and stronger security. Cryptology ePrint Archive (2021)
22. Clet, P.E., Zuber, M., Boudguiga, A., Sirdey, R., Gouy-Pailler, C.: Putting up the swiss army knife of homomorphic calculations by means of TFHE functional bootstrapping. Cryptology ePrint Archive (2022)
23. Chillotti, I., Ligier, D., Orfila, J.-B., Tap, S.: Improved programmable bootstrapping with larger precision and efficient arithmetic circuits for TFHE. In: Tibouchi, M., Wang, H. (eds.) ASIACRYPT 2021, Part III 27. LNCS, vol. 13092, pp. 670–699. Springer, Cham (2021). https://doi.org/10.1007/978-3-030-92078-4_23
24. Lyubashevsky, V., Peikert, C., Regev, O.: On ideal lattices and learning with errors over rings. In: Gilbert, H. (ed.) EUROCRYPT 2010. LNCS, vol. 6110, pp. 1–23. Springer, Heidelberg (2010). https://doi.org/10.1007/978-3-642-13190-5_1

25. Chillotti, I., Gama, N., Georgieva, M., Izabachène, M.: Faster fully homomorphic encryption: bootstrapping in less than 0.1 seconds. In: Cheon, J.H., Takagi, T. (eds.) ASIACRYPT 2016, Part I 22. LNCS, vol. 10031, pp. 3–33. Springer, Heidelberg (2016). https://doi.org/10.1007/978-3-662-53887-6_1
26. Peikert, C., Waters, B.: Lossy trapdoor functions and their applications. In: Proceedings of the Fortieth Annual ACM Symposium on Theory of Computing, pp. 187–196 (2008)

HERatio: Homomorphic Encryption of Rationals Using Laurent Polynomials

Luke Harmon(✉), Gaetan Delavignette, and Hanes Oliveira

Algemetric Inc., Colorado Springs, USA
{lharmon,gdelavignette,holiveira}@algemetric.com

Abstract. In this work we present HERatio, a homomorphic encryption scheme that builds on the scheme of Brakerski, and Fan and Vercauteren. Our scheme naturally accepts Laurent polynomials as inputs, allowing it to work with rationals via their bounded base-b expansions. This eliminates the need for a specialized encoder and streamlines encryption, while maintaining comparable efficiency to BFV. To achieve this, we introduce a new variant of the *Polynomial Learning With Errors* (PLWE) problem which employs Laurent polynomials instead of the usual "classic" polynomials, and provide a reduction to the PLWE problem.

Keywords: homomorphic encryption · Laurent polynomials · rational numbers · polynomial learning with errors

1 Introduction

In 1978, a year after the eponymous RSA cryptosystem was developed by Rivest, Shamir, and Adleman, the idea of "privacy homomorphisms" was proposed in [27]. This work was, in part, based on the observation that the product of two RSA-encrypted secrets would decrypt to the product of the two secrets. From there the idea of an encryption scheme that allows for various operations to be performed on encrypted data gained traction and became focus of many dedicated researchers. Such a scheme that permits both addition and multiplication on encrypted data is called a Homomorphic Encryption (HE) scheme. An HE scheme allows entities to out-source storage of and computations on sensitive information. Until 2009, all known HE schemes could only handle a bounded number of additions *and* multiplications[1] before decryption was required. This changed when Gentry published his seminal paper [18] that introduced a so-called Fully Homomorphic Encryption (FHE) scheme that could handle an unlimited number of additions and multiplications wthout decryption.

Presently, a large portion of the research and development in homomorphic encryption is focused on the usability of HE schemes in real-world applications. To this end, many researchers are working on efficient implementations with

[1] Several early schemes supported an unlimited number of one operation, but none supported both.

T. Zhu and Y. Li (Eds.): ACISP 2024, LNCS 14895, pp. 180–195, 2024.
https://doi.org/10.1007/978-981-97-5025-2_10

suitable software and/or hardware support and developing practically-usable libraries that can support tasks such as machine learning and and business analytics.

Most of the cutting-edge HE schemes are defined to encrypt integers (modulo integers) or polynomials whose coefficients are modulo integers. However, many real-world applications use real numbers or fixed/floating-point rational numbers instead of integers. Typically this is handled by using an encoder that converts the given inputs to a suitable form for homomorphic encryption. Of course this encoder must be homomorphic with respect to both addition and multiplication, and also injective. It is also important for this encoder to be efficient and not hinder the efficiency of the associated HE scheme. Several works, e.g. [2,3,11, 16,22] among others, have investigated such encodings.

The majority of modern HE schemes are based on the Ring Learning with Errors (RLWE) hard problem or one of its variants. In such schemes the plaintext space is the ring $R_t = \mathbb{Z}_t[x]/\Phi_m\mathbb{Z}_t[x]$ where $\Phi_m(x)$ is the m-th cyclotomic polynomial and $\mathbb{Z}_t$ is the ring of integers modulo t. In practice, the elements of R_t are simply viewed as polynomials with bounded degree. Encoding integers to elements of R_t can be straightforward, a common way being to use the base t representation of the integer. Encoding rational numbers, which are normally represented as fixed-point numbers, is more complex and has been discussed in multiple previous works [4,9,10,12,13,19,23,26]. One approach scales the fixed-point numbers to integers and then encodes them to polynomials using an appropriate base. Another simply treats them as fractional numbers. In [12], Costache et al. demonstrated that these two representations are isomorphic, and that the latter approach, although avoiding the overhead of bookkeeping with homomorphic ciphertexts, is difficult to analyze.

Most of these encodings have the same problem, which is discussed in [9,12]. Namely, the modulus t must be sufficiently large for the encoding to work correctly. This, in turn, creates faster noise growth and requires the HE scheme to use larger parameters, which hinders the scheme's efficiency. A few clever solutions have been devised to remedy this, the first one being presented in [9] by borrowing a mathematical technique from [21] and combining it with the HE scheme introduced by Fan and Vercauteren in [17] and by Brakerski in [5]. The main idea of solution is to replace the modulus t by the polynomial $x - b$ for a positive integer b, and then use as the plaintext the quotient ring $\mathbb{Z}/(b^n + 1)\mathbb{Z}$. The second solution, introduced in [8] encodes rationals by computing their base-b expansion, replacing b by an unknown x, and then mapping the resulting Laurent polynomial to an appropriate "classic" polynomial using a novel ring homomorphism. Similar encodings have been considered in [3,10,14,16].

Our Results. We introduce here a homomorphic encryption scheme for rationals. HERatio naturally accepts Laurent polynomials corresponding to bounded base-b expansions of rational numbers without the need of a specialized encoder. While enjoying efficiency comparable to BFV, it is more mathematically streamlined than prior art, and also mitigates the difficulty in choosing parameters to make sure a rational encoding "plays well" with the underlying HE scheme. HERatio may be viewed as a variant of the well-known Brakerski/Fan-Vercauteren (BFV) scheme [5,17], and is obtained (among other modifications) by replacing the rings of "classic" polynomials in BFV by rings of Laurent polynomials. Of course, it must be shown that these changes do not disturb the security of BFV. This is done by introducing a new hardness assumption using Laurent polynomials that can be reduced to the hardness assumption used by BFV. In particular, we introduce a new (decisional) version of the *Polynomial Learning With Errors* (PLWE) problem which uses the Laurent polynomial ring $\mathbb{Z}_q[x^{\pm 1}]/f\mathbb{Z}_q[x^{\pm 1}]$ instead of $\mathbb{Z}_q[x]/f\mathbb{Z}_q[x]$ as in the decisional-PLWE problem [7,15,20,28]. We then use the novel encoding homomorphism from [8] to show that the new problem based on Laurent polynomials is at least as hard as the decisional-PLWE problem under certain conditions, and that modifying the BFV scheme to use the new problem results in comparable efficiency.

2 Notations and Foundations

2.1 Notations

$\mathbb{Z}$ will denote the ring of integers, and $\mathbb{Z}_a$ will denote the ring of integers modulo $a \in \mathbb{Z}$. For $a \in \mathbb{Z}$, we will identify the elements of $\mathbb{Z}_a$ with integer representatives $\big[-\lfloor (a-1)/2 \rfloor, \lceil (a-1)/2 \rceil\big] \cap \mathbb{Z}$. For a ring R, $R[x]$ will denote the ring of polynomials in x with coefficients from R, and $R[x^{\pm 1}]$ will denote the ring of Laurent polynomials. For $r \in R$, R/rR will denote the quotient ring whose elements are the cosets (in R) of the ideal rR. Quotient rings will only arise when R is a ring of (Laurent) polynomials. For non-negative integers ℓ, k we use $\mathbb{Z}[x^{\pm 1}]_{-\ell}^{k}$ (resp. $\mathbb{Z}_a[x^{\pm 1}]_{-\ell}^{k}$) to denote the subset of $\mathbb{Z}[x^{\pm 1}]$ (resp. $\mathbb{Z}_a[x^{\pm 1}]$) with exponents ranging from $-\ell$ to k. For n a power of 2, Φ_{2n} denotes the $2n^{\text{th}}$ cyclotomic polynomial $x^n + 1$. For a distribution χ over a set A and a function $f : A \to A'$, we denote by $f(\chi)$ the distribution over A' induced by χ and f, $x \leftarrow \chi$ will mean that x is chosen from A according to the distribution χ.

2.2 Laurent Polynomials

Naïvely, one thinks of Laurent polynomials in an unknown x as polynomials in which there can be negative integer powers of x. E.g. $2x^{-3} - 5x^{-1} + 1 + x^2$ is a Laurent polynomial with integer coefficients. In general, the ring of Laurent polynomials in x with coefficients from the ring R may be defined as

$$R[x^{\pm 1}] := \left\{ \sum_{i \in \mathbb{Z}} a_i x^i \,\middle|\, a_i \in R, \text{ and only finitely many } a_i \text{ are nonzero} \right\}$$

An important property of $R[x^{\pm 1}]$ that will be used later (e.g. in the proof of Proposition 1) is that the ring of "classic" polynomials $R[x]$ is a subring of the ring of Laurent polynomials $R[x^{\pm 1}]$ for all rings R. Example 2 and Example 3 show how we use Laurent polynomials for our hardness-assumption reduction and base-b encoding of rationals, respectively.

2.3 Polynomial Learning with Errors

We first recall the *Polynomial Learning With Errors* (PLWE) problem [7,25], on which the well-known Brakerski/Fan-Vercauteren (BFV) scheme is based.

Definition 1 (Decisional PLWE). *For all $\kappa \in \mathbb{N}$, let $f(x) = f_\kappa(x)$ be a polynomial of degree $n = n(\kappa)$, and let $q = q(\kappa)$ be a prime integer. Let $R = \mathbb{Z}[x]/f\mathbb{Z}[x]$, $R_q = R/qR$, and χ denote a distribution over R. The decisional-PLWE problem $\mathsf{PLWE}_{f,q,\chi}$ states that for any $\ell = poly(\kappa)$ it holds that*

$$\{(a_i, a_i \cdot s + e_i)\}_{i\in[\ell]} \text{ is computationally indistinguishable from } \{(a_i, u_i)\}_{i\in[\ell]}$$

where s is sampled from the distribution χ, the a_i are uniform in R_q, the error polynomials e_i are sampled from χ, and the ring elements u_i are uniformly random over R_q.

It is also worth noting that for noise growth and performance reasons, it is possible to use a variant in which the coefficients of the secret key are uniformly selected from $\{-1, 0, 1\}$. This was originally suggested as an optimization in [17]. It was also shown in [6] that certain small-secret PLWE variants are as hard as those with $s \leftarrow \chi$ if the degree is sufficiently increased, even though more attacks can be used in this scenario, as shown in [1].

2.4 The BFV Scheme

Since our scheme is a variant of the Brakerski/Fan-Vercauteren scheme, we briefly recall some of the relevant details.

For its security, BFV relies on the hardness of the decisional-PLWE problem with $f(x) = \Phi_{2n}(x)$, and χ a discrete gaussian distribution on R with small standard deviation, normally chosen to be around 3.2 in practice [24].

The following algorithms are the basis of a common variant of the BFV scheme using the ternary distribution for s and u. Let $\Delta = \lfloor q/t \rfloor$ such that $q = \Delta t + r_t(q)$ for some $r_t(q) < t$. It should be assumed that $t \ll q$, which is required for most useful parameters.

$\mathsf{BFV.SecretKeyGen}$: Sample $s \in R$ with coefficients uniformly distributed in $\{-1, 0, 1\}$.
Output

$$\mathsf{sk} = s$$

$\mathsf{BFV.PublicKeyGen(sk)}$: Let $s = \mathsf{sk}$. Sample $a \leftarrow R_q$, and $e \leftarrow \chi$.
Output

$$\mathsf{pk} = \left(\left[-(as + e)\right]_q, a\right) \in R_q \times R_q$$

$\mathsf{BFV.Enc}(\mathsf{pk}, m \in R_t)$: To encrypt a message $m \in R_t$. Let $\mathsf{pk} = (p_0, p_1)$. Sample $u \in R$ with coefficients uniform in $\{-1, 0, 1\}$, and $e_0, e_1 \leftarrow \chi$.
Output

$$\mathsf{ct} = \left([\Delta m + p_0 u + e_0]_q, [p_1 u + e_1]_q\right) \in R_q \times R_q$$

$\mathsf{BFV.Dec}(\mathsf{sk}, \mathsf{ct} \in R_q \times R_q)$. Let $s = \mathsf{sk}$ and $\mathsf{ct} = (c_0, c_1)$.
Output

$$m'(X) = \left[\left\lfloor \frac{t}{q}[c_0 + c_1 s]_q \right\rceil\right]_t \in R_t$$

The remaining protocols, as well as a proof of correctness can be found in [17]. The security is proven in [25] through an indistinguishability argument which relies on the hardness of the decisional-PLWE problem (Definition 1).

3 LWE with Laurent Polynomials

Let $f \in \mathbb{Z}[x]$ and k, ℓ be non-negative integers such that $k + \ell + 1 = \deg f$. Here we introduce a new (decisional) version of the LWE problem which uses the ring $\mathbb{Z}[x^{\pm 1}]/f\mathbb{Z}[x^{\pm 1}]$ with representatives $\mathbb{Z}[x^{\pm 1}]_{-\ell}^{k}$ instead of $\mathbb{Z}[x]/f\mathbb{Z}[x]$ with representatives $\mathbb{Z}[x^{\pm 1}]_0^{k+\ell} = \{\mathsf{p}(x) \in \mathbb{Z}[x] | \deg \mathsf{p} < \deg f\}$ (as in the decisional-PLWE problem). We then show that the new problem based on Laurent polynomials is at least as hard as the decisional-PLWE problem under certain conditions. Throughout this section, $\mathfrak{R} = \mathbb{Z}[x]$ and $\mathfrak{L} = \mathbb{Z}[x^{\pm 1}]$. Also, for $a \in \mathbb{Z}$, $\mathfrak{R}_a = \mathfrak{R}/a\mathfrak{R}$ and $\mathfrak{L}_a = \mathfrak{L}/a\mathfrak{L}$.

3.1 From "classic" Polynomials to Laurent Polynomials

Before introducing the new problem, we build the tools required to show that the new problem is at least as hard as $\mathsf{PLWE}_{f,q,\chi}$ in certain cases. We first show that the rings $\mathbb{Z}[x]/f\mathbb{Z}[x]$ and $\mathbb{Z}[x^{\pm 1}]/f\mathbb{Z}[x^{\pm 1}]$ are isomorphic for certain polynomials f. The isomorphism ends up simply being the map $\mathsf{p}(x)+f\mathbb{Z}[x] \mapsto \mathsf{p}(x)+f\mathbb{Z}[x^{\pm 1}]$, which means that if we are to use proper Laurent polynomials as reprentatives, we need a way to switch representatives in the ring $\mathbb{Z}[x^{\pm 1}]/f\mathbb{Z}[x^{\pm 1}]$.
Recall the following classic theorem from elementary algebra.

Lemma 1 (Second Isomorphism Theorem). *Let R be a ring, S a subring of R, and I an ideal of R. Then $S + I$ is a subring of R, $S \cap I$, and*

$$\varphi : (S + I)/I \to S/(S \cap I) \text{ defined by } x + I \mapsto x + S \cap I$$

is a ring isomorphism.

Proposition 1. *Let $f \in \mathfrak{R}$ with $f(0) \in \mathbb{Z}$ a unit, $L = \mathfrak{L}/f\mathfrak{L}$, and $R = \mathfrak{R}/f\mathfrak{R}$. Then there is a ring isomorphism $L \cong R$.*

Proof. $\mathfrak{R}$ is a subring of $\mathfrak{L}$, and $f\mathfrak{L}$ is an ideal of $\mathfrak{L}$. So by Lemma 1, $(\mathfrak{R} + f\mathfrak{L})/f\mathfrak{L} \cong \mathfrak{R}/(\mathfrak{R} \cap f\mathfrak{L})$. We claim that $\mathfrak{R} + f\mathfrak{L} = \mathfrak{L}$. That the sum is contained in $\mathfrak{L}$ is easy. For the other containment it suffices to show that $x^k \in \mathfrak{R} + f\mathfrak{L}$ for all $k \in \mathbb{Z}$. That this holds for $k \geq 0$ is immediate from the definition of $\mathfrak{R}$. To see that this also holds for negative powers, first observe that $x^{-1}(f(x) - f(0)) \in \mathfrak{R}$. Now, $f(0)x^{-1} = x^{-1}(f(0) - f(x)) + x^{-1}f(x) \in \mathfrak{R} + f\mathfrak{L}$. Whence

$x^{-1} = f(0)^{-1}(f(0)x^{-1}) \in \mathfrak{R} + f\mathfrak{L}$, since $f(0)$ is a unit. An easy induction then shows that $x^k \in \mathfrak{R} + f\mathfrak{L}$ for all $k < 0$. Clearly $\mathfrak{R} \cap f\mathfrak{L} = f\mathfrak{R}$, whence $\mathfrak{L}/f\mathfrak{L} \cong \mathfrak{R}/f\mathfrak{R}$. Equivalently, $L \cong R$, as desired.

Remark 1. The same result holds if we replace everywhere $\mathbb{Z}$ by $\mathbb{Z}_a$, $a \in \mathbb{Z}$, but with the slightly better condition that $f(0)$ is invertible modulo a.

As mentioned at the beginning of this subsection, we need to map representatives in the set $\mathbb{Z}_a[x^{\pm 1}]^k_{-\ell}$ to their equivalents in $\mathbb{Z}_a[x^{\pm 1}]^{k+\ell}_0$ and vice versa. This can be done efficiently (using only matrix multiplications and inversions) using a ring homomorphism introduced by Castryck *et al* in [8]. We pause briefly to describe this homomorphism and give an example.
Let $f(x) \in \mathfrak{R}_a$ such that $f(0) \in \mathbb{Z}_a$ is a unit. The ring homomorphism $\mathfrak{L}_a \to R_a = \mathfrak{R}_a/f\mathfrak{R}_a$ is induced by the correspondences

$$x \mapsto x \quad \text{and} \quad x^{-1} \mapsto -g(x)f(0)^{-1}, \quad \text{where} \quad f(x) = g(x)x + f(0). \tag{1}$$

Let ℓ, k be non-negative integers satisfying $\deg f = k + \ell + 1$. The ring homomorphism in Equation (1) induces a free $\mathbb{Z}_a$-module isomorphism $\eta_{f,(-\ell,k)}$: $\mathbb{Z}_a[x^{\pm 1}]^k_{-\ell} \to \mathbb{Z}_a[x^{\pm 1}]^{k+\ell}_0$. As a free module isomorphism, $\eta_{f,(-\ell,k)}$ can be computed efficiently using a matrix multiplication. If the polynomial f and integers k, ℓ are clear from context, we will simply write η.

Example 1. Consider $f(x) = 1 + x + x^2 - x^3 + x^4 \in \mathbb{Z}_3[x]$. $f(0) \in \mathbb{Z}_3$ is a unit and $f(x) = x(1 + x - x^2 + x^3) + 1$, so η_f is defined by

$$x \mapsto x \quad \text{and} \quad x^{-1} \mapsto -1 - x + x^2 - x^3$$

Taking $\ell = 3$ and $k = 0$ makes the domain of η_f $\mathbb{Z}_3[x^{\pm 1}]^0_{-3} = \{ax^{-3} + bx^{-2} + cx^{-1} + d \mid a, b, c, d \in \mathbb{Z}_3\}$, and the range $\mathbb{Z}_3[x^{\pm 1}]^3_0 = \{a + bx + cx^2 + dx^3 \mid a, b, c, d \in \mathbb{Z}_3\}$. The domain and range being free $\mathbb{Z}_3$-modules allows us to represent η_f (encoding map) and its inverse (decoding map) using matrices:

$$\text{encode matrix} = \begin{bmatrix} -1 & 0 & -1 & 1 \\ 1 & -1 & -1 & 0 \\ 1 & 1 & 1 & 0 \\ 0 & 1 & -1 & 0 \end{bmatrix}, \qquad \text{decode matrix} = \begin{bmatrix} 0 & -1 & -1 & 0 \\ 0 & -1 & 1 & -1 \\ 0 & -1 & 1 & 1 \\ 1 & 1 & 0 & 1 \end{bmatrix}.$$

It is worth mentioning, as noted in [8], that when $f(x) = \Phi_{2n}(x) = x^n + 1$ (a common choice for PLWE) η_f can be computed without any matrix multiplications. In particular, η_f is defined by $x \mapsto x$ and $x^{-1} \mapsto -x^{n-1}$, meaning that a simple "degree shift" may be applied to any negative powers in a Laurent polynomial argument to compute η_f.

Proposition 2. *Let $f(x) = x \cdot g(x) + f(0) \in \mathbb{Z}_a[x]$ such that $f(0)$ is a unit, and $\ell, k \in \mathbb{Z}$ be non-negative. If $\eta_{f,(-\ell,k)} : \mathbb{Z}_a[x^{\pm 1}]^k_{-\ell} \to \mathbb{Z}_a[x^{\pm 1}]^{k+\ell}_0$ is defined by $x \mapsto x$ and $x^{-1} \mapsto -g(x)f(0)^{-1}$, then $\alpha(x) - \eta_{f,(-\ell,k)}(\alpha(x)) = 0 \bmod f(x)$ for all $\alpha \in \mathbb{Z}_a[x^{\pm 1}]^k{}_{\ell}$.*

Proof. Let $\alpha = a_{-\ell}x^{-\ell} + \cdots + a_{-2}x^{-2} + a_{-1}x^{-1} + a_0 + \cdots + a_k x^k \in Z_a[x^{\pm 1}]^k_{-\ell}$. Since the context is clear, we will write η in place of $\eta_{f,(-\ell,k)}$.
By definition,

$$\eta(\alpha) = \sum_{i=1}^{\ell} a_{-i}(-1)^i g(x)^i f(0)^{-i} + \sum_{i=0}^{k} a_i x^i.$$

It follows that

$$\begin{aligned}\alpha - \eta(\alpha) &= \sum_{i=1}^{\ell} a_{-i}x^{-i} - \sum_{i=1}^{\ell} a_{-i}(-1)^i g(x)^i f(0)^{-i} \\ &= \sum_{i=1}^{\ell} a_{-i}\Big(x^{-i} - (-1)^i g(x)^i f(0)^{-i}\Big)\end{aligned} \tag{2}$$

We claim that for all $i \geq 1$ there exists $\beta_i \in \mathbb{Z}_a[x^{\pm 1}]$ such that $x^{-i} - f(x)\beta_i(x) = \big(-g(x)f(0)^{-1}\big)^i$. Proceeding inductively, observe that $f(x) = x \cdot g(x) + f(0) \implies x^{-1} - f(x)\big(x^{-1}f(0)^{-1}\big) = -g(x)f(0)^{-1}$. So $\beta_1(x) = f(0)^{-1}x^{-1}$. Now, suppose that for some $j \geq 1$ there is $\beta_j \in \mathbb{Z}_a[x^{\pm 1}]$ such that $x^{-j} - f(x)\beta_j(x) = \big(-g(x)f(0)^{-1}\big)^j$. It follows that

$$\begin{aligned}\big(-g(x)f(0)^{-1}\big)^{j+1} &= \big(-g(x)f(0)^{-1}\big)^j\big(-g(x)f(0)^{-1}\big) \\ &= \big(x^{-j} - f(x)\beta_j(x)\big)\big(x^{-1} - f(x)\beta_1(x)\big) \\ &= x^{-(j+1)} - f(x)x^{-1}\beta_j(x) - f(x)x^{-j}\beta_1(x) + f(x)^2\beta_j(x)\beta_1(x) \\ &= x^{-(j+1)} - f(x)\Big(x^{-1}\beta_j(x) - x^{-j}\beta_1(x) + f(x)\beta_j(x)\beta_1(x)\Big)\end{aligned}$$

So $\beta_{j+1}(x) = x^{-1}\beta_j(x) - x^{-j}\beta_1(x) + f(x)\beta_j(x)\beta_1(x)$. This proves the claim. Notice that the claim can be rewritten as $f(x)\beta_i(x) = x^{-i} - (-1)^i g(x)^i f(0)^{-i}$. Substituting this into Equation (2), we get

$$\alpha - \eta(\alpha) = \sum_{i=1}^{\ell} a_{-i}\Big(x^{-i} - (-1)^i g(x)^i f(0)^{-i}\Big) = \sum_{i=1}^{\ell} a_{-i} f(x)\beta_i(x) = 0 \bmod f(x).$$

This completes the proof of the proposition.

Lemma 2. *The set $\mathbb{Z}_a[x^{\pm 1}]^k_{-\ell}$ is a complete set of coset representatives of $L_a = L/aL$ as long as $\deg f = k + \ell + 1$.*

Proof. Observe that for any two $\alpha, \beta \in \mathbb{Z}_a[x^{\pm 1}]^k_{-\ell}$, $\alpha \neq \beta \bmod f$. It now suffices to recall from [8] that $\eta_{f,(-\ell,k)} : \mathbb{Z}_a[x^{\pm 1}]^k_{-\ell} \to \mathbb{Z}_a[x^{\pm 1}]^{k+\ell}_0$ is a free module isomorphism and $\mathbb{Z}_a[x^{\pm 1}]^{k+\ell}_0$ is a complete set of coset representatives for R_a when $\deg f = k + \ell + 1$.

Corollary 1. *$\eta = \eta_{f,(-\ell,k)} : \mathbb{Z}_a[x^{\pm 1}]^k_{-\ell} \to \mathbb{Z}_a[x^{\pm 1}]^{k+\ell}_0$ induces the identity homomorphism on L_a. That is, the mapping $L_a \to L_a$ defined by $\alpha + f\mathfrak{L}_a \mapsto \eta(\alpha) + f\mathfrak{L}_a$ is the identity homomorphism.*

The following example illustrates Corollary 1.

Example 2. Using the polynomial $f(x) = 1 + x + x^2 - x^3 + x^4$ from Example 1, along with $\ell = 3$, $k = 0$, and $a = 3$, we have $\eta : \mathbb{Z}_3[x^{\pm 1}]_{-3}^{0} \to \mathbb{Z}_3[x^{\pm 1}]_{0}^{3}$. Let $\mathsf{p} \in L_3 = \mathbb{Z}_3[x^{\pm 1}]/f\mathbb{Z}_3[x^{\pm 1}]$. By Lemma 2, $\mathbb{Z}_3[x^{\pm 1}]_{-3}^{0}$ is a complete set of coset representatives for L_3, so we can let $\mathsf{p}(x) \in \mathbb{Z}_3[x^{\pm 1}]_{-3}^{0}$. Say $\mathsf{p}(x) = x^{-3} + x^{-2} - x^{-1} + 1$. Now, using the encoding matrix of Example 1, we have

$$\begin{aligned}
\eta(\mathsf{p}) - \mathsf{p} &= 1 + x + x^2 - x^3 - (x^{-3} + x^{-2} - x^{-1} + 1) \\
&= -x^{-3} - x^{-2} + x^{-1} + x + x^2 - x^3 \\
&= (-x^{-3} - x^{-1}) f(x) + 3(-1 - x + x^3) \\
&= 0 \bmod f(x), \text{ since coefficients of } f \text{ are from } \mathbb{Z}_3.
\end{aligned}$$

In particular, this shows that $\mathsf{p} + f\mathbb{Z}_3[x^{\pm 1}] = \eta(\mathsf{p}) + f\mathbb{Z}_3[x^{\pm 1}]$.

We now define a version of the Learning With Errors problem over the Laurent polynomial ring modulo the principal ideal generated by f and show that it is at least as hard as its PLWE counterpart.

Definition 2 (Decisional Laurent LWE). *For all $\kappa \in \mathbb{N}$, let $f(x) = f_\kappa(x)$ be a polynomial of degree $n = n(\kappa)$, and let $q = q(\kappa)$ be a prime integer such that $f(0)$ is invertible modulo q. Let $L = \mathbb{Z}[x^{\pm 1}]/f\mathbb{Z}[x^{\pm 1}]$, $L_q = L/qL$, and $\mathcal{X}$ denote a distribution over R. For non-negative integers ℓ, k such that $\deg f = \ell + k + 1$, take the coset representatives of L to be $\mathbb{Z}[x^{\pm 1}]_{-\ell}^{k}$. The decisional-LLWE problem $\mathsf{LLWE}_{f,q,\mathcal{X}}^{(-\ell,k)}$ states that for any $m = poly(\kappa)$ it holds that*

$$\{(a_i, a_i \cdot S + e_i)\}_{i \in [m]} \textit{ is computationally indistinguishable from } \{(a_i, u_i)\}_{i \in [m]}$$

where S is sampled from the distribution $\mathcal{X}$, the a_i are uniform in R_q, the error polynomials e_i are sampled from $\mathcal{X}$, and the ring elements u_i are uniformly random over R_q.

Theorem 1. *If one can solve the $\mathsf{LLWE}_{f,q,\mathcal{X}}^{(-\ell,k)}$ problem with the polynomial f such that $f(0) \in \mathbb{Z}_q$ is a unit, then one can solve the $\mathsf{PLWE}_{f,q,\chi}$ problem with the same polynomial f, $\chi = \eta(\mathcal{X})$, and $s = \eta(S)$.*

Proof. We will use the following pair of mappings from Proposition 1 and [8], respectively:

$$\begin{aligned}
&\gamma : R_q \to L_q \text{ defined by } \alpha + f\mathfrak{R}_a \mapsto \alpha + f\mathfrak{L}_a, \text{ and} \\
&\eta = \eta_{f,(-\ell,k)} : \mathbb{Z}_q[x^{\pm 1}]_{-\ell}^{k} \to \mathbb{Z}_q[x^{\pm 1}]_{0}^{k+\ell}
\end{aligned} \tag{3}$$

The latter mapping simply switches between sets of coset representatives for L_q. Recall that $A_{s,\chi}^{(q)}$ is the PLWE distribution and $\mathcal{L}_{S,\mathcal{X}}^{(q)}$ is the LLWE distribution. We can map the elements $(a, [a \cdot s + e]_q) \in R_q \times R_q$ to their isomorphic images under γ^{-1}

$(\gamma^{-1}(a), [\gamma^{-1}(a) \cdot \gamma^{-1}(s) + \gamma^{-1}(e)]_q) = (a, [a \cdot s + e]_q) \in L_q \times L_q$[2]. We then use η^{-1} to switch from coset representatives in $\mathbb{Z}_t[x^{\pm 1}]_0^{k+\ell}$ to coset representatives in $\mathbb{Z}_q[x^{\pm 1}]_{-\ell}^k$. By Corollary 1, we see that
$\eta^{-1}(a \cdot s + e) = \eta^{-1}(a)\eta^{-1}(s) + \eta^{-1}(e) = \eta^{-1}(a) \cdot S + \eta^{-1}(e)$. Clearly $\eta^{-1}(a)$ is uniform in L_q and, since $\chi = \eta(\mathcal{X})$, there is $e' \leftarrow \mathcal{X}$ such that $e' = \eta^{-1}(e)$. Consequently, $(\eta^{-1}(a), [\eta^{-1}(a) \cdot S + \eta^{-1}(e)]_q)$ is an LLWE instance according to the distribution $\mathcal{L}_{S,\mathcal{X}}^{(q)}$. So, if we can solve the Decisional-LLWE$_{d,q,\mathcal{X}}$ problem with $\chi = \eta(\mathcal{X})$ and $s = \eta(S)$, then we can solve the Decisional-PLWE$_{d,q,\chi}$ problem.

3.2 When Is Laurent LWE Hard?

If $f = \Phi_{2n}$, and χ is an appropriate gaussian error distribution over R with mean 0 for which $\mathsf{PLWE}_{f,q,\chi}$ problem is hard, then $\mathsf{LLWE}_{f,q,\chi}^{(-\ell,k)}$ problem is also hard. This comes down to the fact that, in this case, $\eta_f^{-1}(\chi) = \chi$. We elaborate below.

Proposition 3. *If $f(x) = x^n + 1$ and χ is a spherical discrete gaussian distribution over L with $\mu = 0$ and diagonal covariance matrix $\sigma^2 I$, then $\eta_{f,(-\ell,k)}^{-1}(\chi) = \chi$ for all integers $\ell, k \geq 0$ such that $n = k + \ell + 1$.*

Sketch of proof. First observe that for $f(x) = x^n + 1$, η_f is defined by the correspondences $x \mapsto x$ and $x^{-1} \mapsto -x^{n-1}$. This means, depending on the choice of ℓ, k, that η_f^{-1} can be viewed as a composition of negacyclic permutations of the coefficient vector of its argument. So, for $e \leftarrow \chi$, the set coefficients of $\eta_f^{-1}(e)$ and e are the same up to sign. The covariance matrix of χ being $\sigma^2 I$ implies that sampling $e \leftarrow \chi$ can be done coefficient-wise, sampling each coefficient from a gaussian distribution over $\mathbb{Z}$ with mean 0 and variance σ^2. Since $\eta_f^{-1}(e)$ and e have the same set of coefficients up to sign, the distribution from which each coefficient is selected remains unchanged. Consequently, $\eta_f^{-1}(\chi) = \chi$.

Theorem 2 (corollary of Proposition 3). *If χ is a spherical discrete gaussian distribution over L with mean 0 and diagonal covariance matrix $\sigma^2 I$, then $\mathsf{LLWE}_{f,q,\chi}^{(-\ell,k)}$ is as hard as $\mathsf{PLWE}_{f,q,\chi}$.*

We are unsure whether the LLWE and PLWE problem are actually equivalent, or whether there are polynomials f and distributions $\chi, \mathcal{X}$ with $\eta_f(\mathcal{X}) = \chi$ for which one of $\mathsf{PLWE}_{f,q,\chi}$ and $\mathsf{LLWE}_{f,q,\mathcal{X}}^{(-\ell,k)}$ is hard while the other is not. For now, we relegate this investigation to future work.

4 The New Scheme: HERatio

4.1 Encoding Rationals

Despite removing much of the machinery required by previous works to encode rationals for HE, HERatio does require some pre-processing – one must compute

[2] γ acts like the identity on coset representatives.

a bounded base-b expansion of a rational and then replace b by the unknown x to get a Laurent polynomial. We elaborate below on encode/decode and its correctness conditions.

HERatio.Encode$(m, b, (-\ell, k))$: For a message $m \in \mathbb{Q}$, compute the base-b expansion of m to obtain $\sum_{-\infty}^{\infty} a_i b^i$, where the $a_i \in \mathbb{Z}_b$. Truncate (if necessary) and replace everywhere b by x to obtain the Laurent polynomial $l(x) = \sum_{i=-\ell}^{k} a_i x^i \in \mathbb{Z}[x^{\pm 1}]^k_{-\ell}$ satisfying $l(b) \approx m$. Output $l(x)$.
HERatio.Decode$(l'(x), b)$: Output $l'(b)$.

Example 3. Consider $m = 625/7$, a base $b = 3$, and bounds $\ell = 4$ and $k = 5$. To encode m, first expand it using $b = 3$. Note that the coefficients are from $\mathbb{Z}_3$ whose representatives are $\{-1, 0, 1\}$. This results in:

$$625/7 = O(3^{-5}) + (-1)3^{-4} + (-1)3^{-3} + (1)3^{-1} + (1)3^2 + (1)3^4$$

Truncate using the bounds $\ell = 4$ and $k = 5$ to get:

$$625/7 \approx -3^{-4} - 3^{-3} + 3^{-1} + 3^2 + 3^4$$

Replacing everywhere the base 3 by x to get a Laurent polynomial, gives the encoding of m:

$$l(x) = -x^{-4} - x^{-3} + x^{-1} + x^2 + x^4$$

Observe that $l(3) = 7232/81$, and $|7232/81 - 625/7| < 0.002$, so the approximation is quite close.

Correctness of Decoding. Let $r_1, \ldots, r_n \in \mathbb{Q}$ and $b \in \mathbb{Z}$ such that the base-b expansion of r_i is $l_i(b)$ for $l_i \in \mathbb{Z}[x^{\pm 1}]^k_{-\ell}$. Let $\mathcal{C}$ be an arithmetic circuit, and $l^* = \mathcal{C}(l_1, \ldots, l_n)$ be the evaluation of $\mathcal{C}$ at the l_i. Decoding is correct, i.e. $l^*(b) = \mathcal{C}(l_1(b), \ldots, l_n(b))$, provided l^* remains in the *set* $\mathbb{Z}[x^{\pm 1}]^k_{-\ell}$.

An additional restriction must be imposed for correctness when the encoder is used with HERatio. This is because the plaintext space is a ring of Laurent polynomials whose coefficients come from the ring $\mathbb{Z}_t$. The restriction is that computing the Laurent polynomial l^* above must not result coefficient overflow modulo t. This requires one to choose $b \leq t$ with b and t sufficiently far apart so that the desired circuits can be evaluated.

4.2 HERatio

Let $L = \mathbb{Z}[x^{\pm 1}]/\Phi_{2n}\mathbb{Z}[x^{\pm 1}]$, and for $a \in \mathbb{Z}$ let $L_a = L/aL$. The plaintext space is the ring L_t, for $t \geq 2$, and the ciphertext space is product ring $L_q \times L_q$ for $q \gg t$. The coset representatives of the elements of L are $\mathbb{Z}[x^{\pm 1}]^k_{-\ell}$, where ℓ, k are nonnegative integers satisfying $k + \ell + 1 = n$. We let λ be the security parameter and $\mathcal{X}$ be a discrete Gaussian distribution for which $\mathsf{LLWE}^{(-\ell,k)}_{f,q,\mathcal{X}}$ is hard (see Theorem 2). HERatio is obtained from BFV by replacing everywhere R by L, R_t by L_t, and $R_q \times R_q$ by $L_q \times L_q$.

$\mathsf{HERatio.SecretKeyGen}$: Sample $s \in L$ with coefficients uniformly distributed in $\{-1, 0, 1\}$[3].
Output

$$\mathsf{sk} = s$$

$\mathsf{HERatio.PublicKeyGen(sk)}$: Let $s = \mathsf{sk}$. Sample $a \leftarrow L_q$, and $e \leftarrow \mathcal{X}$.
Output

$$\mathsf{pk} = \left(\left[-\left(as + e\right)\right]_q, a\right) \in L_q \times L_q$$

$\mathsf{HERatio.EvalKeyGen(sk)}$: For $i = 0, \ldots, \ell$, where $w \geq 2$ and $\ell = \lfloor \log_w q \rfloor$, sample $a_i \leftarrow L_q$, and $e_i \leftarrow \mathcal{X}$. Let

$$\mathsf{evk}[i] = ([-(a_i s + e_i) + w^i s^2]_q, a_i) \in L_q \times L_q$$

Output the vector of pairs

$$\mathsf{evk} = \left(\mathsf{evk}[0], \ldots, \mathsf{evk}[\ell]\right)$$

$\mathsf{HERatio.Enc}(\mathsf{pk}, \ell \in L_t)$: Let $\Delta = \left\lfloor \frac{q}{t} \right\rfloor$ and $\mathsf{pk} = (p_0, p_1)$. Sample $u \in L$ with coefficients uniform in $\{-1, 0, 1\}$, and $e_0, e_1 \leftarrow \chi$.
Output

$$\mathsf{ct} = \left([\Delta\ell + p_0 u + e_0]_q, [p_1 u + e_1]_q\right) \in L_q \times L_q$$

$\mathsf{HERatio.Dec}(\mathsf{sk}, \mathsf{ct} \in L_q \times L_q)$: Let $s = \mathsf{sk}$ and $\mathsf{ct} = (c_0, c_1)$.
Output

$$\ell'(X) = \left[\left\lfloor \frac{t}{q} [c_0 + c_1 s]_q \right\rceil\right]_t \in L_t$$

$\mathsf{HERatio.Add}(\mathsf{ct}_0, \mathsf{ct}_1)$:
Output

$$\left(ct_0[0] + ct_1[0], ct_0[1] + ct_1[1]\right) \in L_q \times L_q$$

$\mathsf{HERatio.PartialMult}(\mathsf{ct}_0, \mathsf{ct}_1)$: Denote $(c_0, c_1) = \mathsf{ct}_0$ and $(d_0, d_1) = \mathsf{ct}_1$.
Compute

$$c'_0 = \left[\left\lfloor \frac{t}{q} c_0 d_0 \right\rceil\right]_q, \; c'_1 = \left[\left\lfloor \frac{t}{q} (c_0 d_1 + c_1 d_0) \right\rceil\right]_q, \; c'_2 = \left[\left\lfloor \frac{t}{q} c_1 d_1 \right\rceil\right]_q,$$

Output

$$ct_{\text{prod}} = (c'_0, c'_1, c'_2) \in L_q \times L_q \times L_q$$

$\mathsf{HERatio.Relinearize}(\mathsf{ct}_{\text{prod}} \in L_q \times L_q \times L_q, \mathsf{evk})$: Denote $(c'_0, c'_1, c'_2) = ct_{\text{prod}}$. Express c'_2 in base w, so that $c'_2 = \sum_{i=0}^{\ell} c_2'^{(i)} w^i$. Set

$$c_0 = c'_0 + \sum_{i=0}^{\ell} \mathsf{evk}[i][0] c_2'^{(i)}, \; c_1 = c'_1 + \sum_{i=0}^{\ell} \mathsf{evk}[i][1] c_2'^{(i)},$$

[3] The ability to sample from this set while maintaining security is inherited from BFV.

Output

$$(c_0, c_1) \in L_q \times L_q$$

HERatio.Mult(ct_0, ct_1, evk):
Output

$$\text{HERatio.Relinearize}\big(\text{HERatio.PartialMult}(ct_0, ct_1), \text{evk}\big)$$

4.3 Correctness of HERatio

In [17], Fan and Vercauteren analyze noise growth and derive correctness conditions for BFV using the polynomial infinity norm $\|\cdot\|_\infty$. Since this norm is defined coefficient-wise, switching from "classic" to Laurent polynomials does not change their analysis. This means that HERatio inherits the correctness conditions from [17] (in particular, lemma 1 and theorem 1).

5 Comparison with BFV

Since HERatio is obtained from BFV by simply swapping the ring of polynomials $\mathbb{Z}[x]/\Phi_{2n}\mathbb{Z}[x]$ for the ring of Laurent polynomials $\mathbb{Z}[x^{\pm 1}]/\Phi_{2n}\mathbb{Z}[x^{\pm 1}]$, we expect the performance of the two schemes to be very similar. With this in mind, we compare separately the encodings which allow HERatio and BFV, respectively, to work with rational numbers. The encoding used for BFV is that of [8] (see Equation (1)), and the encoding for HERatio is detailed in Sect. 4.1. We emphasize that the purpose of our implementations is to show that HERatio and BFV have similar performance when implemented the same way.

5.1 Implementation

We divided our results into 4 distinct groups, namely *Codec Operations* (encoding and decoding), *Key Generation*, *Ciphertext Generation*, and *Homomorphic Operations*, although some of them show dependent relationships. We measured the average time execution and memory used on each operation in order to analyze comparatively both HERatio and BFV schemes, along with their respective codecs HERatio.Encode, HERatio.Decode, and the rational encoder and decoder suggested in [8].

We chose messages $m_0 = 12345.678$ and $m_1 = 947.1273$, the additive scalar $s_a = 4$, and multiplicative scalar $s_m = 42.122$ to perform all calculations. The variables were initially defined either as 64-bits float numbers or 64-bits integers. However, during the execution these inputs were converted into arbitrary-size numbers to preserve correctness. For comparison purposes, we used the same parameters for HERatio and BFV, these are polynomials of length 32, with $q = 9,876,523,525, t = 2,131, \sigma = 3.19$, 10 as the expansion base, and $w = 128$ for the relinearization base for both BFV and HERatio.

The experiments were implemented with the Golang programming language version 1.19, on a MacBook Pro 15-inch, MacOS Monterey 12.7.1, 2.7GHz Quad-Core Intel(R) Core(TM) i7, 16GB 2133 MHz LPDDR3, 500GB SSD. Each operation was analyzed by evaluating the mean of 1,000 executions, where the runtime and memory allocated were measured. Note that our implementation of BFV did not include any optimizations, though any existing optimization of BFV could be translated to an equivalent optimization of HERatio. Our implementation of HERatio can be found at https://github.com/Algemetric/HERatio_go.

Codec Operations. For Codec operations, we measured the performance of encoding a message m_0 as c_0 through both codecs, such that $c_0 = \mathsf{Encode}(m_0)$. Furthermore, we also executed the reverse operation to recover the original message from code c_0, such that $m_0' = \mathsf{Decode}(c_0)$ (Table 1).

Table 1. Codec Results.

	Avg. Time (ms)			Memory (MB)	
Operation	[8]	HERatio		[8]	HERatio
Encoding	0.0619	0.0567	+8.40%	0.0227	0.0223
Decoding	0.0270	0.0245	+1.76%	0.0096	0.0083

The HERatio codec was 0.0567 ms faster for encoding a message, and 0.0025 ms faster for decoding.

Key Generation. The key generation analysis measured the functions SecretKeyGen, PublicKeyGen, and EvalKeyGen, where all procedures used the same pseudo-random source. The $sk = \mathsf{SecretKeyGen}()$ function is equivalent for both schemes, whereas $pk = \mathsf{PublicKeyGen}(sk)$, and $evk = \mathsf{EvalKeyGen}(sk)$ differ in their internal polynomial multiplication.

HERatio and BFV had a differential of 0.0073 milliseconds (ms) for the secret key generation, a negligible difference that can be used as a threshold for comparison since the implementation of both functions are nearly equal. For public key generation HERatio was 0.0342 ms faster, and 0.7912 ms ahead for evaluation key.

Ciphertext Generation. On cipher functions we had a small difference, where HERatio encrypted code c_0 as $ct_0 = \mathsf{Encrypt}(pk, c_0)$ 0.0054 ms slower than BFV, and decrypted the same ciphertext as $c_0' = \mathsf{Decrypt}(sk, ct_0)$ 0.0099 faster.

When we consider the difference between equivalent operations for both schemes, such as SecretKeyGen, we notice that the generation of ciphertexts have a negligible margin.

Table 2. Operation Results.

	Avg. Time (ms)			Memory (MB)	
Operation	BFV	HERatio		BFV	HERatio
Secret Key	0.1066	0.1139	−6.40%	0.0324	0.0344
Public Key	5.9967	5.9625	+0.57%	3.2840	3.3453
Evaluation Key	56.2224	55.4312	+1.40%	29.2504	28.0354
Encryption	0.5886	0.5940	−0.90%	0.2867	0.2900
Decryption	0.3501	0.3402	+2.82%	0.1733	0.1756
Scalar Addition	0.0669	0.0719	−6.95%	0.0324	0.0344
Ciphertext Addition	0.1608	0.1598	+0.62%	0.0740	0.0767
Scalar Multiplication	0.1592	0.1551	+2.57%	0.0759	0.0744
Ciphertext Multiplication	3.1545	3.2326	−2.41%	1.6593	1.6857

Homomorphic Operations. We homomorphically tested additive and multiplicative operations (Table 2) between ciphertexts (i.e., ct_0 and ct_1) and additive (i.e., s_a) and multiplicative scalars (i.e., s_m).

HERatio executed the ciphertext addition and scalar multiplication, respectively, 0.001 ms and 0.004 ms faster than BFV, whereas the scalar addition and ciphertext multiplication were 0.005 ms and 0.0781 ms slower. The memory usage did not increase in a rate that was relevant for the experiment.

6 Conclusion and Future Work

We have presented a new variant of the LWE problem based on Laurent polynomials, and constructed a new variant of the BFV scheme based on this problem. The plaintext space of our scheme, HERatio, is a set of Laurent polynomials with exponents ranging from $-\ell$ to k. The exponent range can be chosen to suit the goal application, and does not affect the hardness of the new LWE variant. The main appeal of HERatio is that it can work with rationals via their bounded base-b expansions – all one must do is replace b by the unknown x. While enjoying efficiency comparable to BFV, HERatio is more mathematically streamlined than prior art, and also mitigates the difficulty in choosing parameters to make sure a rational encoding “plays well” with the underlying HE scheme.

Future Work. As mentioned in Sect. 3.2 we are unsure of whether $\mathsf{LLWE}^{(-\ell,k)}_{f,q,\mathcal{X}}$ and $\mathsf{PLWE}_{f,q,\chi}$ are equivalent when $\eta_f(\mathcal{X}) = \chi$. It is our hope that a more detailed analysis of $\mathsf{LLWE}^{(-\ell,k)}_{f,q,\mathcal{X}}$ yields better insight into the Learning With Errors problems. The implementation of HERatio should also be optimized to facilitate meaningful comparison with schemes like CKKS.

Acknowledgements. The authors warmly thank Assistant Professor Arnab Roy, University of Innsbruck, for suggesting the use of Laurent polynomial rings, and for many other helpful comments while this work was in its infancy. Also, a hearty thanks to the three reviewers for their helpful suggestions. This work is fully supported by Algemetric.

Disclosure of Interests. The authors have no competing interests to declare that are relevant to this work.

References

1. Albrecht, M.R.: On dual lattice attacks against small-secret LWE and parameter choices in HElib and SEAL. Cryptology ePrint Archive, Report 2017/047 (2017). https://eprint.iacr.org/2017/047
2. Arita, S., Nakasato, S.: Fully homomorphic encryption for point numbers. Cryptology ePrint Archive, Report 2016/402 (2016). https://ia.cr/2016/402
3. Bonte, C., Bootland, C., Bos, J.W., Castryck, W., Iliashenko, I., Vercauteren, F.: Faster homomorphic function evaluation using non-integral base encoding. In: Fischer, W., Homma, N. (eds.) CHES 2017. LNCS, vol. 10529, pp. 579–600. Springer, Cham (2017). https://doi.org/10.1007/978-3-319-66787-4_28
4. Bos, J.W., Lauter, K.E., Naehrig, M.: Private predictive analysis on encrypted medical data. J. Biomed. Inform. **50**, 234–43 (2014)
5. Brakerski, Z.: Fully homomorphic encryption without modulus switching from classical GapSVP. In: Safavi-Naini, R., Canetti, R. (eds.) CRYPTO 2012. LNCS, vol. 7417, pp. 868–886. Springer, Heidelberg (2012). https://doi.org/10.1007/978-3-642-32009-5_50
6. Brakerski, Z., Langlois, A., Peikert, C., Regev, O., Stehlé, D.: Classical hardness of learning with errors. In: Boneh, D., Roughgarden, T., Feigenbaum, J. (eds.) 45th ACM STOC, pp. 575–584. ACM Press (2013). https://doi.org/10.1145/2488608.2488680
7. Brakerski, Z., Vaikuntanathan, V.: Fully homomorphic encryption from ring-LWE and security for key dependent messages. In: Rogaway, P. (ed.) CRYPTO 2011. LNCS, vol. 6841, pp. 505–524. Springer, Heidelberg (2011). https://doi.org/10.1007/978-3-642-22792-9_29
8. Castryck, W., Iliashenko, I., Vercauteren, F.: Homomorphic SIM^2D operations: single instruction much more data. In: Nielsen, J.B., Rijmen, V. (eds.) EUROCRYPT 2018. LNCS, vol. 10820, pp. 338–359. Springer, Cham (2018). https://doi.org/10.1007/978-3-319-78381-9_13
9. Chen, H., Laine, K., Player, R., Xia, Y.: High-precision arithmetic in homomorphic encryption. In: Smart, N.P. (ed.) CT-RSA 2018. LNCS, vol. 10808, pp. 116–136. Springer, Cham (2018). https://doi.org/10.1007/978-3-319-76953-0_7
10. Cheon, J.H., Jeong, J., Lee, J., Lee, K.: Privacy-preserving computations of predictive medical models with minimax approximation and non-adjacent form. In: Brenner, M., et al. (eds.) FC 2017. LNCS, vol. 10323, pp. 53–74. Springer, Cham (2017). https://doi.org/10.1007/978-3-319-70278-0_4
11. Cheon, J.H., Kim, A., Kim, M., Song, Y.: Homomorphic encryption for arithmetic of approximate numbers. In: Takagi, T., Peyrin, T. (eds.) ASIACRYPT 2017. LNCS, vol. 10624, pp. 409–437. Springer, Cham (2017). https://doi.org/10.1007/978-3-319-70694-8_15

12. Costache, A., Smart, N., Vivek, S., Waller, A.: Fixed point arithmetic in SHE scheme. Cryptology ePrint Archive, Report 2016/250 (2016). https://eprint.iacr.org/2016/250
13. Costache, A., Smart, N.P.: Which ring based somewhat homomorphic encryption scheme is best? In: Sako, K. (ed.) CT-RSA 2016. LNCS, vol. 9610, pp. 325–340. Springer, Cham (2016). https://doi.org/10.1007/978-3-319-29485-8_19
14. Costache, A., Smart, N.P., Vivek, S.: Faster homomorphic evaluation of discrete Fourier transforms. In: Kiayias, A. (ed.) FC 2017. LNCS, vol. 10322, pp. 517–529. Springer, Heidelberg (2017)
15. Devevey, J., Sakzad, A., Stehlé, D., Steinfeld, R.: On the integer polynomial learning with errors problem. In: Garay, J.A. (ed.) PKC 2021. LNCS, vol. 12710, pp. 184–214. Springer, Cham (2021). https://doi.org/10.1007/978-3-030-75245-3_8
16. Dowlin, N., Gilad-Bachrach, R., Laine, K., Lauter, K., Naehrig, M., Wernsing, J.: Manual for using homomorphic encryption for bioinformatics. Proc. IEEE **PP**, 1–16 (2017). https://doi.org/10.1109/JPROC.2016.2622218
17. Fan, J., Vercauteren, F.: Somewhat practical fully homomorphic encryption. IACR Cryptol. ePrint Arch. **2012**, 144 (2012)
18. Gentry, C.: Fully homomorphic encryption using ideal lattices. In: Proceedings of the Forty-First Annual ACM Symposium on Theory of Computing, pp. 169–178 (2009)
19. Harmon, L., Delavignette, G., Roy, A., Silva, D.: PIE: p-adic encoding for high-precision arithmetic in homomorphic encryption. In: Tibouchi, M., Wang, X. (eds.) Applied Cryptography and Network Security. ACNS 2023. LNCS, vol. 13905. Springer, Cham (2023). https://doi.org/10.1007/978-3-031-33488-7_16
20. Håstad, J.: Pseudo-random generators under uniform assumptions. In: 22nd ACM STOC, pp. 395–404. ACM Press (1990). https://doi.org/10.1145/100216.100270
21. Hoffstein, J., Silverman, J.: Optimizations for NTRU. Public-key cryptography and computational number theory (2002)
22. Jäschke, A., Armknecht, F.: Accelerating homomorphic computations on rational numbers. In: Manulis, M., Sadeghi, A.-R., Schneider, S. (eds.) ACNS 2016. LNCS, vol. 9696, pp. 405–423. Springer, Cham (2016). https://doi.org/10.1007/978-3-319-39555-5_22
23. Lauter, K., López-Alt, A., Naehrig, M.: Private computation on encrypted genomic data. In: Aranha, D.F., Menezes, A. (eds.) LATINCRYPT 2014. LNCS, vol. 8895, pp. 3–27. Springer, Cham (2015). https://doi.org/10.1007/978-3-319-16295-9_1
24. Lindner, R., Peikert, C.: Better key sizes (and attacks) for LWE-based encryption. In: Kiayias, A. (ed.) CT-RSA 2011. LNCS, vol. 6558, pp. 319–339. Springer, Heidelberg (2011). https://doi.org/10.1007/978-3-642-19074-2_21
25. Lyubashevsky, V., Peikert, C., Regev, O.: On ideal lattices and learning with errors over rings. In: Gilbert, H. (ed.) EUROCRYPT 2010. LNCS, vol. 6110, pp. 1–23. Springer, Heidelberg (2010). https://doi.org/10.1007/978-3-642-13190-5_1
26. Naehrig, M., Lauter, K., Vaikuntanathan, V.: Can homomorphic encryption be practical? In: Proceedings of the 3rd ACM Workshop on Cloud Computing Security Workshop, pp. 113–124. CCSW '11, Association for Computing Machinery, New York, NY, USA (2011). https://doi.org/10.1145/2046660.2046682
27. Rivest, R.L., Adleman, L., Dertouzos, M.L., et al.: On data banks and privacy homomorphisms. Found. Secure Comput. **4**(11), 169–180 (1978)
28. Stehlé, D., Steinfeld, R., Tanaka, K., Xagawa, K.: Efficient public key encryption based on ideal lattices. In: Matsui, M. (ed.) ASIACRYPT 2009. LNCS, vol. 5912, pp. 617–635. Springer, Heidelberg (2009). https://doi.org/10.1007/978-3-642-10366-7_36

TFHE Bootstrapping: Faster, Smaller and Time-Space Trade-Offs

Ruida Wang[1,2], Benqiang Wei[1,2], Zhihao Li[1,2], Xianhui Lu[1,2](✉), and Kunpeng Wang[1,2](✉)

[1] Key Laboratory of Cyberspace Security Defense, Institute of Information Engineering, Chinese Academy of Sciences, Beijing, China
{wangruida,luxianhui,wangkunpeng}@iie.ac.cn

[2] School of Cyber Security, University of Chinese Academy of Sciences, Beijing, China

Abstract. Fully homomorphic encryption (FHE) allows for computation on encrypted data, providing effective privacy protection in data processing scenarios such as cloud computing and machine learning. However, the efficiency and storage overhead of the FHE *bootstrapping* algorithm remains challenges. The TFHE scheme, proposed by Chillotti et al., offers the fastest bootstrapping among all existing FHE solutions. This scheme has two homomorphic evaluation modes: the LHE mode, which uses circuit bootstrapping; and the FHE mode, which relies on functional bootstrapping. This paper presents four improved bootstrapping algorithms designed to enhance the efficiency and practicality of TFHE applications. Contributions can be summarized as twofold:

1. For the LHE mode, we propose an improved circuit bootstrapping algorithm (ICBS). Compared with the state-of-the-art approach, ICBS reduces the key size by 33.4% and decreases the running time by 36.2%. Furthermore, we present two variants of ICBS trading off computation and storage: the succinct variant (S-ICBS) greatly reduces the key size by 91.2% compared with commonly used implementations, while the efficient variant (E-ICBS) saves the running time by 45% compared with the previous method.
2. For the FHE mode, we propose a succinct functional bootstrapping algorithm (SFBS) that reduces the key size by 49.9% compared with the original TFHE method.

Keywords: Fully Homomorphic Encryption · TFHE · Circuit Bootstrapping · Functional Bootstrapping

1 Introduction

Fully Homomorphic Encryption (FHE) is a powerful cryptographic primitive that enables computations performed on encrypted data without leakage, and gives the potential to achieve privacy-preserving cloud computing [29], machine

T. Zhu and Y. Li (Eds.): ACISP 2024, LNCS 14895, pp. 196–216, 2024.
https://doi.org/10.1007/978-981-97-5025-2_11

learning [2,26] and IoT computing [30], among other applications. But a main drawback of FHE is its inefficiency, to which, possible solutions are researched.

The core operation of any FHE schemes known to date is called bootstrapping, the concept proposed by Gentry in the first FHE solution [17] in 2009. Bootstrapping aims to refresh the ciphertext and reduce the noise level, which prevents error inflation which may lead to a failed decryption. Bootstrapping is always the most time-consuming part of FHE applications. The TFHE scheme (fast fully homomorphic encryption over the torus) [8–10] currently offers the lowest known bootstrapping latency, with *gate bootstrapping* taking only about 7ms [11]. In contrast, the bootstrapping algorithms of BFV [3,16], BGV [4] take minutes [21], and CKKS [7] requires seconds at a cost of precision loss [6,23,24].

TFHE scheme supports two homomorphic evaluation modes. The first is the leveled homomorphic encryption (LHE) mode, which allows for a leveled circuit computation, such as muti-in/1-out lookup table (LUT), torus bit sequence representation (TBSR), weighted finite automata (WFA) [8] before performing a bootstrapping. This makes it possible to speed up large-scale arithmetic functions compared with gate-by-gate evaluation. However, it is important to note that the input and output of the leveled homomorphic operation have different ciphertext types. To achieve composability, *circuit bootstrapping* is required to convert the ciphertext type while reducing the noise. The latency of the circuit bootstrapping is almost 10 times higher than that of the gate bootstrapping [9] and is the main computational overhead in the LHE mode. The other mode is the FHE mode, which builds arbitrary circuits by gate bootstrapping. This technique has been further developed into *functional bootstrapping* [5] (FBS, or programmable bootstrapping [2]), which can evaluate an 1-in/1-out LUT almost for free while refreshing the ciphertext. Optimizing the efficiency of the bootstrapping algorithms is a critical issue that needs to be addressed to improve the practicality of the TFHE scheme.

Hardware acceleration is one approach to tackle this challenge, as it has been demonstrated to significantly improve the performance of TFHE bootstrapping [22,33,39]. However, these solutions are memory-constrained rather than computation-bound. Specifically, TFHE accelerators have limited on-chip storage, but run with a large amount of constant data, especially the bootstrapping key. As a result, the data transfer latency can account for almost 80% of the execution time [33]. These technical challenges indeed provide the motivation for us to design improved bootstrapping algorithms. By reducing the run time and key size, and further exploring a flexible computation-storage trade-off, our aim is to enhance the usefulness and practicality of FHE for real-world applications.

1.1 Our Results

In this paper, we propose four optimized TFHE bootstrapping algorithms. These include an improved circuit bootstrapping algorithm (ICBS), a functional bootstrapping algorithm with a smaller key size, and two variants of the ICBS algorithm that make trade-offs between computation and storage:

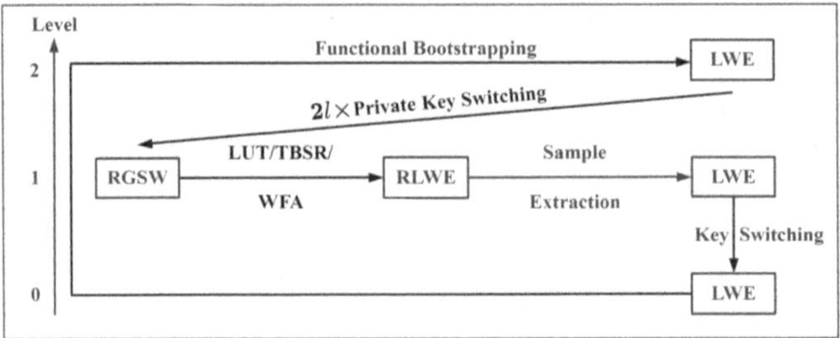

Fig. 1. The LHE evaluation mode of the TFHE scheme. TFHE CBS method contains functional bootstrapping and $2l$ times private key switching. Rectangles hold the types of ciphertexts, arrows hold the conversion algorithms.

- **Improved circuit bootstrapping algorithm:** The ICBS algorithm improves the performance of the original circuit bootstrapping algorithm in terms of both the evaluation key size and running time. Specifically, under a security level of no less than 128-bit (where circuit bootstrapping crosses three parameter levels, and the security parameters are 128, 129 and 153 in our experiment), ICBS reduces the key size by 33.4%, and decreases the running time by 36.2% compared with TFHE circuit bootstrapping.
- **Succinct functional bootstrapping algorithm (SFBS):** The SFBS algorithm significantly reduces the bootstrapping key size compared with previous work [10]. For instance, at a 153-bit target security level, this algorithm reduces the key size by 49.9% (from 158.75 MB to 79.5 MB), while at a cost of increased running time (from 24.6 ms to 33.6 ms). The reduced key size also allows SFBS to provide a means for trading computation for less storage in the design of the succinct circuit bootstrapping algorithm.
- **Succinct-ICBS (S-ICBS):** The S-ICBS variant algorithm uses our proposed SFBS to further reduce the key size. Compared with the current TFHE implementation [31], our approach achieves a significant 91.2% reduction in key size (from 2720.03 MB to 239.52 MB). Moreover, it maintains a 14.3% efficiency advantage under the same parameter settings.
- **Efficient-ICBS (E-ICBS):** The E-ICBS variant algorithm saves the running time of ICBS by pre-calculating and storing. Specifically, it resulted in a further 13.7% reduction in ICBS running time, and a 45% reduction compared with the original TFHE method (from 80 ms to 44 ms).

1.2 Challenges and Techniques

Improved Circuit Bootstrapping Algorithm: The TFHE encryption scheme is based on the learning with errors (LWE) problem and its ring variant (RLWE) on lattice [28,35]. It incorporates three ciphertext types: LWE, RLWE and RGSW. The aim of circuit bootstrapping is to convert LWE ciphertext to RGSW ciphertext, thereby completing the TFHE LHE evaluation mode (Fig. 1).

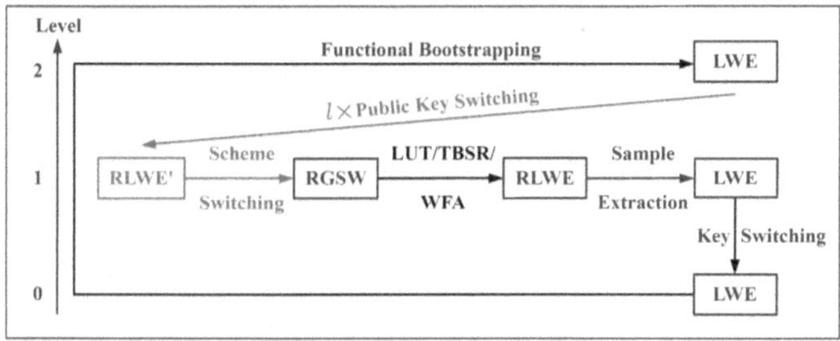

Fig. 2. Our proposed ICBS contains functional bootstrapping, l times public key switching and scheme switching, we highlight the differences with the original method in blue. Rectangles hold the types of ciphertexts, arrows hold the conversion algorithms. (Color figure online)

The primary resource-consuming part of CBS is the LWE-to-RLWE private key switching[1] procedure, both in terms of storage and computation. Our innovation lies in replacing half of the expensive LWE-to-RLWE conversions (which has an $O(N^2)$ multiplication complexity, where N is the RLWE dimension) with an efficient RLWE-to-RLWE conversion (which have an $O(N \log N)$ multiplication complexity). The technique is derived from an observation of the RGSW structure. Specifically, an RGSW ciphertext of message $\mathbf{m}$ can be written as $(\text{RLWE}'(\mathbf{sk} \cdot \mathbf{m}), \text{RLWE}'(\mathbf{m}))$, where $\mathbf{sk}$ is the secret key and RLWE$'$ denotes the gadget RLWE ciphertext [32]. The original TFHE method generates RGSW by computing $\text{RLWE}'(\mathbf{sk} \cdot \mathbf{m})$ and $\text{RLWE}'(\mathbf{m})$ separately, each requiring l times of private key switching, where l is the gadget decomposition length [9]. Our method calculates $\text{RLWE}'(\mathbf{m})$ by l times of public key switching, which is slightly cheaper than private key switching, and converts $\text{RLWE}'(\mathbf{m})$ to $\text{RLWE}'(\mathbf{sk} \cdot \mathbf{m})$ using a scheme switching algorithm, as shown in Fig. 2.

Succinct Functional Bootstrapping Algorithm: The large storage size of the TFHE FBS algorithm is due to the need to store n RGSW ciphertexts as the evaluation key, where n is the LWE dimension. Leveraging the same observation of the RGSW structure, our method is to store only $\text{RLWE}'(\mathbf{m})$ and generate $\text{RLWE}'(\mathbf{sk} \cdot \mathbf{m})$ on-the-fly. This conversion algorithm employs an extra switching key $\text{RLWE}'(\mathbf{sk}^2)$, which is small compared with the savings.

Trade-Off Between Computation and Storage: We provide novel perspectives on the trade-offs between storage and computation in the circuit bootstrapping algorithm. The first method calls our proposed SFBS algorithm in ICBS noise refreshing step instead of FBS, further decreasing the key size; the other method reduces computation with some additional storage inspired by the key switching tricks proposed by Ducas and Micciancio [15], which pre-computes the

[1] We consider the optimization brought by the PBSmanyLUT [12] technique, then the key switching step accounts for about 65% of the time overhead of the circuit bootstrap algorithm.

potential multiplication results as a part of the evaluation key, and replaces scalar product operations with reading operations during the algorithm execution.

1.3 Related Work

Trading off bootstrapping computation for storage is also a valuable approach in potential FHE applications such as privacy-preserving Internet of Things (IoT) [34,37] and mobile devices [18,36]. In these scenarios, client devices often have limited resources, making it crucial to reduce key size in order to minimize local storage and transmission overhead. This is often more important than reducing the computational load on the server side. Additionally, enhancing the functionality of TFHE bootstrapping is also a research focus. Our SFBS method can be also utilized to optimize functional bootstrapping variants, such as multi-value FBS [5,12], high-precision FBS [13,20], and full-domain FBS [12,13,25,27, 38].

1.4 Paper Organization

This paper is organized as follows: Sect. 2 provides a review of the TFHE cryptosystem; Section 3 describes our improved circuit bootstrapping; Section 4 presents a succinct functional bootstrapping; Section 5 considers the time-space trade-off, gives a succinct variant and an efficient variant of ICBS; Section 6 implements all proposed methods, and compares them with prior works; finally, Sect. 7 concludes the paper.

2 Preliminaries

2.1 Notations

Let $\mathbb{A}$ be a set. We denote by $\mathbb{A}_q^n$ the set of vectors with n elements in $\mathbb{A}$ modulo q, and by $\mathbb{A}_N[X]^n$ the set of vectors of n polynomials modulo $X^N + 1$, where N is a power of 2. Let $\mathbb{Z}$ denote the set of integers, and $\mathcal{R}$ denotes $\mathbb{Z}_N[X]$. We use regular letters for (modular) integers, e.g., $a \in \mathbb{Z}_q$, and bold letters for polynomials $\mathbf{a} \in \mathcal{R}_q$ or vectors $\mathbf{a} \in \mathbb{Z}_q^n$. The notation a_i refers to the i-th coefficient/term of $\mathbf{a}$. $\mathbb{B} = \mathbb{Z}_2$ is the set of binary numbers.

2.2 TFHE Cryptosystem

TFHE is an FHE scheme with the fastest bootstrapping algorithm. This section describes ciphertext types and useful algorithms in the TFHE cryptosystem.

2.2.1 (Ring) LWE Ciphertexts

The security of TFHE cryptosystem is based on the (ring) learning with errors problem [28,35]. We summarize the three kinds of ciphertexts as follow:

- **LWE:** Giving positive integers n and q, the LWE encryption of the message $m \in \mathbb{Z}$ is a vector $(\mathbf{a}, b) \in \mathbb{Z}_q^{n+1}$, where $b = -\mathbf{a} \cdot \mathbf{sk} + m + e$. The vector $\mathbf{a}$ is uniformly sampled from $\mathbb{Z}_q^n$, the secret key $\mathbf{sk}$ is sampled from a key distribution χ, the error e is sampled from an error distribution χ'.
- **RLWE:** RLWE is a ring version of LWE on $\mathcal{R}_q$. The RLWE encryption of the message $\mathbf{m} \in \mathcal{R}_q$ is a pair $(\mathbf{a}, \mathbf{b}) \in \mathcal{R}_q^{n+1}$, where $\mathbf{b} = -\mathbf{a} \cdot \mathbf{sk} + \mathbf{m} + \mathbf{e}$. The vector $\mathbf{a}$ is uniformly sampled from $\mathcal{R}_q$, the secret key $\mathbf{sk}$ is sampled from a key distribution χ, and each coefficient of the error e_i is sampled from χ'.
- **RLWE′:** Given a gadget vector $\mathbf{v} = (v_0, v_1, ..., v_{l-1})$, the gadget RLWE written as RLWE′, is defined by:

$$\mathrm{RLWE'}_{\mathbf{sk},q}(\mathbf{m}) = (\mathrm{RLWE}_{\mathbf{sk},q}(v_0 \cdot \mathbf{m}), \cdots, \mathrm{RLWE}_{\mathbf{sk},q}(v_{l-1} \cdot \mathbf{m})).$$

- **RGSW:** The RGSW encryption of the message $\mathbf{m} \in \mathcal{R}_q$ is defined by: $\mathrm{RGSW}_{\mathbf{sk},q}(\mathbf{m}) = (\mathrm{RLWE'}_{\mathbf{sk},q}(\mathbf{sk} \cdot \mathbf{m}), \mathrm{RLWE'}_{\mathbf{sk},q}(\mathbf{m}))$.

Remark 1. The gadget RLWE, which serves as an auxiliary input in algorithms, is a key component of the TFHE cryptosystem. For clarity and ease of understanding, we follow the definition and notation of gadget RLWE as provided by Micciancio and Polyakov [32].

Remark 2. It is important to note that these definitions, which follow Micciancio and Polyakov [32], and Chillotti et.al [12], utilize different notions compared with the original TFHE papers [8–10]. Specifically, TFHE uses real torus $\mathbb{T} = \mathbb{R}/\mathbb{Z}$ and $\mathbb{T}_N[X] = \mathcal{R}/\mathbb{Z}$ to describe the message and ciphertext spaces, but implements $\mathbb{T}$ by $\mathbb{Z}_q$ with $q = 2^{32}$ or $q = 2^{64}$. Thus we straightforwardly use $\mathbb{Z}_q$ instead of $\mathbb{T}$.

2.2.2 Products

Gadget Product $\odot : \mathcal{R} \times \mathrm{RLWE'} \rightarrow \mathrm{RLWE}$ is defined as:

$$\mathbf{t} \odot \mathrm{RLWE}'_{\mathbf{sk}}(\mathbf{m}) = \sum_{i=0}^{l-1} \mathbf{t}_i \cdot \mathrm{RLWE}_{\mathbf{sk}}(v_i \cdot \mathbf{m}) = \mathrm{RLWE}_{\mathbf{sk}}(\mathbf{t} \cdot \mathbf{m}),$$

where $\mathbf{t} \in \mathcal{R}$. The gadget decomposition $(\mathbf{t}_0, ..., \mathbf{t}_{l-1})$ of $\mathbf{t}$ corresponding to the gadget vector $\mathbf{v}$, is defined to minimize the decomposition error $\varepsilon_{\mathsf{gadget}}(\mathbf{t}) = \sum_i v_i \mathbf{t}_i - \mathbf{t}$.

In this paper, we use the canonical gadget decomposition [9], where $\mathbf{v} = (\lceil \frac{q}{B^l} \rceil, \lceil \frac{q}{B^l} \rceil B, ..., \lceil \frac{q}{B^l} \rceil B^{l-1})$, with l denoting the decomposition length, B denoting the decomposition base. Let ϵ denote the infinite norm of the decomposition error, then $\epsilon = \| \sum_i v_i \mathbf{t}_i - \mathbf{t} \|_\infty \leq \frac{1}{2} \lceil \frac{q}{B^l} \rceil$ [10].

Lemma 1. *[12] Let N denote the dimension of the ring polynomial of RLWE ciphertexts, B and l denote the base and the length of the gadget decomposition. Then the error variance of the result of the gadget product is bounded by*

$$\sigma^2_{\odot,input} \leq \frac{B^2}{12} lN\sigma^2_{input} + \frac{1}{3}\|\mathbf{m}\|_2^2\epsilon^2,$$

where σ^2_{input} is the error variance of the input RLWE' *ciphertext, and $\|\mathbf{m}\|_2$ is the 2-norm of the message* $\mathbf{m}$.

External Product $\boxdot$: RLWE × RGSW → RLWE is defined as:

$$\begin{aligned}&\mathrm{RLWE}_{\mathbf{sk}}(\mathbf{m}_1) \boxdot \mathrm{RGSW}_{\mathbf{sk}}(\mathbf{m}_2) := \mathbf{a} \odot \mathrm{RLWE}'_{\mathbf{sk}}(\mathbf{sk} \cdot \mathbf{m}_2) + \mathbf{b} \odot \mathrm{RLWE}'_{\mathbf{sk}}(\mathbf{m}_2)\\ &= \mathrm{RLWE}_{\mathbf{sk}}(\mathbf{m}_1 \cdot \mathbf{m}_2),\end{aligned}$$

where $\mathrm{RLWE}_{\mathbf{sk}}(\mathbf{m}_1) = (\mathbf{a}, \mathbf{b})$.

Lemma 2. *[10] Let N denote the dimension of the ring polynomial of RLWE ciphertexts, B and l denote the base and the length of the gadget decomposition, respectively. Then the error variance of the result of the external product is bounded by*

$$\sigma^2_{\boxdot,\mathrm{RGSW}} \leq \frac{B^2}{6} lN\sigma^2_{\mathrm{RGSW}} + \sigma^2_{\mathrm{RLWE}} + \frac{1}{3}(1 + \|\mathbf{sk}\|_2^2)\|\mathbf{m}_2\|_2^2\epsilon^2,$$

where σ^2_{RGSW} and σ^2_{RLWE} is the error variance of the input RGSW and RLWE ciphertext, respectively, and $\|\mathbf{sk}\|_2$ and $\|\mathbf{m}_2\|_2$ is the 2-norm of the key $\mathbf{sk}$ *and the message* $\mathbf{m}_2$.

Remark 3. The result given by Lemma 2 can be derived from Lemma 1. Readers can also refer to the theorem given by Chillotti et al., [10], Corollary 3.14.). However, there is a coefficient difference because we considered the coefficient 1/12 when calculating the variance of the uniform distribution.

2.2.3 Key Switching

LWE to RLWE Public Functional Key Switching: Given a public Lipschitz morphism $f : \mathbb{Z} \to \mathcal{R}$, a key switching key $\mathsf{PubKS}_{key} = \mathrm{RLWE}'_{\mathbf{sk}'}(sk_i)_{i\in[1,n]}$, where sk_i is the i-th term of the LWE secret key, then the LWE to RLWE public functional key switching algorithm is defined as:

$$\mathsf{PubKS}^f_{\mathbf{sk}\to\mathbf{sk}'}(\mathrm{LWE}_{\mathbf{sk}}(m)) := \sum_{i=1}^{n} f(a_i) \odot \mathrm{RLWE}'_{\mathbf{sk}'}(sk_i) + (0, f(b)).$$

LWE to RLWE Private Functional Key Switching: For a private Lipschitz morphism $f : \mathbb{Z} \to \mathcal{R}$, given a key switching key PrivKS_{key} containing $\mathrm{RLWE}'_{\mathbf{sk}'}(f(sk_i))_{i\in[1,n]}$ and $\mathrm{RLWE}'_{\mathbf{sk}'}(f(1))$, where sk_i is the i-th term of the

Algorithm 1. TFHE's Functional Bootstrapping (FBS) Algorithm [10]

Input: an LWE encryption $\underline{\mathbf{c}} = (\underline{\mathbf{a}}, \underline{b}) \in \mathrm{LWE}_{\underline{\mathbf{sk}}}(m)$,
Input: a test polynomial testP,
Input: a bootstrapping key $\mathsf{FBS}_{key} = \mathrm{RGSW}_{\bar{\mathbf{sk}}}(\underline{sk}_i)_{i \in [1,n]}$.
Output: $\bar{c} \in \mathrm{LWE}_{\bar{\mathbf{sk}}}(f(m))$.
1: $b = \lfloor \frac{2N}{q} \underline{b} \rceil$ and $a_i = \left\lfloor \frac{2N}{q} \underline{a}_i \right\rceil \in \mathbb{Z}_{2N}$ for each $i \in [1, n]$
2: $\mathsf{acc} = (0, X^b \cdot \mathsf{testP})$.
3: **for** $i = 1$ to n **do**:
4: $\quad \mathsf{acc} = ((X^{a_i} - 1) \cdot \mathsf{acc}) \boxdot \mathrm{RGSW}_{\bar{\mathbf{sk}}}(\underline{sk}_i) + \mathsf{acc}$;
5: **end for**
6: $\bar{c} = \mathsf{SampleExtract}^2(\mathsf{acc})$,
7: **return** $\bar{c}$

LWE secret key, then the LWE to RLWE private functional key switching algorithm is defined as:

$$\mathsf{PrivKS}^f_{\mathbf{sk} \to \mathbf{sk}'}(\mathrm{LWE}_{\mathbf{sk}}(m)) := \sum_{i=1}^{n} a_i \odot \mathrm{RLWE}'_{\mathbf{sk}'}(f(sk_i)) + b \odot \mathrm{RLWE}'_{\mathbf{sk}'}(f(1)).$$

RLWE Key Switching: Given a key switching key $\mathsf{KS}_{key} = \mathrm{RLWE}'_{\mathbf{sk}'}(\mathbf{sk})$, the RLWE key switching algorithm is defined as:

$$\mathsf{KS}_{\mathbf{sk} \to \mathbf{sk}'}(\mathrm{RLWE}_{\mathbf{sk}}(\mathbf{m})) := \mathbf{a} \odot \mathrm{RLWE}'_{\mathbf{sk}'}(\mathbf{sk}) + (0, \mathbf{b}).$$

2.2.4 Bootstrapping

Functional Bootstrapping: Given a test polynomial testP encoded with a LUT function f , and a bootstrapping key containing $\mathrm{RGSW}_{\bar{\mathbf{sk}}}(\underline{sk}_i)$, where $\underline{sk}_i$ is the i-th term of the LWE secret key, then the functional bootstrapping algorithm $\mathsf{FBS}^{\mathsf{testP}}_{\mathrm{LWE} \to \mathrm{LWE}} : \mathrm{LWE}_{\underline{\mathbf{sk}}}(m) \to \mathrm{LWE}_{\bar{\mathbf{sk}}}(f(m))$ is shown as Algorithm 1.

Circuit Bootstrapping: The circuit bootstrapping algorithm crosses three levels, each using different keys and parameters. To distinguish different levels, we adopt the convention that the barred represents level 2 variables, the underlined represents level 0 variables, and no extra mark for level 1 variables.

[2] The sample extraction algorithm $\mathsf{SampleExtract}$ receives a RLWE sample $\mathbf{c} = (\mathbf{a}, \mathbf{b}) \in \mathrm{RLWE}_{\mathbf{sk}}(\mathbf{m})$, returns an LWE sample $\bar{c} = (\bar{\mathbf{a}}, \bar{b}) \in \mathrm{LWE}_{\bar{\mathbf{sk}}}(m)$, where m is the constant term of $\mathbf{m}$. This algorithm is efficient without error growth, we omit the algorithm details which can be refered to [10].

Given a circuit bootstrapping key containing a functional bootstrapping key $\text{RGSW}_{\bar{\mathbf{sk}}}(\underline{sk}_i)$, 2 private key switching keys $(\mathsf{PrivKS}_{key}^{f_0}, \mathsf{PrivKS}_{key}^{f_1})$, then the circuit bootstrapping algorithm can be divided into two steps:

1. **LWE to LWEs:** Functional bootstrapping:
   $\mathsf{FBS}_{\text{LWE}\to\text{LWE}} : \text{LWE}_{\underline{\mathbf{sk}}}(m) \to \text{LWE}_{\bar{\mathbf{sk}}}(v_i \cdot m), i \in [0, l-1]$.
2. **LWEs to RGSW:** Private key switching:
   $\mathsf{PrivKS}_{\bar{\mathbf{sk}}\to\mathbf{sk}}^{f_0} : \text{LWE}_{\bar{\mathbf{sk}}}(v_i \cdot m) \to \text{RLWE}_{\mathbf{sk}}(v_i \cdot \mathbf{m})$,
   $\mathsf{PrivKS}_{\bar{\mathbf{sk}}\to\mathbf{sk}}^{f_1} : \text{LWE}_{\bar{\mathbf{sk}}}(v_i \cdot m) \to \text{RLWE}_{\mathbf{sk}}(v_i \cdot \mathbf{sk} \cdot \mathbf{m}), i \in [0, l-1]$,
   $\text{RGSW}_{\mathbf{sk}}(m) = (\text{RLWE}'_{\mathbf{sk}}(\mathbf{sk} \cdot \mathbf{m}), \text{RLWE}'_{\mathbf{sk}}(\mathbf{m}))$.

To clarify the above notations, $\underline{sk}_i$ is the i-th term of the level 0 LWE secret key $\underline{\mathbf{sk}}$, $i \in [1, \underline{n}]$, f_0 is the identity function, $f_1(x) = \mathbf{sk} \cdot x$, $\mathbf{sk}$ is the RLWE secret key for level 1, and $\mathbf{m}$ is a monomial with m in its constant term.

3 Improved Circuit Bootstrapping Algorithm

This section delves into our improved circuit bootstrapping algorithm (ICBS), which offers the dual advantage of reducing key size and decreasing running time compared with previous work. We concentrate on the technical details in this section, while the performance is shown in Sect. 6.

3.1 Scheme Switching

We commence this section by introducing a powerful tool - the scheme switching algorithm [14] proposed by De Micheli et al. to build amortized bootstrapping. Notably, this technique facilitates the transformation of RLWE$'$ ciphertext into RGSW, thereby enhancing the efficiency of our approach.

Given a scheme switching key $\mathsf{SS}_{key} = \text{RLWE}'_{\mathbf{sk}}(\mathbf{sk}^2)$, the scheme switching algorithm $\mathsf{SS}_{\text{RLWE}'\to\text{RGSW}} : \text{RLWE}'_{\mathbf{sk}}(\mathbf{m}) \to \text{RGSW}_{\mathbf{sk}}(\mathbf{m})$ works as follows:

for every $\text{RLWE}_{\mathbf{sk}}(v_i \cdot \mathbf{m}) = (\mathbf{a}_i, \mathbf{b}_i)$ with error $\mathbf{e}_i$, $i \in [0, l-1]$, computing

$$\begin{aligned} \mathbf{a}_i \odot \text{RLWE}'_{\mathbf{sk}}\left(\mathbf{sk}^2\right) + (\mathbf{b}_i, 0) &= \text{RLWE}_{\mathbf{sk}}\left(\mathbf{a}_i \cdot \mathbf{sk}^2 + \mathbf{b}_i \cdot \mathbf{sk}\right) \\ &= \text{RLWE}_{\mathbf{sk}}\left((\mathbf{a}_i \cdot \mathbf{sk} + \mathbf{b}_i) \cdot \mathbf{sk}\right) \\ &= \text{RLWE}_{\mathbf{sk}}\left(v_i \cdot \mathbf{sk} \cdot \mathbf{m} + \mathbf{e}_i \cdot \mathbf{sk}\right), \end{aligned}$$

then we obtain $\text{RLWE}'_{\mathbf{sk}}(\mathbf{sk} \cdot \mathbf{m})$ with an additional error $\mathbf{e}_i \cdot \mathbf{sk}$. This term can remain small by choosing the secret key with a small norm (e.g., binary). Then,

$$\mathsf{SS}_{\text{RLWE}'\to\text{RGSW}}(\text{RLWE}'_{\mathbf{sk}}(\mathbf{m})) := (\text{RLWE}'_{\mathbf{sk}}(\mathbf{sk} \cdot \mathbf{m}), \text{RLWE}'_{\mathbf{sk}}(\mathbf{m})).$$

De Micheli et al. utilized scheme switching on the prime cyclotomic ring. However, for our target power-of-two cyclotomic ring, we must reanalyze the error growth and give Theorem 1.

Theorem 1. *Let N denote the dimension of the ring polynomial of RLWE ciphertexts, B and l denote the base and the length of the gadget decomposition, respectively. Then the error variance of the result of the scheme switching is bounded by*

$$\sigma_{\mathsf{ss}}^2 \leq \frac{B^2}{12} l N \sigma_{\mathsf{SS}_{key}}^2 + \frac{N^2}{12}\epsilon^2 + \frac{N}{2}\sigma_{input}^2,$$

where σ_{input}^2 is the error variance of the input RLWE′ *ciphertext, and $\sigma_{\mathsf{SS}_{key}}^2$ is the error variance of the scheme switching key* $\mathrm{RLWE}'_{\mathbf{sk}}(\mathbf{sk}^2)$.

Proof. The error analysis can be derived from Lemma 1.

3.2 Proposed Algorithm

Our ICBS crosses three levels, each level uses different keys and parameters. We use bar to represent level 2 variables, underline for level 0 variables, and no special representation for level 1 variables.

Given a circuit bootstrapping key containing a level 2 functional bootstrapping key $\mathrm{RGSW}_{\bar{\mathbf{sk}}}(\underline{sk}_i)$, where $\underline{sk}_i$ is the i-th term of the level 0 LWE secret key $\underline{\mathbf{sk}}$, $i \in [1, \underline{n}]$; a level 1 public key switching key $\mathrm{RLWE}'_{\mathbf{sk}}(\bar{sk}_j)$, where $\bar{sk}_j$ is the j-th term of the level 2 LWE secret key $\bar{\mathbf{sk}}$, $j \in [1, \bar{n}]$; and a level 1 scheme switching key $\mathrm{RLWE}'_{\mathbf{sk}}(\mathbf{sk}^2)$, we can then define the improved circuit bootstrapping algorithm $\mathsf{ICBS}_{\mathrm{LWE}\to\mathrm{RGSW}} : \mathrm{LWE}_{\underline{\mathbf{sk}}}(m) \to \mathrm{RGSW}_{\mathbf{sk}}(\mathbf{m})$, where $\mathbf{m}$ is a monomial with m in constant term. This algorithm is shown as Algorithm 2. Here, l denotes the gadget length of the level 1 RGSW ciphertext.

Algorithm 2. Improved Circuit Bootstrapping (ICBS) Algorithm

Input: a level 0 LWE encryption $\underline{\mathbf{c}} = (\underline{\mathbf{a}}, \underline{b}) \in \mathrm{LWE}_{\underline{\mathbf{sk}}}(m)$,
Input: test polynomials testP_k encoding $f_k(m) = v_k \cdot m, k \in [0, l-1]$,
Input: a level 2 $\mathsf{FBS}_{key} = \mathrm{RGSW}_{\bar{\mathbf{sk}}}(\underline{sk}_i)_{i\in[1,\underline{n}]}$,
Input: a level 1 $\mathsf{PubKS}_{key} = \mathrm{RLWE}'_{\mathbf{sk}}(\bar{sk}_j)_{j\in[1,\bar{n}]}$,
Input: a level 1 $\mathsf{SS}_{key} = \mathrm{RLWE}'_{\mathbf{sk}}(\mathbf{sk}^2)$,
Output: $\mathbf{c} \in \mathrm{RGSW}_{\mathbf{sk}}(\mathbf{m})$.

1: **for** $k = 0$ to $l-1$ **do**:
2: $\quad\mathrm{LWE}_{\bar{\mathbf{sk}}}(v_k \cdot m) \leftarrow \mathsf{FBS}^{\mathsf{testP}_k}_{\mathrm{LWE}\to\mathrm{LWE}}(\mathrm{LWE}_{\underline{\mathbf{sk}}}(m))$,
3: $\quad\mathrm{RLWE}_{\mathbf{sk}}(v_k \cdot \mathbf{m}) \leftarrow \mathsf{PubKS}_{\bar{\mathbf{sk}}\to\mathbf{sk}}(\mathrm{LWE}_{\bar{\mathbf{sk}}}(v_k \cdot m))$;
4: **end for**
5: $\mathrm{RLWE}'_{\mathbf{sk}}(m) = (\mathrm{RLWE}_{\mathbf{sk}}(v_0 \cdot \mathbf{m}), ..., \mathrm{RLWE}_{\mathbf{sk}}(v_{l-1} \cdot \mathbf{m}))$,
6: $\mathrm{RGSW}_{\mathbf{sk}}(m) \leftarrow \mathsf{SS}_{\mathrm{RLWE}'\to\mathrm{RGSW}}(\mathrm{RLWE}'_{\mathbf{sk}}(m))$,
7: **return** $\mathbf{c} = \mathrm{RGSW}_{\mathbf{sk}}(m)$

Our approach differs from the original circuit bootstrapping (CBS) algorithm in two significant ways. Firstly, the original CBS employs l times of private key switching to convert $\mathrm{LWE}_{\bar{\mathbf{sk}}}(v_i \cdot m)$ to $\mathrm{RLWE}_{\mathbf{sk}}(v_i \cdot \mathbf{m})$. However, we observed that the evaluation function in this step is an identity function that can be public. As a result, we use l times of public key switching instead (as seen in line 3). This modification slightly reduces both the computation and storage.

Secondly, while CBS generates $\mathrm{RLWE}_{\mathbf{sk}}(v_i \cdot \mathbf{sk} \cdot \mathbf{m})$ through another l times of private key switching algorithms, we use only one scheme switching instead (as seen in line 6). This change significantly reduces the computational complexity and the key size, thereby enhancing the efficiency of our approach.

3.3 Error Analysis

We then analyze the error growth of our proposed ICBS algorithm. The first step of ICBS is to convert $\mathrm{LWE}_{\underline{\mathbf{sk}}}(m)$ to $\mathrm{LWE}_{\bar{\mathbf{sk}}}(v_i \cdot m)$ using FBS algorithm (as seen in line 2). This step is identical to the original CBS algorithm. We denote the error growth of this step as σ^2_{FBS} and consider the public key switching operation as the starting point for our analysis. We then propose the following Theorem 2.

Theorem 2. *Let n, N denote LWE/RLWE level 1 parameters, and the same variables names with underlines/bars for level 0 and level 2 parameters, B_{ks}, l_{ks}, ϵ_{ks} denote the gadget decomposition parameters of the public key switching algorithm, B_{ss}, l_{ss}, ϵ_{ss} denote the gadget decomposition parameters of the scheme switching algorithm. Then the error variance of the result of the improved circuit bootstrapping algorithm is bounded by*

$$\sigma^2_{\mathsf{ICBS}} \leq \frac{N}{2}\sigma^2_{\mathsf{FBS}} + \frac{B^2_{\mathsf{ks}}}{24}N^2\bar{n}l_{\mathsf{ks}}\sigma^2_{\mathsf{PubKS}_{key}} + \frac{1}{6}N^2\bar{n}\epsilon^2_{\mathsf{ks}} + \frac{B^2_{\mathsf{ss}}}{12}l_{\mathsf{ss}}N\sigma^2_{\mathsf{SS}_{key}} + \frac{1}{12}N^2\epsilon^2_{\mathsf{ss}},$$

where $\sigma^2_{\mathsf{PubKS}_{key}}$ is the error variance of the public key switching key $\mathrm{RLWE}'_{\mathbf{sk}}(s\bar{k}_j)$, and $\sigma^2_{\mathsf{SS}_{key}}$ is the error variance of the scheme switching key $\mathrm{RLWE}'_{\bar{\mathbf{sk}}}(\bar{\mathbf{sk}}^2)$.

Proof. The analysis is conducted in accordance with Lemma 1 and Theorem 1.

Remark 4. Indeed, the error rate of the LWE/RLWE ciphertext will significantly influence the decryption failure probability, which can be calculated using the formula $1 - \mathrm{erf}\left(\frac{q}{4t\sigma}\right)$, where t is the modulus of the plaintext space and σ is the standard deviation of the error. Therefore, in each of our proposed algorithms, we carefully analyze the error growth formula, and finally ensure that the algorithm decryption failure rate does not exceed 2^{-32} by selecting appropriate parameters to meet the general application scenarios of FHE. A lower decryption failure rate can be achieved by adjusting parameters.

3.4 Comparison with Previous Work

We proceed to compare the ICBS algorithm with previous work [9,10] in terms of key size and computational complexity. For both the FBS step and the scheme switching step, we use the same gadget decomposition parameters l, B at level 1, and $\bar{l}$, $\bar{B}$ at level 2. For the key switching step, we use the gadget decomposition parameters l_{ks}, B_{ks}. This comparison allows us to evaluate the efficiency and practicality of our proposed ICBS algorithm.

Key Size. The original circuit bootstrapping key contains a level 2 bootstrapping key and two level 1 private key switching keys:

- the bootstrapping key has $\underline{n}$ level 2 RGSW samples, each RGSW sample has $4\bar{l}\bar{n}\log\bar{Q}$ bits,
- the private key switching key has $(\bar{n}+1)l_{\mathsf{ks}}$ level 1 RLWE samples, each RLWE sample has $2N\log Q$ bits.

Thus, the original circuit bootstrapping key size is $4\underline{n}\bar{l}\bar{n}\log\bar{Q} + 4N(\bar{n}+1)l_{\mathsf{ks}}\log Q$ bits. On the other side, our improved circuit bootstrapping key needs a level 2 bootstrapping key, a level 1 public key switching key, and a level 1 scheme switching key:

- the bootstrapping key has $\underline{n}$ level 2 RGSW samples,
- the public key switching key has $\bar{n}l_{\mathsf{ks}}$ level 1 RLWE samples,
- the scheme switching key has l level 1 RLWE samples.

Thus, our improved circuit bootstrapping key size is $4\underline{n}\bar{l}\bar{n}\log\bar{Q} + 2N(\bar{n}l_{\mathsf{ks}} + l)\log Q$ bits. We summarize these comparison results in Sect. 5, see Table 2.

Complexity. Indeed, both our proposed ICBS and the original CBS algorithm include one functional bootstrapping. To facilitate theoretical comparison, we focus on analyzing the computational complexity of the other steps. The overall running time comparison results will be tested in the experimental section. In the TFHE circuit bootstrapping, the private key switching step involves $4l_{\mathsf{ks}}lN(\bar{n}+1)$ modular multiplications. In our ICBS, it involves the public key switching and scheme switching instead:

- the public key switching algorithm involves $2l_{\mathsf{ks}}lN\bar{n}$ modular multiplications,
- the scheme switching step employs l level 2 gadget products, each of which can be translated into $(l+1)$ NTT/FFTs. Therefore, the multiplication complexity of this step is $O(\bar{n}\log\bar{n})$.

Taking into account the difference in order of magnitude, the complexity of our ICBS algorithm is nearly half that of the original CBS algorithm in all other steps, with the exception of the FBS algorithm, which remains unchanged.

4 Succinct Functional Bootstrapping Algorithm

This section describes our succinct functional bootstrapping algorithm (SFBS), which reduces the key size by approximately 50% compared with Algorithm 1.

4.1 Proposed Algorithm

Our algorithm stores half of the traditional bootstrapping key (i.e. $\text{RLWE}'_{\mathbf{sk}}(\underline{sk}_i)$), and generates the whole RGSW on-the-fly. Given that the blind rotation is executed in a serial manner (as seen in Algorithm 1, lines 3–5), it is possible to generate RGSW, compute the external product, and discard it in a loop. This approach can almost halve the bootstrapping storage overhead. In scenarios such as delegated computing, this also reduces the transmission bandwidth by half, as only RLWE′ rather than whole RGSW needs to be transmitted.

Succinct Functional Bootstrapping: Given a test polynomial testP encoded with a LUT function f, a succinct bootstrapping key containing $\text{RLWE}'_{\bar{\mathbf{sk}}}(\underline{sk}_i)$ and $\text{RLWE}'_{\bar{\mathbf{sk}}}(\bar{\mathbf{sk}}^2)$, where $\underline{sk}_i$ is the i-th term of the LWE secret key, the succinct functional bootstrapping algorithm $\mathsf{SFBS}^{\mathsf{testP}}_{\text{LWE}\to\text{LWE}} : \text{LWE}_{\underline{\mathbf{sk}}}(m) \to \text{LWE}_{\bar{\mathbf{sk}}}(f(m))$ is shown as Algorithm 3. We highlight the differences with the original FBS in gray.

Algorithm 3. Succinct Functional Bootstrapping (SFBS) Algorithm

Input: an LWE encryption $\underline{\mathbf{c}} = (\underline{\mathbf{a}}, \underline{b}) \in \text{LWE}_{\underline{\mathbf{sk}}}(m)$,
Input: a test polynomial testP,
Input: a SFBS_{key} contains $\text{RLWE}'_{\bar{\mathbf{sk}}}(\underline{sk}_i)_{i\in[1,n]}$ and $\text{RLWE}'_{\bar{\mathbf{sk}}}(\bar{\mathbf{sk}}^2)$.
Output: $\bar{c} \in \text{LWE}_{\bar{\mathbf{sk}}}(f(m))$.

1: $b = \lfloor \frac{2N}{q}\underline{b} \rceil$ and $a_i = \left\lfloor \frac{2N}{q}\underline{a}_i \right\rceil \in \mathbb{Z}_{2N}$ for each $i \in [1, n]$
2: $\mathsf{acc} = (0, X^b \cdot \mathbf{testP})$.
3: **for** $i = 1$ to n **do**:
4: $\text{RGSW}_{\bar{\mathbf{sk}}}(\underline{sk}_i) = \mathsf{SS}_{\text{RLWE}'\to\text{RGSW}}(\text{RLWE}'_{\bar{\mathbf{sk}}}(\underline{sk}_i))$
5: $\mathsf{acc} = ((X^{a_i} - 1)\cdot \mathsf{acc}) \boxdot \text{RGSW}_{\bar{\mathbf{sk}}}(\underline{sk}_i) + \mathsf{acc}$;
6: **end for**
7: $\bar{c} = \mathsf{SampleExtract}(\mathsf{acc})$,
8: **return** $\bar{c}$

4.2 Correctness and Error Analysis

Since we do not change the calculation of acc in line 5 of Algorithm 3, the correctness of SFBS algorithm can be directly derived from the correctness of Algorithm 1, as detailed in Theorem 6.2 in [10]. We then propose the error analysis.

In previous works, the RGSW ciphertexts for external product with the accumulator are fresh (as seen in line 4 of Algorithm 1). However, in our case, the analysis of error variance is different because we use a scheme switching algorithm to recover RGSW ciphertexts, which introduces additional noise. We thus give Theorem 3.

Theorem 3. *Let n denote the dimension of LWE ciphertexts, N denote the ring polynomials dimension of RLWE ciphertexts, B_{br}, l_{br}, ϵ_{br} denote the gadget decomposition parameters of the blind rotation algorithm, B_{ss}, l_{ss}, ϵ_{ss} denote the gadget decomposition parameters of the scheme switching algorithm. Then the error variance of the result of the SFBS algorithm is bounded by*

$$\sigma^2_{\mathsf{SFBS}} \leq \frac{B^2_{\mathsf{br}} B^2_{\mathsf{ss}}}{72} l_{\mathsf{br}} l_{\mathsf{ss}} N^2 n \sigma^2_{\mathsf{SS}_{key}} + \frac{B^2_{\mathsf{br}}}{72} l_{\mathsf{br}} N^3 n \epsilon^2_{\mathsf{ss}} + \frac{B^2_{\mathsf{br}}}{3} l_{\mathsf{br}} N n \sigma^2_{\mathrm{RLWE}'_{\bar{\mathbf{sk}}}(\underline{sk}_i)} + \frac{1}{6} N(n+1)\epsilon^2_{\mathsf{br}},$$

where $\sigma^2_{\mathsf{SS}_{key}}$ is the error variance of the scheme switching key $\mathrm{RLWE}'_{\bar{\mathbf{sk}}}(\bar{\mathbf{sk}}^2)$.

Proof. The error analysis is conducted following the method in TFHE [10].

4.3 Comparison with Previous Work

To evaluate our approach, we compare the key size of our scheme with those of previous works [9,10]. The bootstrapping key in the original FBS algorithm contains n RGSW ciphertexts, i.e., $\mathrm{RGSW}_{\bar{\mathbf{sk}}}(\underline{sk}_i)$, which equates to $2nl_{\mathsf{br}}$ RLWE ciphertexts. Ours succinct FBS key, however, contains $n+1$ RLWE′ ciphertexts ($\mathrm{RLWE}'_{\bar{\mathbf{sk}}}(\underline{sk}_i)$ and $\mathrm{RLWE}'_{\bar{\mathbf{sk}}}(\bar{\mathbf{sk}}^2)$), which amounts to $(nl_{\mathsf{br}} + l_{\mathsf{ss}})$ RLWE ciphertexts. Since n is always a power-of-two integer greater than 512, and l is consistently set between 2–4, our SFBS algorithm effectively reduces the key size by almost half. The results of this comparison can be found in Table 1.

Table 1. Key size of the FBS algorithm and our SFBS algorithm.

Algorithm	# Keys (in RLWEs)
FBS [10]	$2nl_{\mathsf{br}}$
SFBS [Ours]	$nl_{\mathsf{br}} + l_{\mathsf{ss}}$

5 Trade-Off Between Storage and Computation

This section introduces two variants of our ICBS algorithm, one that trades computation for storage, and another prioritizing storage over computation. Furthermore, we provide a comparison of all existing circuit bootstrapping solutions.

Algorithm 4. Succinct ICBS (S-ICBS) Algorithm

Input: a level 0 LWE encryption $\underline{\mathbf{c}} = (\underline{\mathbf{a}}, \underline{b}) \in \mathrm{LWE}_{\underline{\mathbf{sk}}}(m)$,
Input: test polynomials testP_k encoding $f_k(m) = v_k \cdot m, k \in [0, l-1]$,
Input: a level 2 $\mathsf{SFBS}_{key} = \mathrm{RLWE}'_{\bar{\mathbf{sk}}}(\underline{sk}_i)_{i\in[1,\underline{n}]}$,
Input: a level 1 $\mathsf{PubKS}_{key} = \mathrm{RLWE}'_{\mathbf{sk}}(\bar{sk}_j)_{j\in[1,\bar{n}]}$,
Input: a level 1 $\mathsf{SS}_{key} = \mathrm{RLWE}'_{\mathbf{sk}}(\mathbf{sk}^2)$,
Output: $\mathbf{c} \in \mathrm{RGSW}_{\mathbf{sk}}(\mathbf{m})$.
1: **for** $k = 0$ to $l - 1$ **do**:
2: $\mathrm{LWE}_{\bar{\mathbf{sk}}}(v_k \cdot m) \leftarrow \mathsf{SFBS}^{\mathsf{testP}_k}_{\mathrm{LWE}\to\mathrm{LWE}}(\mathrm{LWE}_{\underline{\mathbf{sk}}}(m))$,
3: $\mathrm{RLWE}_{\mathbf{sk}}(v_k \cdot \mathbf{m}) \leftarrow \mathsf{PubKS}_{\bar{\mathbf{sk}}\to\mathbf{sk}}(\mathrm{LWE}_{\bar{\mathbf{sk}}}(v_k \cdot m))$;
4: **end for**
5: $\mathrm{RLWE}'_{\mathbf{sk}}(m) = (\mathrm{RLWE}_{\mathbf{sk}}(v_0 \cdot \mathbf{m}), ..., \mathrm{RLWE}_{\mathbf{sk}}(v_{l-1} \cdot \mathbf{m}))$,
6: $\mathrm{RGSW}_{\mathbf{sk}}(m) \leftarrow \mathsf{SS}_{\mathrm{RLWE}'\to\mathrm{RGSW}}(\mathrm{RLWE}'_{\mathbf{sk}}(m))$,
7: **return** $\mathbf{c} = \mathrm{RGSW}_{\mathbf{sk}}(m)$

5.1 Succinct ICBS

The first variant is the succinct ICBS (S-ICBS) algorithm, as shown in Algorithm pagination4. This variant use the SFBS algorithm (Algorithm 3) proposed in Sect. 4, instead of the original FBS step. This substitution further reduces the key size at the expense of a slight increase in computational complexity and error growth. The differences between the S-ICBS and the ICBS (Algorithm 2) are highlighted in gray.

The new functional bootstrapping key contains $\mathrm{RLWE}'_{\bar{\mathbf{sk}}}(\underline{sk}_i)$ instead of $\mathrm{RGSW}_{\bar{\mathbf{sk}}}(\underline{sk}_i)$. Then the succinct ICBS algorithm further reduces the key size from $(4\underline{n}\bar{l}\bar{n}\log\bar{Q} + 2N(\bar{n}l_{\mathsf{ks}} + l)\log Q)$ bits to $(2(\underline{n}+1)\bar{l}\bar{n}\log\bar{Q} + 2N(\bar{n}l_{\mathsf{ks}} + l)\log Q)$ bits. The error variance of the algorithm result can be directly derived from Theorem 2 and Theorem 3.

5.2 Efficient ICBS

The second variant is the efficient ICBS (E-ICBS), which uses a pre-computing method proposed by Ducas and Micciancio [15] in the LWE to RLWE key switching algorithm. Specifically, under an observation that:

$$\begin{aligned}\mathsf{PubKS}_{\mathbf{sk}\to\mathbf{sk}'}(\mathrm{LWE}_{\mathbf{sk}}(m)) &:= \sum_{i=1}^{n} a_i \odot \mathrm{RLWE}'_{\mathbf{sk}'}(sk_i) + (0, f(b)) \\ &= \sum_{i=1}^{n}\sum_{j=0}^{t-1} a_{i,j} \cdot \mathrm{RLWE}_{\mathbf{sk}'}(v_j \cdot sk_i) + (0, f(b))\end{aligned}$$

where $a_{i,j}$ is the gadget decomposition of a_i, and since $a_{i,j} \in [0, 1, ..., B_{\mathsf{ks}} - 1]$, we can store all the possible values of $a_{i,j} \cdot \mathrm{RLWE}_{\mathbf{sk}'}(v_j \cdot sk_i)$ as the new public key switching key and then replace the gadget product by reading and adding.

This method increases the public key switching key size by a factor of B_{ks}.

Remark 5. For the sake of engineering simplicity, main existing implementations of the TFHE scheme employ this pre-computing trick to reduce running time. This includes the private/public key switching in the TFHE, Concrete, Palisade, OpenFHE, THFEpp, MOSFHET libraries, even though it is not explicitly mentioned in the corresponding papers [1,8–11,20,31]. We therefore refer to the CBS algorithm using this pre-computing method as E-CBS, which is not part of our contributions, we just regard it, together with the CBS algorithm, as comparison targets for the various types of circuit bootstrapping methods discussed in this paper (see Table 2 and Table 4).

5.3 Comparison

We proceed to compare the circuit bootstrapping key size in previous works and different variants of our ICBS, as shown in Table 2. The key size and algorithm performance under concrete parameters are presented in Sect. 6, Table 4.

Table 2. Evaluation key size of different variances of CBS algorithm.

Algorithm	Key Size (in bit)
CBS [9]	$4\underline{n}\bar{l}\bar{n}\log\bar{Q} + 4N(\bar{n}+1)l_{\mathsf{ks}}\log Q$
E-CBS [31]	$B_{\mathsf{ks}}(4\underline{n}\bar{l}\bar{n}\log\bar{Q} + 4N(\bar{n}+1)l_{\mathsf{ks}}\log Q)$
ICBS [Ours]	$4\underline{n}\bar{l}\bar{n}\log\bar{Q} + 2N(\bar{n}l_{\mathsf{ks}} + l)\log Q$
S-ICBS [Ours]	$2(\underline{n}+1)\bar{l}\bar{n}\log\bar{Q} + 2N(\bar{n}l_{\mathsf{ks}} + l)\log Q$
E-ICBS [Ours]	$B_{\mathsf{ks}}(4\underline{n}\bar{l}\bar{n}\log\bar{Q} + 2N(\bar{n}l_{\mathsf{ks}} + l)\log Q)$

6 Security, Parameters and Experiment

In this section, we conduct experiments on all of our proposed algorithms and compare them with the priors approaches in terms of key size and running time.

6.1 Security and Parameters

Circuit bootstrapping spans across three levels with different parameters. We set these parameters to ensure that the ciphertexts at each level achieve over 128-bit security, and that the decryption failure probability does not exceed 2^{-32}, as per our strict error analysis theorems. The FBS and SFBS algorithms are performed at level 2. The security parameter λ (test by Lattice estimator[3]) and other parameters are shown in Table 3.

[3] https://github.com/malb/lattice-estimator.

Table 3. Security and Parameters.

Level	λ	n	Q	σ^2_{fresh}	l	B	l_{ks}	B_{ks}
2	153	$\bar{n} = 2048$	$\bar{Q} = 2^{64}$	$\bar{\sigma}^2 = 2^{-88}$	$\bar{l} = 4$	$\bar{B} = 2^9$	–	–
1	129	$N = 1024$	$Q = 2^{32}$	$\sigma^2 = 2^{-50}$	$l = 3$	$B = 2^6$	$l_{\mathsf{ks}} = 10$	$B_{\mathsf{ks}} = 2^3$
0	128	$\underline{n} = 635$	$\underline{q} = 2^{32}$	$\underline{\sigma}^2 = 2^{-30}$	–	–	–	–

6.2 Experiments

Our experiment is run on an Intel(R) Core(TM) i5-12500 with 15.3 GB RAM, running the Ubuntu 20.04 LTS operating system. The tests are conducted on the TFHEpp [31] library. The running time and the key size for different bootstrapping methods are shown in Table 4.

Table 4. Run time and key size of different circuit bootstrapping methods.

Algorithm	Run Time (ms)	Key Size (MB)
CBS [9]	80	478.91
E-CBS [31]	70	2720.03
ICBS [Ours]	51	318.77
S-ICBS [Ours]	60	239.52
E-ICBS [Ours]	44	1438.91
FBS [10]	25	158.75
SFBS [Ours]	34	79.5

Analysis. In the LHE mode, our improved circuit bootstrapping algorithm (ICBS) outperforms Chillotti et al.'s method [9] by reducing 33.4% key size and 36.2% running time. Compared with the E-CBS algorithm, which is widely implemented in TFHE libraries, our ICBS still has 27.1% efficiency advantage and over 88.3% size reduction. Furthermore, the succinct-ICBS variant (S-ICBS) saves 49.9% and 91.2% storage, while the efficient variant (E-ICBS) decrease the execution time by 45% and 37.2%, compared with the CBS and the E-CBS.

In the FHE mode of the TFHE evaluation, our succinct functional bootstrapping algorithm (SFBS) reduces the key size by 49.9% compared with Chillotti et al.'s method [10], but increases the running time from 24.6 ms to 33.6 ms. We do not want to oversell this result, but take it as a trade-off method towards practical TFHE applications.

6.3 More Discussions with Instructions Optimizations

AVX-512 Instructions Optimizations. For bootstrapping implementations, enabling SIMD instructions such as advanced vector extensions (AVX) can leverage the vector computation power of the CPU to accelerate the process. For instance, our experiments, as previously discussed, employed the default AVX-2 support provided by the TFHEpp library.

With the release of Version 7 update in May 2023, the TFHEpp library expanded its support to include AVX-512. Given that the AVX-512 has not been updated since 2021 and is only compatible with Intel machines up to the 11-th generation, we replace our experimental environment to an 11-th Gen Intel(R) Core(TM) i5-11500 with 32 GB RAM, running the Ubuntu 22.04.2 LTS operating system. We then conduct comparative experiments using AVX-512, the results of which are presented in Table 5. From the table, it is evident that the performance enhancement achieved by our proposed algorithms aligns closely with the original experimental results.

Table 5. Run time of different circuit bootstrapping methods with AVX-512 instructions.

Algorithms	Run Time (ms)	
	TFHEpp	MOSFHET
CBS [9]	60	201
E-CBS [31]	54	180
ICBS [Ours]	42	137
S-ICBS [Ours]	45	146
E-ICBS [Ours]	37	123
FBS [10]	21	21
SFBS [Ours]	29	29

Implementations in MOSFHET Library. We note that the MOSFHET [19] homomorphic encryption library also implements the circuit bootstrapping algorithm with AVX-512 instructions. Therefore, we conduct comparative experiments under this library as well. The experimental results for FBS and SFBS are consistent with those of TFHEpp due to the fact that AVX-512 spqlios are used in these libraries. However, there is a notable disparity in performance between the circuit bootstrapping algorithms employed in MOSFHET and TFHEpp. This discrepancy emerges from the execution of key switching algorithms in MOSFHET, where they are executed at level 2 instead of original level 1.

7 Conclusion

We propose several improved algorithms of the TFHE bootstrapping, including a succinct functional bootstrapping algorithm, and three circuit bootstrapping variants performing a flexible trade-off between computation and storage. These algorithms can deal with different scenarios that have limitations in terms of computation complexity or memory storage. We take this work as a stepping stone towards more efficient TFHE applications.

References

1. Al Badawi, A., et al.: OpenFHE: open-source fully homomorphic encryption library. In: Proceedings of the 10th Workshop on Encrypted Computing and Applied Homomorphic Cryptography, pp. 53–63 (2022)
2. Boura, C., Gama, N., Georgieva, M., Jetchev, D.: Simulating homomorphic evaluation of deep learning predictions. In: Dolev, S., Hendler, D., Lodha, S., Yung, M. (eds.) CSCML 2019. LNCS, vol. 11527, pp. 212–230. Springer, Cham (2019). https://doi.org/10.1007/978-3-030-20951-3_20
3. Brakerski, Z.: Fully homomorphic encryption without modulus switching from classical GapSVP. In: Safavi-Naini, R., Canetti, R. (eds.) CRYPTO 2012. LNCS, vol. 7417, pp. 868–886. Springer, Heidelberg (2012). https://doi.org/10.1007/978-3-642-32009-5_50
4. Brakerski, Z., Gentry, C., Vaikuntanathan, V.: (Leveled) fully homomorphic encryption without bootstrapping. ACM Trans. Comput. Theory (TOCT) (2014)
5. Carpov, S., Izabachène, M., Mollimard, V.: New techniques for multi-value input homomorphic evaluation and applications. In: Matsui, M. (ed.) CT-RSA 2019. LNCS, vol. 11405, pp. 106–126. Springer, Cham (2019). https://doi.org/10.1007/978-3-030-12612-4_6
6. Cheon, J.H., Han, K., Kim, A., Kim, M., Song, Y.: Bootstrapping for approximate homomorphic encryption. In: Nielsen, J.B., Rijmen, V. (eds.) EUROCRYPT 2018. LNCS, vol. 10820, pp. 360–384. Springer, Cham (2018). https://doi.org/10.1007/978-3-319-78381-9_14
7. Cheon, J.H., Kim, A., Kim, M., Song, Y.: Homomorphic encryption for arithmetic of approximate numbers. In: Takagi, T., Peyrin, T. (eds.) ASIACRYPT 2017. LNCS, vol. 10624, pp. 409–437. Springer, Cham (2017). https://doi.org/10.1007/978-3-319-70694-8_15
8. Chillotti, I., Gama, N., Georgieva, M., Izabachène, M.: Faster fully homomorphic encryption: bootstrapping in less than 0.1 seconds. In: Cheon, J.H., Takagi, T. (eds.) ASIACRYPT 2016. LNCS, vol. 10031, pp. 3–33. Springer, Heidelberg (2016). https://doi.org/10.1007/978-3-662-53887-6_1
9. Chillotti, I., Gama, N., Georgieva, M., Izabachène, M.: Faster packed homomorphic operations and efficient circuit bootstrapping for TFHE. In: Takagi, T., Peyrin, T. (eds.) ASIACRYPT 2017. LNCS, vol. 10624, pp. 377–408. Springer, Cham (2017). https://doi.org/10.1007/978-3-319-70694-8_14
10. Chillotti, I., Gama, N., Georgieva, M., Izabachène, M.: Tfhe: fast fully homomorphic encryption over the torus. J. Cryptol. **33**(1), 34–91 (2020)
11. Chillotti, I., Joye, M., Ligier, D., Orfila, J.B., Tap, S.: CONCRETE: Concrete operates on ciphertexts rapidly by extending TFHE. In: WAHC 2020-8th Workshop on Encrypted Computing and Applied Homomorphic Cryptography (2020)
12. Chillotti, I., Ligier, D., Orfila, J.-B., Tap, S.: Improved programmable bootstrapping with larger precision and efficient arithmetic circuits for TFHE. In: Tibouchi, M., Wang, H. (eds.) ASIACRYPT 2021. LNCS, vol. 13092, pp. 670–699. Springer, Cham (2021). https://doi.org/10.1007/978-3-030-92078-4_23

13. Clet, P.E., Zuber, M., Boudguiga, A., Sirdey, R., Gouy-Pailler, C.: Putting up the swiss army knife of homomorphic calculations by means of TFHE functional bootstrapping. Cryptology ePrint Archive (2022)
14. De Micheli, G., Kim, D., Micciancio, D., Suhl, A.: Faster amortized FHEW bootstrapping using ring automorphisms. Cryptology ePrint Archive (2023)
15. Ducas, L., Micciancio, D.: FHEW: bootstrapping homomorphic encryption in less than a second. In: Oswald, E., Fischlin, M. (eds.) EUROCRYPT 2015. LNCS, vol. 9056, pp. 617–640. Springer, Heidelberg (2015). https://doi.org/10.1007/978-3-662-46800-5_24
16. Fan, J., Vercauteren, F.: Somewhat practical fully homomorphic encryption. Cryptology ePrint Archive (2012)
17. Gentry, C.: A Fully Homomorphic Encryption Scheme. Stanford University, Stanford (2009)
18. Gomes, F.A., de Matos, F., Rego, P., Trinta, F.: Analysis of the impact of homomorphic algorithm on offloading of mobile application tasks. In: 2023 IEEE 20th Consumer Communications and Networking Conference (CCNC), pp. 961–962. IEEE (2023)
19. Guimarães, A., Borin, E., Aranha, D.F.: MOSFHET: optimized software for FHE over the torus. IACR Cryptology ePrint Archive (2022). https://eprint.iacr.org/2022/515
20. Guimarães, A., Borin, E., Aranha, D.F.: Revisiting the functional bootstrap in TFHE. IACR Trans. Cryptogr. Hardw. Embed. Syst. 229–253 (2021)
21. Halevi, S., Shoup, V.: Bootstrapping for Helib. J. Cryptol. **34**(1), 7 (2021)
22. Jiang, L., Lou, Q., Joshi, N.: Matcha: A fast and energy-efficient accelerator for fully homomorphic encryption over the torus. In: Proceedings of the 59th ACM/IEEE Design Automation Conference, pp. 235–240 (2022)
23. Jutla, C.S., Manohar, N.: Modular Lagrange interpolation of the mod function for bootstrapping of approximate HE. Cryptology ePrint Archive (2020)
24. Jutla, C.S., Manohar, N.: Sine series approximation of the mod function for bootstrapping of approximate he. In: Dunkelman, O., Dziembowski, S. (eds.) EUROCRYPT 2022. LNCS, vol. 13275, pp. 491–520. Springer, Cham (2022). https://doi.org/10.1007/978-3-031-06944-4_17
25. Kluczniak, K., Schild, L.: FDFB: full domain functional bootstrapping towards practical fully homomorphic encryption. IACR Trans. Cryptogr. Hardw. Embed. Syst. 501–537 (2023)
26. Lee, J.W., et al.: Privacy-preserving machine learning with fully homomorphic encryption for deep neural network. IEEE Access **10**, 30039–30054 (2022)
27. Liu, Z., Micciancio, D., Polyakov, Y.: Large-precision homomorphic sign evaluation using FHEW/TFHE bootstrapping. In: Agrawal, S., Lin, D. (eds.) ASIACRYPT 2022. LNCS, vol. 13792, pp. 130–160. Springer, Cham (2022). https://doi.org/10.1007/978-3-031-22966-4_5
28. Lyubashevsky, V., Peikert, C., Regev, O.: On ideal lattices and learning with errors over rings. In: Gilbert, H. (ed.) EUROCRYPT 2010. LNCS, vol. 6110, pp. 1–23. Springer, Heidelberg (2010). https://doi.org/10.1007/978-3-642-13190-5_1
29. Marcolla, C., Sucasas, V., Manzano, M., Bassoli, R., Fitzek, F.H., Aaraj, N.: Survey on fully homomorphic encryption, theory, and applications. Proc. IEEE **110**(10), 1572–1609 (2022)
30. Matsumoto, M., Oguchi, M.: Speeding up encryption on IoT devices using homomorphic encryption. In: 2021 IEEE International Conference on Smart Computing (SMARTCOMP), pp. 270–275. IEEE (2021)

31. Matsuoka, K., Banno, R., Matsumoto, N., Sato, T., Bian, S.: Virtual secure platform: a five-stage pipeline processor over TFHE. In: USENIX Security Symposium, pp. 4007–4024 (2021)
32. Micciancio, D., Polyakov, Y.: Bootstrapping in FHEW-like cryptosystems. In: Proceedings of the 9th on Workshop on Encrypted Computing and Applied Homomorphic Cryptography, pp. 17–28 (2021)
33. Nam, K., Oh, H., Moon, H., Paek, Y.: Accelerating n-bit operations over TFHE on commodity CPU-FPGA. In: Proceedings of the 41st IEEE/ACM International Conference on Computer-Aided Design, pp. 1–9 (2022)
34. Peralta, G., Cid-Fuentes, R.G., Bilbao, J., Crespo, P.M.: Homomorphic encryption and network coding in IoT architectures: advantages and future challenges. Electronics **8**(8), 827 (2019)
35. Regev, O.: On lattices, learning with errors, random linear codes, and cryptography. J. ACM **56**(6), 34:1–34:40 (2009)
36. Ren, W., et al.: Privacy-preserving using homomorphic encryption in mobile IoT systems. Comput. Commun. **165**, 105–111 (2021)
37. Shrestha, R., Kim, S.: Integration of IoT with blockchain and homomorphic encryption: challenging issues and opportunities. In: Advances in Computers, vol. 115, pp. 293–331. Elsevier (2019)
38. Yang, Z., Xie, X., Shen, H., Chen, S., Zhou, J.: TOTA: fully homomorphic encryption with smaller parameters and stronger security. Cryptology ePrint Archive (2021)
39. Ye, T., Kannan, R., Prasanna, V.K.: FPGA acceleration of fully homomorphic encryption over the torus. In: 2022 IEEE High Performance Extreme Computing Conference (HPEC), pp. 1–7. IEEE (2022)

Approximate Methods for the Computation of Step Functions in Homomorphic Encryption

Tairong Huang[1], Shihe Ma[2], Anyu Wang[1,3,4](✉), and Xiaoyun Wang[1,3,4,5,6]

[1] Institute for Advanced Study, BNRist, Tsinghua University, Beijing, China
htr19@mails.tsinghua.edu.cn, {anyuwang,xiaoyunwang}@tsinghua.edu.cn
[2] Institute for Network Sciences and Cyberspace, Tsinghua University, Beijing, China
msh21@mails.tsinghua.edu.cn
[3] Zhongguancun Laboratory, Beijing, China
[4] National Financial Cryptography Research Center, Beijing, China
[5] Shandong Institute of Blockchain, Jinan, China
[6] Key Laboratory of Cryptologic Technology and Information Security (Ministry of Education), School of Cyber Science and Technology, Shandong University, Jinan, China

Abstract. This paper proposes two polynomial approximation methods for general step functions to tackle this problem. The first method leverages the fact that any step function can be expressed as a linear combination of shifted sign functions. This connection enables the homomorphic evaluation of any step function using known polynomial approximations of the sign function. The second method boosts computational efficiency by employing a composite polynomial approximation strategy. We present a systematic approach to construct a composite polynomial $f_k \circ f_{k-1} \circ \cdots \circ f_1$ that increasingly approximates the step function as k increases. This method utilizes an adaptive linear programming approach that we developed to optimize the approximation effect of f_i while maintaining the degree and coefficients bounded. We demonstrate the effectiveness of these two methods by applying them to typical step functions such as the round function and encrypted data bucketing, implemented in the HEAAN homomorphic encryption library. Experimental results validate that our methods can effectively address the homomorphic computation of step functions.

Keywords: Step function · Homomorphic encryption · CKKS · Polynomial approximation · Round function · Encrypted data bucketing

1 Introduction

Fully homomorphic encryption (FHE) is a powerful cryptographic primitive which enables performing any computation on encrypted data without having access to the secret key. Since Gentry developed the first FHE scheme [17], various FHE schemes have been proposed following Gentry's blueprint [18,32].

T. Zhu and Y. Li (Eds.): ACISP 2024, LNCS 14895, pp. 217–237, 2024.
https://doi.org/10.1007/978-981-97-5025-2_12

According to the type of computations to which they are suitable, these FHEs can be divided into three categories. The first category contains GSW [19] and its improvements FHEW/TFHE [11,13], which are ideal for evaluating Boolean circuits since they *bit-wisely* encrypt the input data. The second category contains BGV/FV [4,5,16], which pack their input data into finite fields or finite rings and are frequently used to evaluate integer arithmetic with a fixed modulus. The CKKS scheme [6–8], which forms the third category, can process fixed-point input numbers and supports approximate computations over complex and real numbers. In BGV/FV and CKKS, the input data is *word-wisely* encrypted. The operations on these numbers can be performed in a Single Instruction Multiple Data (SIMD) fashion [31], i.e., encrypted numbers are packed in slots such that the operations performed on a single ciphertext are automatically performed on each slot in parallel. Due to the SIMD property, these word-wise FHEs are very efficient in homomorphic addition, multiplication, and, more generally, polynomial evaluation. However, the effective evaluation of non-polynomial functions in word-wise FHEs presents a challenge and has recently garnered significant attention.

For CKKS, a natural way to tackle this issue involves approximating non-polynomial functions with polynomials. This approach has been successfully applied to evaluate a range of non-polynomial functions, such as logistic regression [20,22], inverse [10], square root [10,27], etc. [21,24,25]. Nevertheless, the homomorphic evaluation of discontinuous functions, such as the sign function and the step function, presents a significantly greater challenge.

1.1 Our Results

This paper delves into the polynomial approximation problem for general step functions. Let $\kappa(x)$ be a step function on the interval $[a, b]$ such that

$$\kappa(x) = y_i \text{ for } x \in (a_{i-1}, a_i), 1 \leq i \leq n,$$

where $a = a_0 < a_1 < \cdots < a_n = b$. The main contribution of this paper is two systematical methods for solving the polynomial approximation problem of $\kappa(x)$.

Method I (`SgnToStep`). This method utilizes the fact that a step function $\kappa(x)$ can be expressed as a linear combination of shifted sign functions, i.e.,

$$\kappa(x) = c_1 \mathtt{sgn}(x - a_1) + \cdots + c_{n-1} \mathtt{sgn}(x - a_{n-1}) + c_n,$$

where c_i's are real constants defined in Lemma 1, and $\mathtt{sgn}$ is the sign function defined in Sect. 2. Based on the polynomial approximations of $\mathtt{sgn}(x)$ as provided in [9,10,23], we show that this connection can be used to generate polynomial approximations for any step function $\kappa(x)$. We present a comprehensive analysis of the evaluation complexity and the required homomorphic multiplicative depth of this method. Moreover, we demonstrate that this method can be generalized to address the polynomial approximation problem for any piece-wise polynomial.

Method II (AdaptiveLP). This method reduces the number of multiplications by employing the composite polynomial strategy. Specifically, we construct a composite polynomial $g \circ f_k \circ \cdots \circ f_1$ approximating $\kappa(x)$ in two steps.

The first step aims to construct polynomials $f_1, f_2, \cdots, f_k$ which progressively map the intervals (a_{i-1}, a_i) to smaller intervals around their midpoint $\frac{1}{2}(a_i + a_{i-1})$ for $1 \leq i \leq n$. We demonstrate that the task of determining f_j is equivalent to solving the *weighted minimax polynomial approximation* problem as defined in Problem 1. An additional desirable property of f_j's is that their coefficients can be bounded, thereby allowing for high precision homomorphic evaluation [21]. We introduce an *adaptive linear programming algorithm* (see Algorithm 2), which gives the optimal weighted minimax polynomial approximation for step functions while keeping the coefficients bounded.

The second step involves constructing a polynomial $g(x)$ that maps the midpoints to $y_i, 1 \leq i \leq n$. We demonstrate that the optimal $g(x)$, which has bounded coefficients and minimizes the approximation error, can be derived using the adaptive linear programming algorithm again.

Applications to Concrete Step Functions. We demonstrate the two methods by presenting polynomial approximations for the round function and the bucketing function. Specifically, we give concrete polynomial approximations for the 7-step function $\frac{1}{3}\lfloor 3x \rceil$ and a 5-step function obtained from a bucketing example, and provide explicit error rates and running time for these approximations by evaluating them with the HEAAN library. The source code is available at https://anonymous.4open.science/r/code_upload-131E/.

1.2 Related Works

Numerical Analysis on Piece-Wise Functions. The problem of polynomial approximation for piece-wise functions has been studied for decades in numerical analysis. Some of these works focus on the polynomial approximations of piece-wise smooth functions [3,28–30]. Because step functions have discontinuities and piece-wise smooth functions are continuous, these results are not applicable to step functions. Another portion of works focus on functions with a single discontinuity, such as the sign function [14,15,29]. However, as observed in [9], when the approximation error needs to be exponentially small, the degree of the approximation polynomial becomes quite large, resulting in exponential homomorphic evaluation complexity.

Composite Polynomial Approximation of Sign Function. To improve the homomorphic computational complexity for the sign function, Cheon et al. proposed composite polynomial method that achieves asymptotic computational optimality [9,10]. Later, Lee et al. proposed a minimax composite polynomial method that achieves practical computational optimality [23,24]. However, these methods cannot be extended to handle polynomial approximations for general step functions because the intervals and values of a step function can be intricacy.

1.3 Organization

Section 2 introduces some notations. Section 3 and Sect. 4 propose `SgnToStep` and `AdaptiveLP` respectively. To demonstrate our method, we apply `SgnToStep` and `AdaptiveLP` to the round function and bucketing function in Sect. 5, and present experimental results of evaluating these step functions in HEAAN library in Sect. 6.

2 Preliminary

- We denote $\mathbb{Z}, \mathbb{R}$ and $\mathbb{C}$ to be the ring of integers, the field of real numbers and the field of complex numbers, respectively. $T_0(x) = 1, T_1(x) = x, T_i(x) = 2xT_{i-1}(x) - T_{i-2}(x)$ for $i \geq 2$.
- For a real function f defined over $\mathbb{R}$, let $f^{(d)} := f \circ f \circ \cdots \circ f$ denote the d-time composition of f. We use the infinite norm to measure the accuracy of polynomial approximations as suggested in [9,10]. For a function f and a a compact set $I \subset \mathbb{R}$, the infinite norm is defined by

$$\|f\|_I := \sup_{x \in I} |f(x)|.$$

  Besides, let $C_{max}(f)$ denote the maximum absolute value of f's coefficients (in terms of a polynomial basis, such as the power basis or the Chebyshev basis, depending on the context).
- Let $\log(\cdot)$ denote the logarithm of base 2. For $x \in \mathbb{R}$, let $\lfloor x \rceil = \lfloor x + 1/2 \rfloor$ denote the integer closest to x, and let $\mathbf{sgn}(x)$ denote the sign function

$$\mathbf{sgn}(x) = \begin{cases} 1 & \text{if } x > 0 \\ 0 & \text{if } x = 0 \\ -1 & \text{if } x < 0 \end{cases}. \quad (1)$$

2.1 Step Function

The step functions considered in this paper are piece-wise constant real functions with finitely many pieces, which can be formally defined as follows.

Definition 1 (Step Function). *A real function $\kappa(x)$ defined on the interval $[a, b]$ is a step function if there exists a finite partition $a = a_0 < a_1 < \cdots < a_n = b$ such that $\kappa(x)$ is a constant function on each interval (a_{i-1}, a_i), i.e., there exist $y_i \in \mathbb{R}$ such that*

$$\kappa(x) = y_i \text{ for } x \in (a_{i-1}, a_i), 1 \leq i \leq n.$$

We call $(a_{i-1}, a_i), 1 \leq i \leq n$, the intervals of $\kappa(x)$, and call $y_i, 1 \leq i \leq n$, the values of $\kappa(x)$. For convenience we always assume $\kappa(a_0) = y_1$ and $\kappa(a_n) = y_n$. The value of $\kappa(a_i)$, $1 \leq i \leq n-1$, is not specified in the definition. From the perspective of polynomial approximation, we do not mind the specific value of

$\kappa(a_i)$ for $1 \le i \le n-1$. Besides, we always assume that a_i is a jump discontinuity, i.e., $y_i \ne y_{i+1}$ for $1 \le i \le n-1$.

Since $\kappa(x)$ can not be approximated by polynomials near the discontinuities a_i's, $1 \le i \le n-1$, we follow the approach adopted in [9,10,23] and define the following measurement of approximation error.

Definition 2. *For small real numbers* $2^{-\alpha}, \epsilon > 0$, *we say a polynomial* $f(x)$ *is* (α, ϵ)*-close to a step function* $\kappa(x)$ *with partition* $a = a_0 < a_1 < \cdots < a_n = b$ *if*

$$\|f(x) - \kappa(x)\|_I \le 2^{-\alpha},$$

where $I = [a, b] - \bigcup_{1 \le i < n}(a_i - \epsilon, a_i + \epsilon)$.

Definition 3. *For a step function* $\kappa(x)$ *with partition* $a = a_0 < a_1 < \cdots < a_n = b$ *and constant values* $y_1, \ldots, y_n$. *We say* $\kappa(x)$ *is normalized if* $y_1 = a, y_n = b$ *and* $y_i = \frac{1}{2}(a_{i-1} + a_i)$ *for* $1 < i < n$. *We say* $\tilde{\kappa}(x)$ *is the normalization of* $\kappa(x)$ *if* $\tilde{\kappa}(x)$ *is normalized and* $\tilde{\kappa}(x)$ *has the same partition as* $\kappa(x)$.

2.2 CKKS FHE Scheme

CKKS scheme is introduced by Cheon et al. in [8], which enables approximate homomorphic arithmetic computation over real/complex numbers. We denote λ as the security parameter of CKKS. Let $L \in \mathbb{Z}^+$ denote the bit-length of the initial ciphertext modulus, and define $q_l := 2^l$ for $1 \le l \le L$. We denote χ_s, χ_e, χ_r as the distribution of secret, error, and encryption respectively. Let $N \in \mathbb{Z}$ be a power of 2. Denote $R = \mathbb{Z}[X]/(X^N + 1)$ be the $2N$-th cyclotomic ring, and $R_q := R/qR$ for an integer $q > 1$. The isomorphism $\tau : \mathbb{R}[X]/(X^N + 1) \to \mathbb{C}^{\frac{N}{2}}$ is used for encoding and decoding of plaintexts. Let Δ denote the scaling factor. The CKKS scheme contains the following algorithms.

- $\texttt{KeyGen}(1^\lambda)$:
  Sample $s \leftarrow \chi_s$, $a \leftarrow \mathcal{U}(R_{q_L})$, $e \leftarrow \chi_e$, $e' \leftarrow \chi_e$ and $a' \leftarrow \mathcal{U}(R_{q_L^2})$;
  Output $\texttt{sk} = (1, s)$, $\texttt{pk} = (-as + e, a) \in R_{q_L}^2$ and $\texttt{evk} = (-a' \cdot s + e' + q_L \cdot s^2, a') \in R_{q_L^2}^2$.
- $\texttt{Enc}_{\text{pk}}(m; \Delta)$:
  For a plaintext $m \in \mathbb{C}^{\frac{N}{2}}$, compute $\mathfrak{m} = \lfloor \Delta \cdot \tau^{-1}(m) \rceil$, sample $v \leftarrow \chi_r$ and $e_0, e_1 \leftarrow \chi_e$;
  Output $v \cdot \texttt{pk} + (m + e_0, e_1) \mod q_L$.
- $\texttt{Dec}_{\text{sk}}(\texttt{ct}; \Delta)$:
  For a ciphertext $\texttt{ct} = (c_0, c_1) \in R_{q_l}^2$, compute $\mathfrak{m}' = c_0 + c_1 \cdot s \mod q_l$;
  Output $m' = \frac{1}{\Delta} \cdot \tau(\mathfrak{m}')$.
- $\texttt{Add}(\texttt{ct}_1, \texttt{ct}_2)$:
  For $\texttt{ct}_1, \texttt{ct}_2 \in R_{q_l}^2$, output $\texttt{ct}_{\text{add}} = \texttt{ct}_1 + \texttt{ct}_2 \mod q_l$.
- $\texttt{Mult}_{\text{evk}}(\texttt{ct}_1, \texttt{ct}_2)$:
  For $\texttt{ct}_1 = (b_1, a_1), \texttt{ct}_2 = (b_2, a_2) \in R_{q_l}^2$, let $(d_0, d_1, d_2) = (b_1 b_2, a_1 b_2 + a_2 b_1, a_1 a_2)$, compute $\texttt{ct}'_{\text{mult}} \leftarrow$
  $(d_0, d_1) + \lfloor q_L^{-1} \cdot d_2 \cdot \texttt{evk} \rceil \mod q_l$;
  Output $\texttt{ct}_{\text{mult}} = \lfloor \Delta^{-1} \cdot \texttt{ct}'_{\text{mult}} \rceil \mod (q_l/\Delta)$.

Note that we can deal with $\frac{N}{2}$ encrypted data in a SIMD manner, so the amortized running time is $\frac{2}{N}$ times the total time.

3 SgnToStep: Step Function Approximation by Using the Connection with sgn

In this section, we provide a linear relation between the step function and the sign function. Based on this connection, any step function $\kappa(x)$ can be homomorphically evaluated by using the approximations of $\mathtt{sgn}(x)$ as presented in [9,10,23,26].

3.1 A Connection Between Step Function and Sign Function

It is obvious that a step function $\kappa(x)$ with n intervals can be expressed as a linear combination of at most n indicator functions of intervals. In fact, the following lemma states that $\kappa(x)$ can also be written as a linear combination of $n-1$ shifted sign functions.

Lemma 1. *A step function $\kappa(x)$ with partition $a_0 < a_1 < \cdots < a_n$ and values $y_1, \cdots, y_n$ can be expressed by a linear combination of $n-1$ shifted sign functions, i.e.,*

$$\kappa(x) = \sum_{i=1}^{n-1} c_i \mathtt{sgn}(x - a_i) + c_n, \tag{2}$$

where $c_i = \frac{1}{2}(y_{i+1} - y_i)$ for $1 \leq i \leq n-1$ and $c_n = \frac{1}{2}(y_1 + y_n)$.

Proof. It suffices to check Eq. (2) for the intervals $(a_{i-1}, a_i), 1 \leq i \leq n$. Suppose $x \in (a_{i-1}, a_i)$, then the left hand side of (2) is $\kappa(x) = y_i$. Note that $\mathtt{sgn}(x - a_1) = \cdots = \mathtt{sgn}(x - a_{i-1}) = 1$ and $\mathtt{sgn}(x - a_i) = \cdots = \mathtt{sgn}(x - a_{n-1}) = -1$ for $x \in (a_{i-1}, a_i)$, then the right hand side of (2) is

$$\sum_{j=1}^{i-1} c_j - \sum_{j=i}^{n-1} c_j + c_n = \frac{1}{2}(y_i - y_1) - \frac{1}{2}(y_n - y_i) + \frac{1}{2}(y_1 + y_n) = y_i.$$

Thus Eq. (2) holds. □

Because a linear combination of k shifted sign functions has at most k discontinuities, $n-1$ is the smallest number of shifted sign functions required to represent $\kappa(x)$ linearly.

3.2 Step Function Approximation Based on the Linear Combination

We demonstrate how to use Lemma 1 and a polynomial approximation of $\mathtt{sgn}(x)$ to obtain a polynomial approximation of a step function $\kappa(x)$. Suppose $g(x)$ is a (composite) polynomial approximation of $\mathbf{sgn}(x)$ as constructed in [10,23], such that $g(x)$ is (α, ϵ)-close to $\mathbf{sgn}(x)$ on $[-1,1]$, i.e.

$$\|g(x) - \mathbf{sgn}(x)\|_{[-1,-\epsilon]\cup[\epsilon,1]} \leq 2^{-\alpha}. \tag{3}$$

Then an approximation of $\kappa(x)$ can be constructed as follows.

Theorem 1. *Let $\kappa(x)$ be a step function with partition $-1 = a_0 < a_1 < \cdots < a_n = 1$ and values $y_1, \cdots, y_n$. Suppose $g(x)$ is (α, ϵ)-close to $\mathbf{sgn}(x)$ on $[-1,1]$. Then the function*

$$f(x) = \sum_{i=1}^{n-1} \frac{1}{2}(y_{i+1} - y_i) \cdot g(\frac{x - a_i}{1 + |a_i|}) + \frac{1}{2}(y_1 + y_n)$$

is (α', ϵ')-close to $\kappa(x)$ on $[-1,1]$, where $\alpha' = \alpha - \log(\sum_{i=1}^{n-1} \frac{1}{2}|y_{i+1} - y_i|)$ and $\epsilon' = (1 + \max\{|a_1|, |a_{n-1}|\})\epsilon$.

Proof. We first show that $g(\frac{x-a_i}{1+|a_i|})$ is an approximation of $\mathbf{sgn}(x - a_i)$ on $I := [-1,1] - \bigcup_{1\leq i<n}(a_i - \epsilon', a_i + \epsilon')$. Denote $y = \frac{x-a_i}{1+|a_i|}$, then for $x \in I$ it has $|y| \leq \frac{|x|+|a_i|}{1+|a_i|} \leq 1$ and $|y| = \frac{|x-a_i|}{1+|a_i|} \geq \frac{\epsilon'}{1+|a_i|} \geq \epsilon$, i.e., $y \in [-1,-\epsilon] \cup [\epsilon, 1]$. Thus $\left\|g(\frac{x-a_i}{1+|a_i|}) - \mathbf{sgn}(x - a_i)\right\|_I \leq \|g(y) - \mathbf{sgn}((1 + |a_i|)y)\|_{[-1,-\epsilon]\cup[\epsilon,1]} = \|g(y) - \mathbf{sgn}(y)\|_{[-1,-\epsilon]\cup[\epsilon,1]} \leq 2^{-\alpha}$. Therefore, by Lemma 1 it has

$$\begin{aligned}\|f(x) - \kappa(x)\|_I &= \left\|\sum_{i=1}^{n-1} \frac{1}{2}(y_{i+1} - y_i) \cdot (g(\frac{x - a_i}{1 + |a_i|}) - \mathbf{sgn}(x - a_i))\right\|_I \\ &\leq \sum_{i=1}^{n-1} \frac{1}{2}|y_{i+1} - y_i| \cdot 2^{-\alpha} = 2^{-\alpha'},\end{aligned}$$

which completes the proof. □

Remark 1. Different approximations for $\mathbf{sgn}(x - a_i)$ can be chosen to balance the overall error rate. Specifically, suppose $g_i(x)$ is (α_i, ϵ_i)-close to $\mathbf{sgn}(x)$ on $[-1,1]$ for $1 \leq i < n$. Then it can be similarly proved that the function $f(x) = \sum_{i=1}^{n-1} \frac{1}{2}(y_{i+1} - y_i) \cdot g_i(\frac{x-a_i}{1+|a_i|}) + \frac{1}{2}(y_1 + y_n)$ is (α', ϵ')-close to $\kappa(x)$ on $[a, b]$, where $\alpha' = \log(\sum_{i=1}^{n-1} \frac{1}{2}|y_{i+1} - y_i| \cdot 2^{-\alpha_i})$ and $\epsilon' = \max_{1\leq i<n}\{(1 + |a_i|)\epsilon_i\}$.

Computation Complexity. The polynomial approximation for $\mathbf{sgn}(x)$ is usually given in a composite polynomial form to reduce the number of homomorphic multiplications, i.e., $g = h_k \circ h_{k-1} \circ \cdots \circ h_1$. Then the $g(\frac{x-a_i}{1+|a_i|})$ in Theorem 1 should be evaluated individually before performing the linear combination (Algorithm 1).

The required multiplicative depth for Algorithm 1 is roughly the same as that for `ComputeG` (or $g(x)$), and the number of multiplications is $n - 1$ times as that of `ComputeG` (or $g(x)$). The total running time can be reduced if each $g(\frac{x-a_i}{1+|a_i|})$ can be computed in parallel.

Algorithm 1: Compute step function by using Theorem 1.

Input: A real number $x_0 \in [-1, 1]$
Input: A step function $\kappa(x)$ with partition $a_0 < \cdots < a_n$ and values $y_1, \cdots, y_n$
Input: A sub-algorithm `ComputeG` that computes $g(x)$, where $g(x)$ is a composite polynomial approximation of $\mathbf{sgn}(x)$
Output: Approximate value of $\kappa(x_0)$
1: **for** i from 1 to $n-1$ **do**
2: $\quad z_i = \texttt{ComputeG}(\frac{x_0 - a_i}{1+|a_i|})$
3: **end for**
4: $z = \sum_{i=1}^{n-1} \frac{1}{2}(y_{i+1} - y_i) \cdot z_i + \frac{1}{2}(y_1 + y_n)$
5: **return** z

3.3 Extension to Piece-Wise Polynomials

Suppose $\rho(x)$ is a piece-wise polynomial defined on $[a, b]$ such that

$$\rho(x) = p_i(x) \text{ for } x \in (a_{i-1}, a_i), 1 \le i \le n, \tag{4}$$

where $a = a_0 < a_1 < \cdots < a_n = b$, and $p_i(x)$'s are polynomials defined on $[a, b]$. Similar to Lemma 1, the following lemma can be proved.

Lemma 2. *$\rho(x)$ can be expressed as*

$$\rho(x) = \sum_{i=1}^{n-1} \frac{1}{2}(p_{i+1}(x) - p_i(x)) \cdot \mathbf{sgn}(x - a_i) + \frac{1}{2}(p_1(x) + p_n(x)) \tag{5}$$

for $x \in [a, b]$ other than the singularity points.

Then a polynomial approximation of $\rho(x)$ can be constructed based on the polynomial approximation of $\mathbf{sgn}(x)$ as follows.

Theorem 2. *Suppose $\rho(x)$ is a piece-wise polynomial on $[-1, 1]$ and $g(x)$ is (α, ϵ)-close to $\mathbf{sgn}(x)$ on $[-1, 1]$. Then the function*

$$f(x) = \sum_{i=1}^{n-1} \frac{1}{2}(p_{i+1}(x) - p_i(x)) \cdot g(\frac{x - a_i}{1 + |a_i|}) + \frac{1}{2}(p_1(x) + p_n(x))$$

is (α', ϵ')-close to $\rho(x)$ on $[-1, 1]$, i.e., $\|\rho(x) - f(x)\|_I \le 2^{-\alpha'}$, where $I = [-1, 1] - \bigcup_{1 \le i < n}(a_i - \epsilon', a_i + \epsilon')$ and $\alpha' = \alpha - \log(\sum_{i=1}^{n-1} \frac{1}{2}\|p_{i+1}(x) - p_i(x)\|_I, \epsilon' = (1 + \max\{|a_1|, |a_{n-1}|\})\epsilon$.

Proof. It can be proved as in Theorem 1 that $\|g(\frac{x-a_i}{1+|a_i|}) - \mathbf{sgn}(x - a_i)\|_I \le 2^{-\alpha}$. Then by Lemma 2 it has

$$\begin{aligned} \|f(x) - \rho(x)\|_I &= \left\| \sum_{i=1}^{n-1} \frac{1}{2}(p_{i+1}(x) - p_i(x)) \cdot (g(\frac{x - a_i}{1 + |a_i|}) - \mathbf{sgn}(x - a_i)) \right\|_I \\ &\le \sum_{i=1}^{n-1} \frac{1}{2}\|p_{i+1}(x) - p_i(x)\|_I \cdot 2^{-\alpha} = 2^{-\alpha'}, \end{aligned}$$

which completes the proof. □

4 AdaptiveLP: Step Function Approximation by Polynomial Composition

In this section, we consider the composite polynomial strategy to approximate step functions. For any step function $\kappa(x)$, we aim to construct a composite polynomial $g \circ f_k \circ \cdots \circ f_1$ that approximates $\kappa(x)$. The construction can be divided into two steps, which are specified in Sect. 4.1 and Sect. 4.2 respectively.

Step 1. Construct a composite polynomial $f = f_k \circ \cdots \circ f_1$ approximating $\tilde{\kappa}(x)$, where $\tilde{\kappa}(x)$ is the normalization of $\kappa(x)$.
Step 2. Construct a polynomial $g(x)$ such that $g(\tilde{\kappa}(x)) \approx \kappa(x)$.

4.1 Construction of the Composite Polynomial f

Suppose $\tilde{\kappa}(x) = z_i$ for $x \in (a_{i-1}, a_i)$, where $z_1 = a, z_n = b$ and $z_i = \frac{1}{2}(a_{i-1} + a_i), 1 < i < n$. Our goal is to construct polynomials $f_1, \cdots, f_k$ such that they gradually map the intervals to small intervals. For a small positive real number ϵ, denote $I_{10} = [a_0, a_1 - \epsilon], I_{n0} = [a_{n-1} + \epsilon, a_n]$ and $I_{i0} = [a_{i-1} + \epsilon, a_i - \epsilon]$ for $1 < i < n$. Then the polynomials $f_1, \cdots, f_k$ should satisfy

$$I_{i0} \xrightarrow{f_1} [z_i - t_{i1}, z_i + t_{i1}] \xrightarrow{f_2} \cdots \xrightarrow{f_k} [z_i - t_{ik}, z_i + t_{ik}] \quad (6)$$

for $1 \leq i \leq n$, where $t_{i1} > \cdots t_{ik} > 0$. Denote $I_{ij} = [z_i - t_{ij}, z_i + t_{ij}]$ for $1 \leq i \leq n$ and $1 \leq j \leq k$, and let $t_{10} = a_1 - a_0 - \epsilon$, $t_{n0} = a_n - a_{n-1} - \epsilon$, and $t_{i0} := \frac{1}{2}(a_i - a_{i-1}) - \epsilon$ for $1 < i < n$. Then the optimal polynomial f_{j+1} should minimize the ratio

$$\max_{1 \leq i \leq n} \frac{t_{i,j+1}}{t_{ij}} = \max_{1 \leq i \leq n} \frac{1}{t_{ij}} \cdot \|f_{j+1}(x) - z_i\|_{I_{ij}} = \max_{1 \leq i \leq n} \frac{1}{t_{ij}} \cdot \|f_{j+1}(x) - \tilde{\kappa}(x)\|_{I_{ij}}. \quad (7)$$

On the other hand, we want the coefficients of f_j to be bounded by a real constant number B_j to ensure evaluation precision. In other words, for $0 \leq j \leq k-1$, the polynomial f_{j+1} is a solution to the following optimization problem.

Problem 1 (Weighted Minimax Polynomial Approximation). *For input step function $\tilde{\kappa}(x)$, constant numbers $t_{ij} > 0$, intervals I_{ij} for $1 \leq i \leq n$, find a polynomial $f_{j+1}(x)$ with degree no more than d_{j+1} and coefficients bounded by $\mathcal{C}_{max}(f_{j+1}) \leq B_{j+1}$ that minimizes*

$$\max_{1 \leq i \leq n} \frac{1}{t_{ij}} \cdot \|f_{j+1}(x) - \tilde{\kappa}(x)\|_{I_{ij}}. \quad (8)$$

Solving Problem 1 via Adaptive Linear Programming. Suppose c_{opt} is the minimum value of (8). The adaptive linear programming algorithm iteratively computes a polynomial $\hat{f}_{j+1}$ such that the value $\max_{1 \leq i \leq n}\{\frac{1}{t_{ij}} \cdot \|\hat{f}_{j+1}(x) - \tilde{\kappa}(x)\|_{I_{ij}}\}$ approaches c_{opt}.

To begin with, we choose a set of reference points $\mathcal{X} \subset \cup_{1 \leq i \leq n} I_{ij}$, and consider the conditions

$$\begin{cases} \frac{1}{t_{ij}} \cdot |f_{j+1}(x_l) - \tilde{\kappa}(x_l)| \leq c, \forall 1 \leq i \leq n, \text{ for } x_l \in \mathcal{X}; \\ \mathcal{C}_{max}(f_{j+1}) \leq B_{j+1}, \end{cases} \tag{9}$$

where c is the objective to be minimized. Then (9) provides linear constrains on the coefficients of f_{j+1} and c. As a result, we can obtain a polynomial $\hat{f}_{j+1}$ and a real number $c_l > 0$ by using linear programming to minimize c. Clearly c_l is a lower bound of c_{opt} since the solution $f_{j+1}^{(\text{opt})}$ to Problem 1 must satisfy $\frac{1}{t_{ij}} \cdot |f_{j+1}^{(\text{opt})}(x_l) - \tilde{\kappa}(x_l)| \leq c_{\text{opt}}, \forall x_l \in \mathcal{X}$.

On the other hand, for the polynomial $\hat{f}_{j+1}(x)$ obtained by solving (9), let

$$c_u := \max_{1 \leq i \leq n} \frac{1}{t_{ij}} \cdot \|\hat{f}_{j+1}(x) - \tilde{\kappa}(x)\|_{I_{ij}}.$$

Clearly c_u is an upper bound of c_{opt}. In order to decrease c_u, we collect all the extreme and boundary points $x' \in \cup_{1 \leq i \leq n} I_{ij}$ of the polynomial $\hat{f}_{j+1}(x)$ such that $\frac{1}{t_{ij}} \cdot |\hat{f}_{j+1}(x') - \tilde{\kappa}(x')| > c_l$, add all these points to the set $\mathcal{X}$, and repeat the linear programming process. Algorithm 2 summarizes the above procedure. For the choice of polynomial basis, it was observed that the Chebyshev basis is suitable for minimax polynomial approximation [6,24]. Besides, an efficient homomorphic computation method for the Chebyshev basis has been proposed [25]. Thus we also adopt the Chebyshev basis for polynomial approximation in this paper.

Termination and Runtime of the Algorithm. When performing Algorithm 2, it is clear that the c_l gradually increases because more linear constrains are added to (9). Moreover, through experiments, we find that the c_u quickly approaches c_l and thus approaches c_{opt}. Figure 1 depicts the first two iterations of Algorithm 2 for solving the weighted minimax problem that corresponds to construct f_1 for $\tilde{\kappa}(x) = [x], x \in [-1, 1]$, and $\epsilon = 2^{-16}$.

In each iteration of Algorithm 2, a linear programming algorithm is employed to solve c_l. It is shown in [12] that solving such linear programming takes $\mathcal{O}^*(|\mathcal{X}|^c \log(|\mathcal{X}|/\delta))$ time, where $2 < c < 3$ is a constant determined by the matrix multiplication algorithm, and δ is the relative accuracy. According to our experiment, for a step function with n intervals, and a polynomial degree d, a coefficient bound B, setting $|\mathcal{X}| = \mathcal{O}(nd)$ and $\delta = \mathcal{O}(\epsilon/(dB))$ suffices for the computation (Table 1).

Determine the Composite Polynomial. The polynomials f_{j+1} can be constructed using Algorithm 2 iteratively for $0 \leq j \leq k-1$. Here the $t_{i,j+1}$'s are determined by $t_{i,j+1} = \|f_{j+1}(x) - \tilde{\kappa}(x)\|_{I_{ij}}$ after f_j has been determined. Due

Algorithm 2: Adaptive linear programming

Input: A step function $\tilde{\kappa}(x)$ and an approximation factor $\gamma \in \mathbb{R}^+$
Input: Real numbers $t_{ij} > 0$ and intervals I_{ij} for $1 \leq i \leq n$
Input: Polynomial degree $d_{j+1} \in \mathbb{Z}^+$ and coefficient bound $B_{j+1} > 0$
Input: A polynomial basis $\{p_l(x)\}_{1 \leq l \leq d_{j+1}}$
Output: Approximate polynomial $f_{j+1}(x)$ that minimize (8)
1: Choose a set of reference points $\mathcal{X} \subset \cup_{1 \leq i \leq n} I_{ij}$
2: Solve the following linear programming problem and obtain $\hat{f}_{j+1}$ and c_l
Minimize c
Subject to $\mathcal{C}_{max}(f_{j+1}) \leq B_{j+1}$ and $|f_{j+1}(x_l) - \tilde{\kappa}(x_l)| \leq c t_{ij}, \forall x_l \in \mathcal{X}$
3: Collect the extreme and boundary points $x' \in \cup_{1 \leq i \leq n} I_{ij}$ such that $|\hat{f}_{j+1}(x') - \tilde{\kappa}(x')| > c_l t_{ij}$, and add them to $\mathcal{X}$
4: Compute $c_u = \max_{1 \leq i \leq n} \{\frac{1}{t_{ij}} \cdot \|\hat{f}_{j+1}(x) - \tilde{\kappa}(x)\|_{I_{ij}}\}$.
5: **if** $c_u < (1+\gamma)c_l$ **then**
6: **return** $\hat{f}_{j+1}$
7: **else**
8: Go to line 2
9: **end if**

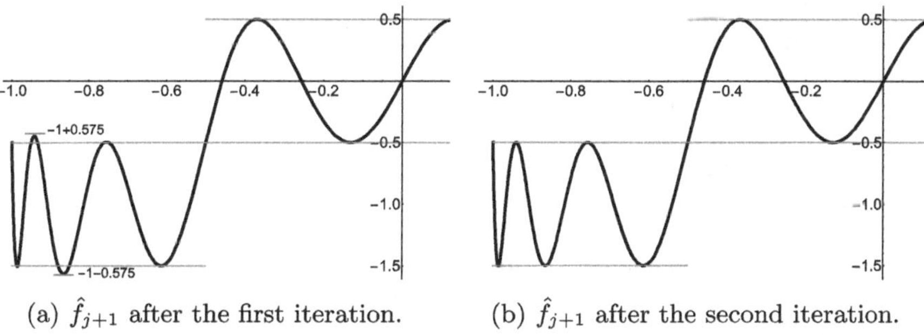

(a) $\hat{f}_{j+1}$ after the first iteration. (b) $\hat{f}_{j+1}$ after the second iteration.

Fig. 1. Illustration of the first two iterations of adaptive linear programming algorithm. The graph of f_{j+1} is symmetric with respect to the origin.

to our choice of $\tilde{\kappa}(x)$ and I_{ij}, it has

$$\max_{1 \leq i \leq n} \frac{1}{t_{ij}} \cdot \|f_{j+1}(x) - \tilde{\kappa}(x)\|_{I_{ij}} < (1+\gamma) \max_{1 \leq i \leq n} \frac{1}{t_{ij}} \cdot \|f_{j+1}^{(\text{opt})} - \tilde{\kappa}(x)\|_{I_{ij}}$$
$$\leq (1+\gamma) \max_{1 \leq i \leq n} \frac{1}{t_{ij}} \cdot \|x - \tilde{\kappa}(x)\|_{I_{ij}} = (1+\gamma),$$

i.e. $t_{i,j+1} < (1+\gamma)t_{i,j}$. In our experiment, it holds $t_{i,j+1} < t_{i,j}$ for an appropriate choice of the factor γ, thus the mapping of intervals in (6) can be guaranteed.

Nevertheless, in the encrypted state, f_{j+1} will be homomorphically evaluated, and $f_{j+1}(I_{ij})$ may not fall into $I_{i,j+1}$ due to the homomorphic computation errors. This can cause an evaluation failure of the composite polynomial $f = f_k \circ \cdots \circ f_1$. To solve this problem, we introduce a parameter η_{j+1} which is an

Table 1. The number of iterations of Algorithm 2 for approximating the step function $\frac{1}{3}\lfloor 3x \rceil$ on $[-1, 1]$, where $\{z_i\}_i = \{0, \pm\frac{1}{3}, \pm\frac{2}{3}, \pm 1\}$ and $\{t_{ij}\}_i$ are roughly equal for the same j. The degrees of f_j are set to be 31.

f_{j+1}	f_1	f_2	f_3	f_4	f_5	f_6	f_7	f_8
$t_{ij} \approx$	$1/6 - 2^{-16}$	$1/6 - 2^{-13.5}$	$1/6 - 2^{-11.0}$	$1/6 - 2^{-8.6}$	$1/6 - 2^{-6.2}$	$1/6 - 2^{-4.0}$	$1/6 - 2^{-2.8}$	$1/6 - 2^{-2.6}$
#Iterations	5	4	4	4	4	3	3	1

upper bound of the homomorphic evaluation error, i.e.,

$$|\texttt{Eval}(f_j)(x) - f_j(x)| \leq \eta_{j+1} \ll 1,$$

for $0 \leq j \leq k-1$. Besides, we set η_0 to be the encryption error. Then we use the intervals $I'_{ij} := [z_i - t_{ij} - \eta_j, z_i + t_{ij} + \eta_j]$ as input to solve f_{j+1} (instead of I_{ij}), which ensures the mapping of intervals in (6) for the encrypted state. The above process is summarized in Algorithm 3.

Algorithm 3: Construct the composite polynomial

Input: A step function $\tilde{\kappa}(x)$ and an approximation factor $\gamma \in \mathbb{R}^+$
Input: $t_{i0} \in \mathbb{R}^+$ and intervals I'_{i0} for $1 \leq i \leq n$
Input: Polynomial degree $d_{j+1} \in \mathbb{Z}^+$, coefficient bound $B_{j+1} \in \mathbb{Z}^+$ and error bound $\eta_j \in \mathbb{R}^+$ for $0 \leq j \leq k-1$
Input: A polynomial basis $\{p_l(x)\}_l$
Output: Composite polynomial $f = f_k \circ \cdots \circ f_1$ approximating $\tilde{\kappa}(x)$
1: **for** j from 0 to $k-1$ **do**
2: Compute a polynomial f_{j+1} by using $\tilde{\kappa}(x)$, γ, t_{ij}, I'_{ij}, d_{j+1}, B_{j+1} and $\{p_l(x)\}_l$ as the inputs of Algorithm 2
3: Compute $t_{i,j+1} = \|f_{j+1}(x) - \tilde{\kappa}(x)\|_{I'_{ij}}$
4: $I'_{i,j+1} := [z_i - t_{i,j+1} - \eta_{j+1}, z_i + t_{i,j+1} + \eta_{j+1}]$ for $1 \leq i \leq n$
5: **end for**
6: **return** $f_k \circ \cdots \circ f_1$

4.2 Construction of the Polynomial $g(x)$

Using Algorithm 3 we obtain a composite polynomial $f = f_k \circ \cdots \circ f_1$ such that $|f(x) - z_i| \leq t_{ik}, x \in I_{i0}$ for all $1 \leq i \leq n$. Then as discussed in Sect. 1.1, the polynomial $g(x)$ is determined by minimizing

$$\max_{1 \leq i \leq n} \|g(z) - y_i\|_{[z_i - t_{ik}, z_i + t_{ik}]}$$

for a given degree $\deg(g) \leq d$ and coefficient bound $\mathcal{C}_{max}(g) \leq B$. Particularly, the following lemma holds.

Lemma 3. *Suppose* $|g(z) - y_i| \leq 2^{-\alpha}$ *for* $z \in [z_i - t_{ik}, z_i + t_{ik}], 1 \leq i \leq n$. *Then the composite polynomial* $g \circ f$ *is an approximation of* $\kappa(x)$ *such that* $\|g \circ f - \kappa\|_I \leq 2^{-\alpha}$, *where* $I = \cup_{i=1}^n I_{i0}$.

Proof. For any $x \in I_{i0}$, it has $z := f(x) \in [z_i - t_{ik}, z_i + t_{ik}]$. Thus $|g(f(x)) - y_i| = |g(z) - y_i| \leq 2^{-\alpha}$. □

Using Algorithm 2, the problem of minimizing $\|g \circ f - \kappa\|_I \leq 2^{-\alpha}$ can be solved. Taking into account the homomorphic computation errors, $g(x)$ should be computed in a similar way as f_{j+1}. Specifically, let η_{k+1} and η_g be the upper bounds of the homomorphic evaluation error of f_k and g, respectively. Then our goal is to find $g(x)$ that minimizes $\max_{1 \leq i \leq n} \|g(z) - y_i\|_{[z_i - t_{ik} - \eta_{k+1}, z_i + t_{ik} + \eta_{k+1}]}$. Moreover, due to the error introduced by homomorphically evaluating $g(x)$, the composite polynomial $g \circ f$ is an approximation of $\kappa(x)$ such that $\|g \circ f - \kappa\|_I \leq 2^{-\alpha} + \eta_g$ in the encrypted state. We specify the computation of $g(x)$ in Algorithm 4.

Algorithm 4: Compute the polynomial $g(x)$.

Input: A step function $\kappa(x)$ and an approximation factor $\gamma \in \mathbb{R}^+$
Input: $z_i \in \mathbb{R}$ and $t_{ik} \in \mathbb{R}^+$ for $1 \leq i \leq n$
Input: Polynomial degree $d \in \mathbb{Z}^+$, coefficient bound $B \in \mathbb{Z}^+$, error bound $\eta_{k+1}, \eta_g \in \mathbb{R}^+$
Input: A polynomial basis $\{p_l(x)\}_{1 \leq l \leq d}$
Output: Polynomial $g(x)$ approximating $\kappa(x)$, and an error rate $2^{-\alpha}$

1: $I'_i = [z_i - t_{ik} - \eta_{k+1}, z_i + t_{ik} + \eta_{k+1}]$ for $1 \leq i \leq n$
2: Compute a polynomial $g(x)$ by using $\kappa(x), \gamma, I'_i, d, B$ and $\{p_l(x)\}_{1 \leq l \leq d}$ as the inputs of Algorithm 2
3: Compute $t = \max_{1 \leq i \leq n} \|g(x) - y_i\|_{I'_i}$
4: Compute $\alpha = -\log(t + \eta_g)$
5: **return** $g(x)$ and $2^{-\alpha}$

Remark 2. To use Algorithms 2, 3, 4 for computing concrete step functions, we need to choose the parameters polynomial degree d, coefficient bound B and error bound η in advance. In our experiments, we set $d \in \{15, 31\}$ and adjust η according to the homomorphic errors, and select B to minimize the number of composite polynomials.

5 Application to Concrete Step Functions

In this section, we apply the two methods `SgnToStep` and `AdaptiveLP` to the round function $\texttt{Round}_m(x)$ and an example of bucketing function in the plaintext state ($\eta = 0$ in Algorithm 2). Suppose a step function $\kappa(x)$ is approximated by a polynomial $f(x)$. Then any step function obtained by applying stretching, shifting and reflecting transformations to $\kappa(x)$ will be approximated by the polynomial obtained by applying the same transformations to $f(x)$. The approximation error rate will change but can be easily predicted. As a result, in this section, we assume all step functions are defined over $[-1,1]$, and their values also fall in $[-1,1]$.

5.1 Application to the Round Function

The round function in this section is a step function with $2m+1$ intervals defined over $[-1,1]$, i.e.,

$$\texttt{Round}_m(x) = \frac{1}{m}\lfloor mx \rceil = \begin{cases} -1, & x \in (-1, -1+\frac{1}{2m}) \\ \frac{i}{m}, & x \in (\frac{i}{m} - \frac{1}{2m}, \frac{i}{m} + \frac{1}{2m}) \text{ for } -m < i < m \\ 1, & x \in (1 - \frac{1}{2m}, 1) \end{cases}$$

where m is a positive integer.

Apply `SgnToStep` **to** $\texttt{Round}_m(x)$. The following corollary directly results from Theorem 1.

Corollary 1. *Suppose* $g(x)$ *is a polynomial that is* (α', ϵ')*-close to* $\texttt{sgn}(x)$ *on* $[-1,1]$*, then the polynomial*

$$f(x) = \frac{1}{2m} \sum_{i=0}^{m-1} \left(g(\frac{mx + i + \frac{1}{2}}{m + i + \frac{1}{2}}) + g(\frac{mx - i - \frac{1}{2}}{m + i + \frac{1}{2}}) \right) \tag{10}$$

is (α, ϵ)*-close to* $\texttt{Round}_m(x)$ *on* $[-1,1]$*, where* $\alpha = \alpha'$ *and* $\epsilon = (2 - \frac{1}{2m})\epsilon'$.

In the following, we focus on $m = 3$ and give concrete polynomial approximations of $\texttt{Round}_3(x)$ based on the constructions in [9,23]. According to Corollary 1, to obtain a polynomial $f(x)$ that is (α, ϵ)-close to $\texttt{Round}_3(x)$ on $[-1,1]$, it suffices to construct a polynomial $g(x)$ that is $(\alpha, \epsilon/(2 - \frac{1}{2m}))$-close to $\texttt{sgn}(x)$. Such $g(x)$ is chosen as follows.

- Using the construction in Sect. 3.1 of [9], $g(x)$ can be defined to be the composite polynomial $h_r^{(k)}$ where $h_r(x) = \sum_{i=0}^{r} \frac{1}{4^i}\binom{2i}{i} x(1-x^2)^i$. It is pointed out in [9] that $r = 4$ is asymptotically optimal concerning the number of multiplications. In our example, using $h_r^{(k)}$ with $r = 4$ for $\epsilon \le 2^{-12}$ results in a large k, which requires very large multiplicative depth, thus very large HEAAN parameters. Therefore we set $r = 4$ for $\epsilon = 2^{-8}$ and $r = 7$ for $\epsilon = 2^{-12}, 2^{-16}, 2^{-20}$. Besides, we set k to be the minimum integer such that $h_r^{(k)}$ is $(\alpha, \epsilon/(2 - \frac{1}{2m}))$-close to $\texttt{sgn}(x)$.
- Using the construction in [23], $g(x)$ is defined to be the composite polynomial $g_k \circ \cdots \circ g_1$, where g_i is constructed by solving the minimax problem to $\texttt{sgn}(x)$. For simplicity, we assume that g_i's have the same degree $d \in \{15, 31\}$, and k is set to be the minimum integer such that $g_k \circ \cdots \circ g_1$ is $(\alpha, \epsilon/(2 - \frac{1}{2m}))$-close to $\texttt{sgn}(x)$.

Based on these $g(x)$'s, we estimate the multiplicative depth and number of multiplications for evaluating $\texttt{Round}_3(x)$ for different (α, ϵ), which are listed in Table 2.

Table 2. The multiplicative depth and number of multiplications for the evaluation of $\texttt{Round}_3(x)$ and $\kappa(x)$ using `SgnToStep` and `AdaptiveLP`. The number of iterations k is minimized such that α satisfies $2^{-\alpha} < \epsilon$.

		Evaluation of $\texttt{Round}_3(x)$				Evaluation of $\kappa(x)$			
		$\epsilon = 2^{-8}$	$\epsilon = 2^{-12}$	$\epsilon = 2^{-16}$	$\epsilon = 2^{-20}$	$\epsilon = 2^{-8}$	$\epsilon = 2^{-12}$	$\epsilon = 2^{-16}$	$\epsilon = 2^{-20}$
`SgnToStep` using [9]	k	8	9	12	14	8	9	12	14
	depth	24	36	48	56	24	36	48	56
	#Mults	240	432	576	672	160	288	384	448
`SgnToStep` using [23] (deg = 15)	k	4	5	6	7	3	5	6	7
	depth	16	20	24	28	12	20	24	28
	#Mults	192	240	288	336	96	160	192	224
`SgnToStep` using [23] (deg = 31)	k	3	4	5	6	3	4	5	6
	depth	15	20	25	30	15	20	25	30
	#Mults	216	288	360	432	144	192	240	288
`AdaptiveLP` (deg = 31)	k	4	6	8	9	3	5	6	8
	depth	20	30	40	45	15	25	30	40
	#Mults	48	72	96	108	36	60	72	96

Apply `AdaptiveLP` to $\texttt{Round}_m(x)$. Again we focus on $m = 3$ and give concrete polynomial approximations of $\texttt{Round}_3(x)$ via Sect. 4, i.e., constructing composite polynomial $f_k \circ \cdots \circ f_1$ that is (α, ϵ)-close to $\texttt{Round}_3(x)$. In this example, the degree of f_i is set as $d = 31$ for $1 \leq i \leq k$, $\epsilon \in \{2^{-8}, 2^{-12}, 2^{-16}, 2^{-20}\}$, and k is set to be the minimum integer such that the approximation error rate $2^{-\alpha} < \epsilon$. We note that choosing smaller d, e.g., $d = 15$, in this example will slow down the convergence thus greatly increase the required multiplicative depth (> 100). As a result, we do not take smaller d into consideration in our implementation. For different ϵ, Table 2 lists the multiplicative depth and number of multiplications required for evaluating $\texttt{Round}_3(x)$.

5.2 Application to the Bucketing Function

In machine learning, bucketing is usually used to map continuous data into discrete categorical values using thresholds, which can be directly viewed as a step function. For example, when training XGBoost, a gradient tree boosting model, the continuous input features can be categorized into buckets (e.g., according to percentiles) to simplify the subtree splitting operation in subsequent training. When a user's data is used by multiple models for training but the models have different granularity for bucketing (i.e. different shape of step function), a user can simply encrypt his data and let the models decide how to perform bucketing.

We consider a bucketing example that maps a latitude data $x \in (-90, 90)$ to discrete data $\{0, 1, 2\}$ for $x \in (-30, 30)$ (low latitude), $x \in (-60, -30) \cup (30, 60)$ (middle latitude), $x \in (-90, -60) \cup (60, 90)$ (high latitude) respectively. By rescaling the data, the corresponding step function can be written as

$$\kappa(x) = \begin{cases} 0, & x \in (-\frac{1}{3}, \frac{1}{3}) \\ \frac{1}{2}, & x \in (-\frac{2}{3}, -\frac{1}{3}) \cup (\frac{1}{3}, \frac{2}{3}) \\ 1, & x \in (-1, -\frac{2}{3}) \cup (\frac{2}{3}, 1) \end{cases}. \tag{11}$$

In the following, we use `SgnToStep` and `AdaptiveLP` to construct polynomial approximations of $\kappa(x)$.

Apply `SgnToStep` to $\kappa(x)$. Suppose $g(x)$ is a polynomial that is (α', ϵ')-close to $\mathtt{sgn}(x)$ on $[-1, 1]$, then by Theorem 1 the polynomial

$$f(x) = \frac{1}{4}\left(g(\frac{3}{5}x - \frac{2}{5}) - g(\frac{3}{5}x + \frac{2}{5}) + g(\frac{3}{4}x - \frac{1}{4}) - g(\frac{3}{4}x + \frac{1}{4})\right) + 1 \tag{12}$$

is (α, ϵ)-close to $\kappa(x)$ on $[-1, 1]$, where $\alpha = \alpha'$ and $\epsilon = \frac{5}{3}\epsilon'$.

Based on the $g(x)$'s constructed in [9,23] (as chosen in Sect. 5.1), we estimate the multiplicative depth and number of multiplications for evaluating $\kappa(x)$ for different ϵ, which are listed in Table 2.

Apply `AdaptiveLP` to $\kappa(x)$. According the Sect. 4, $\kappa(x)$ is approximated by first constructing composite polynomial $f = f_k \circ \cdots \circ f_1$ that approximates the normalization of $\kappa(x)$, i.e.,

$$\tilde{\kappa}(x) = \begin{cases} -1 & x \in (-1, -2/3) \\ -1/2 & x \in (-2/3, -1/3) \\ 0 & x \in (-1/3, 1/3) \\ 1/2 & x \in (1/3, 2/3) \\ 1 & x \in (2/3, 1) \end{cases},$$

then constructing $g(x)$ such that $g \circ f$ approximates $\kappa(x)$. Similarly, we assume that f_i's have the same degree $d = 31$, g has degree $d_g = 31$, $\epsilon \in \{2^{-8}, 2^{-12}, 2^{-16}, 2^{-20}\}$, and k is the minimum integer such that the approximation error rate $2^{-\alpha} < \epsilon$. Table 2 lists the multiplicative depth and number of multiplications required for evaluating $\kappa(x)$ for different ϵ.

We note that the optimal number of multiplications for approximating `sgn` obtained by the dynamic programming approach is given in [23]. We also give a comparison of `SgnToStep` based on these approximations and `AdaptiveLP` with degree $= 31$ in Table 3.

6 Experimental Results

This section presents some experimental results of homomorphically evaluating the step functions in Sect. 5. The computation in this section is performed using the HEAAN library on a Linux PC with an Intel Core i9 CPU at 3.00GHz.

Table 3. The multiplicative depth and number of multiplications for the evaluation of $\mathtt{Round}_3(x)$ and $\kappa(x)$ using `SgnToStep` and `AdaptiveLP`, where `SgnToStep` is based on the approximation of `sgn` in [23] with the optimal number of multiplications.

			$\epsilon = 2^{-8}$	$\epsilon = 2^{-12}$	$\epsilon = 2^{-16}$	$\epsilon = 2^{-20}$
Approximate $\mathtt{Round}_3(x)$	`SgnToStep` using [23] with optimal #Mults.	k	5	8	10	11
		depth	16	23	30	34
		#Mults.	102	132	180	210
	`AdaptiveLP` with degree 31	k	4	6	8	9
		depth	20	30	40	45
		#Mults	48	72	96	108
Approximate $\kappa(x)$ in (11)	`SgnToStep` using [23] with optimal #Mults.	k	5	8	10	11
		depth	16	23	30	34
		#Mults.	68	92	120	140
	`AdaptiveLP` with degree 31	k	3	5	6	8
		depth	15	25	30	40
		#Mults	36	60	72	96

6.1 Parameters Setting

In our experiment, we set $\lambda = 128, N = 2^{17}$, and the highest level modulus q_L upto 2^{1700} to achieve 128-bit security estimated by Albrecht's LWE estimator [1,2]. The scaling factor is set to $\Delta = 2^{40}$. Besides, we expect the final modulus after evaluation to be $\log \Delta + 10$ bits long. Then the initial modulus q_L is given as follows.

- For `SgnToStep`, suppose g is the approximate polynomial of `sgn` used in our construction, then

$$\log q_L = \log \Delta \cdot (\mathtt{dep}(g)) + \log \Delta + 10.$$

- For `AdaptiveLP`, suppose $g \circ f_k \circ \cdots \circ f_1$ is the composite polynomial in our construction, then

$$\log q_L = \log \Delta \cdot (\mathtt{dep}(f_1) + \cdots + \mathtt{dep}(f_k) + \mathtt{dep}(g)) + \log \Delta + 10.$$

Here $\mathtt{dep}(\cdot)$ denotes the depth consumption for evaluating a polynomial, which is usually set to be $\lceil \log(d+1) \rceil$ for a polynomial of degree d.

Table 4. Running time and depth consumption of $\texttt{Round}_3(x)$ and $\kappa(x)$ in HEAAN.

		Evaluation of $\texttt{Round}_3(x)$				Evaluation of $\kappa(x)$			
		$\epsilon = 2^{-8}$	$\epsilon = 2^{-12}$	$\epsilon = 2^{-16}$	$\epsilon = 2^{-20}$	$\epsilon = 2^{-8}$	$\epsilon = 2^{-12}$	$\epsilon = 2^{-16}$	$\epsilon = 2^{-20}$
`SgnToStep` using [9]	k	8	9	12	14	8	9	12	14
	running time	4.63 ms	7.83 ms	13.24 ms	16.32 ms	3.08 ms	4.73 ms	9.38 ms	10.93 ms
	bit consumption	1360	1520	2000*	2320*	1360	1520	2000*	2320*
`SgnToStep` using [23] (deg = 15)	k	4	5	6	8	3	5	6	7
	running time	2.56 ms	3.39 ms	4.17 ms	5.75 ms	1.20 ms	2.26 ms	2.68 ms	3.21 ms
	bit consumption	716	875	1034	1352	557	875	1034	1193
`SgnToStep` using [23] (deg = 31)	k	3	4	5	6	3	4	5	6
	running time	3.01 ms	4.36 ms	6.18 ms	8.09 ms	2.21 ms	3.41 ms	4.11 ms	5.32 ms
	bit consumption	688	891	1094	1297	688	891	1094	1297
`AdaptiveLP` (deg = 31)	k	4	6	8	10	4	6	8	10
	running time	0.74 ms	1.36 ms	1.98 ms	2.81 ms	0.78 ms	1.24 ms	2.15 ms	2.95 ms
	bit consumption	808	1212	1616	2020*	808	1212	1616	2020*

* an asterisk (*) means the parameter set does not achieve 128−bit security for large $\log q_L \geq 1700$ in HEAAN.

6.2 Evaluating $\texttt{Round}_3(x)$

We evaluate the step function $\texttt{Round}_3(x)$ based on the polynomial approximations constructed in Sect. 5. To handle the homomorphic evaluation error, we construct composite polynomials by introducing the error bound η as in Algorithm 2, where η is dynamically determined by the coefficient bound in our implementation. The polynomials are evaluated by the BSGS method as in [25], and the amortized running time and bit consumption $\log(\frac{q_L}{q_l})$ for different approximation error rates are listed in Table 4. In the table, we set $\epsilon = 2^{-8}, 2^{-12}, 2^{-16}, 2^{-20}$ and let the number of iterations k to be minimum such that $2^{-\alpha} < \epsilon$.

Through the table we can see that `SgnToStep` shows an advantage in bit consumption, while `AdaptiveLP` provides better performance in amortized running time. For example, comparing with `SgnToStep` that uses the minimax approximation in [23] with polynomial degree 15, `AdaptiveLP` has roughly 1.5× bit consumption but approximately 0.5× running time. Though each evaluation of `sgn` requires less bit consumption and less running time than `AdaptiveLP`, the evaluation of $\texttt{Round}_3(x)$ based on `SgnToStep` involves 6 evaluations of `sgn` thus requires more running time.

6.3 Evaluating Bucketing Function

We evaluate the bucketing example given in (11) based on the polynomial approximations given in Sect. 5. Again, the dynamic error bound is used to handle the homomorphic evaluation error, and BSGS method is used to evaluate the polynomials. We set $\epsilon = 2^{-8}, 2^{-12}, 2^{-12}, 2^{-16}$ and let the number of iterations k to be minimum such that $2^{-\alpha} < \epsilon$. The amortized running time and bit consumption $\log(\frac{q_L}{q_l})$ for different approximation error rates are listed in Table 4.

Through the table we can see that `AdaptiveLP` still outperforms `SgnToStep` in amortized running time. However, because the bucketing example given in (11) has a less number of intervals $n = 5$, `SgnToStep` requires only 4 evaluations of `sgn`. As a result, `AdaptiveLP` shows less advantage in running time. In general, `SgnToStep` and `AdaptiveLP` provide a trade-off in terms of running time and bit consumption.

References

1. Albrecht, M.R.: On dual lattice attacks against small-secret LWE and parameter choices in HElib and SEAL, pp. 103–129. https://doi.org/10.1007/978-3-319-56614-6_4
2. Albrecht, M.R., Player, R., Scott, S.: On the concrete hardness of learning with errors. J. Math. Cryptol. **9**(3), 169–203 (2015). http://www.degruyter.com/view/j/jmc.2015.9.issue-3/jmc-2015-0016/jmc-2015-0016.xml
3. Andrievskii, V.: Polynomial approximation of piecewise analytic functions on a compact subset of the real line. J. Approx. Theory **161**(2), 634–644 (2009). https://doi.org/10.1016/j.jat.2008.11.015, https://doi.org/10.1016/j.jat.2008.11.015
4. Brakerski, Z.: Fully homomorphic encryption without modulus switching from classical GapSVP, pp. 868–886. https://doi.org/10.1007/978-3-642-32009-5_50
5. Brakerski, Z., Gentry, C., Vaikuntanathan, V.: (Leveled) fully homomorphic encryption without bootstrapping, pp. 309–325. https://doi.org/10.1145/2090236.2090262
6. Chen, H., Chillotti, I., Song, Y.: Improved bootstrapping for approximate homomorphic encryption, pp. 34–54. https://doi.org/10.1007/978-3-030-17656-3_2
7. Cheon, J.H., Han, K., Kim, A., Kim, M., Song, Y.: Bootstrapping for approximate homomorphic encryption, pp. 360–384. https://doi.org/10.1007/978-3-319-78381-9_14
8. Cheon, J.H., Kim, A., Kim, M., Song, Y.S.: Homomorphic encryption for arithmetic of approximate numbers, pp. 409–437. https://doi.org/10.1007/978-3-319-70694-8_15
9. Cheon, J.H., Kim, D., Kim, D.: Efficient homomorphic comparison methods with optimal complexity, pp. 221–256. https://doi.org/10.1007/978-3-030-64834-3_8
10. Cheon, J.H., Kim, D., Kim, D., Lee, H.H., Lee, K.: Numerical method for comparison on homomorphically encrypted numbers, pp. 415–445. https://doi.org/10.1007/978-3-030-34621-8_15
11. Chillotti, I., Gama, N., Georgieva, M., Izabachène, M.: Faster fully homomorphic encryption: Bootstrapping in less than 0.1 seconds, pp. 3–33. https://doi.org/10.1007/978-3-662-53887-6_1
12. Cohen, M.B., Lee, Y.T., Song, Z.: Solving linear programs in the current matrix multiplication time, pp. 938–942. https://doi.org/10.1145/3313276.3316303
13. Ducas, L., Micciancio, D.: FHEW: bootstrapping homomorphic encryption in less than a second, pp. 617–640. https://doi.org/10.1007/978-3-662-46800-5_24
14. Eremenko, A., Yuditskii, P.: Uniform approximation of sgn x by polynomials and entire functions. J. d'Analyse Mathématique **101**(1), 313–324 (2007)

15. Eremenko, A., Yuditskii, P.: Polynomials of the best uniform approximation to sgn (x) on two intervals. J. d'Analyse Mathématique **114**(1), 285–315 (2011)
16. Fan, J., Vercauteren, F.: Somewhat practical fully homomorphic encryption. Cryptology ePrint Archive, Report 2012/144 (2012). https://eprint.iacr.org/2012/144
17. Gentry, C.: Fully homomorphic encryption using ideal lattices, pp. 169–178. https://doi.org/10.1145/1536414.1536440
18. Gentry, C.: Computing arbitrary functions of encrypted data. Commun. ACM **53**(3), 97–105 (2010). https://doi.org/10.1145/1666420.1666444. https://doi.org/10.1145/1666420.1666444
19. Gentry, C., Sahai, A., Waters, B.: Homomorphic encryption from learning with errors: Conceptually-simpler, asymptotically-faster, attribute-based, pp. 75–92. https://doi.org/10.1007/978-3-642-40041-4_5
20. Han, K., Hong, S., Cheon, J.H., Park, D.: Logistic regression on homomorphic encrypted data at scale. In: The Thirty-Third AAAI Conference on Artificial Intelligence, AAAI 2019, The Thirty-First Innovative Applications of Artificial Intelligence Conference, IAAI 2019, The Ninth AAAI Symposium on Educational Advances in Artificial Intelligence, EAAI 2019, Honolulu, Hawaii, USA, January 27 - February 1, 2019, pp. 9466–9471. AAAI Press (2019). https://doi.org/10.1609/aaai.v33i01.33019466. https://doi.org/10.1609/aaai.v33i01.33019466
21. Jutla, C.S., Manohar, N.: Sine series approximation of the mod function for bootstrapping of approximate HE, pp. 491–520. https://doi.org/10.1007/978-3-031-06944-4_17
22. Kim, M., Song, Y., Li, B., Micciancio, D.: Semi-parallel logistic regression for GWAS on encrypted data. Cryptology ePrint Archive, Report 2019/294 (2019). https://eprint.iacr.org/2019/294
23. Lee, E., Lee, J.W., Kim, Y.S., No, J.S.: Minimax approximation of sign function by composite polynomial for homomorphic comparison. IEEE Trans. Dependable Secure Comput. (2021)
24. Lee, J.W., Lee, E., Lee, Y., Kim, Y.S., No, J.S.: High-precision bootstrapping of RNS-CKKS homomorphic encryption using optimal minimax polynomial approximation and inverse sine function, pp. 618–647. https://doi.org/10.1007/978-3-030-77870-5_22
25. Lee, Y., Lee, J.W., Kim, Y.S., Kim, Y., No, J.S., Kang, H.: High-precision bootstrapping for approximate homomorphic encryption by error variance minimization, pp. 551–580. https://doi.org/10.1007/978-3-031-06944-4_19
26. Liu, Z., Micciancio, D., Polyakov, Y.: Large-precision homomorphic sign evaluation using FHEW/TFHE bootstrapping. Cryptology ePrint Archive, Report 2021/1337 (2021). https://eprint.iacr.org/2021/1337
27. Panda, S.: Polynomial approximation of inverse sqrt function for FHE. In: Dolev, S., Katz, J., Meisels, A. (eds.) Cyber Security, Cryptology, and Machine Learning - 6th International Symposium, CSCML 2022, Be'er Sheva, Israel, June 30 - July 1, 2022, Proceedings. Lecture Notes in Computer Science, vol. 13301, pp. 366–376. Springer (2022). https://doi.org/10.1007/978-3-031-07689-3_27, https://doi.org/10.1007/978-3-031-07689-3_27
28. Plaskota, L., Wasilkowski, G.W.: Uniform approximation of piecewise r-smooth and globally continuous functions. SIAM J. Numer. Anal. **47**(1), 762–785 (2008). https://doi.org/10.1137/070708937. https://doi.org/10.1137/070708937
29. Plaskota, L., Wasilkowski, G.W., Zhao, Y.: The power of adaption for approximating functions with singularities. Math. Comput. **77**(264), 2309–2338 (2008). https://doi.org/10.1090/S0025-5718-08-02103-0. https://doi.org/10.1090/S0025-5718-08-02103-0

30. Saff, E.B., Totik, V.: Polynomial approximation of piecewise analytic functions. J. Lond. Math. Soc. **2**(3), 487–498 (1989)
31. Smart, N.P., Vercauteren, F.: Fully homomorphic SIMD operations. Des. Codes Cryptogr. **71**(1), 57–81 (2014). https://doi.org/10.1007/s10623-012-9720-4. https://doi.org/10.1007/s10623-012-9720-4
32. van Dijk, M., Gentry, C., Halevi, S., Vaikuntanathan, V.: Fully homomorphic encryption over the integers, pp. 24–43. https://doi.org/10.1007/978-3-642-13190-5_2

Encryption and Its Applications

Key Cooperative Attribute-Based Encryption

Luqi Huang[1(✉)], Willy Susilo[1], Guomin Yang[2], and Fuchun Guo[1]

[1] Institute of Cybersecurity and Cryptology, School of Computing and Information Technology, University of Wollongong, Wollongong, Australia
lh852@uowmail.edu.au, {wsusilo,fuchun}@uow.edu.au

[2] School of Computing and Information Systems, Singapore Management University, Singapore, Singapore
gmyang@smu.edu.sg

Abstract. Attribute-based encryption (ABE) is an important technology in building access control systems with precise control and scalability. In an ABE system, there exists a private key generator (PKG) that issues all private keys. The PKG has a significant drawback referred to as the huge key management burden in large-scale user systems. To overcome this limitation, we propose a more flexible system that offers users the choice to utilize decryption keys either from the PKG or from trusted users to decrypt the ciphertext, reducing the workload of the PKG. Unfortunately, users are restricted to only receiving private keys from the PKG in most ABE schemes. Thus, our system ABE extends the ability of trusted users to generate and distribute decryption keys. Furthermore, decryption keys from trusted users possess equivalent decryption with a private key from the PKG when satisfying the cooperative access policy set by the encryptor. We define the concept of key cooperative ABE for the first time, presenting a key cooperative ABE scheme.

Keywords: Attribute-based encryption · Key management · Key cooperative · Trust delegation

1 Introduction

ABE plays a key role in realizing access control systems with fine granularity and scalability. It was proposed [24] by Sahai and Waters in 2005. As a promising cryptographic primitive, it has successfully attracted considerable research efforts and comes in two categories: Ciphertext-Policy ABE (CP-ABE) and Key-Policy ABE (KP-ABE). In a CP-ABE system, a party encrypting data can specify access to the data as a boolean formula over a set of attributes. A user will be able to decrypt a ciphertext if the attributes associated with their private key satisfy the boolean formula ascribed to the ciphertext.

W. Susilo—Supported by the ARC Australian Laureate Fellowship FL230100033.
F. Guo—Supported by the ARC Future Fellowship FT220100046.

T. Zhu and Y. Li (Eds.): ACISP 2024, LNCS 14895, pp. 241–260, 2024.
https://doi.org/10.1007/978-981-97-5025-2_13

In an ABE system, private keys were distributed by the PKG that would need to be in a position to verify all the attributes it issued for each user in the system. This approach imposes two key management challenges: key escrow and a huge key management burden. To reduce the workload of the PKG, we propose a more flexible ABE system in terms of generating and distributing decryption keys. Specifically, users within this system can obtain decryption keys either from the PKG or from trusted users.

The common strategy for managing a substantial workload of the PKG is distributing its authority to other attribute authorities. One such approach is Hierarchical ABE (HABE) [27], which enables hierarchical access control by introducing hierarchical attribute authorities. The PKG is only responsible for the generation and distribution of system parameters and domain keys. Similarly, multi-authority ABE [5] introduced parallel attribute authorities to decentralize the PKG's power. These authorities monitor different sets of attributes and issue corresponding private keys to users. However, both HABE and multi-authority ABE still require users to heavily rely on these attribute authorities. Furthermore, the introduction of additional authorities resulted in increased storage costs and communication overheads.

Recently, registered ABE [10] allows users to generate private keys on their own. A key curator (KC) maintains no secrets and replaces the PKG to aggregate public keys from registered users into a master public key and generates helper decryption keys for users. However, removing the central authority (PKG) with secrets results in significant additional updating costs. Since the master public key of the registered ABE scheme changes whenever new users join the system, users must periodically refresh their helper decryption keys over the lifetime of the system. This can be particularly challenging in scenarios with a large number of users who frequently join and leave the system. As a result, registered ABE is limited to applications within fixed, small-scale user systems.

Different from the decentralization of PKG's power, online/offline ABE [11] divides the private key generation computation into two phases. However, the PKG remains responsible for generating all private keys, even when operating as an online server within this framework. Besides the above ABE schemes, some classic ABE [2,9,28] support one-to-one key delegation. Nevertheless, the scope for key delegation among users is restricted to a small number of individuals in the system.

In the Pretty Good Privacy (PGP) encryption program [22] and the GNU Privacy Guard (GNU PG) software suite [1] for secure communication, a web of trust model allows the responsibility for validating public keys is delegated to people a user trusts. In this way, it avoids relying exclusively on a certificate authority and communicating with unknown parties. Inspired by the trust delegation of the decentralized web of trust model, our system considers delegating the generation and distribution of decryption keys to trusted users for reducing the huge workload of the PKG in an ABE system. Therefore, we propose a new ABE system, Key Cooperative ABE.

1.1 Our Contribution

In this paper, we present a novel ABE called Key Cooperative ABE (KC-ABE), aiming at the problem of huge workload for the PKG. Based on trust delegation between users in our system, KC-ABE offers users the option to decrypt ciphertexts using either the private key from the PKG or user private keys from trusted users. This scheme enhances flexibility for users in obtaining decryption keys, thereby reducing the workload burden on the PKG.

In KC-ABE, we extend the capabilities of trusted users, enabling them to generate and distribute user private keys to other users. These cooperative user private keys possess equivalent decryption compared with private keys from the PKG when satisfying the cooperative access policy set by the encryptor. We propose a detailed KC-ABE scheme based on [23] and provide a formal proof of security for our construction within a well-defined security model.

KC-ABE stands out as a distinct and innovative solution, differing from the aforementioned approaches by its practicality, lightweight and extensibility. Specifically, the main contributions of our work are summarized as follows:

- *Practicality.* Our solution avoids the need for additional entities and eliminates the associated overheads of storage, communication, and update. This innovative approach provides key management for dynamic, large-scale user systems, offering enhanced practicality.
- *Lightweight.* In our KC-ABE, we transfer the majority of key generation costs from the PKG to trusted users. As a result, the PKG concentrates on generating private keys for only a small number of users, leading to a substantial improvement in key management.
- *Extensibility.* Our system expands key delegation from a one-to-one model to key cooperation in a many-to-one framework. Leveraging the extended trust among users, the majority of users can obtain decryption keys from cooperative trusted users.

1.2 Related Work

In 1984, Shamir [26] introduced identity-based cryptography, with practical identity-based encryption (IBE) schemes emerging through the pioneering work of Boneh and Franklin [3]. Subsequently, Sahai and Waters proposed fuzzy IBE and ABE in [24]. Since then, numerous variants of ABE have been developed, such as revocable ABE [6,13], accountable ABE [17,19], policy-hiding ABE [15,20], multi-authority ABE [5,16,18], and HABE [27].

HABE can be adopted to realize hierarchical access control by introducing hierarchical attribute authorities. It allows higher-level attribute authorities to delegate private keys to their subordinates and conduct private key transmission locally. Wang et al. [27] proposed hierarchical ABE schemes by combing ABE and hierarchical IBE introduced by Horwitz and Lynn [12]. Chase proposed the first multi-authority scheme in [5]. Lin et al. [18] focused on removing the central authority from the multi-authority ABE scheme. Then Lewko and Waters [16]

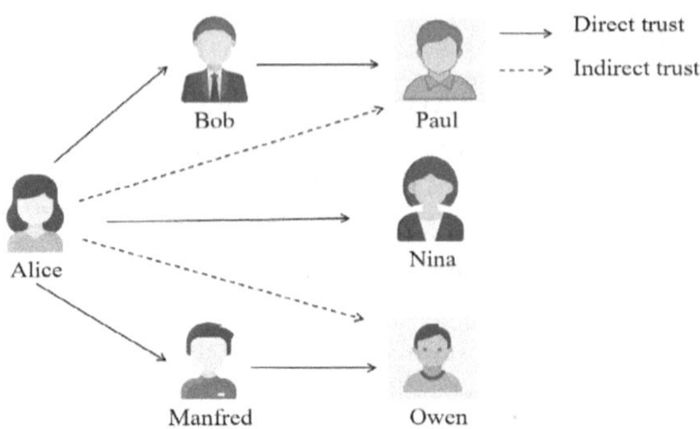

Fig. 1. A web of trust model

proposed a new multi-authority ABE system. Any party can become an authority and there is no requirement for any global coordination.

In 2018, Garg et al. [8] introduced registration-based encryption (RBE) to address the key escrow problem in IBE. In 2023, Hohenberger et al. [10] proposed registered ABE, which removed the need for long-term secret keys in the context of ABE. Moreover, the KC is entirely transparent and maintains no secrets but with inefficient key generations and key updating costs.

In 2014, Hohenberger and Waters proposed online/offline ABE [11] to address the huge encryption and key generation costs scale with the complexity of the access policy or the number of attributes. They developed new techniques for ABE that split the computation into offline and online phases.

When given a private key for attribute set A, a user can generate a new private key for any subset $B \subseteq A$, known as key delegation abuse in ABE schemes [2,9,28]. Additionally, Jiang et al. [14] proposed an ABE scheme against the issue of key delegation abuse. Unlike the one-to-one model of key delegation abuse, our KC-ABE extends to key cooperation from many-to-one. Specifically, these multiple low-level delegated keys process equivalent decryption with a high-level private key when satisfying the cooperative access policy set by the encryptor.

2 Definitions

The web of trust model [1,22] is formed by expanding trust delegations illustrated between users in Fig. 1. For example, Alice places direct trust in her friend Bob, who in turn extends direct trust to stranger Paul. Therefore, Alice can indirectly trust Paul. This allows users to communicate with known parties rather than rely exclusively on a certificate authority. Inspired by its trust delegation and decentralized nature, we consider a novel system in which users can obtain decryption keys from trusted users to distribute PKG authority and reduce its workload.

As shown in Fig. 2, users with attribute set A_i and identity ID_i who obtain the private key $sk_{A_i}^{ID_i}$ directly from the PKG are called high-level users. However,

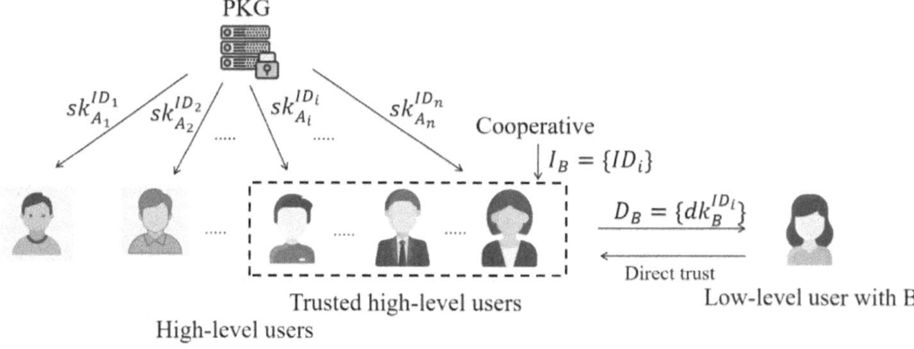

Fig. 2. Key Cooperative ABE system model

in cases where the PKG becomes overwhelmed and is unable to respond to an increasing number of private key requests in time, a user with an attribute set B encounters difficulties in establishing a direct connection with the PKG. In such instances, this user (referred to as a low-level user) has the alternative of obtaining user private keys $dk_B^{ID_i}$ from trusted high-level users based on his/her direct trust of the web of trust model. The specific process unfolds as follows in the right half of Fig. 2.

A low-level user with an attribute set B determines a cooperative group (dotted box) identity set $I_B = \{ID_i\}$ of his/her trusted high-level users based on direct trust. Each trusted high-level user in I_B, with the private key $sk_{A_i}^{ID_i}$ from the PKG, generates a user private key $dk_B^{ID_i}$ under $B \subseteq A_i$. These cooperative trusted high-level users produce $|I_B|$ user private keys, consisting a set $D_B = \{dk_B^{ID_i}\}$ to the low-level user.

Therefore there are two types of decryption keys in our system: the private key $sk_{A_i}^{ID_i}$ generated from the PKG to a high-level user; user private keys $\{dk_B^{ID_i}\}$ issued from trusted high-level users to a low-level user. To enable user private keys for decryption, we introduce a valid cooperation minimum l within the ciphertext. Specifically, successful key cooperation for a low-level user is achieved when the number of user private keys from trusted high-level users, denoted as $|I_B|$, is not less than l. This means that the cooperation set of user private keys $D_B = \{dk_B^{ID_i}\}$ possess commensurate decryption with the private key $sk_{A_i}^{D_i}$.

It is essential to emphasize that the users' identities, represented by ID_i, are only associated with their private keys and they do not directly participate in the decryption process itself. Instead, these identities play a role between trusted high-level users during the generation of cooperative keys.

2.1 Algorithm Definitions

Different from basic ABE, decryption keys in KC-ABE are private keys from the PKG (PKeyGen) or user private keys from trusted high-level users (UKeyGen). Correspondingly, there are two decryption algorithms PDec and UDec. Moreover, the ciphertext with an attribute set $A^{'}$ in our KC-ABE is embedded with the key cooperation minimum l. And the ciphertext can be decrypted using either a private key under $A^{'} \subseteq A_i$ or user private keys under $A^{'} \subseteq B$ and $l \leq |I_B|$. Hence, based on the basic ABE algorithm definitions, we define our key cooperative ABE consisting of six algorithms: Setup, PKeyGen and PDec, Enc, UKeyGen and UDec.

The algorithms in our KC-ABE are defined as follows:

- **Setup**$(\lambda, U, n) \rightarrow (mpk, msk)$. The Setup algorithm is executed by the PKG and takes as input the security parameter λ, the attribute universal set U and the number of users n. It outputs the master public key and master secret key (mpk, msk).
- **PKeyGen**$(mpk, msk, ID_i, A_i) \rightarrow sk_{A_i}^{ID_i}$. The PKeyGen algorithm is executed by the PKG and takes as input the master public key, master secret key mpk, msk, the user identity ID_i and attribute set A_i. It returns the private key $sk_{A_i}^{ID_i}$.
- **UKeyGen**$(mpk, ID_i, sk_{A_i}^{ID_i}, I_B = \{ID_i\}, B \subseteq A_i) \rightarrow dk_B^{ID_i}$. The UKeyGen algorithm is executed by a trusted high-level user with identity ID_i and takes as input identity ID_i, private key $sk_{A_i}^{ID_i}$, the trusted high-level user identity set $I_B, |I_B| < n$. It also inputs the master public key mpk, attribute set $B \subseteq A_i$, and returns the user private key $dk_B^{ID_i}$.
- **Enc**$(M, mpk, A^{'}, l) \rightarrow CT$. The Enc algorithm is executed by the encryptor and takes as input a message M, the master public key mpk, the attribute set $A^{'}$, and the key cooperative minimum $l < n$. It returns the ciphertext CT.
- **PDec**$(sk_{A_i}^{ID_i}, CT) \rightarrow M$. The PDec algorithm is executed by a high-level user and takes as input the private key $sk_{A_i}^{ID_i}$ and the ciphertext CT. It outputs the message M if $A^{'} \subseteq A_i$.
- **UDec**$(D_B = \{dk_B^{ID_i}\}, CT) \rightarrow M$. The UDec algorithm is executed by a low-level user and takes as input a set D_B of user private keys $\{dk_B^{ID_i}\}$ and the ciphertext CT. It outputs the message M if $A^{'} \subseteq B$ and $l \leq | I_B |$.

Correctness: We require the correctness of the KC-ABE are as follows:

- **PDec Correctness:** For any $(mpk, msk) \leftarrow$ Setup(λ, U, n), ciphertext $CT \leftarrow$ Enc$(M, mpk, A^{'}, l)$ and private key $sk_{A_i}^{ID_i} \leftarrow$ PKeyGen(mpk, msk, ID_i, A_i), if $A^{'} \subseteq A$, the PDec algorithm always outputs the correctness message M. Otherwise, the message in CT cannot be decrypted using $sk_{A_i}^{ID_i}$.
- **UDec Correctness:** For any $(mpk, msk) \leftarrow$ Setup(λ, U, n), ciphertext CT $\leftarrow$ Enc$(M, mpk, A^{'}, l)$ and $|I_B|$ user private keys $D_B = \{dk_B^{ID_i}\}$, $dk_B^{ID_i} \leftarrow$ UKeyGen$(mpk, ID_i, sk_{A_i}^{ID_i}, I_B = \{ID_i\}, B \subseteq A_i)$, if $A^{'} \subseteq B$ and $l \leq |I_B|$, that means if $A^{'} \subseteq B$ and the number of trusted high-level users in the group I_B is not less than the key cooperative minimum l decided by the encryptor, the UDec algorithm always outputs the correctness message M. Otherwise, the message in CT cannot be decrypted using $D_B = \{dk_B^{ID_i}\}$.

Discussion. Based on the KC-ABE system model in Fig. 2 and algorithm definitions provided above, we now present a comparison between our KC-ABE proposal and other relevant ABE schemes in three aspects: whether additional entities are introduced, who generates the decryption key and whether there is key delegation. We discuss them as follows.

Table 1. Comparison with related ABE schemes.

Schemes	Additional entities	Key generation	Key delegation
HABE [27]	Domain AAs	Domain AAs	–
Multi-authority ABE [5]	Multiple AAs	Multiple AAs	–
Registered ABE [10]	KC	KC & Users	–
Online/Offline ABE [11]	–	PKG	–
Key delegation ABE [2,9,28]	–	PKG	Single key delegation
KC-ABE	–	PKG or Trusted users	Trust user delegation

As shown in Table 1, one straightforward approach to address the huge workload on the PKG is to introduce additional entities aimed at decentralizing the PKG's power in [5,10,27]. However, introducing other entities still requires users to depend heavily on them to generate decryption keys. In contrast, our approach differs in that we directly enhance the capabilities of existing trusted users and avoid the need for additional entities. It eliminates the associated overheads of storage, communication, and update, which holds more practical significance.

Another method is splitting key generations into two phases in online/offline ABE [11]. However, the PKG in online/offline ABE still generates final private keys for all users. Different from it, we transfer most of the key generation costs from the PKG to trusted users. Consequently, the PKG's role focuses on generating private keys for a small number of users, resulting in lightweight key management of PKG.

Early classic CP-ABE schemes [2,9,28] allowed one-to-one key delegation based on the subset relation of users' attribute sets. Our KC-ABE expands the scope of key delegation beyond its limitations. Based on trust relationships between users, multiple trusted users can generate decryption keys for a user, facilitating extensible key delegation throughout the system.

In conclusion, our KC-ABE represents a unique solution that distinguishes itself from existing approaches by its practicality, lightweight, and extensibility. This innovative approach offers a novel solution to the huge key management of the PKG in ABE, with a summary of the results presented in Table 1.

2.2 Security Model

The security model is built from respective of semantic security against the *Indistinguishability under selective chosen plaintext attacks* (IND-sCPA). More precisely, KC-ABE security is defined by the following game between an adversary $\mathcal{A}$ and a challenger $\mathcal{C}$. $\mathcal{A}$ specifies the target attribute set A^* and the key cooperative minimum l^* in advance. The game says that it is indistinguishable from deciding the message in a ciphertext for any attribute set A^* and any private key $sk_{A_i}^{ID_i}$ if $A^* \nsubseteq A_i$ holds. Additionally, it also says that it is indistinguishable to decide the message in a ciphertext for the attribute set A^*, the key cooperative minimum l^* and any $|I_B|$ user private keys $D_B = \{dk_B^{ID_i}\}$ if $l^* > |I_B|$ holds under $A^* \subseteq B$.

- **Initialization.** $\mathcal{C}$ chooses an attribute universal set U and the number of users n. $\mathcal{A}$ declares the challenge attribute set $A^* \subseteq U$ and the challenge key cooperative minimum $l^* < n$, which it will try to attack, and sends them to $\mathcal{C}$.
- **Setup.** $\mathcal{C}$ runs the Setup algorithm to generate the master public key and master secret key (mpk, msk) with a security parameter λ. It sends mpk to the adversary $\mathcal{A}$. $\mathcal{C}$ keeps msk to respond to private key queries and user private key queries.
- **Query phase 1.**
  1. $\mathcal{A}$ makes q_1 private key queries with attribute sets $A_i \subseteq U$ and user identities ID_i, $i \in \{1, \ldots, q_1\}$ with $A^* \nsubseteq A_1, \ldots, A_{q_1}$.
  2. $\mathcal{A}$ makes q_2 user private key queries with attribute sets $B_j \subseteq U$ and trusted high-level user identity sets I_{B_j}, $j \in \{1, \ldots, q_2\}$ with following restrictions: $l^* > |I_{B_j}|, I_{B_j} = \{ID_i\}, j = 1, \ldots, q_2$ under $A^* \subseteq B_1, \ldots, B_{q_2}$.
- **Challenge.** $\mathcal{A}$ outputs two equal length messages M_0, M_1. $\mathcal{C}$ flips a random coin b, and encrypts M_b. The ciphertext CT^* is given to the adversary $\mathcal{A}$.
- **Query phase 2.** Phase 2 is repeated with the same restrictions as follows:
  1. $\mathcal{A}$ makes q_1' private key queries that attribute sets $A_i \subseteq U$ satisfy $A^* \nsubseteq A_i$ and user identities ID_i, $i \in \{1, \ldots, q_1'\}$.
  2. $\mathcal{A}$ makes q_2' user private key queries that I_{B_j} satisfy $l^* > |I_{B_j}|$ under $A^* \subseteq B_j$, $j \in \{1, \ldots, q_2'\}$.
- **Guess.** $\mathcal{A}$ outputs a guess b' of b and wins the game if $b' = b$.

The advantage of an adversary $\mathcal{A}$ in the IND-sCPA game is defined as follows:

$$\mathsf{Adv}_{\mathsf{KC\text{-}ABE}}^{\mathsf{IND\text{-}sCPA}}(\mathcal{A}) = |\Pr[b' = b] - \frac{1}{2}|$$

Definition 1. *We say the KC-ABE scheme is selectively secure against chosen plaintext attacks if all polynomial-time adversaries $\mathcal{A}$ have at most a negligible advantage in the above security game.*

3 Constructions

In this section, we present our realization of key cooperative scheme of CP-ABE. We begin with an overview of our KC-ABE scheme. Subsequently, we review the first efficient and secure two-party identity-based non-interactive key exchange (IB-NIKE) scheme [21], the Sakai-Ohgishi-Kasahara (SOK) scheme [25]. Lastly, we show the construction of a detailed CP-ABE scheme for our key cooperative ABE, based on the practical CP-ABE scheme proposed in [23] and the SOK scheme.

3.1 Overview

Building on the practical CP-ABE scheme presented in [23], we propose a key cooperative CP-ABE scheme but simplify the access policy of the scheme in [23] from the Linear Secret Sharing Scheme (LSSS) to a subset structure, i.e., $A^{'} \subseteq A_i$ or $A^{'} \subseteq B$. Additionally, in our construction, the encryptor possesses an attribute set of $A^{'}$ and specifies the minimum of user private key cooperation, denoted as l.

To achieve cooperation of user private keys within I_B, each trusted high-level user with A_i and ID_i first generates a cooperative key ck_i^B. We should guarantee that only the combination of $|I_B|$ cooperative keys is a specific and deterministic value for I_B, which we set to $g^{|I_B|}$. Furthermore, $g^{|I_B|}$ can be eliminated when the attribute set B satisfies the access structure ascribed in the ciphertext. It is worth noting that the low-level user will send the identity set I_B to the PKG in the SOK scheme. The PKG will verify the trusted high-level user with identity $ID_i \in I_B$ during the generation of shared keys and prevent unauthorized users from forging any shared key.

Our system prevents collusion attacks by tying a user's attributes together using different random values for different users. This makes it impossible to successfully combine these attributes with another user's attributes during decryption. Collusion attacks won't help since attribute mappings are randomized to a particular user's private key. Furthermore, the user private keys of trusted high-level users are re-randomized, making it impossible to combine different user private keys belonging to different users.

3.2 Sakai-Ohgishi-Kasahara Scheme

We first review the IB-NIKE [21] scheme: SOK scheme [25]. IB-NIKE is an analog of non-interactive key exchange (NIKE) [7] in the identity-based setting, which enables a group of users registered in the same PKG to agree on a unique shared key without any interaction. The SOK scheme is as follows:

- **Setup**$(1^k) \rightarrow (params, msk)$. The Setup takes as input security parameter 1^k and obtains $(\mathbb{G}, \mathbb{G}_T, e)$. The PKG picks a random $\beta \in \mathbb{Z}_p$ as master secret key msk and two hash functions $H_1 : \{0,1\}^* \rightarrow \mathbb{G}$, $H_2 : \mathbb{G}_T \rightarrow \mathbb{Z}_p$. It also outputs system parameters $params = (\mathbb{G}, \mathbb{G}_T, e, g, H_1, H_2)$.
- **Extract**$(params, msk, ID \in \{0,1\}^*) \rightarrow S_{ID}$. The Extract algorithm is run from the PKG with ID and inputs $params$ and msk. The PKG first verifies the validity of identity ID and then outputs the private key $S_{ID} = H_1(ID)^\beta$.
- **SharedKey**$(params, S_{ID_A}, ID_B \in \{0,1\}^*) \rightarrow K_{A,B}$. The SharedKey generation algorithm takes as input identity $ID_B \in \{0,1\}^*$ and $params$. It inputs private key $S_{ID_A} = H_1(ID_A)^\beta$ and returns the shared key $K_{A,B} = H_2(e(H_1(ID_A)^\beta, H_1(ID_B)))$.

Correctness: We require that, for any pair of identities ID_A, ID_B, and corresponding private keys S_{ID_A}, S_{ID_B}, **SharedKey** satisfies the correctness:

$$\textbf{SharedKey}(params, S_{ID_A}, ID_B) = \textbf{SharedKey}(params, S_{ID_B}, ID_A).$$

This ensures that users A and B can indeed generate a shared key $K_{A,B}$ without any interaction.

3.3 Scheme

In this section, we will show the KC-ABE scheme based on the CP-ABE presented in [23] and the SOK scheme. The KC-ABE scheme comprises six algorithms. Specifically, the Setup and PKeyGen algorithms are executed by the PKG. The UKeyGen algorithm is employed by each trusted high-level user to generate a user private key $dk_B^{ID_i}$ for a low-level user, utilizing the private key $sk_{A_i}^{ID_i}$ and the cooperative key associated with identity ID_i, following the execution of the SOK scheme detailed in Sect. 3.2. Users have the option to choose different decryption algorithms based on the type of decryption keys, either PDec or UDec.

Note that users' identities ID_i are only associated with their private keys and are used to generate cooperative keys between trusted high-level users. However, they do not directly partake in the decryption process. The KC-ABE scheme is structured as follows:

- **Setup**$(\lambda, U, n) \rightarrow (mpk, msk)$. The Setup algorithm takes as input the security parameter λ, the universal attribute set U and the number of users n. It randomly chooses $g, w, v, \{h_k\} \in \mathbb{G}, k \in \{1, \ldots, n\}$ for users, $\{u_j\} \in \mathbb{G}, j \in \{1, \ldots, |U|\}$ for attributes and $\alpha \in \mathbb{Z}_p$. The Setup algorithm outputs the master public key and master secret key (mpk, msk) as follows.

$$mpk = (g, w, v, \{h_k\}_{k\in\{1,\ldots,n\}}, \{u_j\}_{j\in\{1,\ldots,|U|\}}, e(g,g)^{\alpha}), msk = \alpha.$$

- **PKeyGen**$(mpk, msk, ID_i, A_i) \rightarrow sk_{A_i}^{ID_i}$. The PKeyGen algorithm takes as input attribute set A_i, user identity ID_i and master public key, master secret key mpk, msk. It randomly chooses $r_i \in \mathbb{Z}_p$. The PKeyGen algorithm returns the private key $sk_{A_i}^{ID_i}$ as follows.

$$sk_{A_i}^{ID_i} = (g^{\alpha} w^{r_i}, g^{r_i}, \{u_j^{-r_i}\}_{j\in\{1,\ldots,|A_i|\}}, \{h_k^{r_i}\}_{k\in\{1,\ldots,n\}}).$$

- **UKeyGen**$(mpk, ID_i, sk_{A_i}^{ID_i}, I_B = \{ID_i\}, B \subseteq A_i) \rightarrow dk_B^{ID_i}$. The UKeyGen algorithm takes as input the master public key mpk, the private key $sk_{A_i}^{ID_i}$ of trusted high-level user with identity ID_i, the low-level user's attribute set $B \subseteq A_i$, the trusted high-level user identity set $I_B = \{ID_i\}$ and $|I_B| < n$. The trusted high-level user with ID_i first generates a cooperative key ck_i^B, through the following process:
  1. Request private key S_{ID_i} for the PKG that inputs the system parameters $params$ and master secret key msk in the SOK scheme and then obtain $S_{ID_i} = H_1(ID_i)^{\beta}$.
  2. Execute the SharedKey algorithm in the SOK scheme by inputting $params$, private key S_{ID_i}, identity $ID_j \in I_B, j \neq i$. The trusted high-level user with ID_i returns shared key $K_{i,j} = H_2(e(H_1(ID_i)^{\beta}, H_1(ID_j)))$.

3. Repeat the SharedKey algorithm $|I_B|-1$ times and then obtain $|I_B|-1$ shared keys $K_{i,j}, j \neq i, j \in \{1,\ldots,|I_B|\}$ between the trusted high-level user with ID_i and every user else in the cooperative group I_B.
4. For $j \in \{1,\ldots,|I_B|\}$ and $j \neq i$, the trusted high-level user with ID_i sets $a_{i,j}$ as follows:

$$a_{i,j} = \begin{cases} -K_{i,j} & i > j \\ K_{i,j} & i < j \end{cases}$$

Trusted high-level user with ID_i compacts these $|I_B|-1$ shared keys into a cooperative key $ck_i^B = g^{x_i} = g^{1+\sum_{j=1,j\neq i}^{|I_B|} a_{i,j}}$.

Then the trusted high-level user with identity ID_i randomly chooses $y_i \in \mathbb{Z}_p$ and $r_i^{'} = r_i + y_i$. The UKeyGen algorithm returns the user private key $dk_B^{ID_i}$ as follows.

$$dk_B^{ID_i} = (g^{\alpha} w^{r_i} \cdot g^{x_i} \cdot (\prod_{k=1}^{n-|I_B|} h_k)^{r_i} \cdot (\prod_{k=1}^{n-|I_B|} h_k w)^{y_i}, g^{r_i} \cdot g^{y_i},$$
$$\{u_j^{-r_i} \cdot u_j^{-y_i}\}_{j\in\{1,\ldots,|B|\}}, \{h_k^{r_i} \cdot h_k^{y_i}\}_{k\in\{n-|I_B|+1,\ldots,n\}}).$$

- **Enc**$(M, mpk, A^{'}, l) \rightarrow CT$. The Enc algorithm takes as input a message M, the master public key mpk, the attribute set $A^{'}$, and the key cooperative minimum $l < n$. It randomly chooses $s, z, z_j, t_j \in \mathbb{Z}_p$ and $\sum_{j=1}^{|A^{'}|} z_j = z$. The Enc algorithm returns the ciphertext CT as follows.

$$CT = (Me(g,g)^{\alpha s}, g^s, w^s v^z, (\prod_{k=1}^{n-l} h_k)^s, \{u_j^{t_j} v^{z_j}, g^{t_j}\}_{j\in\{1,\ldots,|A^{'}|\}}).$$

- **PDec**$(sk_{A_i}^{ID_i}, CT) \rightarrow M$. The PDec algorithm takes as input the private key $sk_{A_i}^{ID_i}$ and the ciphertext CT. It outputs the message if $A^{'} \subseteq A_i$. If $A^{'} \subseteq A_i$, we can computer as follows:

$$\frac{Me(g,g)^{\alpha s} e(w^s v^z, g^{r_i})}{e(g^{\alpha} w^{r_i}, g^s) \prod_{j=1}^{|A^{'}|} e(u_j^{t_j} v^{z_j}, g^{r_i}) e(u_j^{-r_i}, g^{t_j})} = M.$$

- **UDec**$(D_B = \{dk_B^{ID_i}\}, CT) \rightarrow M$. The UDec algorithm takes as input the $|I_B|$ user private keys $D_B = \{dk_B^{ID_i}\}$ and the ciphertext CT. It outputs the message if $A^{'} \subseteq B$ and $l \leq |I_B|$.
First, if $A^{'} \subseteq B$, same as in PDec algorithm, we obtain $e(v,g)^{zr_i^{'}}$. Then if $l \leq |I_B|$, we can computer $e(g,g)^{\alpha s}$ as follows:

$$(\prod_{i=1}^{|I_B|} \frac{e(g^{\alpha+x_i}(\prod_{k=1}^{n-|I_B|} h_k w)^{r_i^{'}}, g^s) \prod_{k=n-|I_B|+1}^{n-l} e(h_k^{r_i^{'}}, g^s) e(v,g)^{zr_i^{'}}}{e(\prod_{k=1}^{n-l} h_k^s \cdot w^s v^z, g^{r_i^{'}}) e(g^{|I_B|}, g^s)})^{\frac{1}{|I_B|}},$$

At last,

$$\frac{Me(g,g)^{\alpha s}}{e(g,g)^{\alpha s}} = M.$$

Correctness:

- **PDec Correctness:** For attribute set A_i in the private key $sk_{A_i}^{ID_i}$, satisfying $A' \subseteq A_i$, we can compute

$$\prod_{j=1}^{|A'|} e(u_j^{t_j} v^{z_j}, g^{r_i}) e(u_j^{-r_i}, g^{t_j}) = \prod_{j=1}^{|A'|} e(v^{z_j}, g^{r_i}) = e(v^z, g^{r_i}).$$

  Thus $e(g,g)^{\alpha s}$ can be computed from

$$\frac{e(g^{\alpha} w^{r_i}, g^s) e(v^z, g^{r_i})}{e(w^s v^z, g^{r_i})} = e(g,g)^{\alpha s}.$$

- **UDec Correctness:** For attribute set B in $|I_B|$ user private keys $D_B = \{dk_B^{ID_i}\}$, satisfying $A' \subseteq B$, for each user private key $dk_B^{ID_i}$, we can compute

$$\prod_{j=1}^{|A'|} e(u_j^{t_j} v^{z_j}, g^{r_i'}) e(u_j^{-r_i'}, g^{t_j}) = \prod_{j=1}^{|A'|} e(v^{z_j}, g^{r_i'}) = e(v^z, g^{r_i'}).$$

  Then $e(g,g)^{\alpha s}$ will be computed under the number of trusted high-level users $|I_B| \geq l$ as following:

$$\prod_{i=1}^{|I_B|} \frac{e(g^{\alpha + x_i} (\prod_{k=1}^{n-|I_B|} h_k w)^{r_i'}, g^s) \prod_{k=n-|I_B|+1}^{n-l} e(h_k^{r_i'}, g^s) e(v,g)^{z r_i'}}{e(\prod_{k=1}^{n-l} h_k^s \cdot w^s v^z, g^{r_i'})}$$

$$= \prod_{i=1}^{|I_B|} \frac{e(g^{\alpha + x_i}, g^s) e(w^{r_i'}, g^s) e((\prod_{k=1}^{n-|I_B|} h_k)^{r_i'}, g^s) \prod_{k=n-|I_B|+1}^{n-l} e(h_k^{r_i'}, g^s) e(v,g)^{z r_i'}}{e(\prod_{k=1}^{n-l} h_k^s \cdot w^s v^z, g^{r_i'})}$$

$$= \prod_{i=1}^{|I_B|} \frac{e(g^{\alpha + x_i}, g^s) e(w^{r_i'}, g^s) \prod_{k=1}^{n-l} e(h_k^{r_i'}, g^s) e(v,g)^{z r_i'}}{e(\prod_{k=1}^{n-l} h_k^s, g^{r_i'}) e(w^s, g^{r_i'}) e(v^z, g^{r_i'})} = \prod_{i=1}^{|I_B|} e(g^{\alpha + x_i}, g^s).$$

Here we show how to achieve key cooperation in the KC-ABE scheme that the multiplication of $|I_B|$ cooperative keys is $\prod_{i=1}^{|I_B|} ck_i^B = g^{|I_B|}$.
From the generation of cooperative key ck_i^B in UKeyGen algorithm, trusted high-level user with ID_i sets $ck_i^B = g^{x_i} = g^{1 + \sum_{j=1, j \neq i}^{|I_B|} a_{i,j}}$. It is important to note the following points:

1. In the key cooperative group I_B, for $i, j \in \{1, \ldots, |I_B|\}, j \neq i$, when $i > j$, $a_{i,j}$ in ck_i^B is $-K_{i,j}$ and $a_{j,i}$ in ck_j^B is $K_{j,i}$; conversely when $i < j$.

2. Because of shared keys in the SOK scheme satisfying $K_{i,j} = K_{j,i}$, thus $-K_{i,j}, K_{j,i}$ can be eliminated to zero when combining cooperative keys ck_i^B, ck_j^B.

Therefore all $a_{i,j}, a_{j,i}$ can be eliminated to zero and achieve $g^{|I_B|}$ when combining $|I_B|$ cooperative keys $ck_i^B, i \in \{1, \ldots, |I_B|\}$. Thus,

$$\prod_{i=1}^{|I_B|} e(g^{\alpha+x_i}, g^s) = \prod_{i=1}^{|I_B|} e(g^{\alpha+1+\sum_{j=1,j\neq i}^{|I_B|} a_{i,j}}, g^s)$$

$$= \prod_{i=1}^{|I_B|} e(g^{\alpha} \cdot g^{1+\sum_{j=1,i>j}^{|I_B|} -K_{i,j}+\sum_{j=1,i<j}^{|I_B|} K_{i,j}}, g^s) = e(g,g)^{|I_B|s(\alpha+1)}$$

Thus $e(g,g)^{\alpha s}$ can be computed from

$$(\frac{(e(g,g)^{|I_B|s(\alpha+1)})}{e(g^{|I_B|}, g^s)})^{\frac{1}{|I_B|}} = e(g,g)^{\alpha s}.$$

Example. Here, we provide a simple example of key cooperation in our KC-ABE scheme. Suppose the low-level user with attribute set B determines an identity set of trusted high-level users $I_B = \{ID_1, ID_2, ID_3\}$. The trusted high-level user with ID_1 generates the shared key $ck_1^B = g^{1+K_{1,2}+K_{1,3}}$. Similarly, trusted high-level users with ID_2, ID_3 also generate $ck_2^B = g^{1-K_{2,1}+K_{2,3}}$, $ck_3^B = g^{1-K_{3,2}-K_{3,1}}$, respectively. Because of $K_{i,j} = K_{j,i}, i,j \in \{1,2,3\}, i \neq j$, it easy to see that $\prod_{i=1}^{3} ck_i^B = g^{|I_B|} = g^3$.

4 Security Analysis

4.1 Complexity Assumption

Decisional $(q-1)$ **Assumption** [23]. For our construction, we will use a $q-1$ type assumption in [23], which is similar to the Decisional Parallel Bilinear Diffle-Hellman Exponent Assumption [28]. This assumption can be proved secure in the generic group model and is defined via the following game between $\mathcal{C}$ and $\mathcal{A}$:

Initially $\mathcal{C}$ calls the group generation algorithm with input the security parameter, picks a random group element $g \in \mathbb{G}$, and $q+2$ random exponents $a, s, \overrightarrow{b} = \{b_1, \ldots, b_q\} \in \mathbb{Z}_p$. Then it sends to $\mathcal{A}$ the group description $(p, \mathbb{G}, \mathbb{G}_p, e)$ and all of the following terms:

$$g, g^s$$

$$g^{a^i}, g^{b_j}, g^{sb_j}, g^{a^i b_j}, g^{\frac{a^i}{(b_j)^2}}, \forall (i,j) \in [q,q]$$

$$g^{\frac{a^i b_j}{(b_{j'})^2}}, \forall (i,j,j') \in [2q,q,q], j \neq j'$$

$$g^{\frac{a^i}{b_j}}, \forall (i,j) \in [2q,q], i \neq q+1$$

$$g^{\frac{sa^i b_j}{b_{j'}}}, g^{\frac{sa^i b_j}{(b_{j'})^2}} \forall (i,j,j') \in [q,q,q], j \neq j'.$$

$\mathcal{C}$ flips a random coin $b \in \{0,1\}$ and if $b = 0$, it gives to $\mathcal{A}$ the term $T = e(g,g)^{sa^{q+1}}$. Otherwise, it gives a random term $T = R \in \mathbb{G}_T$. Finally the $\mathcal{A}$ outputs a guess $b^{'} \in \{0,1\}$.

4.2 Security Proof

We used the SOK scheme in user private key generation of trusted high-level users. Paterson [21] established a new security model for IB-NIKE and then proved the security of the SOK scheme in the extended security model. As mentioned before in Sect. 3.3 scheme, $a_{i,j}$ actually are shared keys between users with ID_i and ID_j in the SOK scheme. The final cooperative key ck_i^B in our KC-ABE scheme for the user with ID_i is just the combination of different $a_{i,j}$. This setting would not influence the security level when using the SOK scheme. Due to the original SOK scheme's indistinguishability of shared key (IND-SK) secure under Bilinear Diffie-Hellman (BDH) assumption in the extended security model in [21], thus our generation of cooperative keys is secure. Therefore, we only prove the security of the KC-ABE scheme in Sect. 3.3 as follows.

Theorem 1. *If the $q-1$ assumption in [23] holds then no polynomial adversary can selectively break our scheme with a challenge attribute set A^* and a challenge key cooperative minimum l^*.*

Proof. Suppose there exists an adversary $\mathcal{A}$ who breaks the proposed KC-ABE scheme with $\mathsf{Adv}_{\mathsf{KC\text{-}ABE}}^{\mathsf{IND\text{-}sCPA}}(\mathcal{A}) = \epsilon$. We built a simulator $\mathcal{B}$ that has advantage $\frac{\epsilon}{2}$ in solving the $q-1$ assumption in [23]. Given as input an assumption instance, $\mathcal{B}$ plays the role of the challenger $\mathcal{C}$ in the game, and interacts with the adversary $\mathcal{A}$ as follows.

- **Initialization.** $\mathcal{B}$ first chooses the attribute universal set U and the number of users n. The selective game with $\mathcal{A}$ outputs a challenge attribute set A^* and a challenge key cooperative minimum $l^*, l^* < n \ll q$ that it intends to attack.
- **Setup.** To generate the system parameters, $\mathcal{B}$ picks randoms $\tilde{\alpha}$ and sets $\alpha = a^{q+1} + \tilde{\alpha}$. Notice that this way α is correctly distributed and information-theoretically hidden from $\mathcal{A}$.
  $\mathcal{B}$ sets $w = g^a, v = g^{\frac{-a}{b_q}} \cdot g^{\frac{a}{b_1}}$ and picks randoms $w_1, \ldots, w_{|U|}$ in $\mathbb{Z}_p$ and $\gamma_1, \ldots, \gamma_n$ in $\mathbb{Z}_p$. Then $\mathcal{B}$ sets $\{u_j\}, \{h_k\}$ as follows.
$$u_j = g^{\frac{a^{q-j+1} - a^2 w_j}{b_q}}, j = 1, \ldots, |U|. \quad h_k = g^{\frac{\gamma_k a^{k+1} b_{n-l^*+1}}{(b_k)^2}}, k = 1, \ldots, n.$$
  Finally, $\mathcal{B}$ gives $\mathcal{A}$ $mpk = (g, w, v, h_1, \ldots, h_n, u_1, \ldots, u_{|U|})$. Observe that all these values are distributed uniformly and independently in $\mathbb{G}$ as required. The master key corresponding to the system parameters is $g^\alpha = g^{a^{q+1} + \tilde{\alpha}}$, which is unknown to $\mathcal{B}$ since $\mathcal{B}$ does not have $g^{a^{q+1}}$.
- **Query phase 1.** The adversary $\mathcal{A}$ makes private key queries and user private key queries in this phase as follows.

1. The adversary $\mathcal{A}$ makes private key queries for non authorized sets A_i and user identity ID_i in this phase, $i \in \{1, \ldots, q_1\}$. For a private key query on A_i and ID_i. $\mathcal{B}$ picks randoms $\theta_j \in \mathbb{Z}_p$ for $a_j \in A^*$, $\sum_{j=1}^{|A^*|} \theta_j = 0$. $\mathcal{B}$ picks randoms $\tilde{r_i} \in \mathbb{Z}_p$ and sets

$$-r_i = \tilde{r_i} + a^q + \frac{(a^{q-1} + \sum_{i=1}^{n} w_i a^i)}{(\sum_{j=1}^{|S|} \theta_j)}$$

Notice that r_i is well-defined only for attribute sets in the specific unauthorized set A_i or unleaded attribute sets. At first, we set $S = A^* \bigcap A_i$ and only $\sum_{j=1}^{|A^*|} \theta_j = 0$. For the unauthorized set A_i, $A^* \not\subseteq A_i$, we obtain $S \neq A^*$ and $\sum_{j=1}^{|S|} \theta_j$ is not zero. Otherwise, for the authorized set A_i, $A^* \subseteq A_i$, we obtain $S = A^*$ and $\sum_{j=1}^{|S|} \theta_j$ is zero. Namely, the set of private keys is distributed in two types: the unauthorized, which the simulator can use the above method, and the authorized, which the simulator cannot create. This is properly distributed due to $\tilde{r_i}$. Then, using the suitable terms from the assumption, it calculates. The private key $sk_{A_i}^{ID_i}$ can be simulated as following:

$$g^{\alpha} w^{r_i} = g^{a^{q+1} + \tilde{\alpha}} \cdot g^{-a^{q+1}} \cdot g^{\frac{a^q + \sum_{i=1}^{n} w_i a^{i+1}}{-(\sum_{j=1}^{|S|} \theta_j)}} \cdot g^{-a\tilde{r_i}}, g^{r_i} = g^{-\tilde{r_i} - a^q + \frac{(a^{q-1} + \sum_{i=1}^{n} w_i a^i)}{-(\sum_{j=1}^{|S|} \theta_j)}},$$

$$\{h_k^{r_i}\} = \{h_k^{-\tilde{r_i}} \cdot g^{\frac{-\gamma_k a^{q+k+1} b_{n-l^*+1}}{(b_k)^2}} \cdot g^{\frac{\gamma_k (a^{q+k} + \sum_{t=1}^{n} \gamma_k w_t a^{t+k+1}) b_{n-l^*+1}}{-(\sum_{j=1}^{|S|} \theta_j)(h_k)^2}}\},$$

$$\begin{aligned}
\{u_j^{-r_i}\} =& \{u_j^{-\tilde{r_i}} \cdot g^{\frac{a^{2q-j} + \sum_{i=1}^{n} w_i a^{q-j+i+1} - w_j a^{q+1} - \sum_{i=1}^{n} w_i w_j a^{i+2}}{b_q (\sum_{j=1}^{|S|} \theta_j)}} \cdot g^{\frac{a^{2q-j+1} - w_j a^{q+2}}{b_q}}\} \\
=& \{u_j^{-\tilde{r_i}} \cdot g^{\frac{a^{2q-j} + \sum_{i=1, i \neq j}^{n} w_i a^{q-j+i+1} - \sum_{i=1}^{n} w_i w_j a^{i+2}}{b_q (\sum_{j=1}^{|S|} \theta_j)}} \cdot g^{\frac{a^{2q-j+1} - w_j a^{q+2}}{b_q}}\},
\end{aligned}$$

$\mathcal{B}$ sets private key $sk_{A_i}^{ID_i} = (g^{\alpha} w^{r_i}, g^{r_i}, \{h_k^{r_i}\}_{k \in \{1, \ldots, n\}}, \{u_j^{-r_i}\}_{j \in \{1, \ldots, |A_i|\}})$ as above and sends them to $\mathcal{A}$.
2. The adversary $\mathcal{A}$ also makes user private key queries in this phase for authorized attribute sets B_j, $A^* \subseteq B_j$ and unauthorized identity sets $I_{B_j} = \{ID_i\}, |I_{B_j}| < l^*$, $j \in \{1, \ldots, q_2\}$. For simplicity, we set B and I_B as attribute sets and identity sets in each user private key query, respectively. Here, for each user private key query on attribute set B and identity $ID_i \in I_B$, $\mathcal{B}$ first generates private keys $sk_{A_i'}^{ID_i}$ with satisfying attribute set A_i', $A^* \subseteq A_i'$ and identity $ID_i \in I_B$. $\mathcal{B}$ generates the user private key $dk_B^{ID_i}$ based on private keys $sk_{A_i'}^{ID_i}$ under $B \subseteq A_i'$. Here we set $\sum_{j=1}^{|S|} \theta_j = 0$ in this phase for each authorized attribute set B under $A^* \subseteq B$ and $S = A^* \cap B$. $\mathcal{B}$ picks a random $\tilde{y_i} \in \mathbb{Z}_p$ and sets

$$-y_i = \tilde{y}_i + a^{q-1}(\sum_{j=1}^{|S|} \theta_j + 1) + \sum_{i=1}^{n} w_i a^i - \frac{a^{q-(n-l^*)-1}}{\gamma_{n-l^*+1}}$$

$\mathcal{B}$ picks a random $\tilde{x}_i$ and sets a cooperative key $ck_i^B = g^{x_i} = g^{\tilde{x}_i - a^{q+1} - \frac{a^{q+1}}{b_{n-l^*+1}}}$.
$\mathcal{B}$ sets

$$\Phi = (\prod_{k=1}^{n-|I_B|} h_k)^{-\tilde{y}_i - a^{q-1}(\sum_{j=1}^{|S|}\theta_j+1) - \sum_{i=1}^{n} w_i a^i}$$

$\mathcal{B}$ computes $(\prod_{k=1}^{n-|I_B|} h_k)^{y_i}$

$$= \Phi \cdot (g^{\frac{\gamma_1 a^2 b_{n-l^*+1}}{(b_1)^2} + \ldots + \frac{\gamma_{n-l^*+1} a^{n-l^*+2} b_{n-l^*+1}}{(b_{n-l^*+1})^2} + \ldots + \frac{\gamma_{n-|I_B|} a^{n-|I_B|+1} b_{n-l^*+1}}{(b_{n-|I_B|})^2}})^{\frac{a^{q-(n-l^*)-1}}{\gamma_{n-l^*+1}}}$$
$$= \Phi \cdot (g^{\frac{\gamma_1 a^2 b_{n-l^*+1}}{(b_1)^2} + \ldots + \frac{\gamma_{n-|I_B|} a^{n-|I_B|+1} b_{n-l^*+1}}{(b_{n-|I_B|})^2}})^{\frac{a^{q-(n-l^*)-1}}{\gamma_{n-l^*+1}}} \cdot g^{\frac{a^{q+1}}{b_{n-l^*+1}}}$$
$$= \Phi \cdot \Psi \cdot g^{\frac{a^{q+1}}{b_{n-l^*+1}}}.$$

Notice that $(\prod_{k=1}^{n-|I_B|} h_k)^y$ is well-defined only for the unauthorized number of trusted high-level users $|I_B|, |I_B| < l^*$. Since if $|I_B| < l^*, n-|I_B| > n-l^*$, then $\prod_{k=1}^{n-|I_B|} h_k$ includes $g^{\frac{\gamma_{n-l^*+1} a^{n-l^*+2} b_{n-l^*+1}}{(b_{n-l^*+1})^2}}$. Otherwise, when $|I_B| > l^*, n - |I_B| < n - l^*$, $\prod_{k=1}^{n-|I_B|} h_k$ will not include it.
Thus the item $g^{\frac{a^{q+1}}{b_{n-l^*+1}}}$ can be computed from $(\prod_{k=1}^{n-|I_B|} h_k)^{y_i}$ only when $|I_B| < l^*$. $\mathcal{B}$ also can calculate the remaining elements in $dk_B^{ID_i}$ as follows, because they do not involve a $g^{\frac{a^{q+1}}{b_{n-l^*+1}}}$ term. Thus, $\mathcal{B}$ computes $\{u_j^{-r_i - y_i}\}$ in user private keys:

$$\{u_j^{-r_i+\tilde{y}_i} \cdot g^{\frac{a^{2q-j} + \sum_{i=1, i\neq j}^{n} w_i a^{q-j+i+1} - \sum_{i=1}^{n} w_i w_j a^{i+2}}{b_q}} \cdot g^{\frac{a^{2q-(n-l^*)-j} - a^{q-(n-l^*)+1} w_j}{-b_q \gamma_{n-l^*+1}}}\},$$

And other parts in user private key $dk_B^{ID_i}$ are as follows:

$$g^{\alpha} w^{r_i} \cdot g^{x_i} \cdot (\prod_{k=1}^{n-|I_B|} h_k)^{r_i} \cdot (\prod_{k=1}^{n-|I_B|} h_k w)^{y_i} = g^{\tilde{\alpha}+\tilde{x}_i} \cdot w^{r_i+y_i} \cdot (\prod_{k=1}^{n-|I_B|} h_k)^{r_i} \cdot \Phi \cdot \Psi,$$

$$\{h_k^{r_i+y_i}\} = \{h_k^{r_i-\tilde{y}_i} \cdot g^{\frac{-\gamma_i (a^{q+k} + \sum_{t=1}^{n} \gamma_k w_t a^{t+k+1}) b_{n-l^*+1}}{(b_k)^2}} \cdot g^{\frac{\gamma_k a^{q-(n-l^*)+k} b_{n-l^*+1}}{(b_k)^2 \gamma_{n-l^*+1}}}\},$$
$$g^{r_i+y_i} = g^{r_i - \tilde{y}_i - a^{q-1} - \sum_{i=1}^{n} w_i a^i + \frac{a^{q-(n-l^*)-1}}{\gamma_{n-l^*+1}}}.$$

$\mathcal{B}$ sets user private key $dk_B^{ID_i} = (g^{\alpha} w^{r_i} \cdot g^{x_i} \cdot (\prod_{k=1}^{n-|I_B|} h_k)^{r_i} \cdot (\prod_{k=1}^{n-|I_B|} h_k w)^{y_i}, g^{r_i} \cdot g^{y_i}, \{u_j^{-r_i} \cdot u_j^{-y_i}\}_{j\in\{1,\ldots,|B|\}}, \{h_k^{r_i} \cdot h_k^{y_i}\}_{k\in\{n-|I_B|+1,\ldots,n\}})$ as above and sends them to $\mathcal{A}$.

- **Challenge.** When $\mathcal{A}$ decides that Phase 1 is over, it outputs two messages $M_0, M_1 \in \mathbb{G}$ on which it wishes to be challenged. $\mathcal{B}$ picks a random bit $b \in \{0,1\}$ and responds with the challenge ciphertext

$$CT = (M_b \cdot T \cdot e(g, g^s)^{\tilde{\alpha}}, g^s, g^{\frac{sab_q}{b_1}}, \prod_{k=1}^{n-l^*} g^{\frac{s\gamma_k a^{k+1} b_{n-l^*+1}}{(b_k)^2}}, \{u_j^{t_j} v^{z_j}, g^{t_j}\}_{j\in\{1,\ldots,|A^*|\}})$$

where g^s and T are from the input tuple given to $\mathcal{B}$. First note that g^s, then $\mathcal{B}$ randomly chooses $z = sb_q, t_j, z_j \in \mathbb{Z}_p$, and calculates the ciphertext.
- **Query phase 2.** $\mathcal{A}$ issues queries not issued in Phase 1 and $\mathcal{B}$ responds as before. Specifically,
  1. $\mathcal{A}$ issues private key queries that attribute sets $A_i \subseteq U$ satisfy $A^* \not\subseteq A_i$ and user identities ID_i, $i \in \{1, \ldots, q_1'\}$.
  2. $\mathcal{A}$ makes q_2' user private key queries that trusted high-level identity sets I_{B_j} satisfy $l^* > |I_{B_j}|$ under $A^* \subseteq B_j$, $j \in \{1, \ldots, q_2'\}$.
- **Guess.** After the query phase 2, $\mathcal{A}$ outputs guess b' for the challenge bit. If $b' = b$, $\mathcal{B}$ outputs 1, it claims that the challenge term is $T = e(g,g)^{sa^{q+1}}$. Otherwise, it outputs 0 meaning $T = R$ is random in $\mathbb{G}_T$.
When $T = R$, the ciphertext will give no information about the simulator's choice of b. In this case, we have $\Pr[b' = b | T = R] = \Pr[b' \neq b | T = R] = \frac{1}{2}$, because the challenge ciphertext will contain only random numbers. Since the simulator outputs a guess $T' = R$ when $b' = b$, we then have $\Pr[T' = T | T = R] = \frac{1}{2}$. When $T = e(g,g)^{sa^{q+1}}$, $\mathcal{A}$ can sees a valid encryption of M_b. In this situation, the $\mathcal{A}$'s advantage is ϵ by definition. The probability is $\Pr[b' = b | T = e(g,g)^{sa^{q+1}}] = \frac{1}{2} + \epsilon$. Since the simulator outputs a guess $T = e(g,g)^{sa^{q+1}}$ of T when $b' = b$, we have $\Pr[T' = T | T = e(g,g)^{sa^{q+1}}] = \frac{1}{2} + \epsilon$. Hence, by putting them all together, $\mathcal{B}$'s advantage in the above game is

$$\begin{aligned}
&\Pr[T' = T] - \frac{1}{2} \\
&= \Pr[T' = T | T = e(g,g)^{sa^{q+1}}] \Pr[T = e(g,g)^{sa^{q+1}}] + \Pr[T' = T | T = R] \Pr[T = R] - \frac{1}{2} \\
&= \frac{1}{2} \Pr[T' = T | T = e(g,g)^{sa^{q+1}}] + \frac{1}{2} \Pr[T' = T | T = R] - \frac{1}{2} \\
&= \frac{1}{2}(\frac{1}{2} + \epsilon + \frac{1}{2}) - \frac{1}{2} \\
&= \frac{\epsilon}{2}
\end{aligned}$$

Therefore, we obtain Theorem 1 and prove the security of our proposed scheme.

5 Discussion and Extensions

Efficiency. Based on the CP-ABE scheme in [23], both the size of the private key and the ciphertext in our scheme scale linearly with the number of attribute sets. For each trusted high-level user in the identity set I_B, the generation of a cooperative key ck_i^B requires $|I_B| - 1$ shared keys with other users.

Specifically, each shared key needs $|I_B| - 1$ pairing operations. After the generation of a cooperative key, each trusted high-level user generates a user private key. Although the user private keys' generation adds extra $O(|I_B|^2)$ computation costs for users, the PKG concentrates on generating private keys for only a small number of high-level users, leading to a substantial improvement in the PKG key management.

Expressive Access Policy. In our KC-ABE scheme, the access policy is that the attribute set of ciphertext is the subset of the users' attribute set. Here we extend the access control to LSSS access policy (M, ρ) in [28]. When M_j, $\overrightarrow{v}$ are the vector corresponding to the jth row of M and a random vector, respectively, then $\lambda_j = \overrightarrow{v} \cdot M_j$ denotes the vector shares of the secret s. As a result, attribute elements in the ciphertext are represented as $\{u_j^{t_j} v^{\lambda_j}, g^{t_j}\}$, and the decryption procedure corresponds to the one in [28].

Chosen-Ciphertext Security. Our security definitions and proofs have been in the chosen-plaintext model. We remark that our construction can be extended to the chosen-ciphertext model by applying the technique of one-time signature scheme in Canettu, Halevi, and Katz [4] by some simple alterations.

6 Conclusion

In this paper, we introduce a new concept called Key Cooperative ABE. It allows trusted users to generate user private keys for other users. This empowers users to decrypt ciphertexts using either the private key from the PKG or user private keys from trusted users. Thus Key Cooperative ABE increases flexibility for users in obtaining decryption keys, solving the huge key management problem. Key Cooperative ABE avoids the introduction of additional attribute authorities and provides key cooperation between multiple trusted users, making it more practical, lightweight, and extensible. We present a basic Key Cooperative ABE scheme that still poses several challenges, such as ensuring security against malicious users, developing more efficient constructions with shorter private keys and user private keys, and exploring its potential applications in scenarios.

References

1. Ashley, M.: The gnu privacy handbook (1999). https://www.gnupg.org/gph/en/manual.html
2. Bethencourt, J., Sahai, A., Waters, B.: Ciphertext-policy attribute-based encryption. In: IEEE Symposium on Security and Privacy 2007, pp. 321–334. IEEE Computer Society (2007)
3. Boneh, D., Franklin, M.K.: Identity based encryption from the weil pairing. IACR Cryptol. ePrint Arch, p. 90 (2001)
4. Canetti, R., Halevi, S., Katz, J.: Chosen-ciphertext security from identity-based encryption. In: Cachin, C., Camenisch, J.L. (eds.) EUROCRYPT 2004. LNCS, vol. 3027, pp. 207–222. Springer, Heidelberg (2004). https://doi.org/10.1007/978-3-540-24676-3_13

5. Chase, M.: Multi-authority attribute based encryption. In: Vadhan, S.P. (ed.) TCC 2007. LNCS, vol. 4392, pp. 515–534. Springer, Heidelberg (2007). https://doi.org/10.1007/978-3-540-70936-7_28
6. Cui, H., Deng, R.H., Li, Y., Qin, B.: Server-aided revocable attribute-based encryption. In: Askoxylakis, I., Ioannidis, S., Katsikas, S., Meadows, C. (eds.) ESORICS 2016. LNCS, vol. 9879, pp. 570–587. Springer, Cham (2016). https://doi.org/10.1007/978-3-319-45741-3_29
7. Freire, E.S.V., Hofheinz, D., Kiltz, E., Paterson, K.G.: Non-interactive key exchange. In: Kurosawa, K., Hanaoka, G. (eds.) PKC 2013. LNCS, vol. 7778, pp. 254–271. Springer, Heidelberg (2013). https://doi.org/10.1007/978-3-642-36362-7_17
8. Garg, S., Hajiabadi, M., Mahmoody, M., Rahimi, A.: Registration-based encryption: removing private-key generator from IBE. In: Beimel, A., Dziembowski, S. (eds.) TCC 2018. LNCS, vol. 11239, pp. 689–718. Springer, Cham (2018). https://doi.org/10.1007/978-3-030-03807-6_25
9. Goyal, V., Jain, A., Pandey, O., Sahai, A.: Bounded ciphertext policy attribute based encryption. In: Aceto, L., Damgård, I., Goldberg, L.A., Halldórsson, M.M., Ingólfsdóttir, A., Walukiewicz, I. (eds.) ICALP 2008. LNCS, vol. 5126, pp. 579–591. Springer, Heidelberg (2008). https://doi.org/10.1007/978-3-540-70583-3_47
10. Hohenberger, S., Lu, G., Waters, B., Wu, D.J.: Registered attribute-based encryption. In: Hazay, C., Stam, M. (eds.) EUROCRYPT 2023. LNCS, vol. 14006, pp. 511–542. Springer (2023). https://doi.org/10.1007/978-3-031-30620-4_17
11. Hohenberger, S., Waters, B.: Online/offline attribute-based encryption. In: Krawczyk, H. (ed.) PKC 2014. LNCS, vol. 8383, pp. 293–310. Springer, Heidelberg (2014). https://doi.org/10.1007/978-3-642-54631-0_17
12. Horwitz, J., Lynn, B.: Toward hierarchical identity-based encryption. In: Knudsen, L.R. (ed.) EUROCRYPT 2002. LNCS, vol. 2332, pp. 466–481. Springer, Heidelberg (2002). https://doi.org/10.1007/3-540-46035-7_31
13. Hur, J., Noh, D.K.: Attribute-based access control with efficient revocation in data outsourcing systems. IEEE Trans. Parallel Distributed Syst. **22**(7), 1214–1221 (2011)
14. Jiang, Y., Susilo, W., Mu, Y., Guo, F.: Ciphertext-policy attribute-based encryption with key-delegation abuse resistance. In: Liu, J.K., Steinfeld, R. (eds.) ACISP 2016. LNCS, vol. 9722, pp. 477–494. Springer, Cham (2016). https://doi.org/10.1007/978-3-319-40253-6_29
15. Lai, J., Deng, R.H., Li, Y.: Fully secure cipertext-policy hiding CP-ABE. In: Bao, F., Weng, J. (eds.) ISPEC 2011. LNCS, vol. 6672, pp. 24–39. Springer, Heidelberg (2011). https://doi.org/10.1007/978-3-642-21031-0_3
16. Lewko, A.B., Waters, B.: Decentralizing attribute-based encryption. IACR Cryptol. ePrint Arch., p. 351 (2010)
17. Li, J., Ren, K., Zhu, B., Wan, Z.: Privacy-aware attribute-based encryption with user accountability. In: Samarati, P., Yung, M., Martinelli, F., Ardagna, C.A. (eds.) ISC 2009. LNCS, vol. 5735, pp. 347–362. Springer, Heidelberg (2009). https://doi.org/10.1007/978-3-642-04474-8_28
18. Lin, H., Cao, Z., Liang, X., Shao, J.: Secure threshold multi authority attribute based encryption without a central authority. Inf. Sci. **180**(13), 2618–2632 (2010)
19. Liu, Z., Cao, Z., Wong, D.S.: White-box traceable ciphertext-policy attribute-based encryption supporting any monotone access structures. IEEE Trans. Inf. Forensics Secur. **8**(1), 76–88 (2013)
20. Nishide, T., Yoneyama, K., Ohta, K.: Abe with partially hidden encryptor-specified access structure (2008)

21. Paterson, K.G., Srinivasan, S.: On the relations between non-interactive key distribution, identity-based encryption and trapdoor discrete log groups. Des. Codes Crypt. **52**, 219–241 (2009)
22. PGP, W.I.W.: Philip zimmermann
23. Rouselakis, Y., Waters, B.: Practical constructions and new proof methods for large universe attribute-based encryption. In: Sadeghi, A., Gligor, V.D., Yung, M. (eds.) CCS 2013, pp. 463–474. ACM (2013)
24. Sahai, A., Waters, B.: Fuzzy identity based encryption. IACR Cryptol. ePrint Arch., p. 86 (2004)
25. Sakai, R., Ohgishi, K., Kasahara, M.: Cryptosystems based on pairing. SCIS 2000 (2000)
26. Shamir, A.: Identity-based cryptosystems and signature schemes. In: Blakley, G.R., Chaum, D. (eds.) CRYPTO 1984. LNCS, vol. 196, pp. 47–53. Springer, Heidelberg (1985). https://doi.org/10.1007/3-540-39568-7_5
27. Wang, G., Liu, Q., Wu, J.: Hierarchical attribute-based encryption for fine-grained access control in cloud storage services. In: Al-Shaer, E., Keromytis, A.D., Shmatikov, V. (eds.) CCS 2010, pp. 735–737. ACM (2010)
28. Waters, B.: Ciphertext-policy attribute-based encryption: an expressive, efficient, and provably secure realization. In: Catalano, D., Fazio, N., Gennaro, R., Nicolosi, A. (eds.) PKC 2011. LNCS, vol. 6571, pp. 53–70. Springer, Heidelberg (2011). https://doi.org/10.1007/978-3-642-19379-8_4

On the Feasibility of Identity-Based Encryption with Equality Test Against Insider Attacks

Keita Emura(✉)

Kanazawa University, Kanazawa, Japan
k-emura@se.kanazawa-u.ac.jp

Abstract. Public key encryption with equality test, proposed by Yang et al. (CT-RSA 2010), allows anyone to check whether two ciphertexts of distinct public keys are encryptions of the same plaintext or not using trapdoors, and identity-based encryption with equality test (IBEET) is its identity-based variant. As a variant of IBEET, IBEET against insider attacks (IBEETIA) was proposed by Wu et al. (ACISP 2017), where a token is defined for each identity and is used for encryption. Lee et al. (ACISP 2018) and Duong et al. (ProvSec 2019) proposed IBEETIA schemes constructed by identity-based encryption (IBE) related complexity assumptions. Later, Emura and Takayasu (IEICE Transactions 2023) demonstrated that symmetric key encryption and pseudo-random permutations are sufficient to construct IBEETIA which is secure in the previous security definition. In this paper, we demonstrate a sufficient condition that IBEETIA implies IBE. We define one-wayness against chosen-plaintext/ciphertext attacks for the token generator (OW-TG-CPA/CCA) and for token holders (OW-TH-CPA/CCA), which were not considered in the previous security definition. We show that OW-TG-CPA secure IBEETIA with additional conditions implies OW-CPA secure IBE. On the other hand, we propose a generic construction of OW-TH-CCA secure IBEETIA from public key encryption. Our results suggest a design principle to efficiently construct IBEETIA without employing IBE-related complexity assumptions.

Keywords: Identity-based encryption with equality test against insider attacks · Searchable Encryption · Generic Construction

1 Introduction

Public key encryption with equality test (PKEET) was proposed in [34], where anyone can check whether two ciphertexts of distinct public keys are encryptions of the same plaintext or not using trapdoors. Unlike to public key encryption with keyword search [5], the equality test works on ciphertexts generated by different public keys. As applications, Duong et al. [15] mentioned that keyword search on

T. Zhu and Y. Li (Eds.): ACISP 2024, LNCS 14895, pp. 261–280, 2024.
https://doi.org/10.1007/978-981-97-5025-2_14

encrypted data, encrypted data partitioning for efficient encrypted data management, personal health record system and spam filtering in encrypted email systems. Identity-based encryption with equality test (IBEET) is its identity-based variant, where anyone can check whether two ciphertexts of distinct identities are encryptions of the same plaintext or not using trapdoors. Wu at al. [32] proposed a lattice-based IBEET scheme based on the Tsabary IBE scheme [27]. Lee et al. [24] demonstrated that IBEET can generically be constructed from 3-level hierarchical identity-based encryption (HIBE).[1] Asano et al. [2] further improved the Lee et al. generic construction to capture state-of-the-art lattice-based IBE schemes [22,33]. Lee [23] gave an attack against the Zhu et al's IBEET scheme [35].

IBEET cannot provide any indistinguishability-based security for insiders who have a trapdoor due to the test functionality of IBEET. Concretely, two security notions are defined against two types of adversaries called Type-I and Type-II. The Type-I adversary is allowed to obtain trapdoors for the challenge identity, and thus the one-wayness against the Type-I adversary was defined. The Type-II adversary is not allowed to obtain trapdoors for the challenge identity, and thus the indistinguishability against the Type-II adversary was defined.[2]

IBEET **Against Insider Attacks.** Since IBEET does not provide any indistinguishability security against insiders who run the test algorithm and have trapdoors (the Type-I adversary above), Wu et al. [31] introduced IBEET against insider attacks (IBEETIA). We briefly introduce IBEETIA as follows. The setup algorithm outputs a master token key MTK, in addition to a master public key MPK and a master secret key MSK. As in IBE, MSK is used for generating a secret key $\mathsf{sk}_{\mathsf{ID}}$ for an identity ID. MTK is used for generating a token $\mathsf{tok}_{\mathsf{ID}}$ for an identity ID. The encryption algorithm takes $\mathsf{tok}_{\mathsf{ID}}$ in addition to MPK, ID, a plaintext M, and outputs a ciphertext ct. The decryption algorithm takes MPK, ct, and both $\mathsf{sk}_{\mathsf{ID}}$ and $\mathsf{tok}_{\mathsf{ID}}$, and outputs M. Anyone can run the test algorithm that takes MPK and two ciphertexts without using trapdoors. Thus, IBEETIA can provide an indistinguishability security against an insider adversary $\mathcal{A}$ when $\mathcal{A}$ is not allowed to obtain a token $\mathsf{tok}_{\mathsf{ID}^*}$ and a secret key $\mathsf{sk}_{\mathsf{ID}^*}$ for the challenge identity ID^* (and it is required that $\mathcal{A}$ did not query (ID, M_0^*) or (ID, M_1^*) for any ID and the challenge plaintexts (M_0^*, M_1^*) as an encryption query).

Lee et al. [25] pointed out a security flaw of the Wu et al. scheme [31], and proposed a pairing-based IBEETIA scheme. Moreover, Duong et al. [15] proposed a lattice-based IBEETIA scheme based on the Agrawal-Boneh-Boyen (ABB) IBE scheme [1]. That is, they employed IBE-related complexity assumptions. According to the implication result shown by Boneh et al. [7], IBE is recognized as a strong cryptographic primitive because no black-box construction of IBE from

[1] Lee et al. [24] also demonstrated that PKEET can generically be constructed from 2-level HIBE.

[2] Chosen-ciphertext attack (CCA) is considered against both types of adversaries. See [2,24] for more details.

trapdoor permutations (TDPs) exist.[3] At the first place, these selections are reasonable because HIBE is employed to construct IBEET [2,24]. However, Emura and Takayasu [18] demonstrated that symmetric key encryption and pseudo-random permutations are sufficient to construct IBEETIA which is secure in the security definition given in [15,25,31]. They paid attention to the syntax that a token $\mathsf{tok}_{\mathsf{ID}}$ is used for both encryption and decryption, as in symmetric key encryption, and they set $\mathsf{tok}_{\mathsf{ID}}$ as a key SKE.sk of a symmetric encryption scheme. Interestingly, $\mathsf{sk}_{\mathsf{ID}} = \perp$ in the Emura-Takayasu construction because $\mathsf{sk}_{\mathsf{ID}}$ is not used for decryption. Emura and Takayasu [18] mentioned that:

> *One may wonder whether our construction provides the same functionality of previous* IBEETIA *schemes since* $\mathsf{sk}_{\mathsf{ID}} = \perp$. *More precisely, because the* Dec *algorithm takes both* $\mathsf{sk}_{\mathsf{ID}}$ *and* $\mathsf{tok}_{\mathsf{ID}}$ *as input, anyone who has* $\mathsf{tok}_{\mathsf{ID}}$ *can decrypt all ciphertexts generated by* ID *in our construction, whereas one who has* $\mathsf{tok}_{\mathsf{ID}}$ *but does not have* $\mathsf{sk}_{\mathsf{ID}}$ *cannot decrypt such ciphertexts in the previous schemes. We emphasize that this difference does not violate not only the correctness but also the wIND-CCA security.*

Here, wIND-CCA stands for weak indistinguishability against chosen ciphertext attacks. According to this observation, it is natural to consider a security notion against token holders who have $\mathsf{tok}_{\mathsf{ID}}$ but do not have $\mathsf{sk}_{\mathsf{ID}}$, as in the Type-I adversary in IBEET. Obviously, the Emura-Takayasu construction does not provide the security because revealing $\mathsf{tok}_{\mathsf{ID}}$ immediately breaks the security of the underlying SKE scheme.

Our Contribution. In this paper, we demonstrate a sufficient condition that IBEETIA implies IBE. We define one-wayness against chosen-plaintext/ciphertext attacks for the token generator (OW-TG-CPA/CCA), where an adversary has MTK, and for token holders (OW-TH-CPA/CCA), where an adversary is allowed to issue token queries for any identity. Our main results are explained as follows.

- We show that OW-TG-CPA secure IBEETIA implies OW-CPA secure IBE. Moreover, we show that the Lee et al. [25] and Duong et al. [15] schemes are OW-TG-CPA secure. Due to the implication result, we can conclude that their IBE-based constructions are reasonable. Here, we assume the following additional conditions hold, where $\mathsf{sk}_{\mathsf{ID}}$ is related to MSK and is independent to MTK, and $\mathsf{tok}_{\mathsf{ID}}$ is related to MTK and is independent to MSK.
  - This structure was employed in the previous constructions [15,18,25,31]. For example, in the Lee et al. scheme [25], $\mathsf{tok}_{\mathsf{ID}}$ is a pseudo-random permutation key and a message authenticated code (MAC) key, and $\mathsf{sk}_{\mathsf{ID}}$ is a secret key of the Boneh-Franklin IBE scheme [6] with the form $H(\mathsf{ID})^{\alpha}$. In the Duong et al. scheme [15], $\mathsf{tok}_{\mathsf{ID}}$ is a trapdoor for a matrix A' of the ABB IBE scheme [1] and $\mathsf{sk}_{\mathsf{ID}}$ is a secret key of the ABB IBE scheme generated by a trapdoor for a matrix A. Here, two matrices A' and A are independently chosen.

[3] As a remark, some techniques can be employed to bypass the impossible result, e.g., [9,14,30].

  - This structure and the OW-TG-CPA security allow us to construct IBE from $\mathsf{IBEETIA}$. Intuitively, $(\mathsf{MPK}, \mathsf{MTK})$ is set as a master public key of IBE. Revealing MTK allows anyone to generate a ciphertext by computing $\mathsf{tok}_{\mathsf{ID}}$ and supports exponentially many identities.

- On the other hand, we propose a generic construction of OW-TH-CCA secure $\mathsf{IBEETIA}$ from CCA-secure public key encryption (PKE), pseudo-random permutations, and a hash function.. Because the encryption algorithm takes a token $\mathsf{tok}_{\mathsf{ID}}$ as input and the number of tokens are bounded by a polynomial of the security parameter, there is room for constructing an OW-TG-CCA secure $\mathsf{IBEETIA}$ scheme from cryptographic primitives which are weaker than IBE.
  - Unlike to the previous constructions, $\mathsf{sk}_{\mathsf{ID}}$ and $\mathsf{tok}_{\mathsf{ID}}$ are related. Let $(\mathsf{PKE.pk}, \mathsf{PKE.dk})$ be a pair of public and decryption key of a PKE scheme. $\mathsf{tok}_{\mathsf{ID}}$ contains a pseudo-random permutation key k and $\mathsf{PKE.pk}$, and $\mathsf{sk}_{\mathsf{ID}} = \mathsf{PKE.dk}$. Since revealing $\mathsf{PKE.pk}$ does not affect the security, this structure provides the OW-TH-CPA security.[4] Moreover, CCA security of the underlying PKE scheme provides OW-TH-CCA security.

Briefly, OW-TG-CPA guarantees that a plaintext is not recovered from a ciphertext even against the token generator that has MTK and issues tokens to users similar to the key generation center of IBE. OW-TG-CPA is a stronger notion because it considers a kind of the key escrow problem of IBE [11,16,17]. In other words, $\mathsf{IBEETIA}$ implies IBE when such a stronger security notion is considered. OW-TH-CPA guarantees that a plaintext is not recovered from a ciphertext even against token holders who have tokens, as in the Type-I adversary in IBEET, and seems sufficient in practice. Our results suggest a design principle to efficiently construct $\mathsf{IBEETIA}$ without employing IBE-related complexity assumptions.

Remark. The term "Insider attackers" refers to a security aspect, and does not refer the functionality of $\mathsf{IBEETIA}$, i.e., a token is defined for each identity and is used for encryption. In this perspective, $\mathsf{IBEETIA}$ could be renamed, e.g., token-controlled identity-based encryption with equality test ($\mathsf{TCIBEET}$) because token-controlled encryption [4,10,19] supports the functionality where plaintexts are encrypted by a public key together with a secret token, and a ciphertext is decrypted by the corresponding private key after the token is released. Nevertheless, we employ the naming $\mathsf{IBEETIA}$ in this paper because changing the name of a cryptographic primitive would obscure its relevance to past researches, and could lead readers to mistakenly decide that a different cryptographic primitive is analyzed in this paper.

---

[4] We inspired the construction of fully anonymous group signatures with verifier-local revocation [21] where $\mathsf{PKE.pk}$ is set as a part of a signing key and $\mathsf{PKE.dk}$ is set as a revocation token. Then, revealing a signing key does not affect the anonymity.

2 Preliminaries

SKE. An symmetric key encryption scheme SKE consists of the three algorithms $(\mathsf{SKE.KeyGen}, \mathsf{SKE.Enc}, \mathsf{SKE.Dec})$. The key generation algorithm $\mathsf{SKE.KeyGen}$ takes a security parameter $\lambda \in \mathbb{N}$, and outputs a secret key $\mathsf{SKE.sk}$. The encryption algorithm $\mathsf{SKE.Enc}$ takes $\mathsf{SKE.sk}$ and a plaintext M as input, and outputs a ciphertext $\mathsf{ct_{SKE}}$. The decryption algorithm $\mathsf{SKE.Dec}$ takes $\mathsf{SKE.sk}$ and $\mathsf{ct_{SKE}}$ as input, and output M or $\perp$.

PKE. An public key encryption scheme PKE consists of the three algorithms $(\mathsf{PKE.KeyGen}, \mathsf{PKE.Enc}, \mathsf{PKE.Dec})$. The key generation algorithm $\mathsf{PKE.KeyGen}$ takes a security parameter $\lambda \in \mathbb{N}$, and outputs a public key $\mathsf{PKE.pk}$ and a decryption key $\mathsf{PKE.dk}$. The encryption algorithm $\mathsf{PKE.Enc}$ takes $\mathsf{PKE.pk}$ and a plaintext M as input, and outputs a ciphertext $\mathsf{ct_{PKE}}$. The decryption algorithm $\mathsf{PKE.Dec}$ takes $\mathsf{PKE.pk}$, $\mathsf{PKE.dk}$, and $\mathsf{ct_{PKE}}$ as input, and output M or $\perp$. For all $\lambda \in \mathbb{N}$, $(\mathsf{PKE.pk}, \mathsf{PKE.dk}) \leftarrow \mathsf{PKE.KeyGen}(1^\lambda)$ and $M \in \mathcal{M}$, it is required that $\Pr[M \leftarrow \mathsf{PKE.Dec}(\mathsf{PKE.pk}, \mathsf{PKE.dk}, \mathsf{PKE.Enc}(\mathsf{PKE.pk}, M))] = 1 - \mathsf{negl}(\lambda)$ holds.

Next, we define the indistinguishability against chosen-ciphertext attacks (IND-CCA) as follows. Let $\mathcal{A}$ be a PPT adversary and $\mathcal{C}$ be the challenger. $\mathcal{C}$ runs $(\mathsf{PKE.pk}, \mathsf{PKE.dk}) \leftarrow \mathsf{PKE.KeyGen}(1^\lambda)$ and sends $\mathsf{PKE.pk}$ to $\mathcal{A}$. $\mathcal{A}$ is allowed to issue decryption queries. $\mathcal{A}$ sends $\mathsf{ct_{PKE}}$ to $\mathcal{C}$. $\mathcal{C}$ returns the result of $\mathsf{PKE.Dec}(\mathsf{PKE.pk}, \mathsf{PKE.dk}, \mathsf{ct_{PKE}})$. In the challenge phase, $\mathcal{A}$ declares two equality-length challenge plaintexts (M_0^*, M_1^*). $\mathcal{C}$ flips a coin $b \in \{0,1\}$, runs $\mathsf{ct^*_{PKE}} \leftarrow \mathsf{PKE.Enc}(\mathsf{PKE.pk}, M_b^*)$, and sends $\mathsf{ct^*_{PKE}}$ to $\mathcal{A}$. $\mathcal{A}$ is allowed to issue decryption queries. $\mathcal{A}$ sends $\mathsf{ct_{PKE}} \neq \mathsf{ct^*_{PKE}}$ to $\mathcal{C}$. $\mathcal{C}$ returns the result of $\mathsf{PKE.Dec}(\mathsf{PKE.pk}, \mathsf{PKE.dk}, \mathsf{ct_{PKE}})$. Finally, $\mathcal{A}$ outputs $b' \in \{0,1\}$. The advantage of $\mathcal{A}$ is defined as $\mathsf{Adv}^{\mathsf{IND\text{-}CCA}}_{\mathsf{PKE},\mathcal{A}}(1^\lambda) := |\Pr[b = b'] - 1/2|$. We say that PKE is IND-CCA secure if $\mathsf{Adv}^{\mathsf{IND\text{-}CCA}}_{\mathsf{PKE},\mathcal{A}}(1^\lambda)$ is negligible.

IBE. An identity-based encryption scheme IBE consists of the four algorithms $(\mathsf{IBE.Setup}, \mathsf{IBE.Extract}, \mathsf{IBE.Enc}, \mathsf{IBE.Dec})$. The setup algorithm $\mathsf{IBE.Setup}$ takes a security parameter $\lambda \in \mathbb{N}$, and outputs a master public key MPK and a master secret key MSK. The key extraction algorithm $\mathsf{IBE.Extract}$ takes MPK, MSK, and an identity $\mathsf{ID} \in \mathcal{ID}$ as input, and outputs a secret key $\mathsf{sk_{ID}}$. Here, $\mathcal{ID}$ is the identity space which is implicitly included in MPK. The encryption algorithm $\mathsf{IBE.Enc}$ takes MPK, ID, and a plaintext $M \in \mathcal{M}$ as input, and outputs a ciphertext $\mathsf{ct_{IBE}}$. Here, $\mathcal{M}$ is the message space which is implicitly included in MPK. The decryption algorithm $\mathsf{IBE.Dec}$ takes MPK, $\mathsf{ct_{IBE}}$, and $\mathsf{sk_{ID}}$, and outputs M or $\perp$. For all $\lambda \in \mathbb{N}$, $(\mathsf{MPK}, \mathsf{MSK}) \leftarrow \mathsf{IBE.Setup}(1^\lambda)$, $\mathsf{ID} \in \mathcal{ID}$, and $M \in \mathcal{M}$, it is required that $\Pr[M \leftarrow \mathsf{IBE.Dec}(\mathsf{MPK}, \mathsf{IBE.Enc}(\mathsf{MPK}, \mathsf{ID}, M), \mathsf{IBE.Extract}(\mathsf{MPK}, \mathsf{MSK}, \mathsf{ID}))] = 1 - \mathsf{negl}(\lambda)$ holds.

Next, we define the one-wayness against chosen-plaintext attacks (OW-CPA) as follows. Let $\mathcal{A}$ be a PPT adversary and $\mathcal{C}$ be the challenger. $\mathcal{C}$ runs $(\mathsf{MPK}, \mathsf{MSK}) \leftarrow \mathsf{IBE.Setup}(1^\lambda)$ and sends MPK to $\mathcal{A}$. $\mathcal{A}$ is allowed to issue key extraction queries. $\mathcal{A}$ sends ID to $\mathcal{C}$. $\mathcal{C}$ runs $\mathsf{sk_{ID}} \leftarrow \mathsf{IBE.Extract}(\mathsf{MPK}, \mathsf{MSK}, \mathsf{ID})$ and sends $\mathsf{sk_{ID}}$ to $\mathcal{A}$. In the challenge phase, $\mathcal{A}$ declares the challenge identity ID^* which was not sent as a key extraction query. $\mathcal{C}$ randomly choose $M^* \leftarrow \mathcal{M}$, runs $\mathsf{ct}^*_{\mathsf{IBE}} \leftarrow \mathsf{IBE.Enc}(\mathsf{MPK}, \mathsf{ID}^*, M^*)$, and sends $\mathsf{ct}^*_{\mathsf{IBE}}$ to $\mathcal{A}$. Finally, $\mathcal{A}$ outputs $\hat{M}$. The advantage of $\mathcal{A}$ is defined as $\mathsf{Adv}^{\text{OW-CPA}}_{\mathsf{IBE},\mathcal{A}}(1^\lambda) := |\Pr[M^* = \hat{M}] - 1/|\mathcal{M}||$. We say that IBE is OW-CPA secure if $\mathsf{Adv}^{\text{OW-CPA}}_{\mathsf{IBE},\mathcal{A}}(1^\lambda)$ is negligible.

IBEETIA [15,18,25,31]. An IBEETIA scheme IBEETIA consists of the following five algorithms (IBEETIA.Setup, IBEETIA.Extract, IBEETIA.Enc, IBEETIA.Dec, IBEETIA.Test) defined below. A master token key MTK is used for generating a token $\mathsf{tok}_{\mathsf{ID}}$ for ID, and $\mathsf{tok}_{\mathsf{ID}}$ is required in the encryption algorithm. Anyone can run the test algorithm against two ciphertexts without using trapdoors. By restricting who can run the encryption algorithm, IBEETIA can provide the indistinguishability security against non-insiders who do not have $\mathsf{tok}_{\mathsf{ID}}$.

IBEETIA.Setup: The setup algorithm takes a security parameter $\lambda \in \mathbb{N}$, and outputs a master public key MPK, a master secret key MSK, and a master token key MTK.

IBEETIA.Extract: The key extraction algorithm IBE.Extract takes MPK, MSK, and MTK, and an identity $\mathsf{ID} \in \mathcal{ID}$ as input, and outputs a secret key $\mathsf{sk}_{\mathsf{ID}}$ and a token $\mathsf{tok}_{\mathsf{ID}}$. Here, $\mathcal{ID}$ is the identity space which is implicitly included in MPK.

IBEETIA.Enc: The encryption algorithm takes MPK, $\mathsf{tok}_{\mathsf{ID}}$, ID, and a plaintext $M \in \mathcal{M}$ as input, and outputs a ciphertext $\mathsf{ct}_{\mathsf{IBEETIA}}$. Here, $\mathcal{M}$ is the message space which is implicitly included in MPK.

IBEETIA.Dec: The decryption algorithm takes MPK, $\mathsf{sk}_{\mathsf{ID}}$, $\mathsf{tok}_{\mathsf{ID}}$, and $\mathsf{ct}_{\mathsf{IBEETIA}}$, and outputs M or $\perp$.

IBEETIA.Test: The test algorithm takes MPK and two ciphertexts $\mathsf{ct}_{\mathsf{IBEETIA}}$ and $\mathsf{ct}'_{\mathsf{IBEETIA}}$, and outputs 1 or 0.

The correctness of an IBEETIA scheme is defined as follows. Here, a probability space is random coins to run IBEETIA.Setup, IBEETIA.Extract, and IBEETIA.Enc. We note that the first condition is employed in our security proof.

1. For all $\lambda \in \mathbb{N}$, $\mathsf{ID} \in \mathcal{ID}$, and $M \in \mathcal{M}$, $M' = M$ holds with overwhelming probability where $(\mathsf{MPK}, \mathsf{MSK}, \mathsf{MTK}) \leftarrow \mathsf{IBEETIA.Setup}(1^\lambda)$, $(\mathsf{sk}_{\mathsf{ID}}, \mathsf{tok}_{\mathsf{ID}}) \leftarrow \mathsf{IBEETIA.Extract}(\mathsf{MPK}, \mathsf{MSK}, \mathsf{MTK}, \mathsf{ID})$, $\mathsf{ct}_{\mathsf{IBEETIA}} \leftarrow \mathsf{IBEETIA.Enc}(\mathsf{MPK}, \mathsf{tok}_{\mathsf{ID}}, \mathsf{ID}, M)$, and $M' \leftarrow \mathsf{IBEETIA.Dec}(\mathsf{MPK}, \mathsf{sk}_{\mathsf{ID}}, \mathsf{tok}_{\mathsf{ID}}, \mathsf{ct}_{\mathsf{IBEETIA}})$.
2. For all $\lambda \in \mathbb{N}$, $\mathsf{ID}, \mathsf{ID}' \in \mathcal{ID}$, and $M \in \mathcal{M}$, $\mathsf{IBEETIA.Test}(\mathsf{MPK}, \mathsf{ct}_{\mathsf{IBEETIA}}, \mathsf{ct}'_{\mathsf{IBEETIA}}) = 1$ holds with overwhelming probability, where $(\mathsf{MPK}, \mathsf{MSK}, \mathsf{MTK}) \leftarrow \mathsf{IBEETIA.Setup}(1^\lambda)$, $(\mathsf{sk}_{\mathsf{ID}}, \mathsf{tok}_{\mathsf{ID}}) \leftarrow \mathsf{IBEETIA.Extract}(\mathsf{MPK}, \mathsf{MSK}, \mathsf{MTK}, \mathsf{ID})$, $(\mathsf{sk}_{\mathsf{ID}'}, \mathsf{tok}_{\mathsf{ID}'}) \leftarrow \mathsf{IBEETIA.Extract}(\mathsf{MPK}, \mathsf{MSK}, \mathsf{MTK}, \mathsf{ID}')$, $\mathsf{ct}_{\mathsf{IBEETIA}} \leftarrow \mathsf{IBEETIA.Enc}(\mathsf{MPK}, \mathsf{tok}_{\mathsf{ID}}, \mathsf{ID}, M)$, and $\mathsf{ct}'_{\mathsf{IBEETIA}} \leftarrow \mathsf{IBEETIA.Enc}(\mathsf{MPK}, \mathsf{tok}_{\mathsf{ID}'}, \mathsf{ID}', M)$.
3. For all $\lambda \in \mathbb{N}$, $\mathsf{ID}, \mathsf{ID}' \in \mathcal{ID}$, and $M, M' \in \mathcal{M}$ such that $M \neq M'$, $\mathsf{IBEETIA.Test}(\mathsf{MPK}, \mathsf{ct}_{\mathsf{IBEETIA}}, \mathsf{ct}'_{\mathsf{IBEETIA}}) = 1$ holds with negligible probability, where $(\mathsf{MPK}, \mathsf{MSK}, \mathsf{MTK}) \leftarrow \mathsf{IBEETIA.Setup}(1^\lambda)$, $(\mathsf{sk}_{\mathsf{ID}}, \mathsf{tok}_{\mathsf{ID}}) \leftarrow \mathsf{IBEETIA.Extract}(\mathsf{MPK}, \mathsf{MSK}, \mathsf{MTK}, \mathsf{ID})$, $(\mathsf{sk}_{\mathsf{ID}'}, \mathsf{tok}_{\mathsf{ID}'}) \leftarrow \mathsf{IBEETIA.Extract}(\mathsf{MPK}, \mathsf{MSK}, \mathsf{MTK}, \mathsf{ID}')$, $\mathsf{ct}_{\mathsf{IBEETIA}} \leftarrow \mathsf{IBEETIA.Enc}(\mathsf{MPK}, \mathsf{tok}_{\mathsf{ID}}, \mathsf{ID}, M)$, and $\mathsf{ct}'_{\mathsf{IBEETIA}} \leftarrow \mathsf{IBEETIA.Enc}(\mathsf{MPK}, \mathsf{tok}_{\mathsf{ID}'}, \mathsf{ID}', M')$.

Next, we define the weak indistinguishability against chosen ciphertext attacks (wIND-CCA). Unlike to IBE, the encryption algorithm takes $\mathsf{tok}_{\mathsf{ID}}$ as input. Thus, an adversary $\mathcal{A}$ is allowed to issue an encryption query. If $\mathcal{A}$ declares the challenge identity ID^* before the setup phase, we call it selectively wIND-CCA secure. Due to the nature of IBEETIA, it is required that $\mathcal{A}$ did not query (ID, M_0^*) or (ID, M_1^*) for any ID and the challenge plaintexts (M_0^*, M_1^*) as an encryption query.

Definition 1 (wIND-CCA [18,25,31]). *Let $\mathcal{A}$ be a PPT adversary and $\mathcal{C}$ be the challenger. $\mathcal{C}$ runs* $(\mathsf{MPK}, \mathsf{MSK}, \mathsf{MTK}) \leftarrow \mathsf{IBEETIA.Setup}(1^\lambda)$ *and sends* MPK *to $\mathcal{A}$. $\mathcal{A}$ is allowed to issue queries below.*

Key Extraction: *$\mathcal{A}$ sends* ID *to $\mathcal{C}$. $\mathcal{C}$ runs* $(\mathsf{sk}_{\mathsf{ID}}, \mathsf{tok}_{\mathsf{ID}}) \leftarrow \mathsf{IBEETIA.Extract}(\mathsf{MPK}, \mathsf{MSK}, \mathsf{MTK}, \mathsf{ID})$ *and sends* $\mathsf{sk}_{\mathsf{ID}}$ *to $\mathcal{A}$. Here $\mathcal{C}$ does not send* $\mathsf{tok}_{\mathsf{ID}}$ *to $\mathcal{A}$.*

Encryption: *$\mathcal{A}$ sends* (ID, M) *to $\mathcal{C}$. $\mathcal{C}$ runs* $(\mathsf{sk}_{\mathsf{ID}}, \mathsf{tok}_{\mathsf{ID}}) \leftarrow \mathsf{IBEETIA.Extract}(\mathsf{MPK}, \mathsf{MSK}, \mathsf{MTK}, \mathsf{ID})$ *and* $\mathsf{ct}_{\mathsf{IBEETIA}} \leftarrow \mathsf{IBEETIA.Enc}(\mathsf{MPK}, \mathsf{tok}_{\mathsf{ID}}, \mathsf{ID}, M)$, *and sends* $\mathsf{ct}_{\mathsf{IBEETIA}}$ *to $\mathcal{A}$.*

Decryption: *$\mathcal{A}$ sends* $(\mathsf{ID}, \mathsf{ct}_{\mathsf{IBEETIA}})$ *to $\mathcal{C}$. $\mathcal{C}$ runs* $(\mathsf{sk}_{\mathsf{ID}}, \mathsf{tok}_{\mathsf{ID}}) \leftarrow \mathsf{IBEETIA.Extract}(\mathsf{MPK}, \mathsf{MSK}, \mathsf{MTK}, \mathsf{ID})$ *and return the result of* $\mathsf{IBEETIA.Dec}(\mathsf{MPK}, \mathsf{sk}_{\mathsf{ID}}, \mathsf{tok}_{\mathsf{ID}}, \mathsf{ct}_{\mathsf{IBEETIA}})$.

In the challenge phase, $\mathcal{A}$ declares two plaintexts (M_0^*, M_1^*) and the challenge identity ID^*. *It is required that* ID^* *was not sent as a key extraction query. Moreover, it is required that $\mathcal{A}$ did not query* (ID, M_0^*) *or* (ID, M_1^*) *for any* ID *as an encryption query. $\mathcal{C}$ randomly flips a coin $b \leftarrow \{0,1\}$, runs* $(\mathsf{sk}_{\mathsf{ID}^*}, \mathsf{tok}_{\mathsf{ID}^*}) \leftarrow \mathsf{IBEETIA.Extract}(\mathsf{MPK}, \mathsf{MSK}, \mathsf{MTK}, \mathsf{ID})$, $\mathsf{ct}^*_{\mathsf{IBEETIA}} \leftarrow \mathsf{IBEETIA.Enc}(\mathsf{MPK}, \mathsf{tok}_{\mathsf{ID}^*}, \mathsf{ID}^*, M_b^*)$, *and sends* $\mathsf{ct}^*_{\mathsf{IBEETIA}}$ *to $\mathcal{A}$. In addition to the restrictions above, $\mathcal{A}$ is not allowed to issue* $(\mathsf{ID}^*, \mathsf{ct}^*_{\mathsf{IBEETIA}})$ *as a decryption query. Finally, $\mathcal{A}$ outputs $b' \in \{0,1\}$. The advantage of $\mathcal{A}$ is defined as*

$$\mathsf{Adv}^{\mathsf{wIND\text{-}CCA}}_{\mathsf{IBEETIA},\mathcal{A}}(1^\lambda) := |\Pr[b = b'] - 1/2|$$

We say that IBEETIA *is wIND-CCA secure if* $\mathsf{Adv}^{\mathsf{wIND\text{-}CCA}}_{\mathsf{IBEETIA},\mathcal{A}}(1^\lambda)$ *is negligible.*

3 OW-CPA Security of IBEETIA

In this section, we define one-wayness against chosen-plaintext attacks for the token generator (OW-TG-CPA) which guarantees that a plaintext is not recovered from a ciphertext even against the token generator. $\mathcal{A}$ is allowed to obtain MTK in addition to MPK. Thus, no encryption oracle is defined. If $\mathcal{A}$ declares the challenge identity ID^* before the setup phase, we call it selectively secure. OW-TG-CCA security can also be defined. We remark that OW-TG-CPA is sufficient to show our implication result.

Definition 2 (OW-TG-CPA). *Let $\mathcal{A}$ be a PPT adversary and $\mathcal{C}$ be the challenger. $\mathcal{C}$ runs* $(\mathsf{MPK}, \mathsf{MSK}, \mathsf{MTK}) \leftarrow \mathsf{IBEETIA.Setup}(1^\lambda)$ *and sends* $(\mathsf{MPK}, \mathsf{MTK})$ *to $\mathcal{A}$. $\mathcal{A}$ is allowed to issue queries below.*

Key Extraction: *$\mathcal{A}$ sends* ID *to* $\mathcal{C}$. $\mathcal{C}$ *runs* $(\mathsf{sk_{ID}}, \mathsf{tok_{ID}}) \leftarrow \mathsf{IBEETIA.Extract}(\mathsf{MPK}, \mathsf{MSK}, \mathsf{MTK}, \mathsf{ID})$ *and sends* $\mathsf{sk_{ID}}$ *to* $\mathcal{A}$. *Here* $\mathcal{C}$ *does not send* $\mathsf{tok_{ID}}$ *to* $\mathcal{A}$.

In the challenge phase, $\mathcal{A}$ *declares the challenge identity* ID^* *which was not sent as a key extraction query.* $\mathcal{C}$ *randomly chooses* $M^* \leftarrow \mathcal{M}$, *runs* $(\mathsf{sk_{ID^*}}, \mathsf{tok_{ID^*}}) \leftarrow \mathsf{IBEETIA.Extract}(\mathsf{MPK}, \mathsf{MSK}, \mathsf{MTK}, \mathsf{ID}^*)$, $\mathsf{ct}^*_{\mathsf{IBEETIA}} \leftarrow \mathsf{IBEETIA.Enc}(\mathsf{MPK}, \mathsf{tok_{ID^*}}, \mathsf{ID}^*, M^*)$, *and sends* $\mathsf{ct}^*_{\mathsf{IBEETIA}}$ *to* $\mathcal{A}$. *Finally,* $\mathcal{A}$ *outputs* $\hat{M}$. *The advantage of* $\mathcal{A}$ *is defined as*

$$\mathsf{Adv}^{\mathsf{OW\text{-}TG\text{-}CPA}}_{\mathsf{IBEETIA},\mathcal{A}}(1^\lambda) := |\Pr[M^* = \hat{M}] - 1/|\mathcal{M}||$$

We say that $\mathsf{IBEETIA}$ *is OW-TG-CPA secure if* $\mathsf{Adv}^{\mathsf{OW\text{-}TG\text{-}CPA}}_{\mathsf{IBEETIA},\mathcal{A}}(1^\lambda)$ *is negligible.*

Next, we define one-wayness against chosen-ciphertext attacks for token holders (OW-TH-CCA), where an adversary is allowed to issue token queries for any identity which guarantees that a plaintext is not recovered from a ciphertext even against token holders who have tokens. OW-TH-CPA is also defined when no decryption oracle is considered. The token extraction oracle can be simulated when MTK is given. Thus, the OW-TG-CPA security implies the OW-TH-CPA security.

Definition 3 (OW-TH-CCA). *Let* $\mathcal{A}$ *be a PPT adversary and* $\mathcal{C}$ *be the challenger.* $\mathcal{C}$ *runs* $(\mathsf{MPK}, \mathsf{MSK}, \mathsf{MTK}) \leftarrow \mathsf{IBEETIA.Setup}(1^\lambda)$ *and sends* MPK *to* $\mathcal{A}$. $\mathcal{A}$ *is allowed to issue queries below.*

Key Extraction: *$\mathcal{A}$ sends* ID *to* $\mathcal{C}$. $\mathcal{C}$ *runs* $(\mathsf{sk_{ID}}, \mathsf{tok_{ID}}) \leftarrow \mathsf{IBEETIA.Extract}(\mathsf{MPK}, \mathsf{MSK}, \mathsf{MTK}, \mathsf{ID})$ *and sends* $\mathsf{sk_{ID}}$ *to* $\mathcal{A}$. *Here* $\mathcal{C}$ *does not send* $\mathsf{tok_{ID}}$ *to* $\mathcal{A}$.

Token Extraction: *$\mathcal{A}$ sends* ID *to* $\mathcal{C}$. $\mathcal{C}$ *runs* $(\mathsf{sk_{ID}}, \mathsf{tok_{ID}}) \leftarrow \mathsf{IBEETIA.Extract}(\mathsf{MPK}, \mathsf{MSK}, \mathsf{MTK}, \mathsf{ID})$ *and sends* $\mathsf{tok_{ID}}$ *to* $\mathcal{A}$. *Here* $\mathcal{C}$ *does not send* $\mathsf{sk_{ID}}$ *to* $\mathcal{A}$.

Decryption: *$\mathcal{A}$ sends* $(\mathsf{ID}, \mathsf{ct_{IBEETIA}})$ *to* $\mathcal{C}$. $\mathcal{C}$ *runs* $(\mathsf{sk_{ID}}, \mathsf{tok_{ID}}) \leftarrow \mathsf{IBEETIA.Extract}(\mathsf{MPK}, \mathsf{MSK}, \mathsf{MTK}, \mathsf{ID})$ *and return the result of* $\mathsf{IBEETIA.Dec}(\mathsf{MPK}, \mathsf{sk_{ID}}, \mathsf{tok_{ID}}, \mathsf{ct_{IBEETIA}})$.

In the challenge phase, $\mathcal{A}$ *declares the challenge identity* ID^* *which was not sent as a key extraction query. Note that* $\mathcal{A}$ *is allowed to obtain* $\mathsf{tok_{ID^*}}$. $\mathcal{C}$ *randomly chooses* $M^* \leftarrow \mathcal{M}$, *runs* $(\mathsf{sk_{ID^*}}, \mathsf{tok_{ID^*}}) \leftarrow \mathsf{IBEETIA.Extract}(\mathsf{MPK}, \mathsf{MSK}, \mathsf{MTK}, \mathsf{ID}^*)$, $\mathsf{ct}^*_{\mathsf{IBEETIA}} \leftarrow \mathsf{IBEETIA.Enc}(\mathsf{MPK}, \mathsf{tok_{ID^*}}, \mathsf{ID}^*, M^*)$, *and sends* $\mathsf{ct}^*_{\mathsf{IBEETIA}}$ *to* $\mathcal{A}$. $\mathcal{A}$ *is not allowed to issue* $(\mathsf{ID}^*, \mathsf{ct}^*_{\mathsf{IBEETIA}})$ *as a decryption query. Finally,* $\mathcal{A}$ *outputs* $\hat{M}$. *The advantage of* $\mathcal{A}$ *is defined as*

$$\mathsf{Adv}^{\mathsf{OW\text{-}TH\text{-}CCA}}_{\mathsf{IBEETIA},\mathcal{A}}(1^\lambda) := |\Pr[M^* = \hat{M}] - 1/|\mathcal{M}||$$

We say that $\mathsf{IBEETIA}$ *is OW-TH-CCA secure if* $\mathsf{Adv}^{\mathsf{OW\text{-}TH\text{-}CCA}}_{\mathsf{IBEETIA},\mathcal{A}}(1^\lambda)$ *is negligible.*

4 Our IBE Construction from IBEETIA

In this section, we construct an OW-CPA secure IBE scheme from an OW-TG-CPA secure $\mathsf{IBEETIA}$ scheme. In addition to the OW-TG-CPA security, we assume the following additional conditions, where $\mathsf{sk_{ID}}$ is related to MSK and is independent to MTK, and $\mathsf{tok_{ID}}$ is related to MTK and is independent to MSK. This is a sufficient condition that $\mathsf{IBEETIA}$ implies IBE. Concretely, the $\mathsf{IBEETIA.Extract}$ algorithm is divided to two sub-algorithms as follows.

$\mathsf{IBEETIA.ExtractSK}$: The secret key extraction algorithm takes as MPK, MSK, and $\mathsf{ID} \in \mathcal{ID}$ as input, and outputs a secret key $\mathsf{sk_{ID}}$.
$\mathsf{IBEETIA.ExtractTK}$: The token extraction algorithm takes as MPK, MTK, and $\mathsf{ID} \in \mathcal{ID}$ as input, and outputs a token $\mathsf{tok_{ID}}$.

Then, the $\mathsf{IBEETIA.Extract}$ algorithm is defined as follows.

$\mathsf{IBEETIA.Extract(MPK, MSK, MTK, ID)}$: Run $\mathsf{sk_{ID}} \leftarrow \mathsf{IBEETIA.ExtractSK(MPK, MSK, ID)}$ and $\mathsf{tok_{ID}} \leftarrow \mathsf{IBEETIA.ExtractTK(MPK, MTK, ID)}$, and output $(\mathsf{sk_{ID}}, \mathsf{tok_{ID}})$.

We give our IBE construction as follows. Here, the test functionality of $\mathsf{IBEETIA}$ is not used. Importantly, IBE must support exponentially many identities, i.e., there are exponentially many public keys.[5] If IBE supports only polynomially many identities, IBE can be constructed from PKE by assigning a public key of PKE to each identity. Due to the OW-TG-CPA security, an adversary is allowed to obtain MTK. This allows us to contain MTK to the master public key of IBE. Then, unlike to $\mathsf{IBEETIA}$, anyone can generate a ciphertext by internally computing $\mathsf{tok_{ID}}$ using MTK and by running the $\mathsf{IBEETIA.Enc}$ algorithm. Thus, as in IBE, exponentially many identities are supported. This publicly-executable encryption algorithm is mandatory to construct IBE.

$\mathsf{IBE.Setup}(1^\lambda)$: Run $(\mathsf{MPK'}, \mathsf{MSK}, \mathsf{MTK}) \leftarrow \mathsf{IBEETIA.Setup}(1^\lambda)$ and output $\mathsf{MPK} = (\mathsf{MPK'}, \mathsf{MTK})$ and MSK.
$\mathsf{IBE.Extract(MPK, MSK, ID)}$: Parse $\mathsf{MPK} = (\mathsf{MPK'}, \mathsf{MTK})$. Run $\mathsf{sk_{ID}} \leftarrow \mathsf{IBEETIA.ExtractSK(MPK', MSK, ID)}$ and output $\mathsf{sk_{ID}}$.
$\mathsf{IBE.Enc}(\mathsf{MPK}, \mathsf{ID}, M)$: Parse $\mathsf{MPK} = (\mathsf{MPK'}, \mathsf{MTK})$. Run $\mathsf{tok_{ID}} \leftarrow \mathsf{IBEETIA.ExtractTK(MPK, MTK, ID)}$ and $\mathsf{ct_{IBEETIA}} \leftarrow \mathsf{IBEETIA.Enc}(\mathsf{MPK'}, \mathsf{tok_{ID}}, \mathsf{ID}, M)$. Output $\mathsf{ct_{IBE}} = \mathsf{ct_{IBEETIA}}$.
$\mathsf{IBE.Dec}(\mathsf{MPK}, \mathsf{ct_{IBE}}, \mathsf{sk_{ID}})$: Parse $\mathsf{MPK} = (\mathsf{MPK'}, \mathsf{MTK})$ and $\mathsf{ct_{IBE}} = \mathsf{ct_{IBEETIA}}$. Run $\mathsf{tok_{ID}} \leftarrow \mathsf{IBEETIA.ExtractTK(MPK, MTK, ID)}$. Output the result of $\mathsf{IBEETIA.Dec}(\mathsf{MPK'}, \mathsf{sk_{ID}}, \mathsf{tok_{ID}}, \mathsf{ct_{IBEETIA}})$.

[5] Boneh et al. [7] wrote that "Our proof brings to light the essential property of IBE systems: an IBE system creates an exponential number of public keys (identities) and compresses all of them into a short string called the public parameters. This ability to represent exponentially many public keys using a short string is not possible with a generic PKE or TDP"

Correctness of the IBE scheme is directly followed by the first condition of the correctness of the underlying $\mathsf{IBEETIA}$ scheme.

Theorem 1. *The proposed* IBE *scheme is OW-CPA secure if the underlying* $\mathsf{IBEETIA}$ *is OW-TG-CPA secure.*

Proof. Let $\mathcal{A}$ be the adversary of the OW-CPA security of IBE and $\mathcal{C}$ be the challenger of the OW-TG-CPA security of $\mathsf{IBEETIA}$. We construct an algorithm $\mathcal{B}$ that breaks the OW-TG-CPA security using $\mathcal{A}$ as follows.

First, $\mathcal{C}$ runs $(\mathsf{MPK}', \mathsf{MSK}, \mathsf{MTK}) \leftarrow \mathsf{IBEETIA.Setup}(1^\lambda)$ and sends $(\mathsf{MPK}', \mathsf{MTK})$ to $\mathcal{B}$. $\mathcal{B}$ sets $\mathsf{MPK} = (\mathsf{MPK}', \mathsf{MTK})$ and sends MPK to $\mathcal{A}$. When $\mathcal{A}$ issues a key extraction query ID, $\mathcal{B}$ forwards ID to $\mathcal{C}$. $\mathcal{C}$ runs $(\mathsf{sk}_{\mathsf{ID}}, \mathsf{tok}_{\mathsf{ID}}) \leftarrow \mathsf{IBEETIA.Extract}(\mathsf{MPK}, \mathsf{MSK}, \mathsf{MTK}, \mathsf{ID})$ and sends $\mathsf{sk}_{\mathsf{ID}}$ to $\mathcal{B}$. In the challenge phase, $\mathcal{A}$ declares ID^*. $\mathcal{B}$ sends ID^* to $\mathcal{C}$. $\mathcal{C}$ randomly chooses $M^* \leftarrow \mathcal{M}$, runs $(\mathsf{sk}_{\mathsf{ID}^*}, \mathsf{tok}_{\mathsf{ID}^*}) \leftarrow \mathsf{IBEETIA.Extract}(\mathsf{MPK}, \mathsf{MSK}, \mathsf{MTK}, \mathsf{ID}^*)$, $\mathsf{ct}^*_{\mathsf{IBEETIA}} \leftarrow \mathsf{IBEETIA.Enc}(\mathsf{MPK}, \mathsf{tok}_{\mathsf{ID}^*}, \mathsf{ID}^*, M^*)$, and sends $\mathsf{ct}^*_{\mathsf{IBEETIA}}$ to $\mathcal{B}$. $\mathcal{B}$ sets $\mathsf{ct}^*_{\mathsf{IBE}} = \mathsf{ct}^*_{\mathsf{IBEETIA}}$ and sends $\mathsf{ct}^*_{\mathsf{IBE}}$ to $\mathcal{A}$. Finally, $\mathcal{A}$ outputs $\hat{M}$ and $\mathcal{B}$ outputs the same $\hat{M}$. Then, $\mathcal{B}$ breaks the OW-TG-CPA security with the advantage at least $\mathsf{Adv}^{\mathsf{OW\text{-}CPA}}_{\mathsf{IBE},\mathcal{A}}(1^\lambda)$. □

5 Proposed Generic Construction of IBEETIA

In this section, we propose a generic construction of $\mathsf{IBEETIA}$ that provides the OW-TH-CCA security, in addition to the wIND-CCA security and correctness. The proposed construction basically follows the Emura-Takayasu construction, except that a PKE scheme is employed. Unlike to the OW-TG-CCA security, an adversary is not allowed to obtain MTK, and is allowed to issue token queries. Because the number of tokens is bounded by a polynomial of the security parameter and the encryption algorithm takes a token $\mathsf{tok}_{\mathsf{ID}}$ as input, the number of identities are also bounded by a polynomial of the security parameter. Thus, there is room for constructing an OW-TG-CCA secure $\mathsf{IBEETIA}$ scheme from cryptographic primitives which are weaker than IBE.

5.1 Emura-Takayasu IBEETIA Construction

Before giving the proposed construction, we revisit the Emura-Takayasu construction [18]. Let π be a pseudo-random permutation with a key space $\mathcal{K}$ and $\mathsf{SKE} = (\mathsf{SKE.KeyGen}, \mathsf{SKE.Enc}, \mathsf{SKE.Dec})$ be a CCA-secure SKE scheme. We denote $\mathcal{R}$ as a randomness space for key generation. That is, we denote $\mathsf{SKE.sk} \leftarrow \mathsf{SKE.KeyGen}(1^\lambda; r)$ for $r \xleftarrow{\$} \mathcal{R}$.

$\mathsf{IBEETIA.Setup}(1^\lambda)$: Choose $k \xleftarrow{\$} \mathcal{K}$. Output $\mathsf{MPK} = \perp$, $\mathsf{MSK} = \perp$, and $\mathsf{MTK} = k$.

$\mathsf{IBEETIA.Extract}(\mathsf{MPK}, \mathsf{MSK}, \mathsf{MTK}, \mathsf{ID})$: Parse $\mathsf{MTK} = k$. Choose $r_{\mathsf{ID}} \xleftarrow{\$} \mathcal{R}$ and $\mathsf{SKE.sk}_{\mathsf{ID}} \leftarrow \mathsf{SKE.KeyGen}(1^\lambda; r_{\mathsf{ID}})$ and output $\mathsf{sk}_{\mathsf{ID}} = \perp$ and $\mathsf{tok}_{\mathsf{ID}} = (\mathsf{SKE.sk}_{\mathsf{ID}}, k)$.

$\mathsf{IBEETIA.Enc}(\mathsf{MPK}, \mathsf{tok}_{\mathsf{ID}}, \mathsf{ID}, M)$: Parse $\mathsf{tok}_{\mathsf{ID}} = (\mathsf{SKE.sk}_{\mathsf{ID}}, k)$. Run $x_M \leftarrow \pi(k, M)$ and $\mathsf{ct}_{\mathsf{SKE}} \leftarrow \mathsf{SKE.Enc}(\mathsf{SKE.sk}_{\mathsf{ID}}, M)$ and outputs $\mathsf{ct}_{\mathsf{IBEETIA}} = (x_M, \mathsf{ct}_{\mathsf{SKE}})$.

$\mathsf{IBEETIA.Dec}(\mathsf{MPK}, \mathsf{sk}_{\mathsf{ID}}, \mathsf{tok}_{\mathsf{ID}}, \mathsf{ct}_{\mathsf{IBEETIA}})$: Parse $\mathsf{tok}_{\mathsf{ID}} = (\mathsf{SKE.sk}_{\mathsf{ID}}, k)$ and $\mathsf{ct}_{\mathsf{IBEETIA}} = (x_M, \mathsf{ct}_{\mathsf{SKE}})$. Run $M' \leftarrow \mathsf{SKE.Dec}(\mathsf{SKE.sk}_{\mathsf{ID}}, \mathsf{ct}_{\mathsf{SKE}})$ and $x_{M'} \leftarrow \pi(k, M')$. If $x_M = x_{M'}$, then output M', and $\perp$ otherwise.

$\mathsf{IBEETIA.Test}(\mathsf{MPK}, \mathsf{ct}_{\mathsf{IBEETIA}}, \mathsf{ct}'_{\mathsf{IBEETIA}})$: Parse $\mathsf{ct}_{\mathsf{IBEETIA}} = (x_M, \mathsf{ct}_{\mathsf{SKE}})$ and $\mathsf{ct}'_{\mathsf{IBEETIA}} = (x_{M'}, \mathsf{ct}'_{\mathsf{SKE}})$. If $x_M = x_{M'}$, then output 1, and 0 otherwise.

Since π is permutation, if $M = M'$, then $x_M = x_{M'}$ holds and if $M \neq M'$, then $x_M \neq x_{M'}$ holds. Moreover, the IBEETIA construction is wIND-CCA secure if π is a pseudo-random permutation and SKE is CCA secure. Intuitively, if $\mathsf{MTK} = k$ is hidden, then x_M is indistinguishable from a value generated by a random function due to the pseudo-randomness. We remark that k is not revealed in the definition of the wIND-CCA security. Moreover, no information of M is revealed from $\mathsf{ct}_{\mathsf{SKE}}$ due to the IND-CCA security of the SKE scheme. We briefly explain how to prove that the construction is wIND-CCA secure as follows. In the security proof, a simulator $\mathcal{B}$ obtains the challenge ciphertext $\mathsf{ct}^*_{\mathsf{SKE}}$ from the challenger of the SKE scheme. Then $\mathcal{B}$ randomly chooses $x_{M_b^*}$ from $\mathcal{M}$ and sets $(x_{M_b^*}, \mathsf{ct}^*_{\mathsf{SKE}})$ as the challenge ciphertext of IBEETIA. One may wonder how to respond a decryption query $(x_M, \mathsf{ct}^*_{\mathsf{SKE}}) \wedge x_M \neq x_{M_b^*}$ since $\mathcal{B}$ is not allowed to issue a decryption query $\mathsf{ct}^*_{\mathsf{SKE}}$ to $\mathcal{C}$. Because π is permutation, there is only one valid $x_{M_b^*}$. Thus, $\mathcal{B}$ simply returns $\perp$ if $(x_M, \mathsf{ct}^*_{\mathsf{SKE}}) \wedge x_M \neq x_{M_b^*}$.

Obviously, the Emura-Takayasu construction does not provide the OW-TH-CPA security because $\mathsf{tok}_{\mathsf{ID}} = (\mathsf{SKE.sk}, k)$ and revealing $\mathsf{tok}_{\mathsf{ID}}$ immediately breaks the security of the underlying SKE scheme. In other words, symmetric key primitives are sufficient if no security against token holders is required.

5.2 Proposed Construction

We tweak the Emura-Takayasu construction to provide the OW-TH-CCA security. Basically, we employ a PKE scheme, and set $\mathsf{sk}_{\mathsf{ID}} = \mathsf{PKE.dk}$ and $\mathsf{tok}_{\mathsf{ID}} = (\mathsf{PKE.pk}, k)$. Then, revealing PKE.pk does not affect the security of PKE. We need to care the fact that no security of π can be assumed if k is revealed. Especially, M may be recovered from $\pi(k, M)$ without contradicting the security of π. For example, when a block cipher is employed as π, revealing the key k immediately recovers M. As the first attempt, we employ a hash function H and set $x_M \leftarrow \pi(k, H(M))$. Due to the collision resistance of H, if $M \neq M'$, then $H(M) \neq H(M')$. Thus, employing H does not affect the correctness. Moreover, due to the one-wayness of H, M is not recovered from $H(M)$. Informally, for $M^* \xleftarrow{\$} \mathcal{M}$, the challenge ciphertext is $(x^*, \mathsf{ct}^*_{\mathsf{PKE}})$ where $x^* = \pi(k, H(M^*))$ and $\mathsf{ct}^*_{\mathsf{PKE}} \leftarrow \mathsf{PKE.Enc}(\mathsf{PKE.pk}, M^*)$. First, we replace $\mathsf{ct}^*_{\mathsf{PKE}}$ such that $\mathsf{ct}^*_{\mathsf{PKE}} \leftarrow \mathsf{PKE.Enc}(\mathsf{PKE.pk}, 0^{|M^*|})$ due to the IND-CCA security of PKE, and second we break the one-wayness of H using the output of the OW-TH-CCA adversary. Here, we need to consider that usually PKE does not hide the plaintext size. Thus, if we set $x^* = \pi(k, H(M))$ where $H(M)$ is given from the challenger of the one-wayness of H, the simulator has no way to generate $\mathsf{ct}^*_{\mathsf{PKE}} \leftarrow \mathsf{PKE.Enc}(\mathsf{PKE.pk}, 0^{|M|})$. Thus, we assume that the size of all plaintexts are the same, e.g., by adding a padding to each plaintext. We further need to consider the case that an adversary sends $(x_M, \mathsf{ct}^*_{\mathsf{PKE}})$ where $x^* \neq x_M$. Since H is deterministic, $H(M)$ is uniquely determined when M is fixed. Thus, we can reject a decryption query $(x_M, \mathsf{ct}^*_{\mathsf{PKE}})$ if $x^* \neq x_M$.

Let $H : \mathcal{M} \to \{0,1\}^\lambda$ be a hash function and $\mathsf{PKE} = (\mathsf{PKE.KeyGen}, \mathsf{PKE.Enc}, \mathsf{PKE.Dec})$ be a CCA-secure PKE scheme. We denote $\mathcal{R}$ as a randomness space for key generation. That is, we denote $(\mathsf{PKE.pk}, \mathsf{PKE.dk}) \leftarrow \mathsf{PKE.KeyGen}(1^\lambda; r)$ for $r \xleftarrow{\$} \mathcal{R}$.

$\mathsf{IBEETIA.Setup}(1^\lambda)$: Specify a hash function H and choose $k \xleftarrow{\$} \mathcal{K}$. Output $\mathsf{MPK} = H$, $\mathsf{MSK} = \perp$, and $\mathsf{MTK} = k$.

$\mathsf{IBEETIA.Extract}(\mathsf{MPK}, \mathsf{MSK}, \mathsf{MTK}, \mathsf{ID})$: Parse $\mathsf{MTK} = k$. Choose $r_{\mathsf{ID}} \xleftarrow{\$} \mathcal{R}$ and $(\mathsf{PKE.pk_{ID}}, \mathsf{PKE.dk_{ID}}) \leftarrow \mathsf{PKE.KeyGen}(1^\lambda; r_{\mathsf{ID}})$ and output $\mathsf{sk_{ID}} = \mathsf{PKE.dk_{ID}}$ and $\mathsf{tok_{ID}} = (\mathsf{PKE.pk_{ID}}, k)$.

$\mathsf{IBEETIA.Enc}(\mathsf{MPK}, \mathsf{tok_{ID}}, \mathsf{ID}, M)$: Parse $\mathsf{MPK} = H$ and $\mathsf{tok_{ID}} = (\mathsf{PKE.pk_{ID}}, k)$. Run $x_M \leftarrow \pi(k, H(M))$ and $\mathsf{ct_{PKE}} \leftarrow \mathsf{PKE.Enc}(\mathsf{PKE.pk_{ID}}, M)$ and outputs $\mathsf{ct_{IBEETIA}} = (x_M, \mathsf{ct_{PKE}})$.

$\mathsf{IBEETIA.Dec}(\mathsf{MPK}, \mathsf{sk_{ID}}, \mathsf{tok_{ID}}, \mathsf{ct_{IBEETIA}})$: Parse $\mathsf{MPK} = H$, $\mathsf{sk_{ID}} = \mathsf{PKE.dk_{ID}}$, $\mathsf{tok_{ID}} = (\mathsf{PKE.pk_{ID}}, k)$, and $\mathsf{ct_{IBEETIA}} = (x_M, \mathsf{ct_{PKE}})$. Run $M' \leftarrow \mathsf{PKE.Dec}(\mathsf{PKE.dk_{ID}}, \mathsf{ct_{PKE}})$ and $x_{M'} \leftarrow \pi(k, H(M'))$. If $x_M = x_{M'}$, then output M', and $\perp$ otherwise.

$\mathsf{IBEETIA.Test}(\mathsf{MPK}, \mathsf{ct_{IBEETIA}}, \mathsf{ct}'_{\mathsf{IBEETIA}})$: Parse $\mathsf{ct_{IBEETIA}} = (x_M, \mathsf{ct_{PKE}})$ and $\mathsf{ct}'_{\mathsf{IBEETIA}} = (x_{M'}, \mathsf{ct}'_{\mathsf{PKE}})$. If $x_M = x_{M'}$, then output 1, and 0 otherwise.

Due to the correctness of the underlying PKE scheme, π is a permutation, and H is a deterministic hash function, the first and second conditions of the correctness hold. Since π is a permutation, and H is collision resistant, $x_M \neq x_{M'}$ holds if $M \neq M'$. Thus, the third condition of the correctness holds.

Theorem 2. *The proposed* IBEETIA *construction is wIND-CCA secure if the underlying* PKE *is IND-CCA secure and π is a pseudo-random permutation.*

Proof. The proof proceeds with the following sequence of games.

Game_0: This game is a real wIND-CCA security game. Queries issued by $\mathcal{A}$ is answered as follows.

Key Extraction: $\mathcal{A}$ sends ID to $\mathcal{C}$. If ID has been issued as a query, $\mathcal{C}$ retrieves r_{ID}. Otherwise, $\mathcal{C}$ chooses $r_{\mathsf{ID}} \xleftarrow{\$} \mathcal{R}$ and preserves $(\mathsf{ID}, r_{\mathsf{ID}})$ locally. $\mathcal{C}$ runs $(\mathsf{PKE.pk_{ID}}, \mathsf{PKE.dk_{ID}}) \leftarrow \mathsf{PKE.KeyGen}(1^\lambda; r_{\mathsf{ID}})$ and returns $\mathsf{sk_{ID}} = \mathsf{PKE.dk_{ID}}$ to $\mathcal{A}$.

Encryption: $\mathcal{A}$ sends (ID, M) to $\mathcal{C}$. If ID has been issued as a query, $\mathcal{C}$ retrieves r_{ID} and preserves $(\mathsf{ID}, r_{\mathsf{ID}})$ locally. Otherwise, $\mathcal{C}$ chooses $r_{\mathsf{ID}} \xleftarrow{\$} \mathcal{R}$. $\mathcal{C}$ runs $(\mathsf{PKE.pk_{ID}}, \mathsf{PKE.dk_{ID}}) \leftarrow \mathsf{PKE.KeyGen}(1^\lambda; r_{\mathsf{ID}})$, runs $x_M \leftarrow \pi(k, H(M))$ and $\mathsf{ct_{PKE}} \leftarrow \mathsf{PKE.Enc}(\mathsf{PKE.pk_{ID}}, M)$, and returns $\mathsf{ct_{IBEETIA}} = (x_M, \mathsf{ct_{PKE}})$ to $\mathcal{A}$.

Decryption: $\mathcal{A}$ sends $(\mathsf{ID}, \mathsf{ct}_{\mathsf{IBEETIA}})$ to $\mathcal{C}$. If ID has been issued as a query, $\mathcal{C}$ retrieves r_{ID} and preserves $(\mathsf{ID}, r_{\mathsf{ID}})$ locally. Otherwise, $\mathcal{C}$ chooses $r_{\mathsf{ID}} \xleftarrow{\$} \mathcal{R}$. $\mathcal{C}$ runs $(\mathsf{PKE.pk}_{\mathsf{ID}}, \mathsf{PKE.dk}_{\mathsf{ID}}) \leftarrow \mathsf{PKE.KeyGen}(1^\lambda; r_{\mathsf{ID}})$. $\mathcal{C}$ runs $M' \leftarrow \mathsf{PKE.Dec}(\mathsf{PKE.dk}_{\mathsf{ID}}, \mathsf{ct}_{\mathsf{PKE}})$ and $x_{M'} \leftarrow \pi(k, H(M'))$. If $x_M = x_{M'}$, then return M', and $\perp$ otherwise.

Game_1: This game is the same as Game_0 except that $\mathcal{C}$ replaces the permutation π with a random function.

Game_2: This game is the same as Game_1 except that $\mathcal{C}$ randomly chooses $x_M \xleftarrow{\$} \mathcal{M}$ instead of using the random function and stores (M, x_M) locally. If the same M is queried, then $\mathcal{C}$ retrieves x_M.

Game_3: Let q_{ID} denote the number of distinct identities that are chosen when the $\mathcal{A}$ queries the oracles, and $r_1, r_2, \ldots, r_{q_{\mathsf{ID}}}$ are random coins for running the $\mathsf{PKE.KeyGen}$ algorithm. This game is the same as Game_2 except that $\mathcal{C}$ randomly chooses $q^* \xleftarrow{\$} \{1, 2, \ldots, q_{\mathsf{ID}}\}$ at the beginning of the game, and aborts the game if there exists $i \in \{1, 2, \ldots, q_{\mathsf{ID}}\} \setminus \{q^*\}$ such that $r_{q^*} = r_i$ holds.

By Lemma 1 in [18], Game_0 and Game_1 are computationally indistinguishable if π is a pseudo-random. By Lemma 2 in [18], Game_1 and Game_2 are statistically indistinguishable. By Lemma 3 in [18], Game_2 and Game_3 are statistically indistinguishable.

Lemma 1 **($\mathcal{A}$'s advantage in Game_3).** *For any PPT adversary $\mathcal{A}$, there exists a reduction algorithm $\mathcal{B}$ for breaking the IND-CCA security of* PKE.

Proof. We show that $\mathcal{A}$'s advantage in Game_3 is negligible. Concretely, we construct an algorithm $\mathcal{B}$ that breaks the IND-CCA security of the PKE scheme. Let $\mathcal{C}$ be the IND-CCA challenger. First, $\mathcal{C}$ runs $(\mathsf{PKE.pk}, \mathsf{PKE.dk}) \leftarrow \mathsf{PKE.KeyGen}(1^\lambda)$ and sends $\mathsf{PKE.pk}$ to $\mathcal{B}$. $\mathcal{B}$ specifies a has function H and chooses $k \xleftarrow{\$} \mathcal{K}$. $\mathcal{B}$ sends $\mathsf{MPK} = H$ to $\mathcal{A}$. During the game, $\mathcal{B}$ aborts if $\mathsf{ID}_{q^*} \neq \mathsf{ID}^*$. From now on, we assume that $\mathcal{B}$'s guess is correct.

$\mathcal{B}$ responds $\mathcal{A}$'s queries as follows.

Key Extraction: $\mathcal{A}$ sends $\mathsf{ID} \neq \mathsf{ID}^*$ to $\mathcal{B}$. If ID has been issued as a query, $\mathcal{B}$ retrieves r_{ID}. Otherwise, $\mathcal{B}$ chooses $r_{\mathsf{ID}} \xleftarrow{\$} \mathcal{R}$ and preserves $(\mathsf{ID}, r_{\mathsf{ID}})$ locally. $\mathcal{B}$ runs $(\mathsf{PKE.pk}_{\mathsf{ID}}, \mathsf{PKE.dk}_{\mathsf{ID}}) \leftarrow \mathsf{PKE.KeyGen}(1^\lambda; r_{\mathsf{ID}})$ and returns $\mathsf{sk}_{\mathsf{ID}} = \mathsf{PKE.dk}_{\mathsf{ID}}$ to $\mathcal{A}$.

Encryption: $\mathcal{A}$ sends (ID, M) to $\mathcal{B}$. $\mathcal{B}$ randomly chooses $x_M \xleftarrow{\$} \mathcal{M}$ if (M, x_M) is not stored locally. Otherwise, $\mathcal{B}$ retrieves x_M and preserves (M, x_M) locally.

- If $\mathsf{ID} \neq \mathsf{ID}^*$ (i.e., this is the i-th query where $i \neq q^*$), if ID has been issued as a query, $\mathcal{B}$ retrieves r_{ID}. Otherwise, $\mathcal{B}$ chooses $r_{\mathsf{ID}} \xleftarrow{\$} \mathcal{R}$ and preserves $(\mathsf{ID}, r_{\mathsf{ID}})$ locally. $\mathcal{B}$ runs $(\mathsf{PKE.pk}_{\mathsf{ID}}, \mathsf{PKE.dk}_{\mathsf{ID}}) \leftarrow \mathsf{PKE.KeyGen}(1^\lambda; r_{\mathsf{ID}})$, runs $\mathsf{ct}_{\mathsf{PKE}} \leftarrow \mathsf{PKE.Enc}(\mathsf{PKE.pk}_{\mathsf{ID}}, M)$, and returns $\mathsf{ct}_{\mathsf{IBEETIA}} = (x_M, \mathsf{ct}_{\mathsf{PKE}})$ to $\mathcal{A}$.

- If $\mathsf{ID} = \mathsf{ID}^*$ (i.e., this is the q^*-th query), then $\mathcal{B}$ runs $\mathsf{ct}_{\mathsf{PKE}} \leftarrow \mathsf{PKE.Enc}(\mathsf{PKE.pk}, M)$, and returns $\mathsf{ct}_{\mathsf{IBEETIA}} = (x_M, \mathsf{ct}_{\mathsf{PKE}})$ to $\mathcal{A}$.

Decryption: $\mathcal{A}$ sends $(\mathsf{ID}, \mathsf{ct}_{\mathsf{IBEETIA}})$ to $\mathcal{B}$. Parse $\mathsf{ct}_{\mathsf{IBEETIA}} = (x_M, \mathsf{ct}_{\mathsf{PKE}})$.

- If $\mathsf{ID} \neq \mathsf{ID}^*$ (i.e., this is the i-th query where $i \neq q^*$), if ID has been issued as a query, $\mathcal{B}$ retrieves r_{ID}. Otherwise, $\mathcal{B}$ chooses $r_{\mathsf{ID}} \xleftarrow{\$} \mathcal{R}$ and preserves $(\mathsf{ID}, r_{\mathsf{ID}})$ locally. $\mathcal{B}$ runs $M' \leftarrow \mathsf{PKE.Dec}(\mathsf{PKE.dk}_{\mathsf{ID}}, \mathsf{ct}_{\mathsf{PKE}})$. $\mathcal{B}$ randomly chooses $x_{M'} \xleftarrow{\$} \mathcal{M}$ and preserves $(M', x_{M'})$ locally if (M', x_M) is not stored locally. Otherwise, $\mathcal{B}$ retrieves x_M, and returns M' if $x_M = x_{M'}$, and $\perp$ otherwise.
- If $\mathsf{ID} = \mathsf{ID}^*$ (i.e., this is the q^*-th query), $\mathcal{B}$ sends $\mathsf{ct}_{\mathsf{PKE}}$ to $\mathcal{C}$ as a decryption query, and obtains M'. $\mathcal{B}$ randomly chooses $x_{M'} \xleftarrow{\$} \mathcal{M}$ and preserves $(M', x_{M'})$ locally if (M', x_M) is not stored locally. Otherwise, $\mathcal{B}$ retrieves x_M and returns M' if $x_M = x_{M'}$, and $\perp$ otherwise.

When $\mathcal{A}$ declares $(\mathsf{ID}^*, M_0^*, M_1^*)$, then $\mathcal{B}$ sends (M_0^*, M_1^*) to $\mathcal{C}$. $\mathcal{C}$ randomly chooses $b \xleftarrow{\$} \{0,1\}$, computes $\mathsf{ct}^*_{\mathsf{PKE}} \leftarrow \mathsf{PKE.Enc}(\mathsf{PKE.pk}, M_b^*)$, and sends $\mathsf{ct}^*_{\mathsf{PKE}}$ to $\mathcal{B}$. $\mathcal{B}$ randomly chooses $x^* \xleftarrow{\$} \mathcal{M}$ and sends $\mathsf{ct}^*_{\mathsf{IBEETIA}} = (x^*, \mathsf{ct}^*_{\mathsf{PKE}})$ to $\mathcal{A}$.

$\mathcal{B}$ responds $\mathcal{A}$'s queries as follows.

Key Extraction: $\mathcal{B}$ responds the query as in the pre-challenge phase.
Encryption: $\mathcal{B}$ responds the query as in the pre-challenge phase.
Decryption: $\mathcal{A}$ sends $(\mathsf{ID}, \mathsf{ct}_{\mathsf{IBEETIA}})$ to $\mathcal{B}$. Parse $\mathsf{ct}_{\mathsf{IBEETIA}} = (x_M, \mathsf{ct}_{\mathsf{PKE}})$.

- If $\mathsf{ID} \neq \mathsf{ID}^*$ (i.e., this is the i-th query where $i \neq q^*$), $\mathcal{B}$ responds the query as in the pre-challenge phase.
- If $\mathsf{ID} = \mathsf{ID}^*$ (i.e., this is the q^*-th query), if $\mathsf{ct}_{\mathsf{PKE}} \neq \mathsf{ct}^*_{\mathsf{PKE}}$, then $\mathcal{B}$ sends $\mathsf{ct}_{\mathsf{PKE}}$ to $\mathcal{C}$ as a decryption query, and obtains M'. $\mathcal{B}$ randomly chooses $x_{M'} \xleftarrow{\$} \mathcal{M}$ and preserves $(M', x_{M'})$ locally if (M', x_M) is not stored locally. Otherwise, $\mathcal{B}$ retrieves x_M and returns M' if $x_M = x_{M'}$, and $\perp$ otherwise. If $\mathsf{ct}_{\mathsf{PKE}} = \mathsf{ct}^*_{\mathsf{PKE}}$ (and then $x_M \neq x^*$), then $\mathcal{B}$ returns $\perp$ because x^* is deterministically fixed when the challenge plaintext is fixed and thus $(x_M, \mathsf{ct}^*_{\mathsf{PKE}})$ is an invalid ciphertext.

Finally, $\mathcal{A}$ outputs a bit b'. $\mathcal{B}$ outputs the same b'. If the guess q^* is correct (with the probability at least $1/(q^{\mathsf{ext}} + q^{\mathsf{enc}} + q^{\mathsf{dec}})$ where q^{ext}, q^{enc}, and q^{dec} are the number of key extraction, encryption, and decryption queries, respectively), then $\mathcal{B}$'s simulation is perfect and $\mathcal{B}$ can break the IND-CCA security. □

This concludes the proof of Theorem 2. □

Theorem 3. *The proposed* IBEETIA *construction is OW-TH-CCA secure if the underlying* PKE *is IND-CCA secure and H is a one-way hash function.*

Proof.

Game_0: This game is a real OW-TH-CCA security game. Queries issued by $\mathcal{A}$ is answered as follows.

Key Extraction: $\mathcal{A}$ sends ID to $\mathcal{C}$. If ID has been issued as a query, $\mathcal{C}$ retrieves r_{ID}. Otherwise, $\mathcal{C}$ chooses $r_{\mathsf{ID}} \xleftarrow{\$} \mathcal{R}$ and preserves $(\mathsf{ID}, r_{\mathsf{ID}})$ locally. $\mathcal{C}$ runs $(\mathsf{PKE.pk}_{\mathsf{ID}}, \mathsf{PKE.dk}_{\mathsf{ID}}) \leftarrow \mathsf{PKE.KeyGen}(1^\lambda; r_{\mathsf{ID}})$ and returns $\mathsf{sk}_{\mathsf{ID}} = \mathsf{PKE.dk}_{\mathsf{ID}}$ to $\mathcal{A}$.

Token Extraction: $\mathcal{A}$ sends ID to $\mathcal{C}$. If ID has been issued as a query, $\mathcal{C}$ retrieves r_{ID}. Otherwise, $\mathcal{C}$ chooses $r_{\mathsf{ID}} \xleftarrow{\$} \mathcal{R}$ and preserves $(\mathsf{ID}, r_{\mathsf{ID}})$ locally. $\mathcal{C}$ runs $(\mathsf{PKE.pk}_{\mathsf{ID}}, \mathsf{PKE.dk}_{\mathsf{ID}}) \leftarrow \mathsf{PKE.KeyGen}(1^\lambda; r_{\mathsf{ID}})$ and returns $\mathsf{tok}_{\mathsf{ID}} = (\mathsf{PKE.pk}_{\mathsf{ID}}, k)$ to $\mathcal{A}$.

Decryption: $\mathcal{A}$ sends $(\mathsf{ID}, \mathsf{ct}_{\mathsf{IBEETIA}})$ to $\mathcal{C}$. If ID has been issued as a query, $\mathcal{C}$ retrieves r_{ID} and preserves $(\mathsf{ID}, r_{\mathsf{ID}})$ locally. Otherwise, $\mathcal{C}$ chooses $r_{\mathsf{ID}} \xleftarrow{\$} \mathcal{R}$. $\mathcal{C}$ runs $(\mathsf{PKE.pk}_{\mathsf{ID}}, \mathsf{PKE.dk}_{\mathsf{ID}}) \leftarrow \mathsf{PKE.KeyGen}(1^\lambda; r_{\mathsf{ID}})$. $\mathcal{C}$ runs $M' \leftarrow \mathsf{PKE.Dec}(\mathsf{PKE.dk}_{\mathsf{ID}}, \mathsf{ct}_{\mathsf{PKE}})$ and $x_{M'} \leftarrow \pi(k, H(M'))$. If $x_M = x_{M'}$, then return M', and $\perp$ otherwise.

Game_1: Let q_{ID} denote the number of distinct identities that are chosen when the $\mathcal{A}$ queries the oracles, and $r_1, r_2, \ldots, r_{q_{\mathsf{ID}}}$ are random coins for running the $\mathsf{PKE.KeyGen}$ algorithm. This game is the same as Game_0 except that $\mathcal{C}$ randomly chooses $q^* \xleftarrow{\$} \{1, 2, \ldots, q_{\mathsf{ID}}\}$ at the beginning of the game, and aborts the game if there exists $i \in \{1, 2, \ldots, q_{\mathsf{ID}}\} \setminus \{q^*\}$ such that $r_{q^*} = r_i$ holds.

Game_2: This game is the same as Game_1 except that $\mathcal{C}$ prepares the challenge ciphertext as follows. In the challenge phase, $\mathcal{A}$ declares ID^*. Let $\mathsf{PKE.pk}$ be the public key used in the q^*-th query. $\mathcal{C}$ randomly chooses $M^* \xleftarrow{\$} \mathcal{M}$, runs $\mathsf{ct}^*_{\mathsf{PKE}} \leftarrow \mathsf{PKE.Enc}(\mathsf{PKE.pk}_{\mathsf{ID}}, 0^{|M^*|})$, computes $x^* \leftarrow \pi(k, H(M^*))$, and sends $\mathsf{ct}^*_{\mathsf{IBEETIA}} = (x^*, \mathsf{ct}^*_{\mathsf{PKE}})$ to $\mathcal{A}$.

By Lemma 3 in [18], Game_0 and Game_1 are statistically indistinguishable.

Lemma 2 (Indistinguishability between Game_1 *and* Game_2). *For any PPT adversary $\mathcal{A}$, there exists a reduction algorithm $\mathcal{B}_1$ for breaking the IND-CCA security of* PKE.

Proof Sketch. The proof is almost the same as that of Lemma 1, except that $\mathcal{A}$ declares ID^* and $\mathcal{B}_1$ sends $(\mathsf{ID}^*, M^*, 0^{|M^*|})$ to the IND-CCA challenger. If M^* is encrypted, then $\mathcal{B}_1$ simulates Game_1 and if $0^{|M^*|}$ is encrypted, then $\mathcal{B}_1$ simulates Game_2. □

Lemma 3 ($\mathcal{A}$'s advantage in Game_2). *For any PPT adversary $\mathcal{A}$, there exists a reduction algorithm $\mathcal{B}_2$ for breaking the one-wayness of H.*

Proof. Let $\mathcal{C}$ be the challenger of the one-wayness of H. We construct the algorithm $\mathcal{B}_2$ as follows. First, $\mathcal{C}$ sends the description of H to $\mathcal{B}_2$. $\mathcal{B}_2$ sets $\mathsf{MPK} = H$ and sends MPK to $\mathcal{A}$. $\mathcal{B}_2$ prepares all public keys and decryption keys of PKE and thus $\mathcal{B}_2$ can respond to all queries. In the challenge phase, $\mathcal{B}_2$ randomly chooses $M^* \xleftarrow{\$} \mathcal{M}$ and runs $\mathsf{ct}^*_{\mathsf{PKE}} \leftarrow \mathsf{PKE.Enc}(\mathsf{PKE.pk}_{\mathsf{ID}}, 0^{|M^*|})$. $\mathcal{C}$ sends $H(M)$ to $\mathcal{B}_2$. If $H(M^*) = H(M)$, then $\mathcal{B}_2$ outputs M^*. Otherwise, $\mathcal{B}_2$ computes $x^* \leftarrow \pi(k, H(M))$ and sends $\mathsf{ct}^*_{\mathsf{IBEETIA}} = (x^*, \mathsf{ct}^*_{\mathsf{PKE}})$ to $\mathcal{A}$. We remark that, for M, which is not known by $\mathcal{B}_2$, $|M| = |M^*|$ holds because we assume that the size of all plaintexts are the same. Thus, $\mathcal{B}_2$ properly simulates the challenge ciphertext in Game_2. Finally, $\mathcal{A}$ outputs $\hat{M}$. $\mathcal{B}_2$ outputs the same $\hat{M}$ and breaks the one-wayness of H with the advantage of $\mathcal{A}$. □

This concludes the proof of Theorem 3. □

Table 1. Comparison: STD stands for standard model and $*$ stands for selective security.

Scheme	Tools/Assumptions	IND	OW	STD
Emura-Takayasu [18]	SKE	wIND-CCA	No	Yes
Lee et al. [25]	Pairing	wIND-CCA	OW-TG-CPA	No
Duong et al. [15]	LWE	wIND-CPA*	OW-TG-CPA*	Yes
Proposed Construction	PKE	wIND-CCA	OW-TH-CCA	Yes

6 Discussion

In this section, we discuss the efficiency and security level of the proposed construction. We give the comparisons in Table 1. We can employ any CCA-secure PKE scheme, e.g., the Cramer-Shoup scheme [12] (which is IND-CCA secure under the decisional Diffie-Hellman assumption) or the CRYSTALS-Kyber key encapsulation mechanism [8] (which is IND-CCA secure under the learning with errors (LWE) assumption over module lattices) with an appropriate data decapsulation mechanism, and so on.

Of course, the Emura-Takayasu construction [18] is more efficient than Lee et al. and Duong et al. schemes and the proposed construction since it employs only symmetric key primitives. However, it does not provide the OW-TH-CPA security.

The Lee et al. scheme [25] is wIND-CCA secure under the bilinear Diffie-Hellman (BDH) assumption in the random oracle model. The form of ciphertext is:

- $C_1 = F(K_1, H_1(M))$, $C_2 = g^r$, $C_3 = (M||r) \oplus H_2(T||C_2||e(P_{\mathsf{pub}}, H(\mathsf{ID}))^r)$

where F is a permutation, T is a MAC (message authentication code) on C_1 such that $T = \mathsf{MAC}(K_2, C_1)$, H, H_1, and H_2 are hash functions modeled as random oracles, e is a bilinear pairing, and $\mathsf{tok}_{\mathsf{ID}} = (K_1, K_2)$. Because the MAC part has the role of rejecting an invalid decryption query, it is not clear whether the Lee et al. scheme provides the OW-TH-CCA security after K_2 is given to the adversary. However, the Lee et al. scheme is at least OW-TH-CPA secure due to the one-wayness of H_1 and the fact that $e(P_{\mathsf{pub}}, H(\mathsf{ID}))^r$ can be regarded as a solution of the BDH problem. In the Lee et al. scheme, $\mathsf{tok}_{\mathsf{ID}} = \mathsf{MTK}$. Thus, the Lee et al. scheme is OW-TG-CPA secure when it is OW-TH-CPA secure. Thus, we state that the Lee et al. scheme is OW-TG-CPA secure in Table 1. Due to our implication result, employing pairings in the Lee et al. scheme is reasonable. On the other hand, proposed construction does not employ pairings, e.g., when the Cramer-Shoup scheme is employed. In this perspective, our constriction is more efficient than the Lee et al. scheme.

The Duong et al. scheme [15] is wIND-CPA secure under the LWE assumption over integer lattices. The form of ciphertext for $M \in \{0,1\}^t$ for some $t \in \mathbb{N}$ is:

- $C_1 = T_{\mathbf{A}'}\mathbf{s}^\top + H(M||T_{\mathbf{A}'})$, C_2, and C_3.

where H is a one-way and collision-resistant hash function, $\mathsf{A}' \in \mathbb{Z}^{n \times m}$ is a matrix (we omit the explanations of n and m here), $T_{\mathbf{A}'}$ is the trapdoor, and $\mathsf{tok}_{\mathsf{ID}} = T_{\mathbf{A}'}$. (C_2, C_3) can be regarded as a ciphertext of the ABB IBE scheme [1], and is independent to A'. In the OW-TH-CPA security definition, the trapdoor $T_{\mathbf{A}'}$ is leaked. Thus, C_1 may leak $H(M||T_{\mathbf{A}'})$. Since the ABB IBE is selectively IND-CPA secure under the LWE assumption, the Duong et al. scheme provides the selective OW-TH-CPA security after $\mathsf{tok}_{\mathsf{ID}} = T_{\mathbf{A}'}$ is given to the adversary. In the Duong et al. scheme, $\mathsf{tok}_{\mathsf{ID}} = \mathsf{MTK}$. Thus, we state that the Duong et al. scheme is selectively OW-TG-CPA secure in Table 1. The proposed construction provides the wIND-CCA security which is stronger than the wIND-CPA security. That is, a post-quantum instantiation of the proposed construction, e.g., from the CRYSTALS-Kyber scheme, provides the first IBEETIA scheme which is wIND-CCA secure under a post-quantum complexity assumption.

7 Conclusion

In this paper, we introduced OW security notions, OW-TG-CPA and OW-TH-CCA/CPA, and demonstrated OW-TG-CPA secure IBEETIA implies IBE, and proposed a generic construction of OW-TH-CCA secure IBEETIA from PKE.

As an extension of IBEET, attribute-based encryption with equality test (ABEET) has been proposed [3,13,26,28,29,36]. However, ABEET against insider attacks (ABEETIA) has not been considered so far, to the best of our knowledge. Because ABE implies IBE [20], it would be interesting to explore whether a similar separation holds or not, i.e., whether OW-TH-CCA/CPA secure ABEETIA can be constructed from PKE/IBE or not.

Acknowledgment. The main part of study was done when the author was with the National Institute of Information and Communications Technology (NICT), Japan. This work was supported by JSPS KAKENHI Grant Number JP21K11897.

References

1. Agrawal, S., Boneh, D., Boyen, X.: Efficient lattice (H)IBE in the standard model. In: Gilbert, H. (ed.) EUROCRYPT 2010. LNCS, vol. 6110, pp. 553–572. Springer, Heidelberg (2010). https://doi.org/10.1007/978-3-642-13190-5_28
2. Asano, K., Emura, K., Takayasu, A.: More efficient adaptively secure lattice-based IBE with equality test in the standard model. In: ISC, pp. 75–83 (2022)
3. Asano, K., Emura, K., Takayasu, A., Watanabe, Y.: A generic construction of CCA-secure attribute-based encryption with equality test. In: Ge, C., Guo, F. (eds.) ProvSec 2022. LNCS, vol. 13600, pp. 3–19. Springer, Cham (2022). https://doi.org/10.1007/978-3-031-20917-8_1
4. Baek, J., Safavi-Naini, R., Susilo, W.: Token-controlled public key encryption. In: Deng, R.H., Bao, F., Pang, H.H., Zhou, J. (eds.) ISPEC 2005. LNCS, vol. 3439, pp. 386–397. Springer, Heidelberg (2005). https://doi.org/10.1007/978-3-540-31979-5_33
5. Boneh, D., Di Crescenzo, G., Ostrovsky, R., Persiano, G.: Public key encryption with keyword search. In: Cachin, C., Camenisch, J.L. (eds.) EUROCRYPT 2004. LNCS, vol. 3027, pp. 506–522. Springer, Heidelberg (2004). https://doi.org/10.1007/978-3-540-24676-3_30
6. Boneh, D., Franklin, M.: Identity-based encryption from the Weil pairing. In: Kilian, J. (ed.) CRYPTO 2001. LNCS, vol. 2139, pp. 213–229. Springer, Heidelberg (2001). https://doi.org/10.1007/3-540-44647-8_13
7. Boneh, D., Papakonstantinou, P.A., Rackoff, C., Vahlis, Y., Waters, B.: On the impossibility of basing identity based encryption on trapdoor permutations. In: IEEE FOCS, pp. 283–292 (2008)
8. Bos, J.W., et al.: CRYSTALS - Kyber: a CCA-secure module-lattice-based KEM. In: EuroS&P, pp. 353–367. IEEE (2018)
9. Brakerski, Z., Lombardi, A., Segev, G., Vaikuntanathan, V.: Anonymous IBE, leakage resilience and circular security from new assumptions. In: Nielsen, J.B., Rijmen, V. (eds.) EUROCRYPT 2018. LNCS, vol. 10820, pp. 535–564. Springer, Cham (2018). https://doi.org/10.1007/978-3-319-78381-9_20
10. Chow, S.S.M.: Token-controlled public key encryption in the standard model. In: Garay, J.A., Lenstra, A.K., Mambo, M., Peralta, R. (eds.) ISC 2007. LNCS, vol. 4779, pp. 315–332. Springer, Heidelberg (2007). https://doi.org/10.1007/978-3-540-75496-1_21
11. Chow, S.S.M.: Removing escrow from identity-based encryption. In: Jarecki, S., Tsudik, G. (eds.) PKC 2009. LNCS, vol. 5443, pp. 256–276. Springer, Heidelberg (2009). https://doi.org/10.1007/978-3-642-00468-1_15
12. Cramer, R., Shoup, V.: A practical public key cryptosystem provably secure against adaptive chosen ciphertext attack. In: Krawczyk, H. (ed.) CRYPTO 1998. LNCS, vol. 1462, pp. 13–25. Springer, Heidelberg (1998). https://doi.org/10.1007/BFb0055717
13. Cui, Y., Huang, Q., Huang, J., Li, H., Yang, G.: Outsourced ciphertext-policy attribute-based encryption with equality test. In: Guo, F., Huang, X., Yung, M. (eds.) Inscrypt 2018. LNCS, vol. 11449, pp. 448–467. Springer, Cham (2019). https://doi.org/10.1007/978-3-030-14234-6_24
14. Döttling, N., Garg, S.: Identity-based encryption from the Diffie-Hellman assumption. J. ACM **68**(3), 14:1–14:46 (2021)

15. Duong, D.H., Le, H.Q., Roy, P.S., Susilo, W.: Lattice-based IBE with equality test in standard model. In: Steinfeld, R., Yuen, T.H. (eds.) ProvSec 2019. LNCS, vol. 11821, pp. 19–40. Springer, Cham (2019). https://doi.org/10.1007/978-3-030-31919-9_2
16. Emura, K., Katsumata, S., Watanabe, Y.: Identity-based encryption with security against the KGC: a formal model and its instantiation from lattices. In: Sako, K., Schneider, S., Ryan, P.Y.A. (eds.) ESORICS 2019. LNCS, vol. 11736, pp. 113–133. Springer, Cham (2019). https://doi.org/10.1007/978-3-030-29962-0_6
17. Emura, K., Katsumata, S., Watanabe, Y.: Identity-based encryption with security against the KGC: a formal model and its instantiations. Theor. Comput. Sci. **900**, 97–119 (2022)
18. Emura, K., Takayasu, A.: A generic construction of CCA-secure identity-based encryption with equality test against insider attacks. IEICE Trans. Fundam. Electron. Commun. Comput. Sci. **106-A**(3), 193–202 (2023)
19. Galindo, D., Herranz, J.: A generic construction for token-controlled public key encryption. In: Di Crescenzo, G., Rubin, A. (eds.) FC 2006. LNCS, vol. 4107, pp. 177–190. Springer, Heidelberg (2006). https://doi.org/10.1007/11889663_16
20. Herranz, J.: Attribute-based encryption implies identity-based encryption. IET Inf. Secur. **11**(6), 332–337 (2017)
21. Ishida, A., Sakai, Y., Emura, K., Hanaoka, G., Tanaka, K.: Fully anonymous group signature with verifier-local revocation. In: Catalano, D., De Prisco, R. (eds.) SCN 2018. LNCS, vol. 11035, pp. 23–42. Springer, Cham (2018). https://doi.org/10.1007/978-3-319-98113-0_2
22. Jager, T., Kurek, R., Niehues, D.: Efficient adaptively-secure IB-KEMs and VRFs via near-collision resistance. In: Garay, J.A. (ed.) PKC 2021. LNCS, vol. 12710, pp. 596–626. Springer, Cham (2021). https://doi.org/10.1007/978-3-030-75245-3_22
23. Lee, H.T.: Cryptanalysis of Zhu et al.'s identity-based encryption with equality test without random oracles. IEEE Access **11**, 84533–84542 (2023)
24. Lee, H.T., Ling, S., Seo, J.H., Wang, H., Youn, T.-Y.: Public key encryption with equality test in the standard model. Inf. Sci. **516**, 89–108 (2020)
25. Lee, H.T., Wang, H., Zhang, K.: Security analysis and modification of ID-based encryption with equality test from ACISP 2017. In: Susilo, W., Yang, G. (eds.) ACISP 2018. LNCS, vol. 10946, pp. 780–786. Springer, Cham (2018). https://doi.org/10.1007/978-3-319-93638-3_46
26. Li, C., Shen, Q., Xie, Z., Feng, X., Fang, Y., Zhonghai, W.: Large universe CCA2 CP-ABE with equality and validity test in the standard model. Comput. J. **64**(4), 509–533 (2021)
27. Tsabary, R.: Fully secure attribute-based encryption for t-CNF from LWE. In: Boldyreva, A., Micciancio, D. (eds.) CRYPTO 2019. LNCS, vol. 11692, pp. 62–85. Springer, Cham (2019). https://doi.org/10.1007/978-3-030-26948-7_3
28. Wang, Q., Li Peng, H., Xiong, J.S., Qin, Z.: Ciphertext-policy attribute-based encryption with delegated equality test in cloud computing. IEEE Access **6**, 760–771 (2018)
29. Wang, Y., Cui, Y., Huang, Q., Li, H., Huang, J., Yang, G.: Attribute-based equality test over encrypted data without random oracles. IEEE Access **8**, 32891–32903 (2020)
30. Wu, H., Chow, S.S.M.: Anonymous (hierarchical) identity-based encryption from broader assumptions. In: Tibouchi, M., Wang, X. (eds.) ACNS 2023. LNCS, vol. 13906, pp. 366–395. Springer, Cham (2023). https://doi.org/10.1007/978-3-031-33491-7_14

31. Wu, T., Ma, S., Mu, Y., Zeng, S.: ID-based encryption with equality test against insider attack. In: Pieprzyk, J., Suriadi, S. (eds.) ACISP 2017. LNCS, vol. 10342, pp. 168–183. Springer, Cham (2017). https://doi.org/10.1007/978-3-319-60055-0_9
32. Wu, Z., et al.: Efficient and fully secure lattice-based IBE with equality test. In: Gao, D., Li, Q., Guan, X., Liao, X. (eds.) ICICS 2021. LNCS, vol. 12919, pp. 301–318. Springer, Cham (2021). https://doi.org/10.1007/978-3-030-88052-1_18
33. Yamada, S.: Asymptotically compact adaptively secure lattice IBEs and verifiable random functions via generalized partitioning techniques. In: Katz, J., Shacham, H. (eds.) CRYPTO 2017. LNCS, vol. 10403, pp. 161–193. Springer, Cham (2017). https://doi.org/10.1007/978-3-319-63697-9_6
34. Yang, G., Tan, C.H., Huang, Q., Wong, D.S.: Probabilistic public key encryption with equality test. In: Pieprzyk, J. (ed.) CT-RSA 2010. LNCS, vol. 5985, pp. 119–131. Springer, Heidelberg (2010). https://doi.org/10.1007/978-3-642-11925-5_9
35. Zhu, H., Ahmad, H., Xue, Q., Li, T., Liu, Z., Liu, A.: New constructions of equality test scheme without random oracles. IEEE Access **11**, 49519–49529 (2023)
36. Zhu, H., Wang, L., Ahmad, H., Niu, X.: Key-policy attribute-based encryption with equality test in cloud computing. IEEE Access **5**, 20428–20439 (2017)

Non-interactive Publicly Verifiable Searchable Encryption with Forward and Backward Privacy

Zhilong Luo[1], Shi-Feng Sun[1,2](✉), Zhedong Wang[1], and Dawu Gu[1,2]

[1] School of Cyber Science and Engineering, Shanghai Jiao Tong University, Shanghai, China
{jerry001,wzdstill,dwgu}@sjtu.edu.cn

[2] (Wuxi) Blockchain Advanced Research Center, Shanghai Jiao Tong University, Wuxi, Jiangsu, China
shifeng.sun@sjtu.edu.cn

Abstract. Publicly Verifiable Symmetric Searchable Encryption (PV-SSE) enables a client to delegate verification process of search results to an auditor without revealing private information. However, most of existing PV-SSE schemes are only designed for static databases, and how to design dynamic PV-SSE (PV-DSSE) schemes with strong security and good efficiency remains an open problem. In this paper, we propose a new dynamic PV-SSE scheme, which is, to the best of our knowledge, the first non-interactive PV-DSSE with forward privacy and Type-II backward privacy. In particular, we achieve both strong backward privacy and public verifiability within one roundtrip, by leveraging a special cryptographic primitive called compressed symmetric revocable encryption (CSRE) and developing a novel verification method based on set hash functions. Moreover, we provide concrete implementation of our scheme, and conduct a comprehensive performance evaluation compared with the state-of-art. In a typical network environment, our result shows that the search process in our scheme is 5x to 7x faster than the state-of-art with 8x to 9x lower communication cost.

Keywords: Searchable encryption · Public verifiability · Forward and backward privacy

1 Introduction

With the rapid development of cloud services, resource-constraint users prefer to outsource their data and related process to cloud servers to reduce local storage and computation overhead. However, cloud servers are not fully trusted, which brings data security concerns. A naive solution is to encrypt local data using classical encryption schemes (e.g. AES) before outsourcing them to cloud servers. When the client wants to query some data, he downloads the entire encrypted database from the cloud server, decrypts it, and then searches on the

T. Zhu and Y. Li (Eds.): ACISP 2024, LNCS 14895, pp. 281–302, 2024.
https://doi.org/10.1007/978-981-97-5025-2_15

decrypted data. It is oblivious that this approach will arise a significant amount of redundant computation and communication overhead, which is unacceptable in practice, especially for resource-constrained clients.

Symmetric Searchable Encryption (SSE) [22] is an efficient solution for above issue, which allows clients searching on encrypted data directly. It enables the client to generate an encrypted index, and send it along with encrypted data to the cloud server. When the client wants to search data containing a certain keyword, he generates a search token and submits it to the server. Then the server uses this token to search on the encrypted index, and returns the matched results. SSE has been well studied in both academic and industrial community [6,11,13,17,29], but most of them adopt the adversarial setting where the cloud server is assumed to be honest-but-curious. That is, the cloud server tries to infer private information but follows the protocol honestly. Nevertheless, the cloud server may be malicious in practical application scenarios, who behaves arbitrarily during the protocol. Verifiable SSE [3,7–9,21,31] are proposed to verify integrity of the search results, fighting against the malicious servers.

Unfortunately, most of existing verifiable SSE schemes suffer from at least one of the following two drawbacks. First, verification process is undertaken by the resource-constraint client who issues the search queries, which contradicts the client's intention to reduce local overhead by outsourcing. Second, the schemes are only designed for static databases and cannot work in the scenarios where the client may update the encrypted database. Recently, Chen [8] et al. proposed the first publicly verifiable DSSE, named BPVSE, which overcomes the above two drawbacks simultaneously. In more details, BPVSE enables the client to perform insertions and deletions on the encrypted database and delegate the verification of search results to an auditor without revealing private information. Moreover, BPVSE achieves forward privacy and Type-II backward privacy. However, the latter is achieved by simply treating deletions as insertions, which results in poor efficiency in real-world applications. First, compared with traditional SSE schemes, the server needs *one extra roundtrip* of interaction with the client to complete search process. Second, deletions induce *significant communication and computation overhead.* Therefore, it remains interesting to design practical, non-interactive PV-DSSE with forward and backward privacy, which is still left open.

1.1 Our Contributions

In this work, we revisit the design of publicly verifiable DSSE scheme, and put forward a new efficient, non-interactive scheme by developing a new lightweight verification method. In detail, our main contributions are summarized as follows.

- Based on a special encryption scheme CSRE and set hash functions, we propose the first non-interactive publicly verifiable DSSE scheme with support for forward privacy and Type-II backward privacy, named Seren. To this end, we develop a novel "indirect" verification method consisting of three phases, through which our scheme can achieve efficient search, update, and public verification simultaneously.

- Our scheme is based on the only existing non-interactive Type-II backward-private scheme Aura [25], thus it inherits all the merits of Aura like strong backward privacy and high-efficiency. Thanks to our novel verification strategy, our scheme achieves public verifiability with almost no extra overhead for the client. Compared with the state-of-art publicly verifiable DSSE BPVSE, our scheme avoids extra roundtrips of interaction during search, and reduces most of communication and computation overhead of deletion operations. In addition, our scheme maintains the same level of security as BPVSE (i.e. forward privacy and Type-II backward privacy). A brief comparison of Seren with the most related works is shown in Table 1.
- We formally prove the security and public verifiability of Seren. Meanwhile, we provide concrete implementation and experimental evaluation for Seren, which further demonstrates that our scheme enjoys a great efficiency under a real-world 201ms-delay network. For deletions, Seren is 3 to 5 orders of magnitude faster than BPVSE with zero communication cost; for search, Seren is 5x to 7x faster than BPVSE, with 8x to 9x lower communication cost. Further, we also show that storage cost of the client and verification time cost of Seren are acceptable under large-scale databases.

Table 1. Comparison with previous works. FS, BS, and PV denote forward-private, backward-private, and publicly verifiable respectively. We denote the number of interaction rounds during search as $\#R$. For a update operation (op, w, in) on keyword w, op denotes operation type (insert or delete), $|in|$ denotes the number of the updated identifiers. For a keyword w, we denote a_w as the total number of inserted identifiers matching w, d_w as the number of deleted identifiers matching w, and $d_{max} = max_w d_w$. W denotes the total number of distinct keywords in database. Moreover, an encryption or decryption operation is denoted by t_{enc} or t_{dec}. An operation of adding an element to a compressed data structure (e.g. Bloom filter) is denoted as t_{com}.

Schemes	FS	BS	PV	#R	Communication			Computation			Client Storage
					Delete	Insert	Search	Delete	Insert	Search	
Aura	✔	✔	✘	1	0	$O(\|in\|)$	$O(a_w - d_w)$	$t_{com}\|in\|$	$t_{enc}\|in\|$	$t_{dec}(a_w - d_w)$	$O(Wd_{max})$
Seren	✔	✔	✔	1	0	$O(\|in\|)$	$O(a_w - d_w)$	$t_{com}\|in\|$	$t_{enc}\|in\|$	$t_{dec}(a_w - d_w)$	$O(Wd_{max})$
BPVSE	✔	✔	✔	2	$O(\|in\|)$	$O(\|in\|)$	$O(2a_w)$	$t_{enc}\|in\|$	$t_{enc}\|in\|$	$t_{dec}(a_w + d_w)$	$O(W)$

1.2 Related Work

Song et al. [22] proposed the first SSE scheme. After that, plenty of schemes have been proposed to improve the security [2,11,17,24], functionality [6,16], and efficiency [1,13,18] of SSE. Especially, dynamic SSE (DSSE), which supports updates on the encrypted database, has drawn strong attention from the community [6,10,18,19,24]. Meanwhile, the community tries to study attacks against SSE [5,15,30], in which the leakage information of update operations can be exploited to recover the client's query. To resist this kind of attacks, many works aim to design schemes with forward and backward privacy, which guarantees higher level of security for DSSE.

Forward privacy ensures that the server can not link an update operation with previous queries. Stefanov et al. [24] first proposed the notion of it. Bost [2] formalized the definition of forward privacy, and proposed a forward-private DSSE based on trapdoor permutation. However, trapdoor permutation has a high computational cost due to the public key operations involved. Song et al. [23] proposed two more practical forward-private DSSE based on only lightweight symmetric primitives, leveraging a state chain method. Backward privacy ensures that the server can not learn the deleted entries from subsequent queries that happen after the deletion. Bost et al. [4] first proposed the formal definition of backward privacy with three variable types of leakage (Type-I to Type-III ordered from most to least secure). Chamani et al. [14] proposed three improved schemes that excel in search time or number of interaction roundtrips. Sun et al. [26] proposed a practical Type-III scheme based on only symmetric primitives. Demertzis et al. [12] proposed several backward-private DSSE with small client storage. In particular, all above Type-I and II schemes rely on either oblivious structures or multi-round of interactions for search. Sun et al. [25] propose a new primitive named as compressed symmetric revocable encryption (CSRE), and built the first non-interactive DSSE with Type-II backward privacy.

Verifiable SSE schemes are proposed to verify the integrity of search results in the adversarial setting where the server is malicious. Chai et al. [7] proposed the first verifiable SSE scheme. Bost et al. [3] proposed the first forward-private verifiable DSSE based on verifiable hash tables. Subsequently, Zhang et al. [31] proposed a more efficient forward-private verifiable DSSE based on multiset hash functions. All the mentioned schemes only achieve private verifiability, which means verification is performed by the client who issues the search query. To further reduce the client's overhead, Cheng et al. [9] proposed the first publicly verifiable SSE based on indistinguishability obfuscation where verification is delegated to a third-party auditor. Afterwards, Soleimanian et al. [21] proposed two publicly verifiable SSE based on lightweight primitives, and extended them to support boolean queries. Most recently, leveraging blockchain and smart contracts, Chen et al. [8] proposed the first dynamic SSE with public verifiability named as BPVSE, which achieves forward privacy and Type-II backward privacy. However, their scheme achieves Type-II backward privacy by a naive idea in [4] of treating deletions as insertions, which induces high communication, and computation overhead.

In this work, we explore the new approaches of designing more practical, non-interactive, and publicly verifiable DSSE with support for forward privacy and backward privacy.

2 Preliminaries

2.1 Compressed Symmetric Revocable Encryption

Compressed Symmetric Revocable Encryption(CSRE) [25] is a special encryption scheme that supports revoking the decryption capability of the secret key on specific elements in message space. In CSRE, messages are encrypted under some tags. When the client wants to revoke the decryption capability of secret key on a set of messages, he adds the corresponding tags into a storage-efficient data structure D. Based on D, the client can generate a revocation key which allows decryption of all ciphertexts except for that are generated under the revoked tags. A CSRE scheme with key space $\mathcal{SK}$, message space $\mathcal{M}$ and tag space $\mathcal{T}$ includes five polynomial-time algorithms:

- $\mathsf{CSRE.KGen}(1^\lambda)$: It takes a security parameter λ as input and outputs a system secret key $sk \leftarrow \mathcal{SK}$, and an initialized revocation list D.
- $\mathsf{CSRE.Enc}((sk, D), m, T)$: It takes a system secret key sk, a revocation list D, and a message $m \in \mathcal{M}$ with a list of tags $T \subseteq \mathcal{T}$ as input, and outputs the ciphertext ct for m under tags T.
- $\mathsf{CSRE.Comp}(D, R)$: On inputting the revocation list D, and a tag list $R = \{t_1, t_2, ..., t_\tau\}$ to be revoked, it compresses every tag in R into D, and outputs the updated D.
- $\mathsf{CSRE.KRev}(sk, D)$: On inputting a secret key sk, a compressed revocation list D, it computes the revoked secret key sk_R based on D.
- $\mathsf{CSRE.Dec}(sk_R, ct, T)$: It takes a revoked secret key sk_R and a ciphertext ct under tags T as input, and outputs a message m or a failure symbol $\perp$.

The semantic security of CSRE is defined via an IND-sREV-CPA experiment, which guarantees that the adversary can not infer any information about the encrypted message, even given access to an Encryption oracle and a KeyRevocation oracle. For the detailed security and correctness definition, please refer to [25], due to the space limit. We remark that in this work, we employ CSRE scheme with a single tag per message, namely $T = \{t\}$.

2.2 Set Hash Function

Wegman and Carter [28] introduced a cryptographic primitive named set hash function, which achieves efficient set representation and set equality test. Set hash functions compress a finite set of strings rather than a single string like traditional hash functions to a byte array with fixed-length l (i.e. hash value). We can test the equality of two sets by comparing their hash values under set hash functions. Secure set hash functions satisfy set-collision resistance defined as follows:

Definition 1. *(Set-collision resistance). Let U denote the set of all finite sets composed of elements selected from $\{0,1\}^*$. Set hash function $\mathcal{H}$ with output length l satisfies set-collision resistance if for all P.P.T. algorithms $\mathcal{A}$*

$$\Pr\left[\begin{matrix}(S_1, S_2) \leftarrow \mathcal{A}(\mathcal{H}) : S_1,\ S_2 \in U \\ \textit{such that } S_1 \neq S_2 \textit{ and } \mathcal{H}(S_1) = \mathcal{H}(S_2)\end{matrix}\right] \leq 2^{-l}.$$

We denote the input set as $B = \{b_1, b_2, ..., b_{|B|}\}$. In [28], Wegman and Carter provided an implementation of set-collision resistant set hash function:

$$\mathcal{H}(B) = \bigoplus_{i=1}^{|B|} h_s(b_i)$$

where h_s is a strongly universal hash function from B to $\{0,1\}^l$. A hash function is called strongly universal if keys are mapped pairwise independently, and more formal definition is shown in [28]. Wegman and Carter proved the above implementation satisfies set-collision resistance. In this work, we adopt above implementation.

2.3 Publicly Verifiable Dynamic SSE

Dynamic SSE enables the client to encrypt his data before outsourcing it to the server, while maintaining the ability to search and update on it. Furthermore, PV-DSSE is an extended DSSE which allows the client to delegate the verification process of search results to a third-party auditor without revealing any private information. As shown in [21], when the client issues a search query, the server performs search, and sends the obtained raw search results to the auditor. Then the auditor verifies the integrity of the raw search results with some auxiliary information from client which reveals no private information, and decides to accept or reject the search results. A PV-DSSE scheme $\Sigma = (\mathsf{Setup}, \mathsf{Update}, \mathsf{Search}, \mathsf{Verify})$consists of the following four protocols:

- $(K, \sigma; \mathsf{EDB}; \mathsf{PF}, \mathsf{AUX}) \leftarrow \mathsf{Setup}(1^\lambda; \perp; \perp)$: In this protocol, the client takes a security parameter λ as input, and outputs (K, σ), where K is a secret key and σ is the client's state. The server outputs the initialized encrypted database EDB, and the auditor outputs initialized tables PF, AUX for storing proofs generated in Update and auxiliary information for subsequent verification respectively.
- $(\sigma'; \mathsf{EDB}'; \mathsf{PF}') \leftarrow \mathsf{Update}(K, (w, op, in), \sigma; \mathsf{EDB}; \mathsf{PF})$: For the update protocol, the client takes secret key K, update query (w, op, in), and his state σ as input where w is the keyword, op is the operation type taken from $\{\mathsf{insert}, \mathsf{delete}\}$, and in is the set of identifiers to be updated. The client encrypts in, and sends the ciphertexts to the server. Moreover, he may generate proofs for updated identifiers and send them to the auditor, which enables the auditor to verify search results. The client outputs his updated state σ'. The server updates EDB and the auditor updates PF with ciphertexts and proofs uploaded from the client.
- $(\sigma'; \mathsf{EDB}', R_{raw}; \mathsf{AUX}') \leftarrow \mathsf{Search}(K, \sigma, w; \mathsf{EDB}; \mathsf{AUX})$: In this protocol, the client's input is secret key K, state σ, and the keyword w, while his output is the updated state σ'. He generates a search token and some auxiliary information for verification based on the searched keyword, and sends them to the server and auditor, respectively. The server receives the search token, takes EDB as input and outputs the raw search results R_{raw} to be verified and the updated EDB. The auditor stores the auxiliary information in AUX, and outputs the updated AUX.
- $(sign, R) \leftarrow \mathsf{Verify}(R_{raw}, \mathsf{PF}, \mathsf{AUX})$: The auditor runs this protocol alone. He takes the raw search results R_{raw} from server, the proof table PF, and the auxiliary information table AUX as input. After verification, he outputs a signal $sign$ taken from $\{\mathsf{Accept}, \mathsf{Reject}\}$ with the verified search results R.

Remark. Our PV-DSSE definition can be seen as a dynamic extension of the PV-SSE definition presented in [21]. Except for adding an update process, we separate the search and verification processes into two distinct protocols, namely Search, Verify, whereas [21] integrates them into a single protocol Search.

2.4 Security Definition of PV-DSSE

PV-DSSE considers an adversarial setting where the client, the auditor and the server are respectively honest, honest-but-curious and malicious, while the auditor and the server may collude to extract information. We adopt the definition in [21] with some modifications, which consists of confidentiality and verifiability.

Confidentiality. Confidentiality ensures the revealed information during interactions between the client, server, and auditor is captured by the pre-defined leakage function. For a PV-DSSE scheme $\Sigma = (\mathsf{Setup}, \mathsf{Update}, \mathsf{Search}, \mathsf{Verify})$, we describe leakage function as $\mathcal{L} = (\mathcal{L}_{Stp}, \mathcal{L}_{Updt}, \mathcal{L}_{Srch}, \mathcal{L}_{Vrfy})$, which corresponds to the information leaked during each protocol respectively. Let $\mathcal{A}$ be an adversary and $\mathcal{S}$ be a simulator, and we define following two experiments:

- $\mathbf{Real}_{\mathcal{A}}^{\Sigma}(\lambda)$: $\mathcal{A}$ chooses a security parameter λ. Then the experiment runs $\mathsf{Setup}(1^\lambda; \perp; \perp)$ and returns initialized data structures. After that, $\mathcal{A}$ adaptively issues a polynomial number $poly(\lambda)$ of queries. For update query (w, op, in), search query w, and verify query R_{raw}, the experiment runs corresponding protocols including Update, Search, Verify, and returns the outputs. Eventually, $\mathcal{A}$ chooses a bit $b \in \{0, 1\}$ as the output of the experiment.
- $\mathbf{Ideal}_{\mathcal{A},\mathcal{S}}^{\Sigma}(\lambda)$: $\mathcal{A}$ chooses a security parameter λ. Then the experiment runs $\mathcal{S}_{Stp}(\mathcal{L}_{stp})$ and returns the outputs. After this, $\mathcal{A}$ adaptively issues a polynomial number $poly(\lambda)$ of queries. For update query (w, op, in), search query w, and verify query R_{raw}, the experiment runs $\mathcal{S}_{Updt}(\mathcal{L}_{Updt})$, $\mathcal{S}_{Srch}(\mathcal{L}_{Srch})$, and $\mathcal{S}_{Vrfy}(\mathcal{L}_{Vrfy})$ respectively, and returns the outputs. Eventually, $\mathcal{A}$ chooses a bit $b \in \{0, 1\}$ as the output of the experiment.

We say that scheme Σ is secure if for any P.P.T adversary $\mathcal{A}$, there exists a probabilistic polynomial time simulator $\mathcal{S}$ such that:

$$|\Pr[\mathbf{Real}_{\mathcal{A}}^{\Sigma}(\lambda) = 1] - \Pr[\mathbf{Ideal}_{\mathcal{A},\mathcal{S}}^{\Sigma}(\lambda) = 1]| \leq \mathbf{negl}(\lambda)$$

Verifiability. Verifiability ensures any incorrect or incomplete result produced by a malicious server who may collude with an honest but curious auditor is unable to pass verification. Formal definition is captured by following experiment:

- $\mathbf{Exp}_{\Sigma,\mathcal{A}}^{\mathsf{verify}}(\lambda)$: $\mathcal{A}$ chooses a security parameter λ. Then, the experiment runs $\mathsf{Setup}(1^\lambda; \perp; \perp)$, and returns the outputs. After this, $\mathcal{A}$ issues a polynomial number $poly(\lambda)$ of queries. For update queries, the experiment runs corresponding protocol Update and returns the outputs. For each search query $q = w$, the experiment:

1) runs $R_{raw,w} \leftarrow \mathsf{Search}(K, \sigma, w; \mathsf{EDB}; \mathsf{AUX})$.
2) runs Search with $\mathcal{A}$ and gets output $R^*_{raw,w}$ from $\mathcal{A}$.
Finally, for all search queries $w_1, ..., w_q$ issued by the adversary, the experiment checks if there exists $i \in [1, q]$ s.t. $\mathsf{Accept} \leftarrow \mathsf{Verify}(R^*_{raw,w_i}, \mathsf{PF}, \mathsf{AUX})$ and $R^*_{raw,w_i} \neq R_{raw,w_i}$. If so, the experiment outputs 1 otherwise 0.

We say that Σ is publicly verifiable if for any P.P.T adversary $\mathcal{A}$:

$$\Pr[\mathbf{Exp}^{\mathsf{verify}}_{\Sigma,\mathcal{A}}(\lambda) = 1] \leq negl(\lambda)$$

2.5 Forward and Backward Privacy

In this part, we recall the definition of forward and backward privacy, which follows the notations from [26] with some modifications to adapt to our PV-DSSE definition. We first revisit some common leakage functions in DSSE, and then present the formal definitions.

The leakage function $\mathcal{L}$ maintains a list Q recording the queries issued so far, whose entry is either (u, w) or (u, op, w, in). (u, w) is a search query on w performed on timestamp u, (u, op, w, in) is an update query on w where $op \in \{\mathsf{insert}, \mathsf{delete}\}$ and in is a set of updated identifiers. Here the timestamp u is initialized with 0 and increases with the coming queries. Since an update query updates a set of identifiers rather than a single identifier, so (u, op, w, in) in Q can be also represented as another form $\{((u, v), op, w, ind)\}_{ind \in in}$, where v is the unique serial number of ind in set in.

- $\mathsf{sp}(w) = \{i \mid (i, w) \in Q\}$: search pattern over keyword w and consists of timestamps of all search queries on w.
- $\mathsf{UpHist}(w) = \{(u, op, in) \mid (u, op, in) \in Q\}$: the history of updates on keyword w. It can be also represented as $\{((u, v), op, ind) \mid ((u, v), op, w, ind) \in Q\}$,.
- UpTime(w): the timestamps of updates on w. Formal definition is defined as:

$$\mathsf{UpTime}(w) = \{(u, v) \mid \exists((u, v), \mathsf{insert}, w, ind) \in Q,\ or$$

$$\exists((u, v), \mathsf{delete}, w, ind) \in Q\}$$

- TimeDB(w): the "extended" access pattern, which consists of the identifiers that are previously inserted and not deleted until now, along with corresponding timestamps.

$$\mathsf{TimeDB}(w) = \{((u, v), ind) \mid \exists((u, v), \mathsf{insert}, w, ind) \in Q,\ and$$

$$\forall\ (u', v'), ((u', v'), \mathsf{delete}, w, ind) \notin Q\}$$

- DelTime(w): timestamps of the inserted identifiers that were deleted later.

$$\mathsf{DelTime}(w) = \{(u,v) \mid \exists((u,v), \mathsf{insert}, w, ind) \in Q,\ \ and$$

$$\exists((u',v'), \mathsf{delete}, w, ind) \in Q\}$$

Definition 2. *(Forward Privacy). An $\mathcal{L}$-adaptively-secure DSSE scheme is forward-private if for an update query (w, op, in), $\mathcal{L}_{Updt}(w, op, in) = \mathcal{L}'(op, in)$.*

Definition 3.
(Type-II Backward Privacy). An $\mathcal{L}$-adaptively-secure DSSE is Type-II backward-private if for an update query (w, op, in) and a search query w, $\mathcal{L}_{Updt}(w, op, in) = \mathcal{L}'(op, w)$ and $\mathcal{L}_{Srch}(w) = \mathcal{L}''(\mathsf{sp}(w), \mathsf{TimeDB}(w), \mathsf{UpTime}(w))$.

3 Constructions

In this section, we propose a non-interactive PV-DSSE scheme, dubbed Seren, with forward and Type-II backward privacy. We first present an overview of our design. Specifically, we first show how to construct publicly verifiable DSSE by treating deletions as insertions, which is a well-known framework for achieving Type-II backward privacy. Then we present how to overcome the drawbacks of the naive solution by leveraging CSRE and a novel "indirect" verification method we propose. After that, we present the concrete construction and security analysis of Seren.

3.1 Overview of Our Construction

A Naive Construction. Our naive construction employs state chain method adopted by [23,31] for efficient forward privacy. When the client inserts a set of identifiers in on keyword w, he generates a new state st_c, and encrypts in along with the state st_{c-1} used in the last insertion on w. The locations of the ciphertexts are derived from st_c. Therefore, EDB behaves as a chain-like structure where the preceding state is embedded in the ciphertexts under subsequent state. When the client performs search, he sends the most recent state st_c to server. Starting from that, the server can retrieve all the previous states $st_1, ..., st_c$ and corresponding inserted identifiers. In this method, forward privacy is achieved by generating states randomly, instead of expensive primitives such as the trapdoor permutation in [2,3]. For deletions, we treat them as same as insertions and encrypt identifiers with an operation type sign. During search, the client decrypts returned ciphertexts and obtains both inserted and deleted identifiers. Then he locally computes the real non-deleted identifiers, and sends them back to server to complete search, which causes one more roundtrip. Further, Type-II backward privacy is achieved by this way of deletions.

To achieve public verifiability, we additionally generate a hash proof π_c for in and upload it to the auditor during each update. The location of π_c is also derived by the state st_c, so that the updated identifiers are linked with a proof by the same state st_c. Notice that the proofs perform on in which is a set of strings instead of a single string, so we employ set hash functions to generate them. During verification, the auditor obtains search results along with corresponding states from the server. Then he verifies if the search result is matched with previous stored hash proof for each state, and if so accepts else rejects. The auditor does not obtain any private information since the proofs do not leak information, guaranteed by the security of set hash function.

BPVSE achieves forward and backward privacy essentially in the same way as the naive construction, that is, by state method chain method and treating deletions as insertions. They differs in the way of generating proofs and performing verification. To generate a proof on a set of identifiers, our naive construction uses set hash function while BPVSE uses standard hash function with the concatenation of all identifiers as input. Our scheme improves the robustness of verification, since we does not need to ensure the order of returned identifiers. Further, BPVSE leverages blockchain and smart contracts to perform verification instead of a third-party auditor in our naive construction. All in all, their backward privacy designs have poor efficiency. First, compared with traditional SSE, the client needs one additional round of interaction with server to complete search. Second, decryption computation on ciphertexts of identifiers is undertaken by the resource-constrained client, which results in significant search latency. Third, deletions induce high communication and computation overhead.

Main Idea of Seren. Following the basic idea of [25], we use CSRE and a novel "indirect" verification idea to overcome the above performance bottleneck, achieving non-interactive backward privacy while maintaining efficient public verifiability. We encrypt identifiers with CSRE during insertions. For deletions, the client only caches corresponding tags of deleted identifiers in the compressed data structure D locally. When issuing a search query, the client sends the revocation key sk_R generated from D, which only allows the server decrypting ciphertexts of identifiers except for deleted ones. By this way, our improved scheme performs deletions locally and obliviously, and doesn't need to upload new proofs, ciphertexts to server as BPVSE and naive construction. Further, the server can directly obtain the non-deleted identifiers without extra interaction and takes on the expensive decryption computation instead of the client.

The *key challenge* lies in how to maintain efficient public verifiability based on above non-interactive backward-private construction. We can not directly follow the verification idea in naive construction, since the search results are inconsistent with the pre-stored proofs due to deletions during verification. To be more specific, the search results for each state $st_k, k \in [1, c]$ are identifiers which are inserted previously and not deleted until this search, denoted by $AddInd_k$. Meanwhile, we denote the deleted identifiers which are inserted before under st_k as $DelInd_k$. Then we know that the pre-stored proof π_k is computed base on the union of the two sets $AddInd_k \cup DelInd_k$, and we can not verify the search result $AddInd_k$ by simply comparing with π_k. A simple solution is that, when some identifiers are deleted, the client additionally fetches corresponding proofs that contain these deleted identifiers from the auditor. Then he adapts these proofs according to the deleted identifiers, and updates new proofs with auditor. However, this solution has two main shortcomings:

- **Shortcoming 1:** proof adaptation leads to high communication cost and makes deletions not oblivious to the server, resulting in security lower than Type-II backward privacy.
- **Shortcoming 2:** the client has to pay huge storage cost for a lookup table to keep corresponding identifiers for each proof, so that he can determine which proofs should be adapted and how to adapt them during deletions.

To overcome the above shortcomings, we develop a novel "indirect" verification method for achieving public verifiability, the high-level idea of which is illustrated as in Fig. 1. Generally, instead of directly verifying the correctness of query results as in the naive solution, we first verify the correctness of the union set $AddInd_k \cup DelInd_k$ by the pre-stored proof π_k, and then separately verify the correctness of $DelInd_k$, thus achieving the verifiability of $AddInd_k$ indirectly. In this way, the auditor completes verification without interactively adapting the previously generated proof π_k during deletion. However, it is impossible to perform verification on $DelInd_k$ directly, since the server can not decrypt the ciphertexts of deleted identifiers using sk_R. But he can learn the tags of deleted identifiers, denoted by $DelTag_k$, so we generate proofs and perform verification on tags rather than identifiers. To further verify correctness of identifiers by correct tags, we generate tags by a secret key which is sent to the auditor during search, enabling him to check if tags and identifiers are consistent.

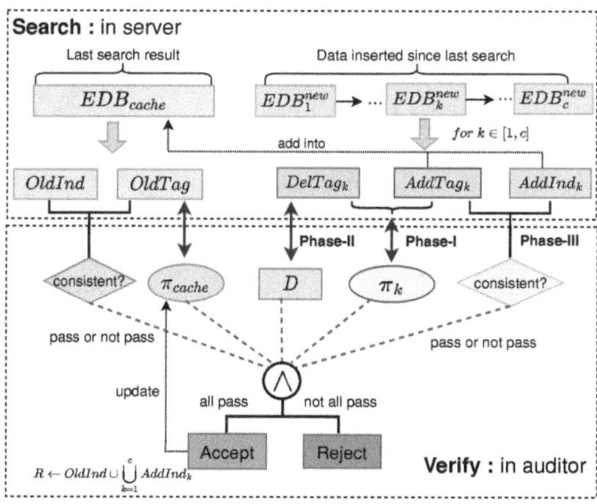

Fig. 1. Main design of Seren.

To be more specific, during search, the client additionally sends some auxiliary information to the auditor, namely key K_w used to generate tags, and some information to verify $DelTag_k$. Fortunately, the compressed deletion list D in CSRE is just capable for that, and we do not need to use extra data structure for this. This auxiliary information will be stored by auditor, and play an important role in subsequent verification. After performing search, except for search results $AddInd_k$, the server additionally sends the tags corresponding to $AddInd_k$, denoted by $AddTag_k$ and $DelTag_k$ to auditor. The auditor then performs the following three steps to verify the correctness of search results:

- **Phase-I**: for each state st_k, the auditor verifies $AddTag_k \cup DelTag_k$ by recomputing a proof based on this union tag set and comparing it with the proof π_k stored in auditor before.
- **Phase-II**: for each state st_k, the auditor verifies $DelTag_k$ by checking if every tag in it belongs to the compressed deletion list D.
- **Phase-III**: for each state st_k, the auditor verifies if $AddInd_k$ and $AddTag_k$ are consistent.

By phase-I and phase-II, the auditor can verify the correctness of $AddTag_k \cup DelTag_k$ and $DelTag_k$. Based on that, correctness of $AddTag_k$ is indirectly ensured. Then the auditor check if $AddTag_k$ and $AddInd_k$ are consistent, and if so, the correctness of $AddInd_k$ is ensured.

At first glance, it seems that phase-II only ensures that $DelTag_k$ returned by the server is a subset of the correct one, and the server may bypass verification by deliberately returning incomplete $DelTag_k$. We argue that phase-I and phase-III prevent this case together. If this case happens, the tags intentionally omitted by the server must belongs to $AddTag_k$ to pass the verification in phase-I. So in phase-III, the server must provide the identifiers corresponding to these omitted tags in $AddInd_k$ to pass verification. However, these identifiers can not be obtained by the server during search due to the security of CSRE. Therefore, we can ensure the correctness of $DelTag_k$ only by performing membership test with D in phase-II. Moreover, we note that Seren, same as [4,25], also requires the client to refresh the encryption key after each search, as the server can use the previous revoked secret key to decrypt the non-deleted identifiers inserted in future. To avoid re-encrypting the search results with the refreshed key, which can be stored in a cache $\mathsf{EDB}_{\mathsf{cache}}$ as processed in [4,25]. As shown in Fig. 1, except for the recent inserted identifiers since last search, denoted as $\mathsf{EDB}_k^{new}, k \in [1, c]$, the server also retrieves the previous search results $(OldInd, OldTag)$ in EDB_{cache} and sends them together to the auditor during search. Then the server will update EDB_{cache} with the recent search results. To support verification on $OldInd$, the auditor maintain a additional proof π_{cache} which is computed on the tags in the previous search results, and perform similar steps as phase-I and phase-III.

3.2 Concrete Construction of Seren

$\mathsf{Setup}(1^\lambda; \perp; \perp)$: With security parameter λ, the client picks the secret key $K \leftarrow \{0,1\}^\lambda$, initializes a counter table C, and tables SK, D, ST for storing the

Algorithm 1. Update

Input: $K, (w, op, in), \sigma$; EDB; PF
Output: updated $\mathsf{EDB}, \mathsf{PF}$
Client:
1: $\mathsf{C}[w] \leftarrow \mathsf{C}[w] + 1$
2: $c \leftarrow \mathsf{C}[w]$, $sk \leftarrow \mathsf{SK}[w]$
3: **if** $sk = \perp$ **then**
4: $\quad sk, D \leftarrow \mathsf{CSRE.KGen}(1^\lambda)$
5: $\quad \mathsf{SK}[w] \leftarrow sk$, $\mathsf{D}[w] \leftarrow D$
6: **end if**
7: $st_c \leftarrow \{0,1\}^\lambda$, $K_w \leftarrow F(K, w)$
8: **if** $op = \mathsf{insert}$ **then**
9: $\quad$ **for** $ind_j \in in$ **do**
10: $\quad\quad idx^{ct}_{c,j} \leftarrow h_0(j, st_c)$
11: $\quad\quad t_j \leftarrow h_1(K_w || ind_j)$
12: $\quad\quad ct_j \leftarrow \mathsf{CSRE.Enc}((sk, D), ind_j, t_j) \oplus h_2(st_c)$
13: $\quad$ **end for**
14: $\quad st_{c-1} \leftarrow \mathsf{ST}[w]$, $\mathsf{ST}[w] \leftarrow st_c$
15: $\quad idx^s_c \leftarrow h_3(st_c)$, $idx^\pi_c \leftarrow h_5(st_c)$
16: $\quad S_c \leftarrow h_4(st_c) \oplus (|in|||st_{c-1})$
17: $\quad \pi_c \leftarrow \mathcal{H}(t_1, t_2, ..., t_{|in|})$
18: $\quad$ send (idx^π_c, π_c) to auditor
19: $\quad$ send (idx^s_c, S_c), to server
20: $\quad$ send $\{(idx^{ct}_{c,j}, ct_j || t_j)\}_{j=1}^{|in|}$ to server
Auditor:
21: $\quad$ set $\mathsf{PF}[idx^\pi_c] \leftarrow \pi_c$
Server:
22: $\quad$ set $\mathsf{EDB}_{add}[idx^s_c] \leftarrow S_c$
23: $\quad$ **for** $j \in [1, |in|]$ **do**
24: $\quad\quad$ set $\mathsf{EDB}_{add}[idx^{ct}_{c,j}] \leftarrow ct_j || t_j$
25: $\quad$ **end for**
26: **else** (i.e. $op = \mathsf{delete}$)
27: $\quad$ **for** $j \in [1, |in|]$ **do**
28: $\quad\quad t_j \leftarrow h_1(K_w || ind_j)$
29: $\quad\quad \mathsf{D}[w] \leftarrow \mathsf{CSRE.Comp}(\mathsf{D}[w], t_j)$
30: $\quad$ **end for**
31: **end if**

encryption key, the compressed deletion list, and the most recent state for each keyword, respectively. The auditor initializes maps PF, AUX for storing proofs of inserted identifiers and the auxiliary information for subsequent verification from client. The server initializes EDB_{add}, EDB_{cache} for storing newly inserted ciphertexts since the last search and the previous search results respectively.

$\mathsf{Update}(K, (w, op, in), \sigma; \mathsf{EDB}; \mathsf{PF})$: As shown in Algorithm 1, when the client inserts a set of identifiers $in = \{ind_1, ..., ind_{|in|}\}$ for keyword w, he obtains the most recent key sk from $\mathsf{SK}[w]$, chooses a new state st_c randomly, and computes key $K_w \leftarrow F(K, w)$ which is used to generate tags and index EDB_{cache}. Then the client generates two types of ciphertexts and a proof, which will be stored as key-value pairs such as $(index, value)$. The first type ciphertexts are $\{(idx^{ct}_{c,j}, ct_j||t_j)\}_{j=1}^{|in|}$, where $t_j \leftarrow h_1(K_w||ind_j)$ is the tag corresponding to ind_j and $ct_j \leftarrow h_2(st_c) \oplus \mathsf{CSRE.Enc}(sk, ind_j, t_j)$ is the masked encryption of ind_j under t_j. The second type ciphertexts are (idx^s_c, S_c), where S_c is the ciphertext of $|in|$ and the last state st_{c-1}. Moreover, the client generates a proof (idx^π_c, π_c) for the tags of inserted identifiers by a set hash function $\mathcal{H}$. Finally, the client sends the two types of ciphertexts and the proof to the server and the auditor respectively, who store them in $\mathsf{EDB}_{\mathsf{add}}$, PF respectively. To delete a set of identifiers in on keyword w, the client computes the corresponding tags of these identifiers, and inserts them into the compressed deletion list D corresponding to w.

$\mathsf{Search}(K, w, \sigma; \mathsf{EDB}; \mathsf{AUX})$: To perform search on w, as shown in Algorithm 2, the client obtains the most recent state st_c, compressed deletion list D, the secret key sk by looking up $\mathsf{ST}[w]$, $\mathsf{D}[w]$, $\mathsf{SK}[w]$, respectively. Then it computes the key $K_w \leftarrow F(K, w)$ and the revoked key $sk_R \leftarrow \mathsf{CSRE.KRev}(sk, D)$. Then he sends

Algorithm 2. Search

Input: K, w, σ; EDB; AUX
Output: R_{raw}; EDB; AUX
Client:
1: $st_c \leftarrow \mathsf{ST}[w],\ D \leftarrow \mathsf{D}[w]$
2: $(sk, D) \leftarrow \mathsf{SK}[w],\ K_w \leftarrow F(K, w)$
3: $sk_R \leftarrow \mathsf{CSRE.KRev}(sk, D)$
4: $idx_c^{aux} \leftarrow h_5(st_c)$
5: send (st_c, sk_R, D, K_w) to server
6: send $(idx_c^{aux}, (D, K_w))$ to auditor
7: $(sk, D) \leftarrow \mathsf{CSRE.KGen}(1^\lambda), \mathsf{SK}[w] \leftarrow sk$
8: $\mathsf{D}[w] \leftarrow D, \mathsf{ST}[w] \leftarrow U_{st}, \mathsf{C}[w] \leftarrow 0$
Auditor:
9: $\mathsf{AUX}[idx_c^{aux}] \leftarrow (D, K_w)$
Server:
10: **while** $st_c \neq U_{st}$ **do**
11: $\quad idx_c^s \leftarrow h_3(st_c)$
12: $\quad |in_c|, st_{c-1} \leftarrow \mathsf{EDB}[idx_c^s] \oplus h_4(st_c)$
13: $\quad$ **for** $j \in [1, |in_c|]$ **do**
14: $\qquad idx_{c,j}^{ct} \leftarrow h_0(j, st_c)$
15: $\qquad (ct_j, t_j) \leftarrow \mathsf{EDB}_{\mathsf{add}}[idx_{c,j}^{ct}]$
16: $\qquad ct_j \leftarrow ct_j \oplus h_2(st_c)$
17: $\qquad ind_j \leftarrow \mathsf{CSRE.Dec}(sk_R, ct_j, t_j)$
18: $\qquad$ **if** $ind_j == \perp$ **then**
19: $\qquad\quad DelTag_c \leftarrow DelTag_c \cup \{t_j\}$
20: $\qquad$ **else**
21: $\qquad\quad AddTag_c \leftarrow Add_c \cup \{t_j\}$
22: $\qquad\quad AddInd_c \leftarrow AddInd_c \cup \{ind_j\}$
23: $\qquad$ **end if**
24: $\quad$ **end for**
25: $\quad st_c \leftarrow st_{c-1}$
26: **end while**
27: $R_{cache} \leftarrow \mathsf{EDB}_{cache}[K_w]$
28: $R_{new} \leftarrow \{(st_k, DelTag_k, AddTag_k, AddInd_k)\}_{k=1}^{c}$
29: $R_{raw} \leftarrow (R_{new}, R_{cache})$
30: send R_{raw} to Auditor
31: add $(AddInd_c, AddTag_c)$ to $\mathsf{EDB}_{cache}[K_w]$

search token (st_c, sk_R, D, K_w), auxiliary information for verification (D, K_w) to the server and the auditor respectively. To retrieve all encrypted data matching search token (st_c, sk_R, D, K_w), the server first retrieves the second type ciphertext S_c by computing idx_c^s using st_c and obtains length of identifier list $|in|_c$ and previous state st_{c-1}. Then it uses a for loop to retrieve and decrypt the encrypted identifiers $(ct_1, t_1), ..., (ct_{|in_c|}, t_{|in_c|})$ corresponding to the state st_c using the revoked key sk_R. If the decryption of (ct_j, t_j) is successful, the obtained identifier ind_j will be added to $AddInd_c$, and the tag t_j will be added to $AddTag_c$. Otherwise, t_j will be added to $DelTag_c$. The server repeats above steps for all the previous state $st_1, ..., st_{c-1}$. The three sets $AddInd_k, AddTag_k, DelTag_k$ along with state st_k form R_{new}, which is the information about the updates since the last search. When that is done, it additionally retrieves the previous search results R_{cache} stored in EDB_{cache} by K_w. After that, the server sends the raw search results $R_{raw} = (\{(st_k, AddInd_k, AddTag_k, DelTag_k)\}_{k=1}^{c}, R_{cache})$ which should be further verified to the auditor. Finally, the server updates EDB_{cache} with above newly obtained search results.

$\mathsf{Verify}(R_{raw}, \mathsf{PF}, \mathsf{AUX})$: As shown in Algorithm 3, the auditor performs a three-phase verification process described in Sect. 3.1. First, it retrieves the auxiliary information (D, K_w) from AUX by st_c in R_{raw}. Then in phase-I, as line 7-line 9, the auditor verifies $AddTag_k \cup DelTag_k$ for each state st_k by recomputing a proof based on them, and then comparing them with the previously stored proof π_k. In phase-II, as line 10-line 12, the auditor verifies $DelTag_k$ for each state st_k by testing if all tags in it belong to D. The above steps ensure the correctness of $AddTag_k$. To further ensure correctness of $AddInd_k$, the auditor checks the consistency between $AddInd_k$ and $AddTag_k$ by K_w as line14-line16. After that, the auditor verifies R_{cache} by repeating phase-I and phase-III. Finally, if all above verification success, the auditor accepts R_{raw}, sends the verified search results R to the client, and updates π_{cache} with this new search results. Otherwise, he rejects R_{raw} and returns an empty set.

Algorithm 3. Verify

Input: R_{raw}, PF, AUX
Output: $(sign, R)$; updated PF, AUX
Auditor:
1: $(D, K_w) \leftarrow \mathsf{AUX}[h_5(st_c)]$, $sign \leftarrow$ Accept
2: **for** $k \in \{1, ..., c\}$ **do**
3: parse $DelTag_k$ as $\{t_j^D\}_{j=1}^{|DelTag_k|}$
4: parse $AddInd_k$ as $\{ind_j^A\}_{j=1}^{|AddInd_k|}$
5: parse $AddTag_k$ as $\{t_j^A\}_{j=1}^{|AddTag_k|}$
6: $idx_k^\pi \leftarrow h_5(st_k)$, $\pi_k \leftarrow \mathsf{PF}[idx_k^\pi]$
7: **if** $\pi_k \neq \mathcal{H}(AddTag_k \cup DelTag_k)$ **then**
8: $sign \leftarrow$ Reject
9: **end if**
10: **if** $\exists\, t_i^D \in DelTag_k$ s.t. $t_i^D \notin D$ **then**
11: $sign \leftarrow$ Reject
12: **end if**
13: **end for**
14: **if** $\exists (ind_k^A, t_k^A)$ $s.t.$ $t_k^A \neq h_1(K_w || ind_k^A)$ **then**
15: $sign \leftarrow$ Reject
16: **end if**
17: parse R_{cache} as $\{(ind_j^O, t_j^O)\}_{j=1}^{|R_{cache}|}$
18: $OldInd \leftarrow \{ind_j^O\}, j \in [1, |R_{cache}|]$
19: $OldTag \leftarrow \{t_j^O\}, j \in [1, |R_{cache}|]$
20: $idx_{cache}^\pi \leftarrow h_5(K_w)$
21: $\pi_{cache} \leftarrow \mathsf{PF}[idx_{cache}^\pi]$
22: **if** $\pi_{cache} \neq \mathcal{H}(OldTag)$ **then**
23: $sign \leftarrow$ Reject
24: **end if**
25: **if** $\exists (ind_k^O, t_k^O)$ $s.t.$ $t_k^O \neq h_1(K_w || ind_k^O)$ **then**
26: $sign \leftarrow$ Reject
27: **end if**
28: **if** $sign =$ Accept **then**
29: $R \leftarrow OldInd \cup (\bigcup_{k=1}^{c} AddInd_k)$
30: $UniTag \leftarrow OldTag \cup (\bigcup_{k=1}^{c} AddTag_k)$
31: $\pi_{cache}^{new} \leftarrow \mathcal{H}(UniTag)$
32: $\mathsf{PF}[idx_{cache}^\pi] \leftarrow \pi_{cache}^{new}$
33: **end if**
34: send $sign, R$ to client

3.3 Security Analysis

Theorem 1. *Let F be a secure PRF, h_0, h_1, h_2, h_3, h_4, h_5, the hash function h_s to implement set hash function $\mathcal{H}$ are secure hash functions modeled by random oracles, and* CSRE *is an IND-sREV-CPA secure compressed symmetric revocable encryption scheme. Define leakage function $\mathcal{L}$ as:*

$$\begin{cases} \mathcal{L}_{Stp} = \perp \\ \mathcal{L}_{Updt}(w, in, op) = (op, |in|) \\ \mathcal{L}_{Srch}(w) = (\mathsf{sp}(w), \mathsf{TimeDB}(w), \mathsf{DelTime}(w)) \\ \mathcal{L}_{Vrfy} = \perp \end{cases}$$

Then the proposed PV-DSSE scheme Seren *is $\mathcal{L}$-adaptively-secure.*

Proof Sketch. The proof proceeds via game hopping, starting from the real-world game and ending with a game simulated with $\mathcal{L}$. The proof is conducted in a similar way as Aura [25], except for additional random oracles programming during simulation.

Theorem 2. *Let F be a secure PRF, h_1 a first-preimage and second-preimage resistant hash function, and $\mathcal{H}$ a set-collision resistant set hash function. Then the proposed PV-DSSE scheme* Seren *is publicly verifiable.*

Proof. To argue the public verifiability of Seren, we show $\mathbf{Exp}_{\Sigma,\mathcal{A}}^{\mathsf{verify}}(\lambda)$ is indistinguishable from one game where adversary $\mathcal{A}$ succeeds with a negligible probability. Such a game is derived by gradually modifying the real game $\mathbf{Exp}_{\Sigma,\mathcal{A}}^{\mathsf{verify}}(\lambda)$.

Game G_1: G_1 replaces PRF F which is used to generate K_w with a truly random function. G_1 and $\mathbf{Exp}_{\Sigma,\mathcal{A}}^{\mathsf{verify}}(\lambda)$ are indistinguishable otherwise we can build a reduction to distinguish between F and a truly random function. Formally, there exists an efficient adversary $\mathcal{B}_1$ s.t.

$$|\Pr[G_1 = 1] - \Pr[\mathbf{Exp}_{\Sigma,\mathcal{A}}^{\mathsf{verify}}(\lambda) = 1]| \leq Adv_{F,\mathcal{B}_1}^{prf}(\lambda)$$

where $Adv_{F,\mathcal{B}_1}^{prf}$ is negligible if F is a secure PRF.

Game G_2: G_2 changes the way of verifying $AddTag_k^* \cup DelTag_k^*$ in phase-I. G_1 verifies $AddTag_k^* \cup DelTag_k^*$ derived from R_{raw,w_i}^* by recomputing a proof $\mathcal{H}(AddTag_k^* \cup DelTag_k^*)$, and comparing it with the pre-stored proof π_k. Instead, G_2 directly compares $AddTag_k^* \cup DelTag_k^*$ with the set $AddTag_k \cup DelTag_k$ derived from the correct search result R_{raw,w_i}. Since R_{raw,w_i} is correct, so we have $\pi_k = \mathcal{H}(AddTag_k \cup DelTag_k)$, so G_2 and G_1 are indistinguishable except for the event that $\mathcal{H}(AddTag_k^* \cup DelTag_k^*) = \mathcal{H}(AddTag_k \cup DelTag_k)$ and $AddTag_k^* \cup DelTag_k^* \neq AddTag_k \cup DelTag_k$ happens, denoted as Bad_1. Then we have:

$$|\Pr[G_2 = 1] - Pr[G_1 = 1]| \leq \Pr[\mathsf{Bad}_1]$$

It is oblivious that $\Pr[\mathsf{Bad}_1]$ is negligible if $\mathcal{H}$ is set-collision resistant. G_2 makes same modification in verifying $OldTag^*$ with the same argument above.

Game G_3: G_3 changes the way of verifying $DelTag_k^*$ in phase-II. G_2 verifies $DelTag_k^*$ derived from R_{raw,w_i}^* by checking if every tag in $DelTag_k^*$ belongs to D. Instead, G_3 directly compares $DelTag_k^*$ with $DelTag_k$ derived from the correct search result R_{raw,w_i}. G_3 and G_2 are indistinguishable except for the event that $DelTag_k^* \neq DelTag_k$ happens but R_{raw,w_i} passes verification, denoted as Bad_2.

$$|\Pr[G_3 = 1] - \Pr[G_2 = 1]| \leq \Pr[\mathsf{Bad}_2]$$

If Bad_2 happens, then there exists some tag $t \notin DelTag_k^*$ but $\in DelTag_k$. Since $AddTag_k^* \cup DelTag_k^* = AddTag_k \cup DelTag_k$ according to phase-I, so t must belong to $AddTag_k^*$. So $\mathcal{A}$ is able to infer corresponding identifier ind from t so that the identifier-tag pair (ind, t) can pass phase-III. That is, given a tag t, the adversary can find ind s.t. $t = h_1(K_w||ind)$. It is straightforward to see that if h_1 is first-preimage resistant, the probability that Bad_2 happens is negligible.

Game G_4: In phase-III, G_4 verifies $AddInd_k^*$ by directly comparing it with $AddInd_k$ derived from R_{raw,w_i}, instead of checking if $AddInd_k^*$ is consistent with correct $AddTag_k^*$. G_4 and G_3 are indistinguishable except for the event that $AddInd_k^*$ is not correct even if it is consistent with correct $AddTag_k^*$, denoted as Bad_3, so we have:

$$|\Pr[G_4 = 1] - \Pr[G_3 = 1] \leq \Pr[\mathsf{Bad}_3]$$

If Bad_3 happens, then there exists two identifier-tag pairs with different identifiers but same tag are both able to pass phase-III, which means:

$$\exists\ t \in AddTag_k^*, ind_a \neq ind_b\ s.t.\ t = h_1(K_w||ind_a) = h_1(K_w||ind_b)$$

It is straightforward that $\Pr[\mathsf{Bad}_3]$ is negligible if h_1 is a second-preimage resistant hash function. Similarly, G_4 makes same modification in verifying of $OldInd^*$ with same argument presented above. It is oblivious that $\Pr[G_4 = 1] = 0$ since G_4 directly compares R^*_{raw,w_i} with R_{raw,w_i} during Verify. So we have:

$$\Pr[\mathbf{Exp}^{\mathsf{verify}}_{\Sigma,\mathcal{A}}(\lambda) = 1] \leq negl(\lambda)$$

4 Implementation and Evaluation

We implement Seren in C++ and use OpenSSL to implement cryptographic primitives, i.e., AES and SHA256. For CSRE, we follow the implementation provided by [25], which is based on GGM tree-based multi-puncturable PRF and Bloom filter (as instantiation of D). We use Rocksdb [27] to store outsourced data in the server and auditor, and leverage Thrift [20] to enable the network communication between the client and server. We use C++ standard library chrono and Wireshark to precisely obtain the latency and communication cost of each operation respectively.

Table 2. Storage cost of D

d	10	100	1000	10000	10000
Storage(KB)	0.293	3.01	29.75	295.32	3054.79

Table 3. Deletion latency and communication cost comparison

Schemes	Latency(ms)				Communication cost (Bytes)			
	d=10	d=100	d=1000	d=10000	d=10	d=100	d=1000	d=10000
Aura	0.048	0.429	3.495	11.507	0	0	0	0
Seren	0.063	0.382	2.574	12.069	0	0	0	0
BPVSE	401.45	413.13	416.70	444.02	1596	11046	105546	574153

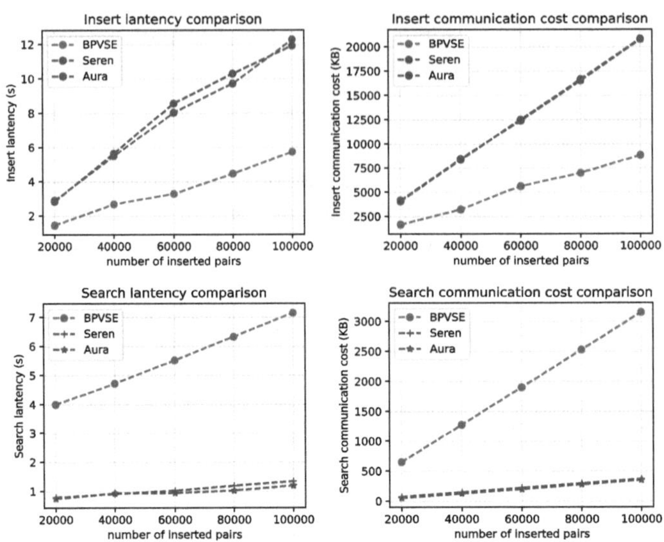

Fig. 2. Latency and communication cost comparison of insertion and search

Client storage cost. We first present our settings for the compressed deletion list D, which is implemented as Bloom filter. Then we present storage cost under different settings of d, which is the number of deletions on a certain keyword, ranging from $\{10, 100, 1000, 10000\}$. We set the false positive $p = 10^{-6}$, and derive the optimal size of Bloom filter as $b = -dlnp/(ln2)^2$, the number of hash functions as $h = \lceil b/dln2 \rceil$. We set the height of GGM tree as $log(b)$, and more concrete information about it is shown in [25]. Detailed space consumption of D is given in Table 2. We can see that the overhead of D is less than 3MB under 10000 deletions, which is always refreshed after each search. That demonstrates the storage overhead for clients in Seren is acceptable in practical scenarios.

Deletion efficiency. As shown in Table 3, we run deletion operations for d times, as $10, 100, 1000, 10000$, and record the latency and communication cost in 201ms-delay network. When performing deletions, BPVSE treats them as insertions, and uploads same amount of data to the server. Differently, Seren doesn't communicate with server, and just locally inserts the tags of deleted identifiers into D. We can see that our scheme Seren is 3 to 5 orders of magnitude faster than BPVSE, and deletion latency of BPVSE under different d changes slightly. These two phenomena are due to the fact that deletion latency of BPVSE is dominated by the huge cost of network delay rather than the computation of protocol. Furthermore, the communication cost induced by deletions is zero in Seren instead of same cost as insertions in BPVSE. In addition, Seren has close deletion latency and communication cost compared with Aura.

Insertion efficiency. We insert 20000, 40000, 60000, 80000, 100000 keyword-identifier pairs to EDB in 201ms-delay network. Figure 2 presents the experimental results, which shows that the insertion latency and communication overhead of Seren are both 3x to 4x lower than BPVSE. The reason is that, to avoid false positive cases, CSRE has to generate and upload h ciphertexts rather than one ciphertext in BPVSE for an inserted identifier, where h is the number of hash functions in Bloom filter settings. It is noteworthy that this additional overhead is caused by CSRE, and we can see Seren and Aura have close insertion performance since they both employ CSRE. However, these h ciphertexts are generated independently, which can be accelerated by multithreading. In this way, the insertion performance gaps of these schemes will be eliminated.

Search efficiency. We update EDB with 20000, 40000, 60000, 80000, 100000 insertions, and 1000 deletions. After that we perform search queries and record the search latency and communication cost, as shown in Fig. 2. Our experimental results show that for search in 201ms-delay network, Seren is 5x to 7x faster than BPVSE, and has 8x to 9x lower communication overhead compared with BPVSE. Moreover, the larger the data scale, the greater the advantage of Seren. The reason is that in BPVSE, the server needs one extra roundtrip of interaction with the client to get search results, and the expensive decryption computation is undertaken by the client locally, which induces large latency and communication cost, e.g., under the setting of 100000 insertions and 1000 deletions in BPVSE, the client additionally receives 101000 encrypted identifiers from the server, decrypts them, and sends the results back. However, Seren achieves non-interactive search and the decryption computation is undertaken by server, so it performs much better in real-world conditions where the network has delay and the server's computing resources are much richer than the client's. Seren and Aura have close search latency and communication cost, and Aura has a little better performance since Seren needs to transfer extra auxiliary information (K_w, D) for subsequent verification to the auditor compared with Aura.

Verification efficiency. We evaluate the time cost of each phase in the verification process under n = 20000, 40000, 60000, 80000, 100000 insertions and 1000 deletions. The detailed evaluation results are shown in Table 4. We can see that the verification process only costs about 50ms under 100000 insertions and 1000 deletions, which demonstrates that our verification scheme maintains good efficiency under large-scale databases.

Table 4. Verification time cost (ms) under different n

Phase	n=20000	n=40000	n=60000	n=80000	n=100000
Phase-I	6.458	12.124	16.910	21.636	25.028
Phase-II	4.225	5.004	4.897	5.136	4.768
Phase-III	6.073	10.272	13.467	16.281	18.293
Total	16.456	27.400	36.460	44.619	50.475

5 Conclusion

In this work, we explore the problem of designing more practical publicly verifiable DSSE schemes. Based on CSRE, set hash functions, and a novel "indirect" verification idea with three phases, we propose the first non-interactive PV-DSSE scheme with support for forward privacy and Type-II backward privacy. We provide formal security proof and comprehensive evaluation in real networks for our scheme, which shows that our scheme outperforms the state-of-art work BPVSE in search, update latency and communication cost. Still, all existing publicly verifiable DSSE schemes support only single-keyword queries, and it is is challenging to design practical PV-DSSE schemes for richer query types.

Acknowledgement. We would like to thank the anonymous reviewers for their constructive comments. This work is supported by the National Natural Science Foundation of China (Grant No. 62272294, 62202305), Young Elite Scientists Sponsorship Program by China Association for Science and Technology (YESS20220150), Shanghai Pujiang Program under Grant 22PJ1407700.

References

1. Bossuat, A., Bost, R., Fouque, P.A., Minaud, B., Reichle, M.: SSE and SSD: page-efficient searchable symmetric encryption. In: Advances in Cryptology–CRYPTO 2021, Part III 41, pp. 157–184 (2021)
2. Bost, R.: $\sigma o\varphi o\varsigma$: Forward secure searchable encryption. In: Proceedings of the 2016 ACM SIGSAC Conference on Computer and Communications Security (2016)
3. Bost, R., Fouque, P.A., Pointcheval, D.: Verifiable dynamic symmetric searchable encryption: Optimality and forward security. Cryptol. ePrint Archive (2016)
4. Bost, R., Minaud, B., Ohrimenko, O.: Forward and backward private searchable encryption from constrained cryptographic primitives. In: Proceedings of the 2017 ACM SIGSAC Conference on Computer and Communications Security (2017)
5. Cash, D., Grubbs, P., Perry, J., Ristenpart, T.: Leakage-abuse attacks against searchable encryption. In: Proceedings of the 22nd ACM SIGSAC Conference on Computer and Communications Security, pp. 668–679 (2015)
6. Cash, D., et al.: Dynamic searchable encryption in very-large databases: data structures and implementation. Cryptol. ePrint Archive (2014)
7. Chai, Q., Gong, G.: Verifiable symmetric searchable encryption for semi-honest-but-curious cloud servers. In: 2012 IEEE International Conference on Communications (ICC), pp. 917–922. IEEE (2012)
8. Chen, B., Xiang, T., He, D., Li, H., Choo, K.K.R.: BPVSE: publicly verifiable searchable encryption for cloud-assisted electronic health records. IEEE Trans. Inf. Forensics Secur. **18**, 3171–3184 (2023)
9. Cheng, R., Yan, J., Guan, C., Zhang, F., Ren, K.: Verifiable searchable symmetric encryption from indistinguishability obfuscation. In: Proceedings of the 10th ACM Symposium on Information, Computer and Communications Security (2015)
10. Chenghao, YUAN, Yong LI, S.R.: Dynamic multi-keyword searchable encryption scheme. Chin. J. Netw. Inf. Secur. **9**(2), 143 (2023)
11. Curtmola, R., Garay, J., Kamara, S., Ostrovsky, R.: Searchable symmetric encryption: improved definitions and efficient constructions. In: Proceedings of the 13th ACM Conference on Computer and Communications security, pp. 79–88 (2006)

12. Demertzis, I., Chamani, J.G., Papadopoulos, D., Papamanthou, C.: Dynamic searchable encryption with small client storage. Cryptol. ePrint Archive (2019)
13. Demertzis, I., Papadopoulos, D., Papamanthou, C.: Searchable encryption with optimal locality: Achieving sublogarithmic read efficiency. In: Advances in Cryptology–CRYPTO 2018, Part I 38, pp. 371–406 (2018)
14. Ghareh Chamani, J., Papadopoulos, D., Papamanthou, C., Jalili, R.: New constructions for forward and backward private symmetric searchable encryption. In: Proceedings of the 2018 ACM SIGSAC Conference on Computer and Communications Security, pp. 1038–1055 (2018)
15. Grubbs, P., Lacharité, M.S., Minaud, B., Paterson, K.G.: Pump up the volume: practical database reconstruction from volume leakage on range queries. In: Proceedings of the 2018 ACM SIGSAC Conference on Computer and Communications Security, pp. 315–331 (2018)
16. Jutla, C., Patranabis, S.: Efficient searchable symmetric encryption for join queries. In: International Conference on the Theory and Application of Cryptology and Information Security, pp. 304–333. Springer (2022). https://doi.org/10.1007/978-3-031-22969-5_11
17. Kamara, S., Moataz, T., Ohrimenko, O.: Structured encryption and leakage suppression. In: Advances in Cryptology–CRYPTO 2018: 38th Annual International Cryptology Conference, pp. 339–370. Springer (2018).https://doi.org/10.1007/978-3-319-96884-1_12
18. Miers, I., Mohassel, P.: Io-DSSE: scaling dynamic searchable encryption to millions of indexes by improving locality. Cryptol. ePrint Archive (2016)
19. Minaud, B., Reichle, M.: Dynamic Local Searchable Symmetric Encryption. In: Dodis, Y., Shrimpton, T. (eds) Advances in Cryptology – CRYPTO 2022. CRYPTO 2022. LNCS, vol. 13510. Springer, Cham (2022). https://doi.org/10.1007/978-3-031-15985-5_4
20. Slee, M., Agarwal, A., Kwiatkowski, M.: Thrift: scalable cross-language services implementation. Facebook white pap. **5**(8), 127 (2007)
21. Soleimanian, A., Khazaei, S.: Publicly verifiable searchable symmetric encryption based on efficient cryptographic components. Des. Codes Crypt. **87**(1), 123–147 (2019)
22. Song, D.X., Wagner, D., Perrig, A.: Practical techniques for searches on encrypted data. In: Proceeding 2000 IEEE Symposium on Security and Privacy. S&P 2000, pp. 44–55 (2000)
23. Song, X., Dong, C., Yuan, D., Xu, Q., Zhao, M.: Forward private searchable symmetric encryption with optimized I/O efficiency. IEEE Trans. Dependable Secure Comput. **17**(5), 912–927 (2018)
24. Stefanov, E., Papamanthou, C., Shi, E.: Practical dynamic searchable encryption with small leakage. Cryptol. ePrint Archive (2013)
25. Sun, S.F., et al.: Practical non-interactive searchable encryption with forward and backward privacy. In: NDSS (2021)
26. Sun, S.F., et al.: Practical backward-secure searchable encryption from symmetric puncturable encryption. In: Proceedings of the 2018 ACM SIGSAC Conference on Computer and Communications Security, pp. 763–780 (2018)
27. Team, F.: RocksDB: a persistent key-value store for flash and ram storage (2021)
28. Wegman, M.N., Carter, J.L.: New hash functions and their use in authentication and set equality. J. Comput. Syst. Sci. **22**(3), 265–279 (1981)
29. Ying, L.I, Chunguang, M.A.: Overview of searchable encryption research. Chin. J. Netw. Inf. Secur. **4**(7), 13 (2018)

30. Zhang, Y., Katz, J., Papamanthou, C.: All your queries are belong to us: the power of file-injection attacks on searchable encryption. In: 25th USENIX Security Symposium (USENIX Security 16), pp. 707–720 (2016)
31. Zhang, Z., Wang, J., Wang, Y., Su, Y., Chen, X.: Towards efficient verifiable forward secure searchable symmetric encryption. In: Computer Security–ESORICS 2019: 24th European Symposium on Research in Computer Security, Luxembourg, September 23–27, 2019, Proceedings, Part II 24, pp. 304–321. Springer (2019).https://doi.org/10.1007/978-3-030-29962-0_15

On the Implications from Updatable Encryption to Public-Key Cryptographic Primitives

Yuichi Tanishita[1,2(✉)], Ryuya Hayashi[1,2], Ryu Ishii[3], Takahiro Matsuda[2], and Kanta Matsuura[1]

[1] The University of Tokyo, Tokyo, Japan
{rhys,kanta}@iis.u-tokyo.ac.jp, y-tani@iis.u-tokyo.ac.jp
[2] National Institute of Advanced Industrial Science and Technology, Tokyo, Japan
t-matsuda@aist.go.jp
[3] Deloitte Tohmatsu Cyber LLC, Tokyo, Japan
ryu.ishii@tohmatsu.co.jp

Abstract. Updatable encryption (UE) is a special type of symmetric-key encryption (SKE) that allows a third party to update ciphertexts while protecting plaintexts. Alamati et al. (CRYPTO 2019) showed a curious connection between UE and public-key encryption (PKE) that PKE can be constructed from UE. This implication result is somewhat surprising since it is well-known that PKE cannot be constructed from (ordinary) SKE in a black-box manner.

In this paper, we continue to study the relationships between UE and other cryptographic primitives to obtain further insights into the existence and power of UE, and assumptions required for it. More specifically, we introduce some security properties that are natural to consider for UE (and are indeed satisfied by some existing UE schemes), and then investigate what types of public-key cryptographic primitives can be constructed from UE with the additional properties. Specifically, we show the following results:

- 2-round oblivious transfer (OT) can be constructed from UE that satisfies the *oblivious samplability (OS)* of original ciphertexts (i.e. those generated by the ordinary encryption algorithm, as opposed to those generated by the ciphertext-update algorithm) and the OS of update tokens (that are used for updating ciphertexts).
- 3-round OT can be constructed from UE with OS of updated ciphertexts (i.e. those generated by the ciphertext-update algorithm).
- Lossy encryption and PKE secure against selective-opening attacks can be constructed from UE if it satisfies what we call *statistical confidentiality of original ciphertexts.*

Keywords: Updatable encryption · Public-key encryption · Oblivious samplability · Lossy encryption

T. Zhu and Y. Li (Eds.): ACISP 2024, LNCS 14895, pp. 303–323, 2024.
https://doi.org/10.1007/978-981-97-5025-2_16

1 Introduction

Compromise of a private key is one of the most serious incidents that can make cryptography insecure. To minimize the impact of such a key compromise, it is important to periodically update the secret key and ciphertexts. *Updatable Encryption* (UE) [4] is a symmetric-key encryption (SKE) scheme that allows a third party to update the ciphertext without revealing the secret information (the key and the plaintexts) to the third party. Specifically, a ciphertext stored on a remote server managed by a third party, such as a cloud server, can be updated without disclosing the key or the plaintext to the administrator by using the so-called update token generated from a secret key. In recent years, there has been a number of works on UE, especially on its security notions [1,4,5,9,10,13–15,17]. UE and its basic security definition called IND-ENC security, were first proposed by Boneh et al. [4] as an extension of the standard IND-CPA security for SKE. Subsequently, Lehmann and Tackmann [15] proposed the security notion called IND-UPD security, which captures so-called post-compromise security, and Boyd et al. [5] proposed the security notion called IND-UE security, which is a stronger security notion than IND-ENC and IND-UPD, and intuitively captures the property that a ciphertext generated by the standard encryption procedure (called an *original ciphertext*) is indistinguishable from a ciphertext obtained by updating an existing ciphertext (called an *updated ciphertext*) encrypting any plaintext.

Depending on how an update token is generated, UE has two major variants: one with *ciphertext-dependent* updates in which an update token is associated with a particular ciphertext, and one with *ciphertext-independent* updates where an update token is not specific to an individual ciphertext, and can be used to update any ciphertext of the same epoch. In terms of functionality, UE with ciphertext-independent updates is more useful than the one with ciphertext-dependent updates. The previous works [4,9] showed that a UE scheme with ciphertext-dependent updates can be constructed from the combination of symmetric-key primitives such as authenticated encryption and thus quite efficient, while no such scheme has been known for UE with ciphertext-independent updates. Alamati et al. [1] showed a curious result that in some sense explains the phenomenon. Specifically, they showed that an IND-CPA secure public-key encryption (PKE) scheme can be constructed from an IND-UPD secure UE scheme with ciphertext-independent updates. In general, it is not known whether it is possible to construct PKE from SKE, and it is even impossible if we consider black-box constructions/reductions [12] which capture the most natural and efficient constructions. The result of Alamati et al. suggests that although UE with ciphertext-independent updates (and post-compromise security) is a symmetric-key primitive, it requires an assumption that is equal to or stronger than those required for public-key primitives.

From here on, by UE, we mean UE with ciphertext-independent updates. The main purpose of our paper is to clarify further relationships between UE and public-key primitives in a similar manner to Alamati et al. [1]. Clarifying such relationships deepens our understanding of UE, and hopefully leads to insights into constructing more efficient schemes than existing ones, or showing some barriers of efficiency or assumptions on which UE can be based. Also, it leads to systematization in the relationships among cryptographic primitives. Our approach is to consider some additional yet natural properties for UE, and then show new implications of UE with such natural additional properties to public-key primitives beyond IND-CPA secure PKE.

1.1 Our Contributions

As mentioned earlier, we introduce some natural additional properties for UE, and then show new implications of UE to public-key primitives beyond IND-CPA secure PKE. Specifically, one of the properties we consider for UE is *oblivious samplability* [6] of original ciphertexts, updated ciphertexts, and update tokens. Recall that oblivious samplability is a property that can be generally considered for any pair of a "public" value P (e.g., a public key or a ciphertext) and a "secret" value S (e.g., a secret key or a plaintext) such that either P is generated from S, or they are generated together, by some legitimate procedure (e.g., the key generation or the encryption algorithm). For such a pair, oblivious samplability means that a public value P (or something that is indistinguishable from a legitimate one) can be sampled without knowing the corresponding secret value S. The formal definitions for oblivious samplability for UE are given in Sect. 3, where we also discuss that several existing UE schemes indeed satisfy oblivious samplability (without introducing additional assumptions). Another property we consider for UE is a certain statistical property of the ciphertext-update algorithm that we call *statistical confidentiality of original ciphertexts*, which roughly means that the ciphertext-update procedure statistically hides the information of the original ciphertext (from which the updated ciphertext is generated), beyond the plaintext encrypted in the original ciphertext. Its formal definition is given in Sect. 5.

Given these additional properties for UE, our main results in this paper are described as follows:

- A UE scheme that satisfies the oblivious samplability of original ciphertexts and that of update tokens, implies a 2-round oblivious transfer (OT).
- A UE scheme that satisfies the oblivious samplability of updated ciphertexts, implies a 3-round OT.
- A UE scheme that satisfies IND-ENC security and statistical confidentiality of original ciphertexts, implies lossy encryption (Sect. 2.2) and PKE secure against selective opening attacks [2].

Technically, the first two results mentioned above are obtained by the following two steps: As the first step, we show how to construct a PKE scheme with oblivious samplability of public keys or ciphertexts from a UE scheme with oblivious samplability (Sect. 3). The construction of PKE from UE used here is a simplified variant of Alamati et al. [1]. Therefore, this step could be seen to clarify a different property of Alamati et al.'s construction based on a different assumption on the building block UE scheme. Then, as the second step, we invoke the results by Gertner et al. [11] who showed how to construct 2-round (resp. 3-round) OT from PKE with the oblivious samplability of public keys (resp. ciphertexts) (Sect. 6). The first and second steps are combined together to obtain the implication of UE with oblivious samplability to OT. Furthermore, for our third result (i.e. the implication of UE to lossy encryption), we also use the above mentioned simplified variant of Alamati et al.'s PKE construction based on UE (Sect. 5).

We note that our results are not implied by the result of Alamati et al. [1], since the result of Gertner et al. [11] shows that OT and PKE are separated with each other in terms of black-box constructions/reductions. Our results suggest further evidence that in terms of primitive-to-primitive construction, UE schemes satisfying the natural additional properties require strictly stronger assumptions than those required for an (ordinary) PKE scheme. For a more detailed discussion, see Sect. 6.

2 Preliminaries

In this section, we review the basic notation and the formal definitions for existing cryptographic primitives and their security properties used in this paper.

Notations. $\mathbb{N}$ denotes the set of all natural numbers. For $n \in \mathbb{N}$, we define $[n] := \{1, \ldots, n\}$. For strings x and y, $|x|$ denotes the bit length of x, and $(x \stackrel{?}{=} y)$ denotes the operation that returns 1 if and only if $x = y$. For a set S, $x \stackrel{\$}{\leftarrow} S$ denotes that an element x is chosen uniformly at random from S. $y \leftarrow x$ denotes that the value x is deterministically assigned to the variable y. If $\mathcal{A}$ is a probabilistic algorithm, then $y \stackrel{\$}{\leftarrow} \mathcal{A}(x)$ denotes that $\mathcal{A}$ receives the input x and outputs y. When we need to make the randomness r used by $\mathcal{A}$ explicit, we denote $y \leftarrow \mathcal{A}(x; r)$ (in which case, the computation by $\mathcal{A}$ is deterministic on input x and r). A function $f(\lambda) : \mathbb{N} \to [0, 1]$ is called *negligible* if for all constants $c \in \mathbb{N}$ and all sufficiently large $\lambda \in \mathbb{N}$, we have $f(\lambda) < 1/\lambda^c$. An unspecified negligible function is denoted by negl. PPT stands for *probabilistic polynomial time.*

2.1 Public-Key Encryption

Here, we review the definitions for PKE.

Definition 2.1 (PKE). *A PKE scheme* PKE *consists of the following three PPT algorithms* $(\mathsf{PKE.KG}, \mathsf{PKE.Enc}, \mathsf{PKE.Dec})$.

$\mathsf{PKE.KG}(1^\lambda) \stackrel{\$}{\to} (\mathsf{pk}, \mathsf{sk})$: *The key generation algorithm* $\mathsf{PKE.KG}$ *takes a security parameter* 1^λ *as input, and outputs a pair of public key and secret key* $(\mathsf{pk}, \mathsf{sk})$.

$\mathsf{PKE.Enc}(\mathsf{pk}, \mathsf{m}) \stackrel{\$}{\to} \mathsf{c}$: *The encryption algorithm* $\mathsf{PKE.Enc}$ *takes a public key* pk *and a plaintext* m *as input, and outputs a ciphertext* c.

$\mathsf{PKE.Dec}(\mathsf{sk}, \mathsf{c}) \to \mathsf{m}'/\bot$: *The decryption algorithm* $\mathsf{PKE.Dec}$ *takes a secret key* sk *and a ciphertext* c *as input, and outputs a decrypted plaintext* m' *(which could be the special symbol* $\bot$ *indicating a decryption error).*

Experiment $\mathsf{Expt}^{\mathsf{IND\text{-}CPA}}_{\mathsf{PKE},\mathcal{A}}(\lambda)$	**Experiment** $\mathsf{Expt}^{\mathsf{key\text{-}OS}}_{\mathsf{PKE},\mathcal{A}}(\lambda)$	**Experiment** $\mathsf{Expt}^{\mathsf{ctx\text{-}OS}}_{\mathsf{PKE},\mathcal{A}}(\lambda)$
$(\mathsf{pk}, \mathsf{sk}) \xleftarrow{\$} \mathsf{PKE.KG}(1^\lambda)$	$(\mathsf{pk}_0, \mathsf{sk}) \xleftarrow{\$} \mathsf{PKE.KG}(1^\lambda)$	$(\mathsf{pk}, \mathsf{sk}) \xleftarrow{\$} \mathsf{PKE.KG}(1^\lambda)$
$(\mathsf{m}_0, \mathsf{m}_1, \mathsf{st}) \xleftarrow{\$} \mathcal{A}_0(\mathsf{pk})$	$s_0 \xleftarrow{\$} \mathsf{rKG}(\mathsf{pk}_0)$	$(\mathsf{m}, \mathsf{st}) \xleftarrow{\$} \mathcal{A}_0(\mathsf{pk})$
$b \xleftarrow{\$} \{0,1\}$	$s_1 \xleftarrow{\$} \{0,1\}^\ell$	$\mathsf{c}_0 \xleftarrow{\$} \mathsf{PKE.Enc}(\mathsf{pk}, \mathsf{m})$
$\mathsf{c} \xleftarrow{\$} \mathsf{PKE.Enc}(\mathsf{pk}, \mathsf{m}_b)$	$\mathsf{pk}_1 \leftarrow \mathsf{oKG}(1^\lambda; s_1)$	$s_0 \xleftarrow{\$} \mathsf{rEnc}_{\mathsf{PKE}}(\mathsf{pk}, \mathsf{c}_0)$
$b' \xleftarrow{\$} \mathcal{A}_1(\mathsf{c}, \mathsf{st})$	$b \xleftarrow{\$} \{0,1\}$	$s_1 \xleftarrow{\$} \{0,1\}^\ell$
Return $(b' \stackrel{?}{=} b)$.	$b' \xleftarrow{\$} \mathcal{A}(\mathsf{pk}_b, s_b)$	$\mathsf{c}_1 \leftarrow \mathsf{oEnc}_{\mathsf{PKE}}(\mathsf{pk}; s_1)$
	Return $(b' \stackrel{?}{=} b)$.	$b \xleftarrow{\$} \{0,1\}$
		$b' \xleftarrow{\$} \mathcal{A}_1(\mathsf{c}_b, s_b, \mathsf{st})$
		Return $(b' \stackrel{?}{=} b)$.

Fig. 1. Security experiments for PKE: The IND-CPA experiment (left), the experiment for oblivious samplability of public keys (center), and that for oblivious samplability of ciphertexts (right). (In the IND-CPA experiment, it is required that $|\mathsf{m}_0| = |\mathsf{m}_1|$.)

Correctness. For all $\lambda \in \mathbb{N}$, m, and $(\mathsf{pk}, \mathsf{sk})$ generated by $\mathsf{PKE.KG}(1^\lambda)$, it always holds that $\mathsf{PKE.Dec}(\mathsf{sk}, \mathsf{PKE.Enc}(\mathsf{pk}, \mathsf{m})) = \mathsf{m}$.

IND-CPA Security. The most basic security notion for PKE is IND-CPA security defined as follows:

Definition 2.2 (IND-CPA Security). *We say that a PKE scheme* $\mathsf{PKE} = (\mathsf{PKE.KG}, \mathsf{PKE.Enc}, \mathsf{PKE.Dec})$ *is* IND-CPA secure, *if for all PPT adversaries* $\mathcal{A} = (\mathcal{A}_0, \mathcal{A}_1)$, *we have* $|\Pr[\mathsf{Expt}^{\mathsf{IND\text{-}CPA}}_{\mathsf{PKE},\mathcal{A}}(\lambda) = 1] - 1/2| = \mathsf{negl}(\lambda)$, *where the experiment* $\mathsf{Expt}^{\mathsf{IND\text{-}CPA}}_{\mathsf{PKE},\mathcal{A}}(\lambda)$ *is defined as in Fig. 1(left).*

Oblivious Samplability. Here, we review the definitions of oblivious samplability for public keys and that for ciphertexts (also called *simulatability* [6]). This property could be considered as a generalization of pseudorandomness that requires indistinguishability from a random string. Informally, in PKE with the oblivious samplability of public keys, via the so-called "oblivious sampling" algorithm, one can generate a public key (or, more precisely, something that looks like a public key) $\widetilde{\mathsf{pk}}$ without knowing the corresponding secret key, in such a way that it is indistinguishable from a legitimate public key pk generated together with a secret key by the ordinary key generation algorithm. It is required that there be the corresponding algorithm that on input a legitimate public key pk, can generate a "fake" randomness s for the oblivious sampling algorithm as if pk was generated by the oblivious sampling algorithm using the randomness s. The oblivious samplability for ciphertexts is similarly defined, and a ciphertext (or something that looks like a ciphertext) can be generated via the oblivious sampling algorithm, without knowing the corresponding plaintext.

Formally, the oblivious samplability of public keys is defined as follows:

Definition 2.3 (Oblivious Samplability of Public Keys). *We say that a PKE scheme* $\mathsf{PKE} = (\mathsf{PKE.KG}, \mathsf{PKE.Enc}, \mathsf{PKE.Dec})$ *satisfies the* oblivious samplability of public keys, *if there exists a pair of the PPT algorithms* $(\mathsf{oKG}, \mathsf{rKG})$ *that satisfies the following:*

- $\mathsf{oKG}(1^\lambda; \widetilde{s}) \rightarrow \widetilde{\mathsf{pk}}$*: This is the algorithm for oblivious sampling: It takes a security parameter* 1^λ *as input, uses a randomness* $\widetilde{s} \in \{0,1\}^\ell$ *(where* $\ell = \ell(\lambda)$ *is some polynomial)*[1]*, and outputs an obliviously sampled public key* $\widetilde{\mathsf{pk}}$.
- $\mathsf{rKG}(\mathsf{pk}) \xrightarrow{\$} s$*: This is the algorithm for explaining that a legitimate public key was generated obliviously: It takes a public key* pk *(supposedly generated by* $\mathsf{PKE.KG}$*) as input, and outputs a randomness* $s \in \{0,1\}^\ell$ *satisfying* $\mathsf{oKG}(1^\lambda; s) = \mathsf{pk}$.
- *For all PPT adversaries* $\mathcal{A}$*, it holds that* $|\Pr[\mathsf{Expt}^{\mathsf{key\text{-}OS}}_{\mathsf{PKE},\mathcal{A}}(\lambda) = 1] - 1/2| = \mathsf{negl}(\lambda)$*, where the experiment* $\mathsf{Expt}^{\mathsf{key\text{-}OS}}_{\mathsf{PKE},\mathcal{A}}(\lambda)$ *is defined as in Fig. 1(center).*

The oblivious samplability of ciphertexts is defined as follows. Note that the oblivious samplability of ciphertexts implies IND-CPA security.

Definition 2.4 (Oblivious Samplability of Ciphertexts). *A PKE scheme* $\mathsf{PKE} = (\mathsf{PKE.KG}, \mathsf{PKE.Enc}, \mathsf{PKE.Dec}$ *satisfies the* oblivious samplability of ciphertexts *if there exists a pair of the PPT algorithms* $(\mathsf{oEnc}_{\mathsf{PKE}}, \mathsf{rEnc}_{\mathsf{PKE}})$ *that satisfies the following:*

- $\mathsf{oEnc}_{\mathsf{PKE}}(\mathsf{pk}; \widetilde{s}) \rightarrow \widetilde{\mathsf{c}}$*: This is the algorithm for oblivious sampling: It takes a public key* pk *as input, uses a randomness* $\widetilde{s} \in \{0,1\}^\ell$ *(where* $\ell = \ell(\lambda)$ *is some polynomial), and outputs an obliviously sampled ciphertext* $\widetilde{\mathsf{c}}$.
- $\mathsf{rEnc}_{\mathsf{PKE}}(\mathsf{pk}, \mathsf{c}) \xrightarrow{\$} s$*: This is the algorithm for explaining that a legitimate ciphertext was generated obliviously: It takes a public key* pk *and a ciphertext* c *(supposedly generated by* $\mathsf{PKE.Enc}$*) as input, and outputs a randomness* $s \in \{0,1\}^\ell$ *satisfying* $\mathsf{oEnc}_{\mathsf{PKE}}(\mathsf{pk}; s) = \mathsf{c}$.
- *For all PPT adversaries* $\mathcal{A} = (\mathcal{A}_0, \mathcal{A}_1)$*, it holds that* $|\Pr[\mathsf{Expt}^{\mathsf{ctx\text{-}OS}}_{\mathsf{PKE},\mathcal{A}}(\lambda) = 1] - 1/2| = \mathsf{negl}(\lambda)$*, where the experiment* $\mathsf{Expt}^{\mathsf{ctx\text{-}OS}}_{\mathsf{PKE},\mathcal{A}}(\lambda)$ *is defined as in Fig. 1(right).*

2.2 Lossy Encryption

Here, we review the definitions for lossy encryption. It is a special kind of PKE that has a procedure for generating a "lossy" public key. A ciphertext generated under a lossy key statistically loses the information of a plaintext.

Definition 2.5 (Lossy Encryption). *A lossy encryption scheme* LE *consists of the following four PPT algorithms* $(\mathsf{LE.KG}, \mathsf{LE.Enc}, \mathsf{LE.Dec}, \mathsf{LE.LKG})$.

[1] Throughout the paper, we will denote the randomness length of oblivious sampling algorithms by ℓ.

Experiment $\mathsf{Expt}^{\mathsf{LE\text{-}KEY\text{-}IND}}_{\mathsf{LE},\mathcal{A}}(\lambda)$	**Experiment** $\mathsf{Expt}^{\mathsf{LE\text{-}conf}}_{\mathsf{LE},\mathcal{A}}(\lambda)$
$(\mathsf{pk}_1, \mathsf{sk}) \xleftarrow{\$} \mathsf{LE.KG}(1^\lambda)$	$\mathsf{pk}_{\mathsf{loss}} \xleftarrow{\$} \mathsf{LE.LKG}(1^\lambda)$
$\mathsf{pk}_0 \xleftarrow{\$} \mathsf{LE.LKG}(1^\lambda)$	$(\mathsf{m}_0, \mathsf{m}_1, \mathsf{st}) \xleftarrow{\$} \mathcal{A}_0(\mathsf{pk}_{\mathsf{loss}})$
$b \xleftarrow{\$} \{0,1\}$	$b \xleftarrow{\$} \{0,1\}$
$b' \xleftarrow{\$} \mathcal{A}(\mathsf{pk}_b)$	$\mathsf{c} \xleftarrow{\$} \mathsf{LE.Enc}(\mathsf{pk}_{\mathsf{loss}}, \mathsf{m}_b)$
Return $(b' \stackrel{?}{=} b)$.	$b' \xleftarrow{\$} \mathcal{A}_1(\mathsf{c}, \mathsf{st})$
	Return $(b' \stackrel{?}{=} b)$.

Fig. 2. Security experiments for lossy encryption: The experiment for indistinguishability between ordinary and lossy public keys (left), the experiment for statistical confidentiality under lossy public keys (right). (In the experiment for statistical confidentiality under lossy public keys, it is required that $|\mathsf{m}_0| = |\mathsf{m}_1|$.)

$(\mathsf{LE.KG}, \mathsf{LE.Enc}, \mathsf{LE.Dec})$: *These algorithms are defined in the same way as the key generation, encryption, and decryption algorithms of a PKE scheme, respectively. We also define correctness in exactly the same way as that for ordinary PKE.*

$\mathsf{LE.LKG}(1^\lambda) \xrightarrow{\$} \mathsf{pk}_{\mathsf{loss}}$: *The lossy key generation algorithm* $\mathsf{LE.LKG}$ *takes a security parameter* 1^λ *as input, and outputs a lossy public key* $\mathsf{pk}_{\mathsf{loss}}$.

Security Notions. We recall the two security properties of lossy encryption: *indistinguishability between ordinary and lossy public keys*, and *statistical confidentiality under lossy keys.*

Definition 2.6 (Indistinguishability between Ordinary and Lossy Public Keys). *We say that a lossy encryption scheme* $\mathsf{LE} = (\mathsf{LE.KG}, \mathsf{LE.Enc}, \mathsf{LE.Dec}, \mathsf{LE.LKG})$ *satisfies* indistinguishability between ordinary and lossy public keys, *if for all PPT adversaries* $\mathcal{A}$, *it holds that* $|\Pr[\mathsf{Expt}^{\mathsf{LE\text{-}KEY\text{-}IND}}_{\mathsf{LE},\mathcal{A}}(\lambda) = 1] - 1/2| = \mathsf{negl}(\lambda)$, *where the experiment* $\mathsf{Expt}^{\mathsf{LE\text{-}KEY\text{-}IND}}_{\mathsf{LE},\mathcal{A}}(\lambda)$ *is defined in Fig. 2(left).*

Definition 2.7 (Statistical Confidentiality under Lossy Keys). *We say that a lossy encryption scheme* $\mathsf{LE} = (\mathsf{LE.KG}, \mathsf{LE.Enc}, \mathsf{LE.Dec}, \mathsf{LE.LKG})$ *satisfies* statistical confidentiality, *if for all computationally unbounded adversaries* $\mathcal{A} = (\mathcal{A}_0, \mathcal{A}_1)$, *it holds that* $|\Pr[\mathsf{Expt}^{\mathsf{LE\text{-}wconf}}_{\mathsf{LE},\mathcal{A}}(\lambda) = 1] - 1/2| = \mathsf{negl}(\lambda)$, *where the experiment* $\mathsf{Expt}^{\mathsf{LE\text{-}wconf}}_{\mathsf{LE},\mathcal{A}}(\lambda)$ *is defined as in Fig. 2(right).*

It is easy to see that a lossy encryption scheme satisfying indistinguishability of ordinary and lossy public keys and statistical confidentiality implies that its first three algorithms constitute an IND-CPA secure PKE scheme.

2.3 Updatable Encryption

Here, we review the definitions for UE.

Definition 2.8 (Updatable Encryption). *A UE scheme* UE *consists of the following five PPT algorithms* $(\mathsf{UE.KG}, \mathsf{UE.Enc}, \mathsf{UE.Dec}, \mathsf{UE.Next}, \mathsf{UE.Upd})$.

$\mathsf{UE.KG}(1^\lambda) \xrightarrow{\$} \mathsf{k}_1$*: The key generation algorithm* $\mathsf{UE.KG}$ *takes a security parameter* 1^λ *as input, and outputs a key* k_1 *(at the initial epoch).*

$\mathsf{UE.Enc}(\mathsf{k}_e, \mathsf{m}) \xrightarrow{\$} \mathsf{c}_e$*: The encryption algorithm* $\mathsf{UE.Enc}$ *takes a key* k_e *(at some epoch* e*) and a plaintext* m *as input, and outputs a ciphertext* c_e.

$\mathsf{UE.Dec}(\mathsf{k}_e, \mathsf{c}_e) \rightarrow \mathsf{m}'/\bot$*: The decryption algorithm* $\mathsf{UE.Dec}$ *takes a key* k_e *and a ciphertext* c_e *(both at some epoch* e*) as input, and outputs a decrypted plaintext* m' *(which could be the special symbol* $\bot$ *indicating a decryption error).*

$\mathsf{UE.Next}(\mathsf{k}_e) \xrightarrow{\$} (\mathsf{k}_{e+1}, \Delta_{e+1})$*: This is the algorithm for updating a key and generating a corresponding token.* $\mathsf{UE.Next}$ *takes a key* k_e *at epoch* e *as input, and outputs a new key* k_{e+1} *and an update token (or just "token" for simplicity)* Δ_{e+1}*, both at the next epoch* $e+1$.

$\mathsf{UE.Upd}(\Delta_{e+1}, c_e) \xrightarrow{\$} \mathsf{c}_{e+1}$*: The ciphertext-update algorithm* $\mathsf{UE.Upd}$ *takes a token* Δ_{e+1} *at some epoch* $e+1$ *and a ciphertext* c_e *at epoch* e *as input, and outputs a ciphertext* c_{e+1} *at epoch* $e+1$.

We refer to a ciphertext generated by $\mathsf{UE.Enc}$ *as an* original ciphertext, *and one generated by* $\mathsf{UE.Upd}$ *as an* updated ciphertext.

Correctness. For all $\lambda \in \mathbb{N}$, all plaintexts m, all initial-epoch keys k_1 generated by $\mathsf{UE.KG}(1^\lambda)$, all epochs $e \geq 2$, all pairs $(\mathsf{k}_2, \Delta_2), \ldots, (\mathsf{k}_e, \Delta_e)$ that are recursively generated by $(\mathsf{k}_i, \Delta_i) \xleftarrow{\$} \mathsf{UE.Next}(\mathsf{k}_{i-1})$ for each $i \in \{2, \ldots, e\}$, all $\widetilde{e} \in [e]$, and all ciphertexts $\mathsf{c}_{\widetilde{e}}$ generated by $\mathsf{UE.Enc}(\mathsf{k}_{\widetilde{e}}, \mathsf{m})$ $(1 \leq \widetilde{e} \leq e)$ and c_j generated by $\mathsf{UE.Upd}(\Delta_j, \mathsf{c}_{j-1})$ $(j = \widetilde{e}+1, \ldots, e)$, it holds that $\mathsf{UE.Dec}(\mathsf{k}_j, \mathsf{c}_j) = \mathsf{m}$.

Security Definitions. Here, we recall the definitions for $\mathsf{IND\text{-}ENC}$ security and $\mathsf{IND\text{-}UPD}$ security. The security definitions in this paper are based on those in Miao et al. [16].[2]

For defining the security notions, we will consider an adversary that has access to the oracles described below, and we will also use the sets listed in Table 1 that are used by the oracles to handle (challenge) ciphertexts and the values leaked to the adversary. In the following, unless stated otherwise, e will denote the current epoch.

[2] For those who are familiar with the security notions for UE, the type of UE we consider in this paper is so-called *uni-directional ciphertext update* and *no-directional key update*. However, our results in Sects. 4 and 5 can be established irrespectively of the directionality of ciphertext update and key update, since we will rely only on the security property of UE in case no key is corrupted. For detailed discussions on the directionality of ciphertext update and key update, see [10,16].

$\mathcal{O}.\mathsf{Enc}(\mathsf{m})$: This is the encryption oracle. It takes a plaintext m as input, and computes and returns $\mathsf{c}_e \xleftarrow{\$} \mathsf{UE.Enc}(\mathsf{k}_e, \mathsf{m})$. It also adds the pair (c_e, e) into $\mathcal{L}$.

$\mathcal{O}.\mathsf{Next}$: This is the oracle to advance the current epoch. When queried (with no input), it increments the current epoch e as $e \leftarrow e+1$, executes $(\mathsf{k}_e, \Delta_e) \xleftarrow{\$} \mathsf{UE.Next}(\mathsf{k}_{e-1})$. If $\mathsf{phase} = 1$ (which guarantees $(\widetilde{\mathsf{c}}_{e-1}, e-1) \in \widetilde{\mathcal{L}}$ for some $\widetilde{\mathsf{c}}_{e-1}$), then it computes $\widetilde{\mathsf{c}}_e \xleftarrow{\$} \mathsf{UE.Upd}(\Delta_e, \widetilde{\mathsf{c}}_{e-1})$ and adds $(\widetilde{\mathsf{c}}_e, e)$ into $\widetilde{\mathcal{L}}$. (This oracle returns nothing.)

$\mathcal{O}.\mathsf{Upd}(\mathsf{c}_{e-1})$: This is the ciphertext-update oracle. It takes a ciphertext c_{e-1} as input, where it is required that $(\mathsf{c}_{e-1}, e-1) \in \mathcal{L}$. Then, it computes $\mathsf{c}_e \xleftarrow{\$} \mathsf{UE.Upd}(\Delta_e, \mathsf{c}_{e-1})$ and returns c_e. It also adds (c_e, e) into $\mathcal{L}$.

$\mathcal{O}.\mathsf{Corr}(\mathsf{inp}, e^*)$: This is an oracle that leaks a token or a key of the current or a past epoch. It takes a command $\mathsf{inp} \in \{\mathsf{token}, \mathsf{key}\}$ and an epoch $e^* \in [e]$ as input.

- If $\mathsf{inp} = \mathsf{token}$, then it returns Δ_{e^*} and adds e^* into $\mathcal{T}$.
- If $\mathsf{inp} = \mathsf{key}$, then it returns k_{e^*} and adds e^* into $\mathcal{K}$.

$\mathcal{O}.\mathsf{Upd}\widetilde{\mathsf{c}}$: This is the oracle for leaking an updated challenge ciphertext to an adversary. If $\mathsf{phase} = 0$ then it returns $\perp$. Otherwise, it returns the current-epoch challenge ciphertext $\tilde{\mathsf{c}}_e$ (retrieved from $\widetilde{\mathcal{L}}$).

$\mathcal{O}.\mathsf{Chall\text{-}IND\text{-}ENC}(\mathsf{m}_0, \mathsf{m}_1)$: This is the challenge oracle for $\mathsf{IND\text{-}ENC}$ security (used only once in the experiment). It takes a plaintext pair $(\mathsf{m}_0, \mathsf{m}_1)$ (of the same length) as input, and computes and returns $\widetilde{\mathsf{c}} \xleftarrow{\$} \mathsf{UE.Enc}(\mathsf{k}_{\widetilde{e}}, \mathsf{m}_b)$ (where b denotes the challenge bit chosen before this oracle is called). It also adds $\widetilde{e}$ and $(\widetilde{\mathsf{c}}, \widetilde{e})$ into $\mathcal{C}$ and $\widetilde{\mathcal{L}}$, respectively.

$\mathcal{O}.\mathsf{Chall\text{-}IND\text{-}UPD}(\bar{\mathsf{c}}_0, \bar{\mathsf{c}}_1)$: This is the challenge oracle for $\mathsf{IND\text{-}UPD}$ security (used only once in the experiment). It takes a pair of ciphertexts $(\bar{\mathsf{c}}_0, \bar{\mathsf{c}}_1)$ as input, where it is required that $(\bar{\mathsf{c}}_0, \widetilde{e}-1), (\bar{\mathsf{c}}_1, \widetilde{e}-1) \in \mathcal{L}$ and $|\bar{\mathsf{c}}_0| = |\bar{\mathsf{c}}_1|$. Then, it computes and returns $\widetilde{\mathsf{c}} \xleftarrow{\$} \mathsf{UE.Upd}(\Delta_{\widetilde{e}}, \bar{\mathsf{c}}_b)$ (where b denotes the challenge bit chosen just before this oracle is called). It also adds $\widetilde{e}$ and $(\widetilde{\mathsf{c}}, \widetilde{e})$ into $\mathcal{C}$ and $\widetilde{\mathcal{L}}$, respectively.

The formal definitions of the sets $\mathcal{K}^*, \mathcal{T}^*, \mathcal{C}^*$ are as follows, where e_{end} denotes the epoch at which an adversary terminates and outputs a guess bit in the security experiment: $\mathcal{K}^* := \mathcal{K}$, $\mathcal{T}^* := \{e \in [e_{\mathsf{end}}] | (e \in \mathcal{T}^*) \vee (e-1 \in \mathcal{K}^* \wedge e \in \mathcal{K}^*)\}$, and $\mathcal{C}^* := \{e \in [e_{\mathsf{end}}] | \mathsf{ChallEq}(e) = \mathsf{true}\}$, where $\mathsf{ChallEq}(e) = \mathsf{ture} \Leftrightarrow (e \in \mathcal{C}) \vee (\mathsf{ChallEq}(e-1) = \mathsf{ture} \wedge e \in \mathcal{T}^*)$.

Definition 2.9 (IND-ENC and IND-UPD Security). *Let* $\mathsf{X} \in \{\mathsf{ENC}, \mathsf{UPD}\}$. *We say that a UE scheme* $\mathsf{UE} = (\mathsf{UE.KG}, \mathsf{UE.Enc}, \mathsf{UE.Dec}, \mathsf{UE.Next}, \mathsf{UE.Upd})$ *is* $\mathsf{IND\text{-}X}$ *secure, if for all PPT adversaries* $\mathcal{A} = (\mathcal{A}_0, \mathcal{A}_1)$, *it holds that* $|\Pr[\mathsf{Expt}^{\mathsf{IND\text{-}X}}_{\mathsf{UE},\mathcal{A}}(\lambda) = 1] - 1/2| = \mathsf{negl}(\lambda)$, *where the experiment* $\mathsf{Expt}^{\mathsf{IND\text{-}X}}_{\mathsf{UE},\mathcal{A}}(\lambda)$ *is defined as in Fig. 3.*

Table 1. The roles of the sets used in the security experiments for UE.

$\mathcal{L}$	The set containing the pairs (c_e, e) of a (non-challenge) ciphertext and its epoch that is generated by $\mathcal{O}.\mathsf{Enc}$ or $\mathcal{O}.\mathsf{Upd}$
$\widetilde{\mathcal{L}}$	The set containing the pairs $(\widetilde{\mathsf{c}}_e, e)$ of an (updated) challenge ciphertext and its epoch that is generated by the challenge oracle or $\mathcal{O}.\mathsf{Upd}\widetilde{\mathsf{c}}$
$\mathcal{K}$	The set of epochs e at which a key k_e is leaked to an adversary via $\mathcal{O}.\mathsf{Corr}$
$\mathcal{T}$	The set of epochs e at which a token Δ_e is leaked to an adversary via $\mathcal{O}.\mathsf{Corr}$
$\mathcal{C}$	The set of epochs e at which the challenge ciphertext (or its updated version) at epoch e is given to an adversary via the challenge oracle or $\mathcal{O}.\mathsf{Upd}\widetilde{\mathsf{c}}$
$\mathcal{K}^*$	The set of epochs e (extended from $\mathcal{K}$) at which an adversary knows a key (or its equivalent) via the information collected during the experiment
$\mathcal{T}^*$	The of epochs e (extended from $\mathcal{T}$) at which an adversary knows a token (or its equivalent) via the information collected during the experiment
$\mathcal{C}^*$	The set of epochs e (extended from $\mathcal{C}$) at which an adversary knows the challenge ciphertext or its updated version (or its equivalent) via the information collected during the experiment

Experiment $\mathsf{Expt}^{\mathsf{IND\text{-}X}}_{\mathsf{UE},\mathcal{A}}$

$e \leftarrow 1;\ \ \mathsf{phase} \leftarrow 0;\ \ \mathcal{L}, \widetilde{\mathcal{L}}, \mathcal{K}, \mathcal{T}, \mathcal{C}, \mathcal{K}^*, \mathcal{T}^*, \mathcal{C}^* \leftarrow \emptyset$

$\mathsf{k}_1 \xleftarrow{\$} \mathsf{UE.KG}(1^\lambda)$

$(\mathbf{Chall}_0, \mathbf{Chall}_1, \mathsf{st}) \xleftarrow{\$} \mathcal{A}_0^{\mathcal{O}.\mathsf{Enc}, \mathcal{O}.\mathsf{Next}, \mathcal{O}.\mathsf{Upd}, \mathcal{O}.\mathsf{Corr}}(1^\lambda)$

$\widetilde{e} \leftarrow e;\ \ \mathsf{phase} \leftarrow 1;\ \ b \xleftarrow{\$} \{0,1\}$

$\widetilde{\mathsf{c}} \xleftarrow{\$} \mathcal{O}.\mathsf{Chall\text{-}IND\text{-}X}(\mathbf{Chall}_0, \mathbf{Chall}_1)$

$b' \xleftarrow{\$} \mathcal{A}_1^{\mathcal{O}.\mathsf{Enc}, \mathcal{O}.\mathsf{Next}, \mathcal{O}.\mathsf{Upd}, \mathcal{O}.\mathsf{Corr}, \mathcal{O}.\mathsf{Upd}\widetilde{\mathsf{c}}}(\widetilde{\mathsf{c}}, \mathsf{st})$

If $\mathcal{C}^* \cap \mathcal{K}^* \neq \emptyset$ then return a random bit and halt.

Return $(b' \stackrel{?}{=} b)$.

Fig. 3. The IND-ENC and IND-UPD experiments. (In the IND-ENC experiment, $(\mathbf{Chall}_0, \mathbf{Chall}_1)$ is a pair of plaintexts $(\mathsf{m}_0, \mathsf{m}_1)$; In the IND-UPD experiment, $(\mathbf{Chall}_0, \mathbf{Chall}_1)$ is a pair of ciphertexts $(\bar{\mathsf{c}}_0, \bar{\mathsf{c}}_1)$.)

3 Oblivious Samplability for UE

In this section, we introduce the formal definitions for oblivious samplability for UE. We consider oblivious samplability for three kinds of values in UE: the oblivious samplability of original ciphertexts (in Sect. 3.1), that for updated ciphertexts (in Sect. 3.2), and that for update tokens (in Sect. 3.3). We also explain concrete UE schemes that satisfy the definitions of oblivious samplability in Sect. 3.4.

3.1 Oblivious Samplability of Original Ciphertexts

Here, we define the oblivious samplability of original ciphertexts. Like the oblivious samplability of ciphertexts for PKE, this definition should capture the prop-

Experiment $\mathsf{Expt}^{\mathsf{OS\text{-}X}}_{\mathsf{UE},\mathcal{A}}(\lambda)$

$e \leftarrow 1$; $\mathsf{phase} \leftarrow 0$; $\mathcal{L}, \widetilde{\mathcal{L}}, \mathcal{K}, \mathcal{T}, \mathcal{C}, \mathcal{K}^*, \mathcal{T}^*, \mathcal{C}^* \leftarrow \emptyset$
$\mathsf{k}_1 \xleftarrow{\$} \mathsf{UE.KG}(1^\lambda)$
$(\mathbf{Chall}, \mathsf{st}) \xleftarrow{\$} \mathcal{A}_0^{\mathcal{O}.\mathsf{Enc}, \mathcal{O}.\mathsf{Next}, \mathcal{O}.\mathsf{Upd}, \mathcal{O}.\mathsf{Corr}}(1^\lambda)$
$\widetilde{e} \leftarrow e$; $\mathsf{phase} \leftarrow 1$; $b \xleftarrow{\$} \{0,1\}$
$(\widetilde{\mathsf{c}}_b, s_b) \xleftarrow{\$} \mathcal{O}.\mathsf{Chall\text{-}OS\text{-}X}(\mathbf{Chall})$
$b' \xleftarrow{\$} \mathcal{A}_1^{\mathcal{O}.\mathsf{Enc}, \mathcal{O}.\mathsf{Next}, \mathcal{O}.\mathsf{Upd}, \mathcal{O}.\mathsf{Corr}, \mathcal{O}.\mathsf{Upd}\widetilde{\mathsf{c}}}(\widetilde{\mathsf{c}}_b, s_b, \mathsf{st})$
If $\mathcal{C}^* \cap \mathcal{K}^* \neq \emptyset$ then return a random bit and halt.
Return $(b' \stackrel{?}{=} b)$.

Experiment $\mathsf{Expt}^{\mathsf{OS\text{-}token}}_{\mathsf{UE},\mathcal{A}}(\lambda)$

$\mathsf{k} \xleftarrow{\$} \mathsf{UE.KG}(1^\lambda)$
$(\mathsf{k}', \Delta_0) \xleftarrow{\$} \mathsf{UE.Next}(\mathsf{k})$
$s_0 \xleftarrow{\$} \mathsf{rTkn}(\Delta_0)$
$s_1 \xleftarrow{\$} \{0,1\}^\ell$
$\Delta_1 \leftarrow \mathsf{oTkn}(1^\lambda, 2; s_1)$
$b \xleftarrow{\$} \{0,1\}$
$b' \xleftarrow{\$} \mathcal{A}(\Delta_b, s_b)$
Return $(b' \stackrel{?}{=} b)$.

Fig. 4. Security experiments for defining oblivious samplability for UE: The experiment for the oblivious samplability of original/updated ciphertexts (left), and the experiment for the oblivious samplability of update tokens (right). (In the OS-ENC experiment, **Chall** is a plaintext m; in the OS-UPD experiment, **Chall** is a ciphertext $\bar{\mathsf{c}}$ (that is supposedly at epoch $\widetilde{e} - 1$).)

erty that an original ciphertext can be obliviously sampled (without knowing the corresponding plaintext) so that it is indistinguishable from an ordinary original ciphertext generated by UE.Enc, and it is necessary that there is a corresponding mechanism with which it can be explained that an ordinary original ciphertext was obliviously sampled.

We note that there is some freedom as to how we define such an intuitive property expected for the oblivious samplability of original ciphertexts for UE. Our definitional approach is to simply modify the challenge part of the IND-ENC security experiment, envisioning a natural relation that our oblivious samplability implies IND-ENC security for UE, very much like the oblivious samplability of ciphertexts for PKE that implies IND-CPA security.

Definition 3.1 (Oblivious Samplability of Original Ciphertexts). *We say that a UE scheme* $\mathsf{UE} = (\mathsf{UE.KG}, \mathsf{UE.Enc}, \mathsf{UE.Dec}, \mathsf{UE.Next}, \mathsf{UE.Upd})$ *satisfies the* oblivious samplability of original ciphertexts, *if there exists a pair of the PPT algorithms* $(\mathsf{oEnc}_{\mathsf{UE}}, \mathsf{rEnc}_{\mathsf{UE}})$ *that satisfies the following:*

- $\mathsf{oEnc}_{\mathsf{UE}}(1^\lambda, e; \widetilde{s}) \to \widetilde{\mathsf{c}}$: *This is the algorithm for oblivious sampling: It takes a security parameter* 1^λ *and an epoch* e *as input, uses a randomness* $\widetilde{s} \in \{0,1\}^\ell$ *(where* $\ell = \ell(\lambda)$ *is some polynomial), and outputs an obliviously sampled original ciphertext* $\widetilde{\mathsf{c}}$.
- $\mathsf{rEnc}_{\mathsf{UE}}(\mathsf{c}_e) \xrightarrow{\$} s$: *This is the algorithm for explaining that a legitimate original ciphertext was generated obliviously: It takes a ciphertext* c_e *at epoch* e *as input (supposedly generated by* UE.Enc*), and outputs* $s \in \{0,1\}^\ell$ *satisfying* $\mathsf{oEnc}_{\mathsf{UE}}(1^\lambda, e; s) = \mathsf{c}_e$.
- *For all PPT adversaries* $\mathcal{A} = (\mathcal{A}_0, \mathcal{A}_1)$, *it holds that* $|\Pr[\mathsf{Expt}^{\mathsf{OS\text{-}ENC}}_{\mathsf{UE},\mathcal{A}}(\lambda) = 1] - 1/2| = \mathsf{negl}(\lambda)$, *where the experiment* $\mathsf{Expt}^{\mathsf{OS\text{-}ENC}}_{\mathsf{UE},\mathcal{A}}(\lambda)$ *is defined in Fig. 4(left) with* $\mathsf{X} = \mathsf{ENC}$. *There, the oracles that* $\mathcal{A}$ *has access to (other than the chal-*

lenge oracle) are the same as in the IND-ENC *experiment, and the challenge oracle* $\mathcal{O}$.Chall-OS-ENC *is defined as follows:*

- $\mathcal{O}$.Chall-OS-ENC(m)*: It computes* $\widetilde{c}_0 \xleftarrow{\$} \mathsf{UE.Enc}(\mathsf{k}_{\widetilde{e}}, \mathsf{m})$, $s_0 \xleftarrow{\$} \mathsf{rEnc}_{\mathsf{UE}}(\widetilde{c}_0)$, $s_1 \xleftarrow{\$} \{0,1\}^{\ell}$, *and* $\widetilde{c}_1 \leftarrow \mathsf{oEnc}_{\mathsf{UE}}(1^{\lambda}; s_1)$. *Then, it returns the pair* $(\widetilde{c}_b, s_b)$ *(where b denotes the challenge bit chosen before this oracle is called). It also adds* $\widetilde{e}$ *and* $(\widetilde{c}_b, \widetilde{e})$ *into* $\mathcal{C}$ *and* $\widetilde{\mathcal{L}}$*, respectively.*

Due to how we define the oblivious samplability of original ciphertexts, it is not hard to see that it implies IND-ENC security.

3.2 Oblivious Samplability of Updated Ciphertexts

Here, we define the oblivious samplability of updated ciphertexts. This is a property analogous to the oblivious samplability of original ciphertexts, except that here what is sampled obliviously is an updated ciphertext. Since the definitional approach is essentially the same (except for the difference between UE.Enc and UE.Upd), we directly proceed to the formal definition.

Definition 3.2 (Oblivious Samplability of Updated Ciphertexts). *We say that a UE scheme* UE = (UE.KG, UE.Enc, UE.Dec, UE.Next, UE.Upd) *satisfies the* oblivious samplability of updated ciphertexts, *if there exists a pair of the PPT algorithms* (oUpd, rUpd) *that satisfies the following:*

- $\mathsf{oUpd}(\Delta_e; \widetilde{s}) \rightarrow \widetilde{c}$*: This is the algorithm for oblivious sampling: It takes a token* Δ_e *at some epoch e as input, uses a randomness* $\widetilde{s} \in \{0,1\}^{\ell}$ *(where* $\ell = \ell(\lambda)$ *is some polynomial), and outputs an obliviously sampled updated ciphertext at epoch e.*
- $\mathsf{rUpd}(\mathsf{c}_e, \Delta_e) \xrightarrow{\$} s$*: This is the algorithm for explaining that a legitimate updated ciphertext was generated obliviously. It takes a ciphertext* c_e *and token* Δ_e *(both at some epoch e) as input, and outputs a randomness* $s \in \{0,1\}^{\ell}$ *satisfying* $\mathsf{oUpd}(\Delta_e; s) = \mathsf{c}_e$.
- *For all PPT adversaries* $\mathcal{A} = (\mathcal{A}_0, \mathcal{A}_1)$*, it holds that* $|\Pr[\mathsf{Expt}^{\mathsf{OS\text{-}UPD}}_{\mathsf{UE},\mathcal{A}}(\lambda) = 1] - 1/2| = \mathsf{negl}(\lambda)$*, where the experiment* $\mathsf{Expt}^{\mathsf{OS\text{-}UPD}}_{\mathsf{UE},\mathcal{A}}(\lambda)$ *is defined as in Fig. 4(left) with* X = UPD*. There, the oracles that* $\mathcal{A}$ *has access to (other than the challenge oracle) are the same as in the* IND-ENC *experiment, and the challenge oracle* $\mathcal{O}$.Chall-OS-UPD *is defined as follows:*
  - $\mathcal{O}$.Chall-OS-UPD($\bar{c}$)*: (The input ciphertext must satisfy* $(\bar{c}, \widetilde{e} - 1) \in \widetilde{\mathcal{L}}$.) *It computes* $\widetilde{c}_0 \xleftarrow{\$} \mathsf{UE.Upd}(\Delta_{\widetilde{e}}, \bar{c})$, $s_0 \xleftarrow{\$} \mathsf{rUpd}(\widetilde{c}_0, \Delta_{\widetilde{e}})$, $s_1 \xleftarrow{\$} \{0,1\}^{\ell}$, *and* $\widetilde{c}_1 \leftarrow \mathsf{oUpd}(\Delta_{\widetilde{e}}; s_1)$. *Then, it returns the pair* $(\widetilde{c}_b, s_b)$ *(where b denotes the challenge bit chosen before this oracle is called). It also adds* $\widetilde{e}$ *and* $(\widetilde{c}_b, \widetilde{e})$ *into* $\mathcal{C}$ *and* $\widetilde{\mathcal{L}}$*, respectively.*

Similarly to the implication of oblivious samplability of original ciphertexts to IND-ENC security, it is not hard to see that the oblivious samplability of updated ciphertexts implies IND-UPD security.

3.3 Oblivious Samplability of Update Tokens

Here, we define the oblivious samplability of update tokens for UE. The property we would like to capture by this notion is that an update token Δ_e at some epoch e can be sampled obliviously, without knowing the corresponding previous-epoch key k_{e-1} (and an ordinary update token should be able to be explained as having been sampled obliviously).

There should be many possible definitions capturing such an intuitive property. We choose a simple definition in the sense that it only considers a token at the epoch next to the initial epoch. Unlike the oblivious samplability of original/updated ciphertexts given in the previous subsections (which implies IND-ENC and IND-UPD, respectively), it is not clear whether the oblivious samplability for tokens given below is useful in general, while it is sufficient for our purpose.

Definition 3.3 (Oblivious Samplability of Update Tokens). *We say that a UE scheme* UE *satisfies the oblivious samplability of update tokens if there exists a pair of the PPT algorithms* (oTkn, rTkn) *that satisfies the following:*

- $\mathsf{oTkn}(1^\lambda, e; \widetilde{s}) \rightarrow \widetilde{\Delta}_e$: *This is the algorithm for oblivious sampling. It takes a security parameter 1^λ and an epoch e as input, uses a randomness $\widetilde{s} \in \{0,1\}^\ell$ (where $\ell = \ell(\lambda)$ is some polynomial), and outputs an obliviously sampled token $\widetilde{\Delta}_e$ at epoch e.*
- $\mathsf{rTkn}(\Delta_e) \xrightarrow{\$} s$: *This is the algorithm for explaining that a legitimate token was generated obliviously. It takes a token Δ_e as input, and outputs $s \in \{0,1\}^\ell$ satisfying $\mathsf{oTkn}(1^\lambda, e; s) = \Delta_e$.*
- *For all PPT adversaries $\mathcal{A} = (\mathcal{A}_0, \mathcal{A}_1)$, it holds that $|\Pr[\mathsf{Expt}^{\mathsf{OS\text{-}token}}_{\mathsf{UE},\mathcal{A}}(\lambda) = 1] - 1/2| = \mathsf{negl}(\lambda)$, where the experiment $\mathsf{Expt}^{\mathsf{OS\text{-}token}}_{\mathsf{UE},\mathcal{A}}(\lambda)$ is defined as in Fig. 4 (right).*

3.4 Examples of UE Schemes Satisfying the Oblivious Samplability

The UE schemes proposed by Galteland and Pan [10, Sect. 5] and Miao et al. [16, Sect. 3.3], which are both based on a certain type of homomorphic PKE scheme supporting homomorphisms for keys and plaintexts, which can be instantiated from either the decisional Diffie-Hellman (DDH) or the learning-with-errors (LWE) assumption.[3] More specifically, in the UE schemes in [10] and [16], an original ciphertext, an updated ciphertext, and an update token essentially consist only of public keys and/or ciphertexts of the underlying homomorphic PKE scheme, and an update token consists of public key and ciphertext in the homomorphic encryption scheme, respectively. The homomorphic encryption schemes

[3] The definitions of homomorphic PKE supporting homomorphisms for keys and plaintexts used in [10] and [16] are different. In the case of the DDH-based instantiation, the one in [10] is simply the ElGamal PKE scheme, while the one in [16] is based on the PKE scheme by Boneh et al. [3]. For the details of the formal definitions as well as the instantiations, see the original papers.

instantiated based on DDH and LWE satisfy the pseudorandom property (indistinguishability from a uniform random element in the corresponding space) both for a public key and a ciphertext, and a random element in the public key/ciphertext space can be obliviously sampled when properly instantiated.[4] Hence, the UE schemes in [10] and [16] naturally succeed the pseudorandom property for an original/updated ciphertext and an update token, and in turn the oblivious samplability of these values.

4 Constructing PKE with Oblivious Samplability

In this section, we show our main technical result that a UE scheme with the oblivious samplability can be used to construct a PKE scheme with the oblivious samplability. The construction we use is, in fact, a simplified variant of the construction of PKE based on UE by Alamati et al. [1].

More specifically, in Sect. 4.1, we present a simplified variant of the PKE construction based on UE by Alamati et al. [1]. In Sect. 4.2, we show that this PKE scheme satisfies the oblivious samplability of public keys in case the underlying UE scheme satisfies the oblivious samplability of original ciphertexts and update tokens. In Sect. 4.3, we show that this PKE scheme satisfies the oblivious samplability of ciphertexts in case the underlying UE scheme satisfies the oblivious samplability of updated ciphertexts.

4.1 Simplified Version of Alamati et al.'s PKE Scheme

Alamati et al. [1] presented a PKE scheme based on a UE scheme, but their scheme is somewhat redundant in its treatment of multiple epochs of the underlying UE scheme, which is actually not necessary for the implication of UE to PKE. Hence, for simplicity, here we present a simplified variant of Alamati et al.'s PKE scheme, which uses only the initial and next epochs.

Let $\mathsf{UE} = (\mathsf{UE.KG}, \mathsf{UE.Enc}, \mathsf{UE.Dec}, \mathsf{UE.Next}, \mathsf{UE.Upd})$ be a UE scheme. Using it as a building block, we construct a PKE scheme $\mathsf{PKE} = (\mathsf{PKE.KG}, \mathsf{PKE.Enc}, \mathsf{PKE.Dec})$ whose plaintext space is $\{0,1\}$, as in Fig. 5.

The correctness of PKE follows immediately from the correctness of UE. The following theorem is proved in the same way as in the proof by Alamati et al. [1, Theorem 4.3] and thus its proof is omitted.

Theorem 4.1. *If* UE *is* $\mathsf{IND\text{-}UPD}$ *secure, then* PKE *is* $\mathsf{IND\text{-}CPA}$ *secure.*

4.2 Oblivious Samplability of Public Keys

Here, we show the following theorem that guarantees the oblivious samplability of public keys of PKE in Sect. 4.1.

[4] In the case of LWE-based instantiations, the public key/ciphertext space is just the set of integer vectors, and thus obliviously sampling an element from the spaces is trivial. For the DDH-based instantiations, we can use the simulatable group as formalized by Dent [7].

PKE.KG(1^λ):	PKE.Enc(pk, m $\in \{0,1\}$):	PKE.Dec(sk, c$'$):
$\mathsf{k} \xleftarrow{\$} \mathsf{UE.KG}(1^\lambda)$ $\mathsf{c}_0 \xleftarrow{\$} \mathsf{UE.Enc}(\mathsf{k}, 0)$ $\mathsf{c}_1 \xleftarrow{\$} \mathsf{UE.Enc}(\mathsf{k}, 1)$ $(\mathsf{k}', \Delta) \xleftarrow{\$} \mathsf{UE.Next}(\mathsf{k})$ $\mathsf{pk} \leftarrow (\mathsf{c}_0, \mathsf{c}_1, \Delta)$ $\mathsf{sk} \leftarrow \mathsf{k}'$ Return $(\mathsf{pk}, \mathsf{sk})$.	Parse pk as $(\mathsf{c}_0, \mathsf{c}_1, \Delta)$. $\mathsf{c}' \xleftarrow{\$} \mathsf{UE.Upd}(\Delta, \mathsf{c}_\mathsf{m})$ Return c'.	Parse sk as k'. $\mathsf{m}' \leftarrow \mathsf{UE.Dec}(\mathsf{k}', \mathsf{c}')$ Return m'.

Fig. 5. Simplified variant of Alamati et al.'s construction of PKE based on UE.

Theorem 4.2. *If* UE *satisfies the oblivious samplability of original ciphertexts and update tokens, then* PKE *satisfies the oblivious samplability of public keys.*

Proof. Let $(\mathsf{oEnc}_{\mathsf{UE}}, \mathsf{rEnc}_{\mathsf{UE}})$ and $(\mathsf{oTkn}, \mathsf{rTkn})$ be the algorithms that are guaranteed to exist due to the oblivious samplability of original ciphertexts and that for update tokes of UE, respectively. Let ℓ_{enc} be the randomness length of $\mathsf{oEnc}_{\mathsf{UE}}$ and ℓ_{tkn} be the randomness length of oTkn. Consider the following pair of the two PPT algorithms (oKG, rKG). (The randomness length of oKG is $2\ell_{\mathsf{enc}} + \ell_{\mathsf{tkn}}$.)

$\mathsf{oKG}(1^\lambda; s)$: On input 1^λ, and $s = (s_0, s_1, s_{\mathsf{tkn}}) \in (\{0,1\}^{\ell_{\mathsf{enc}}})^2 \times \{0,1\}^{\ell_{\mathsf{tkn}}}$, it runs $c_0 \leftarrow \mathsf{oEnc}_{\mathsf{UE}}(1^\lambda, 1; s_0)$, $c_1 \leftarrow \mathsf{oEnc}_{\mathsf{UE}}(1^\lambda, 1; s_1)$, and $\Delta \leftarrow \mathsf{oTkn}(1^\lambda, 2; s_{\mathsf{tkn}})$. Then, it returns an obliviously sampled public key $\widetilde{\mathsf{pk}} := (\mathsf{c}_0, \mathsf{c}_1, \Delta)$.

$\mathsf{rKG}(\mathsf{pk})$: On input $\mathsf{pk} = (\mathsf{c}_0, \mathsf{c}_1, \Delta)$, it runs $s_0 \xleftarrow{\$} \mathsf{rEnc}_{\mathsf{UE}}(\mathsf{c}_0)$, $s_1 \xleftarrow{\$} \mathsf{rEnc}_{\mathsf{UE}}(\mathsf{c}_1)$, and $s_{\mathsf{tkn}} \xleftarrow{\$} \mathsf{rTkn}(\Delta)$. Then, it returns a randomness $s := (s_0, s_1, s_{\mathsf{tkn}})$.

Let $\mathcal{A}$ be any PPT adversary that attacks the oblivious samplability of public keys of PKE, and consider the following sequence of games. (The steps that are not mentioned are not changed from the previous game.)

Game 1: This is the experiment for the oblivious samplability of public keys of PKE with the challenge bit $b = 0$. That is, execute $(\mathsf{pk} = (\mathsf{c}_0, \mathsf{c}_1, \mathsf{tkn}), \mathsf{sk}) \xleftarrow{\$} \mathsf{PKE.KG}(1^\lambda)$, $s = (s_0, s_1, s_{\mathsf{tkn}}) \xleftarrow{\$} \mathsf{rKG}(\mathsf{pk})$, and $b' \xleftarrow{\$} \mathcal{A}(\mathsf{pk}, s)$.

Game 2: Generate c_0 in $\mathsf{pk} = (\mathsf{c}_0, \mathsf{c}_1, \Delta)$ and s_0 in $s = (s_0, s_1, s_{\mathsf{tkn}})$ in Game 1 as follows: $s_0 \xleftarrow{\$} \{0,1\}^{\ell_{\mathsf{enc}}}$, $\mathsf{c}_0 \leftarrow \mathsf{oEnc}_{\mathsf{UE}}(1^\lambda, 1; s_0)$.

Game 3: Generate $\mathsf{c}_1 \in \mathsf{pk} = (\mathsf{c}_0, \mathsf{c}_1, \Delta)$ and s_1 in $s = (s_0, s_1, s_{\mathsf{tkn}})$ in Game 2 as follows: $s_1 \xleftarrow{\$} \{0,1\}^{\ell_{\mathsf{enc}}}$, $\mathsf{c}_1 \leftarrow \mathsf{oEnc}_{\mathsf{UE}}(1^\lambda, 1; s_1)$.

Game 4: Replace Δ in $\mathsf{pk} = (\mathsf{c}_0, \mathsf{c}_1, \Delta)$ and s_{tkn} in $s = (s_0, s_1, s_{\mathsf{tkn}})$ in Game 3 as follows: $s_{\mathsf{tkn}} \xleftarrow{\$} \{0,1\}^{\ell_{\mathsf{tkn}}}$, $\Delta \leftarrow \mathsf{oTkn}(1^\lambda, 1; s_{\mathsf{tkn}})$. Note that pk and s in this game are generated exactly as $s = (s_0, s_1, s_{\mathsf{tkn}}) \xleftarrow{\$} (\{0,1\}^{\ell_{\mathsf{enc}}})^2 \times \{0,1\}^{\ell_{\mathsf{tkn}}}$ and $\mathsf{pk} = (\mathsf{c}_0, \mathsf{c}_1, \Delta) \leftarrow \mathsf{oKG}(1^\lambda; s)$.

For each $i \in [4]$, let T_i be the event that $\mathcal{A}$ outputs $b' = 1$ in Game i. From the definition of Game 1, $\Pr[\mathsf{T}_1] = \Pr[b' = 1 \mid b = 0]$. By the definition of Game

4 and the discussion above, $\Pr[\mathsf{T}_4] = \Pr[b' = 1 \mid b = 1]$. In addition, we have that $|\Pr[\mathsf{Expt}^{\mathsf{key\text{-}OS}}_{\mathsf{PKE},\mathcal{A}}(\lambda) = 1] - 1/2| = (1/2)|\Pr[b' = 1 \mid b = 1] - \Pr[b' = 1 \mid b = 0]| = (1/2)|\Pr[\mathsf{T}_1] - \Pr[\mathsf{T}_4]|$. Therefore, by the triangular inequality, it is sufficient to show that all of $\{|\Pr[\mathsf{T}_i] - \Pr[\mathsf{T}_{i+1}]|\}_{i\in[3]}$ are negligible, in order to show that PKE satisfies the oblivious samplability of public keys.

We first show $|\Pr[\mathsf{T}_1]-\Pr[\mathsf{T}_2]| = \mathsf{negl}(\lambda)$. Using $\mathcal{A}$, we construct the following adversary $\mathcal{B} = (\mathcal{B}_0, \mathcal{B}_1)$ against the oblivious samplability of original ciphertexts of UE.

$\mathcal{B}_0^{\mathcal{O}.\mathsf{Enc},\mathcal{O}.\mathsf{Next},\mathcal{O}.\mathsf{Upd},\mathcal{O}.\mathsf{Corr}}(1^\lambda)$: $\mathcal{B}_0$ sets $\mathsf{m} = 0$ as its challenge plaintext, set $\mathsf{st}_\mathcal{B} = \mathsf{m}$, and terminates with the output $(\mathsf{m}, \mathsf{st}_\mathcal{B})$.

$\mathcal{B}_1^{\mathcal{O}.\mathsf{Enc},\mathcal{O}.\mathsf{Next},\mathcal{O}.\mathsf{Upd},\mathcal{O}.\mathsf{Corr},\mathcal{O}.\mathsf{Upd}\widetilde{\mathsf{c}}}(\widetilde{\mathsf{c}}, \widetilde{s}, \mathsf{st}_\mathcal{B})$: $\mathcal{B}_1$ calls $\mathcal{O}.\mathsf{Enc}(1)$ to get c_1. Next, $\mathcal{B}_1$ calls $\mathcal{O}.\mathsf{Next}$ to advance the epoch, and calls $\mathcal{O}.\mathsf{Corr}(\mathsf{token}, 2)$ to get a token Δ (at epoch $e = 2$). Then, $\mathcal{B}_1$ executes $s_1 \xleftarrow{\$} \mathsf{rEnc}_{\mathsf{UE}}(\mathsf{c}_1)$ and $s_{\mathsf{tkn}} \xleftarrow{\$} \mathsf{rTkn}(\Delta)$, sets $\mathsf{pk} := (\widetilde{\mathsf{c}}, \mathsf{c}_1, \Delta)$ and $s := (\widetilde{s}, s_1, s_{\mathsf{tkn}})$. Then, $\mathcal{B}_1$ runs $b' \xleftarrow{\$} \mathcal{A}(\mathsf{pk}, s)$. Finally, $\mathcal{B}_1$ sets $\beta' \leftarrow b'$, and terminates with output β'.

Let $\beta \in \{0, 1\}$ be $\mathcal{B}$'s challenge bit in $\mathsf{Expt}^{\mathsf{OS\text{-}ENC}}_{\mathsf{UE},\mathcal{B}}(\lambda)$. We see that $\mathcal{B}$ perfectly simulates Game 1 (resp. Game 2) for $\mathcal{A}$ when $\beta = 0$ (resp. $\beta = 1$), and $\mathcal{B}$ uses $\mathcal{A}$'s output as its output. Thus, we have $\Pr[\beta' = 1|\beta = 0] = \Pr[\mathsf{T}_1]$ and $\Pr[\beta' = 1|\beta = 1] = \Pr[\mathsf{T}_2]$. Hence, we have $|\Pr[\mathsf{T}_1] - \Pr[\mathsf{T}_2]| = |\Pr[\beta' = 1|\beta = 0] - \Pr[\beta' = 1|\beta = 1]| = 2 \cdot |\Pr[\mathsf{Expt}^{\mathsf{OS\text{-}ENC}}_{\mathsf{UE},\mathcal{B}}(\lambda)] - \frac{1}{2}| = \mathsf{negl}(\lambda)$.

The proof for showing $|\Pr[\mathsf{T}_2] - \Pr[\mathsf{T}_3]| = \mathsf{negl}(\lambda)$ is very similar to the one we showed above, and thus omitted due to the lack of space.

Finally, we show that $|\Pr[\mathsf{T}_3] - \Pr[\mathsf{T}_4]| = \mathsf{negl}(\lambda)$. Using $\mathcal{A}$, we construct the following adversary $\mathcal{B}''$ against the oblivious samplability of update tokens of UE.

$\mathcal{B}''(\widetilde{\Delta}, \widetilde{s})$: $\mathcal{B}''$ executes $s_0, s_1 \xleftarrow{\$} \{0,1\}^{\ell_{\mathsf{enc}}}$, $\mathsf{c}_0 \leftarrow \mathsf{oEnc}_{\mathsf{UE}}(1^\lambda, 1; s_0)$, and $\mathsf{c}_1 \leftarrow \mathsf{oEnc}_{\mathsf{UE}}(1^\lambda, 1; s_1)$. Then, $\mathcal{B}''$ sets $\mathsf{pk} := (\mathsf{c}_0, \mathsf{c}_1, \widetilde{\Delta})$ and $s := (s_0, s_1, \widetilde{s})$, and runs $b' \xleftarrow{\$} \mathcal{A}(\mathsf{pk}, s)$. Finally, $\mathcal{B}''$ sets $\beta' \leftarrow b'$, and terminates with output β'.

Let $\beta \in \{0, 1\}$ be $\mathcal{B}''$'s challenge bit in $\mathsf{Expt}^{\mathsf{OS\text{-}token}}_{\mathsf{UE},\mathcal{B}''}(\lambda)$. We see that $\mathcal{B}''$ perfectly simulates Game 3 (resp. Game 4) for $\mathcal{A}$ when $\beta = 0$ (resp. $\beta = 1$), and $\mathcal{B}''$ uses $\mathcal{A}$'s output as its output. Thus, we have $\Pr[\beta' = 1|\beta = 0] = \Pr[\mathsf{T}_3]$ and $\Pr[\beta' = 1|\beta = 1] = \Pr[\mathsf{T}_4]$. Hence, we have $|\Pr[\mathsf{T}_3] - \Pr[\mathsf{T}_4]| = |\Pr[\beta' = 1|\beta = 0] - \Pr[\beta' = 1|\beta = 1]| = 2 \cdot |\Pr[\mathsf{Expt}^{\mathsf{OS\text{-}token}}_{\mathsf{UE},\mathcal{B}''}(\lambda)] - \frac{1}{2}| = \mathsf{negl}(\lambda)$.

We have shown that all of $\{|\Pr[\mathsf{T}_i] - \Pr[\mathsf{T}_{i+1}]|\}_{i\in[3]}$ are negligible, which in turn implies that $\mathcal{A}$'s advantage in breaking the oblivious samplability of public keys of PKE is also negligible. This completes the proof of Theorem 4.2. □

4.3 Oblivious Samplability of Ciphertexts

Here, we show the following theorem that guarantees the oblivious samplability of ciphertexts of PKE in Sect. 4.1.

Theorem 4.3. *If* UE *satisfies the oblivious samplability of updated ciphertexts, then* PKE *satisfies the oblivious samplability of ciphertexts.*

Proof. Let (oUpd, rUpd) be the pair of the algorithms that is guaranteed to exist due to the oblivious samplability of updated ciphertexts of UE. Then, consider the following pair of the PPT algorithm ($\mathsf{oEnc}_{\mathsf{PKE}}, \mathsf{rEnc}_{\mathsf{PKE}}$).

$\mathsf{oEnc}_{\mathsf{PKE}}(\mathsf{pk}; s)$: On input $\mathsf{pk} = (\mathsf{c}_0, \mathsf{c}_1, \Delta)$, and $s \in \{0,1\}^\ell$, it runs $\mathsf{c}' \leftarrow \mathsf{oUpd}(\Delta; s)$, and returns c' as an obliviously sampled ciphertext.

$\mathsf{rEnc}_{\mathsf{PKE}}(\mathsf{pk}, \mathsf{c}')$: On input $\mathsf{pk} = (\mathsf{c}_0, \mathsf{c}_1, \Delta)$ and c', it runs $s \xleftarrow{\$} \mathsf{rUpd}(\mathsf{c}', \Delta)$ (where c' and Δ are in the epoch $e = 2$), and then returns s as a randomness.

Let $\mathcal{A} = (\mathcal{A}_0, \mathcal{A}_1)$ be any PPT adversary that attacks the oblivious samplability of ciphertexts of PKE. Then, consider the following adversary $\mathcal{B} = (\mathcal{B}_0, \mathcal{B}_1)$ against the oblivious samplability of updated ciphertexts of UE:

$\mathcal{B}_0^{\mathcal{O}.\mathsf{Enc}, \mathcal{O}.\mathsf{Next}, \mathcal{O}.\mathsf{Upd}, \mathcal{O}.\mathsf{Corr}}(1^\lambda)$: $\mathcal{B}_0$ calls $\mathcal{O}.\mathsf{Enc}(0)$ and $\mathcal{O}.\mathsf{Enc}(1)$ to get c_0 and c_1, respectively. Then, $\mathcal{B}_0$ calls $\mathcal{O}.\mathsf{Next}$ to advance the epoch, and calls $\mathcal{O}.\mathsf{Corr}(\mathsf{token}, 2)$ to get a token Δ (at epoch $e = 2$). Then, $\mathcal{B}_0$ sets $\mathsf{pk} := (\mathsf{c}_0, \mathsf{c}_1, \Delta)$, and runs $(\mathsf{m} \in \{0,1\}, \mathsf{st}) \leftarrow \mathcal{A}_0(\mathsf{pk})$. Finally, $\mathcal{B}_0$ sets $\bar{\mathsf{c}} := \mathsf{c}_\mathsf{m}$ as its challenge and $\mathsf{st}_\mathcal{B}$ as all the values known to $\mathcal{B}_0$, and terminates with the output $(\bar{\mathsf{c}}, \mathsf{st}_\mathcal{B})$.

$\mathcal{B}_1^{\mathcal{O}.\mathsf{Enc}, \mathcal{O}.\mathsf{Next}, \mathcal{O}.\mathsf{Upd}, \mathcal{O}.\mathsf{Corr}, \mathcal{O}.\mathsf{Upd}\tilde{\mathsf{c}}}(\tilde{\mathsf{c}}, \tilde{s}, \mathsf{st}_\mathcal{B})$: $\mathcal{B}_1$ runs $b' \leftarrow \mathcal{A}_1(\tilde{\mathsf{c}}, \tilde{s}, \mathsf{st})$. Finally, $\mathcal{B}_1$ sets $\beta' \leftarrow b'$, and terminates with output β'.

It is easy to see that $\mathcal{B}$ perfectly simulates the experiment for the oblivious samplability of ciphertexts of PKE for $\mathcal{A}$ so that the challenge bit in $\mathcal{B}$'s experiment is that of $\mathcal{A}$. In addition, $\mathcal{B}$ uses the output of $\mathcal{A}$ as is. Therefore, the probability that $\mathcal{B}$ guesses the challenge bit correctly is identical to the probability that $\mathcal{A}$ does so in $\mathsf{Expt}_{\mathsf{PKE},\mathcal{A}}^{\mathsf{ctx\text{-}OS}}(\lambda)$. Therefore, $\mathcal{B}$ and $\mathcal{A}$ have exactly the same advantages, and thus we have $|\Pr[\mathsf{Expt}_{\mathsf{PKE},\mathcal{A}}^{\mathsf{ctx\text{-}OS}}(\lambda) = 1] - 1/2| = |\Pr[\mathsf{Expt}_{\mathsf{UE},\mathcal{B}}^{\mathsf{OS\text{-}UPD}}(\lambda) = 1] - 1/2| = \mathsf{negl}(\lambda)$. This completes the proof of Theorem 4.3. □

5 Constructing Lossy Encryption

In this section, we show a connection of UE to lossy encryption (Sect. 2.2). Specifically, we introduce a new property that we call *statistical confidentiality of original ciphertexts* for UE, and then show that if the underlying UE scheme in the PKE construction PKE in Sect. 4.1 satisfies IND-ENC security and our new property, then the PKE scheme is in fact a lossy encryption scheme.

The statistical confidentiality of original ciphertexts guarantees that an updated ciphertext, which is updated from some original ciphertext encrypting a plaintext m, statistically hides the information of the original ciphertext other than the fact that it is encrypting m. Technically, we require that even a computationally unbounded adversary who is given the challenge updated ciphertext and two candidate original ciphertexts that encrypt the same plaintext, cannot tell which of the original ciphertexts is used for generating the challenge.

$$\begin{array}{l}
\underline{\textbf{Experiment } \mathsf{Expt}^{\mathsf{UE\text{-}conf}}_{\mathsf{UE},\mathcal{A}}(\lambda)} \\
(\mathsf{m}, \mathsf{st}) \xleftarrow{\$} \mathcal{A}_0(1^\lambda) \\
\mathsf{k} \xleftarrow{\$} \mathsf{UE.KG}(1^\lambda) \\
\mathsf{c}_0, \mathsf{c}_1 \xleftarrow{\$} \mathsf{UE.Enc}(\mathsf{k}, \mathsf{m}) \\
(\mathsf{k}', \Delta) \xleftarrow{\$} \mathsf{UE.Next}(\mathsf{k}) \\
b \xleftarrow{\$} \{0, 1\} \\
\widetilde{\mathsf{c}} \xleftarrow{\$} \mathsf{UE.Upd}(\Delta, \mathsf{c}_b) \\
b' \xleftarrow{\$} \mathcal{A}_1(\widetilde{\mathsf{c}}, \Delta, \mathsf{c}_0, \mathsf{c}_1, \mathsf{st}) \\
\text{Return } (b' \stackrel{?}{=} b).
\end{array}$$

Fig. 6. The experiment for defining statistical confidentiality of original ciphertexts.

Definition 5.1 (Statistical Confidentiality of Original Ciphertexts). *We say that a UE scheme* $\mathsf{UE} = (\mathsf{UE.KG}, \mathsf{UE.Enc}, \mathsf{UE.Dec}, \mathsf{UE.Next}, \mathsf{UE.Upd})$ *satisfies* statistical confidentiality of original ciphertexts *if for all computationally unbounded adversaries* $\mathcal{A} = (\mathcal{A}_0, \mathcal{A}_1)$, *it holds that* $|Pr[\mathsf{Expt}^{\mathsf{UE\text{-}conf}}_{\mathsf{UE},\mathcal{A}}(\lambda) = 1] - 1/2| = \mathsf{negl}(\lambda)$, *where the experiment* $\mathsf{Expt}^{\mathsf{UE\text{-}conf}}_{\mathsf{UE},\mathcal{A}}(\lambda)$ *is defined as in Fig. 6.*

The UE scheme called RISE proposed by Lehman and Tackmann [15], based on the ElGamal PKE scheme, satisfies the property unconditionally. Specifically, in this scheme, an original ciphertext and an updated ciphertext are just a fresh ciphertext of the ElGamal PKE scheme, and an updated ciphertext has no information about the randomness used in the "source" ciphertext (i.e. either an original ciphertext, or another updated ciphertext at the previous epoch). The UE scheme based on iO by Nishimaki [17] has exactly the same property, and hence satisfies this property.[5]

Now, using a UE scheme $\mathsf{UE} = (\mathsf{UE.KG}, \mathsf{UE.Enc}, \mathsf{UE.Dec}, \mathsf{UE.Next}, \mathsf{UE.Upd})$ as a building block, consider the following lossy encryption scheme $\mathsf{LE} = (\mathsf{LE.KG}, \mathsf{LE.Enc}, \mathsf{LE.Dec}, \mathsf{LE.LKG})$:

$(\mathsf{LE.KG}, \mathsf{LE.Enc}, \mathsf{LE.Dec})$: These are exactly the same algorithms as $(\mathsf{PKE.KG}, \mathsf{PKE.Enc}, \mathsf{PKE.Dec})$ in PKE in Sect. 4.1, respectively.

$\mathsf{LE.LKG}(1^\lambda)$: It executes $\mathsf{k} \xleftarrow{\$} \mathsf{UE.KG}(1^\lambda)$, $\mathsf{c}_0 \xleftarrow{\$} \mathsf{UE.Enc}(\mathsf{k}, 0)$, $\mathsf{c}_1 \xleftarrow{\$} \mathsf{UE.Enc}(\mathsf{k}, 0)$, and $(\mathsf{k}', \Delta) \xleftarrow{\$} \mathsf{UE.Next}(\mathsf{k})$. Then, it returns a lossy public key $\mathsf{pk}_{\mathrm{loss}} := (\mathsf{c}_0, \mathsf{c}_1, \Delta)$.

The security properties of LE are guaranteed by the following theorem.

[5] If this UE scheme based on iO had no problem, then the combination of it with our result in this section would imply a construction of lossy encryption based on iO, which to the best of our knowledge is not known before. Unfortunately, however, the iO-based UE scheme had an error and was retracted recently [18].

Theorem 5.1. *If* UE *is* $\mathsf{IND\text{-}ENC}$ *secure, then* LE *satisfies indistinguishability of ordinary and lossy public keys. Furthermore, if* UE *satisfies statistical confidentiality of original ciphertexts, then* LE *satisfies statistical confidentiality under lossy keys.*

Proof. Since the implication of the indistinguishability of ordinary and lossy public keys of LE based on the $\mathsf{IND\text{-}ENC}$ security of the underlying UE scheme is fairly straightforward (just replace c_1 in pk with an encryption of 0 by relying on the power of $\mathsf{IND\text{-}ENC}$ security), we omit the formal proof.

We thus show that LE satisfies statistical confidentiality under lossy keys if UE satisfies statistical confidentiality of original ciphertexts. Let $\mathcal{A} = (\mathcal{A}_0, \mathcal{A}_1)$ be any (unbounded) adversary against the statistical confidentiality under lossy keys of LE. For simplicity, we assume that $\mathcal{A}$'s challenge plaintexts $(\mathsf{m}_0, \mathsf{m}_1)$ satisfies $\mathsf{m}_0 = 0$ and $\mathsf{m}_1 = 1$. (This is without loss of generality since the plaintext space of LE is $\{0,1\}$.) Using $\mathcal{A}$, we construct another adversary $\mathcal{B} = (\mathcal{B}_0, \mathcal{B}_1)$ against the statistical confidentiality of original ciphertexts of UE, as follows:

$\mathcal{B}_0(1^\lambda)$: $\mathcal{B}_0$ sets $\mathsf{m} = 0$ as its challenge plaintext, set $\mathsf{st}_\mathcal{B} = \mathsf{m}$, and terminates with the output $(\mathsf{m}, \mathsf{st}_\mathcal{B})$.

$\mathcal{B}_1(\widetilde{\mathsf{c}}, \Delta, \mathsf{c}_0, \mathsf{c}_1, \mathsf{st}_\mathcal{B})$: $\mathcal{B}_1$ sets $\mathsf{pk}_{\mathsf{loss}} := (\mathsf{c}_0, \mathsf{c}_1, \Delta)$, and then executes $(\mathsf{m}_0 = 0, \mathsf{m}_1 = 1, \mathsf{st}) \xleftarrow{\$} \mathcal{A}_0(\mathsf{pk}_{\mathsf{loss}})$, and $b' \xleftarrow{\$} \mathcal{A}_1(\widetilde{\mathsf{c}}, \mathsf{st})$ Finally, $\mathcal{B}_1$ sets $\beta' \leftarrow b'$, and terminates with output β'.

It is easy to see that $\mathcal{B}$ perfectly simulates the experiment for the statistical confidentiality under lossy keys of LE for $\mathcal{A}$ so that the challenge bit in $\mathcal{B}$'s experiment is that of $\mathcal{A}$. In addition, $\mathcal{B}$ uses the output of $\mathcal{A}$ as is. Therefore, the probability that $\mathcal{B}$ guesses the challenge bit correctly is identical to that of $\mathcal{A}$ in $\mathsf{Expt}^{\mathsf{LE\text{-}conf}}_{\mathsf{LE},\mathcal{A}}(\lambda)$. Therefore, $\mathcal{B}$ and $\mathcal{A}$ have exactly the same advantages, and thus we have $|\Pr[\mathsf{Expt}^{\mathsf{LE\text{-}conf}}_{\mathsf{LE},\mathcal{A}}(\lambda) = 1] - 1/2| = |\Pr[\mathsf{Expt}^{\mathsf{UE\text{-}conf}}_{\mathsf{UE},\mathcal{B}}(\lambda) = 1] - 1/2| = \mathsf{negl}(\lambda)$. This completes the proof of Theorem 5.1. □

6 Implications of Our Results

Oblivious Transfer (OT) [8] is one of the fundamental public-key cryptographic primitives that is a basis for secure two-party and multi-party computation [19]. Gertner et al. [11] showed that it is possible to construct a 2-round OT from a PKE scheme satisfying oblivious samplability of public keys, and a 3-round OT from a PKE scheme satisfying oblivious samplability of ciphertexts. Combining these results with the results in Sect. 4, we obtain the following theorems.

Theorem 6.1. *If there exists a UE scheme that satisfies the oblivious samplability of original ciphertexts and update tokens, then a 2-round OT exists.*

Theorem 6.2. *If there exists a UE scheme that satisfies the oblivious samplability of updated ciphertexts, then a 3-round OT exists.*

Furthermore, it was shown by Gertner et al. [11] that there is no black-box construction of OT (with any round) from a PKE scheme that does not satisfy oblivious sampling properties, and that there is no black-box construction of PKE from OT with more than two rounds. In other words, Theorems 6.1 and 6.2 do not follow from the results of Alamati et al. (i.e., Theorem 4.1). These results suggest further evidence that in terms of primitive-to-primitive construction, UE schemes satisfying various kinds of oblivious samplability require strictly stronger assumptions than those required for an (ordinary) PKE scheme.

It was shown that lossy encryption implies a PKE scheme secure against selective opening attacks [2]. Therefore, combining these results with the results in Sect. 5, we obtain the following theorem.

Theorem 6.3. *If there exists a UE scheme that satisfies* $\mathsf{IND\text{-}ENC}$ *security and statistical confidentiality of original ciphertexts, then there exists a lossy encryption scheme and a PKE scheme secure against selective opening attacks.*

Acknowledgement. This work was partially supported by JST CREST Grant Number JPMJCR22M1 and JSPS KAKENHI Grant Number JP23KJ0548.

References

1. Alamati, N., Montgomery, H., Patranabis, S.: Symmetric primitives with structured secrets. In: Boldyreva, A., Micciancio, D. (eds.) CRYPTO 2019, Part I. LNCS, vol. 11692, pp. 650–679. Springer, Cham (2019). https://doi.org/10.1007/978-3-030-26948-7_23
2. Bellare, M., Hofheinz, D., Yilek, S.: Possibility and impossibility results for encryption and commitment secure under selective opening. In: Joux, A. (ed.) EUROCRYPT 2009. LNCS, vol. 5479, pp. 1–35. Springer, Heidelberg (2009). https://doi.org/10.1007/978-3-642-01001-9_1
3. Boneh, D., Halevi, S., Hamburg, M., Ostrovsky, R.: Circular-secure encryption from decision Diffie-Hellman. In: Wagner, D. (ed.) CRYPTO 2008. LNCS, vol. 5157, pp. 108–125. Springer, Heidelberg (2008). https://doi.org/10.1007/978-3-540-85174-5_7
4. Boneh, D., Lewi, K., Montgomery, H., Raghunathan, A.: Key homomorphic PRFs and their applications. In: Canetti, R., Garay, J.A. (eds.) CRYPTO 2013, Part I. LNCS, vol. 8042, pp. 410–428. Springer, Heidelberg (2013). https://doi.org/10.1007/978-3-642-40041-4_23
5. Boyd, C., Davies, G.T., Gjøsteen, K., Jiang, Y.: Fast and secure updatable encryption. In: Micciancio, D., Ristenpart, T. (eds.) CRYPTO 2020, Part I. LNCS, vol. 12170, pp. 464–493. Springer, Cham (2020). https://doi.org/10.1007/978-3-030-56784-2_16
6. Damgård, I., Nielsen, J.B.: Improved non-committing encryption schemes based on a general complexity assumption. In: Bellare, M. (ed.) CRYPTO 2000. LNCS, vol. 1880, pp. 432–450. Springer, Heidelberg (2000). https://doi.org/10.1007/3-540-44598-6_27
7. Dent, A.W.: The Cramer-Shoup encryption scheme is plaintext aware in the standard model. In: Vaudenay, S. (ed.) EUROCRYPT 2006. LNCS, vol. 4004, pp. 289–307. Springer, Heidelberg (2006). https://doi.org/10.1007/11761679_18

8. Even, S., Goldreich, O., Lempel, A.: A randomized protocol for signing contracts. In: Chaum, D., Rivest, R.L., Sherman, A.T. (eds.) Advances in Cryptology, pp. 205–210. Springer, Boston, MA (1983). https://doi.org/10.1007/978-1-4757-0602-4_19
9. Everspaugh, A., Paterson, K., Ristenpart, T., Scott, S.: Key rotation for authenticated encryption. In: Katz, J., Shacham, H. (eds.) CRYPTO 2017, Part III. LNCS, vol. 10403, pp. 98–129. Springer, Cham (2017). https://doi.org/10.1007/978-3-319-63697-9_4
10. Jiang Galteland, Y., Pan, J.: Backward-leak uni-directional updatable encryption from (homomorphic) public key encryption. In: Boldyreva, A., Kolesnikov, V. (eds.) Public-Key Cryptography, PKC 2023, Part II. LNCS, vol. 13941, pp. 399–428. Springer, Cham (2023). https://doi.org/10.1007/978-3-031-31371-4_14
11. Gertner, Y., Kannan, S., Malkin, T., Reingold, O., Viswanathan, M.: The relationship between public key encryption and oblivious transfer. In: FOCS 2000, pp. 325–335 (2000)
12. Impagliazzo, R., Rudich, S.: Limits on the provable consequences of one-way permutations. In: STOC 1989, pp. 44–61 (1989)
13. Jiang, Y.: The direction of updatable encryption does not matter much. In: Moriai, S., Wang, H. (eds.) ASIACRYPT 2020, Part III. LNCS, vol. 12493, pp. 529–558. Springer, Cham (2020). https://doi.org/10.1007/978-3-030-64840-4_18
14. Klooß, M., Lehmann, A., Rupp, A.: (R)CCA secure updatable encryption with integrity protection. In: Ishai, Y., Rijmen, V. (eds.) EUROCRYPT 2019, Part I. LNCS, vol. 11476, pp. 68–99. Springer, Cham (2019). https://doi.org/10.1007/978-3-030-17653-2_3
15. Lehmann, A., Tackmann, B.: Updatable encryption with post-compromise security. In: Nielsen, J.B., Rijmen, V. (eds.) EUROCRYPT 2018, Part III. LNCS, vol. 10822, pp. 685–716. Springer, Cham (2018). https://doi.org/10.1007/978-3-319-78372-7_22
16. Miao, P., Patranabis, S., Watson, G.: Unidirectional updatable encryption and proxy re-encryption from DDH. In: Boldyreva, A., Kolesnikov, V. (eds.) Public-Key Cryptography, PKC 2023, Part II. LNCS, vol. 13941, pp. 368–398. Springer, Cham (2023). https://doi.org/10.1007/978-3-031-31371-4_13
17. Nishimaki, R.: The direction of updatable encryption does matter. In: Hanaoka, G., Shikata, J., Watanabe, Y. (eds.) Public-Key Cryptography, PKC 2022, Part II. LNCS, vol. 13178, pp. 194–224. Springer, Cham (2022). https://doi.org/10.1007/978-3-030-97131-1_7
18. Nishimaki, R.: The direction of updatable encryption does matter. Cryptology ePrint Archive, Report 2021/221, The latest version as of April 23, 2024. https://eprint.iacr.org/archive/2021/221/20240419:044657
19. Yao, A.C.C.: Protocols for secure computations (extended abstract). In: FOCS 1982, pp. 160–164 (1982)

Continuous Version of Non-malleable Codes from Authenticated Encryption

Anit Kumar Ghosal(✉) and Dipanwita Roychowdhury

Department of Computer Science and Engineering, IIT Kharagpur, Kharagpur, India
anit.ghosa@gmail.com

Abstract. Non-malleable codes are designed to provide security of highly sensitive data against tampering attacks where traditional error correction and error detection codes fail. An attacker can perform tampering experiment on the codeword but *non-malleability* property ensures that outcome is either *completely unrelated to the original data or the original message*, in case of unsuccessful tampering, i.e., tampering has no effect on the codeword at all. Usually, standard non-malleable codes provide security against *one-time* tampering attack. In literature, it is shown that authenticated encryption can be used in the design of such codeword [22,31]. The security of such construction breaks when an adversary tampers the codeword more than once. To overcome the situation, *continuously non-malleable codes* are proposed where an adversary is able to tamper the codeword for *polynomial number of times* and non-malleability property is preserved. We show a computationally secure construction of continuously non-malleable code from *encrypt then MAC* based authenticated encryption in 2-split-state model. Earlier codewords are designed using heavy cryptographic primitives like *non-interactive zero knowledge proof* (NIZK). Our construction is based on *non-malleable non-interactive commitment scheme* of [32] along with *authenticated encryption* only. This is the first construction that achieves strong continuous non-malleability without using NIZK, but only relying on non-malleable non-interactive commitment, authenticated encryption and leakage resilient storage. Whenever the tampering experiment triggers *self-destruct*, the security of continuously non-malleable code is reduced to the security of underlying *leakage resilient storage*.

Keywords: Authenticated encryption · Non-malleable commitment · Non-malleable codes · 2-Split-State model · Tamper-resilient cryptography

1 Introduction

The extensive growth in the area of digital device introduces various new attacks. The main challenging task is to find the prevention mechanism against such attacks. One example of such an attack is *hardware attack or active physical attack* where an adversary is able to tamper the device through some faults. In software modules or platforms, an adversary can tamper the secret information by corrupting some regions of the memory using viruses or malware's.

T. Zhu and Y. Li (Eds.): ACISP 2024, LNCS 14895, pp. 324–344, 2024.
https://doi.org/10.1007/978-981-97-5025-2_17

The ultimate goal of an adversary is to find out the secret data so that they can destroy the complete security of the device. A devastating attack is shown by Boneh et al. [3] where an adversary can make a minor modification in the crypto module and the secret key can be recovered completely. A line of research works have shown to protect any cryptographic implementation from tampering attacks [5,11,12,14,16,31].

Non-malleable codes, introduced by Dziembowski et al. [8], are used to protect secret data against adversarial tampering. It is useful in the scenario when an adversary tampers the data and ensuring the security of data is beyond the scope of *error correction and error detection codes.* Consider a tampering experiment. A secret message m is encoded via encoding function $\mathsf{Encode}_k(m)$ that generates the codeword M_{code}, i.e., $M_{code} = \mathsf{Encode}_k(m)$. However, an attacker can modify the codeword M_{code} using some tampering function $f \in \mathcal{F}$ as $M^{'}_{code} = f(M_{code})$. Further, it can perform decoding through some deterministic function Decode in the following way $m^{'} = \mathsf{Decode}_k(M^{'}_{code})$. In terms of practical perspective, we can expect the reliability, i.e., original secret m should be same as $m^{'}$. Usually, an error correcting code provides reliable communication with respect to the tampering function $f \in \mathcal{F}$ when the Hamming distance d between original codeword M_{code} and tampered codeword $M^{'}_{code}$ is upper bounded by $\lfloor (d-1)/2 \rfloor$. However, it is not possible to achieve reliability when the tampering function set $\mathcal{F}$ is comparatively larger. To deal with such scenario, *non-malleable codes* are introduced to encode the secret information. It ensures with high probability that either the decoding of tampered codeword is possible in *a correct way* or the decoded message is *completely unrelated* or $\perp$. Moreover, it is well known that non-malleable codes exist only for some specific classes of tampering functions. One of the widely used tampering model is *2-split-state model* where the codeword is partitioned into *two shares* and an adversary can arbitrarily tamper the codeword in each part independently. More precisely, the codeword M_{code} is split into M_1, M_2 and the adversary uses the tampering function f_1, f_2 in the following way $f_1(M_1)$, $f_2(M_2)$. The main limitation of *standard non-malleable code* is that it can provide security against *one-time* attack. But in some cases, an adversary can tamper the codeword *more than once or polynomial number of times. Continuous non-malleable codes* ($CNMC$) are proposed to overcome such limitation and it can tolerate *polynomial number of tampering attacks* and preserves the *non-malleability* as discussed in [17].

In literature, continuous version of non-malleable codes have different flavours. To describe it, we define some notations for a particular continuous tampering experiment. The secret message is denoted as m. After encoding the message using non-malleable code, the codeword $M_{code} = \mathsf{Encode}_k(m)$ is generated. An adversary can tamper the codeword to produce $M^{'}_{code}$. Then decoding algorithm generates the tampered message $m^{'} = \mathsf{Decode}_k(M^{'}_{code})$. In *standard* version of continuous non-malleability, the decoded tampered message $m^{'}$ is independent of the original message m but the adversary can modify the codeword M_{code} to generate an another codeword $M^{'}_{code}$. The code $M^{'}_{code}$ is different from M_{code} but decoding of $M^{'}_{code}$ can be highly related to m as described in [8]. On the other hand, *strong* continuous non-malleability ensures that when $M^{'}_{code}$ is different from M_{code}, m and $m^{'}$ are completely unrelated [17,19,20]. Further, *super-strong* continuous non-malleability condition says that two codewords M_{code}, $M^{'}_{code}$ are unrelated implies that $m^{'}$ and m are independent [17,19,20]. Moreover, the way tampering is performed on the codeword, continuous version of non-malleable codes has two versions as shown in [20]. The first one is *non-persistent* tampering where an adversary tampers the initial version of the codeword always. Another flavour is *persistent* tampering. Here, the codeword of previous experiment is tampered by $f \in \mathcal{F}$, not the initial version. The experiment continues until decoding error is triggered. The construction of continuously non-malleable codes are broadly categorized into computational [17,26,28] and information-theoretic [29] domain. Due to some generic attack, information-theoretic continuous non-malleability is not achievable in 2-splt-state model [17]. Further research work shows that in case of persistent tampering function, information-theoretic continuous non-malleability is achievable [24]. All of the constructions of codeword are based on *common reference string* (CRS) based setup. Later, Ostrovsky et al. [28] propose a construction of $CNMC$ in the plain model but their security notion is not stronger. In [30], authors describe that CRS based setup is necessary for stronger security requirement. Hence, our proposed construction is based on authenticated encryption and a non-malleable commitment scheme with CRS based setup.

Table 1. Comparison of various $CNMC$ in 2-split-state model. Comp. stands for *computational* and I.T. denotes *information-theoretic*. *Poly*(n) means polynomial function with security parameter n.

Scheme	Model	Assumption	Types of Tampering	Security against Attacks
[17]	Comp., CRS	NIZK, Collision resistant hash, Leakage resilient storage	Non-persistent with self-destruct	*Poly*(n) number of tampering attacks, Bounded leakage attacks
[24]	I.T	N.A	Persistent with self-destruct	Unbounded adversary with *Poly*(n) number of tampering attacks, Bounded leakage attacks
[26]	Comp., CRS	NIZK, Non-interactive commitment, Leakage resilient public key encryption	Non-persistent with self-destruct	*Poly*(n) number of tampering attacks, Bounded leakage attacks
[28]	Comp	Only one-to-one one-way function	Non-persistent with self-destruct	Unbounded adversary with *Poly*(n) number of tampering attacks, Bounded leakage attacks
[32]	Comp., CRS	1-more weakly extractable leakage-resilient hash, Collision resistant hash	Non-persistent with self-destruct	*Poly*(n) number of tampering attacks, Bounded leakage attacks
[33]	Comp., CRS	NIZK, Non-interactive commitment, Leakage resilient storage	Non-persistent with self-destruct	*Poly*(n) number of tampering attacks, Bounded leakage attacks
This work	**Comp., CRS**	**Non-malleable non-interactive commitment, Authenticated encryption**	**Non-persistent with self-destruct**	***Poly*(n) number of tampering attacks, bounded leakage attacks**

Limitations of Existing Scheme and Motivation of the design. In literature, non-malleable codes are shown as *keyless* encoding scheme [8]. Later, various symmetric-key primitives, i.e., authenticated encryption (AE) [18,22,23,31], related-key secure ciphers [27] etc. are used to construct standard version of non-malleable codes. Unfortunately, the codeword of [18,23] and [22,27,31] do not provide security for polynomial number of tampering attacks. Moreover, an adversary can tamper the right part of a codeword M_2 to generate a related codeword $M_2^{'}$. The attack creates two valid codewords (M_1, M_2) and $(M_1, M_2^{'})$ such that their decoding does not return $\perp$, i.e., $\perp \neq \mathsf{Decode}_k(\alpha, (M_1, M_2)) \neq \mathsf{Decode}_k(\alpha, (M_1, M_2^{'})) \neq \perp$, where $M_2 \neq M_2^{'}$. The adversary generates two valid messages m_0, m_1. Further, the adversary may not activate the *self-destruct* feature and it can leak all the bits of M_2 with the assumption that the underlying tampering function is *non-persistent*. In general, for any continuously non-malleable codes, finding two valid codewords (M_1, M_2) and $(M_1, M_2^{'})$ such that $\mathsf{Decode}_k(\alpha, (M_1, M_2)) \neq \mathsf{Decode}_k(\alpha, (M_1, M_2^{'}))$ should be computationally hard to the adversary. This property is called *message uniqueness* as described in [17]. Our goal is to design non-malleable codes from *encrypt then MAC* based authenticated encryption that is secure against *polynomial number of tampering attacks*. Table 1 shows various constructions of continuous non-malleable codes in 2-split-state model as available in the literature.

Comparison with Kiayias et al. [32]**.** In [32], authors show a construction of computationally secure continuous version of non-malleable codes from *1-more weakly extractable leakage-resilient* hash function (wECRH) along with *collision resistant hash function* and *leakage resilient storage* in the CRS model. On the other hand, we propose a construction of continuous non-malleable code from *encrypt then MAC* based authenticated encryption along with *non-malleable non-interactive commitment scheme* which is instantiated with 1-more wECRH. The main motivation of our work is to secure the various constructions of non-malleable codes from authenticated encryption [18,22,23,31] against *polynomial number of tampering attacks.*

Our Contribution. In this work, we propose computationally secure, a non-persistent version of continuously non-malleable code in 2-split-state model from *encrypt then MAC* based authenticated encryption (AE) along with *non-malleable non-interactive commitment scheme*, instantiated with *1-more wECRH* [32]. We remove the bottleneck [17,33] of *robust non-interactive zero knowledge* (NIZK) proof [4]. Initially, a leakage resilient storage is used to encode the message. Further, it is encrypted with an authenticated encryption along with a non-malleable commitment scheme. The underlying AE should satisfy the following assumptions:

a) Whenever the decryption of an authenticated encryption algorithm with a secret key k works in a proper way, it should return $\perp$ if a different key $k^{'}$ is used, where $k \neq k^{'}$.
b) The encryption scheme of the underlying authenticated encryption should be IND-CPA secure and MAC is strongly unforgeable (Subsection 2.4).

Organization. The paper is structured as follows: basic notations, results and preliminaries are described in Sect. 2. Codeword description is illustrated in Sect. 3. Further, security analysis of the construction is depicted in Sect. 4. Finally, we conclude in Sect. 5.

2 Preliminaries

2.1 Basic Notations and Results

In 2-split-state model, $m \in M$ denotes the plaintext whereas M_{code} is the codeword. M_1 and M_2 are the left and right part of a codeword which is stored into the memory $\mathcal{M}_L$, $\mathcal{M}_R$ respectively. Two tampering functions are f_1^i and f_2^i, where i represents the round of tampering experiment. Oracle $\mathcal{O}^T_{CNMC_\pi}(.,.)$ is the tampering oracle that tampers the codeword. When k is uniformly chosen at random from $\mathcal{K}$, we write $k \xleftarrow{\$} \mathcal{K}$. k_{enc} and k_{mac} are the keys of an authenticated encryption. k_{enc} is used to encrypt the message whereas k_{mac} uses to generate the tag. n is the security parameter. $\mathcal{O}^l(s,.)$ is the leakage oracle that takes string s and leakage function τ as input and it returns at most l bits. $r \in \{0,1\}^n$ denotes the randomness. α represents an untamperable *common reference string* (CRS). A function $\epsilon(n)$ is said to be *negligible* in n when it vanishes faster than the inverse of any polynomial in n. Let $\mathfrak{E} = \{E_k\}_{k \in N}$, $\mathfrak{F} = \{F_k\}_{k \in N}$ be two ensembles and $\mathfrak{E} \underset{c}{\approx} \mathfrak{F}$ represents the *computational indistinguishability* that for every PPT distinguisher D, $|Pr[D(E_k) = 1] - Pr[D(F_k) = 1]| \leq \epsilon(n)$. In a similar way, $\mathfrak{E} \underset{s}{\approx} \mathfrak{F}$ denotes the *statistical indistinguishability* for computationally unbounded scenario. $\mathcal{H}_\infty(X)$ and $\tilde{\mathcal{H}}_\infty(X|Y)$ are the min-entropy and conditional average min-entropy of a random variable X. $\delta_1[i]$, $\delta_2[i]$ denote the two arrays used to store the result of tampering queries whereas $\mu_0[i]$, $\mu_1[i]$ stores leakage queries data at each invocation ($i \in q \wedge q \in poly(n)$) in Algorithm 5 and Algorithm 6. Table 2 illustrates a summary of notations. We now define some definitions and lemmas (**2.1.1** to **2.1.6**) [6] used to construct the codeword.

Definition 2.1.1 (2-Split-State Model). *For any $n \in \{0,1\}^*$ and for any efficiently computable function $f : \{0,1\}^n \rightarrow \{0,1\}^n$, $f \in \mathcal{F}$, there exists two efficiently computable functions $f_1 : \{0,1\}^{n/2} \rightarrow \{0,1\}^{n/2}$, $f_2 : \{0,1\}^{n/2} \rightarrow \{0,1\}^{n/2}$ such that for every $M_1, M_2 \in \{0,1\}^{n/2} \times \{0,1\}^{n/2}$, $f(M_1, M_2) = f_1(M_1)||f_2(M_2)$. Alternatively, we can say for a codeword $M = (M_1, M_2)$ with two tampering functions f_1 and f_2 on the codeword M_1, M_2, an adversary A performs the tampering experiment in an arbitrary and independent way. Such tampering model is said to be 2-split-state model.*

Definition 2.1.2 (Non-persistent 2-Split-State Tampering). *For any efficiently computable function $f = (f_1, f_2)$, the function f always tampers the initial version of the codeword $M = (M_1, M_2)$ in the following way $f(M_1, M_2) = f_1(M_1)||f_2(M_2)$. Moreover, an adversary has access to an n-bit auxiliary memory beyond the active memory, and it copies the initial version of the codeword to the auxiliary memory. Later, the subsequent tampering experiment can be performed on the auxiliary memory and the tampered codeword can be placed to the original memory. The model that satisfies the above property is said to be non-persistent 2-split-state tampering model.*

Definition 2.1.3 (Non-Malleable Codes in 2-Split-State Model). *Let (CRSGen, Encode_k, Decode_k) be a split-state coding scheme in the CRS model. Encode_k takes input a CRS, message m and produces codeword M_{code} which is split into two parts as M_1 and M_2. On the other hand, Decode_k takes the codeword M_1, M_2 and a CRS. Finally, it generates the message m. For any efficiently computable function $f \in \mathcal{F}$, the coding scheme (CRSGen, Encode_k, Decode_k) is said to be non-malleable if $\mathsf{Decode}_k(M_{code})$ outputs m', where m' can be m, $\perp$ or completely unrelated.*

Definition 2.1.4 (Continuously Non-Malleable Codes in 2-Split-State Model). *The coding scheme* ($\mathsf{CRSGen}, \mathsf{Encode}_k, \mathsf{Decode}_k$) *is said to be* **split-state continuously non-malleable code** *with respect to the efficiently computable function* $f = \{f_1, f_2\} \in \mathcal{F}$, *if the* **non-malleability** *property still holds after polynomial number of tampering attempts (Subsect. 2.7).*

Lemma 2.1.1. *Let* X *be a random variable that has min-entropy over the set* $\mathcal{X}$, *written as* $\mathcal{H}_\infty(X) = -log\ max_{x\in\mathcal{X}} Pr[X = x]$. *It represents the probability to guess the variable* X *by an unbounded adversary* A *[6].*

Table 2. Summary of notations

Notation	Terminology
M	Message space
M_{code}	Codeword
$\mathcal{K}$	Key space
$\mathcal{C}$	Ciphertext space
c	A particular ciphertext
m	Original message
M_1, M_2	Left and right half of a codeword
$\mathcal{M}_L, \mathcal{M}_R$	Left and right half of the memory
$\mathcal{O}^T_{CNMC_\pi}(.,.)$	Tampering oracle
f_1, f_2	Tampering functions
$\mathcal{K}$	Key set
$k \xleftarrow{\$} \mathcal{K}$	A particular key is selected
k_{enc}, k_{mac}	Encryption key and MAC key
$Sim = (Sim_0, Sim_1)$	Two simulators
n	Security parameter
$\mathcal{O}^l(s,.)$	Leakage oracle with string s and leakage function as input
$\mathsf{O}(n)$	Big O of n
α	Common reference string
H_n	1-more weakly extractable leakage-resilient hash function
$View_{\mathcal{L}^i_A}$	Stores the view of tampering and leakage operation
S_0, S_1	Two simulators
$\epsilon(n)$	A negligible function
$\mathsf{Commit}(\alpha_1,.)$	Commitment scheme
$\mathsf{Open}(\alpha_1,.)$	Opening of a commitment scheme
$\mathbb{E} \underset{s}{\approx} \mathbb{F}$	Statistical indistinguishability
$\tau()$	Leakage function
r	Randomness
A	Adversary
E_{h_1}	Extractor
Enc_{lrs}, Dec_{lrs}	LRS encoding and decoding
$\widetilde{h}$	Universal hash function
$\mathfrak{M}$	External memory
η	Leakage bits supported
$\|\|$	Concatenation of two strings
$\mathcal{LR}$	Left to right oracle
$\mathcal{E}_k$	Encryption oracle
$\mathcal{D}_k$	Decryption oracle
$\mathcal{T}$	The class that contains all split-state tampering functions

Lemma 2.1.2. *Let X be a random variable that has conditional average min-entropy given some extra information Y over the set* $\mathcal{X}$, $\mathcal{Y}$, *denoted as* $\tilde{\mathcal{H}}_\infty(X|Y) = -log\mathbb{E}_{y\in\mathcal{Y}}\ max_{x\in\mathcal{X}} Pr[X = x|Y = y]$. *The equation represents the probability of guessing X when some related information of X is available to the adversary A through side channel information.*

Lemma 2.1.3. *Let X and Y be two random variables. The equation* $\tilde{\mathcal{H}}_\infty(X|Y) \geq \mathcal{H}_\infty(X) - l$ *can be equated with respect to X and Y, where Y takes* 2^l *possible values.*

Lemma 2.1.4. *Let X be a random variable and other two correlated random variables are* Y_1, Y_2, *we get* $\tilde{\mathcal{H}}_\infty(X|Y_1, Y_2) \geq \tilde{\mathcal{H}}_\infty(X|Y_1) - l$, *where* Y_2 *takes* 2^l *possible values.*

Lemma 2.1.5. *For the leakage function* τ, *(possibly randomized) used by an adversary A on variable X, we get* $\tilde{\mathcal{H}}_\infty(X|\tau(X)) \geq \mathcal{H}_\infty(X) - l$, *where* $\tau(X)$ *generates l bits of leakage through the side channel.*

Lemma 2.1.6. *Let* X, Y *be the correlated random variables and* τ *be the leakage function used by an adversary A. Then,* $\tilde{\mathcal{H}}_\infty(X|\tau(Y)) \geq \tilde{\mathcal{H}}_\infty(X|Y)$.

2.2 Leakage Resilient Storage

Let (Enc_{lrs}, Dec_{lrs}) be a leakage resilient storage (lrs) scheme that encodes the message m. For any algorithm A, message m, and $\beta \in \{0, 1\}$, we define (Enc_{lrs}, Dec_{lrs}) in the following way:

- *Enc_{lrs} encodes the message m into p_0, p_1.*
- *Dec_{lrs} takes p_0, p_1 and generates the original message m.*

The leakage experiment can be described as follows:

$$leak^{\beta}_{A,m} = \left\{ \begin{array}{c} (p_0, p_1) \leftarrow Enc_{lrs}(m); \mathcal{L} \leftarrow A^{\mathcal{O}^l(p_0,.),\mathcal{O}^l(p_1,.)} \\ output : (p_\beta, \mathcal{L}_A), \beta \in \{0, 1\} \end{array} \right\}$$

Usually, lrs provides security of the message until some bounded information leaks to the adversary [17]. Initially, a counter ctr is used to calculate the total leakage supported by the lrs. In the oracle $\mathcal{O}^l(p_0, .)$, $\mathcal{O}^l(p_1, .)$, strings are passed along with arbitrary leakage function τ to calculate the value and finally, it is added to the ctr, until $ctr \leq l$ from each part. Whenever the condition $ctr > l$ is reached, Oracle terminates and further query would return $\perp$. It is said to be *strong* lrs if an adversary fails to distinguish the leakage experiment between two arbitrarily chosen messages m and m' except with negligible probability, i.e., $leak^{\beta}_{A,m} \underset{c}{\approx} leak^{\beta}_{A,m'}$.

2.3 Authenticated Encryption

An *authenticated encryption* (AE) scheme consists of following algorithms ($k = \{k_{enc}||k_{mac}\}, Encypt, Decrypt$). There are three generic compositions of an AE.

- In *Encrypt and MAC* scheme, the message $m \in M$ is encrypted by a key $k_{enc} \in \mathcal{K}$ to produce the ciphertext $c \in \mathcal{C}$ as $c \leftarrow Encrypt(k_{enc}, m)$. The tag $\mathsf{tag} \leftarrow tag(k_{mac}, m)$ is calculated on the message m. The output is generated as $(c||\mathsf{tag})$. Alternatively, $Decrypt + verify$ is performed by first decrypting the ciphertext to get the plaintext and subsequently, the tag is verified or returns $\perp$ if decryption/verification fails.
- In *Encrypt then MAC* scheme, an encryption algorithm takes a key $k_{enc} \in \mathcal{K}$, a message $m \in M$ and produces a ciphertext $c \in \mathcal{C}$ as $c \leftarrow Encrypt(k_{enc}, m)$. The tag $\mathsf{tag} \leftarrow tag(k_{mac}, c)$ is calculated on the ciphertext c. The output is generated as $(c||\mathsf{tag})$. Alternatively, $Decrypt + verify$ is performed by first verifying the tag and if it is successful, the plaintext $m \leftarrow Decrypt(k_{enc}, c)$ is generated or returns $\perp$ if decryption/verification fails.
- In *MAC then Encrypt* scheme, the tag $\mathsf{tag} \leftarrow tag(k_{mac}, m)$ is calculated first on the message m. Then, an encryption algorithm takes a key $k_{enc} \in \mathcal{K}$, a message $m \in M$ and produces a ciphertext $c \in \mathcal{C}$ as $c \leftarrow Encrypt(k_{enc}, m||\mathsf{tag})$. The output is generated as (c). Alternatively, $Decrypt + verify$ is performed by first decrypting the ciphertext to get the plaintext and the candidate tag, and finally, verify the tag or returns $\perp$ if decryption/verification fails.

Moreover, the *correctness* property $Decrypt(k, Encrypt(k, m)) = m$, for all $k \in \mathcal{K}$, $m \in M$ and $c \in \mathcal{C}$ should be satisfied. In [25], authors show that only *Encrypt then MAC* scheme is secure against leakage. Hence, we use only *encrypt then MAC* based AE in our construction of $CNMC$. The underlying AE should satisfy IND-CPA security and INT-CTXT security as described in the below subsection. We incorporate property (a) in the *encrypt then MAC* based AE to construct our codeword as described the Sect. 1 in the following way: When the original key k and the tampered key k' are different, the tag of AE should not match and we return $\perp$ in this case.

2.4 IND-CPA Security and INT-CTXT Security

A symmetric encryption scheme $se = (k = \{k_{enc}||k_{mac}\}, Encypt, Decrypt)$ is said to satisfy *indistinguishablity under chosen plaintext attack* (IND-CPA) security if an adversary A_{cpa} with access to the encryption oracle $\mathcal{E}_k(\mathcal{LR}(.,.,b))$[1] performs the below experiment in Algorithm 1 and the following advantage holds with negligible probability:

[1] $\mathcal{LR}(.,.,b)$ stands for left-to-right oracle, where $b \in \{0,1\}$ with input m_0, m_1. When $b = 0$, it calculates $c \leftarrow \mathcal{E}_k(m_0)$. Otherwise, it sets $c \leftarrow \mathcal{E}_k(m_1)$ [2]. The adversary queries with two equal length message and it can guess the bit b.

Algorithm 1. $Exp_{se,A_{cpa}}^{ind-cpa-b}(.)$

1: Select $\{m_0, m_1\} \leftarrow M$
2: Set $c \leftarrow A_{cpa}^{\mathcal{E}_k(\mathcal{LR}(.,.,b)}(.))$
3: Return c

$\mathbf{Adv}_{se,A_{cpa}}^{ind-cpa}(.) = Pr[Exp_{se,A_{cpa}}^{ind-cpa-0}(.) = 1]$ - $Pr[Exp_{se,A_{cpa}}^{ind-cpa-1}(.) = 1] \leq \epsilon(n)$.

Let $se = (k = \{k_{enc}||k_{mac}\}, Encypt, Decrypt)$ be a symmetric encryption scheme. Let A_{ctxt} be an adversary with access to two oracles $\mathcal{E}_k$ and $\mathcal{D}_k^*$[2]. The adversary performs the experiment in Algorithm 2 and the following advantage holds:

Algorithm 2. $Exp_{se,A_{ctxt}}^{int-ctxt}(.)$

1: Select $m \leftarrow M$
2: If $A_{ctxt}^{\mathcal{E}_k(m),\mathcal{D}_k^*(c)}(.)$ makes a ciphertext query c to the $\mathcal{D}_k^*(c)$ such that $\mathcal{D}_k^*(c)$ returns 1 and c was never a response of $\mathcal{E}_k(.)$, then return 1
3: Else, return 0

$\mathbf{Adv}_{se,A_{ctxt}}^{int-ctxt}(.) = Pr[Exp_{se,A_{ctxt}}^{int-ctxt}(.) = 1] \leq \epsilon(n)$.

The se scheme is said to be *INT-CTXT* secure (*integrity of ciphertext*) if the advantage of $\mathbf{Adv}_{se,A_{ctxt}}^{int-ctxt}(.)$ is negligible with respect to A_{ctxt} with time complexity polynomial in n. For detailed description, we request to refer [2].

2.5 1-More Weakly Extractable Leakage Resilient Hash

The hash function H_n is said to be 1-more weakly extractable leakage resilient hash (wECRH) [32] against η bits of leakage with respect to the leakage function $\tau : \{0,1\}^n \rightarrow \{0,1\}^\eta$ if for any probabilistic polynomial time algorithm algorithm A_{h_1} and any $p_{h_1} \in \{0,1\}^n$, there exists a probabilistic polynomial time extractor E_{h_1} and $p_{\mathsf{E}} \in \{0,1\}^n$, such that for all PPT algorithm A_k, $n \in N$, for any input message $k \in \{0,1\}^n$ and a negligible function $\epsilon(n)$:

$$Pr_{h_z \leftarrow H_n}[\mathbf{Exp}_{A_{h_1},A_k,\mathsf{E}_{h_1}}^{k,h_z}(1, p_{h_1}, p_{\mathsf{E}}) = 1] \leq \epsilon(n),$$

where $\boldsymbol{Exp}_{A_{h_1},A_k,\mathsf{E}_{h_1}}^{k,h_z}(1, p_{h_1}, p_{\mathsf{E}})$ *consists of four properties* [32] *as follows:*

- (**Hash computation**) : $(h_1) \leftarrow h_z(r,k)$, *randomness* $r \leftarrow \{0,1\}^n$
- (**Hash tampering**) : $(h_1^{'}, t) \leftarrow A_{h_1}(h_z, h_1, \tau(r), p_{h_1})$
- (**Preimage extraction**) : $(\hat{k}, \hat{r}) \leftarrow \mathsf{E}_{h_1}(h_z, h_1^{'}, p_{\mathsf{E}})$
- (**Preimage tampering**) : $(k', r^{'}) \leftarrow A_k(h_z, r, k, t)$

[2] $\mathcal{D}_k^*(.)$ works as follows: If $\mathcal{D}_k(c) \neq \perp$ return 1, else return 0. It is called verification oracle, an attacker adversary is allowed to perform chosen-message attack on the scheme, modeled by giving it access to an encryption oracle $\mathcal{E}_k$. The adversary is successful when the verification oracle accepts a ciphertext that is not legitimately produced [2].

If $h_z(k^{'},r^{'}) = h_1^{'} \wedge h_z(\hat{k},\hat{r}) \neq h_1^{'}$, *return 1*
else, return 0

The experiment $\mathbf{Exp}^{k,h_z}_{A_{h_1},A_k,\mathsf{E}_{h_1}}(1,p_{h_1},p_{\mathsf{E}})$ works in the following way: we use the randomized hash function. An attacker A_{h_1} wants to generate a tampered hash $h_1^{'}$, given some additional information, i.e., a hash h_1 with auxiliary information p_{h_1} and leakage function $\tau()$ on randomness r. Further, the extractor E_{h_1} produces the preimage, given h_1 and auxiliary input p_{E}. Finally, the attacker A_k tries to generate the preimage of $h_1^{'}$ with the help of additional information gathered during the experiment. The output of the experiment is 1, if A_{h_1} is able to successfully produce a valid hash $h_1^{'}$, and a valid preimage $h_1^{'}$ is generated by A_k, while the extractor algorithm is unable to generate the output properly.

Let $h_z \leftarrow \mathsf{H}_n$ is sampled from the set of collision resistant, efficiently samplable, 1-more wECRH function family as described in [32]. The adversary selects $\widetilde{h} \leftarrow \widetilde{H_{\eta-1}}$ from the family of universal hash function where $\widetilde{h}: \{0,1\}^n \rightarrow \{0,1\}^{\eta-1}$. The output generated by the hash function consists of $\eta - 1$ bits. The leakage function $\tau^{\widetilde{h},h_z}$ is defined in the following way:

$\tau^{\widetilde{h},h_z}(k,r) = (0,\widetilde{h}(f_1(k,r),h_z(k,r))$ if $(f_1(k,r) = (k,r))$.

else, $\tau^{\widetilde{h},h_z}(k,r) = (1,\widetilde{h}(f_1(k,r),h_z(k,r)), (f_1(k,r) \neq (k,r))$.

The maximum leakages are η bits where $\eta = \omega(logn) + \beta(n)$, where $\beta(n) = poly(n)$. The experiment finds the leakage values rather than the output generated by f_1. We refer our reader to [32] for more details.

2.6 Non-malleable Non-interactive Commitment Scheme

Let $(\mathsf{CRSGen}(1^n), \mathsf{Commit})$ be a *non-interactive commitment scheme* that satisfies the following properties:

- **Computationally binding.** It is computationally infeasible to find two messages $m_0, m_1 \in \{0,1\}^n$ (where $m_0 \neq m_1$) with randomness $r_0, r_1 \in \{0,1\}^n$ that satisfies the equation $\mathsf{Commit}(\alpha, m_0, r_0) = \mathsf{Commit}(\alpha, m_1, r_1)$.
- **Statistically hiding.** For any two different messages $m_0, m_1 \in \{0,1\}^n$ with commitment key $\alpha \leftarrow \mathsf{CRSGen}(1^n)$, the equation $\mathsf{Commit}(\alpha, m_0) \underset{s}{\approx} \mathsf{Commit}(\alpha, m_1)$ should be satisfied.

We now define the notion of non-malleable non-interactive commitment scheme as illustrated in [7] with some simplification for the non-interactive setting.

Algorithm 3. Leakage Oracle $\mathcal{O}^l(m,.)$

```
1: Initialize counter = 0
2: Calculate leakage using the function τ() on m
3: Update counter = counter + |τ(m)|
4: if counter ≤ l then
5:     return counter
6: else
7:     return ⊥
8: end if
```

Algorithm 4. Tampering Oracle $\mathcal{O}^T_{CNMC_\pi}((M_1, M_2), (f_1, f_2))$

```
1: Set state = alive
2: Use CRS α of Encode_k
3: if state = self-destruct then
4:     return ⊥
5: end if
6: (M1', M2') = (f1(M1), f2(M2))
7: if (M1, M2) = (M1', M2') then
8:     return same*
9: end if
10: if Decode_k(α, (M1', M2')) = ⊥ then
11:     set state = self-destruct and return ⊥
12: else
13:     return Decode_k(α, (M1', M2'))
14: end if
```

Let H_n be 1-more wECRH that is used to instantiate the *non-malleable non-interactive commitment scheme* as described in [32]. The $(\mathsf{Init}, \mathsf{Commit}, \mathsf{Open})$ algorithms are illustrated as follows:

- $\mathsf{Init}(1^n)$. Sample $h_z \leftarrow \mathsf{H}_n$ and set $\alpha = h_z$.
- $\mathsf{Commit}(\alpha,.)$. It takes input $m \in \{0,1\}^n$ and $r \in \{0,1\}^n$. The output is generated as $h_z(m, r)$.
- $\mathsf{Open}(\alpha,.)$. It takes input a commitment c and algorithm outputs m, r. The receiver accepts if $h_z(m, r) = c$.

Moreover, it should satisfy the *computationally binding* and *statistically hiding* properties as described above.

2.7 Continuously Non-malleable Codes

Leakage Oracle. Let $\mathcal{O}^l(.,.)$ be the leakage oracle that takes a string and arbitrary leakage function as input. The oracle calculates total amount of leakage using the leakage function $\tau(.)$. In Algorithm 3, we illustrate the leakage experiment where l is the maximum leakage bound. A counter variable *counter* is initialized to 0 and it is updated at each invocation of the oracle call with $\tau(m)$. Whenever $counter > l$, experiment returns $\perp$.

Tampering Oracle. Let $\mathcal{O}^T_{CNMC_\pi}(.,.)$ be a stateful oracle in 2-split-state model that takes input two codewords M_1, M_2 with the tampering function $f = (f_1, f_2) \in \mathcal{F}$ and initial *state* is set to *alive*. In Algorithm 4, we define the tampering experiment.

Coding Scheme. Let $CNMC_\pi = (\mathsf{CRSGen}, \mathsf{Encode}_k, \mathsf{Decode}_k)$ be a split-state coding scheme in the CRS model.

- *CRSGen algorithm inputs a security parameter* 1^n *and produces the common reference string* $\alpha \in \{0,1\}^n$.
- *Encode*$_k$ *algorithm inputs a CRS* α, *message* $m \in M$, *key* $k \in \mathcal{K}$ *and generates the codeword* (M_1, M_2).

– *Decode_k algorithm inputs the codeword (M_1, M_2), key $k \in \mathcal{K}$ and CRS α. Finally, it generates the message m or special symbol $\perp$.*

Continuously Non-malleable Codes. The coding scheme $CNMC_\pi$ is said to be (l, q) continuously non-malleable code in 2-split-state model if for all messages $m_0, m_1 \in \{0,1\}^n$ and for all *probabilistic polynomial-time* adversaries A with at most q tampering queries, $\mathbf{Tamper}^{A,m_0}_{CNMC_\pi}$ and $\mathbf{Tamper}^{A,m_1}_{CNMC_\pi}$ are computationally indistinguishable, i.e.,
$\mathbf{Adv}^{Strong}_{Tamper^A_{CNMC_\pi}}(A) = [Pr[A(\mathbf{Tamper}^{A,m_0}_{CNMC_\pi}) = 1]$ - $Pr[A\,(\mathbf{Tamper}^{A,m_1}_{CNMC_\pi})$ $= 1]] \leq \epsilon(n)$, where m_0, $m_1 \in \{0,1\}^n$ and

$$\mathbf{Tamper}^{A,m_0}_{cnmc_\pi} = \left\{ \begin{array}{c} \alpha \leftarrow \mathsf{CRSGen}(1^n); i = 0; (M_1, M_2) \leftarrow \mathsf{Encode}_k(\alpha, m_0) \\ while \quad i \leq q \\ \mathcal{L}^i_A \leftarrow A^{\mathcal{O}^l(M^i_1), \mathcal{O}^l(M^i_2), \mathcal{O}^T_{CNMC_\pi}(M^i_1, M^i_2)} \\ i = i + 1 \\ end \quad while \\ output : \mathcal{L}^i_A. \end{array} \right\},$$

The view of the tampering experiment is captured into $View_{\mathcal{L}^i_A}$ with two one dimensional arrays μ and δ, where i represents the number of tampering queries ($i \leq q$) performed by an adversary. During the tampering experiment, μ stores the view of leakage experiment with the result of all leakage queries ($\mu \leq 2l$). On the other hand, δ array captures the view of tampering queries ($\delta \leq q$) from $\mathcal{O}^T_{CNMC_\pi}()$ oracle. The value $i = 1$ represents that the codeword can handle *one-time* tampering attack only whereas the $i = 0$ means that no tampering is occurred in the codeword and the underlying codeword is capable of handling *leakage* attacks [9].

Message Uniqueness. A continuously non-malleable code $CNMC_\pi$ = $(\mathsf{CRSGen}, \mathsf{Encode}_k, \mathsf{Decode}_k)$ is said to satisfy *message uniqueness* property if an adversary fails to create two valid pairs (M_1, M_2), $(M_1, M_2^{'})$ such that $\perp \neq$ $\mathsf{Decode}_k(\alpha, (M_1, M_2)) \neq \mathsf{Decode}_k(\alpha, (M_1, M_2^{'})) \neq \perp$, where $M_2 \neq M_2^{'}$ and the experiment produces two valid messages m_0, m_1. Any continuously non-malleable code should maintain the uniqueness property as mentioned in [17].

3 Code Construction

To construct the continuously non-malleable codes, an authenticated encryption[3] along with a non-malleable non-interactive commitment scheme is used. We describe the code construction below:

1. $\underline{\mathsf{CRSGen}(1^n)}$. The CRS generation algorithm samples a hash from 1-more weakly extractable leakage-resilient hash function H_n as $h_z \leftarrow \mathsf{H}_n$ and it sets $\alpha = h_z$. Finally, it outputs the CRS α as a security parameter.

[3] We show *Encrypt then MAC* scheme for the code construction.

2. <u>*Encode*$_k(\alpha, m)$.</u> To encode the message $m \in M$, a uniformly random key $k \in \mathcal{K}$ (where $k = \{k_{enc}||k_{mac}\}$) is selected with common reference string α. The algorithm first computes $(p_0, p_1) \leftarrow Enc^{lrs}(m||r)$ with some randomness $r \leftarrow \{0,1\}^n$. Further, p_0, p_1 are encrypted by an encryption algorithm of the authenticated encryption (i.e., $c_0 \leftarrow Encrypt(k_{enc}, p_0)$, $c_1 \leftarrow Encrypt(k_{enc}, p_1)$). The *tag* tag_{p_0} and tag_{p_1} are generated as $\mathsf{tag}_{p_0} \leftarrow tag(k_{mac}, c_0)$, $\mathsf{tag}_{p_1} \leftarrow tag(k_{mac}, c_1)$. The commitment scheme is used to check uniqueness of the key $k = \{k_{enc}, k_{mac}\}$, i.e., $com_0 = \mathsf{commit}(\alpha, k; r)$ (where $\alpha = h_z$). Further, we store the commitment value of the encrypted message and the tag as $com_1 = \mathsf{commit}(\alpha, c_0; \mathsf{tag}_{p_0})$, $com_2 = \mathsf{commit}(\alpha, c_1; \mathsf{tag}_{p_1})$. Finally, the codeword (M_1, M_2) is set to $M_1 = (k, p_0, (com_0, c_1, \mathsf{tag}_{p_1}), com_1, com_2)$, $M_2 = (k, p_1, (com_0, c_0, \mathsf{tag}_{p_0}), com_1, com_2)$ and subsequently, it is stored into the memory $\mathcal{M}_L$ and $\mathcal{M}_R$ respectively.
3. <u>*Decode*$_k(\alpha, (M_1, M_2))$.</u> To decode the codeword, com_0, com_1, com_2 are parsed and the following steps are performed:
   (a) <u>*Left & Right verification.*</u> If the verification of commitment does not satisfy, i.e., $com_1 \neq \mathsf{commit}(\alpha, c_0; \mathsf{tag}_{p_0})$ in M_2 and $com_2 \neq$ $\mathsf{commit}(\alpha, c_1; \mathsf{tag}_{p_1})$ in M_1, return $false$ and output $\perp$. Otherwise, go to the next step.
   (b) <u>*Uniquenesscheck.*</u> If $com_0 = \mathsf{commit}(\alpha, k; r)$, go to the next step. Otherwise, return $\perp$.
   (c) <u>*Cross check & Decode.*</u> tag_{p_0} and tag_{p_1} are retrieved using $\mathsf{Open}()$ operation of the commitment scheme. Then, tag_{p_0} and tag_{p_1} are verified, if it is not successful, return $\perp$. The key k of an authenticated encryption is obtained by $\mathsf{Open}()$ operation to perform the decryption. When the $\mathsf{Open}()$ is executed properly, compare $p_0 = Decrypt(k_{enc}, c_0)$, $p_1 = Decrypt(k_{enc}, c_1)$, if it is satisfied, invoke the decoding function $Dec^{lrs}(p_0, p_1)$.

Lemma 1. $CNMC_\pi$ = (*CRSGen*, *Encode*$_k$, *Decode*$_k$) satisfies message uniqueness property if implemented with a non-malleable commitment scheme.

Proof. To ensure message uniqueness, the binding property of a commitment scheme is considered. The instantiation of the $\mathsf{commit}()$ is based on 1-more wECRH and the underlying hash is collision resistant. Let us consider an adversary A which has the capability to generate a pair (M_1, M_2), $(M_1, M_2^{'})$ such that both are valid and $M_2 \neq M_2^{'}$. Therefore, the adversary is able to generate the following equation: $\perp \neq \mathsf{Decode}_k(\alpha, (M_1, M_2)) \neq \mathsf{Decode}_k(\alpha, (M_1, M_2^{'})) \neq \perp$. It is only possible if an adversary generates a valid key pair $(k, k^{'})$ in such a way that satisfies $\mathsf{commit}(\alpha, k, r) = com_0 = \mathsf{commit}(\alpha, k^{'}, r)$, where $k \neq k^{'}$. Unfortunately, such equation violates the binding property of the commitment scheme. Moreover, the violation of binding property implies that 1-more wECRH H_n is not collision resistant which is the contradiction. Hence, $\mathsf{commit}(\alpha, k, r) = com_0 \neq \mathsf{commit}(\alpha, k^{'}, r)$. Therefore, we can conclude that integrity of the key (property a) is violated and the decoding function should return $\perp$.

Algorithm 5. Tampering Experiment of Left Share $T_1(M_1, f_1^i, r, i)$

1: Parse $M_1 = (k, p_0, (com_0, c_1, \mathsf{tag}_{p_1}), com_1, com_2)$
2: Apply leakage function τ_1^i on M_1, i.e., $\mu_1[i] = \tau_1^i(M_1)$
3: Apply f_1^i on M_1. $M_1'' = f_1^i(M_1) = (k'', p_0'', (com_0'', c_1'', \mathsf{tag}_{p_1}''), com_1'', com_2'')$
4: Sample $h \leftarrow H_k$
5: Set $\alpha = h$
6: **if** $M_1 = M_1''$ **then**
7: set $\delta_1[i] = same^*$
8: **end if**
9: **if** $com_2 \neq \mathsf{commit}(\alpha, c_1''; \mathsf{tag}_{p_1}'')$ **then**
10: set $\delta_1[i] = \perp$ and return $\perp$
11: **end if**
12: **if** $com_2 = com_2''$ **then**
13: set $\delta_1[i] = \perp$ and return $\perp$
14: **end if**
15: Retrieve key k'' of AE by $k'' \leftarrow \mathsf{Open}(\alpha, com_0'')$
16: **if** $k'' = \perp$ **then**
17: set $\delta_1[i] = \perp$ and return $\perp$
18: **end if**
19: **if** $k'' \neq k$ **then**
20: set $\delta_1[i] = \perp$ and return $\perp$
21: **end if**
22: set $k'' = \{k_{enc}'' || k_{mac}''\})$
23: **if** $\mathsf{tag}_{p_1} \neq tag(k_{mac}'', c_1'')$ **then**
24: set $\delta_1[i] = \perp$ and return $\perp$
25: **end if**
26: $p_1'' \leftarrow Decrypt(k_{enc}'', c_1'')$,
27: Call decode $\mathfrak{Dec}_{lrs}(p_0'', p_1'')$ to retrieve m''
28: Set $\delta_1[i] = M_1''$ and return m''

3.1 Proof Idea of the Construction $CNMC_\pi$

We show a construction of continuously non-malleable codes from *encrypt then MAC* based authenticated encryption in 2-split-state model. To show the security of the design, we develop a simulator $\mathbf{SimTamper}^{A,0^n}_{CNMC_\pi}$ that takes $0^n||r$ ($r \leftarrow \{0,1\}^n$) as input instead of m in case of $\mathbf{Tamper}^{A,m}_{CNMC_\pi}$ experiment. An adversary can not distinguish between $\mathbf{Tamper}^{A,m}_{CNMC_\pi}$, the real experiment and $\mathbf{SimTamper}^{A,0^n}_{CNMC_\pi}$, the ideal experiment except with negligible probability $\epsilon(n)$, i.e., $[Pr[D(\mathbf{Tamper}^{A,m}_{CNMC_\pi}) = 1] - Pr[D(\mathbf{SimTamper}^{A,0^n}_{CNMC_\pi}=1]$ $] < \epsilon(n)$. At each invocation of $\mathbf{SimTamper}^{A,0^n}_{CNMC_\pi}$, a negligible amount of error is introduced to change the distribution. In simulated experiment, $Sim = (Sim_0, Sim_1)$ executes the operation where Sim_1 invokes two algorithms T_1 (Algorithm 5) and $T2$ (Algorithm 6) in the interleaved way. The experiment checks the message m from each part of the algorithm. Whenever, the outputs are not matched, execution is stopped by returning $\perp$ and the state is set to *self-destruct*. Due to non-persistent tampering attack model, a separate memory $\mathfrak{M}$ stores the results of all tampered and leakage data.

In the experiment, the difficult task is to find *self-destruct* point. Any further query performed by an adversary A from that point would return $\perp$ and the *state* is set to *self-destruct*. The adversary uses the leakage function $\tau(.)$ on the codeword M. Let $\mathcal{H}_\infty(M|\tau(M))$ be the conditional average entropy of the codeword M when some side-channel information is available to A, i.e., the best way to guess the message m from the codeword M with some side-channel information. An adversary uses the leakage function $\tau(.)$ on the codeword (M_1, M_2) in the following way: $\tau_1^0(M_1)$, $\tau_2^0(M_2)$, $\tau_1^1(M_1)$, $\tau_2^1(M_2)$, ... $\tau_1^{i-1}(M_1)$, $\tau_2^{i-1}(M_2)$. The simulation proceeds until the output of Algorithm 5 and Algorithm 6 are same. The left part of a codeword M_1 is simulated by Algorithm 5 with tampering function f_1^i whereas Algorithm 6 simulates the right part of a codeword M_2 with tampering function f_2^i. Information theoretically, we can see that $\tilde{\mathcal{H}}_\infty(M_1|\tau_1^i(M_1))$ $= \tilde{\mathcal{H}}_\infty(M_2|\tau_2^i(M_2))$, i.e., the best way to predict the message m from the codeword $M = (M_0, M_1)$ is same when some additional information is available through side-channel leakage to the adversary A. During the query phase, simulated outputs are checked, if it matches leak the entire part so that the total amount of leakages are upper bounded by $\mathsf{O}(n)$, where n represents the security parameter. The simulator $Sim = (Sim_0, Sim_1)$ uses Sim_0 to select the CRS, i.e., the hash $h_z \leftarrow \mathsf{H}_n$ in our case whereas Sim_1 performs the actual game with the help to two algorithms T_1 and T_2. In this construction, non-malleable non-interactive commitment scheme is used and it is instantiated with 1-more wECRH. Continuous non-malleability in 2-split-state model implies commitment scheme. It helps to maintain the message uniqueness property. Since the hash function is collision resistant and 1-more extractable as well as leakage resilient, we can achieve the uniqueness property from this primitive. The integrity of key k is maintained by binding property of the commitment scheme. Whenever the key is tampered and changed to other key $k^{'}$, the commitment value should be different than original one. Moreover, the commitment open operation is performed using the extractable hash to retrieve the key later. Further, the message is encrypted using AE and tag is stored along with the commitment value as part of codeword. In the simulated algorithm, we check the uniqueness of the key and the tag. If any alteration is present, experiment is stopped at that point. The complete view of simulated experiment is persevered into the $\mu_b[i]$ and $\delta_b[i]$ array. When all the checks are passed, both algorithms return the message m. Finally, they are compared with each other to check whether any modification is occurred or not. The total leakages of the experiment are upper bounded to $2l$.

Now, it is interesting to show that some attacks are impossible to perform on the codeword. Initially, an attacker can tamper c_1 to change it $c_1^{''}$ in the codeword M_1. But the binding property of commitment scheme says that it is hard to open a given commitment com_2 in two different ways. So, it implies that any modification in c_1 should produce a different com_2 value though the tag is not altered. Moreover, the attacker also has to make the same change in com_2 of codeword M_2 to generate the commitment value which is hard as the commitment scheme is non-malleable in nature. Further, if any alteration of key is performed, the authenticated encryption algorithm should return $\perp$ (property a). Such property is ensured by com_0 value. Hence, the codeword is capable of tolerating continuous tampering attacks and it is secure. The next section shows the security proof in detail.

4 Proof of Security

Theorem 1. *Let* $(k = \{k_{enc}||k_{mac}\}, Encypt, Decrypt)$ *be the authenticated encryption with message space* M*, key space* $\mathcal{K}$ *and ciphertext space* $\mathcal{C}$*,*

(Enc_{lrs}, Dec_{lrs}) be l'' leakage resilient storage, H_n be 1-more weakly extractable leakage-resilient hash function that is used to instantiate the non-malleable non-interactive commitment scheme (Init, Commit, Open). Then $CNMC_\pi$ = (CRSGen, $Encode_k$, $Decode_k$) is $((l + \gamma + \theta), q)$ continuously non-malleable and l leakage resilient code under non-persistent tampering when instantiated with all the above primitives, where $q = poly(n)$, $\gamma = log(M)$, $\theta = log(\mathcal{K})$, $l'' \geq (2l + n)$ and n denotes the security parameter.

Proof. The proof of the theorem is illustrated in the following way: In Subsect. 3.1, we show the $\mathbf{Tamper}^{A,m}_{CNMC_\pi}$ experiment where an adversary A tampers the codeword for polynomial number of times. The main motivation of our proof is show that after polynomial number of tampering attacks the security of the codeword preserves and it maintains non-malleability. We denote $\mathbf{Tamper}^{A,m}_{CNMC_\pi}$ as real game. A simulator is constructed that simulates the tampering experiment in an ideal world and it is denoted as ideal game. The simulated game $\mathbf{SimTamper}^{A,0^n}_{CNMC_\pi}$ takes 0^n to the input instead of original message m and performs the simulation. An adversary should not be able to distinguish between real game and ideal game except with negligible probability ϵ, i.e., $|Pr[\mathbf{Tamper}^{A,m}_{CNMC_\pi} = 1]$ - $Pr[\mathbf{SimTamper}^{A,0^n}_{CNMC_\pi} = 1]| \leq \epsilon(n)$. A simulator $Sim = (Sim_0, Sim_1)$ is developed to simulate the ideal game. Sim_0 generates an untamperable common reference string α using $\mathsf{CRSGen}(1^n)$ algorithm. The algorithm selects a hash $h_z \leftarrow \mathsf{H}_n$, where h_z is a *1-more weakly extractable leakage-resilient hash* function and it sets $\alpha = h_z$ $((\alpha) \leftarrow Sim_0)$. The goal of Sim_1 is to perform the simulated tampering experiment by invoking two algorithms (T_1, T_2) with tampering functions f_1^i and f_2^i $(i \leq q \wedge q \in poly(n))$. In Algorithm 5, we show that T_1 works on the left part of a codeword M_1 whereas in Algorithm 6, T_2 simulates the right part of a codeword M_2 in 2-split-state model. To work the simulation properly, distribution of simulated experiment $\mathbf{SimTamper}^{A,0^n}_{CNMC_\pi}$ is changed in a incremental way until we reach to the real tampering experiment $\mathbf{Tamper}^{A,m}_{CNMC_\pi}$. At each invocation of the experiment, a negligible error is inserted and the change is not noticeable due to the security of lrs scheme. This process changes the encryption of 0^n to the codeword M, i.e., encoding of message m. Sim_0 invokes the two algorithms (T_1, T_2) in the interleaved way and the experiment stops when the output generated from the algorithms T_1 and T_2 are different, i.e., $T_1(M, f_1^i, r, i) \neq T_2(M_2, f_2^i, r, i)$. From this point, any tampering query would return $\perp$ and experiment leads to *$self$-$destruct$* in $\mathbf{SimTamper}^{A,0^n}_{CNMC_\pi}$. Whenever the experiment triggers *$self$-$destruct$*, security of continuous non-malleability reduces to the security of underlying lrs scheme. Alternatively, the reduction can be described as follows: Let us assume that there exist an adversary A, a pair of messages $(m, 0^n)$, a negligible function ϵ, a distinguisher D, such that

$$[Pr[D(\mathbf{Tamper}^{A,m}_{CNMC_\pi}) = 1] - Pr[D(\mathbf{SimTamper}^{A,0^n}_{CNMC_\pi} = 1]] > \epsilon(n)$$

Moreover, we can construct another distinguisher D' against the lrs scheme with an adversary A' such that

$$[Pr[D^{'}(leak^{\beta}_{A,m}) = 1] - Pr[D^{'}(leak^{\beta}_{A,0^n}) = 1]] > \epsilon^{'}(n)$$

It leads to the contradiction. We run $D^{'}$ and $A^{'}$ for many n to break the lrs security with non-negligible probability. At the end of experiment, they are given access to D and A that win in the game $\mathbf{Tamper}^{A,m}_{CNMC_{\pi}}$ with non-negligible advantage. Algorithm 5 and Algorithm 6 requires $l^{'} \geq (2l + n)$ bits of leakage to find the self-destruct point i (where $i = q$ during self-destruct). From which any tampering query would return $\perp$ and self-destruct would trigger. $A^{'}$ runs A as subroutine to find the index i (where $i = q$ during self-destruct). $\delta_1[i]$ and $\delta_2[i]$ contain the result of tampering queries whereas $\mu_1[i]$ and $\mu_2[i]$ contain the result of leakage queries. Upto $(q-1)^{th}$ query, views are same for the real and the ideal game. As we assume that A, D has non-negligible advantage, $D^{'}$, $A^{'}$ also have non-negligible advantage that contradicts the fact that underlying lrs scheme is secure. Hence, the security of codeword is reduced to the security of leakage resilient storage.

Algorithm 5 describes the simulated experiment T_1 on the codeword M_1. Initially, the codeword is parsed and the leakage function $\tau^i_1()$ is applied on it. As per the underlying lrs security, it can tolerate maximum l bits of leakage. The complete view of leakage experiment is stored into the array $\mu_1[i]$. Next, the tampering function f^i_1 is applied the codeword M_1, and the modified codeword is set to $M^{''}_1 = f^i_1(M_1) = (k^{''}, p^{''}_0, (com^{''}_0, c^{''}_1, \mathsf{tag}^{''}_{p_1}), com^{''}_1, com^{''}_2)$. If the modified codeword $M^{''}_1$ is same as original one M_1, $\delta_1[i]$ array is set to *same**. Now, the non-malleable non-interactive commitment scheme is used to verify whether com_2 is *same* with the new modified $c^{''}_1$, and $\mathsf{tag}^{''}_{p_1}$, i.e., $com_2 = \mathsf{commit}(\alpha, c^{''}_1; \mathsf{tag}^{''}_{p_1})$. If, the condition returns *true*, the algorithm further checks whether $com_2 = com^{''}_2$ is *true or false*. Next, a hash function is sampled from $h_z \leftarrow \mathsf{H}_n$, where H_n is 1-more wECRH. It is used to instantiate the non-malleable commitment scheme. When the $com_0 = \mathsf{commit}(\alpha, k; r)$ is invoked during the codeword construction, $h_z(k, r)$ is called and the hash is preserved in com_0 (where, $\alpha = h_z$). Since the hash function is leakage resilient and extractable, such key k can be retrieved using $\mathsf{open}(\alpha, com_0)$ operation of the commitment scheme. Hence, the modified key is generated by $\mathsf{open}(\alpha, com^{''}_0)$. Further, the modified key $k^{''}$ is compared with the key k and the algorithm proceeds. The key $k^{''}$ is parsed to select the encryption key $k^{''}_{enc}$ and the MAC key $k^{''}_{mac}$. The *tag* of the underlying authenticated encryption is verified by $\mathsf{tag}_{p_0} \neq tag(k^{''}_{mac}, c^{''}_0)$. When the verification algorithm returns *true*, the decryption function is called $p^{''}_0 \leftarrow Decrypt(k^{''}_{enc}, c^{''}_0)$ to get $p^{''}_0$. Finally, the decoding of lrs $\mathfrak{Dec}_{lrs}(p^{''}_0, p^{''}_1)$ returns the message $m^{''}$. $\delta_1[i]$ and $\mu_1[i]$ stores the complete view of the algorithm T_1.

Algorithm 6. Tampering Experiment of Right Share $T_2(M_2, f_2^i, r, i)$

1: Parse $M_2 = (k, p_1, (com_0, c_0, \mathsf{tag}_{p_0}), com_1, com_2)$
2: Apply leakage function τ_2^i on M_2, i.e., $\mu_2[i] = \tau_2^i(M_2)$
3: Apply f_2^i on M_2. $M_2'' = f_2^i(M_2) = (k'', p_1'', (com_0'', c_0'', \mathsf{tag}_{p_0}), com_1'', com_2'')$
4: Sample $h \leftarrow H_k$
5: Set $\alpha = h$
6: **if** $M_2 = M_2''$ **then**
7:     set $\delta_2[i] = same^*$
8: **end if**
9: **if** $com_1 \neq \mathsf{commit}(\alpha, c_0''; \mathsf{tag}_{p_0}'')$ **then**
10:     set $\delta_2[i] = \perp$ and return $\perp$
11: **end if**
12: **if** $com_1 = com_1''$ **then**
13:     set $\delta_2[i] = \perp$ and return $\perp$
14: **end if**
15: Retrieve key k'' of AE by $k'' \leftarrow \mathsf{Open}(\alpha, com_0'')$
16: **if** $k'' = \perp$ **then**
17:     set $\delta_2[i] = \perp$ and return $\perp$
18: **end if**
19: **if** $k'' \neq k$ **then**
20:     set $\delta_2[i] = \perp$ and return $\perp$
21: **end if**
22: set $k'' = \{k''_{enc} || k''_{mac}\})$
23: **if** $\mathsf{tag}_{p_0} \neq tag(k''_{mac}, c_0'')$ **then**
24:     set $\delta_2[i] = \perp$ and return $\perp$
25: **end if**
26: $p_0'' \leftarrow Decrypt(k''_{enc}, c_0'')$,
27: Call decode $\mathfrak{Dec}_{lrs}(p_0'', p_1'')$ to retrieve m''
28: Set $\delta_2[i] = M_2''$ and return m''

Algorithm 6 performs the simulated tampering experiment T_2 on the codeword M_2. The leakage function τ_2^i on M_2 stores the result in $\mu_2[i]$. The maximum leakage of T_2 is upper bounded to l. During each tampering attack, the tampering function f_2^i is used on the codeword M_2 with a different value of i and subsequently, the codeword is modified in the following way: $M_2'' = f_2^i(M_2) = (k'', p_1'', (com_0'', c_0'', \mathsf{tag}_{p_0}''), com_1'', com_2'')$. Further, the experiment checks whether $com_1 = \mathsf{commit}(\alpha, c_0''; \mathsf{tag}_{p_0}'')$ is *true or false*. If so, the condition $com_1 = com_1''$ is invoked and the corresponding views are stored into $\delta_2[i]$. The extractable hash h_z is sampled from H_n which is used to retrieve key k'' of AE as $k'' \leftarrow \mathsf{Open}(\alpha, com_0'')$. Such key is again compared with the key k and the tag verification of the AE is invoked. If it is successful, p_0'' is restored as $p_0'' \leftarrow Decrypt(k''_{enc}, c_0'')$. Finally, $\mathfrak{Dec}_{lrs}(p_0'', p_1'')$ returns the message m''.

The main goal of simulator Sim_1 is to run the algorithm T_1 and T_2 alternatively as long as their outputs are same. Let the average conditional entropy of codeword M_1 be denoted as $\tilde{\mathcal{H}}_\infty(M_1|\tau_1^i(M_1))$. Here, entropy represents the randomness of the codeword when the adversary A has side channel leakage access using $\tau_0^i(M_0)$ to guess some bits of information. In information theoretic observation, it is written in the following way: $\tilde{\mathcal{H}}_\infty(M_1|\tau_1^i(M_1)) = \tilde{\mathcal{H}}_\infty(M_2|\tau_1^i(M_2))$ from the simulation strategy. We can write $\tilde{\mathcal{H}}_\infty(M_1|\tau_1^i(M_1))$ as follows (**Lemma 2.1.3**).

$$\tilde{\mathcal{H}}_\infty(M_1|\tau_1^i(M_1)) = \mathcal{H}_\infty(M_1) - l$$

Similarly,

$$\tilde{\mathcal{H}}_\infty(M_2|\tau_2^i(M_2)) = \mathcal{H}_\infty(M_2) - l$$

In the experiment, $\tau_1^i(M_1)$ or $\tau_2^i(M_2)$ can tolerate leakage at most l bits as per security of the lrs scheme. The simulator Sim_1 executes the $\mathbf{SimTamper}_{CNMC_\pi}^{A,0^n}$ experiment until *self-destruct* is invoked or returns $\perp$. Let an adversary invokes q number of tampering queries in $\mathbf{Tamper}_{CNMC_\pi}^{A,m}$. We assume that the experiment stops at q^{th} query invocation. In the $\mathbf{SimTamper}_{CNMC_\pi}^{A,0^n}$ experiment, same number of queries are performed and the experiment stops by returning $\perp$ when outputs from T_1 and T_2 are unequal. The algorithm $T_1(M_1, f_1^q, r, q)$ and $T_2(M_2, f_2^q, r, q)$ are capable of sustaining l bits of leakage. For 1 to $(q-1)^{th}$ query, we get the below equation by assuming that function output cannot be more informative than its own input and last inequality comes from **Lemma 2.1.4**. Moreover, we consider that M_2, $T_1(M_1, f_1^q, r, q))$ do not give much useful information about M_1 to guess the message m and the min-entropy of M_0 is decreased by $\mathsf{O}(n)$, i.e., its size

$$\tilde{\mathcal{H}}_\infty(M_1|T_1(M_1, f_1^1, r, 1), .., T_1(M_1, f_1^q, r, q))$$
$$= \tilde{\mathcal{H}}_\infty(M_1|T_2(M_2, f_2^1, r, 1), .., T_2(M_2, f_2^{q-1}, r, q-1), T_1(M_1, f_1^q\ , r, q)).$$
$$\Longrightarrow \tilde{\mathcal{H}}_\infty(M_1|T_2(M_2, f_2^1, r, 1), .., T_2(M_2, f_2^{q-1}, r, q-1), T_1(M_1,\ f_1^q, r, q))$$
$$\geq \tilde{\mathcal{H}}_\infty(M_1|M_2, q, T_1(M_1, f_1^q, r, q)).$$

The simulated experiment performed by Sim_1 compares the tampered output from both sides of (M_1, M_2) and if it matches, the entire codeword is leaked. At q^{th} query execution, when the outputs of T_1 and T_2 are different and $\tau_1^q(M_1) \neq \tau_2^q(M_2)$, leak the entire tampered codeword so that total amount of leakage is bounded by $\mathsf{O}(n)$. The lrs scheme used in this construction is able to tolerate maximum $2l$ (l bits from each side) bits. Combining all the parameters, the value of l' has to be at least greater than $(2l+n)$ to work the simulator Sim_1 properly.

5 Conclusion

In this work, we show that continuously non-malleable codes can be constructed from *encrypt then MAC* based authenticated encryption in *common reference string* model. Our codeword does not require *robust* NIZK proof since NIZKs are the bottleneck. Instead of using NIZK, we use non-malleable non-interactive commitment scheme instantiated with a *1-more weakly extactable leakage-resilient hash function.* An adversary can perform *non-persistent* tampering attacks for polynomial number of times until *self-destruct* state is triggered. Further research work can be pursued to construct *super-strong* continuously non-malleable codes from authenticated encryption.

References

1. Goldreich, O., Micali, S., Wigderson, A.: Proofs that yield nothing but their validity for all languages in NP have zero-knowledge proof systems. J. ACM **38**(3), 691–729 (1991)
2. Bellare, M., Namprempre, C.: Authenticated encryption: relations among notions and analysis of the generic composition paradigm. In: Okamoto, T. (ed.) ASIACRYPT 2000. LNCS, vol. 1976, pp. 531–545. Springer, Heidelberg (2000)

3. Boneh, D., DeMillo, R.A., Lipton, R.J.: On the importance of eliminating errors in cryptographic computations. J. Cryptology **14**(2), 101–119 (2001)
4. De Santis, A., Di Crescenzo, G., Ostrovsky, R., Persiano, G., Sahai, A.: Robust non-interactive zero knowledge. In: Kilian, J. (ed.) CRYPTO 2001. LNCS, vol. 2139, pp. 566–598. Springer, Heidelberg (2001)
5. Bellare, M., Kohno, T.: A theoretical treatment of related-key attacks: RKA-PRPS, RKA-PRFS, and applications. In: Biham, E. (ed.) EUROCRYPT 2003. LNCS, vol. 2656, pp. 491–506. Springer, Heidelberg (2003)
6. Dodis, Y., Reyzin, L., Smith, A.: Fuzzy extractors: how to generate strong keys from biometrics and other noisy data. In: Cachin, C., Camenisch, J.L. (eds.) EUROCRYPT 2004. LNCS, vol. 3027, pp. 523–540. Springer, Heidelberg (2004)
7. Pass, R., Rosen, A.: New and improved constructions of non-malleable cryptographic protocols. In Gabow H. N. Fagin, R. Eds, 37th ACM STOC, pp. 533-542. ACM Press (2005)
8. Dziembowski, S., Pietrzak, K., Wichs, D.: Non-malleable codes. In: Yao, A.C.-C. (ed.) ICS 2010, Beijing, China, January 5-7, pp. 434-452. Tsinghua University Press (2010)
9. Davì, F., Dziembowski, S., Venturi, D.: Leakage-resilient storage. In: Garay, J.A., De Prisco, R. (eds.) SCN 2010. LNCS, vol. 6280, pp. 121–137. Springer, Heidelberg (2010)
10. Dziembowski, S., Faust, S.: Leakage-Resilient Cryptography from the Inner-Product Extractor. In: Lee, D.H., Wang, X. (eds.) ASIACRYPT 2011. LNCS, vol. 7073, pp. 702–721. Springer, Heidelberg (2011). https://doi.org/10.1007/978-3-642-25385-0_38
11. Bellare, M., Cash, D., Miller, R.: Cryptography secure against related-key attacks and tampering. In: Lee, D.H., Wang, X. (eds.) ASIACRYPT 2011. LNCS, vol. 7073, pp. 486–503. Springer, Heidelberg (2011)
12. Kalai, Y.T., Kanukurthi, B., Sahai, A.: Cryptography with tamperable and leaky memory. In: Rogaway, P. (ed.) CRYPTO 2011. LNCS, vol. 6841, pp. 373–390. Springer, Heidelberg (2011)
13. Liu, F.-H., Lysyanskaya, A.: Tamper and leakage resilience in the split-state model. In: Safavi-Naini, R., Canetti, R. (eds.) CRYPTO 2012. LNCS, vol. 7417, pp. 517–532. Springer, Heidelberg (2012)
14. Bellare, M., Paterson, K.G., Thomson, S.: RKA security beyond the linear barrier: IBE, encryption and signatures. In: Wang, X., Sako, K. (eds.) ASIACRYPT 2012. LNCS, vol. 7658, pp. 331–348. Springer, Heidelberg (2012)
15. Dziembowski, S., Kazana, T., Obremski, M.: Non-malleable codes from two-source extractors. In: Canetti, R., Garay, J.A. (eds.) CRYPTO 2013. LNCS, vol. 8043, pp. 239–257. Springer, Heidelberg (2013)
16. Damgård, I., Faust, S., Mukherjee, P., Venturi, D.: Bounded tamper resilience: how to go beyond the algebraic barrier. In: Sako, K., Sarkar, P. (eds.) ASIACRYPT 2013, Part II. LNCS, vol. 8270, pp. 140–160. Springer, Heidelberg (2013)
17. Faust, S., Mukherjee, P., Nielsen, J.B., Venturi, D.: Continuous non-malleable codes. In: Lindell, Y. (ed.) TCC 2014. LNCS, vol. 8349, pp. 465–488. Springer, Heidelberg (2014)
18. Aggarwal, D., Dodis, Y., Lovett, S.: Non-malleable codes from additive combinatorics. In: STOC, pp. 774-783 (2014)
19. Faust, S., Mukherjee, P., Venturi, D., Wichs, D.: Efficient non-malleable codes and key-derivation for poly-size tampering circuits. In: EUROCRYPT, pp. 111-128 (2014)

20. Jafargholi, Z., Wichs, D.: Tamper detection and continuous non-malleable codes. In: Dodis, Y., Nielsen, J.B. (eds.) TCC 2015. LNCS, vol. 9014, pp. 451–480. Springer, Heidelberg (2015)
21. Aggarwal, D., Dodis, Y., Kazana, T., Obremski, M.: Non-malleable reductions and applications. In: Proceedings of the Forty-Seventh Annual ACM on Symposium on Theory of Computing, pp. 459-468. ACM (2015)
22. Kiayias, A., Liu, F.H., Tselekounis, Y.: Practical non-malleable codes from l-more extractable hash functions. In: Weippl, E.R., Katzenbeisser, S., Kruegel, C., Myers, A.C., Halevi, S. (eds.) ACM CCS 2016, pp. 1317-1328. ACM Press, October (2016)
23. Aggarwal, D., Agrawal, S., Gupta, D., Maji, H.K., Pandey, O., Prabhakaran, M.: Optimal computational split-state non-malleable codes. In: Kushilevitz, E., Malkin, T. (eds.) TCC 2016. LNCS, vol. 9563, pp. 393–417. Springer, Heidelberg (2016)
24. Aggarwal, D., Kazana, T., Obremski, M.: Inception makes non-malleable codes stronger. In: Kalai, Y., Reyzin, L. (eds.) TCC 2017. LNCS, vol. 10678, pp. 319–343. Springer, Cham (2017)
25. Barwell, G., Martin, D.P., Oswald, E., Stam, M.: Authenticated encryption in the face of protocol and side channel leakage. In: Takagi, T., Peyrin, T. (eds.) ASIACRYPT 2017. LNCS, vol. 10624, pp. 693–723. Springer, Cham (2017)
26. Faonio, A., Nielsen, J.B., Simkin, M., Venturi, D.: Continuously non-malleable codes with split-state refresh. In: Preneel, B., Vercauteren, F. (eds.) ACNS 2018. LNCS, vol. 10892, pp. 1–19. Springer, Cham (2018)
27. Fehr, S., Karpman, P., Mennink, B.: Short Non-Malleable Codes from Related-Key Secure Block Ciphers. IACR Trans Symmetric Cryptology, 336-352, (2018)
28. Ostrovsky, R., Persiano, G., Venturi, D., Visconti, I.: Continuously non-malleable codes in the split-state model from minimal assumptions. In: Shacham, H., Boldyreva, A. (eds.) CRYPTO 2018, Part III. LNCS, vol. 10993, pp. 608–639. Springer, Cham (2018)
29. Aggarwal, D., Döttling, N., Nielsen, J.B., Obremski, M., Purwanto, E.: Continuous non-malleable codes in the 8-split-state model. In: Ishai, Y., Rijmen, V. (eds.) EUROCRYPT 2019, Part I. LNCS, vol. 11476, pp. 531–561. Springer, Cham (2019)
30. Dachman-Soled, D. Kulkarni, M.: Upper and lower bounds for continuous non-malleable codes, in, pp. 519-548 PKC (2019)
31. Ghosal, A.K., Ghosh, S., Roychowdhury, D.: Practical Non-malleable Codes from Symmetric-Key Primitives in 2-Split-State Model. In: Ge, C., Guo, F. (eds) Provable and Practical Security (2022)
32. Kiayias, A., Liu, F.H., Tselekounis, Y.: Leakage Resilient l-more Extractable Hash and Applications to Non-Malleable Cryptography. Cryptology ePrint Archive, Report2022/1745 (2022)
33. Ghosal, A.K., Roychowdhury, D.: Continuously Non-malleable Codes from Authenticated Encryptions in 2-Split-State Model. In: Prabhu, S., Pokhrel, S.R., Li, G. (eds) Applications and Techniques in Information Security, (2022)

Digital Signatures

Pairing-Free ID-Based Signatures as Secure as Discrete Logarithm in AGM

Jia-Chng Loh(✉), Fuchun Guo, and Willy Susilo

School of Computing and Information Technology, Institute of Cybersecurity and Cryptology, University of Wollongong, Wollongong, Australia
{jial,fuchun,wsusilo}@uow.edu.au

Abstract. Identity-based signatures (IBS) allow the signer's identity information to be used as the public key for signature verification, eliminating the need for managing certificates to establish ownership of the corresponding public key. The Schnorr-like IBS due to Galindo and Garcia is known as the most efficient IBS based on the discrete logarithm (DL) problem, without the need for computationally expensive pairing operations. This makes it a lightweight and efficient solution for signature generation and verification. Unfortunately, the security reduction of Schnorr-like IBS is not tight under the standard EUF-CMA in the ID-based setting. Recently, by using the algebraic group model (AGM), where adversary computation is algebraic, the EUF-CMA security of ordinary Schnorr signatures has been proven tightly secure under DL assumption with random oracles. However, one could not trivially apply the reduction of Schnorr signatures in AGM to achieve tight security for the Schnorr-like IBS scheme because of the inability to capture the chosen identity-and-message attacks. In this work, we show that, with the adoption of AGM, it is feasible to tighten the EUF-CMA security for IBS without pairing under DL assumption with random oracles. We resolve the chosen identity-and-message attacks by adopting the OR-proof technique to generate the user's private key containing the DL of either one of the two random group elements, leading to a new pairing-free IBS scheme. We provide a concrete security analysis for the scheme in AGM showing that by embedding the DL problem instance into one of the randomness, the algebraic adversary could only return a non-reducible forgery and representations with half of the success probability.

Keywords: Pairing-free · Identity-based signatures · Tight security reduction · Algebraic group model

1 Introduction

Digital signatures ensure data authenticity and integrity in digital communication. A signature is valid if the verification algorithm holds, which requires the

W. Susilo—Supported by the ARC Australian Laureate Fellowship FL230100033.
F. Guo—Supported by the ARC Future Fellowship FT220100046.

T. Zhu and Y. Li (Eds.): ACISP 2024, LNCS 14895, pp. 347–367, 2024.
https://doi.org/10.1007/978-981-97-5025-2_18

signer's public key as input. The identity-based signatures (IBS), introduced by Shamir [39], replace the signer's public key with a user's identity (e.g., email and identity card number), eliminating the need for a Public Key Infrastructure (PKI) whose responsible to bind between the public key and the signer, thereby substantially reducing costs associated with key management and certificate issuance in practice.

In modern cryptography, constructing a provably secure signature scheme usually comes with a reduction algorithm in some pre-defined security models, which turns algorithm $\mathcal{A}$ breaking the scheme with probability ϵ_A into algorithm $\mathcal{B}$ breaks some hard problems with probability $\epsilon_B \geq \frac{\epsilon_A}{L}$, such that $L \geq 1$ is the loss factor, and if L is small constant, we say the reduction is tight. A tight reduction has a meaningful feature as one could choose the optimal parameter size for the scheme when deploying it into practice, hence the scheme preserves the efficiency as proposed. However, it is a challenging task to achieve tight security for the IBS scheme under its standard security model, namely the existential unforgeability against chosen identity-and-message attacks (EUF-CMA). In particular, such a tight reduction algorithm must have a constant success probability near one that can (i) respond to all adversary's queries and (ii) extract problem solutions based on any given forgery.

With the emergence of pairing-based cryptography, several IBS schemes [3,5,9,14,21,26,34,35] have been proposed over time, each aiming to strike a balance between security and efficiency. For example, Waters-IBS [35] and BBG-IBS [26] are proven in the standard model without relying the random oracles [6]. Pairing-based cryptography, although powerful, may not be desirable under certain applications, especially in resource-constrained environments where computational power is limited, e.g., Internet of Things devices [1,31] and wireless sensor networks [32,37,41]. The Schnorr-like IBS scheme due to Galindo and Garcia [19] at AfricaCrypt'09 has been known as the most efficient IBS scheme due to its simple construction, using well-known Schnorr signatures [38] based on the discrete logarithm (DL) problem, without costly pairing operations.

The EUF-CMA security of Schnorr-like IBS was first proven under the DL assumption [19]. Although its security was revised and improved by Chatterjee et al. [8], the security loss of the scheme is still loosely reduced to the DL problem because of the need for the reset lemma [36], and schemes proven with reset lemma are known to suffer from the tightness barrier [25].

Since the introduction of the algebraic group model (AGM) [15], which idealizes the adversary's computation to be algebraic, such that any returned group element from the algebraic adversary must be described along with the representation based on all received group elements, the EUF-CMA security of ordinary Schnorr signatures [16] has been proven tightly secure under DL assumption in AGM. Recently, Loh et al. [28] proposed an extended BLS signatures [15] in an ID-based setting, namely the BLS-IBS scheme. With the help of the AGM, its EUF-CMA security can be proven tightly reduced to the DL problem. In addition, they mentioned that the Schnorr-like IBS scheme may not be tightly EUF-CMA secure even with the help of the algebraic adversary in the AGM because of the difficulty of designing a reduction algorithm that can capture all kinds of adversary queries and forgeries. While there is no further discussion of this, it is, therefore, an interesting challenge to investigate if there exists any potential way to achieve a tightly secure IBS scheme without pairing.

1.1 Our Contribution

In this work, with the adoption of the AGM, we show the possibility of improving security reductions for pairing-free IBS schemes under DL assumption and the EUF-CMA security model. We first study the main reduction issue encountered in Schnorr-like IBS, for which the simulator could not respond to all adversary's queries, i.e., the inability to simulate the user's private key and extract problem solutions based on any given forgery. We resolve the simulator aborts issue by using the OR-proof technique [10,20] to generate the user's private key. In particular, the user's private key contains two group elements and a Schnorr's signature on the user's identity that binds with one of the randomness. While computing signatures, the user proves the possession of the key belongs to one of the two randomnesses. This leads to a new pairing-free IBS scheme.

We then prove the EUF-CMA security of the proposed IBS scheme can be tightly reduced to the DL problem in the AGM. Given a DL problem instance, we design a reduction algorithm in which there exists a simulator that can randomly embed the problem instance into one of the two group elements of the user's secret key, hence the simulator implicitly manages to respond to all adversary's queries, i.e., signing queries and key extraction queries. In the forgery phase, the algebraic adversary has no advantage in revealing which one is the embedded randomness, hence with the returned forgery and representations, our simulator can successfully extract the DL solution with a success probability of $\frac{1}{2}$.

Table 1 summarizes EUF-CMA secure IBS schemes in cyclic groups. Some IBS schemes with pairing [26,35] offer security in the standard model (SM) but lack tight reductions under the DL assumption. While most pairing-free IBS schemes under DL assumption are proven in the random oracle model (ROM) with loose reductions [5,7,19], except for [17,18] proven under decisional problems. Achieving tight reductions under DL assumption involves leveraging the AGM.

1.2 EUF-CMA Security of Schnorr in AGM

We first take a look at how the EUF-CMA security of Schnorr signatures [38] is proven tightly secure in the algebraic group model (AGM) [15]. Let g be a group generator of a group $\mathbb{G}$ in the prime order of p, $H : \{0,1\}^* \rightarrow \mathbb{Z}_p^*$ be a cryptography hash function, and $Z = g^z$ be the user's public key for some $z \in \mathbb{Z}_p^*$ and $r \in \mathbb{Z}_p^*$ be some randomness. The Schnorr signature $\sigma_m = (R, y)$ on message m is computed as

$$R = g^r, \quad y = r + z \cdot H(m, R).$$

Table 1. Summary of EUF-CMA Secure IBS Schemes

	Pairing-free	Hardness Assumption	Tight Reduction	SM/ROM/ AGM
ChCh-IBS [9]	×	CDH	×	ROM
BBMQ-IBS [3]	×	q-SDH	×	ROM
Waters-IBS [35]	×	CDH	×	SM
BBG-IBS [26]	×	mCDH	×	SM
BNN-IBS [5], Beth-IBS [7], Schnorr-like IBS [19]	✓	DL	×	ROM
FH-IBS [17,18]	✓	DDH	✓	ROM
BLS-IBS [28]	×	DL	✓	ROM+AGM
Ours	✓	DL	✓	ROM+AGM

At EuroCrypt'20, Fuchsbauer et al. [16] showed that its EUF-CMA security can be proven tightly secure in AGM. The security proof in AGM requires the adversary to be algebraic, such that an adversary that outputs a group element $X \in \mathbb{G}$ also outputs a vector $\vec{c} = (c_0, ..., c_n)$ representing how X is the linear combination $X = g^{c_0} C_1^{c_1} \cdots C_n^{c_n}$ of all group elements $g, C_0, ..., C_n \in \mathbb{G}$ that the adversary has obtained.

Suppose the algebraic adversary, in the forgery phase of the EUF-CMA game, returns a forgery $\sigma_{m^*} = (R^*, y^*)$ on chosen message m^* (without query any signature) and a vector $\vec{u}$, such that $R^* = g^{u_0} Z^{u_1}$. While forgery σ_{m^*} is valid, i.e. $g^{y^*} = R^* \cdot Z^{H(m^*,R^*)}$ holds, we obtain

$$R^* = g^{u_0} Z^{u_1} = g^{y^*} (Z)^{-H(m^*,R^*)}.$$

Now, assume that the DL problem instance is embedded into the master public key, i.e. $Z = g^{\alpha}$ where α is unknown. By observing the above equation, one may solve for $\alpha = \frac{y^* - u_0}{u_1 + H(m^*,R^*)}$ where the algebraic adversary has negligible success probability to set $u_1 = -H(m^*, R^*)$. We defer the full proof to [16]. A question arises as to why it is not trivial to adopt the same proof technique for the EUF-CMA security of the Schnorr-like IBS scheme [19].

1.3 Challenge to Achieve Tight Reduction for Schnorr-Like IBS

We first recall the Schnorr-like IBS scheme by Galindo and Garcia [19], which is constructed using two concatenated Schnorr signatures [38], and we discuss the main fact of designing a tight reduction.

Let g be a group generator of a group $\mathbb{G}$ in the prime order of p and $H_1, H_2 : \{0,1\}^* \rightarrow \mathbb{Z}_p^*$ be two cryptography hash functions. In the setup phase of the

IBS scheme, the TTP selects $z \in \mathbb{Z}_p^*$ and the master public and secret key pair as $(mpk = Z = g^z, msk = z)$. To generate a user's private key d_{ID}, the TTP produces a Schnorr signature on the identity of the user by using the master secret key. Let $r \in \mathbb{Z}_p^*$ be some randomness, and the user's private key is defined as $d_{ID} = (R = g^r, y = r + z \cdot H_1(ID, R))$. Next, the user uses its private key $d_{ID} = (R, y)$ to compute a Schnorr-like signature $\sigma_{ID,m} = (R, A, s)$. Let $a \in \mathbb{Z}_p^*$ be some randomness,

$$A = g^a, \quad s = a + y \cdot H_2(ID, m, A).$$

Non-tight Reduction. One notable challenge to design a tight reduction algorithm for Schnorr-like IBS in the EUF-CMA security model under DL assumption is the ability to respond to all adversary's queries and extract problem solution α on any given forgery. In particular, it is hard to design a reduction that the simulator can first respond to signature query $\mathcal{O}_S(ID, m) \rightarrow \sigma_{ID,m} = (R, A, s)$ then user's private key query $\mathcal{O}_E(ID) \rightarrow d_{ID} = (R, y)$ with the same R.

Given a DL problem instance tuple (g, g^α) where $\alpha \in \mathbb{Z}_p^*$, the simulator may embed g^α into either master public key $mpk = Z$, user's private key randomness R, or signature randomness A.

Non-simulatable User's Private Key. Suppose Z, A are embedded with g^α. This yields a critical issue when the adversary queries $\mathcal{O}_S(ID', m)$, then $\mathcal{O}_E(ID')$. Upon receiving the extraction query, the simulator must abort as it cannot simulate $d_{ID'} = (R', y')$, such that R' that was used to compute $\sigma_{ID',m} = (R', A, s)$.

Simulatable User's Private Keys But Non-reducible. As noted in [28], the query phase of the EUF-CMA security game for IBS in AGM can be successful with overwhelming success probability if the simulator can simulate the user's private key $d_{ID'}$ and the corresponding signatures $\sigma_{ID',m}$. That is said, if Z, R' are embedded with g^α, the simulator can randomly choose $A = g^a$ for some $a \in \mathbb{Z}_p^*$ to simulate $\sigma_{ID',m} = (R', A, s)$ on (ID, m). Unfortunately, this is not enough to achieve tight reduction for the Schnorr-like IBS scheme. The following discussion explains the fact.

Suppose $ID' = ID^*$ is the forgery identity and $d_{ID^*} = (R^*, y^*)$ is simulatable. Note that the query phase is perfect if the simulator sets $Z = g^\alpha$ and $R^* = g^{r'^*} Z^{-h_{ID^*}}$, where $r'^*, h_{ID^*} \in \mathbb{Z}_p^*$ are randomly chosen, such that $h_{ID^*} = H_1(ID^*, R^*)$ is programmable due to random oracles. Therefore, upon receiving a queried signature $\sigma_{ID^*,m} = (R^*, A, s) \leftarrow \mathcal{O}_S(ID^*, m)$, we assume the algebraic adversary can return a valid forgery $\sigma_{ID^*,m^*} = (R^*, A^*, s^*)$ on (ID^*, m^*) and two vectors $\vec{u}, \vec{x}$ that representing how R^*, A^* are computed, respectively, such that $R^* = g^{u_0} Z^{u_1} R^{*u_2} A^{u_3}$ and $A^* = g^{x_0} Z^{x_1} R^{*x_2} A^{x_3}$. Since the forgery σ_{ID^*,m^*} is valid, the verification equation holds, such that

$$g^{s^*} = A^*(R^* Z^{h_{ID^*}})^{h_{ID^*,m^*}} \tag{1}$$

where $h_{ID^*} = H_1(ID^*, R^*)$ and $h_{ID^*,m^*} = H_2(ID^*, m^*, A^*)$.

Notably, the above forgery and representations can be non-reducible by simply reusing the obtained $R^* = g^{r'^*} Z^{-h_{ID^*}}$ from the query phase. Therefore, based on Eq. (1), it becomes

$$g^{s^*} = A^* (g^{r'^*})^{h_{ID^*,m^*}}$$

The algebraic adversary can easily compute A^* using representations $\vec{x}$ as the linear combination without any embedded elements, i.e. Z, R^*. Hence, the simulator does not have sufficient knowledge to extract the DL solution α, i.e. coefficients of α are zero.

1.4 Our Techniques

Based on the above observations, we identify that the main challenge is to enable the reduction (i) to answer all adversary's queries; and (ii) to ensure the forgery on (ID^*, m^*) is reducible even though the adversary has queried for signatures on ID^*, i.e. d_{ID^*} was simulated. We hence attempt to use an OR-proof technique to generate the user's private key in the Schnorr-like IBS scheme. The idea is similar to the Twin signatures in [29] that sign with two possible secret keys, but here we adopt this technique to enable simulating the user's private key. In particular, the TTP now selects two randomness $r_0, r_1 \in \mathbb{Z}_p^*$ and computes the user's private key $d_{ID} = (R_0, R_1, b, y)$ as follows. Let $b \in \{0, 1\}$ be a random bit,

$$y = r_b + z \cdot H_1(ID, R_0, R_1), \quad R_0 = g^{r_0}, \quad R_1 = g^{r_1}.$$

Next, the user computes signature $\sigma_{ID,m}$, which can be viewed as a proof of knowledge of y with either the DL of randomness R_0, R_1. The user first retrieves $b \in \{0, 1\}$, y, R_0, R_1, and select new randomness $a'_0, a'_1, c_{1-b} \in \mathbb{Z}_p^*$ to compute A_0, A_1, s_{1-b} as follows

$$\begin{cases} b = 0, & A_0 = g^{a'_0}, \quad A_1 = g^{a'_1}(R_1 \cdot Z^{h_{ID}})^{-c_{1-b}}, \quad s_1 = a'_1, \\ b = 1, & A_0 = g^{a'_0}(R_0 \cdot Z^{h_{ID}})^{-c_{1-b}}, \quad A_1 = g^{a'_1}, \quad s_0 = a'_0. \end{cases}$$

Lastly, the user computes c_b, s_b such that

$$c_b = H_2(ID, m, A_0, A_1) - c_{1-b}, \quad s_b = a'_b + y \cdot c_b.$$

The resulted IBS is returned as $\sigma_{ID,m} = ((R_0,\ A_0,\ s_0,\ c_0), (R_1,\ A_1,\ s_1,\ c_1))$.

1.5 Overview of the Security Proof

Now we elaborate on how the security proof works at a high level. Given a DL problem instance tuple (g, g^α), we construct a simulation that embeds g^α to master public key Z and, based on random bit $b \in \{0, 1\}$, to either randomness (R_0, A_1) or (R_1, A_0). The user's private key and corresponding signature can be simulated as follows.

Simulating User's Private Keys. Let L_E be a list of users' private keys. Upon receiving the ID_i query, the simulator checks if d_{ID_i} exits in L_E and returns as it. Otherwise, the simulator embeds g^α into either one of the randomness $R_{i,0}, R_{i,1}$, which are programmed as follows. It first randomly selects $r'_{i,0}, r'_{i,1}, h_{ID_i} \in \mathbb{Z}_p$. Then, based on a chosen random bit $b_i \in \{0,1\}$, it sets $y_i = r'_{i,b_i}$,

$$\begin{cases} b=0, & R_{i,0} = g^{r'_{i,0}} \cdot (g^\alpha)^{-h_{ID_i}},\ R_{i,1} = g^{r'_{i,1}}, \\ b=1, & R_{i,0} = g^{r'_{i,0}},\ R_{i,1} = g^{r'_{i,1}} \cdot (g^\alpha)^{-h_{ID_i}}, \end{cases}$$

and hash oracle is programmed as $H_1(ID_i, R_{i,0}, R_{i,1}) = h_{ID_i}$. The simulator returns user's private key $d_{ID_i} = (R_{i,0}, R_{i,1}, b_i, y_i)$ and stores $\{ID_i, d_{ID_i}\}$ to the list of users' private keys L_E.

Simulating Signatures. The simulator can simulate the signature σ_{ID_i,m_i} on any user identity and message pair (ID_i, m_i) as any user's private key d_{ID_i} is simulatable – including d_{ID^*} for which ID^* is the chosen identity in the forgery phase. In particular, this resolves the issue that the simulator aborts while the adversary first asks for signatures σ_{ID_i,m_i} on ID_i then its corresponding user's private key d_{ID_i} because the inability to simulate d_{ID_i} with the same randomness generated in the signature query.

Our simulation allows such a query and additionally, the user's private key d_{ID^*} on the chosen identity ID^*, that the adversary is going to forge, is also simulatable. We hence need to enable the corresponding forgery σ_{ID^*,m^*} on (ID^*, m^*) is reducible.

Suppose d_{ID_i} does not exist in L_E. The simulator calls $\mathcal{O}_E(ID_i)$ and retrieves $\{ID_i, d_{ID_i}\}$ from L_E. Let $h_{ID_i} = H_1(ID_i, R_{i,0}, R_{i,1})$. If $b_i = 0$, the simulator obtains $R_{i,0} = g^{r'_{i,0} - \alpha h_{ID_i}}, R_{i,1} = g^{r'_{i,1}}$. It randomly selects $a'_{i,0}, a'_{i,1}, c_{i,1-b} \in \mathbb{Z}_p^*$, and based on $b_i \in \{0,1\}$, it sets

$$\begin{cases} b_i = 0, & A_{i,0} = g^{a'_{i,0}}, \quad A_{i,1} = g^{a'_{i,1}}(R_{i,1} \cdot Z^{h_{ID_i}})^{-c_{i,1-b}}, \\ & s_{i,1} = a'_{i,1}, \\ b_i = 1, & A_{i,0} = g^{a'_{i,0}}(R_{i,0} \cdot Z^{h_{ID_i}})^{-c_{i,1-b}}, \quad A_{i,1} = g^{a'_{i,1}}, \\ & s_{i,0} = a'_{i,0}. \end{cases}$$

It then calls for hash oracle and obtains $h_{ID_i,m_i} = H_2(ID_i, m_i, A_{i,0}, A_{i,1})$ and sets $c_{b_i} = h_{ID_i,m_i} - c_{i,1-b_i}$ and $s_{b_i} = a'_{b_i} + y_i \cdot c_{b_i}$. It is worth noting that in either case, the signature is simulatable and indistinguishable. The signature is returned as $\sigma_{ID_i,m_i} = ((R_{i,0}, A_{i,0}, s_{i,0}, c_{i,0}), (R_{i,1}, A_{i,1}, s_{i,1}, c_{i,1}))$, and the following verification algorithms hold

$$\begin{aligned} g^{s_0} &= A_0(R_0 \cdot Z^{h_{ID}})^{c_0}, \text{ and} \\ g^{s_1} &= A_1(R_1 \cdot Z^{h_{ID}})^{c_1}, \text{ and} \\ h_{ID,m} &= c_0 + c_1. \end{aligned}$$

Forgery. During the forgery phase, the algebraic adversary returns a forged signature $\sigma_{ID^*,m^*} = ((R_0^*, A_0^*, s_0^*, c_0^*), (R_1^*, A_1^*, s_1^*, c_1^*))$ on (ID^*, m^*) that passes the

Let $Z = g^{\alpha}$ be the master public key.
Given a forgery $\sigma_{ID^*,m^*} = ((R_0^*, A_0^*, s_0^*, c_0^*), (R_1^*, A_1^*, s_1^*, c_1^*))$ on (ID^*, m^*)
and representations $\pi_{u,0}, \pi_{u,1}, \pi_{v,0}, \pi_{v,1}, \pi_{x,0}, \pi_{x,1}, \pi_{y,0}, \pi_{y,1} \in \mathbb{Z}_p$, such that

$$R_0^* = g^{\alpha\pi_{u,1}+\pi_{u,0}}, \quad R_1^* = g^{\alpha\pi_{v,1}+\pi_{v,0}},$$
$$A_0^* = g^{\alpha\pi_{x,1}+\pi_{x,0}}, \quad A_1^* = g^{\alpha\pi_{y,1}+\pi_{y,0}}.$$

The validity of the forgery holds, such that

$$g^{s_0^*} = A_0^*(R_0^* Z^{h_{ID^*}})^{c_0^*}, \quad g^{s_1^*} = A_1^*(R_1^* Z^{h_{ID^*}})^{c_1^*},$$
$$c_0^* + c_1^* = H_2(ID^*, m^*, A_0^*, A_1^*).$$

Hence, the simulator obtains following two modular equations:

$$\boxed{s_0^* = \alpha\pi_{x,1} + \pi_{x,0} + c_0^*(\alpha\pi_{u,1} + \pi_{u,0} + \alpha h_{ID^*})}, \text{ and}$$
$$\boxed{s_1^* = \alpha\pi_{y,1} + \pi_{y,0} + c_1^*(\alpha\pi_{y,1} + \pi_{y,0} + \alpha h_{ID^*})}$$

The simulator solves for

$$\alpha = \frac{s_0^* - (\pi_{x,0} + c_0^*\pi_{u,0})}{\pi_{x,1} + c_0^*(\pi_{u,1} + h_{ID^*})}, \text{ or}$$
$$\alpha = \frac{s_1^* - (\pi_{y,0} + c_1^*\pi_{v,0})}{\pi_{y,1} + c_1^*(\pi_{v,1} + h_{ID^*})}.$$

The reduction fails if the following **abort condition** holds, i.e. coefficients of α are zero:

$$\pi_{x,1} + c_0^*(\pi_{u,1} + h_{ID^*}) = 0, \text{ and } \pi_{y,1} + c_1^*(\pi_{v,1} + h_{ID^*}) = 0.$$

Since coefficients can be freely crafted by the adversary, we, therefore, analyze the **abort condition** under the following three cases.

Case 1: $\pi_{u,1} + h_{ID^*} = \pi_{v,1} + h_{ID^*} = 0$

It implies that $\pi_{x,1} = \pi_{y,1} = 0$ is set.
While $h_{ID^*} = H_1(ID^*, R_0^*, R_1^*)$ is randomly chosen after $R_0^* = g^{\alpha\pi_{u,1}+\pi_{u,0}}$ and $R_1^* = g^{\alpha\pi_{v,1}+\pi_{v,0}}$ were queried to hash oracles, the adversary has negligible success probability to set both $\pi_{u,1} = \pi_{v,1} = -h_{ID^*}$, such that

$$\Pr[\pi_{u,1} = \pi_{v,1} = -h_{ID^*}] = \frac{1}{p}.$$

Case 2: $\pi_{u,1} + h_{ID^*} \neq 0$ and $\pi_{v,1} + h_{ID^*} \neq 0$

Since $c_0^* + c_1^* = h_{ID^*,m^*} = H_2(ID^*, m^*, A_0^*, A_1^*)$ holds. The **abort condition** holds if the following behaviors are set:

$$\mathcal{F}_0\colon \ \pi_{x,1} = -(h_{ID^*,m^*} - c_1^*)(\pi_{u,1} + h_{ID^*}), \text{ and}$$
$$\mathcal{F}_1\colon \ \pi_{y,1} = -(h_{ID^*,m^*} - c_0^*)(\pi_{v,1} + h_{ID^*}).$$

While $h_{ID^*,m^*} = H_2(ID^*, m^*, A_0^*, A_1^*)$ is randomly chosen after $A_0^* = g^{\alpha\pi_{x,1}+\pi_{x,0}}$ and $A_1^* = g^{\alpha\pi_{y,1}+\pi_{y,0}}$ were queried to hash oracles, the adversary has negligible success probability to launch $\mathcal{F}_0$ and $\mathcal{F}_1$, such that

$$\Pr[\mathcal{F}_1 \wedge \mathcal{F}_2] = \frac{1}{p}.$$

Case 3: $\pi_{u,1} + h_{ID^*} = 0$ or $\pi_{v,1} + h_{ID^*} = 0$.

Based on simulation definition, the adversary may query for signatures by calling $\mathcal{O}_S(ID^*,\cdot)$.
This means that the simulator simulated the corresponding user's private key $d_{ID^*} = (R_0^*, R_1^*, b^*, y^*)$, which implies that

$$\begin{cases} b^* = 0, \ R_0^* = g^{\alpha(-h_{ID^*})+r_0'^*}, R_1^* = g^{r_1'^*}, \text{ such that } \pi_{u,1} = -h_{ID^*} \text{ and } \pi_{v,1} = 0. \\ \qquad \pi_{x,1} + c_0^*(-h_{ID^*} + h_{ID^*}) = 0 \text{ and } \pi_{y,1} + c_1^*(0 + h_{ID^*}) = 0. \\ b^* = 1, \ R_0^* = g^{r_0'^*}, R_1^* = g^{\alpha(-h_{ID^*})+r_1'^*}, \text{ such that } \pi_{u,1} = 0 \text{ and } \pi_{v,1} = -h_{ID^*}. \\ \qquad \pi_{x,1} + c_0^*(0 + h_{ID^*}) = 0 \text{ and } \pi_{y,1} + c_1^*(-h_{ID^*} + h_{ID^*}) = 0. \end{cases}$$

The **abort condition** holds if either of the two behaviors is set:

$$\begin{cases} b^* = 0, & \mathcal{G}_0\colon \ \pi_{x,1} = 0, \quad \pi_{y,1} = -c_1^* h_{ID^*}; \\ b^* = 1, & \mathcal{G}_1\colon \ \pi_{x,1} = -c_0^* h_{ID^*}, \quad \pi_{y,1} = 0. \end{cases}$$

While $b^* \in \{0,1\}$ is perfectly hidden from the view of the adversary, it is depending on random guessing $b' \in \{0,1\}$ to launch $\mathcal{G}_{b'}$, hence

$$\Pr[b^* = b'] = \frac{1}{2}.$$

Fig. 1. An Overview of Successful Reduction

verification algorithms. While the adversary is algebraic, the simulator obtains representations $\vec{u}, \vec{v}, \vec{x}, \vec{y}$ that describe how $R_0^*, R_1^*, A_0^*, A_1^*$ are linearly combined, respectively. The overview of the successful reduction can be found in Fig. 1.

Suppose $Z = g^{\alpha}$, $R_0^* = g^{\alpha\pi_{u,1}+\pi_{u,0}}$, $R_1^* = g^{\alpha\pi_{v,1}+\pi_{v,0}}$, $A_0^* = g^{\alpha\pi_{x,1}+\pi_{x,0}}$, and $A_1^* = g^{\alpha\pi_{y,1}+\pi_{y,0}}$, where $\pi_{u,0}, \pi_{u,1}, \pi_{v,0}, \pi_{v,1}, \pi_{x,0}, \pi_{x,1}, \pi_{y,0}, \pi_{y,1}$ are computable coefficients according to the simulator's randomness and algebraic adversary's representations. Based on all the above definitions and verification algorithms, the simulator can derive the following modular equations

$$\begin{aligned} s_0^* &= \alpha\pi_{x,1} + \pi_{x,0} + c_0^*(\alpha\pi_{u,1} + \pi_{u,0} + \alpha h_{ID^*}), \\ s_1^* &= \alpha\pi_{y,1} + \pi_{y,0} + c_1^*(\alpha\pi_{v,1} + \pi_{v,0} + \alpha h_{ID^*}), \\ h_{ID^*,m^*} &= c_0^* + c_1^*, \end{aligned}$$

such that $h_{ID^*} = H_1(ID^*, R_0^*, R_1^*)$ and $h_{ID^*,m^*} = H_2(ID^*, m^*, A_0^*, A_1^*)$ were made. This allows the simulator to solve for DL solution α as long as its coefficients are non-zero.

It now remains to analyze the overall successful reduction. The reduction fails if the simulator aborts when the following abort condition holds, i.e. all coefficients of α are zero, such that $\pi_{x,1}+c_0^*(\pi_{u,1}+h_{ID^*}) = \pi_{y,1}+c_1^*(\pi_{v,1}+h_{ID^*}) = 0$. While $\pi_{u,1}, \pi_{v,1}, \pi_{x,1}, \pi_{y,1} \in \mathbb{Z}_p$ can be crafted by the algebraic adversary, we classify them into three potential cases. In particular, $\pi_{u,1}+h_{ID^*} = \pi_{v,1}+h_{ID^*} = 0$ in **Case 1**; and $\pi_{u,1} + h_{ID^*} \neq 0$ and $\pi_{v,1} + h_{ID^*} \neq 0$ in **Case 2**; and either $\pi_{u,1}+h_{ID^*} = 0$ or $\pi_{v,1}+h_{ID^*} = 0$ in **Case 3**. Fortunately, the overall probability of forgery and representations are non-reducible is $\frac{1}{2}$ because the algebraic adversary has no advantage to reveal the indistinguishable randomness, but a random guessing over a random bit $b^* \in \{0,1\}$. We defer this analysis to the full proof in Sect. 3.2.

1.6 Related Work

The first EUF-CMA secure IBS scheme under DL assumption, known as Beth-IBS, was derived from Bellare et al.'s framework [5] transformed Beth's identification scheme [7] to IBS. However, due to the lack of security analysis for the Beth-IBS scheme, Galindo and Garcia [19] proposed a new IBS scheme based on the well-known Schnorr signature [38] in the random oracle model, namely Schnorr-like IBS scheme. The Schnorr-like IBS scheme is considered the most efficient IBS to date, due to its pairing-free setting. Although its security was improved by Chatterjee et al. [8], it is still loosely reduced to the DL problem due to the need for the reset lemma [36], which has been proven that the resulting security loss must suffer from the tightness barrier – the reduction loss cannot be tight [25].

Recently, Fukumitsu and Hasegawa [17,18] proposed enhancements to the Galindo and Garcia's Schnorr-like IBS [19] to design pairing-free IBS schemes with a tight security reduction. However, these two schemes are only proven under a strong assumption, i.e., the decisional Diffie-Hellman assumption. While in the weak-EUF-CMA security model [42], where the adversary is restricted from asking for a user's private key $\mathcal{O}_E(ID')$ if the identity has already been requested for signatures $\mathcal{O}_S(ID',\cdot)$, making the reduction easier but non-standard as the challenge to respond to any adversary's query is omitted.

Other Signature Schemes in AGM. The AGM has been widely adopted in variants of signature schemes to achieve better efficiency and enhanced properties. For instance, multi-signature schemes [4,11,24,27,30] allow multiple signers to collaborate on signing a single message. Blind signature schemes [16,22,23,40] enable a signer to create a valid signature without revealing the message's content. Threshold signature schemes [2,11,12] necessitate at least 't-out-of-n' signers to jointly produce a valid signature. Notably, a threshold blind signature scheme was recently proposed in [13].

To the extent of our current understanding, in the AGM setting that focuses on achieving tight reductions under the DL assumption and standard EUF-CMA security, only Schnorr signatures [16], BLS signatures [15], and its identity-based adaptation [28] have been investigated.

2 Preliminaries

2.1 Definition of Identity-Based Signatures (IBS)

An identity-based signature (IBS) scheme consists of the following algorithms:

- *Setup*(1^k): On input security parameters 1^k, it returns the master public and secret key pair (mpk, msk).
- *Extract*(mpk, msk, ID): On input master public and secret key pair (mpk, msk) and a user identity ID, it returns a user private key d_{ID}.
- *Sign*(mpk, d_{ID}, m): On input (mpk, d_{ID}) and message m, it returns a signature $\sigma_{ID,m}$.
- *Verify*$(mpk, ID, m, \sigma_{ID,m})$: On input $(mpk, ID, m, \sigma_{ID,m})$, it returns 1 or 0 which indicates accept or reject respectively.

Correctness. For user identity ID and user private key pair (ID, d_{ID}) that is extracted based on the master public and secret key pair (mpk, msk), signing a message m with d_{ID} must return a correct signature $\sigma_{ID,m}$ that is valid on user ID, i.e. *Verify*$(mpk, ID, m, \sigma_{ID,m}) = 1$.

2.2 The EUF-CMA Security for IBS

The notion of existential unforgeability against chosen identity-and-message attacks (EUF-CMA) [5] security model is defined between a challenger and an adversary as follows.

- **Setup Phase:** Let L_E, L_S be two sets of extract queries and signing queries, respectively. The challenger runs *Setup* algorithm to compute a master public and secret key pair (mpk, msk). The challenger forwards mpk to the adversary.

- **Query Phase:** The adversary may adaptively ask for user private key d_{ID_i} on any chosen identity ID_i to the extraction oracle $\mathcal{O}_E(ID_i)$ and signatures σ_{ID_i,m_i} on any chosen identity-and-message pair (ID_i, m_i) to the signing oracle $\mathcal{O}_S(ID_i, m_i)$.
  - Extraction oracle $\mathcal{O}_E(ID_i)$: On input i-th query for ID_i, it first checks whether $\langle ID_i, \cdot \rangle \in L_E$. The challenger returns d_{ID_i} if it finds $\langle ID_i, d_{ID_i} \rangle \in L_E$. Otherwise, the challenger calls $Extract(mpk, msk, ID_i) \rightarrow d_{ID_i}$ algorithm and stores $\langle ID_i, d_{ID_i} \rangle$ to L_E. The challenger returns d_{ID_i}.
  - Signing oracle $\mathcal{O}_S(ID_i, m_i)$: On input i-th query for (ID_i, m_i), it first checks and retrieves d_{ID_i} if $\langle ID_i, d_{ID_i} \rangle \in L_E$ exists. If $\langle ID_i, d_{ID_i} \rangle \notin L_E$, the challenger calls $Extract(mpk, msk, ID_i) \rightarrow d_{ID_i}$ algorithm and stores (ID_i, d_{ID_i}) to L_E. The challenger calls $Sign(mpk, d_{ID_i}, m_i) \rightarrow \sigma_{ID_i,m_i}$ and stores $\langle ID_i, m_i, \sigma_{ID_i,m_i} \rangle$ to L_S. The challenger returns σ_{ID_i,m_i} that is valid and verifiable by running $Verify(mpk, ID_i, m_i, \sigma_{ID_i,m_i}) = 1$.
- **Forgery Phase:** The adversary returns challenge identity, message and signature pair $(ID^*, m^*, \sigma_{ID^*,m^*})$. The adversary wins if the verification holds, such that $Verify(mpk, ID^*, m^*, \sigma_{ID^*,m^*}) = 1$, where ID^* has not been queried to the extraction oracle $\mathcal{O}_E(ID^*)$ and (ID^*, m^*) has not been queried to the signing oracle $\mathcal{O}_S(ID^*, m^*)$.

Definition 1. *An IBS scheme is (ϵ, q_e, q_s, t)-secure in the EUF-CMA security model if it is infeasible that any probabilistic polynomial-time adversary who runs in t polynomial time and makes at most q_e extraction queries and q_s signing queries has advantage at most ϵ in winning the game, where ϵ is negligible function of the input security parameter.*

2.3 Discrete Logarithm Problem

Let $\mathbb{G}$ be a group of prime order p with generator $g \in \mathbb{G}$. The discrete logarithm (DL) problem is to compute $\alpha \in \mathbb{Z}_p$, given a random group element $g^\alpha \in \mathbb{G}$.

Definition 2. *The (ϵ, t)-DL assumption holds in $\mathbb{G}$ if there is no probabilistic polynomial-time adversary who runs in t polynomial time has the advantage at most ϵ to solve the DL problem in $\mathbb{G}$.*

2.4 The Algebraic Adversary

The algebraic adversary was studied in [33], and was formalized in the AGM [15]. The security reduction in the AGM is like the standard model, but we say the computation of the adversary is algebraic. Informally, suppose algebraic adversary is given $(g, C_1, ..., C_n) \in \mathbb{G}^{n+1}$ group elements. For any group element it outputs $X \in \mathbb{G}$, it is also required to return a representation vector $\vec{c} = (c_0, c_1, ..., c_n) \in \mathbb{Z}_p^{n+1}$, that indicates how a returned group element X been generated based on the received group elements, such that

$$X = g^{c_0} \cdot C_1^{c_1} \cdots C_n^{c_n} = g^{c_0} \prod_{i=1}^{n} C_i^{c_i}.$$

3 The Proposed Pairing-Free Identity-Based Signatures

3.1 Scheme

- `Setup`: On input security parameter 1^k, it selects $g \in \mathbb{G}$, where g is the generator of group $\mathbb{G}$ in the prime order of p, and $H_1 : \{0,1\}^* \to \mathbb{Z}_p, H_2 : \{0,1\}^* \to \mathbb{Z}_p$ be two cryptography hash functions. It then selects $z \in \mathbb{Z}_p$ and computes $Z = g^z$. The master public and secret key pair (mpk, msk) is returned to a trusted third party (TTP) as follows, such that

$$mpk = (g, p, \mathbb{G}, H_1, H_2, Z), \quad msk = z.$$

- `Extract`: On input (mpk, msk) and user identity $ID \in \{0,1\}^*$. It selects $r_0, r_1 \in \mathbb{Z}_p^*$ and sets $R_0 = g^{r_0}, R_1 = g^{r_1}$. Let $h_{ID} = H_1(ID, R_0, R_1)$, it then selects a random bit $b \in \{0,1\}$ and returns user private key $d_{ID} = (R_0, R_1, b, y)$, such that

$$y = r_b + z \cdot h_{ID}.$$

  In practice, TTP would securely distribute d_{ID} to the user. Hence one may assume that TTP stores a list of user private keys $L_E = \langle ID, d_{ID} \rangle$, which reflects the security proof that returns the same d_{ID} upon receiving extraction query on ID.
- `Sign`: On input (mpk, d_{ID}, m). Let $h_{ID} = H_1(ID, R_0, R_1)$. It retrieves $b \in \{0,1\}$ and randomly selects $a'_0, a'_1, c_{1-b} \in \mathbb{Z}_p^*$. It then computes A_0, A_1, s_{1-b}, such that

$$\begin{aligned} A_0 &= g^{a'_0}(R_0 \cdot Z^{h_{ID}})^{(-c_{1-b})\cdot b}, \\ A_1 &= g^{a'_1}(R_1 \cdot Z^{h_{ID}})^{(-c_{1-b})\cdot(1-b)}, \\ s_{1-b} &= a'_{1-b}. \end{aligned}$$

  Let $h_{ID,m} = H_2(ID, m, A_0, A_1)$, it then computes c_b and s_b, such that

$$c_b = h_{ID,m} - c_{1-b}, \quad s_b = a'_b + y \cdot c_b.$$

  The signature $\sigma_{ID,m}$ is returned as follows, where

$$\sigma_{ID,m} = ((R_0,\ A_0,\ s_0,\ c_0), (R_1,\ A_1,\ s_1,\ c_1)).$$

- `Verify`: On input $(mpk, ID, m, \sigma_{ID,m})$. Let $h_{ID} = H_1(ID, R_0, R_1)$ and $h_{ID,m} = H_2(ID, m, A_0, A_1)$, it checks the following equations, such that

$$g^{s_0} = A_0(R_0 \cdot Z^{h_{ID}})^{c_0}, \tag{2}$$
$$g^{s_1} = A_1(R_1 \cdot Z^{h_{ID}})^{c_1}, \tag{3}$$
$$h_{ID,m} = c_0 + c_1. \tag{4}$$

  It returns 1 if Eqs. (2), (3), and (4) hold. Otherwise, it returns 0.

Correctness. Recall that $Z = g^z$ is from mpk. The scheme is correct if the signature $\sigma_{ID,m} = ((R_0, A_0, s_0, c_0), (R_1, A_1, s_1, c_1))$ on message m generated by the user ID with user private key $d_{ID} = (R_0, R_1, b, y)$ holds. Let $h_{ID} = H_1(ID, R_0, R_1)$ and $h_{ID,m} = H_2(ID, m, A_0, A_1)$. Suppose $b = 0$, we have

$$\begin{aligned} R_0 &= g^{r_0}, & R_1 &= g^{r_1}, \\ A_0 &= g^{a_0}, & A_1 &= g^{a_1}(R_1 \cdot Z^{h_{ID}})^{-c_1}, \\ s_0 &= a_0 + y \cdot c_0, & s_1 &= a_1. \\ y &= r_0 + z \cdot h_{ID}, & c_1 &= h_{ID,m} - c_0. \end{aligned}$$

Otherwise, if $b = 1$, we have

$$\begin{aligned} R_0 &= g^{r_0}, & R_1 &= g^{r_1}, \\ A_0 &= g^{a_0}(R_0 \cdot Z^{h_{ID}})^{-c_0}, & A_1 &= g^{a_1}, \\ s_0 &= a_0, & s_1 &= a_1 + y \cdot c_1. \\ y &= r_1 + z \cdot h_{ID}, & c_0 &= h_{ID,m} - c_1. \end{aligned}$$

It is easy to verify Eqs. (2), (3), and (4) hold in either $b \in \{0, 1\}$.

3.2 Security Analysis

Theorem 1. *If an adversary can break the EUF-CMA of the proposed IBS scheme in the AGM, then there exists an adversary that can solve the DL problem with a success probability of* $\frac{1}{2}$.

Proof. Suppose there exists an algebraic adversary who can (t, q_e, q_s, ϵ)-break the proposed IBS scheme under EUF-CMA in the AGM. Given as input a DL problem instance (g, g^α) over the cyclic group tuple $(p, g, \mathbb{G})$. We design a simulator that can solve the DL problem with the forgery and representations given by the algebraic adversary.

- **Setup.** During the setup phase, the simulator embeds the DL problem instance to $Z = g^\alpha$ and returns the master public mpk, such that

$$mpk = (p, g, \mathbb{G}, Z).$$

- **Hash Oracles.** At the beginning, the simulator prepares two empty sets L_{H1}, L_{H2} to record all hash queries and responses as follows.
  $\mathcal{H}_1(ID_i, R_{i,0}, R_{i,1})$: On an i-th random oracle query, i.e. on input $(ID_i, R_{i,0}, R_{i,1})$. If the tuple exists, such that $\langle ID_i, R_{i,0}, R_{i,1}, h_{ID_i} \rangle \in L_{H1}$, the simulator responds to this query following the record, such that

$$\mathcal{H}_1(ID_i, R_{i,0}, R_{i,1},) = h_{ID_i}.$$

  Otherwise, the simulator randomly selects $h_{ID_i} \in \mathbb{Z}_p^*$ and stores $\langle ID_i, R_{i,0}, R_{i,1}, h_{ID_i} \rangle$ into L_{H1}. The simulator returns $\mathcal{H}_1(ID_i, R_{i,0}, R_{i,1},) = h_{ID_i}$.

$\mathcal{H}_2(ID_i, m_i, A_{i,0}, A_{i,1})$: On an i-th random oracle query, i.e. on input $(ID_i,\ m_i, A_{i,0}, A_{i,1})$. If the tuple exists, such that $\langle ID_i, m_i, A_{i,0}, A_{i,1}, h_{m_i}\rangle \in L_{H2}$, the simulator responds to this query following the record, such that

$$\mathcal{H}_2(ID_i, m_i, A_{i,0}, A_{i,1}) = h_{ID_i,m_i}$$

Otherwise, the simulator randomly selects $h_{ID_i,m_i} \in \mathbb{Z}_p^*$ and stores $\langle ID_i, m_i, A_{i,0}, A_{i,1}, h_{m_i}\rangle$ into L_{H2}. The simulator returns $\mathcal{H}_2(ID_i, m_i, A_{i,0}, A_{i,1}) = h_{ID_i,m_i}$.

- **Query Phase.** During the query phase, the adversary may adaptively make extraction queries and signing queries. The simulator prepares an empty set L_E to record the simulated user's private keys.
  - $\mathcal{O}_E(ID_i)$: On an i-th extraction query (ID_i), the simulator checks whether $\langle ID_i, d_{ID_i}\rangle \in L_E$ exits. If there is, the simulator retrieves the user's private key $d_{ID_i} = (R_{i,0}, R_{i,1}, b_i, y_i)$, which is based on the definition as follows. Otherwise, it first randomly selects $b_i \in \{0,1\}$ and $r'_{i,0}, r'_{i,1}, h_{ID_i} \in \mathbb{Z}_p^*$. It computes $R_{i,0}, R_{i,1}$ in either cases, such that

$$\begin{cases} b=0, & R_{i,0} = g^{r'_{i,0}} \cdot (g^\alpha)^{-h_{ID_i}},\ R_{i,1} = g^{r'_{i,1}}, \\ b=1, & R_{i,0} = g^{r'_{i,0}},\ R_{i,1} = g^{r'_{i,1}} \cdot (g^\alpha)^{-h_{ID_i}}. \end{cases}$$

    It then sets $y_i = r'_{i,b_i}$ and programs hash oracle as follows

$$H_1(ID_i, R_{i,0}, R_{i,1}) = h_{ID_i}.$$

    The simulator returns user private key $d_{ID_i} = (R_{i,0}, R_{i,1}, b_i, y_i)$ and stores $\{ID_i, d_{ID_i}\}$ into a list of user's private key L_E and $\{ID_i, R_{i,0}, R_{i,1}, h_{ID_i}\}$ into L_{H1}. It is worth noting that the extraction query may fail if the hash value h_{ID_i} was called by the adversary. However, since $R_{i,0}, R_{i,1}$ are randomly selected by the simulator, this abort probability is negligible.
  - $\mathcal{O}_S(ID_i, m_i)$: On an i-th signing query (ID_i, m_i). It checks whether $\langle ID_i, d_{ID_i}\rangle \in L_E$. It calls $\mathcal{O}_E(ID_i) \to d_{ID_i}$ to generate a fresh user private key if it does not exist. Otherwise, it retrieves $d_{ID_i} = (R_{i,0}, R_{i,1}, b_i, y_i)$ from L_E. Let $h_{ID_i} = H_1(ID_i, R_{i,0}, R_{i,1})$. It randomly selects $a'_{i,0}, a'_{i,1}, c_{i,1-b} \in \mathbb{Z}_p^*$, and based on $b_i \in \{0,1\}$, it sets

$$\begin{cases} b_i = 0, & A_{i,0} = g^{a'_{i,0}}, \quad A_{i,1} = g^{a'_{i,1}}(R_{i,1} \cdot Z^{h_{ID_i}})^{-c_{i,1-b}}, \\ & s_{i,1} = a'_{i,1}, \\ b_i = 1, & A_{i,0} = g^{a'_{i,0}}(R_{i,0} \cdot Z^{h_{ID_i}})^{-c_{i,1-b}}, \quad A_{i,1} = g^{a'_{i,1}}, \\ & s_{i,0} = a'_{i,0}. \end{cases}$$

    It then calls for hash oracle and obtains $h_{ID_i,m_i} = H_2(ID_i, m_i, A_{i,0}, A_{i,1})$ and sets

$$c_{b_i} = h_{ID_i,m_i} - c_{i,1-b_i}, \quad s_{b_i} = a'_{b_i} + y_i \cdot c_{b_i}.$$

The signature σ_{ID_i,m_i} is returned as follows.

$$\sigma_{ID_i,m_i} = ((R_{i,0},\ A_{i,0},\ s_{i,0},\ c_{i,0}),\ (R_{i,1},\ A_{i,1},\ s_{i,1},\ c_{i,1})).$$

- **Forgery Phase.** Assume the algebraic adversary returns a forged signature $\sigma_{ID^*,m^*} = ((R_0^*, A_0^*, s_0^*, c_0^*), (R_1^*, A_1^*, s_1^*, c_1^*))$ on (ID^*, m^*), such that Eqs. (2), (3), and (4) hold, and representations $\vec{u}, \vec{v}, \vec{x}, \vec{y}$, i.e., for $\pi \in [u, v, x, y]$

$$\vec{\pi} = (\pi_0, \pi_1, \pi_{2,1}, ..., \pi_{2,q_e}, \pi_{3,1}, \pi_{3,q_e}, \pi_{4,1}, ..., \pi_{4,q_s}, \\ \pi_{5,1}, ..., \pi_{5,q_s}, \pi_{6,1}, ..., \pi_{6,q_s}, \pi_{7,1}, ..., \pi_{7,q_s}),$$

that describe the linear combination of $R_0^*, R_1^*, A_0^*, A_1^*$, respectively, such that

$$R_0^* = g^{u_0} Z^{u_1} \prod_{i=1}^{q_e} (R_{i,0})^{u_{2,i}} (R_{i,1})^{u_{3,i}} \prod_{i=1}^{q_s} (R_0^*)^{u_{4,i}} (R_1^*)^{u_{5,i}} (A_{i,0})^{u_{6,i}} (A_{i,1})^{u_{7,i}},$$

$$R_1^* = g^{v_0} Z^{v_1} \prod_{i=1}^{q_e} (R_{i,0})^{v_{2,i}} (R_{i,1})^{v_{3,i}} \prod_{i=1}^{q_s} (R_0^*)^{v_{4,i}} (R_1^*)^{v_{5,i}} (A_{i,0})^{v_{6,i}} (A_{i,1})^{v_{7,i}},$$

$$A_0^* = g^{x_0} Z^{x_1} \prod_{i=1}^{q_e} (R_{i,0})^{x_{2,i}} (R_{i,1})^{x_{3,i}} \prod_{i=1}^{q_s} (R_0^*)^{x_{4,i}} (R_1^*)^{x_{5,i}} (A_{i,0})^{x_{6,i}} (A_{i,1})^{x_{7,i}},$$

$$A_1^* = g^{y_0} Z^{y_1} \prod_{i=1}^{q_e} (R_{i,0})^{y_{2,i}} (R_{i,1})^{y_{3,i}} \prod_{i=1}^{q_s} (R_0^*)^{y_{4,i}} (R_1^*)^{y_{5,i}} (A_{i,0})^{y_{6,i}} (A_{i,1})^{y_{7,i}},$$

where q_e is the number of extraction queries and q_s is the number is signing queries been made. Note that for simplicity, it is reasonable to assume the adversary obtains signatures on chosen ID^* from the signing query $\sigma_{ID^*,m_i} \leftarrow \mathcal{O}_S(ID^*, m_i)$, and for signatures on other ID_i, the adversary may query user's private key $d_{ID_i} \leftarrow \mathcal{O}_E(ID_i)$ and compute the corresponding valid signatures.

To simplify the above definition, suppose representations $\vec{u}, \vec{v}, \vec{x}, \vec{y}$ and chosen randomness $r'_{i,0}, r'_{i,1}, r'^*_0, r'^*_1, a'_{i,0}, a'_{i,1}, c_{i,0}, c_{i,1}, h_{ID_i}, h_{ID^*} \in \mathbb{Z}_p^*$ are sorted by unknown α, such that $\pi_{u,1}, \pi_{v,1}, \pi_{x,1}, \pi_{y,1} \in \mathbb{Z}_p$ are defined as coefficients of α, and $\pi_{u,0}, \pi_{v,0}, \pi_{x,0}, \pi_{y,0} \in \mathbb{Z}_p$ are the remaining coefficients. Therefore, $R_0^*, R_1^*, A_0^*, A_1^*$ can be simplified as follows

$$R_0^* = g^{\alpha\pi_{u,1} + \pi_{u,0}}, \quad R_1^* = g^{\alpha\pi_{v,1} + \pi_{v,0}},$$
$$A_0^* = g^{\alpha\pi_{x,1} + \pi_{x,0}}, \quad A_1^* = g^{\alpha\pi_{y,1} + \pi_{y,0}}.$$

By combining the forgery and representations and verification Eqs. (2), (3) and (4), such that

$$g^{s_0} = A_0(R_0 \cdot Z^{h_{ID}})^{c_0},$$
$$g^{s_1} = A_1(R_1 \cdot Z^{h_{ID}})^{c_1},$$
$$h_{ID,m} = c_0 + c_1$$

The simulator can obtain the following three modular equations

$$\begin{aligned} s_0^* &= \alpha\pi_{x,1} + \pi_{x,0} + c_0^*(\alpha\pi_{u,1} + \pi_{u,0} + \alpha h_{ID^*}), \\ s_1^* &= \alpha\pi_{y,1} + \pi_{y,0} + c_1^*(\alpha\pi_{v,1} + \pi_{v,0} + \alpha h_{ID^*}), \\ h_{ID^*,m^*} &= c_0^* + c_1^*, \end{aligned}$$

where $h_{ID^*} = H_1(ID^*, R_0^*, R_1^*)$ and $h_{ID^*,m^*} = H_2(ID^*, m^*, A_0^*, A_1^*)$.

Abort. The simulator aborts if the following abort condition holds, such that coefficients of α are zero, i.e.

$$\begin{aligned} \pi_{x,1} + c_0^*(\pi_{u,1} + h_{ID^*}) = 0, \text{ and} \\ \pi_{y,1} + c_1^*(\pi_{v,1} + h_{ID^*}) = 0. \end{aligned}$$

Otherwise, the simulator can solve for DL solution α via either of the following equations, such that

$$\alpha = \frac{s_0^* - (\pi_{x,0} + c_0^*\pi_{u,0})}{\pi_{x,1} + c_0^*(\pi_{u,1} + h_{ID^*})}, \text{ or} \tag{5}$$

$$\alpha = \frac{s_1^* - (\pi_{y,0} + c_1^*\pi_{v,0})}{\pi_{y,1} + c_1^*(\pi_{v,1} + h_{ID^*})}. \tag{6}$$

Indistinguishable Simulation. The simulation is correct as the correctness of every simulated user's private key and signature hold, which has been described in the query phase. The simulator sets all elements with perfectly hidden randomness. For example, master public key $Z = g^z$, user's private key randomness $R_{i,0} = g^{r_{i,0}}, R_{i,1} = g^{r_{i,1}}$ and signature randomnesses $R_0^* = g^{r_0^*}, R_1^* = g^{r_1^*}, A_{i,0} = g^{a_{i,0}}, A_{i,1} = g^{a_{i,1}}$. They are, respectively, simulated based on random bit $b_i, b^* \in \{0,1\}$ as

$$\begin{aligned} z = \alpha, \qquad & r_{i,0} = r'_{i,0} - \alpha h_{ID_i}(1 - b_i), \qquad r_{i,1} = r'_{i,1} - \alpha h_{ID_i}(b_i), \\ & r_0^* = r_0'^* - \alpha h_{ID^*} \cdot (1 - b^*), \qquad r_1^* = r_1'^* - \alpha h_{ID^*}(b^*), \\ a_{i,0} = a'_{i,0} - c_{i,1-b^*}&(r_0^* + \alpha h_{ID^*})(b^*), \quad a_{i,1} = a'_{i,1} - c_{i,1-b^*}(r_1^* + \alpha h_{ID^*})(1 - b^*). \end{aligned}$$

While $b_i, b^* \in \{0,1\}$ and $\alpha, r'_{i,0}, r'_{i,1}, r_0'^*, r_1'^*, a'_{i,0}, a'_{i,1}, c_{i,0}, c_{i,1}, h_{ID_i}, h_{ID^*} \in \mathbb{Z}_p^*$ are all randomly chosen by the simulator, group elements $Z, R_{i,0}, R_{i,1}, R_0^*, R_1^*, A_{i,0}, A_{i,1}$ are random and indistinguishable from the view of the adversary. Hence, the simulation is indistinguishable from the proposed IBS scheme.

Probability of Aborts. Based on Eqs. (5) and (6), we note that the simulator aborts if the following abort condition holds, such that

$$\pi_{x,1} + c_0^*(\pi_{u,1} + h_{ID^*}) = \pi_{y,1} + c_1^*(\pi_{v,1} + h_{ID^*}) = 0. \tag{7}$$

Therefore, we classify the forgery and representations into three cases as follows to analyze the probability of setting Eqs. (7) holds.

Case 1. The first case is to analyze the probability that both $\pi_{u,1} + h_{ID^*} = 0$ and $\pi_{v,1} + h_{ID^*} = 0$ hold, which would also implies that setting $\pi_{x,1} = \pi_{y,1} = 0$ based on Eq. (7). Recall that $R_0^* = g^{\alpha\pi_{u,1}+\pi_{u,0}}$ and $R_1^* = g^{\alpha\pi_{v,1}+\pi_{v,0}}$ are algebraically computed, and they must be submitted to hash oracles $H_1(ID^*, R_0^*, R_1^*) = h_{ID^*}$ in advance. While $h_{ID^*} \in \mathbb{Z}_p^*$ is randomly chosen after R_0^*, R_1^* and $\vec{u}, \vec{v}$ were submitted to hash oracles, the success probability of setting both $\pi_{u,1} = \pi_{v,1} = -h_{ID^*}$ is negligible, such that $\Pr[\pi_{u,1} = \pi_{v,1} = -h_{ID^*}] = \frac{1}{p}$.

Case 2. Now, we analyze the case that setting $\pi_{u,1}+h_{ID^*} \neq 0$ and $\pi_{v,1}+h_{ID^*} \neq 0$ hold. Based on Eq. (7), we then need to analyze

$$\pi_{x,1} + c_0^*(\pi_{u,1} + h_{ID^*}) = 0, \quad \text{and} \quad \pi_{y,1} + c_1^*(\pi_{v,1} + h_{ID^*}) = 0.$$

The above equations hold if $\pi_{x,1} = -c_0^*(\pi_{u,1} + h_{ID^*})$ and $\pi_{y,1} = -c_1^*(\pi_{v,1} + h_{ID^*})$ are set. Recall that $H_2(ID^*, m^*, A_0^*, A_1^*) = h_{ID^*,m^*} = c_0^* + c_1^*$, we can observe that the above equations hold if the adversary could launch the following behaviours $\mathcal{F}_0, \mathcal{F}_1$, such that

$$\begin{aligned} \mathcal{F}_0: \quad \pi_{x,1} &= -(h_{ID^*,m^*} - c_1^*)(\pi_{u,1} + h_{ID^*}), \text{ and} \\ \mathcal{F}_1: \quad \pi_{y,1} &= -(h_{ID^*,m^*} - c_0^*)(\pi_{v,1} + h_{ID^*}). \end{aligned}$$

Recall that $A_0^* = g^{\alpha\pi_{x,1}+\pi_{x,0}}$ and $A_1^* = g^{\alpha\pi_{y,1}+\pi_{y,0}}$ are algebraically computed, and they must be submitted to hash oracles $H_2(ID^*, m^*, A_0^*, A_1^*) = h_{ID^*,m^*}$ in advance. While $h_{ID^*,m^*} \in \mathbb{Z}_p^*$ is randomly chosen after A_0^*, A_1^* and $\vec{x}, \vec{y}$ were submitted to random oracles, the success probability of launching both behaviours $\mathcal{F}_0, \mathcal{F}_1$ is negligible, such that $\Pr[\mathcal{F}_0 \wedge \mathcal{F}_1] = \frac{1}{p}$.

Case 3. Lastly, we analyze the case that setting either $\pi_{u,1} + h_{ID^*} = 0$ or $\pi_{v,1} + h_{ID^*} = 0$ holds. This would be possible when $\mathcal{O}_S(ID^*, m_i)$ has been made, such that $d_{ID^*} = (R_0^*, R_1^*, b^*, y^*)$ was simulated by the simulator. So R_0^*, R_1^* must become part of the forgery. It is worth noting that both $b^* \in \{0, 1\}$ and $y^* \in \mathbb{Z}_p^*$ are information-theoretically hidden from the view of the adversary. This means that for some $r_0'^*, r_1'^*, h_{ID^*} \in \mathbb{Z}_p^*$,

$$\begin{cases} b^* = 0, \quad R_0^* = g^{r_0'^*} \cdot (g^\alpha)^{-h_{ID^*}}, \; R_1^* = g^{r_1'^*}, \\ b^* = 1, \quad R_0^* = g^{r_0'^*}, \; R_1^* = g^{r_1'^*} \cdot (g^\alpha)^{-h_{ID^*}}. \end{cases}$$

The hash oracle is programmed, i.e. $h_{ID^*} = H_1(ID^*, R_0^*, R_1^*)$ is set. Based on previous definition, $R_0^* = g^{\alpha\pi_{u,1}+\pi_{u,0}}$ and $R_1^* = g^{\alpha\pi_{v,1}+\pi_{v,0}}$, this indicates that $\pi_{u,1} = -h_{ID^*} \cdot (1-b^*)$ and $\pi_{v,1} = -h_{ID^*} \cdot b^*$ hold. We see Eq. (7) can be rewritten as follows, such that

$$\begin{aligned} \pi_{x,1} + c_0^*(-h_{ID^*} \cdot (1 - b^*) + h_{ID^*}) = 0, \text{ and} \\ \pi_{y,1} + c_1^*(-h_{ID^*} \cdot b^* + h_{ID^*}) = 0. \end{aligned}$$

Recall that $A_0^* = g^{\alpha\pi_{x,1}+\pi_{x,0}}$ and $A_1^* = g^{\alpha\pi_{y,1}+\pi_{y,0}}$. Therefore, the above equations hold if the adversary could launch either one of the following behaviours $\mathcal{G}_0, \mathcal{G}_1$ correctly, such that

$$\begin{cases} b^* = 0, \quad \mathcal{G}_0: \quad \pi_{x,1} = 0, \qquad \pi_{y,1} = -c_1^* h_{ID^*}, \\ b^* = 1, \quad \mathcal{G}_1: \quad \pi_{x,1} = -c_0^* h_{ID^*}, \quad \pi_{y,1} = 0. \end{cases}$$

Fortunately, the adversary succeeds depending on the random guessing $b' \in \{0,1\}$, i.e. either launching $\mathcal{G}_0$ or $\mathcal{G}_1$. Since b^* is information-theoretically hidden from the view of the adversary, the success probability is half, such that $\Pr[b^* = b'] = \frac{1}{2}$.

The Security Loss. The security loss is calculated based on the success probability of the simulation. We describe the success probability of extracting the DL solution, according to the above probability of abort, as follows

$$\begin{aligned}\Pr[\text{Success}] &= 1 - (\Pr[\textbf{Case 1}] + \Pr[\textbf{Case 2}] + \Pr[\textbf{Case 3}]) \\ &= 1 - (\frac{1}{p} + \frac{1}{p} + \frac{1}{2}) \approx \frac{1}{2}.\end{aligned}$$

Therefore, the success probability of extracting the DL problem solution is at least $\frac{1}{2}$, which concludes that the security loss of the proposed IBS scheme under EUF-CMA and DL assumption in AGM is 2. This completes the proof of the Theorem 1. □

4 Conclusion

In this work, through the adoption of the AGM, we addressed the challenge of achieving tight security reductions for pairing-free IBS under EUF-CMA and DL assumption. We obtained the first pairing-free IBS scheme, which is inspired by using an OR-proof technique to generate the user's private keys for the Schnorr-like IBS. We also proposed a tight reduction algorithm that shows how the obtained pairing-free IBS can achieve tight EUF-CMA security under the DL assumption in the AGM.

This work provided insight that one must carefully analyze the security of a scheme when proving security in the AGM. For example, the Schnorr-like IBS scheme is constructed based on Schnorr signatures, although both schemes are proven under DL assumption, the additional algorithm and chosen-identity attacks in the ID-based setting may lead to a loose reduction. It is noteworthy to mention that, as indicated in the summary provided in Table 1, among the current theoretical studies, adopting the AGM may be necessary to achieve IBS schemes under the DL assumption and EUF-CMA security with tight reductions.

Acknowledgement. We extend our gratitude to the anonymous reviewers for their valuable feedback.

References

1. Ahed, K., Benamar, M., El Ouazzani, R.: Content delivery in named data networking based internet of things. In: 2019 15th International Wireless Communications and Mobile Computing Conference (IWCMC), pp. 1397–1402. IEEE (2019)
2. Bacho, R., Loss, J.: On the adaptive security of the threshold BLS signature scheme. In: Proceedings of the 2022 ACM SIGSAC Conference on Computer and Communications Security, pp. 193–207 (2022)

3. Barreto, P.S.L.M., Libert, B., McCullagh, N., Quisquater, J.-J.: Efficient and provably-secure identity-based signatures and signcryption from bilinear maps. In: Roy, B. (ed.) ASIACRYPT 2005. LNCS, vol. 3788, pp. 515–532. Springer, Heidelberg (2005). https://doi.org/10.1007/11593447_28
4. Bellare, M., Dai, W.: Chain reductions for multi-signatures and the HBMS scheme. In: Tibouchi, M., Wang, H. (eds.) ASIACRYPT 2021. LNCS, vol. 13093, pp. 650–678. Springer, Cham (2021). https://doi.org/10.1007/978-3-030-92068-5_22
5. Bellare, M., Namprempre, C., Neven, G.: Security proofs for identity-based identification and signature schemes. J. Cryptol. **22**(1), 1–61 (2009)
6. Bellare, M., Rogaway, P.: Random oracles are practical: a paradigm for designing efficient protocols. In: Proceedings of the 1st ACM Conference on Computer and Communications Security, pp. 62–73 (1993)
7. Beth, T.: Efficient zero-knowledge identification scheme for smart cards. In: Barstow, D., Brauer, W., Brinch Hansen, P., Gries, D., Luckham, D., Moler, C., Pnueli, A., Seegmüller, G., Stoer, J., Wirth, N., Günther, C.G. (eds.) EUROCRYPT 1988. LNCS, vol. 330, pp. 77–84. Springer, Heidelberg (1988). https://doi.org/10.1007/3-540-45961-8_7
8. Chatterjee, S., Kamath, C., Kumar, V.: Galindo-Garcia identity-based signature revisited. In: Kwon, T., Lee, M.-K., Kwon, D. (eds.) ICISC 2012. LNCS, vol. 7839, pp. 456–471. Springer, Heidelberg (2013). https://doi.org/10.1007/978-3-642-37682-5_32
9. Choon, J.C., Hee Cheon, J.: An identity-based signature from gap Diffie-Hellman groups. In: Desmedt, Y.G. (ed.) PKC 2003. LNCS, vol. 2567, pp. 18–30. Springer, Heidelberg (2003). https://doi.org/10.1007/3-540-36288-6_2
10. Cramer, R., Damgård, I., Schoenmakers, B.: Proofs of partial knowledge and simplified design of witness hiding protocols. In: Desmedt, Y.G. (ed.) CRYPTO 1994. LNCS, vol. 839, pp. 174–187. Springer, Heidelberg (1994). https://doi.org/10.1007/3-540-48658-5_19
11. Crites, E., Komlo, C., Maller, M.: How to prove Schnorr assuming Schnorr: Security of multi-and threshold signatures. Cryptology ePrint Archive (2021)
12. Crites, E., Komlo, C., Maller, M.: Fully adaptive Schnorr threshold signatures. In: Handschuh, H., Lysyanskaya, A. (eds.) CRYPTO 2023. LNCS, vol. 14081, pp. 678–709. Springer, Cham (2023). https://doi.org/10.1007/978-3-031-38557-5_22
13. Crites, E., Komlo, C., Maller, M., Tessaro, S., Zhu, C.: Snowblind: a threshold blind signature in pairing-free groups. In: Handschuh, H., Lysyanskaya, A. (eds.) CRYPTO 2023. LNCS, vol. 14081, pp. 710–742. Springer, Cham (2023). https://doi.org/10.1007/978-3-031-38557-5_23
14. Du, H., Wen, Q.: An efficient identity-based short signature scheme from bilinear pairings. In: 2007 International Conference on Computational Intelligence and Security (CIS 2007), pp. 725–729. IEEE (2007)
15. Fuchsbauer, G., Kiltz, E., Loss, J.: The algebraic group model and its applications. In: Shacham, H., Boldyreva, A. (eds.) CRYPTO 2018. LNCS, vol. 10992, pp. 33–62. Springer, Cham (2018). https://doi.org/10.1007/978-3-319-96881-0_2
16. Fuchsbauer, G., Plouviez, A., Seurin, Y.: Blind Schnorr signatures and signed ElGamal encryption in the algebraic group model. In: Canteaut, A., Ishai, Y. (eds.) EUROCRYPT 2020. LNCS, vol. 12106, pp. 63–95. Springer, Cham (2020). https://doi.org/10.1007/978-3-030-45724-2_3
17. Fukumitsu, M., Hasegawa, S.: A Galindo-Garcia-like identity-based signature with tight security reduction. In: 2017 Fifth International Symposium on Computing and Networking (CANDAR), pp. 87–93. IEEE (2017)

18. Fukumitsu, M., Hasegawa, S.: A Galindo-Garcia-like identity-based signature with tight security reduction, revisited. In: 2018 Sixth International Symposium on Computing and Networking (CANDAR), pp. 92–98. IEEE (2018)
19. Galindo, D., Garcia, F.D.: A Schnorr-like lightweight identity-based signature scheme. In: Preneel, B. (ed.) AFRICACRYPT 2009. LNCS, vol. 5580, pp. 135–148. Springer, Heidelberg (2009). https://doi.org/10.1007/978-3-642-02384-2_9
20. Gjøsteen, K., Jager, T.: Practical and tightly-secure digital signatures and authenticated key exchange. In: Shacham, H., Boldyreva, A. (eds.) CRYPTO 2018. LNCS, vol. 10992, pp. 95–125. Springer, Cham (2018). https://doi.org/10.1007/978-3-319-96881-0_4
21. Hess, F.: Efficient identity based signature schemes based on pairings. In: Nyberg, K., Heys, H. (eds.) SAC 2002. LNCS, vol. 2595, pp. 310–324. Springer, Heidelberg (2003). https://doi.org/10.1007/3-540-36492-7_20
22. Kastner, J., Loss, J., Xu, J.: The Abe-Okamoto partially blind signature scheme revisited. In: Agrawal, S., Lin, D. (eds.) ASIACRYPT 2022. LNCS, vol. 13794, pp. 279–309. Springer, Cham (2022). https://doi.org/10.1007/978-3-031-22972-5_10
23. Kastner, J., Loss, J., Xu, J.: On pairing-free blind signature schemes in the algebraic group model. In: Hanaoka, G., Shikata, J., Watanabe, Y. (eds.) PKC 2022. LNCS, vol. 13178, pp. 468–497. Springer, Cham (2022). https://doi.org/10.1007/978-3-030-97131-1_16
24. Kılınç Alper, H., Burdges, J.: Two-round trip Schnorr multi-signatures via delinearized witnesses. In: Malkin, T., Peikert, C. (eds.) CRYPTO 2021. LNCS, vol. 12825, pp. 157–188. Springer, Cham (2021). https://doi.org/10.1007/978-3-030-84242-0_7
25. Kiltz, E., Masny, D., Pan, J.: Optimal security proofs for signatures from identification schemes. In: Robshaw, M., Katz, J. (eds.) CRYPTO 2016. LNCS, vol. 9815, pp. 33–61. Springer, Heidelberg (2016). https://doi.org/10.1007/978-3-662-53008-5_2
26. Kiltz, E., Neven, G.: Identity-based signatures. In: Joye, M., Neven, G. (eds.) Identity-Based Cryptography, Cryptology and Information Security Series, vol. 2, pp. 31–44. IOS Press (2009). https://doi.org/10.3233/978-1-58603-947-9-31
27. Lee, K., Kim, H.: Two-round multi-signature from Okamoto signature. Cryptology ePrint Archive (2022)
28. Loh, J.C., Guo, F., Susilo, W., Yang, G.: A tightly secure id-based signature scheme under dl assumption in AGM. In: Simpson, L., RezazadehBaee, M.A. (eds.) ACISP 2023. LNCS, vol. 13915, pp. 199–219. Springer, Cham (2023). https://doi.org/10.1007/978-3-031-35486-1_10
29. Naccache, D., Pointcheval, D., Stern, J.: Twin signatures: an alternative to the hash-and-sign paradigm. In: Proceedings of the 8th ACM Conference on Computer and Communications Security, pp. 20–27 (2001)
30. Nick, J., Ruffing, T., Seurin, Y.: MuSig2: simple two-round Schnorr multi-signatures. In: Malkin, T., Peikert, C. (eds.) CRYPTO 2021. LNCS, vol. 12825, pp. 189–221. Springer, Cham (2021). https://doi.org/10.1007/978-3-030-84242-0_8
31. Nour, B., et al.: Internet of things mobility over information-centric/named-data networking. IEEE Internet Comput. **24**(1), 14–24 (2019)
32. Oliveira, L.B., et al.: TinyPBC: pairings for authenticated identity-based non-interactive key distribution in sensor networks. Comput. Commun. **34**(3), 485–493 (2011)
33. Paillier, P., Vergnaud, D.: Discrete-log-based signatures may not be equivalent to discrete log. In: Roy, B. (ed.) ASIACRYPT 2005. LNCS, vol. 3788, pp. 1–20. Springer, Heidelberg (2005). https://doi.org/10.1007/11593447_1

34. Paterson, K.G.: Id-based signatures from pairings on elliptic curves. Electron. Lett. **38**(18), 1025–1026 (2002)
35. Paterson, K.G., Schuldt, J.C.N.: Efficient identity-based signatures secure in the standard model. In: Batten, L.M., Safavi-Naini, R. (eds.) ACISP 2006. LNCS, vol. 4058, pp. 207–222. Springer, Heidelberg (2006). https://doi.org/10.1007/11780656_18
36. Pointcheval, D., Stern, J.: Security arguments for digital signatures and blind signatures. J. Cryptol. **13**(3), 361–396 (2000)
37. Rahman, S.M.M., El-Khatib, K.: Private key agreement and secure communication for heterogeneous sensor networks. J. Parallel Distrib. Computi. **70**(8), 858–870 (2010)
38. Schnorr, C.P.: Efficient identification and signatures for smart cards. In: Brassard, G. (ed.) CRYPTO 1989. LNCS, vol. 435, pp. 239–252. Springer, New York (1990). https://doi.org/10.1007/0-387-34805-0_22
39. Shamir, A.: Identity-based cryptosystems and signature schemes. In: Blakley, G.R., Chaum, D. (eds.) CRYPTO 1984. LNCS, vol. 196, pp. 47–53. Springer, Heidelberg (1985). https://doi.org/10.1007/3-540-39568-7_5
40. Tessaro, S., Zhu, C.: Short pairing-free blind signatures with exponential security. Cryptology ePrint Archive (2022)
41. Xiong, W., Wang, R., Wang, Y., Zhou, F., Luo, X.: CPPA-D: efficient conditional privacy-preserving authentication scheme with double-insurance in VANETs. IEEE Trans. Veh. Technol. **70**(4), 3456–3468 (2021)
42. Zhang, X., Liu, S., Gu, D., Liu, J.K.: A generic construction of tightly secure signatures in the multi-user setting. Theoret. Comput. Sci. **775**, 32–52 (2019)

Threshold Ring Signatures with Accountability

Xuan Thanh Khuc[1](✉), Willy Susilo[1], Dung Hoang Duong[1], Fuchun Guo[1], Kazuhide Fukushima[2], and Shinsaku Kiyomoto[2]

[1] Institute of Cybersecurity and Cryptology, School of Computing and Information Technology, University of Wollongong,Wollongong, Australia
xtk929@uowmail.edu.au, {wsusilo,hduong,fuchun}@uow.edu.au

[2] Information Security Laboratory, KDDI Research, Inc., Saitama, Japan
{ka-fukushima,kiyomoto}@kddi-research.jp

Abstract. Threshold ring signatures (TRS) allow several signers to sign the same message on behalf of a group. This scheme is fully anonymity in that a signature reveals the number of signers who created the signature but tells nothing about the identity of all signers who participated in generating the signature. However, in several scenarios, it is required that signers should take responsibility for their actions and hence traceability of the signer's identity is needed. In this paper, we introduce a new type of threshold ring signature, called *threshold ring signature with accountability (ATRS)*. It is a balance of anonymity and accountability. In addition to achieving privacy as in the TRS scheme, an ATRS scheme also allows a designated opener to learn the identity of all signers who participated in creating the signature when necessary. Notably, the signers can decide whether they want to disclose their identity if they have any concerns about the designated opener. Moreover, the opener must provide a verifiable opening to ensure that he/she cannot hold a signer responsibility for a signature that was not created by him/her. Our new type of threshold ring signature combines the notion of tracing in traditional anonymity-oriented signature primitives with threshold ring signatures and can enable various new and appealing privacy-preserving applications.

We formalise the definition and security requirements for ATRS. We present a generic construction to demonstrate the feasibility of designing ATRS in a modular manner from commonly used cryptographic building blocks (verifiable random function, public-key encryption, preconstrained encryption and NIWI). Our scheme is secure in the plain model that does not require the heuristic of random oracle or trusted setup assumptions.

Keywords: ring signatures · threshold ring signatures · accountability · anonymity · plain model · standard model · NIWI

This work is partially supported by the ARC Linkage Project LP190100984. Willy Susilo is supported by the ARC Australian Laureate Fellowship FL230100033. Fuchun Guo is supported by the ARC Future Fellowship FT220100046. Dung Hoang Duong is partially supported by AEGiS 2023 grant from University of Wollongong.

T. Zhu and Y. Li (Eds.): ACISP 2024, LNCS 14895, pp. 368–388, 2024.
https://doi.org/10.1007/978-981-97-5025-2_19

1 Introduction

Anonymity has become a requirement in many real-world implementations of cryptographic systems and privacy-enhancing technologies, including electronic voting and cryptocurrencies [22]. Ring signatures [24] allow signers to remain anonymous inside a group of users. While anonymity is desirable, in certain circumstances, we also want to be assured of signature traceability, which implies that anonymity can be revealed. Group signatures [14] rely on a trust manager to achieve both anonymity and traceability of the signature. The manager identifies the group members and assigns each member a key pair. The signer is anonymous to other team members, but the manager can revoke the anonymity if desired.

Group and ring signatures limit the signature to be created by only one person and do not allow multiple people to generate the same signature. Threshold ring signatures (TRS) [10] are a variant of ring signatures, which allows some t signers to sign a message on behalf of a ring R of size larger than t. The signature reveals t members of the ring signed the message but not the identities of those members. In some practical cases, the full anonymisation of the TRS scheme would be too strong and facilitate illegitimate activity. For example, in money transfer transactions, participating parties will collude to create money laundering transactions. However, being able to completely retrieve the signer's identity will also create concerns among participating parties in creating a joint signature. Therefore, striking a balance between anonymity and accountability is always an open question when building a TRS scheme, which we will resolve in this paper.

1.1 Our Contribution

The notable contributions and innovations of our work can be summarised as follows:

- We introduce the definition and security requirements for the new primitive construction of threshold ring signatures with accountability (ATRS). It is not only private from the public but also from a set of signers. However, it also offers a designated opportunity to uncover the identity of the signer. Importantly, signers can choose whether or not to reveal their identities if they are concerned about the opening process. Moreover, the opener provides a verifiable opening to ensure that the opener cannot hold the non-signer responsible. Hence, an ATRS achieves a balance between privacy and accountability.
- We present a generic construction of ATRS. Our construction is secure in the plain model, i.e., it can be built from falsifiable standard assumptions without the random oracle heuristic or trusted setup assumptions.
- Our proposed generic scheme achieves *anonymity* against adversarially chosen keys: it ensures that a signature does not disclose the identity of signers who have individual thresholds less than or equal to the opener threshold unless the opener reveals it. Moreover, it ensures that if the signer's thresholds are greater than the opener's threshold, then the identity of the signer is not exposed, even if the opener wants to release it.

Our scheme achieves *unforgeability* with respect to insider corruption and control semi-honest opener: it ensures that an adversary cannot generate a valid signature for threshold $\mathsf{t}+1$, even if the adversary is able to corrupt up to t signers and can determine the identity of honest signers via the opening functionality.
Our scheme is also *tracing soundness* with respect to insider corruption and control semi-honest opener: it ensures that an adversary cannot open two distinct subsets of signers from only one ring signature even if the adversary can control the opener and corrupt all signers.

1.2 Overview of Our Techniques

A series of work inspire our construction in [1,2,6,7,19,21]. Our three biggest challenges in building the ATRS scheme are *(i) ensuring a balance between anonymity and accountability (ii) providing the opener can issue a verifiable opening*, and *(iii) ensuring the above settings are secure in the plain model.*

Our construction uses several cryptographic tools, including a one-way function, commitment scheme (CS), public-key encryption (PKE), verifiable random function (VRF), pre-constrained encryption (PCE), somewhere perfectly binding hash function (SPB), and non-interactive witness-indistinguishable (NIWI). At a high level, our construction is based on the sign-then-encrypt-then-prove paradigm that is typically used for building group/ring signatures. However, we also need to use some novel techniques to solve the three challenges mentioned above.

To address the challenge (i), we replace the use of a public key encryption function with the use of a PCE function. The opener will use PCE's key generation algorithm to generate a public and secret key pair, with input being a threshold at which the opener wants to be able to decrypt the ciphertext. The opener will publish his threshold along with the public key. Our signature threshold is a set of individual signatures of signers on the same message. Before participating in generating a signature, each signer will choose an individual threshold t_i based on the published opener's threshold. In the signing algorithm, the signer encrypts their identity along with their threshold. Due to the correctness of the PCE scheme, the opener can only decrypt the ciphertexts with which the message is encrypted with a threshold smaller than the opener's threshold.

The use of thresholds has great implications for both the opener and the signer. On the signer side, it ensures that their signature will only be added to the signature threshold when at least t_i other people agree to sign the message. Besides, it ensures that the signer has the right to choose whether they want to reveal their identity to the opener or not. If the signer chooses a threshold smaller than the opener's threshold, there is a high probability that their signature will be contributed to the threshold signature. However, the opener will know the identity through their signature. Therefore, the signer will have to be more cautious when deciding to sign. On the contrary, if the signer chooses a threshold larger than the opener's threshold, their identity will not be revealed. However, this will also cause a high probability that their signature will not contribute to the threshold signature because the signer has asked a large number of people to agree to sign. On the opener side, signers who choose a low threshold will be those who have the highest agreement with the message. Therefore, these signers will also have to be the ones with the greatest responsibility for the signed message. If the signature has been contributed to the threshold signature but the opener can not know the signer's identity, then these signers have also created a certain trust because their signature is valid, along with many other signers. In our construction, we require the opener to be semi-honest. Actually, this is reasonable since many previous schemes [6,20,21] required the opener to be an honest party. Besides, no signer would like to choose a designated opener as malicious.

To solve the challenge (ii), we require that in addition to providing the signer's identity, the opener also needs to provide proof that can only be computed by the signer's private key. This proof is hidden in the threshold signature, and only the opener holding the secret key can open it. Specifically, we require the opener to present the signature of the signer on the message and the user's ring. Then, using the signer's public key, anyone can check whether this signature was created by the signer revealed by the opener. During the signing process, the signer will also encrypt this signature with their public key to prevent the opener from using this advantage to generate fake signatures later. In our construction, we use the VRF scheme instead of the usual signature scheme. In addition to serving as a traditional signature, the VRF scheme provides a deterministic signature that avoids the signer creating a threshold signature by signing multiple times on the same message and ring.

As for challenge (iii), all of the above would be easy if we construct security proof in the ROM, CRS model, where we can use the zero-knowledge property of NIZK to prove anonymity. Unfortunately, we do not have this property in a plain model. Thus, anonymity becomes a major challenge. Anonymity ensures that the signature does not reveal any information about the identity of the signer. In the anonymity experiment, we will have to construct a transformation to convert the signature computed by signer VK_0 to the signature computed by signer VK_1 via a hybrid sequence. The difference between the hybrids is indistinguishable from the adversary. However, in our construction, we use the VRF function as an alternative to the usual signature. The VRF scheme provides a deterministic signature, so if an adversary requests the same message to be signed twice, this will create a difference in the sequence of hybrids. To overcome this problem, we will add one more relation in the NIWI proof. This relationship would be unlikely in a real key generation algorithm. However, we need the above relationship to happen as proof of anonymity. Now, each user will choose a random value nonce and add it to the signer secret key in the key generation algorithm. Then, the user computes the commitment value on nonce and adds commitment value to the signer public key. The additional relation said that there exist two users in the ring whose nonce values satisfy the condition $F(\mathsf{nonce}_0) = \mathsf{nonce}_1$, where F are one-way functions. We can easily see that this relation is unlikely in a real key generation algorithm because the nonce values are uniformly random. We can choose two users who satisfy the above relation in the anonymity experiment. However, adding a new relation also has a slight effect on proving unforgeability and anonymity. Specifically, we must restrict the adversary from adding his generated public key to the ring of users in the unforgeability experiment. Indeed, assume that the adversary can add his public keys. In that case, the adversary trivially generates a valid forged signature by choosing two public keys whose nonce values satisfy the newly added relation. In the anonymity experiment, we restrict the adversary to querying each message only once to the signing oracle. We note that previous works such as [19] also had to overcome similar situations. However, they add one more relation by using public key encryption instead of using a commitment scheme as our construction.

1.3 Related Work

In [6], Dan Boneh and Chelsea Komlo introduce a new type of primitive by mixing private threshold signature and accountable threshold signature. This scheme reveals nothing about t and quorum of t users and can reveal the identity of all t users when needed. However, the accountable and anonymous properties now totally depend on the combiner. It looks like a threshold group signature. Moreover, the number of users is fixed. These schemes achieve security in the random oracle mode.

A traceable ring signature (TRS) [9,12,16,17] is a variant of ring signatures that allows public traceability of the user's identity if he/she has produced two signatures with respect to the same issue. An accountable ring signature (ARS) [7,11,25] fills the gap between ring signatures and group signatures. They give you the freedom to choose the ring of users when making a signature, and they also make sure people are responsible by having an opener who can open a signature and show who signed it. The mix of flexibility and accountability makes it useful in situations where ring or group signatures are not as good. However, both TRS and ARS schemes are limited to a single signer, as opposed to a threshold of signers within the ring. Moreover, these schemes fail to strike a balance between anonymity and accountability. Specifically, the signer's identity is completely revealed by the opener in the ARS scheme. In the TRS scheme, the signer's identity is revealed only when signing two different messages on the same tag.

Recently, Bifurcated Anonymous Signatures (BiAS) [20] and Multimodal Private Signatures (MPS) [21] strike a balance between anonymity and accountability. However, these schemes are also limited to only one signer and use an honest opener.

1.4 Application

We briefly describe some potential applications for ATRS. Let's consider the scenario of an anonymous financial transaction (such as the privacy-preserving cryptocurrency Monero [22]). When it is necessary to transfer a certain amount of money, the system requires the consent of a number of users in a group of users by signing the money transfer message. We can use the ATRS scheme to detect fraudulent transactions by requiring signers to be responsible for their actions. The opener publics his public key as well as his opening threshold. The signers generate signatures with their threshold t. Signers are guaranteed that their signature is responsible for the payment transaction when at least t other people also co-sign the message. Signers have the right to decide whether their signature can be traced by the opener or not. Choosing a small threshold means that the signer strongly supports this transfer. Therefore, they will also be responsible for this money transfer transaction.

On the contrary, choosing a large threshold means they will only agree when a large number of people also agree. In other words, openers can trust people with large thresholds more. However, choosing a large threshold also means there is a high possibility that their signature is not a valid signature for the transaction. Transactions are only completed successfully when there are enough signers, and each signer has a threshold greater than the verification threshold.

2 Preliminaries

This section will recall some cryptographic primitives that are building blocks for our construction. We will use λ as a main security parameter. We denote by $[N]$ the set $\{0, 1, \ldots, N + 1\}$. Let $r \leftarrow_\$ \mathsf{S}$ denote r be chosen uniformly random from the set S. Denote $\mathsf{S}[i], |\mathsf{S}|$ respectively are i-th element and size of S.

Definition 1 (Public Key Encryption, [2,19]**).** *A public key encryption scheme* PKE *consists of the algorithms* $(\mathsf{PKE.KeyGen}, \mathsf{PKE.Enc}, \mathsf{PKE.Dec})$ *with the following syntax:*

- $(\mathsf{epk}, \mathsf{esk}) \leftarrow \mathsf{PKE.KeyGen}(1^\lambda)$. *On input a security parameter* 1^λ*, this PPT algorithm outputs a pair of public and secret keys* $(\mathsf{epk}, \mathsf{esk})$.
- $\mathsf{ct} \leftarrow \mathsf{PKE.Enc}(\mathsf{epk}, \mathsf{m}, \mathsf{r})$. *On input a public key* epk*, a message* m *and a randomness* r*, this PPT algorithm outputs a ciphertext* ct.
- $\mathsf{m} \leftarrow \mathsf{PKE.Dec}(\mathsf{esk}, \mathsf{ct})$. *On input a secret key* esk *and a ciphertext* ct*, this deterministic algorithm outputs a message* m.

We require the following properties of a PKE scheme. Firstly, the property of *perfect correctness* guarantees that the original plaintext can be obtained through decryption if the keys and ciphertexts were generated honestly. Secondly, *key privacy* guarantees that the adversary, without knowledge about the secret keys, cannot determine for which public key a ciphertext has been computed. Thirdly, *perfect binding* ensures that every unbounded adversary cannot output two pairs of distinct messages and randomness such that it computes the same ciphertext. PKE schemes that satisfy pseudorandom ciphertexts and perfect binding can be instantiated based on DDH [15].

We consider a notion of threshold-based pre-constrained encryption (TB-PCE) that is defined with respect to a constraint family $\mathcal{C}$ and a threshold universe $\mathcal{T}$. In a TB-PCE scheme, messages are encrypted together with a threshold. It has syntax nearly the same as attribute-based (identity-based) pre-constrained encryption [1]. However, we adapted it to fit with our notation. Crucially, the constraint $C \in \mathcal{T}$ is a set of thresholds, i.e. the authority can only decrypt ciphertexts computed using thresholds $\mathsf{t} \in \mathcal{T}$ such that $\mathsf{t} \in C$ (denoted $C(\mathsf{t}) = 1$).

Definition 2 (Threshold-Based Pre-Constrained Encryption, Adapted from [1]). *A threshold-based pre-constrained encryption* TB-PCE *scheme for a constraint family $\mathcal{C}$ and an threshold universe $\mathcal{T}$ consists of three algorithms* (TB-PCE.KeyGen, TB-PCE.Enc, TB-PCE.Dec) *described below.*

- $(\mathsf{ppk}, \mathsf{psk}) \leftarrow \mathsf{TB\text{-}PCE.KeyGen}(1^\lambda, C)$: *On input a security parameter 1^λ, a constraint predicate $C \in \mathcal{C}$, this algorithm outputs a pair of public and secret keys* $(\mathsf{ppk}, \mathsf{psk})$.
- $\mathsf{ct} \leftarrow \mathsf{TB\text{-}PCE.Enc}(\mathsf{ppk}, \mathsf{m}, \mathsf{t}, \mathsf{r})$: *On input a public key* ppk, *a message* m, *a threshold* $\mathsf{t} \in \mathcal{T}$ *and a randomness* r, *this algorithm outputs a ciphertext* ct.
- $\mathsf{m} \leftarrow \mathsf{TB\text{-}PCE.Enc}(\mathsf{psk}, \mathsf{ct})$: *On input a secret key* psk *and a ciphertext* ct, *this algorithm outputs a plaintext* m *iff* $C(\mathsf{t}) = 1$. *Otherwise, it returns* $\perp$.

We consider several following properties for TB-PCE. Firstly, *correctness* ensures that if ciphertexts are computed using thresholds $\mathsf{t} \in \mathcal{T}$ such that $C(\mathsf{t}) = 1$, then the authority can decrypt it. Secondly, *security against semi-honest authority* SH-SAA ensure that an authority who honestly runs the TB-PCE.KeyGen algorithm on input C should not be able to distinguish between encryptions of any two equal length messages $\mathsf{m}_0, \mathsf{m}_1$ and threshold t_0 and t_1 such that for all threshold $\mathsf{t}_0, \mathsf{t}_1 \in \mathcal{T}$ satisfying $C(\mathsf{t}_0) = C(\mathsf{t}_1) = 1$.

- **Correctness:** We say a threshold-based pre-constrained encryption TB-PCE is correctness if it holds for all security parameters $\lambda \in \mathbb{N}$, for all $C \in \mathcal{C}$ and all $\mathsf{t} \in \mathcal{T}$ such that $C(\mathsf{t}) = 1$ and all message m that given $(\mathsf{ppk}, \mathsf{psk}) \leftarrow \mathsf{TB\text{-}PCE.KeyGen}(1^\lambda, C)$, $\mathsf{ct} \leftarrow \mathsf{TB\text{-}PCE.Enc}(\mathsf{ppk}, \mathsf{m}, \mathsf{t}, \mathsf{r})$, then it holds that $\mathsf{PKE.Dec}(\mathsf{psk}, \mathsf{ct}) = \mathsf{m}$.
- **Security against Semi-Honest Authority**: A TB-PCE scheme for constraint family $\mathcal{C}$ and a threshold universe $\mathcal{T}$ satisfies security against semi-honest authority if for any non-uniform PPT adversary $\mathcal{A}$, any $C \in \mathcal{C}$, and every pair $(\mathsf{m}_0, \mathsf{t}_0)$ and $(\mathsf{m}_1, \mathsf{t}_1)$ such that $|\mathsf{m}_0| = |\mathsf{m}_1|$ and $C(\mathsf{t}_0) = C(\mathsf{t}_1)$ has at most has at most negligible advantage in the following experiment:
  $\mathsf{Exp}_{\mathsf{SH\text{-}SAA}}(\mathcal{A})$:
  1. The adversary $\mathcal{A}$ generates $(\mathsf{ppk}, \mathsf{psk}) \leftarrow \mathsf{TB\text{-}PCE.KeyGen}(1^\lambda, C)$ and public ppk.
  2. $\mathcal{A}$ choose $(\mathsf{m}_0, \mathsf{t}_0)$, $(\mathsf{m}_1, \mathsf{t}_1)$ and send them to experiment.

3. The experiment chooses $b \leftarrow \$ \{0,1\}$, computes $\mathsf{ct} \leftarrow \mathsf{TB\text{-}PCE.Enc}(1^\lambda, \mathsf{m}_b, \mathsf{t}_b)$ and provides ct to $\mathcal{A}$.
4. $\mathcal{A}$ outputs a guess b'. If $b' = b$, the experiment outputs 1, otherwise 0.

TB-PCE can be constructed as attribute-based (identity-based) pre-constrained [1], set pre-constrained encryption [4].

Definition 3 (Commitment Schemes, [2]). *A commitment scheme* CS *consists of two algorithms* (CS.Com, CS.Verify) *defined as the following:*

- $(\mathsf{com}, \mathsf{u}) \leftarrow \mathsf{CS.Com}(1^\lambda, \mathsf{m})$*: On input a security parameter* 1^λ*, a message* m*, this PPT algorithm outputs a commitment value* com *and unveil information* u*.*
- $b \leftarrow \mathsf{CS.Verify}(\mathsf{com}, \mathsf{m}, \mathsf{u})$*: On input a commitment value* com*, a message* m*, and unveil information* u*, this deterministic algorithm outputs a single bit* $b \in \{0,1\}$*.*

We consider the commitment scheme CS to be *correctness*, *perfect binding* and *computational hiding*. A commitment scheme is perfectly binding if every unbounded adversary cannot change the committed message after the commitment phase. It is hiding if any computational adversary cannot distinguish the commitments of two different messages.

Definition 4 (Somewhere Perfectly Binding Hashing, [2]). *A somewhere perfectly binding hash family with private local opening* SPB *consists of four algorithms* (SPB.KeyGen, SPB.Hash, SPB.Open, SPB.Verify) *defined as the following:*

- $(\mathsf{hk}, \mathsf{shk}) \leftarrow \mathsf{SPB.KeyGen}(1^\lambda, n, \mathsf{ind})$*. On input a security parameter* 1^λ*, a database size* n *and an index* ind*, this PPT algorithm outputs public hashing key* hk *and corresponding secret hashing key* shk*.*
- $\mathsf{h} \leftarrow \mathsf{SPB.Hash}(\mathsf{hk}, \mathsf{db})$*. On input a hashing key* hk*, a database* db *of size* n*, this deterministic algorithm outputs a hash value* h*.*
- $\tau \leftarrow \mathsf{SPB.Open}(\mathsf{hk}, \mathsf{shk}, \mathsf{db}, \mathsf{ind})$*. On input a hashing key* hk*, private hashing key* shk*, a database* db *of size* n*, and index* ind*, this algorithm outputs a witness* τ*.*
- $b \leftarrow \mathsf{SPB.Verify}(\mathsf{hk}, \mathsf{h}, \mathsf{ind}, x, \tau)$*. On input a hashing key* hk*, a hash value* h*, and an index* ind*, this algorithm outputs a single bit* $b \in \{0,1\}$*.*

The somewhere perfect binding satisfies the following properties. Firstly, *correctness* ensures that if the keys, hashes, and openings are generated truthfully, the verification process will be successful. Secondly, *efficiency* ensures that the size of public hashing hk and witnesses τ is logarithmic with the number of ring users. Moreover, the Verify algorithm can be computed by a circuit of size $\log(n) \cdot \mathsf{poly}(\lambda)$. Thirdly, *somewhere perfectly binding* ensures that if a particular index i and value x pass verification, all valid openings for that index must result in x. Lastly, *index hiding* means that an efficient adversary cannot determine the index i from the public hashing key. SPB scheme can be constructed from somewhere statistically binding hashing (SSB). In [Appendix A.2, [2]], they present how to transform SSB into an SPB scheme. Moreover, SBB schemes can be constructed from the hardness of assumptions such as DDH, QR, and LWE [23].

Let $\mathcal{R}$ be an efficiently computable binary relation, where for $(x, w) \in \mathcal{R}$ (x is a statement and w is the witness). The language $\mathcal{L}$ is defined as all statements that have a valid witness in $\mathcal{R}$, i.e. $\mathcal{L} := \{x | \exists w : (x, w) \in \mathcal{R}\}$.

Definition 5 (Non-Interactive Proof System, [2]**).** *Let $\mathcal{R}$ be an efficiently computable witness relation, and $\mathcal{L}$ be the language accepted by $\mathcal{R}$. A non-interactive witness-indistinguishable proof system* NIWI *for $\mathcal{L}$ includes* (NIWI.Prove, NIWI.Verify) *with the following syntax:*

- $\pi \leftarrow \mathsf{NIWI.Prove}(1^\lambda, x, w)$. *On input a security parameter 1^λ, a statement x and a witness w, it outputs a proof π.*
- $b \leftarrow \mathsf{NIWI.Verify}(x, \pi)$. *Given a statement x and a proof π, it outputs a bit b.*

We require the following four properties for a Non-Interactive Witness Indistinguishable (NIWI) proof system. Firstly, *perfect completeness* ensures that correct statements can always be proven. Secondly, *perfect soundness* prevents the generation of valid proofs for false statements. Thirdly, *witness indistinguishability* states that given two valid witnesses for a statement, no efficient adversary can determine which witness was used to compute the proof. Last one, the *proof-size* requires that π is polynomial size with security parameter λ.

NIWI scheme can be constructed from pairing assumptions [18] as pointed out in [2,8] for classical instantiation. For post-quantum instantiation, we can replace our NIWI by using ZAP (two-message public coin argument system) based on LWE [3], which can be found in [13].

Definition 6 (Verifiable Random Function, [19]**).** *A verifiable random function is 4-tuple* (VRF.Gen, VRF.Eval, VRF.Prove, VRF.Verfify) *with the following syntax:*

- $(\mathsf{vpk}, \mathsf{vsk}) \leftarrow \mathsf{VRF.Gen}(1^\lambda)$: *On input a security parameter 1^λ, this PPT algorithm outputs a public verification key* vpk *and corresponding secret key* vsk.
- $\mathsf{v} \leftarrow \mathsf{VRF.Eval}(\mathsf{vsk}, x)$: *On input the secret key* vsk *and an input value $x \in \{0,1\}^{a(\lambda)}$, this deterministic algorithm outputs a value $\mathsf{v} \in \{0,1\}^{b(\lambda)}$ ($a(\lambda)$ and $b(\lambda)$ are polynomial bounded and efficiently computable function in λ).*
- $\mathsf{p} \leftarrow \mathsf{VRF.Prove}(\mathsf{vsk}, x)$: *On input a secret key* vsk *and an input value x, this PPT algorithm outputs a proof* p.
- $b \leftarrow \mathsf{VRF.Verify}(\mathsf{vpk}, x, \mathsf{v}, \mathsf{p})$. *On input a verification key* vpk*, an input value x, a value* v*, and a proof* p*, this deterministic algorithm outputs a single bit b.*

We consider several following properties for VRFs. Firstly, *complete provability* ensures that outputs and proofs generated from consistent inputs will verify each other. Secondly, *unique provability* provides only one valid proof for each input. Thirdly, *residual unpredictability* ensures that outputs cannot be predicted by an adversary. Fourthly, *key privacy* ensures that the public key cannot be determined from the output. Fifthly, *key collision resistance* guarantees that the same output will not be produced with different secret keys.

3 Threshold Ring Signatures with Accountability

In this section, we introduce the definition and security model for threshold ring signatures with accountability in Sect. 3.1. Our construction is presented in Sect. 3.2.

3.1 Definition and Security Model

Definition 7 (Threshold Ring Signatures with Accountability). *An threshold ring signatures with accountability* ATRS *scheme is given by 6-tuple algorithm* ($\mathsf{ATRS.OKeyGen}$, $\mathsf{ATRS.UKeyGen}$, $\mathsf{ATRS.Sign}$, $\mathsf{ATRS.Verify}$, $\mathsf{ATRS.Open}$, $\mathsf{ATRS.VerifyOpen}$) *defined as follows:*

- $(\mathsf{OPK}, \mathsf{OSK}) \leftarrow \mathsf{ATRS.OKeyGen}(1^\lambda, \mathsf{t_O})$. *On input a security parameter* 1^λ *and threshold* $\mathsf{t_O}$, *this PPT algorithm generates a public encryption key* OPK *and a corresponding secret key* OSK *for an opener.*
- $(\mathsf{VK}, \mathsf{SK}) \leftarrow \mathsf{ATRS.UKeyGen}(1^\lambda)$. *On input a security parameter* 1^λ, *this PPT algorithm generates a public verification key* VK *and a corresponding secret key* SK *for a user.*
- $\Sigma \leftarrow \mathsf{ATRS.Sign}(\mathsf{R}, \mathsf{OPK}, \mathsf{SSK}, \mathsf{T}, \mathsf{m})$. *On input* $\mathsf{R} = (\mathsf{VK}_0, \mathsf{VK}_1, \ldots, \mathsf{VK}_N)$, *a opener's public key* OPK, *a set of secret keys* $\mathsf{SSK} = \{\mathsf{SK}_{i_0}, \ldots, \mathsf{SK}_{i_L}\}$, *where* $|\mathsf{SSK}| = L + 1$ *and* $L < N$ *corresponding to a subset of public keys* $\{\mathsf{VK}_{i_0}, \ldots, \mathsf{VK}_{i_L}\} \subseteq \mathsf{R}$, *a set of thresholds* $\mathsf{T} = \{\mathsf{t}_{i_j}\}_{j \in [L]}$ *and a message* $\mathsf{m} \in \{0,1\}^*$, *this PPT algorithm outputs a signature* Σ *on* $\mathsf{m} \in \{0,1\}^*$.
- $b \leftarrow \mathsf{ATRS.Verify}(\mathsf{R}, \mathsf{OPK}, \Sigma, \mathsf{t_V}, \mathsf{m})$. *On input a ring* R, *an opener public key* OPK, *signature* Σ, *a message* $\mathsf{m} \in \{0,1\}^*$ *and a verification threshold* $\mathsf{t_V}$, *this deterministic algorithm outputs a bit* b.
- $(\mathsf{OS}, \mathsf{AUX}) \leftarrow \mathsf{ATRS.Open}(\mathsf{R}, \mathsf{OSK}, \Sigma, \mathsf{t_V}, \mathsf{m})$. *On input a ring* R, *a secret opener key* OSK, *a signature* Σ, *a verification threshold* $\mathsf{t_V}$ *and a message* $\mathsf{m} \in \{0,1\}^*$, *this deterministic algorithm outputs a set of public keys* OS, *and a set of auxiliary information* AUX.
- $b \leftarrow \mathsf{ATRS.VerifyOpen}(\mathsf{R}, \mathsf{OPK}, \Sigma, \mathsf{t_V}, \mathsf{m}, \mathsf{OS}, \mathsf{AUX})$. *On input a ring* R, *an opener public key* OPK, *a signature* Σ, *a verification threshold* $\mathsf{t_V}$, *a message* $\mathsf{m} \in \{0,1\}^*$, *a set of public keys* OS *and a set of auxiliary information* AUX, *this deterministic algorithm outputs a bit* b.

Firstly, the *correctness* guarantees that a signature generated by honest signers will always pass the verification algorithm.

Definition 8 (Correctness). *A threshold ring signatures with accountability* ATRS *is correct, if for all* $\lambda \in \mathbb{N}$, *all* $N = \mathsf{poly}(\lambda)$, *all size of signer subset* $L < N$, *all set of individual thresholds* $\mathsf{T} = \{\mathsf{t}_{i_j}\}_{j \in [L]}$, *all opener threshold* $\mathsf{t_O}$, *all verifier threshold* $\mathsf{t_V}$ *such that* $\mathsf{t_V} \leq |i_j : \mathsf{t}_{i_j} \leq \mathsf{t_V}|$ *for all* $j \in [L]$ *and all* *message* $\mathsf{m} \in \{0,1\}^*$ *that if* $(\mathsf{OPK}, \mathsf{OSK}) \leftarrow \mathsf{ATRS.OKeyGen}(1^\lambda, \mathsf{t_O})$*, for* $i \in [N]$*,* $(\mathsf{VK}_i, \mathsf{SK}_i) \leftarrow \mathsf{ATRS.UKeyGen}(1^\lambda)$*,* $\mathsf{R} = \{\mathsf{VK}_i\}_{i\in[N]}$*,* $\mathsf{SSK} = \{\mathsf{SK}_{i_j}\}_{i\in[L]}$*,* $\Sigma \leftarrow \mathsf{ATRS.Sign}(\mathsf{R}, \mathsf{OPK}, \mathsf{SSK}, \mathsf{T}, \mathsf{m})$*, then it holds that*

$$\Pr[\mathsf{ATRS.Verify}(\mathsf{R}, \mathsf{OPK}, \Sigma, \mathsf{t_V}, \mathsf{m}) = 1] = 1 - \mathsf{negl}(\lambda).$$

Secondly, the *threshold traceability* guarantees that an honestly generated signature with individual thresholds less than or equal to the verification threshold must be traceable to the correct signer.

Definition 9 (Threshold Traceability). *A threshold ring signatures with accountability* ATRS *is threshold traceable, if for all* $\lambda \in \mathbb{N}$*, all* $N = \mathsf{poly}(\lambda)$*, all size of signer subset* $L < N$*, all set of individual thresholds* $\mathsf{T} = \{\mathsf{t}_{i_j}\}_{j\in[L]}$*, all opener threshold* $\mathsf{t_O}$*, all verifier threshold* $\mathsf{t_V}$ *such that* $\mathsf{t_V} \leq |i_j : \mathsf{t}_{i_j} \leq \mathsf{t_V}|$ *for all* $j \in [L]$ *and all message* $\mathsf{m} \in \{0,1\}^*$ *that if* $(\mathsf{OPK}, \mathsf{OSK}) \leftarrow \mathsf{ATRS.OKeyGen}(1^\lambda, \mathsf{t_O})$*, for* $i \in [N]$*,* $(\mathsf{VK}_i, \mathsf{SK}_i) \leftarrow \mathsf{ATRS.UKeyGen}(1^\lambda)$*,* $\mathsf{R} = (\mathsf{VK}_i)_{i\in[N]}$*,* $\Sigma \leftarrow \mathsf{ATRS.Sign}(\mathsf{R}, \mathsf{OPK}, \mathsf{SSK}, \mathsf{T}, \mathsf{m})$*,* $\mathsf{SSK} = \{\mathsf{SK}_{i_j}\}_{j\in[L]}$*,* $(\mathsf{OS}, \mathsf{AUX}) \leftarrow \mathsf{ATRS.Open}(\mathsf{R}, \mathsf{OSK}, \Sigma, \mathsf{t_V}, \mathsf{m})$*, then it holds that*

$$\Pr[\mathsf{ATRS.VerifyOpen}(\mathsf{R}, \mathsf{OPK}, \Sigma, \mathsf{t_V}, \mathsf{m}, \mathsf{OS}, \mathsf{AUX}) = 1] = 1 - \mathsf{negl}(\lambda).$$

Thirdly, the *anonymity* ensures that a signature does not disclose the identity of signers with individual thresholds less than or equal to the opener threshold unless the opener reveals it. Moreover, it ensures that if the signer's thresholds are greater than the opener's threshold, then the identity of the signers is not exposed even if the opener wants to release it. Our formal definition captures anonymity against adversarially chosen keys as defined in [5]. This model allows an adversary to join the ring with maliciously generated keys and corrupt users. When presented with a challenge signature, we require that the adversary cannot determine which of two potential sets of honest signers generated the signature.

Definition 10 (Anonymity). *A threshold ring signature with accountability* ATRS *is anonymous if every PPT adversary* $\mathcal{A}$*, holds that* $\mathcal{A}$ *has a negligible advantage in the following experiment.*
$\mathsf{Exp}_{\mathsf{ATRS\text{-}Anon}}(\mathcal{A})$:

1. *The experiment generates* $(\mathsf{OPK}, \mathsf{OSK}) \leftarrow \mathsf{ATRS.OKeyGen}(1^\lambda, \mathsf{t_O})$*. Next, it generates two same-size subsets of signers and thresholds* $\mathsf{S}_0 = \{\mathsf{VK}_{i_j}\}_{j\in[L]}$*,* $\mathsf{SSK}_0 = \{\mathsf{SK}_{i_j}\}_{j\in[L]}$*,* $\mathsf{T}_0 = \{\mathsf{t}_{i_j}\}_{j\in[L]}$*,* $\mathsf{S}_1 = \{\mathsf{VK}_{l_k}\}_{k\in[L]}$*,* $\mathsf{SSK}_1 = \{\mathsf{SK}_{l_k}\}_{k\in[L]}$*,* $\mathsf{T}_1 = \{\mathsf{t}_{l_k}\}_{k\in[L]}$ *where* $\mathsf{T}_0 = \mathsf{T}_1$*. The adversary provides* $\mathsf{S}_0, \mathsf{T}_0, \mathsf{S}_1, \mathsf{T}_1$ *to the experiment. Let* $\mathcal{Q} = \emptyset$ *be a set of signing queries.*
2. *The experiment samples* $b \leftarrow\$ \{0,1\}$*.*
3. *The adversary* $\mathcal{A}$ *can append new public keys to the global public keys, and threshold lists* $\mathsf{R} = \{\mathsf{VK}_0, \ldots, \mathsf{VK}_N\}$ *and* $\mathsf{T} = \{\mathsf{t}_0, \ldots, \mathsf{t}_N\}$*, where* $N > L$*. The experiment requires* $\mathsf{S}_0, \mathsf{S}_1 \in \mathsf{R}$ *and* $\mathsf{T}_0, \mathsf{T}_1 \in \mathsf{R}$*.*
4. *The adversary* $\mathcal{A}$ *can access signing three oracles* $\mathsf{ATRS.Sign}_{\mathsf{S}_b}$*,* $\mathsf{ATRS.Sign}_{\mathsf{S}_0}$*,* $\mathsf{ATRS.Sign}_{\mathsf{S}_1}$ *and an opening oracle* $\mathsf{ATRS.Open}$*, where*

i. $\mathsf{ATRS.Sign}_{\mathsf{S}_b}$ *is challenge signing oracle with respect to* S_b *for signing* $(\mathsf{R}, \mathsf{OPK}, \mathsf{T}_b, \mathsf{m})$.
   * *If* $(\mathsf{R}, \mathsf{OPK}, \mathsf{T}_b, \mathsf{m}) \in \mathcal{Q}$ *then the return* $\perp$.
   * *Otherwise, it computes* $\Sigma \leftarrow \mathsf{ATRS.Sign}(\mathsf{R}, \mathsf{OPK}, \ \mathsf{SSK}_b, \mathsf{T}_b, m)$ *and returns* Σ.

ii. $\mathsf{ATRS.Sign}_{\mathsf{S}_0}$ *(resp.* $\mathsf{ATRS.Sign}_{\mathsf{S}_1}$*) is the signing oracle with respect to* S_0 *(resp.* S_1*) for signing* $(\mathsf{R}, \mathsf{OPK}, \mathsf{T}_0, \mathsf{m})$ *(resp.* $(\mathsf{R}, \mathsf{OPK}, \mathsf{T}_1, \mathsf{m})$*).*
   * *If* $(\mathsf{R}, \mathsf{OPK}, \mathsf{T}_0, \mathsf{m}) \in \mathcal{Q}$ *(resp.* $(\mathsf{R}, \mathsf{OPK}, \ \mathsf{T}_1, \mathsf{m}) \in \mathcal{Q}$*) then the return* $\perp$.
   * *Otherwise, it computes* $\Sigma \leftarrow \mathsf{ATRS.Sign}(\mathsf{R}, \mathsf{OPK}, \mathsf{SSK}_0, \mathsf{T}_0, \mathsf{m})$ *(resp.* $\Sigma \leftarrow \mathsf{ATRS.Sign}(\mathsf{R}, \mathsf{OPK}, \mathsf{SSK}_1, \mathsf{T}_1, \mathsf{m})$*) and returns* Σ.

iii. $\mathsf{ATRS.Open}$ *is an opening oracle with respect to* $(\mathsf{R}, \mathsf{OPK}, \mathsf{t_V}, \mathsf{m})$. *If* Σ *is obtained by calling challenge signing oracle* $\mathsf{ATRS.Sign}_{\mathsf{S}_b}$, *then return* $\perp$. *Otherwise, the experiment returns* $\mathsf{ATRS.Open}(\mathsf{R}, \mathsf{OSK}, \Sigma, \mathsf{t_V}, \mathsf{m})$.

5. *$\mathcal{A}$ outputs a guess $b' \in \{0, 1\}$. The experiment outputs 1 if $b' = b$. Otherwise, it outputs 0.*

Fourth, the *unforgeability* ensures that an adversary cannot generate a valid signature for threshold $(\mathsf{t}+1)$, even if the adversary can corrupt up to t signers and determine the identity of honest signers via the opening functionality. In our unforgeability experiment, we presume that the adversary controls the opener and can corrupt users. However, the adversary does not allow adding new public keys to ring R. We require that an adversary cannot generate a valid signature for a ring of honest users if the signature is not the output of the signing oracle.

Definition 11 (Unforgeability). *A threshold ring signature with accountability* ATRS *is unforgeability if, for every PPT adversary* $\mathcal{A}$*, it holds that* $\mathcal{A}$ *has a negligible advantage in the following experiment.*
$\mathsf{Exp}_{\mathsf{ATRS\text{-}Unf}}$:

1. *The adversary generates* $(\mathsf{OPK}^*, \mathsf{OSK}^*) \leftarrow \mathsf{ATRS.OKeyGen}(1^\lambda, \mathsf{t_O})$ *and provides* OPK^* *to the experiment.*
2. *For all* $i = 0, \ldots, N$*, it generates* $(\mathsf{VK}_i, \mathsf{SK}_i) \leftarrow \mathsf{ATRS.UKeyGen}(1^\lambda, \mathsf{r}_i)$ *by using random coins* r_i*. The experiment sets* $\mathsf{R} = (\mathsf{VK}_0, \ldots, \mathsf{VK}_N)$ *and provides* R *to* $\mathcal{A}$*. Let* $\mathcal{Q}, \mathcal{C} = \emptyset$ *be sets of signing queries and corrupted signers.*
3. *The adversary* $\mathcal{A}$ *now can make some queries as the following*

   i. $\mathsf{ATRS.Sign}_{\mathsf{S}_i}$ *is a signing oracle with respect to subset* $\mathsf{S}_i \in \mathsf{R}$ *for signing* $(\mathsf{R}, \mathsf{OPK}^*, \mathsf{T}_i, \mathsf{m})$*, where* T_i *and* m *are chosen by adversary. The experiment checks if* $\mathsf{S}_i \in \mathsf{R}$*, if so, it computes* $\Sigma \leftarrow \mathsf{ATRS.Sign}(\mathsf{R}, \mathsf{OPK}^*, \mathsf{SSK}_i, \mathsf{T}_i, \mathsf{m})$ *and returns* Σ *to* $\mathcal{A}$*. Besides, the experiment sets* $\mathcal{Q} = \mathcal{Q} \cup (\mathsf{R}, \mathsf{OPK}^*, \mathsf{T}_i, \mathsf{m}, \Sigma)$.

   ii. $\mathsf{Corrupt}_{\mathsf{S}_i}$ *is a corruption oracle with respect to subset of signers* $\mathsf{S}_i \subset \mathsf{R}$*. The experiment returns* $\{r_{i_j}\}_{j \in [L_i]}$*, where* $L_i = |\mathsf{S}_i|$ *and adds* $\mathcal{C} = \mathcal{C} \cup \mathsf{S}_i$.
4. *In the end, the adversary* $\mathcal{A}$ *outputs a tuple* $(\mathsf{R}^*, \mathsf{OPK}^*, \mathsf{T}_i^*, \mathsf{m}^*, \Sigma^*, \mathsf{t}_\mathsf{V}^*)$.
5. *The experiment outputs 1 if the following conditions hold:*

   i. $\mathsf{R}^* \subset \mathsf{R} \setminus \mathcal{C}$,

   ii. $(\mathsf{R}^*, \mathsf{OPK}^*, \mathsf{T}_i^*, \mathsf{m}^*, \Sigma^*) \notin \mathcal{Q}$, *iii. Set* $\mathsf{U}^* = \{\mathsf{VK}^*_{i_j} \in \mathsf{R}^* : \mathsf{t}^*_{i_j} \leq \mathsf{t}^*_{\mathsf{V}}\}$, $|\mathcal{C} \cap \mathsf{U}^*| < \mathsf{t}^*_{\mathsf{V}}$

iii. $\mathsf{ATRS.Verify}(\mathsf{R}^*, \mathsf{OPK}^*, \Sigma^*, \mathsf{t}^*_{\mathsf{V}}, \mathsf{m}^*) = 1$.

Fifth, the *tracing soundness* ensures that an adversary cannot open two distinct subsets of signers from only one signature, even if the adversary can control an opener and corrupted signers. Our formal definition achieves strong tracing soundness as defined in [7], where all users of two distinct subsets can be corrupted.

Definition 12 (Tracing Soundness). *A threshold ring signature with accountability* ATRS *is tracing soundness if, for every PPT adversary* $\mathcal{A}$*, it holds that* $\mathcal{A}$ *has a negligible advantage in the following experiment.*
$\mathsf{Exp}_{\mathsf{ATRS\text{-}TrS}}$:

1. *The adversary generates* $(\mathsf{OPK}^*, \mathsf{OSK}^*) \leftarrow \mathsf{ATRS.OKeyGen}(1^\lambda, \mathsf{t_O})$ *and provides* OPK^* *to the experiment.*
2. *For all* $i = 0, \ldots, N$*, it generates* $(\mathsf{VK}_i, \mathsf{SK}_i) \leftarrow \mathsf{ATRS.UKeyGen}(1^\lambda, r_i)$ *by using random coins* r_i*. The experiment sets* $\mathsf{R} = (\mathsf{VK}_0, \ldots, \mathsf{VK}_N)$ *and provides* R, r_i *for all* $i \in [N]$ *to the adversary.*
3. *The adversary* $\mathcal{A}$ *outputs a tuple* $(\mathsf{R}, \mathsf{OPK}^*, \mathsf{m}^*, \Sigma^*, \mathsf{OS}_i, \mathsf{AUX}_i, \mathsf{OS}_j, \mathsf{AUX}_j, \mathsf{t}^*_{\mathsf{V}})$.
4. *The experiment outputs 1 if the following conditions hold:*
   i. $\mathsf{OS}_i, \mathsf{OS}_j \in \mathsf{R}$, $\mathsf{OS}_i \cap \mathsf{OS}_i = \emptyset$,
   ii. $\mathsf{ATRS.VerifyOpen}(\mathsf{R}^*, \mathsf{OPK}^*, \Sigma^*, \mathsf{t}^*_{\mathsf{V}}, \mathsf{m}^*, \mathsf{OS}_i, \mathsf{AUX}_i) = 1$.
   iii. $\mathsf{ATRS.VerifyOpen}(\mathsf{R}^*, \mathsf{OPK}^*, \Sigma^*, \mathsf{t}^*_{\mathsf{V}}, \mathsf{m}^*, \mathsf{OS}_j, \mathsf{AUX}_j) = 1$.

3.2 Our Threshold Ring Signatures with Accountability Construction

This section will present a generic construction of threshold ring signatures with accountability. Our construction employs the following technical building blocks.

- $\mathsf{VRF} = (\mathsf{VRF.KeyGen}, \mathsf{VRF.Sign}, \mathsf{VRF.Verfify})$ be a verifiable random function.
- $\mathsf{PKE} = (\mathsf{PKE.KeyGen}, \mathsf{PKE.Enc}, \mathsf{PKE.Dec})$ be a public encryption scheme.
- $\mathsf{CS} = (\mathsf{CS.Com}, \mathsf{CS.Verfify})$ be a commitment scheme.
- $\mathsf{TB\text{-}PKE} = (\mathsf{TB\text{-}PCE.KeyGen}, \mathsf{TB\text{-}PCE.Enc}, \mathsf{TB\text{-}PCE.Dec})$ be a threshold based pre-constrained encryption scheme.
- $\mathsf{SPB} = (\mathsf{SPB.KeyGen}, \mathsf{SPB.Hash}, \mathsf{SPB.Open}, \mathsf{SPB.Verify})$ be a somewhere perfectly binding hash function.
- F be a one-way function.
- $\mathsf{NIWI} = (\mathsf{NIWI.Prove}, \mathsf{NIWI.Verify})$ be a NIWI proof system for the language $\mathcal{L}$ defined as in (1).

Firstly, we define a relation $\mathcal{R}_0$ consisting of $(\mathbf{x}_0, \mathbf{w}_0)$, where

$$\mathbf{x}_0 = (\mathsf{m}, \mathsf{R}, \mathsf{OPK}, \mathsf{t}, \mathsf{v}_0, \mathsf{v}_1, \{\mathsf{oct}_k\}_{k\in[2]}, \mathsf{ct}_0, \mathsf{ct}_1, \mathsf{h}_0, \mathsf{hk}_0), \text{ where } \mathsf{OPK} = (\mathsf{ppk}, \mathsf{t_O})$$
$$\mathbf{w}_0 = (i, \mathsf{VK}_i, \mathsf{p}_0, \mathsf{p}_1, \tau_0, \mathsf{r}_0, \mathsf{r}_1, \{\mathsf{or}_k\}_{k\in[2]}), \text{ where } \mathsf{VK}_i = (\mathsf{vpk}_i, \mathsf{epk}_i, \mathsf{com}_i)$$

satisfying the following:

$$\begin{aligned}
&\mathsf{SPB.Verfify}(\mathsf{hk}_0, \mathsf{h}_0, i, \mathsf{VK}_i, \tau_0) = 1\\
&\text{and } \mathsf{TB\text{-}PCE.Enc}(\mathsf{ppk}, \mathsf{p}_0, \mathsf{t}, \mathsf{or}_0) = \mathsf{oct}_0 \text{ and } \mathsf{PKE.Enc}(\mathsf{epk}_i, \mathsf{p}_0, \mathsf{r}_0) = \mathsf{ct}_0\\
&\text{and } \mathsf{VRF.Verify}(\mathsf{vpk}_i, \mathsf{m}||\mathsf{R}, \mathsf{v}_0, \mathsf{p}_0) = 1 \text{ and } \mathsf{TB\text{-}PCE.Enc}(\mathsf{ppk}, \mathsf{VK}_i, \mathsf{t}, \mathsf{or}_1) = \mathsf{oct}_1\\
&\text{and } \mathsf{TB\text{-}PCE.Enc}(\mathsf{ppk}, \mathsf{p}_1, \mathsf{t}, \mathsf{or}_2) = \mathsf{oct}_2 \text{ and } \mathsf{PKE.Enc}(\mathsf{epk}_i, \mathsf{p}_1, \mathsf{r}_1) = \mathsf{ct}_1\\
&\text{and } \mathsf{VRF.Verify}(\mathsf{vpk}_i, \mathsf{t}||\mathsf{oct}_0||\mathsf{oct}_1||\mathsf{ct}_0||\mathsf{m}||\mathsf{R}, \mathsf{v}_1, \mathsf{p}_1) = 1.
\end{aligned}$$

The relation $\mathcal{R}_1$ is the same $\mathcal{R}_0$ except that it is true for index j and $j \neq i$.

Now, we define the relation $\mathcal{R}_2$ consisting of $(\mathbf{x}_2, \mathbf{w}_2)$, where

$$\begin{aligned}
\mathbf{x}_2 &= (\mathsf{R}, \mathsf{h}_0, \mathsf{h}_1, \mathsf{hk}_0, \mathsf{hk}_1)\\
\mathbf{w}_2 &= (i, j, \mathsf{VK}_i, \mathsf{VK}_j, \tau_0, \tau_1, \mathsf{nonce}_i, \mathsf{nonce}_j, u_i, u_j)
\end{aligned}$$

satisfying the following:

$$\begin{aligned}
&\mathsf{SPB.Verfify}(\mathsf{hk}_0, \mathsf{h}_0, i, \mathsf{VK}_i, \tau_0) = 1\\
&\text{and } \mathsf{SPB.Verfify}(\mathsf{hk}_1, \mathsf{h}_1, j, \mathsf{VK}_j, \tau_1) = 1\\
&\text{and } \mathsf{CS.Verify}(\mathsf{com}_i, \mathsf{nonce}_i, \mathsf{u}_i) = 1\\
&\text{and } \mathsf{CS.Verify}(\mathsf{com}_j, \mathsf{nonce}_j, \mathsf{u}_j) = 1\\
&\text{and } \mathsf{F}(\mathsf{nonce}_i) = \mathsf{nonce}_j.
\end{aligned}$$

Let $\mathcal{L}_0, \mathcal{L}_1, \mathcal{L}_2$ be languages accepted by $\mathcal{R}_0, \mathcal{R}_1, \mathcal{R}_3$, respectively. We define the language $\mathcal{L}$ as the following:

$$\mathcal{L} := \mathcal{L}_0 \vee \mathcal{L}_1 \vee \mathcal{L}_2 \tag{1}$$

The statement and witness for $\mathcal{L}$ are defined as the following:

$$\begin{aligned}
\mathbf{x} &= (\mathsf{m}, \mathsf{R}, \mathsf{OPK}, \mathsf{t}, \mathsf{v}_0, \mathsf{v}_1, \{\mathsf{oct}_k\}_{k\in[2]}, \mathsf{ct}_0, \mathsf{ct}_1, \mathsf{h}_0, \mathsf{hk}_0, \mathsf{h}_1, \mathsf{hk}_1),\\
\mathbf{w} &= (i, j, \mathsf{VK}_i, \mathsf{VK}_j, \mathsf{p}_0, \mathsf{p}_1, \tau_0, \tau_1, \mathsf{nonce}_i, \mathsf{nonce}_j, \mathsf{r}_0, \mathsf{r}_1, \{\mathsf{or}_k\}_{k\in[2]}, \mathsf{u}_i, \mathsf{u}_j)
\end{aligned}$$

Our threshold ring signature with accountability $\mathsf{ATRS} = (\mathsf{ATRS.OKeyGen}, \mathsf{ATRS.UKeyGen}, \mathsf{ATRS.Sign}, \mathsf{ATRS.Verify}, \mathsf{ATRS.Open}, \mathsf{ATRS.VerifyOpen})$ consists of the algorithms defined as the following:

ATRS.OKeyGen$(1^\lambda, \mathsf{t_O})$: On input a security parameter 1^λ and opener threshold $\mathsf{t_O}$, it works as follows.

- Compute $(\mathsf{ppk}, \mathsf{psk}) \leftarrow \mathsf{TB\text{-}PCE.KeyGen}(1^\lambda, C)$, where $C = \{0, 1, \ldots, \mathsf{t_O}\}$
- Set $\mathsf{OPK} := (\mathsf{ppk}, \mathsf{t_O})$ and $\mathsf{OSK} := (\mathsf{psk}, \mathsf{t_O})$
- Return $(\mathsf{OPK}, \mathsf{OSK})$.

ATRS.UKeyGen(1^λ) : On input a security parameter 1^λ, it does as follows.

- Compute $(\mathsf{vpk}, \mathsf{vsk}) \leftarrow \mathsf{VRF.KeyGen}(1^\lambda)$ and $(\mathsf{epk}, \mathsf{esk}) \leftarrow \mathsf{PKE.KeyGen}(1^\lambda)$.
- Choose $\mathsf{nonce} \leftarrow_\$ \{0,1\}^\lambda$, and compute $(\mathsf{com}, \mathsf{u}) \leftarrow \mathsf{CS.Com}(1^\lambda, \mathsf{nonce})$
- Set $\mathsf{VK} := (\mathsf{vpk}, \mathsf{epk}, \mathsf{com})$ and $\mathsf{SK} := (\mathsf{vsk}, \mathsf{esk}, \mathsf{nonce}, \mathsf{u})$.
- Return $(\mathsf{VK}, \mathsf{SK})$.

ATRS.Sign$(\mathsf{R}, \mathsf{OPK}, \mathsf{SSK}, \mathsf{T}, \mathsf{m})$: On input a ring $\mathsf{R} = (\mathsf{VK}_0, \mathsf{VK}_1, \ldots, \mathsf{VK}_N)$, a opener's public key OPK, a set of secret keys $\mathsf{SSK} = \{\mathsf{SK}_{i_j}\}_{j \in [L]}$, where $|\mathsf{SSK}| = L+1$ and $L < N$ corresponding to a subset of public keys $\{\mathsf{VK}_{i_j}\}_{j \in [L]} \subseteq \mathsf{R}$, a set of thresholds $\mathsf{T} = \{\mathsf{t}_{i_j}\}_{j \in [L]}$ and a message $\mathsf{m} \in \{0,1\}^*$, this algorithm processes as the following.
- To simplify notation, we sort $\{\mathsf{VK}_{i_0}, \ldots, \mathsf{VK}_{i_L}\} = \{\mathsf{VK}_i\}_{i \in [L]}$ by use t_i. Each signer $i \in [L]$ who has secret key SK_i and threshold t_i does as the following:

- Parse $\mathsf{VK}_i = (\mathsf{vpk}_i, \mathsf{epk}_i, \mathsf{com}_i)$, $\mathsf{SK}_i = (\mathsf{vsk}_i, \mathsf{esk}_i, \mathsf{nonce}_i, \mathsf{u}_i)$ and $\mathsf{OPK} = (\mathsf{ppk}, \mathsf{t_O})$.
- Compute $\mathsf{v}_{i,0} \leftarrow \mathsf{VRF.Eval}(\mathsf{vsk}_i, \mathsf{m}||\mathsf{R})$ and $\mathsf{p}_{i,0} \leftarrow \mathsf{VRF.Prove}(\mathsf{vsk}_i, \mathsf{m}||\mathsf{R})$.
- Compute $\mathsf{oct}_{i,0} \leftarrow \mathsf{TB\text{-}PCE.Enc}(\mathsf{ppk}, \mathsf{p}_{i,0}, \mathsf{t}_i, \mathsf{or}_{i,0})$, where $\mathsf{or}_{i,0}$ is a randomness.
- Compute $\mathsf{ct}_{i,0} \leftarrow \mathsf{PKE.Enc}(\mathsf{epk}_i, \mathsf{p}_{i,0}, \mathsf{r}_{i,0})$, $\mathsf{oct}_{i,1} \leftarrow \mathsf{TB\text{-}PCE.Enc}(\mathsf{ppk}, \mathsf{VK}_i, \mathsf{t}_i, \mathsf{or}_{i,1})$,
- Compute $\mathsf{v}_{i,1} \leftarrow \mathsf{VRF.Eval}(\mathsf{vsk}_i, \mathsf{t}_i||\mathsf{oct}_{i,0}||\mathsf{oct}_{i,1}||\mathsf{ct}_{i,0}||\mathsf{m}||\mathsf{R})$.
- Compute $\mathsf{p}_{i,1} \leftarrow \mathsf{VRF.Prove}(\mathsf{vsk}_i, \mathsf{t}_i||\mathsf{oct}_{i,0}||\mathsf{oct}_{i,1}||\mathsf{ct}_{i,0}||\mathsf{m}||\mathsf{R})$.
- Compute $\mathsf{oct}_{i,2} \leftarrow \mathsf{TB\text{-}PCE.Enc}(\mathsf{ppk}, \mathsf{p}_{i,1}, \mathsf{t}_i, \mathsf{or}_{i,2})$,
- Compute $\mathsf{ct}_{i,1} \leftarrow \mathsf{PKE.Enc}(\mathsf{epk}_i, \mathsf{p}_{i,1}, \mathsf{r}_{i,1})$
- Compute $(\mathsf{hk}_{i,0}, \mathsf{skh}_{i,0}) \leftarrow \mathsf{SPB.KeyGen}(1^\lambda, |\mathsf{R}|, i)$.
- Compute $\mathsf{h}_{i,0} \leftarrow \mathsf{SPB.Hash}(\mathsf{hk}_{i,0}, \mathsf{R})$, $\tau_{i,0} \leftarrow \mathsf{SPB.Open}(\mathsf{hk}_{i,0}, \mathsf{shk}_{i,0}, \mathsf{R}, i)$.
- Choose uniformly at random index $j \in [L]$ such that $j \neq i$.
- Choose uniformly at random $\mathsf{u}_j, \mathsf{nonce}_j$.
- Compute $(\mathsf{hk}_{i,1}, \mathsf{skh}_{i,1}) \leftarrow \mathsf{SPB.KeyGen}(1^\lambda, |\mathsf{R}|, j)$.
- Compute $\mathsf{h}_{i,1} \leftarrow \mathsf{SPB.Hash}(\mathsf{hk}_{i,1}, \mathsf{R})$, $\tau_{i,1} \leftarrow \mathsf{SPB.Open}(\mathsf{hk}_{i,1}, \mathsf{shk}_{i,1}, \mathsf{R}, j)$.
- Set $\mathbf{x}_i = (\mathsf{m}, \mathsf{R}, \mathsf{OPK}, \mathsf{t}_i, \mathsf{v}_{i,0}, \mathsf{v}_{i,1}, \{\mathsf{oct}_{i,k}\}_{k \in [2]}, \mathsf{ct}_{i,0}, \mathsf{ct}_{i,1}, \mathsf{h}_{i,1}, \mathsf{h}_{i,0}, \mathsf{hk}_{i,0}, \mathsf{hk}_{i,1})$.
- Set $\mathbf{w}_i = (i, j, \mathsf{VK}_i, \mathsf{VK}_j, \mathsf{p}_{i,0}, \mathsf{p}_{i,1}, \tau_{i,0}, \tau_{i,1}, \mathsf{nonce}_i, \mathsf{nonce}_j, \mathsf{r}_{i,0}, \mathsf{r}_{i,1}, \{\mathsf{or}_{i,k}\}_{k \in [2]}, \mathsf{u}_i, \mathsf{u}_j)$ and compute $\pi_i \leftarrow \mathsf{NIWI.Prove}(\mathbf{x}_i, \mathbf{w}_i)$.
- Broadcast $\Sigma_i = (\mathsf{v}_{i,0}, \mathsf{v}_{i,1}, \{\mathsf{oct}_{i,k}\}_{k \in [2]}, \mathsf{ct}_{i,0}, \mathsf{ct}_{i,1}, \mathsf{hk}_{i,0}, \mathsf{hk}_{i,1}, \pi_i, \mathsf{t}_i)$.

- Return $\Sigma = \{\Sigma_i\}_{i \in [L]}$

ATRS.Verify$(\mathsf{R}, \mathsf{OPK}, \Sigma, \mathsf{t_V}, \mathsf{m})$: On input a ring R, an opener's public key OPK, a signature Σ, a threshold verification $\mathsf{t_V}$, and a message m do the following.

- Parse $\mathsf{OPK} = (\mathsf{ppk}, \mathsf{t_O})$. Sort list $\Sigma = \{\Sigma_i\}_{i \in L}$ by t_i , where $\Sigma_i = (\mathsf{v}_{i,0}, \mathsf{v}_{i,1}, \{\mathsf{oct}_{i,k}\}_{k \in [2]}, \mathsf{ct}_{i,0}, \mathsf{ct}_{i,1}, \mathsf{hk}_{i,0}, \mathsf{hk}_{i,1}, \pi_i, \mathsf{t}_i)$
- Let $\mathsf{VTS} = \emptyset$ be a verification threshold set.
- For $i \in [L]$, do as the following:
  - Parse $\Sigma_i = (\mathsf{v}_{i,0}, \mathsf{v}_{i,1}, \{\mathsf{oct}_{i,k}\}_{k \in [2]}, \mathsf{ct}_{i,0}, \mathsf{ct}_{i,1}, \mathsf{hk}_{i,0}, \mathsf{hk}_{i,1}, \pi_i, \mathsf{t}_i)$.
  - Compute $\mathsf{h}'_{i,0} = \mathsf{SPB.Hash}(\mathsf{hk}_{i,0}, \mathsf{R})$ and $\mathsf{h}'_{i,1} = \mathsf{SPB.Hash}(\mathsf{hk}_{i,1}, \mathsf{R})$.
  - Set $\mathbf{x}_i = (\mathsf{m}, \mathsf{OPK}, \mathsf{v}_{i,0}, \mathsf{v}_{i,1}, \{\mathsf{oct}_{i,k}\}_{k \in [2]}, \mathsf{ct}_{i,0}, \mathsf{ct}_{i,1}, \mathsf{h}'_{i,1}, \mathsf{h}'_{i,0}, \mathsf{hk}_{i,0}, \mathsf{hk}_{i,1})$.
  - If $\mathsf{NIWI.Verify}(\mathbf{x}_i, \pi_i) = 1$ and $\mathsf{v}_{i,0} \neq \mathsf{v}_{j,0}, \forall j \in [i-1]$ then append t_i to VTS.
- If there exists an index $i \in [|\mathsf{VTS}|]$ such that $i \geq \mathsf{t_V}$ and $\mathsf{VTS}[i] \leq \mathsf{t_V}$ then return 1. Otherwise, return 0.

ATRS.Open$(\mathsf{R}, \mathsf{OSK}, \Sigma, \mathsf{t_V}, \mathsf{m})$: On input ring R, an opener's secret key OSK, a signature Σ, a threshold verification $\mathsf{t_V}$, and a message m, do the following.

- If $\mathsf{ATRS.Verify}(\mathsf{R}, \mathsf{OPK}, \Sigma, \mathsf{t_V}, \mathsf{m}) = 0$, then return $\perp$.
- Parse $\mathsf{OSK} = (\mathsf{psk}, \mathsf{t_O})$. Sort list $\Sigma = \{\Sigma_i\}_{i \in L}$ by t_i , where $\Sigma_i = (\mathsf{v}_{i,0}, \mathsf{v}_{i,1}, \{\mathsf{oct}_{i,k}\}_{k\in[2]}, \mathsf{ct}_{i,0}, \mathsf{ct}_{i,1}, \mathsf{hk}_{i,0}, \mathsf{hk}_{i,1}, \pi_i, \mathsf{t}_i)$
- Set $\mathsf{OS} = \emptyset$ and $\mathsf{AUX} = \emptyset$
- For $i \in [L]$ and $\mathsf{t}_i \le \mathsf{t_O}$, do as the following:
  - Parse $\Sigma_i = (\mathsf{v}_{i,0}, \mathsf{v}_{i,1}, \{\mathsf{oct}_{i,k}\}_{k\in[2]}, \mathsf{ct}_{i,0}, \mathsf{ct}_{i,1}, \mathsf{hk}_{i,0}, \mathsf{hk}_{i,1}, \pi_i, \mathsf{t}_i)$
  - Compute $\mathsf{VK}'_i = \mathsf{TB\text{-}PCE.Dec}(\mathsf{psk}, \mathsf{oct}_{i,1})$
  - Compute $\mathsf{p}'_{i,0} = \mathsf{TB\text{-}PCE.Dec}(\mathsf{psk}, \mathsf{oct}_{i,0})$, $\mathsf{p}'_{i,1} = \mathsf{TB\text{-}PCE.Dec}(\mathsf{psk}, \mathsf{oct}_{i,2})$.
  - Set $\mathsf{AUX}_i = (\mathsf{p}'_{i,0}, \mathsf{p}'_{i,1})$.
  - Add VK'_i to OS and AUX_i to AUX.
- Return OS and AUX.

ATRS.VerifyOpen$(\mathsf{R}, \mathsf{OPK}, \Sigma, \mathsf{t_V}, \mathsf{m}, \mathsf{OS}, \mathsf{AUX})$: On input ring R, an opener's public key OPK, a signature Σ, a message m, verification threshold $\mathsf{t_V}$, a set of signer public key OS, and a set of auxiliary information AUX, this algorithm works as the following.

- If $\mathsf{ATRS.Verify}(\mathsf{R}, \mathsf{OPK}, \Sigma, \mathsf{t_V}, \mathsf{m}) = 0$, then return 0.
- For $i \in |\mathsf{OS}|$, do as the following:
  - Parse $\Sigma_i = (\mathsf{v}_{i,0}, \mathsf{v}_{i,1}, \{\mathsf{oct}_{i,k}\}_{k\in[2]}, \mathsf{ct}_{i,0}, \mathsf{ct}_{i,1}, \mathsf{hk}_{i,0}, \mathsf{hk}_{i,1}, \pi_i, \mathsf{t}_i)$
  - Parse $\mathsf{VK}'_i = (\mathsf{vpk}'_i, \mathsf{epk}'_i, \mathsf{com}'_i)$ and $\mathsf{AUX}_i = (\mathsf{p}'_{i,0}, \mathsf{p}'_{i,1})$.
  - If $\mathsf{VRF.Verify}(\mathsf{vpk}'_i, \mathsf{m}||\mathsf{R}, \mathsf{v}_{i,0}, \mathsf{p}'_{i,0}) \neq 1$, then return 0.
  - If $\mathsf{VRF.Verify}(\mathsf{vpk}'_i, \mathsf{t}_i||\mathsf{oct}_{i,0}||\mathsf{oct}_{i,1}||\mathsf{ct}_{i,0}||\mathsf{m}||\mathsf{R}, \mathsf{v}_{i,1}, \mathsf{p}'_{i,1}) \neq 1$, then return 0.
- Return 1.

Theorem 1 (Correctness and Threshold Traceability). *If the* NIWI, VRF, PKE, TB-PCE, CS *and* SPB *schemes are correct and the* VRF *is key collision free then* ATRS *is correct and threshold traceable. Moreover, our* ATRS *scheme has a compact signature size.*

Proof. Correctness: Suppose that $(\mathsf{OPK}, \mathsf{OSK}) \leftarrow \mathsf{ATRS.OKeyGen}(1^\lambda, \mathsf{t_O})$ and for all $i \in [N]$, $\mathsf{VK}_i = (\mathsf{vpk}_i, \mathsf{epk}_i, \mathsf{com}_i)$, $\mathsf{SK}_i = (\mathsf{vsk}_i, \mathsf{esk}_i, \mathsf{nonce}_i, \mathsf{u}_i)$ were generated by $\mathsf{ATRS.UKeyGen}(1^\lambda)$ algorithm. The signature $\Sigma = \{\Sigma_i\}_{i\in[L]}$, sorted by t_i, is an output of $\mathsf{ATRS.Sign}(\mathsf{R}, \mathsf{OPK}, \mathsf{SSK}, \mathsf{T}, \mathsf{m})$ algorithm, where $\mathsf{R} = \{\mathsf{VK}_i\}_{i\in[N]}$, $L < N$, $\{\mathsf{SSK}_i\}_{i\in[L]}$, $\mathsf{T} = \{\mathsf{t}_i\}_{i\in[L]}$ and $\Sigma_i = (\mathsf{v}_{i,0}, \mathsf{v}_{i,1}, \{\mathsf{oct}_{i,k}\}_{k\in[2]}, \mathsf{ct}_{i,0}, \mathsf{ct}_{i,1}, \mathsf{hk}_{i,0}, \mathsf{hk}_{i,1}, \pi_i, \mathsf{t}_i)$.

For each signer $i \in [L]$, since $\mathsf{SPB.Hash}$ is a deterministic algorithm, we have $\mathsf{h}'_{i,0} = \mathsf{h}_{i,0}$ and $h'_{i,1} = \mathsf{h}_{i,1}$. It also holds that one index corresponds to VK_i and the other to VK_j. By the correctness of the SPB function, we have $\mathsf{SPB.Verfify}(\mathsf{hk}_{i,0}, \mathsf{h}'_{i,0}, i, \mathsf{VK}_i, \tau_{i,0}) = 1$.

Since PKE is correctness, $\mathsf{PKE.Enc}(\mathsf{epk}_i, \mathsf{p}_{i,0}, \mathsf{r}_{i,0}) = \mathsf{ct}_{i,0}$, and $\mathsf{PKE.Enc}(\mathsf{epk}_i, \mathsf{p}_{i,1}, \mathsf{r}_{i,1}) = \mathsf{ct}_{i,1}$. By the correctness of TB-PCE, we have $\mathsf{TB\text{-}PCE.Enc}(\mathsf{ppk}, \mathsf{p}_{i,0}, \mathsf{t}_i, \mathsf{or}_{i,0}) = \mathsf{oct}_{i,0}$, $\mathsf{TB\text{-}PCE.Enc}(\mathsf{ppk}, \mathsf{VK}_i, \mathsf{t}_i, \mathsf{or}_{i,1}) = \mathsf{oct}_{i,1}$, $\mathsf{TB\text{-}PCE.Enc}(\mathsf{ppk},$

$\mathsf{p}_{i,1}, \mathsf{t}_i, \mathsf{or}_{i,2}) = \mathsf{oct}_{i,2}$. Since VRF is correctness, $\mathsf{VRF.Verify}(\mathsf{vpk}_i, \mathsf{m}||\mathsf{R}, \mathsf{v}_{i,0}, \mathsf{p}_{i,0}) =$ 1 and $\mathsf{VRF.Verify}(\mathsf{vpk}_i, \mathsf{t}||\mathsf{oct}_{i,0}||\mathsf{oct}_{i,1}||\mathsf{ct}_{i,0}\ ||\mathsf{m}||\mathsf{R}, \mathsf{v}_{i,1}, \mathsf{p}_{i,1}) = 1$.

Therefore, $(\mathbf{x}_i, \mathbf{w}_i) \in \mathcal{L}$, where $\mathbf{x}_i = (\mathsf{m}, \mathsf{R}, \mathsf{OPK}, \mathsf{t}_i, \mathsf{v}_{i,0}, \mathsf{v}_{i,1}, \{\mathsf{oct}_{i,k}\}_{k\in[2]}, \mathsf{ct}_{i,0}, \mathsf{ct}_{i,1}, \mathsf{h}_{i,1}, \mathsf{h}_{i,0}, \mathsf{hk}_{i,0}, \mathsf{hk}_{i,1})$, and $\mathbf{w}_i = (i, j, \mathsf{VK}_i, \mathsf{VK}_j, \mathsf{p}_{i,0}, \mathsf{p}_{i,1}, \tau_{i,0}, \tau_{i,1}, \mathsf{nonce}_i, \mathsf{nonce}_j, \mathsf{r}_{i,0}, \mathsf{r}_{i,1}, \{\mathsf{or}_{i,k}\}_{k\in[2]}, \mathsf{u}_i, \mathsf{u}_j)$. By the correctness of NIWI scheme, it holds that $\mathsf{NIWI.Verify}(x_i, \pi) = 1$.

Since VRF is key collision free, we have $\mathsf{v}_{i,0} \neq \mathsf{v}_{j,0}$ for all $i \neq j \in [L]$. The ordered set VTS contains all individual thresholds of signers who pass the verification algorithm in the NIWI scheme. It is a valid signature if there exists at least $\mathsf{t_V}$ signer and each signer has an individual threshold less than or equal to $\mathsf{t_V}$. Namely, if there exits an index $i \in [|\mathsf{VTS}|]$ such that $i \geq \mathsf{t_V}$ and $\mathsf{VTS}[i] \leq \mathsf{t}_V$. Note that, since VTS is an ordered set, we have $\mathsf{VTS}[j] \leq \mathsf{t_V}$ for all $j \leq i$. Therefore, our ARS scheme is *correct.*

Threshold traceability: By the correctness of TB-PCE, the opener can only decrypt the ciphertexts that are encrypted by individual thresholds less than or equal to the opener threshold. Namely, for each index $i \in [L]$ satisfies $\mathsf{t}_i \leq \mathsf{t_O}$, we have $\mathsf{VK}'_i = \mathsf{TB\text{-}PCE.Dec}(\mathsf{psk}, \mathsf{oct}_{i,1})$, $\mathsf{p}'_{i,0} = \mathsf{TB\text{-}PCE.Dec}(\mathsf{psk}, \mathsf{oct}_{i,0})$ and $\mathsf{p}'_{i,1} = \mathsf{TB\text{-}PCE.Dec}(\mathsf{psk}, \mathsf{oct}_{i,2})$. By the correctness of VRF, we have $\mathsf{VRF.Verify}(\mathsf{vpk}'_i, \mathsf{m}||\mathsf{R}, \mathsf{v}_{i,0}, \mathsf{p}'_{i,0})$ and $\mathsf{VRF.Verify}(\mathsf{vpk}'_i, \mathsf{t}_i||\mathsf{oct}_{i,0}||oct_{i,1}||\mathsf{ct}_{i,0}||\mathsf{m}||\mathsf{R},) = 1$. Hence, our ARS scheme is traceable.

Signature size: Let consider a signature $\Sigma = \{\Sigma_i\}_{i\in L}$, where $\Sigma_i = (\mathsf{v}_{i,0}, \mathsf{v}_{i,1}, \{\mathsf{oct}_{i,k}\}_{k\in[2]}, \mathsf{ct}_{i,0}, \mathsf{ct}_{i,1}, \mathsf{hk}_{i,0}, \mathsf{hk}_{i,1}, \pi, \mathsf{t}_i)$. The size of elements $\mathsf{v}_{i,0}, \mathsf{v}_{i,1}, \{\mathsf{oct}_{i,k}\}_{k\in[2]}, \mathsf{ct}_{i,0}, \mathsf{ct}_{i,1}, \mathsf{t}_i)$ are $\mathcal{O}(\mathsf{poly}(\lambda))$ and independent of the number of users N. By the efficiency property of SPB, the sizes of the hashing keys $\mathsf{hk}_0, \mathsf{hk}_1$ are bounded by $\mathcal{O}(\log(N) \cdot \mathsf{poly}(\lambda))$. By the proof-size property of NIWI proof, it holds that the size of proof π is $\mathcal{O}(\log(N) \cdot \mathsf{poly}(\lambda))$. Therefore, the size of signatures Σ is $L \cdot \mathcal{O}(\log(N) \cdot \mathsf{poly}(\lambda))$. The output of the opener includes $(\mathsf{OS}, \mathsf{AUX})$, which also is linear with a number of subset signers.

4 Security Proofs

In the anonymity experiment, the challenger chooses two honestly generated verification keys S_0 and S_1. The adversary can add new public keys to obtain a ring of signers R, where $\mathsf{S}_0, \mathsf{S}_1 \subset \mathsf{R}$. The challenger picks one of the signing sets S_b and computes a signature Σ. Based on signature Σ, the adversary guesses which of S_0, S_1 signed the message. In our construction, the signature Σ is a collection of L individual signatures. Moreover, they are independent of each other. Hence, it suffices to show anonymity for a single signer ($|\mathsf{S}_0| = |\mathsf{S}_1| = 1$) and one can use a hybrid argument to show anonymity for larger sets. If the probability of distinguishing between two signatures is negligible, then distinguishing between two sets of signatures also is negligible. To prove anonymity, we will build a sequence of hybrids to prove that a signature created under VK_{i_0} is computationally indistinguishable from a signature created under VK_{i_1}, where i_0, i_1 are some indexes in $[N]$.

Theorem 2. *If* NIWI *is computationally witness-indistinguishable,* PKE *has key privacy,* CS *is computational hiding,* TB-PCE *is security against semi-honest authority,* VRF *has key privacy,* F *is one-way function,* SPB *is index hiding, then* ATRS *is anonymous.*

Proof. The proof is presented in a full version of the paper. □

To prove unforgeability, we must show that an adversary $\mathcal{A}$ who knows up to t secret keys cannot forge a signature that verifies for $\mathsf{t}+1$ signers, even if the adversary controls an opener. Because our signature Σ is a collection of L, individual signatures are independent. Hence, it suffices to show unforgeability for a single signer with threshold $\mathsf{t}_i = \mathsf{t}_\mathsf{V} = 1$. Informal speaking, to generate a valid message-signature pair, $\mathcal{A}$ needs to provide a valid NIWI proof in the signature. Because NIWI is perfect soundness, the claimed statement must be true. Since t secret keys are generated by the challenger, F is a one-way function, the probability that there exists a pair nonce_i and nonce_j such that $\mathsf{F}(\mathsf{nonce}_i) = \mathsf{nonce}_j$ is negligible. Even if $\mathcal{A}$ can know t secret key, $\mathcal{A}$ is impossible to find a pair nonce_i and nonce_j in relation $\mathcal{R}_2$.

Now, the adversary $\mathcal{A}$ must find another strategy. Namely, $\mathcal{A}$ needs to know a witness of one in two first relations. At a high level, we can see that an adversary has only two ways. First, the adversary can give a valid forgery signature by using useful information in the opening algorithm and aiming to output a valid forgery signature. Second, the adversary can give a valid signature for any message without using information from the opening algorithm. In the first method, since PKE has perfect correctness and perfect binding, we will prove that the probability that the adversary wins is negligible. In the latter case, by the residual unpredictability of VRF, the adversary also has negligible success. In other words, we have a reduction to the security of VRF. This reduction receives a verification key vpk_{i^*} from the challenger, the residual unpredictability of VRF scheme. Then, it generates vpk_i for all $i \neq i^*$ and $\mathsf{epk}_i, \mathsf{com}_i$ for all $i \in [N]$. Upon receiving a signature Σ^* from the adversary, the reduction decrypts both ct_0^* and ct_1^*, yielding the output of residual unpredictability experiment. Due to the somewhere perfect binding of SPB hashing and the perfect soundness of the NIWI scheme, the reduction produces a valid breaking with a probability non-negligible.

Theorem 3. *If* NIWI *has perfect soundness,* SPB *is somewhere perfectly binding,* VRF *has residual unpredictability and unique provability,* PKE *has perfect correctness and perfect binding,* F *is one-way function then* ATRS *is unforgeable.*

Proof. The proof is presented in a full version of the paper. □

The same argument is in the properties mentioned above since our signature Σ is a non-interactive protocol that collects independent individual signatures. Thus, it suffices to show tracing soundness for a single signer with $\mathsf{t}_\mathsf{V}^* = \mathsf{t}_i = 1$. We show that if the adversary can be able to open two different subsets of signers with only one signature, then we can break key collision resistance properties of VRF scheme.

Theorem 4. *If* VRF *has key collision resistance secure and* F *is a one-way function, then* ATRS *is tracing soundness.*

Proof. The proof is presented in a full version of the paper. □

5 Conclusion and Future Work

A threshold ring signature is a well-studied cryptographic primitive with many applications. An ATRS is a stronger cryptographic primitive than a TRS while retaining the TRS's flexibility. It strikes a balance between anonymity and accountability. We provide a formal definition of ATRS and its security requirements. A generic construction is provided to demonstrate the feasibility of designing ATRS in a modular manner from commonly used cryptographic building blocks.

As a future work, we aim to construct efficient concrete instantiations of ATRS in the plain model. Since our construction is generic based on several underlying primitives, a concrete instantiation may request several choices and possible primitive optimisations. Another direction for our future research is to expand the opener set. Specifically, each signer can choose an opener that he trusts for himself instead of every signer having to select a common opener. This will further enhance user privacy.

References

1. Ananth, P., Jain, A., Jin, Z., Malavolta, G.: Pre-constrained encryption. In: 13th Innovations in Theoretical Computer Science Conference (ITCS 2022). Schloss Dagstuhl-Leibniz-Zentrum für Informatik (2022)
2. Backes, M., Döttling, N., Hanzlik, L., Kluczniak, K., Schneider, J.: Ring signatures: logarithmic-size, no setup—from standard assumptions. In: Ishai, Y., Rijmen, V. (eds.) EUROCRYPT 2019. LNCS, vol. 11478, pp. 281–311. Springer, Cham (2019). https://doi.org/10.1007/978-3-030-17659-4_10
3. Badrinarayanan, S., Fernando, R., Jain, A., Khurana, D., Sahai, A.: Statistical ZAP arguments. In: Canteaut, A., Ishai, Y. (eds.) EUROCRYPT 2020. LNCS, vol. 12107, pp. 642–667. Springer, Cham (2020). https://doi.org/10.1007/978-3-030-45727-3_22
4. Bartusek, J., Garg, S., Jain, A., Policharla, G.V.: End-to-end secure messaging with traceability only for illegal content. In: Hazay, C., Stam, M. (eds.) EUROCRYPT 2023. LNCS, vol. 14008, pp. 35–66. Springer, Cham (2023). https://doi.org/10.1007/978-3-031-30589-4_2
5. Bender, A., Katz, J., Morselli, R.: Ring signatures: stronger definitions, and constructions without random oracles. In: Halevi, S., Rabin, T. (eds.) TCC 2006. LNCS, vol. 3876, pp. 60–79. Springer, Heidelberg (2006). https://doi.org/10.1007/11681878_4
6. Boneh, D., and C. Komlo. Threshold signatures with private accountability. In Annual International Cryptology Conference, pp. 551–581. Springer, Cham (2022). https://doi.org/10.1007/978-3-031-15985-5_19

7. Bootle, J., Cerulli, A., Chaidos, P., Ghadafi, E., Groth, J., Petit, C.: Short accountable ring signatures based on DDH. In: Pernul, G., Ryan, P.Y.A., Weippl, E. (eds.) ESORICS 2015. LNCS, vol. 9326, pp. 243–265. Springer, Cham (2015). https://doi.org/10.1007/978-3-319-24174-6_13
8. Branco, P., Döttling, N., Wohnig, S.: Universal ring signatures in the standard model. In: Agrawal, S., Lin, D. (eds.) ASIACRYPT 2022. LNCS, vol. 13794, pp. 689–718. Springer, Cham. doi:https://doi.org/10.1007/978-3-031-22972-5_9
9. Branco, P., Mateus, P.: A traceable ring signature scheme based on coding theory. In: Ding, J., Steinwandt, R. (eds.) PQCrypto 2019. LNCS, vol. 11505, pp. 387–403. Springer, Cham (2019). https://doi.org/10.1007/978-3-030-25510-7_21
10. Bresson, E., Stern, J., Szydlo, M.: Threshold ring signatures and applications to ad-hoc groups. In: Yung, M. (ed.) CRYPTO 2002. LNCS, vol. 2442, pp. 465–480. Springer, Heidelberg (2002). https://doi.org/10.1007/3-540-45708-9_30
11. Bultel, X., Fraser, A., Quaglia, E.A.: Improving the efficiency of report and trace ring signatures. In: International Symposium on Stabilizing, Safety, and Security of Distributed Systems, pp. 130–145. Springer (2022). https://doi.org/10.1007/978-3-031-21017-4_9
12. Bultel, X., Lafourcade, P.: k-times full traceable ring signature. In: 2016 11th International Conference on Availability, Reliability and Security (ARES), pp. 39–48. IEEE (2016)
13. Chatterjee, R., Garg, S., Hajiabadi, M., Khurana, D., Liang, X., Malavolta, G., Pandey, O., Shiehian, S.: Compact ring signatures from learning with errors. In: Malkin, T., Peikert, C. (eds.) CRYPTO 2021. LNCS, vol. 12825, pp. 282–312. Springer, Cham (2021). https://doi.org/10.1007/978-3-030-84242-0_11
14. Chaum, D., van Heyst, E.: Group signatures. In: Davies, D.W. (ed.) EUROCRYPT 1991. LNCS, vol. 547, pp. 257–265. Springer, Heidelberg (1991). https://doi.org/10.1007/3-540-46416-6_22
15. ElGamal, T.: A public key cryptosystem and a signature scheme based on discrete logarithms. IEEE Trans. Inf. Theory **31**(4), 469–472 (1985)
16. Fujisaki, E.: Sub-linear size traceable ring signatures without random oracles. In: Kiayias, A. (ed.) CT-RSA 2011. LNCS, vol. 6558, pp. 393–415. Springer, Heidelberg (2011). https://doi.org/10.1007/978-3-642-19074-2_25
17. Fujisaki, E., Suzuki, K.: Traceable ring signature. In: Okamoto, T., Wang, X. (eds.) PKC 2007. LNCS, vol. 4450, pp. 181–200. Springer, Heidelberg (2007). https://doi.org/10.1007/978-3-540-71677-8_13
18. Groth, J., Ostrovsky, R., Sahai, A.: Non-interactive zaps and new techniques for NIZK. In: Dwork, C. (ed.) CRYPTO 2006. LNCS, vol. 4117, pp. 97–111. Springer, Heidelberg (2006). https://doi.org/10.1007/11818175_6
19. Haque, A., Krenn, S., Slamanig, D., Striecks, C.: Logarithmic-size (linkable) threshold ring signatures in the plain model. In: Public-Key Cryptography–PKC 2022: 25th IACR International Conference on Practice and Theory of Public-Key Cryptography, Virtual Event, March 8–11, 2022, Proceedings, Part II, pp. 437–467. Springer (2022). https://doi.org/10.1007/978-3-030-97131-1_15
20. Libert, B., Nguyen, K., Peters, T., Yung, M.: Bifurcated signatures: folding the accountability vs. anonymity dilemma into a single private signing scheme. In: Canteaut, A., Standaert, F.-X. (eds.) EUROCRYPT 2021. LNCS, vol. 12698, pp. 521–552. Springer, Cham (2021). https://doi.org/10.1007/978-3-030-77883-5_18
21. Nguyen, K., Guo, F., Susilo, W., Yang, G.: Multimodal private signatures. In: Annual International Cryptology Conference, pp. 792–822. Springer (2022). doi:https://doi.org/10.1007/978-3-031-15979-4_27

22. Noether, S.: Ring signature condential transactions for monero. Cryptology ePrint Archive, Report 2015/1098 (2015)
23. Okamoto, T., Pietrzak, K., Waters, B., Wichs, D.: New realizations of somewhere statistically binding hashing and positional accumulators. In: Iwata, T., Cheon, J.H. (eds.) ASIACRYPT 2015. LNCS, vol. 9452, pp. 121–145. Springer, Heidelberg (2015). https://doi.org/10.1007/978-3-662-48797-6_6
24. Rivest, R.L., Shamir, A., Tauman, Y.: How to leak a secret. In: Boyd, C. (ed.) ASIACRYPT 2001. LNCS, vol. 2248, pp. 552–565. Springer, Heidelberg (2001). https://doi.org/10.1007/3-540-45682-1_32
25. Xu, S., Yung, M.: Accountable ring signatures: a smart card approach. In: Quisquater, J.-J., Paradinas, P., Deswarte, Y., El Kalam, A.A. (eds.) CARDIS 2004. IIFIP, vol. 153, pp. 271–286. Springer, Boston, MA (2004). https://doi.org/10.1007/1-4020-8147-2_18

Threshold Signatures with Private Accountability via Secretly Designated Witnesses

Meng Li[1,2,3], Hanni Ding[1,2,3], Qing Wang[1,2,3], Zijian Zhang[4](✉), and Mauro Conti[5,6]

[1] Key Laboratory of Knowledge Engineering with Big Data, Hefei University of Technology,Hefei, China
mengli@hfut.edu.cn, {hanniding,qingwang}@mail.hfut.edu.cn
[2] Ministry of Education; School of Computer Science and Information Engineering, Hefei University of Technology,Hefei, China
[3] Anhui Province Key Laboratory of Industry Safety and Emergency Technology; and Intelligent Interconnected Systems Laboratory of Anhui Province, Hefei University of Technology,Hefei, China
[4] School of Cyberspace Science and Technology, Beijing Institute of Technology, Beijing, China
zhangzijian@bit.edu.cn
[5] Department of Mathematics and HIT Center, University of Padua, Padua, Italy
mauro.conti@math.unipd.it
[6] Department of Intelligent Systems, CyberSecurity Group, Delft University of Technology, Delft, The Netherlands

Abstract. Threshold signature is a powerful cryptographic technique with a large number of real-life applications. As designed by Boneh and Komlo (CRYPTO'22), TAPS is a new threshold signature integrating privacy and accountability. It allows a combiner to combine t signature shares while protecting t and the signing group from the public. It also enables a tracer to trace a threshold signature to its original signing group. Despite being valuable, TAPS neglects the witnessing of tracing, i.e., leaves the tracing activity unrestrained.

In this paper, we introduce Accountable and Private Threshold Signature with Hidden Witnesses (HiTAPS) that not only provides privacy and accountability, but also incorporates witnessed tracing. In specific, we first utilize Dynamic Threshold Public-Key Encryption (DTPKE) and ElGamal encryption to designate a set of t' witnesses for endorsing the tracing activity. We then compute a keyed-hash tag for the t' witnesses to initiate the tracing activity secretly. Moreover, we present an optimized protocol HiTAPS2 to reduce communication overhead of the combiner. We formalize the definitions, security, and privacy for HiTAPS. We formally prove its security and privacy. To evaluate the performance of HiTAPS and HiTAPS2, we build a prototype based on pypbc. Experimental results show that HiTAPS takes 217 (370) ms to combine (track) a threshold signature of 5 signers (witnesses). The optimized HiTAPS2 only takes 137 ms to combine a threshold signature of 5 signers.

Keywords: Threshold Signatures · Privacy · Accountability · Witness

T. Zhu and Y. Li (Eds.): ACISP 2024, LNCS 14895, pp. 389–407, 2024.
https://doi.org/10.1007/978-981-97-5025-2_20

1 Introduction

1.1 Background

A Threshold Signature Scheme (TSS) [1] allows a set of n people to sign a message via a combiner when no less than t people join in the signing process. The system model is sketched in Fig. 1. Among its multiple variants, Private Threshold Signature (PTS) [2–4] and Accountable Threshold Signature (ATS) [5–8] stand out. A PTS signature σ on a message m tells nothing about the group of t signers who generated σ, which is useful since security and privacy are increasingly gaining importance [9,10]. An ATS signature σ on a message m can disclose the identity of all t people who co-generate σ via a tracer. For this reason, ATS is also considered as traceable secret sharing [11]. ATS can be used in real-world applications where accountability is required. For instance, if five of nine cooperative manufacturers prepare to authorize a product transfer, and all of them expect accountability in case a fraudulent transfer is consented. By using an ATS scheme, a Threshold Signature (TS) on a fraudulent transfer can disclose the five manufacturers who approved of it.

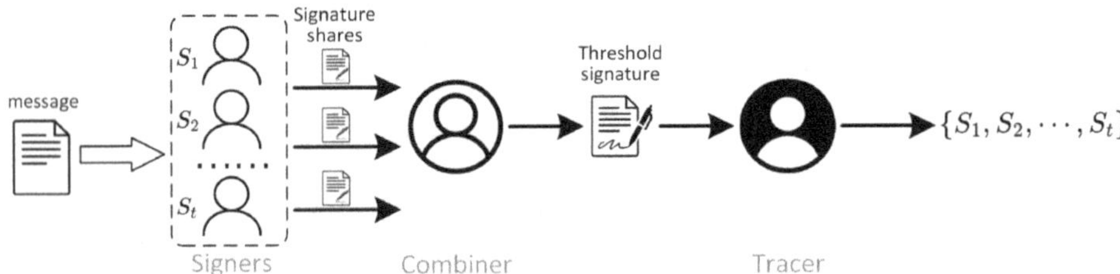

Fig. 1. System Model of ATS.

1.2 Motivation and Goal

Recently, Boneh and Komlo proposed a Threshold, Accountable, and Private Signature (TAPS) [12] to achieve both accountability and privacy for threshold signatures. It works as below: It generates a public key pk and n private keys $\{sk_1, sk_2, \cdots, sk_n\}$ for n signers; t signers $\mathcal{S} = \{S_1, S_2, \cdots, S_t\}$ generate t signature shares for a combiner with a combining key ck to generate a complete signature; and a tracer with a tracing key tk can identify the signing group of a complete signature. Here, accountability guarantees that a TS that is related to a misbehavior can be traced to its t signers. Meanwhile, privacy refers to the fact that any signing group and t are kept secret from the public.

Based on the observations on TAPS, our motivation arises from two aspects. **M1. Unwitnessed Tracing.** Tracing a TS σ to its signing quorum is a formidable capability that should be "kept in a box" at ordinary times. If not properly restrained, the "almighty" tracer can trace any σ to its t signers. Followed by the motivation above, we are driven to achieve **M2. Secretly Designated Witnessing**, i.e., designate another group of witnesses to sanction the tracing while keeping their identities secret.

1.3 Possible Solutions and Technical Challenges

Intuitively, one can first ask the combiner to encrypt the TS σ by using the t' witnesses' public keys and then ask each of them to decrypt the ciphertext right before tracing. However, this will expose σ to all witnesses and increase the risk of leakage. In this work, we resort to the Dynamic Threshold Public-Key Encryption (DTPKE) [13], which allows (1) a sender (combiner) to dynamically choose the authorized group of recipients (witnesses) for a ciphertext (TS), and (2) a set of witnesses to decrypt a ciphertext when a threshold of authorized witnesses collaborate. Such two properties shed light on a promising approach. Still, we have to tackle two technical challenges:

C1. How to awake the t' witnesses to share-decrypt the encrypted TS without exposing their identities? We assume that neither the combiner nor the tracer is aware of the t' witnesses at any phase of the protocol given the identity privacy of witnesses. It is not feasible to ask the combiner or the tracer to "contract" the pertinent witnesses during tracing. Thus, it leaves us to design a method for the ($\leq t$) signers to designate the t' witnesses secretly.

C2. How to prove the validity of a TS that is already encrypted by the combiner via layered encryption? In TAPS, the combiner generates a Non-Interactive Zero Knowledge Proof (NIZKP) π that the output signature σ_m is a valid ΛTS signature on m. However, we intend to protect σ_m from public including the tracer via DTPKE as well as a second layer of encryption due to designation. This in turn makes it challenging for the combiner to prove the signature validity.

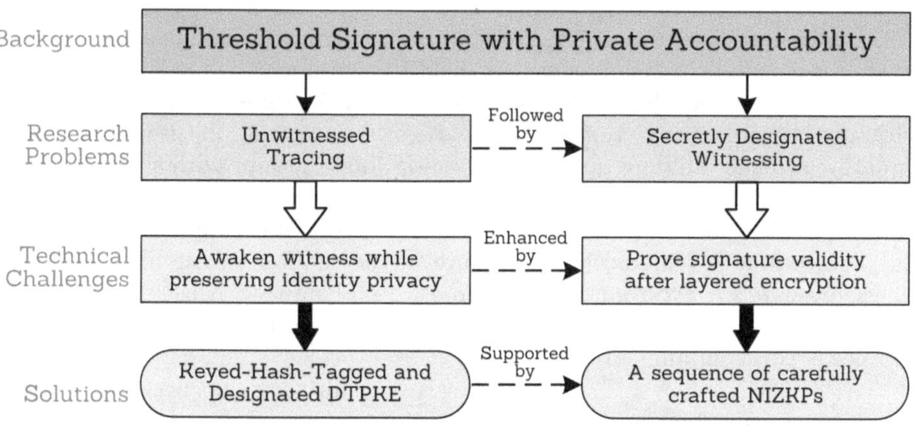

Fig. 2. Technical Roadmap of HiTAPS.

1.4 Our Solutions

To solve the first challenge, we design a Keyed-Hash-Tagged and Designated DTPKE using three carefully structured cryptographic primitives. Specifically, we first designate t' witnesses during signing, encrypt the TS σ_m with DTPKE to obtain $\hat{\sigma}_m$ during combining to protect σ_m from arbitrary tracing, encrypt $\hat{\sigma}_m$ with PKE to obtain t' versions of encrypted $\hat{\sigma}_m$ for t' witnesses, and compte t' keyed-hash tags of the message and witness identities. In this way, we lay the groundwork for awakening secret awakening while preserving the identity privacy of the witnesses.

To solve the second challenge, we carefully craft a sequence of NIZKPs for the combiner to prove the validity of t' ciphertexts sent by the combiner. Specifically, the combiner has to prove the validity of σ_m, the encryption of σ_m, the encryption of $\hat{\sigma}_m$, and the possession of the combining key. By doing so, we pay the way for systematically proving the validity of σ_m after σ_m is encrypted.

We portray our technical roadmap in Fig. 2. Note that, given the high communication overhead of the t' ciphertexts, we further design an optimized version of HiTAPS by computing an aggregated ciphertext at the combiner.

2 Related Work

In this section, we first revisit some related work, including PTS, ATS, and TAPS. Then we discuss how our work advances the state-of-the-art.

Shoup [2] proposed an RSA threshold signature scheme that provided unforgeability and robustness in the random oracle model and made the generation and verification non-interactive. Stern et al. [3] presented new techniques to fully distribute RSA, i.e., generated RSA moduli for Shoup's scheme without a trusted dealer. Koprowski et al. [4] designed a threshold RSA scheme as efficient as Shoup's scheme while not depending on two previously used assumptions and building its robustness on an intractability assumption.

Micali et al. [5] formalized Accountable-Subgroup Multisignatures (ASM) where a subgroup $\mathcal{S}$ of signers was enabled to sign a message m and the resulting signature σ provably discloses the identities of the signers in $\mathcal{S}$ to any verifier without the help of any trusted third parties. Boneh et al. [6] proposed a short signature scheme building upon Gap-Diffie-Hellman groups with small representations. Bellare et al. [7] removed the key-setup requirements (it is not necessary for a signer to have a secret key) and proposed a multi-signature scheme in the plain public-key model that is secure in the random-oracle model. Nick et al. [8] presented a two-round multi-signature scheme that is secure under concurrent signing sessions, supports key aggregation, produces Schnorr signatures, and needs two communication rounds.

TAPS [12] is a novel TS that achieves both privacy and accountability. It protects not only the threshold t, but also the t signers via keeping the tracking key tk to the sole tracer. Based on TAPS, our proposed HiTAPS concentrates on the *secure authorization of tracing for TSs*. It constitutes an important piece of the puzzle. To the best of our knowledge, this problem is not very understood yet, and we aim to fill the gap.

3 Problem Formulation

In this section, we formalize the notion of Accountable and Private Threshold Signature with Hidden Witnesses (HiTAPS), including the system model, definition, unforgeability and accountability, and privacy. We use n for the total number of signers, t for the threshold number of required signers in combining, t' for the threshold number of required witnesses in tracing.

3.1 System Model

The system model of HiTAPS is depicted in Fig. 3. It consists of n signers $\{S_1, S_2, \cdots, S_n\}$, t' witnesses $\{W_1, W_2, \cdots, W_{t'}\}$, a combiner C, and a tracer T. Next, we describe how they work in the system.

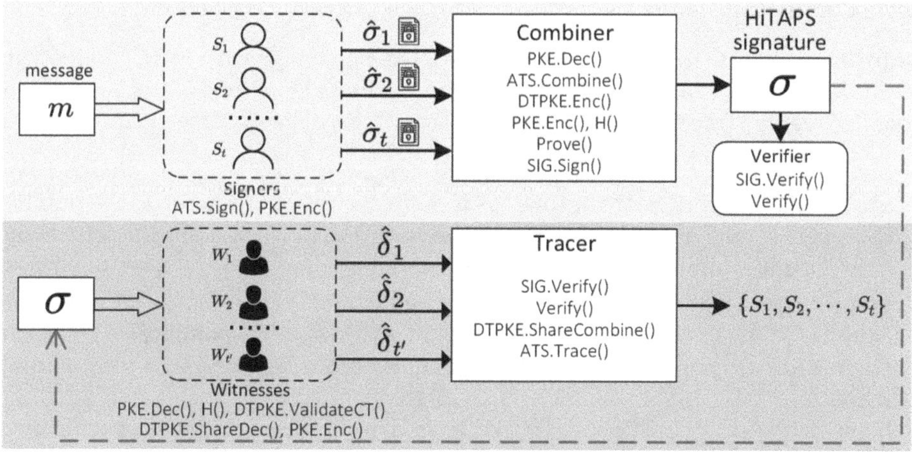

Fig. 3. System Model of HiTAPS.

Signer. A set of t signers $\mathcal{S} = \{S_1, S_2, \cdots, S_t\}$ belonging to a bigger set of signers $\{S_1, S_2, \cdots, S_n\}$ $(t \leq n)$ decides to cooperatively generate a signature on a message m. Each signer S_i has a private key sk_i^s. First, they select a set of t' witness $\mathcal{W} = \{W_1, W_2, \cdots, W_{t'}\}$. Then, each S_i computes a signature share σ_i on m and encrypts σ_i with the combiner C's public key pk^c to obtain $\widehat{\sigma}_i$. Next, each S_i sends $\widehat{\sigma}_i$ and an encrypted list $\widehat{\mathcal{W}}$ of $\mathcal{W}$ to C.

Witness. The selected set $\mathcal{W}$ belongs to a bigger set $\{W_1, W_2, \cdots, W_w\}$ $(t' \leq w)$. Each witness W_i has a public key (pk_i^w, upk_i) as his identity. Upon the call of tracing, W_i computes a keyed hash tag and compares it with the ones on a public bulletin. If a match is found, W_i will decrypt it to have a HiTAPS signature σ, verifies it, and share-decrypts it to obtain a decryption share δ_i. Next, W_i sends $\widehat{\delta}_i$, i.e., the encryption of δ_i under the tracer T's public key pk^t, to C.

Combiner. The combiner C receives the encrypted signature shares and encrypted lists from the signers. C decrypts them to obtain the signature shares $\{\sigma_i\}_{i=1}^t$ in plaintext. C combines t signature shares to compute an ATS signature σ_m and encrypts it via DTPKE and an encryption key ek^c. C also decrypts the encrypted list $\widehat{\mathcal{W}}$ to know which witnesses to send to. After checking all the lists, C encrypts σ_m with the pk_i^w of t' witnesses to get a triad (C_{i1}, C_{i2}, C_{i3}) and computed a keyed hash tag ht_i. We assume that C and each W_i shares a secret key ssk_i. Furthermore, C generates a proof π and a signature $\hat{\sigma}$, and submits them to the public bulletin.

Tracer. The tracer T receives the encrypted decryption shares from witnesses. T decrypts them to get decryption shares and then combines them to obtain σ_m. Lastly, T traces σ_m to its original signing group if σ_m is a valid ATS signature.

3.2 Definition

Definition 1. An accountable and private threshold signature with hidden witness, or HiTAPS, is a tuple of five polynomial time algorithms $\Pi = (\mathsf{Setup}, \mathsf{Sign}, \mathsf{Combine}, \mathsf{Verify}, \mathsf{Trace})$ where

- $\mathsf{Setup}(1^\lambda, n, w, t, t') \rightarrow (PK, \{sk_i^s\}_{i=1}^n, \{wsk_j, ssk_j, sk_j^{\mathrm{wit}}\}_{j=1}^w, ck, tk, SK)$ is a probabilistic algorithm that takes as input a security parameter λ, the number of signers n, the number of witnesses w, a threshold t, and another threshold t' to output a public key PK, a set of n signing keys $\{sk_i^s\}_{i=1}^n$, a set of witness private keys $\{wsk_j\}_{j=1}^w$, a set of shared secret keys $\{ssk_j\}_{j=1}^w$, a set of private keys $\{sk_j^{\mathrm{wit}}\}_{j=1}^w$, three private keys sk_c^{enc}, sk_c^{sig}, sk_t^{enc}, a combining key ck, a tracing key tk, and a secret key SK, an encryption key EK, and a combining key CK.
- $\mathsf{Sign}(m, \mathcal{S}, \{sk_i^s\}_{i=1}^t, \mathcal{W}) \rightarrow (\{\widehat{\sigma}_i, \widehat{\mathcal{W}}_i\})$ is a probabilistic algorithm run by a set of signers that takes as input a set of signers $\mathcal{S}$ and a corresponding set of t signing keys $\{sk_i^s\}_{i=1}^t$, a message m in message space $\mathcal{M}$, and a set of t' witnesses, to output a set of encrypted signature shares $\{\widehat{\sigma}_i\}$ on m and a set of encrypted list $\{\widehat{\mathcal{W}}_i\}$. For simplicity, we denote $\{sk_i^s\}_{i=1}^t$ as the set of t secret keys from any set of t signers in $\{S_1, S_2, \cdots, S_n\}$.
- $\mathsf{Combine}(m, t, t', PK, \mathcal{S}, \{\widehat{\sigma}_i\}, \{\widehat{\mathcal{W}}_i\}, ck, EK) \rightarrow \sigma$ is a probabilistic algorithm run by the combiner that takes as input a message m, two thresholds t and t', the public key PK, a set of signers $\mathcal{S}$, a set of encrypted signature shares $\{\widehat{\sigma}_i\}$ on m and a set of encrypted list $\{\widehat{\mathcal{W}}_i\}$, a combining key ck, and an encryption key EK to output a HiTAPS signature σ.
- $\mathsf{Verify}(PK, m, \sigma) \rightarrow \{0, 1\}$ is a deterministic algorithm run by the public and the tracer that verifies a HiTPAS signature σ on a message m with respect to the public key PK.

- $\mathsf{Trace}(m, PK, tk, \sigma) \rightarrow \mathcal{S}$ is a deterministic algorithm run by the tracer that takes as input a message m, the public key PK, a tracing key tk, and a HiTAPS signature σ to output a set $\mathcal{S}$ who previously generated σ or a failure symbol $\perp$.

For *correctness*, we require that for all $t \in [n]$, all t-size sets $\mathcal{S}$, all $t' \in [w]$, all t'-size set $\mathcal{W}$, all $m \in \mathcal{M}$, and all the $(PK, \{sk_i^s\}_{i=1}^n, \{ssk_j\}_{j=1}^w, \{sk_j\}_{i=1}^w, ck$ $tk, SK) \leftarrow \mathsf{Setup}(1^\lambda, n, t)$, the following two conditions hold:

$$\Pr[\mathsf{Verify}(m, PK, \mathsf{Combine}(ck, sk^s, m, \mathcal{S}, \mathsf{Sign}(m, \mathcal{S}, \{sk_i^s\}_{i=1}^t, \mathcal{W})\})) = 1] = 1,$$
$$\Pr[\mathsf{Trace}(m, tk, \mathsf{Combine}(ck, sk^s, m, \mathcal{S}, \mathsf{Sign}(m, \mathcal{S}, \{sk_i^s\}_{i=1}^t, \mathcal{W})\})) = \mathcal{S}] = 1.$$

3.3 Unforgeability and Accountability

Before we dive into privacy, HiTAPS must satisfy the standard notion of existential unforgeability against a chosen message attack (EUF-CMA) [14,15] like any signature scheme, i.e., an adversary compromising fewer than t signers cannot generate a valid message-signature pair. Meanwhile, HiTAPS should be accountable, i.e., a tracer holding a tracing key tk can output the correct set of signers for a correct message-signature pair. We formalize these two properties in the adversarial experiment in Fig. 4. Let $\mathbf{Adv}_{\mathcal{A},\Pi}^{\mathrm{forg}}(\lambda)$ be the probability that $\mathcal{A}$ wins the experiment $\mathbf{Exp}^{\text{unf-acc}}$ against the HiTAPS scheme Π.

Definition 2 (Unforgeability and Accountability). A HiTAPS scheme Π is **unforgeable and accountable** if for all Probabilistic Polynomial Time (PPT) adversaries $\mathcal{A}$, there is a negligible function negl such that $\mathbf{Adv}_{\mathcal{A},\Pi}^{\mathrm{forg}}(\lambda) \leq \mathsf{negl}(\lambda)$.

$\mathbf{Exp}^{\text{unf-acc}}$

1. $(n, w, t, t', \mathcal{S}, \mathsf{state}) \xleftarrow{\$} \mathcal{A}(1^\lambda)$ where $t \in [n], t' \in [w], \mathcal{S} \subseteq [n]$
2. $(PK, \{sk_i^s\}_{i=1}^n, \{wsk_j, ssk_j, sk_j^{\mathrm{wit}}\}_{i=1}^w, ck, tk, SK) \leftarrow \mathsf{Setup}(1^\lambda, n, w, t, t')$
3. $(m', \sigma') \xleftarrow{\$} \mathcal{A}^{\mathcal{O}(\cdot,\cdot)}(PK, \{sk_i^s\}_{S_i \in \mathcal{S}}, \mathcal{W}, \{wsk_j, ssk_j, sk_j^{\mathrm{wit}}\}_{i=1}^w, ck, tk, SK, \mathsf{state})$ where $\mathcal{O}(\mathcal{S}_j, m_i)$ returns $\mathsf{Sign}(m_i, \mathcal{S}_j, \{sk_i^s\}_{S_i \in \mathcal{S}_j}, \mathcal{W}_j)$

$\mathcal{A}$'s winning conditions:
Let $(\mathcal{S}_1, m_1), (\mathcal{S}_2, m_2), \cdots$ be $\mathcal{A}$'s queries to $\mathcal{O}$
Let $\mathcal{S} \leftarrow \cup \mathcal{S}_i$, union over all queries to $\mathcal{O}(\mathcal{S}_j, m')$,
  if no such queries, set $\mathcal{S}_j = \emptyset$
let $\mathcal{S}_t \leftarrow \mathsf{Trace}(m', tk, PK, \sigma', \hat{\sigma}')$
Output 1 if $\mathsf{Verify}(m', PK, \sigma', \hat{\sigma}') = 1$ and either $\mathcal{S}_t \not\subseteq \mathcal{S} \cup \mathcal{S}'$ or if $\mathcal{S}_t = \mathsf{fail}$

Fig. 4. Experiment of Unforgeability and Accountability.

3.4 Privacy

Now we define privacy for HiTAPS. Usually, the privacy for a TSS is formalized by requiring that a TS on a message m be computationally indistinguishable from a standard signature on m [16]. It ensures that a TS tells nothing about the threshold t or the set of signers that generated the TS.

We give three informal privacy requirements as below and they are further captured by the experiments in Fig. 5 and Fig. 6, respectively.

- **Privacy against public:** A party who only has (PK, n, w) and observes a series of $(m, \sigma, \hat{\sigma})$ triads, learns nothing about t, t' or the set of signers that generated the observed $(\sigma, \hat{\sigma})$.
- **Privacy against signer:** A set of all signers who only has (PK, n, w, t, t') and observes a series of $(m, \sigma, \hat{\sigma})$ triads, learns nothing about the set of signers that generated the observed $(\sigma, \hat{\sigma})$.

Let E be the event that the experiment $\mathbf{Exp}^{\text{privP}}$ in Fig. 5 outputs 1 and E' be the event that the experiment $\mathbf{Exp}^{\text{privS}}$ in Fig. 6 outputs 1. We define the two advantage functions for an adversary $\mathcal{A}$ against the HiTAPS scheme Π, as a function of the security parameter λ:

$$\mathbf{Adv}^{\text{privP}}_{\mathcal{A},\Pi}(\lambda) = |2\Pr[E] - 1|, \ \mathbf{Adv}^{\text{privS}}_{\mathcal{A},\Pi}(\lambda) = |2\Pr[E'] - 1|.$$

Definition 3 (Privacy). A HiTAPS scheme is **private** if for all PPT adversaries $\mathcal{A}$, $\mathbf{Adv}^{\text{privP}}_{\mathcal{A},\Pi}(\lambda)$ and $\mathbf{Adv}^{\text{privS}}_{\mathcal{A},\Pi}(\lambda)$ are negligible functions of λ.

$b_0 \xleftarrow{\$} \{0,1\}$, $b_1 \xleftarrow{\$} \{0,1\}$, $(t_0, t_1, t'_0, t'_1, \mathsf{state}) \xleftarrow{\$} \mathcal{A}(1^\lambda)$ where $t_0, t_1 \in [n], t'_0, t'_1 \in [w]$
$(PK, \{sk_i^s\}_{i=1}^n, \{wsk_j, ssk_j, sk_j^{\text{wit}}\}_{i=1}^w, ck, tk, SK) \leftarrow \mathsf{Setup}(1^\lambda, n, w, t_{b_0}, t'_{b_1})$
$(b'_0, b'_1) \leftarrow \mathcal{A}^{\mathcal{O}_1(\cdot,\cdot,\cdot,\cdot,\cdot),\mathcal{O}_2(\cdot,\cdot,\cdot)}(PK, \mathsf{state})$
Output $(b'_0 = b_0) \wedge (b'_1 = b_1)$.
where $\mathcal{O}_1(m, \mathcal{S}_0, \mathcal{S}_1, \mathcal{W}_0, \mathcal{W}_1)$ returns
  $(\sigma, \hat{\sigma}) \leftarrow \mathsf{Combine}(m, \mathcal{S}_{b_0}, \mathsf{Sign}(m, \mathcal{S}_{b_0}, \{sk_i^s\}_{S_i \in \mathcal{S}_{b_0}}, \mathcal{W}_{b_1}), ck, EK)$
  for $\mathcal{S}_0, \mathcal{S}_1 \subseteq [n]$, $|\mathcal{S}_0| = t_0$ and $|\mathcal{S}_1| = t_1$, $\mathcal{W}_0, \mathcal{W}_1 \subseteq [w]$, $|\mathcal{W}_0| = t'_0$ and $|\mathcal{W}_1| = t'_1$
$\mathcal{O}_2(m, \sigma, \hat{\sigma})$ returns $\mathsf{Trace}(m, tk, PK, \sigma, \hat{\sigma})$.
Restriction: if $(\sigma, \hat{\sigma})$ is returned by $\mathcal{O}_2$, then $(m, \sigma, \hat{\sigma})$ will not be sent to $\mathcal{O}_3$.

Fig. 5. $\mathbf{Exp}^{\text{privP}}$: Experiment of Privacy against the Public.

In $\text{Exp}^{\text{privP}}$, $\mathcal{A}$ generates four thresholds t_0, t_1, t'_0 and t'_1 in $[n]$ and is given PK. $\mathcal{A}$ submits a string of signature queries to a signing oracle $\mathcal{O}_1$, where each query contains a message m and four sets $\mathcal{S}_0$, $\mathcal{S}_1$, $\mathcal{W}_0$, and $\mathcal{W}_1$. Then, $\mathcal{A}$ receives a signature generated using either $\mathcal{S}_0$ or $\mathcal{S}_1$ (same for $\mathcal{W}_0$ or $\mathcal{W}_1$). $\mathcal{A}$ can access a tracing oracle $\mathcal{O}_2$ while not being able to determine whether the string of signatures it observed related to the left or the right sequence of sets.

In $\text{Exp}^{\text{privS}}$, $\mathcal{A}$ generates (t, t'), and is given all the signing keys. Same as $\text{Exp}^{\text{privP}}$, $\mathcal{A}$ cannot determine whether $\mathcal{O}_1$ that takes four sets $\mathcal{S}_0$, $\mathcal{S}_1$, $\mathcal{W}_0$, and $\mathcal{W}_1$ responds using either $\mathcal{S}_0$ or $\mathcal{S}_1$ (same for $\mathcal{W}_0$ or $\mathcal{W}_1$).

$b_0 \xleftarrow{\$} \{0,1\}$, $b_1 \xleftarrow{\$} \{0,1\}$, $(t, t', \mathsf{state}) \xleftarrow{\$} \mathcal{A}(1^\lambda)$ where $t_0, t_1 \in [n], t'_0, t'_1 \in [w]$
$(PK, \{sk_i\}_{i=1}^n, \{wsk_j, ssk_j, sk_j^{\mathsf{wit}}\}_{i=1}^w, ck, tk, SK) \leftarrow \mathsf{Setup}(1^\lambda, n, w, t, t')$
$(b'_0, b'_1) \leftarrow \mathcal{A}^{\mathcal{O}_1(\cdot,\cdot,\cdot,\cdot,\cdot), \mathcal{O}_2(\cdot,\cdot,\cdot,\cdot,\cdot), \mathcal{O}_3(\cdot,\cdot,\cdot)}(PK, \{sk_i^s\}_{i=1}^n, \mathsf{state})$
Output $(b'_0 = b_0) \wedge (b'_1 = b_1)$.
where $\mathcal{O}_1(m, \mathcal{S}_0, \mathcal{S}_1, \mathcal{W}_0, \mathcal{W}_1)$ returns $\widehat{\sigma}_i \leftarrow \mathsf{PKE.Enc}(pk_c^{\mathsf{enc}}, m||\sigma_i)$ and $\widehat{\mathcal{W}}_i \leftarrow \mathsf{PKE.Enc}(pk_c^{\mathsf{enc}}, \mathcal{W}_{b_1})$ for $S_i \in \mathcal{S}_{b_0}$ where $\sigma_i \leftarrow \mathsf{ATS.Sign}(sk_i^s, m)$
$\mathcal{O}_2(m, \mathcal{S}_0, \mathcal{S}_1, \mathcal{W}_0, \mathcal{W}_1)$ returns
$(\sigma, \hat{\sigma}) \leftarrow \mathsf{Combine}(m, \mathcal{S}_{b_0}, \mathsf{Sign}(m, \mathcal{S}_{b_0}, \{sk_i^s\}_{S_i \in \mathcal{S}_{b_0}}, \mathcal{W}_{b_1}), ck, EK)$
for $\mathcal{S}_0, \mathcal{S}_1 \subseteq [n]$, $|\mathcal{S}_0| = |\mathcal{S}_1| = t$, $\mathcal{W}_0, \mathcal{W}_1 \subseteq [w]$, $|\mathcal{W}_0| = |\mathcal{W}_1| = t'$
$\mathcal{O}_3(m, \sigma, \hat{\sigma})$ returns $\mathsf{Trace}(m, tk, PK, \sigma, \hat{\sigma})$.
Restriction: if $(\sigma, \hat{\sigma})$ is returned by $\mathcal{O}_1$, then $(m, \sigma, \hat{\sigma})$ will not be sent to $\mathcal{O}_2$.

Fig. 6. $\mathbf{Exp}^{\mathrm{privS}}$: Experiment of Privacy against the Signers.

4 Preliminaries

In this section, we revisit and review some preliminaries, including notations, ATS, DTPKE, ElGamal encryption, and hash function.

4.1 Notations

Since there are many notations used in this work, we list the key notations in Table 1, including number of all signers n, number of required signers t, etc.

4.2 ATS

An accountable threshold signature is a tuple of five polynomial time algorithms (Setup, Sign, Combine, Verify, Trace) invoked as

$$(\{pk_i^s, sk_i^s\}_{i=1}^n) \leftarrow \mathsf{ATS.Setup}(1^\lambda, n, t), \;\; \sigma_i \leftarrow \mathsf{ATS.Sign}(sk_i^s, m),$$
$$\sigma_m \leftarrow \mathsf{ATS.Combine}(t, \{pk_i^s\}_{i=1}^n, m, \mathcal{S}, \{\sigma_i\}_{i \in \mathcal{S}}),$$
$$\{0,1\} \leftarrow \mathsf{ATS.Verify}(t, \{pk_i^s\}_{i=1}^n, m, \sigma_m), \;\; \mathcal{S} \leftarrow \mathsf{ATS.Trace}(t, \{pk_i^s\}_{i=1}^n, m, \sigma_m).$$

An ATS is secure if it is unforgeable and accountable, i.e., if for every PPT adversary $\mathcal{A}$, the function $\mathbf{Adv}_{\mathcal{A},\mathsf{ATS}}^{\mathrm{forg}}$ of winning an unforgeability and accountability attack game is a negligible function of λ [12,17].

4.3 DTPKE

A dynamic threshold public-key encryption is a tuple of six polynomial algorithms (Setup, Join, Encrypt, ValidateCT, ShareDecrypt, Combine) invoked as

$$(SK, EK, CK) \leftarrow \mathsf{Setup}(1^\lambda), \;\; (wpk, wsk) \leftarrow \mathsf{Join}(SK, id),$$
$$(C_0, C_1, C_2, C_3) \leftarrow \mathsf{Enc}(Ek, t', m), \;\; \{0,1\} \leftarrow \mathsf{ValidateCT}(EK, t', C_0, C_1),$$
$$\sigma_{id} \leftarrow \mathsf{ShareDecrypt}(id, wsk, C_0, C_1),$$
$$\sigma_m \leftarrow \mathsf{Combine}(CK, t', C_0, C_1, C_2, C_3, \mathcal{W}, \{\sigma_m^j\}_{W_j \in \mathcal{W}}).$$

Table 1. Key Notations

Notation	Meaning	Notation	Meaning
TSS	Threshold signature scheme	TS	Threshold signature
PTS	Private threshold signature	ATS	Accountable threshold signature
n	Number of all signers	t	Number of required signers
w	Number of all witnesses	t'	Number of required witnesses
S_i	Signer i	W_j	Witness j
$\mathcal{S}$	Set of required signers	$\mathcal{W}$	Set of required witnesses
C	Combiner	T	Tracer
λ	Security parameter	PK	Public key
pk_j^w, wpk_j	W_j's public key	sk_j^{wit}, wsk_j	W_j's private key
$pk_c^{\text{enc}}, pk_c^{\text{sig}}$	C's Public key	$sk_c^{\text{enc}}, sk_c^{\text{sig}}$	C's private key
pk_t^{enc}	T's public key	sk_t^{enc}	T's private key
(sk_i^s, pk_i^s)	S_i's signing key pair	ssk	Shared secret key
SK	Secret key	EK	Encryption key
CK	Combining key	m	Message
ck	Combining key	tk	Tracing key
σ_i	S_i's signature	$\widehat{\sigma}_i$	Encrypted signature
σ_m	ATS signature	$\widehat{\sigma}_m$	Encrypted ATS signature
$\widehat{\mathcal{W}}$	Encrypted witness list	ht	Keyed hash tag
δ_i	W_i's share decryption	$\widehat{\delta}_i$	Encrypted share decryption
π	NIZKP	σ	HiTAPS signature

Its non-adaptive adversary, non-adaptive corruption, chosen-plaintext attacks (IND-NAA-NAC-CPA) security is based on the Multi-sequence of Exponents Diffie-Hellman (MSE-DDH) assumption, where $\mathbf{Adv}^{\text{ind-cpa}}_{\mathcal{A},\mathsf{DTPKE}}(l, m, t') \leq \mathbf{Adv}^{\text{mse-ddh}}(l, m, t')$ [13,18,19].

4.4 ElGamal Encryption

The ElGamal encryption scheme PKE is a tuple of three polynomial algorithms $(\mathsf{KeyGen}, \mathsf{Enc}, \mathsf{Dec})$ invoked as

$$(sk, pk) \leftarrow \mathsf{KeyGen}(1^\lambda),\ (C_0, C_1) \leftarrow \mathsf{Enc}(pk, m),\ m \leftarrow \mathsf{Dec}(sk, (C_0, C_1)).$$

The ElGamal encryption scheme is secure against Chosen Plaintext Attack (CPA) if the decisional Diffie-Hellman (DDH) problem [17].

4.5 Hash Function

A cryptographic hash function with output length $l(\lambda)$ is a tuple of two polynomial algorithms $(\mathsf{Gen}, \mathsf{H})$ invoked as

$$k \leftarrow \mathsf{Gen}(\lambda), \{0,1\}^{l(\lambda)} \leftarrow \mathsf{H}(k, m),\ m \in \{0,1\}^*.$$

A hash function H is preimage resistant if for every PPT adversary $\mathcal{A}$, there is a negligible function negl such that $\Pr[\mathsf{Hash\text{-}pre}_{\mathcal{A},\mathsf{H}}(\lambda) = 1] \leq \mathsf{negl}(\lambda)$ [17].

4.6 PKE, COM, and SIG

A commitment scheme COM is a pair of algorithms (Comm, Verify) invoked as

$$com \leftarrow \mathsf{Comm}(x, r),\ \{0,1\} \leftarrow \mathsf{Verify}(x, r, com).$$

A COM scheme is secure if it is unconditionally hiding and computationally binding, i.e., for every PPT adversary $\mathcal{A}$, $\mathbf{Adv}^{\text{blind}}_{\mathcal{A},\text{COM}}(\lambda)$ is negligible.

A signature scheme SIG is a triple of algorithms (KeyGen, Sign, Verify) invoked as

$$(pk, sk) \leftarrow \mathsf{KeyGen}(1^\lambda),\ \sigma \leftarrow \mathsf{Sign}(sk, m),\ \{0,1\} \leftarrow \mathsf{Verify}(pk, m, \sigma).$$

A SIG scheme is strongly unforgeable if for every PPT adversary $\mathcal{A}$, $\mathbf{Adv}^{\text{euf-cma}}_{\mathcal{A},\text{SIG}}(\lambda)$ is negligible.

5 The Proposed Scheme HiTAPS

We show the construction of HiTAPS in Fig. 7. In Setup, all system parameters are generated for the ATS, DTPKE, commitment scheme COM, ElGamal encryption PKE, signature scheme SIG, hash function H, including the secret keys, public keys, shared secret keys, and a commitment.

In Sign, t signers from one signer set $\mathcal{S}$ generate t signature shares $\{\sigma_i\}_{S_i \in \mathcal{S}}$ on a message m. The signature shares are further encrypted into $\{\widehat{\sigma}_i\}$ with C's public key pk_c^{enc} and then sent to C along with encrypted witness lists $\widehat{\mathcal{W}}$.

In Combine, C first decrypts $\widehat{\mathcal{W}}$ to collect encrypted signature shares belonging to some signer set $\mathcal{S}$ and then decrypts them to recover $\{\sigma_i\}_{S_i \in \mathcal{S}}$. Next, C combines them to obtain an ATS signature σ_m and encrypts it into $\widehat{\sigma}_m$ with DTPKE. Then, C encrypts (C_0, C_1) in $\widehat{\sigma}_m$ with the public keys of witnesses in $\mathcal{W}$. Here, (C_0, C_1) is encrypted because they can be used to recover $C_2 = k$ to decrypt C_3. For instance, $\mathsf{ElGamal.Enc}(pk_1^w, C_0) = (C_{10}, C_{11}) = (C_0 g^{sk_1^w r_1}, g^{r_1})$ where $pk_1^w = g^{sk_1^w}$. C also computes t' hash tag $ht_j = \mathsf{H}(ssk_j, m||W_j)$ $(W_j \in \mathcal{W})$ as a secret token to awake W_j into share-decrypt $\widehat{\sigma}_m$. Here, the t' witnesses do not actively participate in the process. C generates a NIZPK π to prove the validity of a TS. For example, to prove $C_{10} = C_0 g^{sk_1^w r_1}$ while keeping C_0, sk_1^w, r_1 secret, we first transfer it into $C_{10} = g^x g^{sk_1^w r_1} = g^{x + sk_1^w r_1}$ for simplicity. Now the combiner (Prover) and a verifier proceed as follows. *Prover*: (1) set $A = C_0 g^{sk_1^w r_1}$ where $C_0 = g^x$; (2) compute $B = g$, $\delta = \mathsf{H}(g||A)$, $C = g^\delta$, and $D = g^{(x + sk_1^w r_1)/\delta}$. *Verifier*: check $e(A, B) \stackrel{?}{=} e(C, D)$.

In Verify, a verifier verifies the validity of σ by checking the signature η and the proof π. If they are both valid, then the validity of σ is guaranteed.

In Trace, each W_j computes a hash tag ht'_j and compares it with the ones in $\{ht_j\}_{W_j \in \mathcal{S}}$. If one tag is found to be the same, W_j continues to verify the validity of σ to decide whether this signature is worthy of tracing. If so, W_j share-decrypts (C_0, C_1) and sends an encrypted decryption share $\widehat{\delta}_j$ to T. After decrypting $\{\widehat{\delta}_j\} \in \mathcal{S}$, T verifies whether (C_0, C_1) is a valid ciphertext w.r.t. EK and t'. If so, T combines all decryption shares to obtain σ_m and traces $\mathcal{S}$.

$\mathsf{Setup}(1^\lambda, n, w, t, t') \rightarrow (PK, \{sk_i^s\}_{i=1}^n, \{wsk_j, ssk_j, sk_j^{\mathrm{wit}}\}_{j=1}^w, ck, tk, SK)$

1. $(\{pk_i^s, sk_i^s\}_{i=1}^n) \leftarrow \mathsf{ATS.KeyGen}(1^\lambda, n, t)$, set $pk = (t, \{pk_i^s\}_{i=1}^n)$
2. $r_{pk} \leftarrow \mathcal{R}_\lambda$, $\mathsf{com}_{pk} \leftarrow \mathsf{COM.Comm}(\{pk_i^s\}_{i=1}^n, r_{pk})$
3. $(SK, EK, CK) \leftarrow \mathsf{DTPKE.Setup}(1^\lambda)$
4. $(pk_c^{\mathsf{enc}}, sk_c^{\mathsf{enc}}) \leftarrow \mathsf{PKE.KeyGen}(1^\lambda)$, $(pk_t^{\mathsf{enc}}, sk_t^{\mathsf{enc}}) \leftarrow \mathsf{PKE.KeyGen}(1^\lambda)$
5. $(pk_c^{\mathsf{sig}}, sk_c^{\mathsf{sig}}) \leftarrow \mathsf{SIG.KeyGen}(1^\lambda)$, $(pk_j^{\mathrm{wit}}, sk_j^{\mathrm{wit}}) \leftarrow \mathsf{PKE.KeyGen}(1^\lambda)$, $j \in [w]$
6. $(wpk_j, wsk_j) \leftarrow \mathsf{DTPKE.Join}(SK, id_j)$, $id_j \in [w]$
7. $ck \leftarrow (\{pk_i^s\}_{i=1}^n, sk_c^{\mathsf{enc}}, sk_c^{\mathsf{sig}}, t', \mathsf{com}_{pk}, r_{pk})$, $tk \leftarrow (\{pk_i^s\}_{i=1}^n, sk_t^{\mathsf{enc}}, pk_c^{\mathsf{sig}})$
8. $ssk_j \leftarrow \mathsf{Gen}(\lambda)$, for $W_j \in \mathcal{W}$
9. $PK \leftarrow (\mathsf{com}_{pk}, pk_c^{\mathsf{sig}}, pk_c^{\mathsf{enc}}, pk_t^{\mathsf{enc}}, \{wpk_j, pk_j^{\mathrm{wit}}\}_{j=1}^w)$
10. Output $(PK, \{sk_i^s\}_{i=1}^n, \{wsk_j, ssk_j, sk_j^{\mathrm{wit}}\}_{i=1}^w, ck, tk, SK)$

$\mathsf{Sign}(m, \mathcal{S}, \{sk_i^s\}_{S_i \in \mathcal{S}}, \mathcal{W}) \rightarrow (\widehat{\sigma}_i, \widehat{\mathcal{W}}_i)$:

1. $\sigma_i \leftarrow \mathsf{ATS.Sign}(sk_i^s, m)$ for $S_i \in \mathcal{S}$
2. $\widehat{\sigma}_i \leftarrow \mathsf{PKE.Enc}(pk_c^{\mathsf{enc}}, m||\sigma_i)$, $\widehat{\mathcal{W}}_i \leftarrow \mathsf{PKE.Enc}(pk_c^{\mathsf{enc}}, \mathcal{W})$ for $S_i \in \mathcal{S}$

$\mathsf{Combine}(m, t, t', \mathcal{S}, PK, \{\widehat{\sigma}_i\}_{S_i \in \mathcal{S}}, \{\widehat{\mathcal{W}}_i\}_{S_i \in \mathcal{S}}, ck, EK) \rightarrow \sigma$

1. $\mathcal{W} \leftarrow \mathcal{W}_i \leftarrow \mathsf{PKE.Dec}(\widehat{\mathcal{W}}_i, sk_c^{\mathsf{enc}})$ for $S_i \in \mathcal{S}$
2. $(m||\sigma_i) \leftarrow \mathsf{PKE.Dec}(\widehat{\sigma}_i, sk_c^{\mathsf{enc}})$ for $S_i \in \mathcal{S}$
3. $\sigma_m \leftarrow \mathsf{ATS.Combine}(\{pk_i^s\}_{i=1}^n, m, \mathcal{S}, \{\sigma_i\}_{S_i \in \mathcal{S}})$
4. $(C_0, C_1, k, \mathsf{AES.Enc}(k, \sigma_m)) \leftarrow \mathsf{DTPKE.Enc}(EK, \{wpk_j\}_{W_j \in \mathcal{W}}, \sigma_m)$
5. $(C_{j0}, C_{j1}) = \mathsf{ElGamal.Enc}(pk_j^{\mathrm{wit}}, C_0)$, $(C_{j2}, C_{j3}) = \mathsf{ElGamal.Enc}(pk_j^{\mathrm{wit}}, C_1)$, $ht_j = \mathsf{H}(ssk_j, m||W_j))$ for $W_j \in \mathcal{W}$
6. Generate a proof by using Prove for the relation:

$$\mathcal{R}((\mathsf{com}_{pk}, m, C_0, C_1); (pk, r_{pk}, \sigma_m, \{r_j\}_{j=1}^{t'})) = 1 \text{ iff}$$

$$\begin{Bmatrix} \mathsf{COM.Verify}(pk, r_{pk}, \mathsf{com}_{pk}) = 1, \ \mathsf{ATS.Verify}(pk, m, \sigma_m) = 1 \\ \mathsf{ElGamal.Enc}(pk_j^{\mathrm{wit}}, C_0) = (C_{j0}, C_{j1}), \text{ for } W_j \in \mathcal{W} \\ \mathsf{ElGamal.Enc}(pk_j^{\mathrm{wit}}, C_1) = (C_{j2}, C_{j3}), \text{ for } W_j \in \mathcal{W} \end{Bmatrix}$$

7. $\widehat{\sigma}_m = (\{C_{j0}, C_{j1}, C_{j2}, C_{j3}, ht_j\}, \mathsf{AES.Enc}(k, \sigma_m))$ for $W_j \in \mathcal{W}$
8. $\eta \leftarrow \mathsf{SIG.Sign}(sk_c^{\mathsf{sig}}, (m, \widehat{\sigma}_m, \pi))$
9. Output $\sigma \leftarrow (\widehat{\sigma}_m, \pi, \eta)$

$\mathsf{Verify}(PK, m, \sigma = (\widehat{\sigma}_m, \pi, \eta)) \rightarrow \{0, 1\}$

1. Accept σ if $\mathsf{SIG.Verify}(pk_c^{\mathsf{sig}}, m, \widehat{\sigma}_m, \pi, \eta) = 1$ and $\mathsf{Verify}(\pi) = 1$; reject otherwise.

$\mathsf{Trace}(m, PK, tk = (\{pk_i^s\}_{i=1}^n, sk_t^{\mathsf{enc}}, pk_c^{\mathsf{sig}}), \sigma = (\widehat{\sigma}_m, \pi, \eta)) \rightarrow \mathcal{S}$

1. If $\mathsf{Verify}(PK, m, \sigma) \neq 1$, output fail and return.
2. $ht'_j = \mathsf{H}(ssk_j, m||W_j)$ by $W_j \in \mathcal{W}$ and compare with $\{ht_j\}_{W_j \in \mathcal{W}}$
3. (C_0, C_1)=$(\mathsf{ElGamal.Dec}(sk_j^{\mathrm{wit}}, C_{j0}, C_{j1}), \mathsf{ElGamal.Dec}(sk_j^{\mathrm{wit}}, C_{j2}, C_{j3}))$ for each j
4. $\delta_j \leftarrow \mathsf{DTPKE.ShareDecrypt}(W_j, wsk_j, C_0, C_1)$ for $W_j \in \mathcal{W}$
5. $\widehat{\delta}_j \leftarrow \mathsf{PKE.Enc}(pk_t^{\mathsf{enc}}, \delta_j)$ for $W_j \in \mathcal{W}$

- -

6. $\delta_j \leftarrow \mathsf{PKE.Dec}(sk_t^{\mathsf{enc}}, \widehat{\delta}_j)$ for $W_j \in \mathcal{W}$
7. $\{0, 1\} \leftarrow \mathsf{DTPKE.ValidateCT}(EK, t', C_0, C_1)$
8. $\sigma_m \leftarrow \mathsf{DTPKE.Combine}(CK, t', C_0, C_1, C_2, C_3, \mathcal{W}, \{\sigma_m^j\}_{W_j \in \mathcal{W}})$
9. $\mathcal{S} \leftarrow \mathsf{ATS.Trace}(\{pk_i^s\}_{i=1}^n, m, \sigma_m)$.

Fig. 7. The HiTAPS scheme

6 The Optimized Scheme HiTAPS2

For each encrypted ATS signature $\widehat{\sigma}_m$, C has to encrypt it $2t'$ times via ElGamal encryption, resulting in much time on encryption and produces $4t'$ ciphertexts. Intuitively, such a process is optimizable and we improve it by leveraging the homomorphic feature of ElGamal encryption. Specifically, C encrypts $\widehat{\sigma}_m$ with the public keys of all t' witnesses and $2t'$ random numbers:

$$\begin{aligned}\mathsf{ElGamal.Enc}(pk_j^{\text{wit}}, C_0 \| C_1) =& (C_0 g^{pk_1^{\text{wit}} r_{11}} \cdots g^{pk_{t'}^{\text{wit}} r_{1t'}}, C_1 g^{pk_1^{\text{wit}} r_{21}} \cdots g^{pk_{t'}^{\text{wit}} r_{2t'}},\\ & g^{r_{11}}, \cdots, g^{r_{1t'}}, g^{r_{21}}, \cdots, g^{r_{2t'}}).\end{aligned}$$

In this way, the number of ciphertexts is reduced to $2t' + 2$. For tracing, each witness W_j is awakened by the hash tag, but partially decrypts the first two ciphertexts in the above equation by sending $((g^{r_{1j}})^{-pk_j^{\text{wit}}}, (g^{r_{2j}})^{-pk_j^{\text{wit}}}) = (g^{-r_j pk_j^{\text{wit}}}, g^{-r_{2j} pk_j^{\text{wit}}})$ to T. Next, T recovers (C_0, C_1) by removing $g^{sk_1^{\text{wit}} r_{11}} \cdots g^{pk_{t'}^{\text{wit}} r_{1t'}}$ and $g^{pk_1^{\text{wit}} r_{21}} \cdots g^{pk_w^{\text{wit}} r_{2t'}}$ from $C_0 g^{pk_1^{\text{wit}} r_{11}} \cdots g^{pk_{t'}^{\text{wit}} r_{1t'}}$ and $C_1 g^{pk_1^{\text{wit}} r_{21}} \cdots g^{pk_w^{\text{wit}} r_{2t'}}$ step by step, respectively.

7 Security and Privacy Analysis

Theorem 1. The HiTAPS scheme Π in Fig. 7 is unforgeable, accountable, and private, assuming that the underlying ATS is secure, the DTPKE is IND-NAA-NAC-CPA secure, the PKE is semantically secure, the hash function is preimage resistant, the (Prove, Verify) is an argument of knowledge and Honest Verifier Zero Knowledge (HVZK), the COM is hiding and binding, and the SIG is strongly unforgeable.

The proof of Theorem 1 is captured in the following three lemmas.

Lemma 1. The HiTAPS scheme Π is unforgeable and accountable if the ATS is secure, the (Prove, Verify) is an argument of knowledge, and COM is blinding, i.e., for all PPT adversaries $\mathcal{A}$, $\mathcal{A}_1$, and $\mathcal{A}_2$, such that

$$\mathbf{Adv}_{\mathcal{A},\Pi}^{\text{forg}}(\lambda) \leq \left(\mathbf{Adv}_{\mathcal{A}_1,\mathsf{ATS}}^{\text{forg}}(\lambda) + \mathbf{Adv}_{\mathcal{A}_2,\mathsf{COM}}^{\text{bind}}(\lambda)\right) \cdot \alpha(\lambda) + \beta(\lambda), \tag{1}$$

where α and β are the knowledge error and tightness of the proof system.

Proof. Proof. We prove Lemma 1 by defining three experiments.

Ept 0. It is the unforgeability and accountability experiment in Fig. 4 applied to Π. If Evt_0 is the event that $\mathcal{A}$ wins in Ept 0, then

$$\mathbf{Adv}_{\mathcal{A},\Pi}^{\text{forg}}(\lambda) = \Pr[Evt_0]. \tag{2}$$

Ept 1. Let $PK = (\mathsf{com}_{pk}, pk_c^{\mathsf{sig}}, pk_c^{\mathsf{enc}}, pk_t^{\mathsf{enc}}, \{wpk_j, pk_j^{\text{wit}}\}_{j=1}^{w})$ be the public key and let $tk \leftarrow (\{pk_i^s\}_{i=1}^n, sk_t^{\mathsf{enc}}, pk_c^{\mathsf{sig}})$ be he tracing key that are given to $\mathcal{A}$.

Ept 1 is identical to Ept 0 except that we strengthen the winning condition over Exp 0: to win Exp 1, $\mathcal{A}$ has to produce a valid forgery (m', σ''_m) where $\sigma'_m = (\widehat{\sigma}'_m, \{ht'_j\}, \pi', \eta')$, along with a witness $(pk'', r''_{pk}, \sigma''_m, \{r''_j\}_{j=1}^{t'})$ such that

$$\mathcal{R}((\mathsf{com}_{pk}, m', C'_1, C'_2); (pk'', r''_{pk}, \sigma''_m, \{r''_j\}_{j=1}^{t'})) = 1,$$

We construct an adversary $\mathcal{A}'$ from $\mathcal{A}$ in Exp 0. It invokes $\mathcal{A}$ and answers to all $\mathcal{A}$'s queries until it receives from $\mathcal{A}$ the $(m', \widehat{\sigma}'_m)$ to provide a statement $(\mathsf{com}_{pk}, m', C'_1, C'_2)$. $\mathcal{A}'$ executes an extractor Ext for (P, V) on $\mathcal{A}$'s remaining execution. Ext produces a witness $wt = (pk'', r''_{pk}, \sigma''_m, \{r''_j\}_{j=1}^{t'})$. $\mathcal{A}'$ uses wt and sk_c^{Sig} to generate π' and η' such that $\sigma'_m = (\widehat{\sigma}'_m, \{ht'_j\}, \pi', \eta')$ is a valid signature on m'. $\mathcal{A}'$ outputs (m', σ'_m) and wt. If Evt_1 stands for $\mathcal{A}'$ wins Exp 1, then

$$\Pr[Evt_1] \geq (\Pr[Evt_0] - \alpha(\lambda))/\beta(\lambda). \tag{3}$$

Ept 2. We strengthen the winning condition by requiring $pk'' = (t, \{pk_i^s\}_{i=1}^n)$. Given the binding property of COM, we have $\mathsf{COM.Verify}(pk, r_{pk}, \mathsf{com}_{pk}) = \mathsf{COM.Verify}(pk'', r''_{pk}, \mathsf{com}_{pk}) = 1$. If $pk'' \neq pk$, we find an attack on the binding property of COM. Specifically, let Evt_2 be the event that $\mathcal{A}'$ wins Ept 2 and E be the event that $pk \neq pk''$, then $\Pr[Evt_2] = \Pr[Evt_1 \wedge \neg E] \geq \Pr[Evt_1] - \Pr[E]$. We assume that there is an adversary $\mathcal{A}_2$ such that $\Pr[E] = \mathbf{Adv}^{\text{forg}}_{\mathcal{A}_2,\mathsf{COM}}(\lambda)$

We construct an adversary $\mathcal{A}_1$ that invokes $\mathcal{A}$ and answers to $\mathcal{A}$'s queries. When $\mathcal{A}$ outputs a forgery (m', σ'_m) and a witness $(pk'', r''_{pk}, \sigma''_m, \{r''_j\}_{j=1}^{t'})$ that meet the winning condition of Exp 1 and Exp 2, $\mathcal{A}_1$ outputs (m', σ''^{m}). By $\mathcal{R}$, we have σ''_m is a valid signature on m' with respect to pk''. By Exp 2, we have $pk = pk''$. Therefore, if $\mathcal{A}$ wins Exp 2, then (m', σ''_m) is a valid forgery for the ATS scheme. Since the ATS is secure, we have

$$\mathbf{Adv}^{\text{forg}}_{\mathcal{A}_1,\mathsf{ATS}}(\lambda) \geq \Pr[Evt_2]. \tag{4}$$

Conclusively, combining (2), (3), (4), and (5) proves (1). □

Lemma 2. The HiTAPS scheme Π is private against the public if the PKE is semantically secure, the SIG is strongly unforgeable the (Prove, Veriy) is an argument of knowledge and HVZK, the COM is hiding, the $\mathcal{H}$ is preimage resistant, and the DTPKE is IND-NAA-NAC-CPA secure, i.e., for all PPT adversaries $\mathcal{A}$, there exists adversaries $\mathcal{A}_1$, $\mathcal{A}_2$, $\mathcal{A}_3$, $\mathcal{A}_4$, $\mathcal{A}_5$, and $\mathcal{A}_6$ such that

$$\begin{aligned}\mathbf{Adv}^{\text{priP}}_{\mathcal{A},\Pi}(\lambda) \leq 2 \Big(& 3\mathbf{Adv}^{\text{ind-cpa}}_{\mathcal{A}_1,\mathsf{PKE}}(\lambda) + \mathbf{Adv}^{\text{euf-cma}}_{\mathcal{A}_2,\mathsf{SIG}}(\lambda) + Q \cdot \mathbf{Adv}^{\text{hvzk}}_{\mathcal{A}_3,(\mathsf{P},\mathsf{V})}(\lambda) \\ & + \epsilon_{\mathcal{A}_4}(\lambda) + \mathbf{Adv}^{\text{hash-pre}}_{\mathcal{A}_5,\mathsf{H}}(\lambda) + \mathbf{Adv}^{\text{ind-cpa}}_{\mathcal{A}_6,\mathsf{DTPKE}}(\lambda) \Big)\end{aligned} \tag{5}$$

where $\epsilon(\lambda)_{\mathcal{A}_4}$ is hiding statistical distance of COM and Q is query number.

Proof. We prove Lemma 2 by defining seven experiments.

Exp 0. It is the experiment of privacy against the public $\mathbf{Exp}^{\mathrm{priP}}$ defined in Fig. 2 applied to Π. If Evt_0 stands for $\mathcal{A}$ wins Exp 0, then

$$\mathbf{Adv}^{\mathrm{priP}}_{\mathcal{A},\Pi}(\lambda) = |2\Pr[Evt_0] - 1|. \tag{6}$$

Exp 1. It is identical to Exp 0 except that the signing oracle $\mathcal{O}_1(\mathcal{S}_0, \mathcal{S}_1, \mathcal{N}_0, \mathcal{N}_1, m)$ is modified such that step 2 of Sign in Fig. 5 now returns $\widehat{\sigma}_i \leftarrow \mathsf{PKE.Enc}(pk^e_j, 0)$, where 0 is encrypted instead of $(m||\sigma_i)$. Since PKE is semantically secure, $\mathcal{A}_1$'s **Adv** in Exp 1 is indistinguishable from its **Adv** in Exp 0, i.e., say Evt_1 stands for $\mathcal{A}_1$ wins Exp 1,

$$|\Pr[Evt_1] - \Pr[Evt_0]| \leq \mathbf{Adv}^{\text{ind-cpa}}_{\mathcal{A}_1,\mathsf{PKE}}(\lambda). \tag{7}$$

Exp 2. It is identical to Exp 0 except that responses to $\mathcal{O}_2(m, \sigma, \hat{\sigma})$ are fail. If SIG is strongly unforgeable, $\mathcal{A}_1$'s **Adv** in Exp 2 is indistinguishable from its **Adv** in Exp 1, i.e., say Evt_2 stands for $\mathcal{A}_2$ wins Exp 2,

$$|\Pr[Evt_2] - \Pr[Evt_1]| \leq \mathbf{Adv}^{\text{euf-cma}}_{\mathcal{A}_2,\mathsf{SIG}}(\lambda). \tag{8}$$

Exp 3. It is identical to Exp 2 except that $\mathcal{O}_1(m, \mathcal{S}_0, \mathcal{S}_1, \mathcal{N}_0, \mathcal{N}_1)$ is modified such that step 6 of $\mathsf{Combine}$ now generates a proof π by using the simulator, which is given $(\mathsf{com}_{pk}, m, C_1, C_2)$ as input. Since the simulated proofs are computationally indistinguishable from real proofs, $\mathcal{A}_3$'s **Adv** in Exp 3 is indistinguishable from its **Adv** in Exp 2, i.e., say Evt_3 stands for $\mathcal{A}_2$ wins Exp 3,

$$|\Pr[Evt_3] - \Pr[Evt_2]| \leq Q \cdot \mathbf{Adv}^{\mathrm{hvzk}}_{\mathcal{A}_3,(\mathsf{P},\mathsf{V})}(\lambda). \tag{9}$$

Exp 4. It is identical to Exp 3 except that step 2 of Setup in Fig. 5 is modified such that $r_{pk} \leftarrow \mathcal{R}_\lambda$, $\mathsf{com}_{pk} \leftarrow \mathsf{COM.Comm}(0, r_{pk})$, where 0 is committed instead of pk. Since COM is hiding, the adversary's **Adv** in Exp 4 is indistinguishable from its **Adv** in Exp 3, i.e., say Evt_3 stands for $\mathcal{A}_4$ wins Exp 4,

$$|\Pr[Evt_4] - \Pr[Evt_3]| \leq \epsilon_{\mathcal{A}_4}(\lambda). \tag{10}$$

Exp 5. This step resembles Exp 1, but the first part step 5 of $\mathsf{Combine}$ now returns $\mathsf{ElGamal.Enc}(pk^{\mathrm{wit}}_j, 0)$, $\mathsf{ElGamal.Enc}(pk^{\mathrm{wit}}_j, 0)$, and

$$|\Pr[Evt_5] - \Pr[Evt_4]| \leq 2\mathbf{Adv}^{\text{ind-cpa}}_{\mathcal{A}_5,\mathsf{PKE}}(\lambda). \tag{11}$$

Exp 6. It is identical to Exp 5 except that the signing oracle $\mathcal{O}_1(m, \mathcal{S}_0, \mathcal{S}_1, \mathcal{N}_0, \mathcal{N}_1)$ is modified such that the second part of step 5 of $\mathsf{Combine}$ now returns $\mathcal{H}(0)$. Since the $\mathcal{H}$ is preimage resistant, $\mathcal{A}_6$'s **Adv** in Exp 6 is indistinguishable from its **Adv** in Exp 5, i.e., say Evt_6 stands for $\mathcal{A}_6$ wins Exp 6,

$$|\Pr[Evt_6] - \Pr[Evt_5]| \leq \mathbf{Adv}^{\text{hash-pre}}_{\mathcal{A}_6,\mathsf{H}}(\lambda). \tag{12}$$

Exp 7. It is identical to Exp 6 except that $\mathcal{O}_1(m, \mathcal{S}_0, \mathcal{S}_1, \mathcal{N}_0, \mathcal{N}_1)$ is modified such that step 4 of $\mathsf{Combine}$ now returns $\hat{\sigma} \leftarrow \mathsf{DTPKE.Enc}(EK, 0, \sigma_m)$. Since

DTPKE is secure, $\mathcal{A}_7$'s **Adv** in Exp 7 is indistinguishable from its **Adv** in Exp 6, i.e., say Evt_7 stands for $\mathcal{A}_7$ wins Exp 7,

$$|\Pr[Evt_7] - \Pr[Evt_6]| \leq \mathbf{Adv}^{\text{ind-cpa}}_{\mathcal{A}_7,\mathsf{DTPKE}}(\lambda). \tag{13}$$

In Exp 7, $\mathcal{A}_7$'s view is independent of b, i.e.,

$$\Pr[Evt_7] = 1/2. \tag{14}$$

Lastly, combining (6)-(14) proves (5). This completes the proof of lemma 2. □

Lemma 3. The HiTAPS scheme Π is private against the signer.

The proof of Lemma 3 is almost identical to the proof of Lemma 2 and is omitted. Combining Lemma 1, Lemma 2, and Lemma 3, we prove Theorem 1.

Theorem 2. The HiTAPS2 scheme is unforgeable, accountable, and private.

The proof is Theorem 2 is almost identical to the proof of Theorem 1 except that the step 5 and step 6 of Combine are altered, leading to changes in corresponding experiments.

8 Performance Analysis

8.1 Experiment Settings

Dataset and Parameters. Considering the validity and observability of the results, we comprehensively provide the parameters. We change the number of signers n and the maximum number of witnesses w from 20 to 100, the length of the signer message m from 50 KB to 400 KB, the threshold t from 5 to 15, and the number of participating witnesses t' from 5 to 40, and finally change t and t' from 3 to 5.
Setup. We implemented HiTAPS using Intel(R) Core(TM) i5-10210U CPU @ 1.60GHz 2.11 GHz on a Linux server running Ubuntu 18.04. We use HMAC-SHA256 as the pseudo-random function to implement the hash function and AES as the symmetric encryption.

8.2 Computational Cost

HiTAPS consists of (Setup, Sign, Combine, Verify, Trace). We measured the time required for each phase under the given parameters.

In Setup, HiTAPS generates all keys. As shown in Fig. 8(a), Setup takes 314 ms when $n = 100$ and $w = 100$. In Sign, signers calculate signature shares, and in Fig. 8(b) we can see that it takes 154 ms to sign a message with a length of 400 KB. In Combine, the combiner combines signatures from t signature shares and constructs zero-knowledge proofs. By observing Fig. 8(c), we can find that as t increases, the time required for Combine also increases. When $t' = 5$ and $t = 5$, 100 groups of signers takes 22.3 s, while HiTAPS2 only takes 13.6 s. In

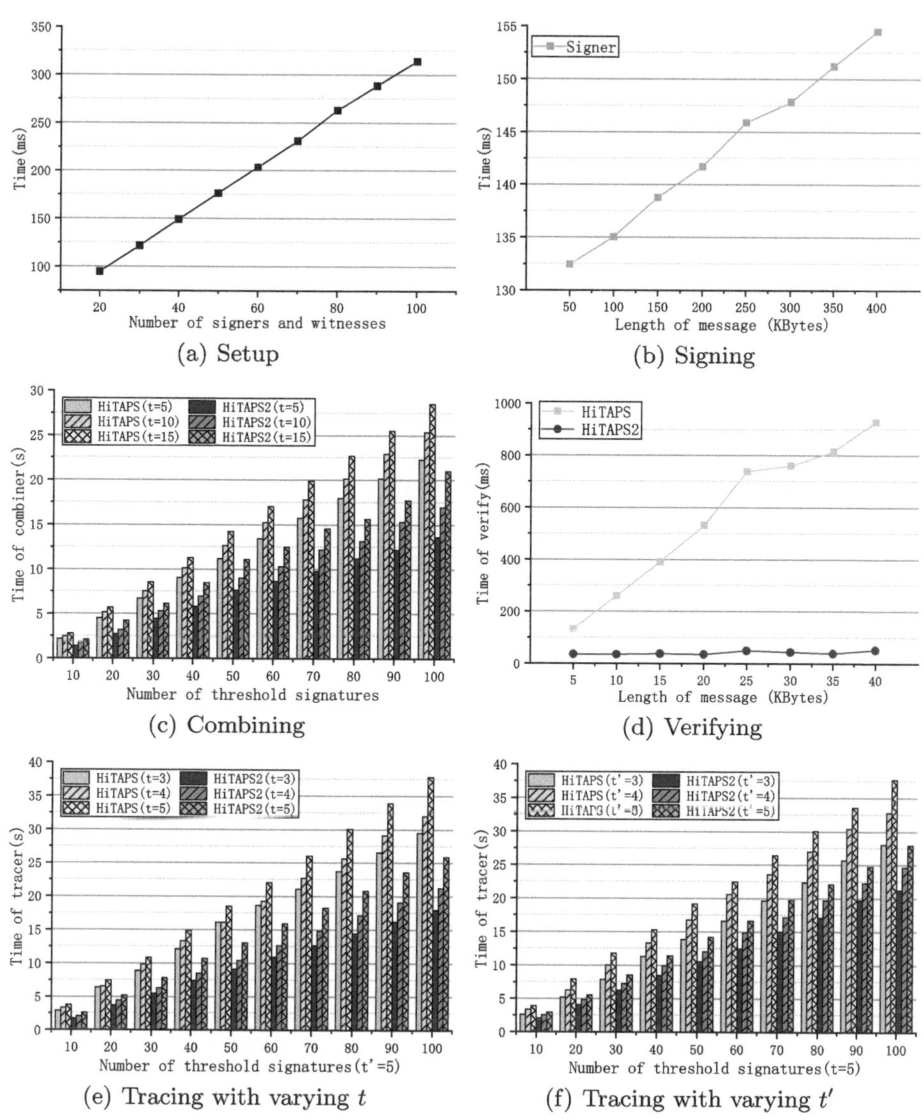

Fig. 8. Computational Costs.

Verify, the verifier verifies the threshold signature and the zero-knowledge proof given by combiner. As shown in Fig. 8(d), as t' increases, the time required for Verify changes from 131 ms to 928 ms, while the verification time of HiTAPS2 stabilizes at 41 ms. In Trace, the tracer traces a signing group. In Fig. 8(e) and Fig. 8(f), we can see that tracking 100 groups of signers at $t = 5$ and $t' = 5$ takes 37.8 s (HiTAPS) and 25.9 s (HiTAPS2), respectively.

8.3 Communication Overhead

We analyze communication overhead by calculating the length of messages transmitted by each party in a signing group.

In Sign, the signer sends plaintext m, signature $\widehat{\sigma}_i$, and encrypted witness sequence $\widehat{\mathcal{W}}_i$. In Combine, the combiner outputs a message m, an encrypted threshold signature σ, and the threshold signature contains $(\widehat{\sigma}_m, \{ht_j\}, \pi, \eta)$. In Verify, the verifier outputs 1 bit. In Trace, the witness gives his partial decryption key $\widehat{\delta}_j$, and the tracer combines the keys to trace and gives the signer sequence S. We record the communication overhead in Table 2.

Table 2. Communication Overhead

Phase	Signing	Combining	Verifying	Tracing	
Party	Signer	Combiner	Verifier	Witness	Tracer
Theory	$\lvert m\rvert$, $\lvert\widehat{\sigma}_i\rvert$, $\lvert\widehat{\mathcal{W}}_i\rvert$	$\lvert m\rvert$, $\lvert\sigma\rvert$	$\lvert b\rvert$	$\lvert\widehat{\delta}_j\rvert$	$\lvert\mathcal{S}\rvert$
Practice	1.969 KB	1.165 KB	1 bit	0.461 KB	0.445 KB

9 Conclusions

In this work, we have proposed HiTAPS, a new threshold signature scheme that achieves unforgeability, accountability, and privacy (against the public and signers). HiTAPS builds on TAPS and moves forward by facilitating witnessed tracing and securely designating witnesses. We formally prove the security and privacy of HiTAPS. Experimental results obtained from running a prototype show that HiTAPS is practical and efficient, HiTAPS2 is even better in Combine and Verify, e.g., HiTAPS2 stabilizes Verify at around 41 ms.

Acknowledgment. This work is supported by National Natural Science Foundation of China (NSFC) under the grant No. 62372149, No. U23A20303, and National Natural Science Foundation of China (NSFC) under the grant No. 62172040. It is partially supported by EU LOCARD Project under Grant H2020-SU-SEC-2018-832735.

References

1. Desmedt, Y., Frankel, Y.: Threshold cryptosystems. In: Proceedings of 6th Annual International Cryptology Conference (CRYPTO), pp. 307–315. Santa Barbara, USA (1989)
2. Shoup, V.: Practical threshold signatures. In: International Conference on the Theory and Application of Cryptographic Techniques(EUROCRYPT), pp. 14–18. Bruges, Belgium (2000)
3. Fouque, P., Stern, J.: Fully distributed threshold RSA under standard assumptions. International Conference on the Theory and Application of Cryptology and Information Security(ASIACRYPT), pp. 310–330. Gold Coast, Australia (2001)

4. Damgård, I., Koprowski, M.: Practical threshold RSA signatures without a trusted dealer. In: International Conference on the Theory and Applications of Cryptographic Techniques(EUROCRYPT), pp. 152–165. Innsbruck, Austria (2001)
5. Micali, S., Ohta, K., Reyzin, L.: Accountable-subgroup multi signatures: extended abstract. In: Proceedings of 8th ACM Conference on Computer and Communications Security (CCS), pp. 245–254. Philadelphia, USA (2001)
6. Boneh, D., Lynn, B., Shacham, H.: Short signatures from the weil pairing. In: Proceedings of 7th International Conference on the Theory and Application of Cryptology and Information Security (ASIACRYPT), pp. 514–532. Gold Coast, Australia (2001)
7. Bellare, M., Neven, G.: Multi-signatures in the plain public-key model and a general forking lemma. In: Proceedings of 13th ACM Conference on Computer and Communications Security (CCS), pp. 390–399. Alexandria, USA (2006)
8. Nick, J., Ruffing, T., Seurin, Y.: MuSig2: simple two-round Schnorr multi-signatures. In: Proceedings of 41st Annual International Cryptology Conference (CRYPTO), pp. 189–221, Virtual (2021)
9. Tang, G.: On tightly-secure (linkable) ring signatures. In: Proceedings of 23rd International Conference on Information and Communications Security (ICICS), pp. 375–393. Chongqing, China (2021)
10. Ji, Y., Tao, Y., Rui, Z.: More efficient construction of anonymous signatures. In: Proceedings of 23rd International Conference on Information and Communications Security (ICICS), pp. 394–411. Chongqing, China (2021)
11. Goyal, V., Song, Y., Srinivasan, A.: Traceable secret sharing and applications. In: Proceedings of 41st Annual International Cryptology Conference (CRYPTO), pp. 718–747, Virtual (2021)
12. Boneh, D., Komlo, C.: Threshold signatures with private accountability. In: Proceedings of 42nd Annual International Cryptology Conference (CRYPTO), pp. 551–581. Santa Barbara, USA (2022)
13. Delerablée, C., Pointcheval, D.: Dynamic threshold public-key encryption. In: Proceedings of 28th Annual International Cryptology Conference (CRYPTO), pp. 317–334. Santa Barbara, USA (2008)
14. Goldwasser, S., Micali, S., Rivest, R.L.: A digital signature scheme secure against adaptive chosen-message attack. SIAM J. Comput. **17**(2), 281–308 (1988)
15. Loh, J.-C., Guo, F., Susilo, W., Yang, G.: A tightly secure ID-based signature scheme under DL assumption in AGM. In: Proceedings of 28th Australasian Conference on Information Security and Privacy (ACISP), pp. 199–219. Brisbane, Australia (2023)
16. Gennaro, R., Jarecki, S., Krawczyk, H., Rabin, T.: Robust threshold DSS signatures. Inf. Comput. **164**(1), 54–84 (2001)
17. Katz, J., Lindell, Y.: Introduction to Modern Cryptography (Third edition), pp. 1–598. CRC Press (2021)
18. Boneh, D., Boyen, X.: Hierarchical identity based encryption with constant size ciphertext. Proceedings of 24th International Conference on the Theory and Application of Cryptographic Technique (EUROCRYPT), pp. 440–456 (2005)
19. Delerablé, C., Paillier, P., Pointcheval, D.: Fully collusion secure dynamic broadcast encryption with constant-size ciphertexts or decryption keys. In: Proceedings of first International Conference on Pairing-Based Cryptography (Pairing), pp. 39–59 (2007)

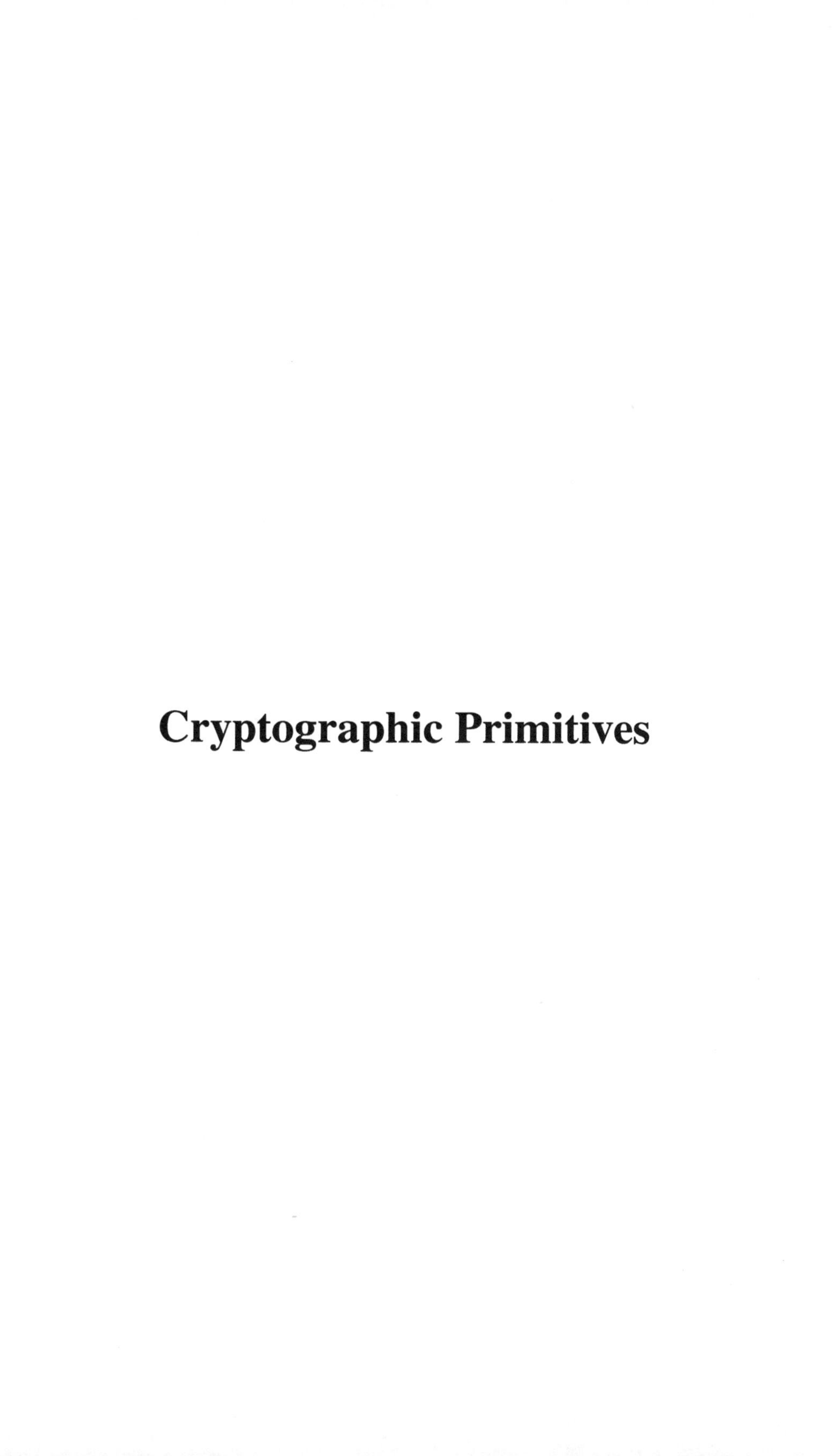

Cryptographic Primitives

A Novel Window τNAF on Koblitz Curves

Xiuxiu Li[1,2,3], Wei Yu[1,2,3(✉)], and Kunpeng Wang[1,3]

[1] Key Laboratory of Cyberspace Security Defense, Institute of Information Engineering, CAS, Beijing 100085, China
yuwei@iie.ac.cn, yuwei_1_yw@163.com
[2] State Key Laboratory of Cryptology, P. O. Box 5159, Beijing 100878, China
[3] School of Cyberspace Security, University of Chinese Academy of Sciences, Beijing, China

Abstract. The window τ-adic non-adjacent form (window τNAF) was initially proposed by Solinas in 2000 as a method to calculate scalar multiplication on the Koblitz curves. To ensure the correctness of the window τNAF, Blake, Murty, and Xu demonstrated that the pre-computation scheme of this method must be convergent. To date, the standard representation system for scalar multiplication on the Koblitz curve $E_a/\mathbb{F}_{2^m}$ using the Frobenius map τ remains the window τNAF. In this paper, we present a novel algorithm that extends the window τNAF approach and improves its pre-computation scheme. Our algorithm integrates several window τNAFs, where the initial few are non-convergent and the final one is the standard window τNAF. Compared with the previous state-of-the-art method, our approach achieves a 5% reduction in the time for implementing scalar multiplication when using the μ_4-Koblitz curves and LD coordinates. This work offers significant advancements in optimizing scalar multiplication on Koblitz curves.

Keywords: Elliptic curve cryptography · Koblitz curve · Scalar multiplication · Window τNAF · Pre-computation

1 Introduction

Elliptic curve cryptography (ECC) is a popular cryptographic technique [20,26] that relies on scalar multiplication as its core operation. The scalar multiplication operation is multiplying a point P on the elliptic curves by a non-negative integer n, denoted as nP. However, the scalar multiplication is computationally expensive, particularly for large scalars n. To accelerate scalar multiplication, several approaches have been devised, including techniques based on windowing, recoding, and projective coordinates [2,8,10,17,21,26].

In 1991, Koblitz [11] introduced a family of non-supersingular curves called Koblitz curves, defined over the binary field $\mathbb{F}_2$. These curves are designed to optimize the scalar multiplication, making them an attractive choice for ECC applications. Koblitz curves E_a are represented by the equation

$$y^2 + xy = x^3 + ax^2 + 1,$$

T. Zhu and Y. Li (Eds.): ACISP 2024, LNCS 14895, pp. 411–428, 2024.
https://doi.org/10.1007/978-981-97-5025-2_21

where $a \in \{0, 1\}$. Koblitz curves are typically used over binary extension fields $\mathbb{F}_{2^m}$, where m is an integer. One of the significant advantage of Koblitz curves is using the Frobenius map τ to replace the point doubling operation. The Frobenius map τ is an inexpensive operation as squaring in $\mathbb{F}_{2^m}$ can be done efficiently, resulting in efficient scalar multiplication.

The Frobenius endomorphism τ takes a point $P(x, y)$ on the curve to another point $Q(x^2, y^2)$, replacing every doubling operation in the standard double-and-add algorithm. This leads to a scalar multiplication algorithm with lower computational cost. In particular, the cost of scalar multiplication using the Frobenius endomorphism method on Koblitz curves is $O(klog(k))\mathbf{M} + O(k)\mathbf{S}$, where k represents the bit length of the scalar n, $\mathbf{M}$ is the cost of a multiplication, and $\mathbf{S}$ is the cost of a squaring in $\mathbb{F}_{2^m}$. In contrast, the standard double-and-add method has a cost of $O(k^2)\mathbf{M} + O(k^2)\mathbf{S}$. The set of all endomorphisms τ on E_a constitutes the Euclidean domain $\mathbb{Z}[\tau] = \mathbb{Z} + \tau\mathbb{Z}$.

Several approaches have been designed to speed up scalar multiplication on Koblitz curves. One approach is the window τNAF method, which was first proposed by Solinas [21]. This method involves precomputing $2^{\omega-2} - 1$ points for a given window width ω. Pre-computation techniques can be employed to further optimize the window τNAF method. For example, Blake, Murty, and Xu [4] proposed a more flexible framework that adapts the pre-computed points based on the scalar values, which is accompanied by a termination proof of the window τNAF. Additionally, Trost and Xu [24] introduced a highly efficient pre-computation scheme that requires only $2^{\omega-2} - 1$ point additions and two τ evaluations for a given window width ω. Regarding software implementation, Oliveira et al. [22] achieved a fast single-core constant-time point multiplication by using a regular τ-adic recoding approach for Koblitz curves. Furthermore, in [23], Oliveira et al. reported constant-time scalar multiplications of Koblitz curves defined over $\mathbb{F}_4$, achieving very competitive timings compared to other libraries that implement scalar multiplication over binary and prime field elliptic curves.

Another promising method to speed up scalar multiplication on Koblitz curves is the use twisted μ_4-normal form elliptic curves over binary field, as introduced by Kohel [13]. Kohel showed that all the Koblitz curves over binary fields can be transformed into μ_4-Koblitz curves. The use of μ_4-Koblitz curves in the window τNAF, particularly for the pre-computation, is therefore of significant interest due to its potential computational advantages. Recent developments in $\mu\bar{\tau}$-operations by Yu and Xu [27] reduced the cost of pre-computation and enabled to use larger widths in the window τNAF where $\mu = (-1)^{1-a}$ and $\bar{\tau}$ is the complex conjugate of τ, resulting in faster scalar multiplication.

Our Contributions. Although many efforts have been made to improve the efficiency of scalar multiplication on Koblitz curves, it remains the most computationally intensive part in corresponding cryptographic schemes. Our purpose is to further accelerate scalar multiplication on Koblitz curves by following key contributions:

1. We introduce a non-convergent window τNAF method that initially expands the integer using bigger window widths and subsequently decreases the window width. While the initial steps of the novel window τNAF are non-convergent, the final step employing the convergent τ-4 NAF ensures the correctness of the full algorithm. The convergent portion of the window τNAF utilizes a width of 4, while the non-convergent part is optimized for speed.
2. We incorporate a non-convergent pre-computation scheme into the novel non-convergent window τNAF, further enhancing the efficiency of scalar multiplication. The pre-computation process is explained in Sect. 3, and we give its cost on three different coordinates of Koblitz curves with $a = 0$ and $a = 1$ in Table 1. The algorithm is designed to operate specifically on Koblitz curves and aims to reduce the number of point additions required for scalar multiplication.
3. This paper highlights that μ_4-Koblitz curves and LD coordinates represent the state-of-the-art in scalar multiplication on Koblitz curves, achieving a 5% reduction in time for scalar multiplication compared to previous literature results.

Table 1. Cost of our pre-computations on Koblitz curves

		$w = 4$	$w = 5$	$w = 6$	$w = 7$	$w = 8$	$w = 9$
$a = 0$	LD	5M+6S	13M+14S	29M+30S	61M+62S	125M+126S	253M+254S
	λ	7M+5S	19M+13S	43M+29S	91M+61S	187M+125S	379M+253S
	μ_4	6M+6S	14M+14S	30M+30S	62M+62S	126M+126S	254M+254S
$a = 1$	LD	5M+3S	13M+7S	29M+15S	61M+31S	125M+63S	253M+127S
	λ	7M+5S	19M+13S	43M+29S	91M+61S	187M+125S	379M+253S
	μ_4	6M+6S	14M+14S	30M+30S	62M+62S	126M+126S	254M+254S

The paper is organized into six sections. Section 1 is the introduction. In Sect. 2, we present the fundamentals that helps to comprehend the rest of this article. Section 3 puts forth a new pre-computation schemes for our window τNAF. Section 4 focuses on the design of the novel window τNAF method. Section 5 analyzes the experimental results of these scalar multiplications using different pre-computation schemes. Finally, Sect. 6 concludes this paper.

2 Preliminary

2.1 Testing $\tau^w|(g + h\tau)$

The criterion $\tau^w|(g + h\tau) \Leftrightarrow 2^w|(g + hs_w)$ for determining whether $\tau^w|(g + h\tau)$ holds in $\mathbb{Z}[\tau]$. This criterion has been established in previous works [5,21], and is done using either the Lucas sequence [21] or Hensel's lift algorithm [9] to compute the value of s_w for a given $w \in \mathbb{Z}^+$. Specifically, the values of s_w for $w = 1, 2, ..., 10$ are explicitly given as $s_1 = 0$, $s_2 = 2\mu$, $s_3 = 6\mu$, $s_4 = 6\mu$, $s_5 = 6\mu$, $s_6 = 38\mu$, $s_7 = 38\mu$, $s_8 = 166\mu$, $s_9 = 422\mu$, and $s_{10} = 934\mu$, where $\mu = (-1)^{1-a}$

and a is parameter of the Koblitz curves. It follows that when $\omega \geq 2$, $s_w/2$ is odd and $s_w \equiv 0 \pmod 2$. Therefore, the criterion can be used to test whether $\tau^w|(g + h\tau)$ holds in $\mathbb{Z}[\tau]$.

2.2 Costs of Point Operations on Koblitz Curves

Let M be the main subgroup of the elliptic curves point group $E_a(\mathbb{F}_{2^m})$, namely the subgroup of order p. We consider rational points P and Q in the M and denote the τ-affine operation of P by $\tau(P)$ or the mixed addition of P and Q by $P + Q$ when the Z-coordinate of P is 1 using LD coordinates, which are inverse-free and commonly used in elliptic curve systems over binary fields. We summarize the costs of point operations on Koblitz curves using LD coordinates [16], λ-coordinates [18], and on μ_4-Koblitz curves [13] in Table 2, neglecting the cost of field addition as it only involves bitwise XORs. These costs are important in the context of scalar multiplication on Koblitz curves, as they impact the overall performance of elliptic curve schemes.

Table 2. Costs of point operations on Koblitz curves

Coordinates	$\tau(P)$	τ-affine operation	addition	mixed addition*
LD coordinates [15,16]	3S	2S	13M+4S	8M+5S
λ-coordinates [18]	3S	2S	11M+2S	8M+2S
μ_4-Koblitz curve ($a = 0$) [12]	4S	3S	7M+2S	6M+2S
μ_4-Koblitz curve ($a = 1$) [13,14]	4S	3S	8M+2S	7M+2S

*

We can translate a Koblitz curve $y^2+xy = x^3+ax^2+1$ into a μ_4-Koblitz curve $X_0^2+X_2^2 = X_1X_3+aX_0X_2$, $X_1^2+X_3^2 = X_0X_2$ by the map $(x, y) \mapsto (x^2 : x^2+y : 1 : x^2 + y + x)$ and the inverse is $(X_0 : X_1 : X_2 : X_3) \mapsto (X_1 + X_3 : X_0 + X_1 : X_2)$. Furthermore, there are $\tau(X_0 : X_1 : X_2 : X_3) = (X_0^2 : X_1^2 : X_2^2 : X_3^2)$ and $\tau^2(P) + 2P = \mu\tau(P)$.

This paper considers four Koblitz curves, namely K-233($a = 0$), K-283($a = 0$), K-409($a = 0$) and K-571($a = 0$), and four other curves, namely K1-163, K1-283, K1-359 and K1-701 [3]. For $a = 0$, adding one affine point to another requires 5M+2S [12], while for $a = 1$, it requires 6M+2S [13,14].

2.3 The Window τNAF Method

In 1991, Koblitz [11] proposed a method for computing scalar multiplication that represents the scalar as a sum of powers of complex number τ. The efficiency of this method was improved by Solinas [21], who introduced the "τNAF" method for computing scalar multiplication using a sparse representation of the scalar. [4] proofs the termination of Solinas algorithm. Later, Blake, Murty, and Xu [5] further refined and extended the τNAF method to every integer in a Euclidean imaginary quadratic number field.

The main subgroup M mentioned above is an annihilating subgroup of $\delta = \frac{\tau^m - 1}{\tau - 1}$ in the sense that $\delta(P) = O$ for every $P \in M$. The window τNAF representation of n is the sparse τ expansion of n's reduction ρ. More specifically, we first choose a suitable $\rho \in \mathbb{Z}[\tau]$ satisfying $\rho \equiv n \pmod{\delta}$. Then, we denote the window τNAF of ρ as $\sum_{i=0}^{l-1} \epsilon_i c_i \tau^i$, where $\epsilon_i \in \{-1, 1\}$ and each coefficient c_i is chosen from the set $R_i = \{g + h\tau | g + h\tau \equiv i \pmod{\tau^w}, N(g + h\tau) < 2^w\}$, where ω is the window width and N denotes the norm function. The set R_i is chosen such that any subset $\{c_k, c_{k+1}, \ldots, c_{k+w-1}\}$ contains at most one nonzero element c_j.

Finally, we pre-compute each $Q_i = c_i P$ and compute nP using the pre-computed values and the window τNAF expansion of ρ. This method allows for efficient scalar multiplication on Koblitz curves, and its effectiveness has been demonstrated in [4,24].

2.4 Previous Pre-computation Schemes

The window τNAF is a popular method for computing scalar multiplication on Koblitz curves that includes pre-computation. Specially, the convergence of window τNAF requires that the norm $N(c_i)$ of pre-computed elements c_i satisfying $N(c_i) < 2^w$. When this condition is met, the window τNAF is said to be convergent; otherwise, it is non-convergent. Previous pre-computation schemes in the literature are all convergent, due to the pre-computed elements c_i satisfying the norm condition. The main operation we need for the pre-computation part is the $\mu\bar{\tau}$-operations. Table 3 gives the cost of $\mu\bar{\tau}$-operations on multiple coordinates, that will be used in our pre-computation scheme.

Table 3. Cost of $\mu\bar{\tau}$-operations on Koblitz curves

Coordinates	$\mu\bar{\tau}(P)$ $Z = 1$	$\mu\bar{\tau}(P)$	re-$\mu\bar{\tau}(P)$	$\mu\bar{\tau}P\&(\mu\bar{\tau})^2P$ with $Z = 1$ *
LD($a = 0$)	**M**+2**S** [7]	2**M**+2**S** [7]	–	–
LD($a = 1$)	**M**+**S** [7]	2**M**+**S** [7]	–	–
λ	2**M**+2**S** [27]	5**M**+3**S** [27]	3**M**+2**S** [27]	4**M**+3**S** [27]
μ_4	–	2**M**+2**S** [27]	–	–

3 A New Pre-computation Scheme for Window τNAF

Firstly, we have two equations $\tau^2 - \mu\tau + 2 = 0$, $\mu\bar{\tau} = 1 - \mu\tau$ and several properties of $\mu\bar{\tau}$ that will be used in our new window τNAF.

Lemma 1 *[1] Let* $\mathrm{ord}_{\tau^w}(\bar{\tau})$ *denote the multiplicative order of* $\bar{\tau} \pmod{\tau^w}$. *Then*

$$\mathrm{ord}_{\tau^w}(\mu\bar{\tau}) = 2^{w-2}, \ w \geq 3.$$

Proof. As per [1], we have

$$\mathrm{ord}_{\tau^w}(\bar{\tau}) = 2^{w-2}$$

Lemma 2 *1. For* $0 \le i < j < 2^{w-2}$, $(\mu\bar{\tau})^i \not\equiv \pm(\mu\bar{\tau})^j \pmod{2^w}$.

2. *Given that* $(\mu\bar{\tau})^i$ *is odd, it must belong to one of the congruence classes of* $-2^{w-1}+1, -2^{w-1}+3, \ldots, -3, -1, 1, 3, \ldots, 2^{w-1}-3, 2^{w-1}-1$ *modulo* 2^w.
3. *The set* $\pm(\mu\bar{\tau})^0, \pm(\mu\bar{\tau})^1, \ldots, \pm(\mu\bar{\tau})^{2^{w-2}-1}$ *covers all congruence classes* $R_i = \{g + h\tau | g + h\tau \equiv i \pmod{\tau^w}, N(g + h\tau) < 2^w\}$, *where* $i \in I_w = \{1, 3, \cdots, 2^{\omega-1}-1\}$.

Proof. 1. Assume that $(\mu\bar{\tau})^i \equiv (\mu\bar{\tau})^j \pmod{2^\omega}$ with $0 \le i < j < 2^{\omega-2}$. According to Corollary 1 in [1], the order of $\bar{\tau}$ modulo τ^ω is $2^{\omega-2}$, which implies that $i = j$. This leads to a contradiction.
   Suppose that $(\mu\bar{\tau})^i \equiv -(\mu\bar{\tau})^j \pmod{2^\omega}$ with $0 \le i < j < 2^{\omega-2}$. Reducing modulo τ^3, we get that the coprime resides modulo τ^3 are ± 1, $\pm\bar{\tau}$. Thus hypothesis is not valid.
2. From (1), we observe that each $(\mu\bar{\tau})^i$ is different when i is different. Since $(\mu\bar{\tau})^i$ is odd and the order of $(\mu\bar{\tau})^i$ is 2^{w-2}. We obtain the result.
3. Combining above (1) and (2), the lemma can be obtained.

The results hitherto proved allow us to designs a novel pre-computation scheme for the window τNAF method. Let $P(X_0 : X_1 : X_2 : X_3)$ and $Q = (Y_0 : Y_1 : Y_2 : Y_3)$ be rational points on Koblitz curves. As shown below, the scheme [27] that employed the $\mu\bar{\tau}$-operations reduces the cost of computing $\mu\bar{\tau}P$ from 7**M**+2**S** to 2**M**+2**S** for both $a = 0$ and $a = 1$ on the μ_4-Koblitz curves.

$$\begin{cases} Z_0 = & (X_0Y_0 + X_2Y_2)^2 \\ Z_1 = & X_0Y_0X_1Y_1 + X_2Y_2X_3Y_3 \\ Z_2 = & (X_1Y_1 + X_3Y_3)^2 \\ Z_3 = & (X_0Y_0 + X_2Y_2)(X_1Y_1 + X_3Y_3) + Z_1 \end{cases} \Longrightarrow \begin{cases} Z_0 = & (X_0 + X_2)^2 \\ Z_1 = & X_0X_3 + X_1X_2 \\ Z_2 = & (X_1 + X_3)^2 \\ Z_3 = & X_0X_1 + X_2X_3 \end{cases}$$

$$\mathbf{7M} + \mathbf{2S} \quad \Longrightarrow \quad \mathbf{2M} + \mathbf{2S}$$

As well as for LD coordinates, their $\mu\bar{\tau}$-operations' costs are 2**M**+2**S** and 2**M**+**S** for $a = 0$ and $a = 1$, respectively.

For window widths from 4 to 9, our new pre-computation requires 6**M**+6**S**, 14**M**+14**S**, 30**M**+30**S**, 62**M**+62**S**, 126**M**+126**S**, and 254**M**+254**S** on the μ_4-Koblitz curves. The explicit computing procedure and costs of the pre-computation using μ_4-Koblitz curves, LD coordinates, and λ-coordinate for window widths
from 4 to 6, 7, 8, and 9 are shown in Tables 4, 5, 9, 10, and 11, with Tables 9, 10, and 11 available in Appendix A. We also employs $\mu\bar{\tau}(P)$ and $(P \pm Q)$-operations to replace point addition for each Q_i $(i = 3, 5, \ldots, 2^{w-1}-1)$, leading to a more speedup for the pre-computation part.

For window τNAF with widths from 4 to 6, Table 6 compares the cost of our pre-computation with other existing pre-computation schemes on the μ_4-Koblitz curves. The results demonstrate that our method exhibiting the fastest performance among the four pre-computation methods considered.

Pre-computation is increasingly important in the computation of scalar multiplication, especially for unfixed point scalar multiplication. Our work offers a promising approach for optimizing scalar multiplication performance on Koblitz curves and can be applied in various cryptographic applications. To evaluate the performance of our window τNAF method, four different cases are mentioned in subsequent experiments, where $\mathbf{I}/\mathbf{M} = 10$ is fixed and the ratio of $\mathbf{S}/\mathbf{M}$ varies among 0, 0.1, 0.2, and 0.5.

Table 4. Pre-computation for widths from 4 to 6

	c_i	Q_i	$LD(a=0/a=1)$	λ	μ_4
$w=3$	$c_3 = 1 - \mu\tau = \mu\bar{\tau}$	$Q_3 = \mu\bar{\tau}P$	M+2S/M+S	2M+2S	2M+2S
$w=4$	$c_5 = -1 + \mu\tau = -\mu\bar{\tau}$	$Q_5 = -\mu\bar{\tau}P$	M+2S/M+S		2M+2S
	$c_7 = 1 + \mu\tau = -(\mu\bar{\tau})^2$	$Q_7 = \mu\bar{\tau}Q_5$	2M+2S/2M+S	4M+3S	2M+2S
	$c_3 = -3 + \mu\tau = (\mu\bar{\tau})^3$	$Q_3 = -\mu\bar{\tau}Q_7$	2M+2S/2M+S	3M+2S	2M+2S
$w=5$	$c_5 = -1 + \mu\tau = -\mu\bar{\tau}$	$Q_5 = -\mu\bar{\tau}P$	M+2S/M+S		2M+2S
	$c_7 = 1 + \mu\tau = -(\mu\bar{\tau})^2$	$Q_7 = \mu\bar{\tau}Q_5$	2M+2S/2M+S	4M+3S	2M+2S
	$c_3 = -3 + \mu\tau = (\mu\bar{\tau})^3$	$Q_3 = -\mu\bar{\tau}Q_7$	2M+2S/2M+S	3M+2S	2M+2S
	$c_{15} = 1 - 3\mu\tau = -(\mu\bar{\tau})^4$	$Q_{15} = -\mu\bar{\tau}Q_3$	2M+2S/2M+S	3M+2S	2M+2S
	$c_{11} = 5 + \mu\tau = (\mu\bar{\tau})^5$	$Q_{11} = -\mu\bar{\tau}Q_{15}$	2M+2S/2M+S	3M+2S	2M+2S
	$c_9 = 7 - 5\mu\tau = (\mu\bar{\tau})^6$	$Q_9 = \mu\bar{\tau}Q_{11}$	2M+2S/2M+S	3M+2S	2M+2S
	$c_{13} = 3 + 7\mu\tau = -(\mu\bar{\tau})^7$	$Q_{13} = -\mu\bar{\tau}Q_9$	2M+2S/2M+S	3M+2S	2M+2S
$w=6$	$c_{27} = 1 - \mu\tau = \mu\bar{\tau}$	$Q_{27} = \mu\bar{\tau}P$	M+2S/M+S		2M+2S
	$c_{25} = -1 - \mu\tau = (\mu\bar{\tau})^2$	$Q_{25} = \mu\bar{\tau}Q_{27}$	2M+2S/2M+S	4M+3S	2M+2S
	$c_{29} = 3 - \mu\tau = -(\mu\bar{\tau})^3$	$Q_{29} = -\mu\bar{\tau}Q_{25}$	2M+2S/2M+S	3M+2S	2M+2S
	$c_{15} = 1 - 3\mu\tau = -(\mu\bar{\tau})^4$	$Q_{15} = \mu\bar{\tau}Q_{29}$	2M+2S/2M+S	3M+2S	2M+2S
	$c_{21} = -5 - \mu\tau = -(\mu\bar{\tau})^5$	$Q_{21} = \mu\bar{\tau}Q_{15}$	2M+2S/2M+S	3M+2S	2M+2S
	$c_9 = 7 - 5\mu\tau = (\mu\bar{\tau})^6$	$Q_9 = -\mu\bar{\tau}Q_{21}$	2M+2S/2M+S	3M+2S	2M+2S
	$c_{13} = 3 + 7\mu\tau = -(\mu\bar{\tau})^7$	$Q_{13} = -\mu\bar{\tau}Q_9$	2M+2S/2M+S	3M+2S	2M+2S
	$c_{31} = 17 - 3\mu\tau = -(\mu\bar{\tau})^8$	$Q_{31} = \mu\bar{\tau}Q_{13}$	2M+2S/2M+S	3M+2S	2M+2S
	$c_5 = 11 - 17\mu\tau = -(\mu\bar{\tau})^9$	$Q_5 = \mu\bar{\tau}Q_{31}$	2M+2S/2M+S	3M+2S	2M+2S
	$c_7 = -23 - 11\mu\tau = -(\mu\bar{\tau})^{10}$	$Q_7 = \mu\bar{\tau}Q_5$	2M+2S/2M+S	3M+2S	2M+2S
	$c_3 = 45 - 23\mu\tau = (\mu\bar{\tau})^{11}$	$Q_3 = -\mu\bar{\tau}Q_7$	2M+2S/2M+S	3M+2S	2M+2S
	$c_{17} = -1 - 45\mu\tau = (\mu\bar{\tau})^{12}$	$Q_{17} = \mu\bar{\tau}Q_3$	2M+2S/2M+S	3M+2S	2M+2S
	$c_{11} = -91 + \mu\tau = (\mu\bar{\tau})^{13}$	$Q_{11} = \mu\bar{\tau}Q_{17}$	2M+2S/2M+S	3M+2S	2M+2S
	$c_{23} = 89 - 91\mu\tau = -(\mu\bar{\tau})^{14}$	$Q_{23} = -\mu\bar{\tau}Q_{11}$	2M+2S/2M+S	3M+2S	2M+2S
	$c_{19} = 93 + 89\mu\tau = (\mu\bar{\tau})^{15}$	$Q_{19} = -\mu\bar{\tau}Q_{23}$	2M+2S/2M+S	3M+2S	2M+2S

Table 5. Pre-computation for width 7

c_i	Q_i	LD(a=0/a=1)	λ	μ_4
$c_{37} = -1 + \mu\tau = -\mu\bar{\tau}$	$Q_{37} = -\mu\bar{\tau}P$	M+2S/M+S		2M+2S
$c_{39} = 1 + \mu\tau = -(\mu\bar{\tau})^2$	$Q_{39} = \mu\bar{\tau}Q_{37}$	2M+2S/2M+S	4M+3S	2M+2S
$c_{35} = -3 + \mu\tau = (\mu\bar{\tau})^3$	$Q_{35} = -\mu\bar{\tau}Q_{39}$	2M+2S/2M+S	3M+2S	2M+2S
$c_{15} = 1 - 3\mu\tau = -(\mu\bar{\tau})^4$	$Q_{15} = -\mu\bar{\tau}Q_{35}$	2M+2S/2M+S	3M+2S	2M+2S
$c_{43} = 5 + \mu\tau = (\mu\bar{\tau})^5$	$Q_{43} = -\mu\bar{\tau}Q_{15}$	2M+2S/2M+S	3M+2S	2M+2S
$c_{55} = -7 + 5\mu\tau = -(\mu\bar{\tau})^6$	$Q_{55} = -\mu\bar{\tau}Q_{43}$	2M+2S/2M+S	3M+2S	2M+2S
$c_{13} = 3 + 7\mu\tau = -(\mu\bar{\tau})^7$	$Q_{13} = \mu\bar{\tau}Q_{55}$	2M+2S/2M+S	3M+2S	2M+2S
$c_{31} = 17 - 3\mu\tau = -(\mu\bar{\tau})^8$	$Q_{31} = \mu\bar{\tau}Q_{13}$	2M+2S/2M+S	3M+2S	2M+2S
$c_5 = 11 - 17\mu\tau = -(\mu\bar{\tau})^9$	$Q_5 = \mu\bar{\tau}Q_{31}$	2M+2S/2M+S	3M+2S	2M+2S
$c_{57} = 23 + 11\mu\tau = (\mu\bar{\tau})^{10}$	$Q_{57} = -\mu\bar{\tau}Q_5$	2M+2S/2M+S	3M+2S	2M+2S
$c_{61} = -45 + 23\mu\tau = -(\mu\bar{\tau})^{11}$	$Q_{61} = -\mu\bar{\tau}Q_{57}$	2M+2S/2M+S	3M+2S	2M+2S
$c_{47} = 1 + 45\mu\tau = -(\mu\bar{\tau})^{12}$	$Q_{47} = \mu\bar{\tau}Q_{61}$	2M+2S/2M+S	3M+2S	2M+2S
$c_{53} = 91 - \mu\tau = -(\mu\bar{\tau})^{13}$	$Q_{53} = \mu\bar{\tau}Q_{47}$	2M+2S/2M+S	3M+2S	2M+2S
$c_{41} = -89 + 91\mu\tau = (\mu\bar{\tau})^{14}$	$Q_{41} = -\mu\bar{\tau}Q_{53}$	2M+2S/2M+S	3M+2S	2M+2S
$c_{19} = 93 + 89\mu\tau = (\mu\bar{\tau})^{15}$	$Q_{19} = \mu\bar{\tau}Q_{41}$	2M+2S/2M+S	3M+2S	2M+2S
$c_{63} = -271 + 93\mu\tau = -(\mu\bar{\tau})^{16}$	$Q_{63} = -\mu\bar{\tau}Q_{19}$	2M+2S/2M+S	3M+2S	2M+2S
$c_{27} = 85 - 271\mu\tau = (\mu\bar{\tau})^{17}$	$Q_{27} = -\mu\bar{\tau}Q_{63}$	2M+2S/2M+S	3M+2S	2M+2S
$c_{25} = -457 - 85\mu\tau = (\mu\bar{\tau})^{18}$	$Q_{25} = \mu\bar{\tau}Q_{27}$	2M+2S/2M+S	3M+2S	2M+2S
$c_{29} = 627 - 457\mu\tau = -(\mu\bar{\tau})^{19}$	$Q_{29} = -\mu\bar{\tau}Q_{25}$	2M+2S/2M+S	3M+2S	2M+2S
$c_{49} = 287 + 627\mu\tau = (\mu\bar{\tau})^{20}$	$Q_{49} = -\mu\bar{\tau}Q_{29}$	2M+2S/2M+S	3M+2S	2M+2S
$c_{21} = -1541 + 287\mu\tau = -(\mu\bar{\tau})^{21}$	$Q_{21} = -\mu\bar{\tau}Q_{49}$	2M+2S/2M+S	3M+2S	2M+2S
$c_9 = 967 - 1541\mu\tau = (\mu\bar{\tau})^{22}$	$Q_9 = -\mu\bar{\tau}Q_{21}$	2M+2S/2M+S	3M+2S	2M+2S
$c_{51} = -2115 - 967\mu\tau = (\mu\bar{\tau})^{23}$	$Q_{51} = \mu\bar{\tau}Q_9$	2M+2S/2M+S	3M+2S	2M+2S
$c_{33} = -4049 + 2115\mu\tau = (\mu\bar{\tau})^{24}$	$Q_{33} = \mu\bar{\tau}Q_{51}$	2M+2S/2M+S	3M+2S	2M+2S
$c_{59} = 181 + 4049\mu\tau = (\mu\bar{\tau})^{25}$	$Q_{59} = \mu\bar{\tau}Q_{33}$	2M+2S/2M+S	3M+2S	2M+2S
$c_7 = -8279 + 181\mu\tau = -(\mu\bar{\tau})^{26}$	$Q_7 = -\mu\bar{\tau}Q_{59}$	2M+2S/2M+S	3M+2S	2M+2S
$c_3 = 7917 - 8279\mu\tau = (\mu\bar{\tau})^{27}$	$Q_3 = -\mu\bar{\tau}Q_7$	2M+2S/2M+S	3M+2S	2M+2S
$c_{17} = -8641 - 7917\mu\tau = (\mu\bar{\tau})^{28}$	$Q_{17} = \mu\bar{\tau}Q_3$	2M+2S/2M+S	3M+2S	2M+2S
$c_{11} = -24475 + 8641\mu\tau = (\mu\bar{\tau})^{29}$	$Q_{11} = \mu\bar{\tau}Q_{17}$	2M+2S/2M+S	3M+2S	2M+2S
$c_{23} = 7193 - 24475\mu\tau = -(\mu\bar{\tau})^{30}$	$Q_{23} = -\mu\bar{\tau}Q_{11}$	2M+2S/2M+S	3M+2S	2M+2S
$c_{45} = -41757 - 7193\mu\tau = -(\mu\bar{\tau})^{31}$	$Q_{45} = \mu\bar{\tau}Q_{23}$	2M+2S/2M+S	3M+2S	2M+2S

Table 6. Cost of pre-computations on a μ_4-Koblitz curve

		$w = 4$	$w = 5$	$w = 6$
$a = 0$	Solinas [21]	15M+15S	38M+38S	–
	Trost, Xu [24]	15M+12S	39M+20S	87M+36S
	Yu, Xu [27]	6M+6S	18M+17S	44M+32S
	Ours	6M+6S	14M+14S	30M+30S
$a = 1$	Solinas [21]	18M+15S	45M+38S	–
	Trost, Xu [24]	18M+12S	46M+20S	102M+36S
	Yu, Xu [27]	6M+6S	19M+17S	47M+32S
	Ours	6M+6S	14M+14S	30M+30S

4 Novel Window τNAF

In this section, we present a novel non-convergent window τNAF method, which will be widely employed when decomposing integers on Koblitz curves. The method begins by expanding the initial integer n_0 using a large window width ω_0 (e.g., 8 or 9), as illustrated in lines 3 to 9 of Algorithm 2 in [1]. Once the norm of the new value of n_0 satisfies $N(n_0) < 2^{2^{\omega_0-2}}$, the method proceeds to the next stage, where a smaller window width $\omega_0 - 1$ (e.g., 7 or 8) is employed to decompose integer n_1 (is the new value n_0 at the end of the previous phase) until the norm of the new value n_1 is less than $N(n_1) < 2^{2^{\omega_0-3}}$. This process is repeated with decreasing window widths until the norm of the new value n_i is less than 2^{2^2}. At this point, the τ-4 NAF method, which has been proven to be convergent in [4], is utilized to complete the decomposition.

The algorithm for decomposing scalar n_0 using the gradient descent approach is depicted in Fig. 1. It is important to note that the initial steps of the window τNAF are non-convergent, as demonstrated in [4]. However, the final

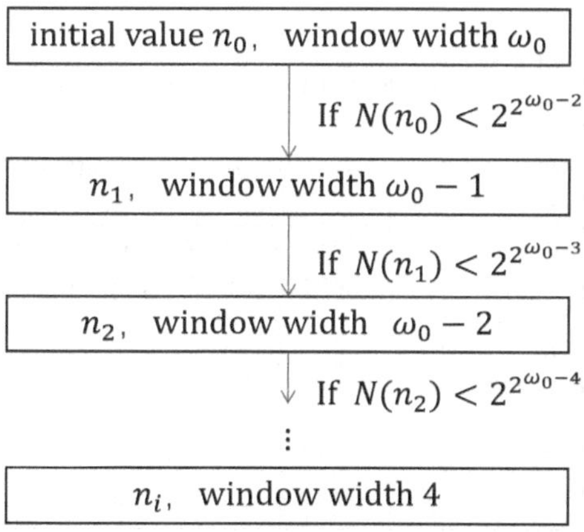

Fig. 1. Novel window τNAF to decompose scalar n_0

step employing the convergent τ-4 NAF method ensures the correctness of the overall algorithm. The non-convergent window τNAF method is particularly advantageous on Koblitz curves, as it enables efficient point multiplication and scalar multiplication.

The window width in the τNAF method can be adjusted to keep balance between the number of non-zero terms in the τ expansion and the pre-computation cost, which significantly impacts the overall performance of scalar multiplication. Previous research suggests that larger window widths (e.g., 7 or 8) can accelerate scalar multiplication, although excessively large windows may result in increased pre-computation and overall scalar multiplication costs. Therefore, it is crucial to find a trade-off between the Hamming weight of τNAF expansion and the pre-computation costs for computing scalar multiplication on Koblitz curves.

5 Scalar Multiplications Using Novel Window τNAF on Koblitz Curves

Scalar multiplication on μ_4-Koblitz curves using our window τNAF can be categorized into two situations: projective coordinates and affine coordinates.

The implementation of our scalar multiplication including pre-computation, m_0 τ-operations (m_0 is the m of $\mathbb{F}_{2^m}$, and to fit the idea of our approach, we use m_0 to denote the initial sequence length), $\frac{m_0-2^{\omega_0-2}+1}{\omega_0+1}\cdot\frac{2^{\omega_0-2}-1}{2^{\omega_0-2}}$ point additions A (ω_0 is the initial window width), and $\frac{m_0-2^{\omega_0-2}+1}{\omega_0+1}\cdot\frac{1}{2^{\omega_0-2}}$ mixed additions mA. The calculation of $\overline{\tau}$ is included in the pre-computation part. In projective coordinates, by summing the costs of the aforementioned four components with $a\in\{0,1\}$, we derive the cost of scalar multiplication:

$$\begin{aligned} &m_0\tau+\frac{m_0-2^{\omega_0-2}+1}{\omega_0+1}\left(\frac{1}{2^{\omega_0-2}}mA+\frac{2^{\omega_0-2}-1}{2^{\omega_0-2}}A\right)\\ &+\frac{2^{\omega_0-3}}{\omega_0}\left(\frac{1}{2^{\omega_0-3}}mA+\frac{2^{\omega_0-3}-1}{2^{\omega_0-3}}A\right)+\cdots+\frac{8}{6}\left(\frac{1}{2^3}mA+\frac{7}{2^3}A\right)\\ &+\frac{7}{5}\left(\frac{1}{4}mA+\frac{3}{4}A\right)+\text{Precomputation} \end{aligned} \tag{1}$$

The scalar multiplication in affine coordinates entirely use mixed additions mA. In this case, we need Montgomery trick [6] to translate points in projective coordinates to those in affine coordinates, which needs $3(n-1)$ multiplication and one inversion. The total cost of scalar multiplication in affine coordinates is:

$$\begin{aligned} &m_0\tau+\left(\frac{m_0-2^{\omega_0-2}+1}{\omega_0+1}+\frac{2^{\omega_0-3}}{\omega_0}+\cdots+\frac{8}{6}+\frac{7}{5}\right)mA\\ &+\text{Precomputation}+\text{Montgomerytrick} \end{aligned} \tag{2}$$

Above two equations, Eq. 1 and Eq. 2, can be used to calculate the cost of scalar multiplication using window τNAF. The choice of equation is independent of pre-computation efficiency. For $a = 0$, Table 7 presents a comparison of the costs of scalar multiplications on four Koblitz curves K-233, K-283, K-409, and K-571 using μ_4-Koblitz curves. For $a = 1$, Table 8 compares the costs of scalar multiplications on K1-163, K1-283, K1-359,and K1-701 using μ_4-Koblitz curves. Through two tables we can see that our method outperforms Yu and Xu's method by up to 5% on μ_4-Koblitz curves.

Table 7. The expected costs of scalar multiplications on K-233, K-283, K-409, and K-571 using μ_4-Koblitz curves in **M**

		K-233(w)	K-283(w)	K-409(w)	K-571(w)
S=0M	τNAF	466	566	818	1142
	Trost, Xu [24]	306.0(5)	363.3(5)	492.3(6)	652.9(6)
	Yu,Xu [27]	274.9(6)	324.5(6)	444.3(7)	585.4(7)
	Ours	264.7(6)	314.2(6)	425.7(7)	566.8(7)
S=0.1M	τNAF	574.7	698.0	1008.8	1408.5
	Trost, Xu [24]	408.9(5)	487.9(5)	671.2(6)	901.2(6)
	Yu,Xu [27]	377.9(6)	448.9(6)	624.3(7)	834.3(7)
	Ours	367.6(6)	438.6(6)	606.0(7)	815.9(7)
S=0.2M	τNAF	683.5	830.1	1199.7	1674.9
	Trost, Xu [24]	511.9(5)	612.5(5)	850.1(6)	1149.5(6)
	Yu,Xu [27]	481(6)	573.4(6)	804.3(7)	1083.1(7)
	Ours	470.6(6)	563.0(6)	786.3(7)	1065.1(7)
S=0.5M	τNAF	1009.7	1226.3	1772.3	2474.3
	Trost, Xu [24]	835.2(6)	991.9(6)	1386.8(6)	1894.5(6)
	Yu,Xu [27]	790.2(6)	946.9(6)	1341.8(6)	1829.8(7)
	Ours	779.5(6)	936.2(6)	1327.0(7)	1812.4(7)

For $a = 0$ and $a = 1$, we compare the costs of our scalar multiplication with other various works using μ_4-Koblitz curves, LD coordinates, and λ-coordinates with different ratios of **S/M** in Figs. 2 and 3, respectively. Our observations indicate that LD coordinates perform comparably to μ_4-Koblitz curves for $a = 0$. Specifically, for $a = 1$, LD coordinates outperform μ_4-Koblitz curves due to their lower pre-computation costs. By the results in the tables and figures mentioned above, we can see that our window τNAF method gives the lowest costs for scalar multiplication on Koblitz curves and is a promising choice for efficient implementation of Koblitz curves cryptosystem.

Table 8. The expected costs of scalar multiplications on K1-163, K1-283, K1-359, and K1-701 using μ_4-Koblitz curves in **M**

		K1-163(w)	K1-283(w)	K1-359(w)	K1-701(w)
S=0M	τNAF	380.3	660.3	837.7	1635.7
	Trost, Xu [24]	259.9(5)	417.4(5)	509.1(6)	896.9(6)
	Yu,Xu [27]	231.8(6)	367.9(6)	450.6(7)	791.3(7)
	Ours	219.2(6)	352.5(7)	428.2(7)	763.4(8)
S=0.1M	τNAF	456.4	792.4	1005.2	1962.8
	Trost, Xu [24]	332.6(5)	542.1(5)	666.6(6)	1200.9(6)
	Yu,Xu [27]	304.9(6)	492.4(6)	609.4(7)	1095.4(7)
	Ours	292.1(6)	479.2(7)	587.2(7)	1072.4(8)
S=0.2M	τNAF	532.5	924.5	1172.7	2289.9
	Trost, Xu [24]	405.2(5)	666.7(5)	824(6)	1505(6)
	Yu,Xu [27]	377.9(6)	616.9(6)	768.1(7)	1399.5(7)
	Ours	365.1(6)	604.1(6)	746.2(7)	1377.6(7)
S=0.5M	τNAF	760.7	1320.7	1675.3	3271.3
	Trost, Xu [24]	623.1(5)	1040.6(5)	1296.4(6)	2417.3(6)
	Yu,Xu [27]	594.6(5)	990.3(6)	1239.4(6)	2311.9(7)
	Ours	584.0(6)	977.3(6)	1223.2(7)	2290.6(7)

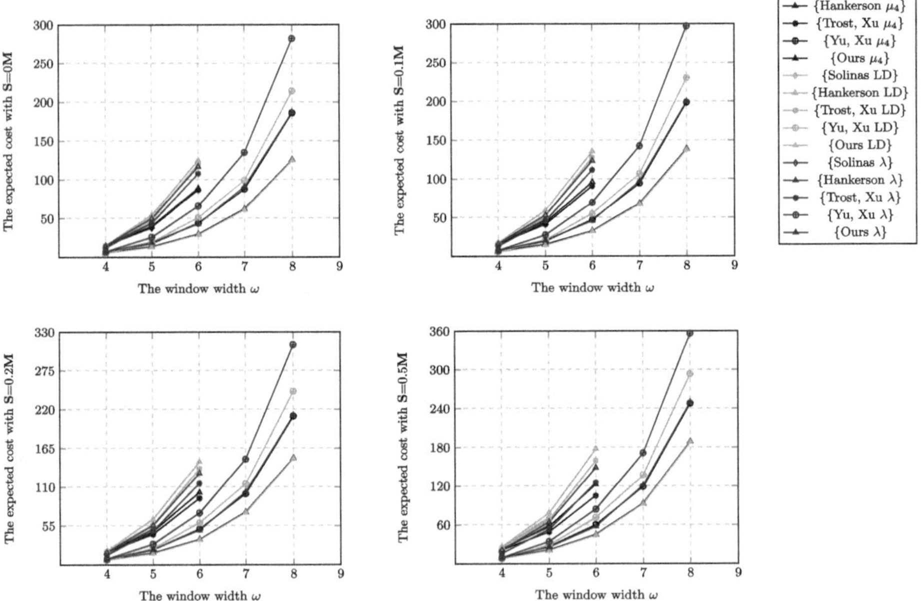

Fig. 2. When $a = 0$, the cost of scalar multiplication using three coordinates with different ratios of **S**/**M** in **M**

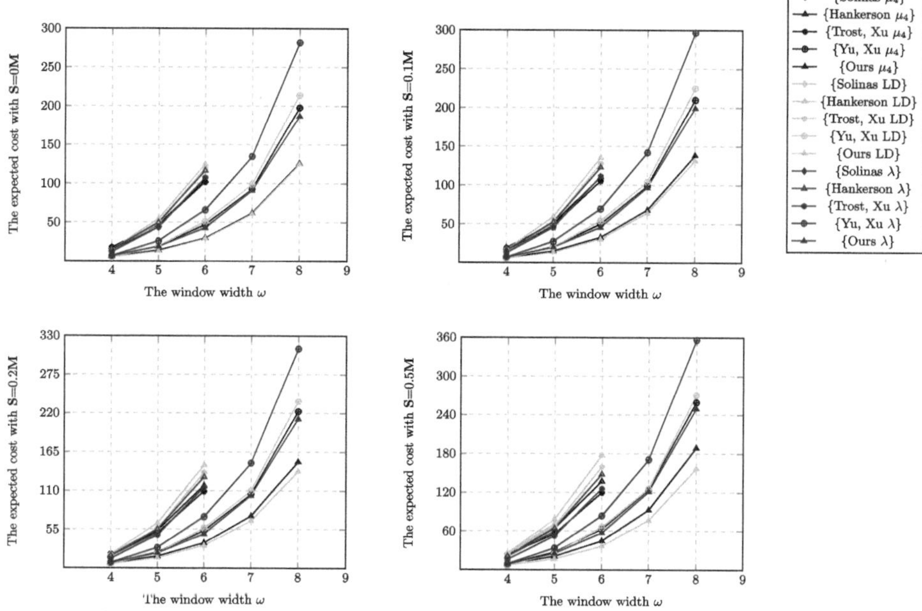

Fig. 3. When $a = 1$, the cost of scalar multiplication using three coordinates with different ratios of $\mathbf{S}/\mathbf{M}$ in $\mathbf{M}$

6 Conclusion

This paper proposes new optimization for scalar multiplication on Koblitz curves, including a non-convergent window τNAF method and a unified pre-computation scheme. The window τNAF method commences with larger window widths and gradually decreases the window width. Although the initial steps are non-convergent, the final step employs a convergent τ-4 NAF to ensure the correctness of the whole decomposition algorithm. The non-convergent part is optimized for speed. Meanwhile, our proposed pre-computation scheme is particularly advantageous for point multiplication on Koblitz curves. Finally, the results of this work demonstrate that μ_4-Koblitz curves and LD coordinates represent the state-of-the-art for Koblitz curves, providing a 5% reduction for scalar multiplication time on Koblitz curves compared to prior literature.

Although the contributions presented in this paper offer some advancements in optimizing scalar multiplication on Koblitz curves, further research is needed to explore the efficiency of these methods in various settings. Another point is that our method may not be constant time, which needs further improvement.

Acknowledgments. The authors would like to thank the anonymous reviewers for many helpful comments and their helpful suggestions. This work was supported by National Key Research and Development Program of China (Grant No. 2022YFB3102500), the National Natural Science Foundation of China (Grant No. 62272453), and the Key Research Program of the Chinese Academy of Sciences (Grant No. ZDRW-XX-2022-1).

A Pre-computation for Window τNAF with Widths 8 and 9

Our pre-computation on a Koblitz curve for the window τNAF with widths 8 and 9 are shown in Tables 9, 10 and 11 respectively.

Table 9. Pre-computation for width 8

c_i	Q_i	LD(a=0/a=1)	λ	μ_4
$c_{91} = 1 - \mu\tau = \mu\bar{\tau}$	$Q_{91} = \mu\bar{\tau}P$	M+2S/M+S		2M+2S
$c_{89} = -1 - \mu\tau = (\mu\bar{\tau})^2$	$Q_{89} = \mu\bar{\tau}Q_{91}$	2M+2S/2M+S	4M+3S	2M+2S
$c_{93} = 3 - \mu\tau = -(\mu\bar{\tau})^3$	$Q_{93} = -\mu\bar{\tau}Q_{89}$	2M+2S/2M+S	3M+2S	2M+2S
$c_{15} = 1 - 3\mu\tau = -(\mu\bar{\tau})^4$	$Q_{15} = \mu\bar{\tau}Q_{93}$	2M+2S/2M+S	3M+2S	2M+2S
$c_{85} = -5 - \mu\tau = -(\mu\bar{\tau})^5$	$Q_{85} = \mu\bar{\tau}Q_{15}$	2M+2S/2M+S	3M+2S	2M+2S
$c_{55} = -7 + 5\mu\tau = -(\mu\bar{\tau})^6$	$Q_{55} = \mu\bar{\tau}Q_{85}$	2M+2S/2M+S	3M+2S	2M+2S
$c_{115} = -3 - 7\mu\tau = (\mu\bar{\tau})^7$	$Q_{115} = -\mu\bar{\tau}Q_{55}$	2M+2S/2M+S	3M+2S	2M+2S
$c_{31} = 17 - 3\mu\tau = -(\mu\bar{\tau})^8$	$Q_{31} = -\mu\bar{\tau}Q_{115}$	2M+2S/2M+S	3M+2S	2M+2S
$c_5 = 11 - 17\mu\tau = -(\mu\bar{\tau})^9$	$Q_5 = \mu\bar{\tau}Q_{31}$	2M+2S/2M+S	3M+2S	2M+2S
$c_{57} = 23 + 11\mu\tau = (\mu\bar{\tau})^{10}$	$Q_{57} = -\mu\bar{\tau}Q_5$	2M+2S/2M+S	3M+2S	2M+2S
$c_{67} = 45 - 23\mu\tau = (\mu\bar{\tau})^{11}$	$Q_{67} = \mu\bar{\tau}Q_{57}$	2M+2S/2M+S	3M+2S	2M+2S
$c_{47} = 1 + 45\mu\tau = -(\mu\bar{\tau})^{12}$	$Q_{47} = -\mu\bar{\tau}Q_{67}$	2M+2S/2M+S	3M+2S	2M+2S
$c_{75} = -91 + \mu\tau = (\mu\bar{\tau})^{13}$	$Q_{75} = -\mu\bar{\tau}Q_{47}$	2M+2S/2M+S	3M+2S	2M+2S
$c_{87} = 89 - 91\mu\tau = -(\mu\bar{\tau})^{14}$	$Q_{87} = -\mu\bar{\tau}Q_{75}$	2M+2S/2M+S	3M+2S	2M+2S
$c_{19} = 93 + 89\mu\tau = (\mu\bar{\tau})^{15}$	$Q_{19} = -\mu\bar{\tau}Q_{87}$	2M+2S/2M+S	3M+2S	2M+2S
$c_{63} = -271 + 93\mu\tau = -(\mu\bar{\tau})^{16}$	$Q_{63} = -\mu\bar{\tau}Q_{19}$	2M+2S/2M+S	3M+2S	2M+2S
$c_{101} = -85 + 271\mu\tau = -(\mu\bar{\tau})^{17}$	$Q_{101} = \mu\bar{\tau}Q_{63}$	2M+2S/2M+S	3M+2S	2M+2S
$c_{25} = -457 - 85\mu\tau = (\mu\bar{\tau})^{18}$	$Q_{25} = -\mu\bar{\tau}Q_{101}$	2M+2S/2M+S	3M+2S	2M+2S
$c_{29} = 627 - 457\mu\tau = -(\mu\bar{\tau})^{19}$	$Q_{29} = -\mu\bar{\tau}Q_{25}$	2M+2S/2M+S	3M+2S	2M+2S
$c_{79} = -287 - 627\mu\tau = -(\mu\bar{\tau})^{20}$	$Q_{79} = \mu\bar{\tau}Q_{29}$	2M+2S/2M+S	3M+2S	2M+2S
$c_{21} = -1541 + 287\mu\tau = -(\mu\bar{\tau})^{21}$	$Q_{21} = \mu\bar{\tau}Q_{79}$	2M+2S/2M+S	3M+2S	2M+2S
$c_{119} = -967 + 1541\mu\tau = -(\mu\bar{\tau})^{22}$	$Q_{119} = \mu\bar{\tau}Q_{21}$	2M+2S/2M+S	3M+2S	2M+2S
$c_{77} = 2115 + 967\mu\tau = -(\mu\bar{\tau})^{23}$	$Q_{77} = \mu\bar{\tau}Q_{119}$	2M+2S/2M+S	3M+2S	2M+2S
$c_{95} = 4049 - 2115\mu\tau = -(\mu\bar{\tau})^{24}$	$Q_{95} = \mu\bar{\tau}Q_{77}$	2M+2S/2M+S	3M+2S	2M+2S
$c_{59} = 181 + 4049\mu\tau = (\mu\bar{\tau})^{25}$	$Q_{59} = -\mu\bar{\tau}Q_{95}$	2M+2S/2M+S	3M+2S	2M+2S
$c_7 = -8279 + 181\mu\tau = -(\mu\bar{\tau})^{26}$	$Q_7 = -\mu\bar{\tau}Q_{59}$	2M+2S/2M+S	3M+2S	2M+2S
$c_{125} = -7917 + 8279\mu\tau = -(\mu\bar{\tau})^{27}$	$Q_{125} = \mu\bar{\tau}Q_7$	2M+2S/2M+S	3M+2S	2M+2S
$c_{111} = 8641 + 7917\mu\tau = -(\mu\bar{\tau})^{28}$	$Q_{111} = \mu\bar{\tau}Q_{125}$	2M+2S/2M+S	3M+2S	2M+2S
$c_{117} = 24475 - 8641\mu\tau = -(\mu\bar{\tau})^{29}$	$Q_{117} = \mu\bar{\tau}Q_{111}$	2M+2S/2M+S	3M+2S	2M+2S
$c_{105} = -7193 + 24475\mu\tau = (\mu\bar{\tau})^{30}$	$Q_{105} = -\mu\bar{\tau}Q_{117}$	2M+2S/2M+S	3M+2S	2M+2S
$c_{83} = 41757 + 7193\mu\tau = (\mu\bar{\tau})^{31}$	$Q_{83} = \mu\bar{\tau}Q_{105}$	2M+2S/2M+S	3M+2S	2M+2S
$c_{127} = -56143 + 41757\mu\tau = -(\mu\bar{\tau})^{32}$	$Q_{127} = -\mu\bar{\tau}Q_{83}$	2M+2S/2M+S	3M+2S	2M+2S
$c_{37} = 27371 + 56143\mu\tau = -(\mu\bar{\tau})^{33}$	$Q_{37} = \mu\bar{\tau}Q_{127}$	2M+2S/2M+S	3M+2S	2M+2S
$c_{39} = 139657 - 27371\mu\tau = -(\mu\bar{\tau})^{34}$	$Q_{39} = \mu\bar{\tau}Q_{37}$	2M+2S/2M+S	3M+2S	2M+2S
$c_{35} = -84915 + 139657\mu\tau = (\mu\bar{\tau})^{35}$	$Q_{35} = -\mu\bar{\tau}Q_{39}$	2M+2S/2M+S	3M+2S	2M+2S
$c_{113} = 194399 + 84915\mu\tau = (\mu\bar{\tau})^{36}$	$Q_{113} = \mu\bar{\tau}Q_{35}$	2M+2S/2M+S	3M+2S	2M+2S
$c_{43} = 364229 - 194399\mu\tau = (\mu\bar{\tau})^{37}$	$Q_{43} = \mu\bar{\tau}Q_{113}$	2M+2S/2M+S	3M+2S	2M+2S
$c_{73} = -24569 - 364229\mu\tau = (\mu\bar{\tau})^{38}$	$Q_{73} = \mu\bar{\tau}Q_{43}$	2M+2S/2M+S	3M+2S	2M+2S
$c_{13} = 753027 - 24569\mu\tau = -(\mu\bar{\tau})^{39}$	$Q_{13} = -\mu\bar{\tau}Q_{73}$	2M+2S/2M+S	3M+2S	2M+2S
$c_{97} = -703889 + 753027\mu\tau = (\mu\bar{\tau})^{40}$	$Q_{97} = -\mu\bar{\tau}Q_{13}$	2M+2S/2M+S	3M+2S	2M+2S
$c_{123} = 802165 + 703889\mu\tau = (\mu\bar{\tau})^{41}$	$Q_{123} = \mu\bar{\tau}Q_{97}$	2M+2S/2M+S	3M+2S	2M+2S
$c_{71} = -2209943 + 802165\mu\tau = -(\mu\bar{\tau})^{42}$	$Q_{71} = -\mu\bar{\tau}Q_{123}$	2M+2S/2M+S	3M+2S	2M+2S
$c_{61} = -605613 + 2209943\mu\tau = -(\mu\bar{\tau})^{43}$	$Q_{61} = \mu\bar{\tau}Q_{71}$	2M+2S/2M+S	3M+2S	2M+2S
$c_{81} = -3814273 - 605613\mu\tau = (\mu\bar{\tau})^{44}$	$Q_{81} = -\mu\bar{\tau}Q_{61}$	2M+2S/2M+S	3M+2S	2M+2S
$c_{53} = 5025499 - 3814273\mu\tau = -(\mu\bar{\tau})^{45}$	$Q_{53} = -\mu\bar{\tau}Q_{81}$	2M+2S/2M+S	3M+2S	2M+2S
$c_{41} = 2603047 + 5025499\mu\tau = (\mu\bar{\tau})^{46}$	$Q_{41} = -\mu\bar{\tau}Q_{53}$	2M+2S/2M+S	3M+2S	2M+2S
$c_{109} = -12654045 + 2603047\mu\tau = -(\mu\bar{\tau})^{47}$	$Q_{109} = -\mu\bar{\tau}Q_{41}$	2M+2S/2M+S	3M+2S	2M+2S
$c_{65} = 7447951 - 12654045\mu\tau = (\mu\bar{\tau})^{48}$	$Q_{65} = -\mu\bar{\tau}Q_{109}$	2M+2S/2M+S	3M+2S	2M+2S
$c_{27} = -17860139 - 7447951\mu\tau = (\mu\bar{\tau})^{49}$	$Q_{27} = \mu\bar{\tau}Q_{65}$	2M+2S/2M+S	3M+2S	2M+2S
$c_{103} = 32756041 - 17860139\mu\tau = -(\mu\bar{\tau})^{50}$	$Q_{103} = -\mu\bar{\tau}Q_{27}$	2M+2S/2M+S	3M+2S	2M+2S
$c_{99} = 2964237 + 32756041\mu\tau = (\mu\bar{\tau})^{51}$	$Q_{99} = -\mu\bar{\tau}Q_{103}$	2M+2S/2M+S	3M+2S	2M+2S
$c_{49} = 68476319 - 2964237\mu\tau = (\mu\bar{\tau})^{52}$	$Q_{49} = \mu\bar{\tau}Q_{99}$	2M+2S/2M+S	3M+2S	2M+2S
$c_{107} = 62547845 - 68476319\mu\tau = (\mu\bar{\tau})^{53}$	$Q_{107} = \mu\bar{\tau}Q_{49}$	2M+2S/2M+S	3M+2S	2M+2S
$c_9 = -74404793 - 62547845\mu\tau = (\mu\bar{\tau})^{54}$	$Q_9 = \mu\bar{\tau}Q_{107}$	2M+2S/2M+S	3M+2S	2M+2S
$c_{51} = -199500483 + 74404793\mu\tau = (\mu\bar{\tau})^{55}$	$Q_{51} = \mu\bar{\tau}Q_9$	2M+2S/2M+S	3M+2S	2M+2S
$c_{33} = -50690897 + 199500483\mu\tau = (\mu\bar{\tau})^{56}$	$Q_{33} = \mu\bar{\tau}Q_{51}$	2M+2S/2M+S	3M+2S	2M+2S
$c_{69} = -348310069 - 50690897\mu\tau = -(\mu\bar{\tau})^{57}$	$Q_{69} = -\mu\bar{\tau}Q_{33}$	2M+2S/2M+S	3M+2S	2M+2S
$c_{121} = 449691863 - 348310069\mu\tau = (\mu\bar{\tau})^{58}$	$Q_{121} = -\mu\bar{\tau}Q_{69}$	2M+2S/2M+S	3M+2S	2M+2S
$c_3 = -246928275 - 449691863\mu\tau = (\mu\bar{\tau})^{59}$	$Q_3 = \mu\bar{\tau}Q_{121}$	2M+2S/2M+S	3M+2S	2M+2S
$c_{17} = -1146312001 + 246928275\mu\tau = (\mu\bar{\tau})^{60}$	$Q_{17} = \mu\bar{\tau}Q_3$	2M+2S/2M+S	3M+2S	2M+2S
$c_{11} = -652455451 + 1146312001\mu\tau = (\mu\bar{\tau})^{61}$	$Q_{11} = \mu\bar{\tau}Q_{17}$	2M+2S/2M+S	3M+2S	2M+2S
$c_{23} = -1640168551 - 652455451\mu\tau = -(\mu\bar{\tau})^{62}$	$Q_{23} = -\mu\bar{\tau}Q_{11}$	2M+2S/2M+S	3M+2S	2M+2S
$c_{45} = 1349887843 + 1640168551\mu\tau = -(\mu\bar{\tau})^{63}$	$Q_{45} = \mu\bar{\tau}Q_{23}$	2M+2S/2M+S	3M+2S	2M+2S

Table 10. Pre-computation for width 9

c_i	Q_i	LD(a=0/a=1)	λ	μ_4
$c_{91} = 1 - \mu\tau = \mu\bar{\tau}$	$Q_{91} = \mu\bar{\tau}P$	M+2S/M+S		2M+2S
$c_{89} = -1 - \mu\tau = (\mu\bar{\tau})^2$	$Q_{89} = \mu\bar{\tau}Q_{91}$	2M+2S/2M+S	4M+3S	2M+2S
$c_{93} = 3 - \mu\tau = -(\mu\bar{\tau})^3$	$Q_{93} = -\mu\bar{\tau}Q_{89}$	2M+2S/2M+S	3M+2S	2M+2S
$c_{241} = -1 + 3\mu\tau = (\mu\bar{\tau})^4$	$Q_{241} = -\mu\bar{\tau}Q_{93}$	2M+2S/2M+S	3M+2S	2M+2S
$c_{85} = -5 - \mu\tau = -(\mu\bar{\tau})^5$	$Q_{85} = -\mu\bar{\tau}Q_{241}$	2M+2S/2M+S	3M+2S	2M+2S
$c_{55} = -7 + 5\mu\tau = -(\mu\bar{\tau})^6$	$Q_{55} = \mu\bar{\tau}Q_{85}$	2M+2S/2M+S	3M+2S	2M+2S
$c_{115} = -3 - 7\mu\tau = (\mu\bar{\tau})^7$	$Q_{115} = -\mu\bar{\tau}Q_{55}$	2M+2S/2M+S	3M+2S	2M+2S
$c_{225} = -17 + 3\mu\tau = (\mu\bar{\tau})^8$	$Q_{225} = \mu\bar{\tau}Q_{115}$	2M+2S/2M+S	3M+2S	2M+2S
$c_5 = 11 - 17\mu\tau = -(\mu\bar{\tau})^9$	$Q_5 = -\mu\bar{\tau}Q_{225}$	2M+2S/2M+S	3M+2S	2M+2S
$c_{57} = 23 + 11\mu\tau = (\mu\bar{\tau})^{10}$	$Q_{57} = -\mu\bar{\tau}Q_5$	2M+2S/2M+S	3M+2S	2M+2S
$c_{67} = 45 - 23\mu\tau = (\mu\bar{\tau})^{11}$	$Q_{67} = \mu\bar{\tau}Q_{57}$	2M+2S/2M+S	3M+2S	2M+2S
$c_{47} = 1 + 45\mu\tau = -(\mu\bar{\tau})^{12}$	$Q_{47} = -\mu\bar{\tau}Q_{67}$	2M+2S/2M+S	3M+2S	2M+2S
$c_{181} = 91 - \mu\tau = -(\mu\bar{\tau})^{13}$	$Q_{181} = \mu\bar{\tau}Q_{47}$	2M+2S/2M+S	3M+2S	2M+2S
$c_{87} = 89 - 91\mu\tau = -(\mu\bar{\tau})^{14}$	$Q_{87} = \mu\bar{\tau}Q_{181}$	2M+2S/2M+S	3M+2S	2M+2S
$c_{237} = -93 - 89\mu\tau = -(\mu\bar{\tau})^{15}$	$Q_{237} = \mu\bar{\tau}Q_{87}$	2M+2S/2M+S	3M+2S	2M+2S
$c_{63} = -271 + 93\mu\tau = -(\mu\bar{\tau})^{16}$	$Q_{63} = \mu\bar{\tau}Q_{237}$	2M+2S/2M+S	3M+2S	2M+2S
$c_{101} = -85 + 271\mu\tau = -(\mu\bar{\tau})^{17}$	$Q_{101} = \mu\bar{\tau}Q_{63}$	2M+2S/2M+S	3M+2S	2M+2S
$c_{25} = -457 - 85\mu\tau = (\mu\bar{\tau})^{18}$	$Q_{25} = -\mu\bar{\tau}Q_{101}$	2M+2S/2M+S	3M+2S	2M+2S
$c_{227} = -627 + 457\mu\tau = (\mu\bar{\tau})^{19}$	$Q_{227} = \mu\bar{\tau}Q_{25}$	2M+2S/2M+S	3M+2S	2M+2S
$c_{177} = 287 + 627\mu\tau = (\mu\bar{\tau})^{20}$	$Q_{177} = \mu\bar{\tau}Q_{227}$	2M+2S/2M+S	3M+2S	2M+2S
$c_{235} = 1541 - 287\mu\tau = (\mu\bar{\tau})^{21}$	$Q_{235} = \mu\bar{\tau}Q_{177}$	2M+2S/2M+S	3M+2S	2M+2S
$c_{119} = -967 + 1541\mu\tau = -(\mu\bar{\tau})^{22}$	$Q_{119} = -\mu\bar{\tau}Q_{235}$	2M+2S/2M+S	3M+2S	2M+2S
$c_{77} = 2115 + 967\mu\tau = -(\mu\bar{\tau})^{23}$	$Q_{77} = \mu\bar{\tau}Q_{119}$	2M+2S/2M+S	3M+2S	2M+2S
$c_{161} = -4049 + 2115\mu\tau = (\mu\bar{\tau})^{24}$	$Q_{161} = -\mu\bar{\tau}Q_{77}$	2M+2S/2M+S	3M+2S	2M+2S
$c_{197} = -181 - 4049\mu\tau = -(\mu\bar{\tau})^{25}$	$Q_{197} = -\mu\bar{\tau}Q_{101}$	2M+2S/2M+S	3M+2S	2M+2S
$c_7 = -8279 + 181\mu\tau = -(\mu\bar{\tau})^{26}$	$Q_7 = \mu\bar{\tau}Q_{197}$	2M+2S/2M+S	3M+2S	2M+2S
$c_{125} = -7917 + 8279\mu\tau = -(\mu\bar{\tau})^{27}$	$Q_{125} = \mu\bar{\tau}Q_7$	2M+2S/2M+S	3M+2S	2M+2S
$c_{111} = 8641 + 7917\mu\tau = -(\mu\bar{\tau})^{28}$	$Q_{111} = \mu\bar{\tau}Q_{125}$	2M+2S/2M+S	3M+2S	2M+2S
$c_{139} = -24475 + 8641\mu\tau = (\mu\bar{\tau})^{29}$	$Q_{139} = -\mu\bar{\tau}Q_{111}$	2M+2S/2M+S	3M+2S	2M+2S
$c_{151} = 7193 - 24475\mu\tau = -(\mu\bar{\tau})^{30}$	$Q_{151} = -\mu\bar{\tau}Q_{139}$	2M+2S/2M+S	3M+2S	2M+2S
$c_{83} = 41757 + 7193\mu\tau = (\mu\bar{\tau})^{31}$	$Q_{83} = -\mu\bar{\tau}Q_{151}$	2M+2S/2M+S	3M+2S	2M+2S
$c_{127} = -56143 + 41757\mu\tau = -(\mu\bar{\tau})^{32}$	$Q_{127} = -\mu\bar{\tau}Q_{83}$	2M+2S/2M+S	3M+2S	2M+2S
$c_{219} = -27371 - 56143\mu\tau = (\mu\bar{\tau})^{33}$	$Q_{219} = -\mu\bar{\tau}Q_{127}$	2M+2S/2M+S	3M+2S	2M+2S
$c_{39} = 139657 - 27371\mu\tau = -(\mu\bar{\tau})^{34}$	$Q_{39} = -\mu\bar{\tau}Q_{219}$	2M+2S/2M+S	3M+2S	2M+2S
$c_{35} = -84915 + 139657\mu\tau = (\mu\bar{\tau})^{35}$	$Q_{35} = -\mu\bar{\tau}Q_{39}$	2M+2S/2M+S	3M+2S	2M+2S
$c_{113} = 194399 + 84915\mu\tau = (\mu\bar{\tau})^{36}$	$Q_{113} = \mu\bar{\tau}Q_{35}$	2M+2S/2M+S	3M+2S	2M+2S
$c_{43} = 364229 - 194399\mu\tau = (\mu\bar{\tau})^{37}$	$Q_{43} = \mu\bar{\tau}Q_{113}$	2M+2S/2M+S	3M+2S	2M+2S
$c_{183} = 24569 + 364229\mu\tau = -(\mu\bar{\tau})^{38}$	$Q_{183} = -\mu\bar{\tau}Q_{43}$	2M+2S/2M+S	3M+2S	2M+2S
$c_{243} = -753027 + 24569\mu\tau = (\mu\bar{\tau})^{39}$	$Q_{243} = -\mu\bar{\tau}Q_{183}$	2M+2S/2M+S	3M+2S	2M+2S
$c_{97} = -703889 + 753027\mu\tau = (\mu\bar{\tau})^{40}$	$Q_{97} = \mu\bar{\tau}Q_{243}$	2M+2S/2M+S	3M+2S	2M+2S
$c_{123} = 802165 + 703889\mu\tau = (\mu\bar{\tau})^{41}$	$Q_{123} = \mu\bar{\tau}Q_{97}$	2M+2S/2M+S	3M+2S	2M+2S
$c_{71} = -2209943 + 802165\mu\tau = -(\mu\bar{\tau})^{42}$	$Q_{71} = -\mu\bar{\tau}Q_{123}$	2M+2S/2M+S	3M+2S	2M+2S
$c_{195} = 605613 - 2209943\mu\tau = (\mu\bar{\tau})^{43}$	$Q_{195} = -\mu\bar{\tau}Q_{71}$	2M+2S/2M+S	3M+2S	2M+2S
$c_{175} = 3814273 + 605613\mu\tau = -(\mu\bar{\tau})^{44}$	$Q_{175} = -\mu\bar{\tau}Q_{195}$	2M+2S/2M+S	3M+2S	2M+2S
$c_{53} = 5025499 - 3814273\mu\tau = -(\mu\bar{\tau})^{45}$	$Q_{53} = \mu\bar{\tau}Q_{175}$	2M+2S/2M+S	3M+2S	2M+2S
$c_{215} = -2603047 - 5025499\mu\tau = -(\mu\bar{\tau})^{46}$	$Q_{215} = \mu\bar{\tau}Q_{53}$	2M+2S/2M+S	3M+2S	2M+2S
$c_{109} = -12654045 + 2603047\mu\tau = -(\mu\bar{\tau})^{47}$	$Q_{109} = \mu\bar{\tau}Q_{215}$	2M+2S/2M+S	3M+2S	2M+2S
$c_{191} = -7447951 + 12654045\mu\tau = -(\mu\bar{\tau})^{48}$	$Q_{191} = \mu\bar{\tau}Q_{109}$	2M+2S/2M+S	3M+2S	2M+2S
$c_{27} = -17860139 - 7447951\mu\tau = (\mu\bar{\tau})^{49}$	$Q_{27} = -\mu\bar{\tau}Q_{191}$	2M+2S/2M+S	3M+2S	2M+2S
$c_{103} = 32756041 - 17860139\mu\tau = -(\mu\bar{\tau})^{50}$	$Q_{103} = -\mu\bar{\tau}Q_{27}$	2M+2S/2M+S	3M+2S	2M+2S
$c_{157} = -2964237 - 32756041\mu\tau = -(\mu\bar{\tau})^{51}$	$Q_{157} = \mu\bar{\tau}Q_{103}$	2M+2S/2M+S	3M+2S	2M+2S
$c_{49} = 68476319 - 2964237\mu\tau = (\mu\bar{\tau})^{52}$	$Q_{49} = -\mu\bar{\tau}Q_{157}$	2M+2S/2M+S	3M+2S	2M+2S
$c_{149} = -62547845 + 68476319\mu\tau = -(\mu\bar{\tau})^{53}$	$Q_{149} = -\mu\bar{\tau}Q_{49}$	2M+2S/2M+S	3M+2S	2M+2S
$c_{247} = 74404793 + 62547845\mu\tau = -(\mu\bar{\tau})^{54}$	$Q_{247} = \mu\bar{\tau}Q_{149}$	2M+2S/2M+S	3M+2S	2M+2S
$c_{51} = -199500483 + 74404793\mu\tau = (\mu\bar{\tau})^{55}$	$Q_{51} = -\mu\bar{\tau}Q_{247}$	2M+2S/2M+S	3M+2S	2M+2S
$c_{33} = -50690897 + 199500483\mu\tau = (\mu\bar{\tau})^{56}$	$Q_{33} = \mu\bar{\tau}Q_{51}$	2M+2S/2M+S	3M+2S	2M+2S
$c_{69} = -348310069 - 50690897\mu\tau = -(\mu\bar{\tau})^{57}$	$Q_{69} = -\mu\bar{\tau}Q_{33}$	2M+2S/2M+S	3M+2S	2M+2S
$c_{135} = -449691863 + 348310069\mu\tau = -(\mu\bar{\tau})^{58}$	$Q_{135} = \mu\bar{\tau}Q_{69}$	2M+2S/2M+S	3M+2S	2M+2S
$c_3 = -246928275 - 449691863\mu\tau = (\mu\bar{\tau})^{59}$	$Q_3 = -\mu\bar{\tau}Q_{135}$	2M+2S/2M+S	3M+2S	2M+2S
$c_{239} = 1146312001 - 246928275\mu\tau = -(\mu\bar{\tau})^{60}$	$Q_{239} = -\mu\bar{\tau}Q_3$	2M+2S/2M+S	3M+2S	2M+2S
$c_{245} = 652455451 - 1146312001\mu\tau = -(\mu\bar{\tau})^{61}$	$Q_{245} = \mu\bar{\tau}Q_{239}$	2M+2S/2M+S	3M+2S	2M+2S
$c_{233} = 1640168551 + 652455451\mu\tau = (\mu\bar{\tau})^{62}$	$Q_{233} = -\mu\bar{\tau}Q_{245}$	2M+2S/2M+S	3M+2S	2M+2S
$c_{211} = -1349887843 - 1640168551\mu\tau = (\mu\bar{\tau})^{63}$	$Q_{211} = \mu\bar{\tau}Q_{233}$	2M+2S/2M+S	3M+2S	2M+2S
$c_{255} = 335257649 - 1349887843\mu\tau = -(\mu\bar{\tau})^{64}$	$Q_{255} = -\mu\bar{\tau}Q_{211}$	2M+2S/2M+S	3M+2S	2M+2S

Table 11. Pre-computation for width 9 (continued)

c_i	Q_i	LD(a=0/a=1)	λ	μ_4
$c_{165} = 1930449259 - 335257649\mu\tau = -(\mu\bar{\tau})^{65}$	$Q_{165} = \mu\bar{\tau}Q_{255}$	2M+2S/2M+S	3M+2S	2M+2S
$c_{167} = 1259933961 - 1930449259\mu\tau = -(\mu\bar{\tau})^{66}$	$Q_{167} = \mu\bar{\tau}Q_{165}$	2M+2S/2M+S	3M+2S	2M+2S
$c_{163} = -1694002739 + 1259933961\mu\tau = (\mu\bar{\tau})^{67}$	$Q_{163} = -\mu\bar{\tau}Q_{167}$	2M+2S/2M+S	3M+2S	2M+2S
$c_{15} = -825865183 - 1694002739\mu\tau = -(\mu\bar{\tau})^{68}$	$Q_{15} = -\mu\bar{\tau}Q_{163}$	2M+2S/2M+S	3M+2S	2M+2S
$c_{171} = -81096635 - 825865183\mu\tau = (\mu\bar{\tau})^{69}$	$Q_{171} = -\mu\bar{\tau}Q_{15}$	2M+2S/2M+S	3M+2S	2M+2S
$c_{201} = -1732827001 + 81096635\mu\tau = (\mu\bar{\tau})^{70}$	$Q_{201} = \mu\bar{\tau}Q_{171}$	2M+2S/2M+S	3M+2S	2M+2S
$c_{141} = 1570633731 - 1732827001\mu\tau = -(\mu\bar{\tau})^{71}$	$Q_{141} = -\mu\bar{\tau}Q_{201}$	2M+2S/2M+S	3M+2S	2M+2S
$c_{31} = -1895020271 - 1570633731\mu\tau = -(\mu\bar{\tau})^{72}$	$Q_{31} = \mu\bar{\tau}Q_{141}$	2M+2S/2M+S	3M+2S	2M+2S
$c_{251} = 741320437 - 1895020271\mu\tau = (\mu\bar{\tau})^{73}$	$Q_{251} = -\mu\bar{\tau}Q_{31}$	2M+2S/2M+S	3M+2S	2M+2S
$c_{199} = -1246247191 + 741320437\mu\tau = -(\mu\bar{\tau})^{74}$	$Q_{199} = -\mu\bar{\tau}Q_{251}$	2M+2S/2M+S	3M+2S	2M+2S
$c_{189} = 236393683 + 1246247191\mu\tau = -(\mu\bar{\tau})^{75}$	$Q_{189} = \mu\bar{\tau}Q_{199}$	2M+2S/2M+S	3M+2S	2M+2S
$c_{209} = 1566079231 + 236393683\mu\tau = (\mu\bar{\tau})^{76}$	$Q_{209} = -\mu\bar{\tau}Q_{189}$	2M+2S/2M+S	3M+2S	2M+2S
$c_{75} = 2038866597 - 1566079231\mu\tau = (\mu\bar{\tau})^{77}$	$Q_{75} = \mu\bar{\tau}Q_{209}$	2M+2S/2M+S	3M+2S	2M+2S
$c_{169} = -1093291865 - 2038866597\mu\tau = (\mu\bar{\tau})^{78}$	$Q_{169} = \mu\bar{\tau}Q_{75}$	2M+2S/2M+S	3M+2S	2M+2S
$c_{19} = -876057763 + 1093291865\mu\tau = (\mu\bar{\tau})^{79}$	$Q_{19} = \mu\bar{\tau}Q_{169}$	2M+2S/2M+S	3M+2S	2M+2S
$c_{193} = 1310525967 + 876057763\mu\tau = (\mu\bar{\tau})^{80}$	$Q_{193} = \mu\bar{\tau}Q_{19}$	2M+2S/2M+S	3M+2S	2M+2S
$c_{155} = -1232325803 - 1310525967\mu\tau = (\mu\bar{\tau})^{81}$	$Q_{155} = \mu\bar{\tau}Q_{193}$	2M+2S/2M+S	3M+2S	2M+2S
$c_{231} = -441589559 - 1232325803\mu\tau = -(\mu\bar{\tau})^{82}$	$Q_{231} = -\mu\bar{\tau}Q_{155}$	2M+2S/2M+S	3M+2S	2M+2S
$c_{29} = 1388726131 + 441589559\mu\tau = -(\mu\bar{\tau})^{83}$	$Q_{29} = \mu\bar{\tau}Q_{231}$	2M+2S/2M+S	3M+2S	2M+2S
$c_{79} = -2023062047 - 1388726131\mu\tau = -(\mu\bar{\tau})^{84}$	$Q_{79} = \mu\bar{\tau}Q_{29}$	2M+2S/2M+S	3M+2S	2M+2S
$c_{21} = -505547013 + 2023062047\mu\tau = -(\mu\bar{\tau})^{85}$	$Q_{21} = \mu\bar{\tau}Q_{79}$	2M+2S/2M+S	3M+2S	2M+2S
$c_{137} = 754390215 - 505547013\mu\tau = (\mu\bar{\tau})^{86}$	$Q_{137} = -\mu\bar{\tau}Q_{21}$	2M+2S/2M+S	3M+2S	2M+2S
$c_{179} = -256703811 - 754390215\mu\tau = (\mu\bar{\tau})^{87}$	$Q_{179} = \mu\bar{\tau}Q_{137}$	2M+2S/2M+S	3M+2S	2M+2S
$c_{95} = 1765484241 - 256703811\mu\tau = -(\mu\bar{\tau})^{88}$	$Q_{95} = -\mu\bar{\tau}Q_{179}$	2M+2S/2M+S	3M+2S	2M+2S
$c_{59} = -1252076619 + 1765484241\mu\tau = (\mu\bar{\tau})^{89}$	$Q_{59} = -\mu\bar{\tau}Q_{95}$	2M+2S/2M+S	3M+2S	2M+2S
$c_{249} = -2016075433 + 1252076619\mu\tau = (\mu\bar{\tau})^{90}$	$Q_{249} = \mu\bar{\tau}Q_{59}$	2M+2S/2M+S	3M+2S	2M+2S
$c_{131} = 488077805 + 2016075433\mu\tau = (\mu\bar{\tau})^{91}$	$Q_{131} = \mu\bar{\tau}Q_{249}$	2M+2S/2M+S	3M+2S	2M+2S
$c_{145} = 225261375 - 488077805\mu\tau = (\mu\bar{\tau})^{92}$	$Q_{145} = \mu\bar{\tau}Q_{131}$	2M+2S/2M+S	3M+2S	2M+2S
$c_{117} = 750894235 + 225261375\mu\tau = -(\mu\bar{\tau})^{93}$	$Q_{117} = -\mu\bar{\tau}Q_{145}$	2M+2S/2M+S	3M+2S	2M+2S
$c_{105} = -1201416985 + 750894235\mu\tau = (\mu\bar{\tau})^{94}$	$Q_{105} = -\mu\bar{\tau}Q_{117}$	2M+2S/2M+S	3M+2S	2M+2S
$c_{173} = -300371485 - 1201416985\mu\tau = -(\mu\bar{\tau})^{95}$	$Q_{173} = -\mu\bar{\tau}Q_{105}$	2M+2S/2M+S	3M+2S	2M+2S
$c_{129} = -1591761841 - 300371485\mu\tau = (\mu\bar{\tau})^{96}$	$Q_{129} = -\mu\bar{\tau}Q_{173}$	2M+2S/2M+S	3M+2S	2M+2S
$c_{37} = -2102462485 - 1591761841\mu\tau = -(\mu\bar{\tau})^{97}$	$Q_{37} = -\mu\bar{\tau}Q_{129}$	2M+2S/2M+S	3M+2S	2M+2S
$c_{217} = 991018871 - 2102462485\mu\tau = (\mu\bar{\tau})^{98}$	$Q_{217} = -\mu\bar{\tau}Q_{37}$	2M+2S/2M+S	3M+2S	2M+2S
$c_{221} = -1081061197 + 991018871\mu\tau = -(\mu\bar{\tau})^{99}$	$Q_{221} = -\mu\bar{\tau}Q_{217}$	2M+2S/2M+S	3M+2S	2M+2S
$c_{143} = 900976545 + 1081061197\mu\tau = -(\mu\bar{\tau})^{100}$	$Q_{143} = \mu\bar{\tau}Q_{221}$	2M+2S/2M+S	3M+2S	2M+2S
$c_{213} = -1231868357 - 900976545\mu\tau = -(\mu\bar{\tau})^{101}$	$Q_{213} = \mu\bar{\tau}Q_{143}$	2M+2S/2M+S	3M+2S	2M+2S
$c_{73} = -1261145849 - 1231868357\mu\tau = (\mu\bar{\tau})^{102}$	$Q_{73} = -\mu\bar{\tau}Q_{213}$	2M+2S/2M+S	3M+2S	2M+2S
$c_{13} = -570084733 - 1261145849\mu\tau = -(\mu\bar{\tau})^{103}$	$Q_{13} = -\mu\bar{\tau}Q_{73}$	2M+2S/2M+S	3M+2S	2M+2S
$c_{159} = 1202590865 + 570084733\mu\tau = -(\mu\bar{\tau})^{104}$	$Q_{159} = \mu\bar{\tau}Q_{13}$	2M+2S/2M+S	3M+2S	2M+2S
$c_{133} = -1952206965 - 1202590865\mu\tau = -(\mu\bar{\tau})^{105}$	$Q_{133} = \mu\bar{\tau}Q_{159}$	2M+2S/2M+S	3M+2S	2M+2S
$c_{185} = 62421399 - 1952206965\mu\tau(\mu\bar{\tau})^{106}$	$Q_{185} = -\mu\bar{\tau}Q_{133}$	2M+2S/2M+S	3M+2S	2M+2S
$c_{61} = -452974765 + 62421399\mu\tau = -(\mu\bar{\tau})^{107}$	$Q_{61} = -\mu\bar{\tau}Q_{185}$	2M+2S/2M+S	3M+2S	2M+2S
$c_{81} = 328131967 - 452974765\mu\tau = (\mu\bar{\tau})^{108}$	$Q_{81} = -\mu\bar{\tau}Q_{61}$	2M+2S/2M+S	3M+2S	2M+2S
$c_{203} = -577817563 - 328131967\mu\tau = (\mu\bar{\tau})^{109}$	$Q_{203} = \mu\bar{\tau}Q_{81}$	2M+2S/2M+S	3M+2S	2M+2S
$c_{41} = -1234081497 + 577817563\mu\tau = (\mu\bar{\tau})^{110}$	$Q_{41} = \mu\bar{\tau}Q_{203}$	2M+2S/2M+S	3M+2S	2M+2S
$c_{147} = -78446371 + 1234081497\mu\tau = (\mu\bar{\tau})^{111}$	$Q_{147} = \mu\bar{\tau}Q_{41}$	2M+2S/2M+S	3M+2S	2M+2S
$c_{65} = -1905250673 + 78446371\mu\tau = (\mu\bar{\tau})^{112}$	$Q_{65} = \mu\bar{\tau}Q_{147}$	2M+2S/2M+S	3M+2S	2M+2S
$c_{229} = 1748357931 - 1905250673\mu\tau = -(\mu\bar{\tau})^{113}$	$Q_{229} = -\mu\bar{\tau}Q_{65}$	2M+2S/2M+S	3M+2S	2M+2S
$c_{153} = 2062143415 + 1748357931\mu\tau = (\mu\bar{\tau})^{114}$	$Q_{153} = -\mu\bar{\tau}Q_{229}$	2M+2S/2M+S	3M+2S	2M+2S
$c_{99} = 1263891981 - 2062143415\mu\tau = (\mu\bar{\tau})^{115}$	$Q_{99} = \mu\bar{\tau}Q_{153}$	2M+2S/2M+S	3M+2S	2M+2S
$c_{207} = -1434572447 + 1263891981\mu\tau = -(\mu\bar{\tau})^{116}$	$Q_{207} = -\mu\bar{\tau}Q_{99}$	2M+2S/2M+S	3M+2S	2M+2S
$c_{107} = -1093211515 - 1434572447\mu\tau = (\mu\bar{\tau})^{117}$	$Q_{107} = -\mu\bar{\tau}Q_{207}$	2M+2S/2M+S	3M+2S	2M+2S
$c_{9} = 332610887 + 1093211515\mu\tau = (\mu\bar{\tau})^{118}$	$Q_{9} = \mu\bar{\tau}Q_{107}$	2M+2S/2M+S	3M+2S	2M+2S
$c_{205} = 1775933379 + 332610887\mu\tau = -(\mu\bar{\tau})^{119}$	$Q_{205} = -\mu\bar{\tau}Q_{9}$	2M+2S/2M+S	3M+2S	2M+2S
$c_{223} = -1853812143 - 1775933379\mu\tau = -(\mu\bar{\tau})^{120}$	$Q_{223} = \mu\bar{\tau}Q_{205}$	2M+2S/2M+S	3M+2S	2M+2S
$c_{187} = 1110711605 - 1853812143\mu\tau = (\mu\bar{\tau})^{121}$	$Q_{187} = -\mu\bar{\tau}Q_{223}$	2M+2S/2M+S	3M+2S	2M+2S
$c_{121} = 1698054615 - 1110711605\mu\tau = (\mu\bar{\tau})^{122}$	$Q_{121} = \mu\bar{\tau}Q_{187}$	2M+2S/2M+S	3M+2S	2M+2S
$c_{253} = 523368595 + 1698054615\mu\tau = -(\mu\bar{\tau})^{123}$	$Q_{253} = -\mu\bar{\tau}Q_{121}$	2M+2S/2M+S	3M+2S	2M+2S
$c_{17} = 375489471 + 523368595\mu\tau = (\mu\bar{\tau})^{124}$	$Q_{17} = -\mu\bar{\tau}Q_{253}$	2M+2S/2M+S	3M+2S	2M+2S
$c_{11} = 1422226661 - 375489471\mu\tau = (\mu\bar{\tau})^{125}$	$Q_{11} = \mu\bar{\tau}Q_{17}$	2M+2S/2M+S	3M+2S	2M+2S
$c_{23} = -671247719 + 1422226661\mu\tau = -(\mu\bar{\tau})^{126}$	$Q_{23} = -\mu\bar{\tau}Q_{11}$	2M+2S/2M+S	3M+2S	2M+2S
$c_{45} = -2121761693 + 671247719\mu\tau = -(\mu\bar{\tau})^{127}$	$Q_{45} = \mu\bar{\tau}Q_{23}$	2M+2S/2M+S	3M+2S	2M+2S

References

1. Avanzi, R.M., Dimitrov, V.S., Doche, C., Sica, F.: Extending scalar multiplication using double bases. In: Lai, X., Chen, K. (eds.) ASIACRYPT 2006. LNCS, vol. 4284, pp. 130–144. Springer, Heidelberg (2006). https://doi.org/10.1007/11935230_9
2. Bernstein, D.J.: Multidigit multiplication for mathematicians. Adv. Appl. Math., 1–19 (2001)
3. Barker, E.: Recommendation for Key Management, Part 1: General. National Institute of Standards and Technology (2016). http://doi.org/10.6028/NIST.SP.800-57pt1r4
4. Blake, I., Murty, V., Xu, G.: A note on window τ-NAF algorithm. Inf. Process. Lett. **95**(5), 496–502 (2005)
5. Blake, I., Murty, V., Xu, G.: Nonadjacent radix-τ expansions of integers in Euclidean imaginary quadratic number fields. Canadian J. Math. **60**, 1267–1282 (2008)
6. Bos J., Lenstra A., Te Riele H., Shumow, D.: Introduction. In: Bos, J., Lenstra, A.,(Eds.), Topics in Computational Number Theory Inspired by Peter L. Montgomery (pp. 1-9). Cambridge University Press, Cambridge (2017). https://doi.org/10.1017/9781316271575.002
7. Doche, C., Kohel, D.R., Sica, F.: Double base number system for multi scalar multiplications. In: Joux, A. (ed.) EUROCRYPT 2009. LNCS, vol. 5479, pp. 502–517. Springer, Heidelberg (2009). https://doi.org/10.1007/978-3-642-01001-9_29
8. Dimitrov, V., Imbert, L., Mishra, P.K.: Efficient and secure elliptic curve point multiplication using double-base chains. In: Roy, B. (ed.) ASIACRYPT 2005. LNCS, vol. 3788, pp. 59–78. Springer, Heidelberg (2005). https://doi.org/10.1007/11593447_4
9. Koblitz, N.: p-adic Numbers, p-adic Analysis, and Zeta-Functions. GTM, vol. 58, New York, Springer, Heidelberg (1984). https://doi.org/10.1007/978-1-4612-1112-9
10. Koblitz, N.: Elliptic curve cryptosystems. Math. Comput. **48**(177), 203–209 (1987)
11. Koblitz N.: CM-curves with good cryptographic properties. In: Proceedings of the 11th Annual International Cryptology Conference on Advances in Cryptology, pp. 279–287 (1992)
12. Kohel D.: Efficient arithmetic on elliptic curves in characteristic 2. https://arxiv.org/abs/1601.03669
13. Kohel, D.: Twisted μ_4-normal form for elliptic curves. In: Coron, J.-S., Nielsen, J.B. (eds.) EUROCRYPT 2017. LNCS, vol. 10210, pp. 659–678. Springer, Cham (2017). https://doi.org/10.1007/978-3-319-56620-7_23
14. Li, W., Yu, W., Li, B., Fan, X.: Speeding up scalar multiplication on Koblitz curves using μ_4 coordinates. In: Jang-Jaccard, J., Guo, F. (eds.) ACISP 2019. LNCS, vol. 11547, pp. 620–629. Springer, Cham (2019). https://doi.org/10.1007/978-3-030-21548-4_34
15. Lange, T.: A note on Lopez-Dahab coordinates. Cryptology ePrint Archive, Report 2004/323 (2004). https://eprint.iacr.org/2004/323.pdf
16. López, J., Dahab, R.: Improved algorithms for elliptic curve arithmetic in $GF(2^n)$. In: Proceedings of Selected Areas Cryptography, pp. 201–212 (1998)
17. Montgomery, P.L.: Speeding the Pollard and elliptic curve methods of factorization. Math. Comput. **48**(177), 243–264 (1987)

18. Oliveira, T., López, J., Aranha, D.F., Rodríguez-Henríquez, F.: Two is the fastest prime: lambda coordinates for binary elliptic curves. J. Crypt. Eng. **4**(1), 3–7 (2014)
19. Scott, M.: MIRACL-Multiprecision integer and rational arithmetic cryptographic library, C/C++ Library. https://github.com/miracl/MIRACL
20. Renes, J., Costello, C., Batina, L.: Complete addition formulas for prime order elliptic curves. In: Fischlin, M., Coron, J.-S. (eds.) EUROCRYPT 2016. LNCS, vol. 9665, pp. 403C428. Springer, Heidelberg (2016). https://doi.org/10.1007/978-3-662-49890-316
21. Solinas J.: Efficient arithmetic on Koblitz curves, Des., Codes Crypt. **19**, 195–249 (2000)
22. Oliveira, T., Aranha, D.F., Lpez, J., Rodrguez-Henrquez, F.: Fast point multiplication algorithms for binary elliptic curves with and without precomputation. Sel. Areas Crypt., 324–344 (2014)
23. Oliveira, T., Lpez-Hernndez, J.C., Cervantes-Vzquez, D., Rodrguez-Henrquez, F.: Koblitz curves over quadratic fields. J. Cryptol. **32**(3), 867–894 (2019)
24. Trost, W., Xu, G.: On the optimal pre-computation of window τNAF for Koblitz curves. IEEE Trans. Comput. **65**(9), 2918–2924 (2016)
25. Vuillaume, C., Okeya, K., Takagi, T.: Defeating simple power analysis on Koblitz curves. In: IEICE Transactions on Fundamentals of Electronics Communications and Computer Sciences, E89-A(5), pp. 1362–1369 (2006)
26. Yu, W., Musa, S.A., Li, B.: Double-base chains for scalar multiplications on elliptic curves. In: Canteaut, A., Ishai, Y. (eds.) EUROCRYPT 2020. LNCS, vol. 12107, pp. 538–565. Springer, Cham (2020). https://doi.org/10.1007/978-3-030-45727-3_18
27. Yu, W., Xu, G.: Pre-computation scheme of window τNAF for Koblitz curves revisited. In: Canteaut, A., Standaert, F.-X. (eds.) EUROCRYPT 2021. LNCS, vol. 12697, pp. 187–218. Springer, Cham (2021). https://doi.org/10.1007/978-3-030-77886-6_7

Parallel Algorithms on Hyperelliptic Pairings Using Hyperelliptic Nets

Chao Chen[1,2] and Fangguo Zhang[1,2](✉)

[1] School of Computer Science and Engineering, Sun Yat-sen University, Guangzhou 510006, China
chench533@mail2.sysu.edu.cn, isszhfg@mail.sysu.edu.cn
[2] Guangdong Province Key Laboratory of Information Security Technology, Guangzhou 510006, China

Abstract. Pairings are useful tools in cryptography and efficient implementations play a critical role in their usage, where Miller's algorithms are the main method for all pairings. As an alternative approach, elliptic nets were first employed to evaluate Tate pairings and generalized to the hyperelliptic nets for Tate pairings on hyperelliptic curves. In this work, for hyperelliptic pairings derived from rational functions, we establish the unitary formulae in terms of hyperelliptic nets. Afterwards, for genus-2 hyperelliptic pairings, we construct a parallel Double-and-Add algorithm on the minimal block. In particular, all terms in new blocks, having irrelevant formulae on current blocks, can be evaluated with 12 processors in parallel, thus the explicit loop cost reduces to $4\mathsf{M}'$ (multiplications in extension fields) with 276 parallel processors. As an additional merit, Double and Double-Add algorithms invoke analogous operations such that our method avoids extra additions in Miller's algorithms.

Keywords: Hyperelliptic Nets · Hyperelliptic Curves · Tate Pairings · Ate Pairings · Parallel Algorithms

1 Introduction

Bilinear pairings on algebraic curves have been studied for decades, especially for Weil pairings and Tate pairings on elliptic curves [34]. It was introduced into cryptography until the prominent work [27] was established, while the first positive application was the one-round tripartite Diffie-Hellman protocol [22]. Nowadays, pairings are one powerful tool in cryptography such that numerous notable protocols, e.g., identity-based encryptions [5], short signatures [6], public key encryptions with keyword search [4] and functional encryptions [1], have been constructed using these non-degenerate maps. Recently, pairings were also utilized for verifiable delay functions [13] and delay encryptions [9].

The efficiency of pairing implementations is critical for cryptographic applications, where Miller [28] established the first practical algorithm for Weil pairings and Tate pairings. Namely, for two r-torsion points $P \in \mathbb{G}_1$ and

T. Zhu and Y. Li (Eds.): ACISP 2024, LNCS 14895, pp. 429–449, 2024.
https://doi.org/10.1007/978-981-97-5025-2_22

$Q \in \mathbb{G}_2$, the Tate pairing $\langle P, Q \rangle_r$ is computed via $f_{r,P}(Q)$, where div $f_{r,P} = r(P) - ([r]P) - [r-1](\mathcal{O})$. In particular, this algorithm utilized an iterated algorithm for rational functions with $\lceil \log r \rceil$ loops, where Zhao et al. [43] utilized the double-based chain for optimization. To obtain speedup, the main approach is to reduce the loops in Miller's algorithms, where Barreto et al. [2] first introduced Eta pairings by modifying the idea in [14] to the supersingular abelian varieties, then Hess et al. [20] generalized Eta pairings to the Ate pairings over ordinary curves, followed by optimized Ate pairings in [26]. After that, more optimized pairings, e.g., Ate_i pairings [44], R-ate pairings [23], and optimal pairings [40], have been constructed for shorter loops.

To enrich the choice of embedding degrees, Galbraith [15] introduced the pairings on hyperelliptic curves, then Duursma and Lee [14] established the first implementation of Tate pairings on the hyperelliptic curves of genus $g \geq 2$, yet it is less practical than elliptic curve pairings [16]. The above methods for the optimized pairings can be applied to hyperelliptic pairings directly, where Galbraith and Zhao [18] constructed self-pairings to shorten the loop length in Miller's algorithms. Besides, motivated by Ate pairings on elliptic curves, Granger et al. [19] proposed the Ate pairings from ordinary hyperelliptic curves, where the loop length is only $1/g$ times of that of Tate pairings and no final exponentiation is required, then Zhang [42] presented twist Ate pairings on hyperelliptic curves with several tricks. As a result, hyperelliptic pairings over small fields have been an alternative to elliptic pairings for cryptographic applications.

In 2010, Lubicz and Robert [24] discovered the intrinsic connection between Theta functions on abelian varieties and Tate pairings such that Tate pairings and Weil pairings can be evaluated through Riemann forms. Consequently, they obtained a unitary algorithm for Tate pairings on abelian varieties of arbitrary genera, moreover, it is facile to apply to Tate pairings on Kummer varieties [17,24]. Unfortunately, this method cannot work for other optimized pairings without the analytic property of abelian varieties. To broaden the usage of Theta functions, Miller's algorithms with Theta functions [25] were generalized to all abelian varieties and Kummer varieties such that all efforts on shortening the loop length would still work.

Another notable breakthrough was established in 2007, when the elliptic net, i.e., a map from a finitely generated free abelian group to an integral domain with the specific recurrence relation as the generalization of elliptic divisibility sequences [41], was first employed for Tate pairings [35]. More precisely, Stange built the Double-and-Add algorithm on the well-selected blocks, containing several elements of elliptic nets, such that Tate pairings can be evaluated in terms of elliptic nets. From then on, several works have been established, where [31,36] presented the formulae of optimized pairings with elliptic nets. Besides, the famous Barreto-Naehrig (BN) curves and supersingular curves [32,33] were employed for optimized pairings and symmetric pairings, respectively, moreover, to obtain accelerations, the parallel computation on elliptic nets was taken into account while the block size grew. If the Hamming weight of the parameter r is low, the dimension of blocks can be reduced [11] at the price of one inversion for the DoubleAdd step, then a recent work [10] eliminated the inversion and employed lazy reductions for better performances.

The above works mainly considered the pairings on elliptic curves, thus it is natural to construct hyperelliptic nets for hyperelliptic pairings with analogous methods. To this issue, Uchida and Uchiyama [39] first introduced hyperelliptic nets, generalized from elliptic nets, and gave the formulae of Tate pairings on hyperelliptic curves in terms of hyperelliptic nets, then Tran [37] extended the above methods to all genera with several optimizations. Nevertheless, the formulae of optimized hyperelliptic pairings in terms of hyperelliptic nets are still unexploited and it is significant to search for more optimizations.

Our Contributions. This work aims to provide the unitary parallel algorithm for computing arbitrary hyperelliptic pairings with hyperelliptic nets.

Firstly, after recalling basic definitions, we regard $\Psi_{1,v_2,\cdots,v_m}(P_1, P_2, \cdots, P_m)$ as a function of P_1 such that

$$\text{div } \Psi_{\mathbf{v}} = t^*_{-\sum_{i=2}^m v_i P_i}\Theta - \sum_{i=2}^m v_i t^*_{-P_i}\Theta - (1 - \sum_{i=2}^m v_i)\Theta,$$

where $t_P : Q \mapsto P + Q$ is a translation map and Θ is the Theta divisor on J_C. Besides, fixed some m and v_i, the divisor of $\Psi_{\mathbf{v}}$ coincides with that of the rational function for pairings, such that the computation only relies on hyperelliptic nets. Afterwards, given several lemmas, we obtain the theorem for all the pairings defined by the rational functions $f_{a,P}$ for $a \in \mathbb{Z}$ and $P \in J_C$.

Theorem 1. *Let C be a hyperelliptic curve of genus g, and $P, Q \in J_C[r]$ with $P \notin \langle Q \rangle$. Let $W_{P,Q}$ be the hyperelliptic nets associated to C, P, Q. Then we have*

$$f_{a,P}(D_Q) = \frac{W_{P,Q}(a+1,1)W_{P,Q}(2,0)^a}{W_{P,Q}(a+1,0)W_{P,Q}(2,1)^a}, \tag{1}$$

where $D_Q = (-P) - (-P - Q)$ satisfies $supp(div\ f_{a,P}) \cap supp(D_Q) = \emptyset$.

Moreover, the above theorem is consistent with all previous works [35–37,39], so it can be leveraged for all pairings on hyperelliptic curves of all genera (including genus-1 elliptic curves).

Our second contribution is an efficient parallel algorithm for hyperelliptic nets. More specifically, for $m > 2^g$, the matrix

$$A = (W_{P,Q}(u_i + u_j, v_i + v_j)W_{P,Q}(u_i - u_j, v_i - v_j))_{1 \le i,j \le m}$$

satisfies $\det A = 0$, leading to a recurrence formula for hyperelliptic nets. For $g = 2$, motivated by the Double-and-Add algorithm in [39], we reach the same optimal block $B(\ell)$ with 23 elements.

For the implementation of hyperelliptic nets, the gist is to employ Eq. (10) to compute the terms in the block $B(2\ell)$ or $B(2\ell + 1)$. A crucial discovery is that the formulae of different terms in new blocks are irrelevant, then parallel computations can be applied. Furthermore, the evaluations of $W_{P,Q}(i,0)$ and $W_{P,Q}(j,1)$ can be accelerated by leveraging more parallel resources. As an additional merit, this method employs unified algorithms for $B(2\ell)$ and $B(2\ell+1)$, so it can avoid possible extra additions in Miller loops. In a word, yet every loop of Double-and-Add algorithms invokes more multiplications, the explicit minimal cost for a new block is only $4\mathsf{M}'$ (M' is the cost of multiplications in extension fields). Compared with previous approaches (c.f. Table 4), this work provides an efficient algorithm for pairing computations at the expense of parallel resources.

Organization. The rest of this paper is organized as follows. Section 2 provides some necessary mathematical backgrounds and the definitions of hyperelliptic nets. Section 3 depicts the unitary formula of hyperelliptic pairings with hyperelliptic nets. In Sect. 4, the parallel algorithms on the minimal blocks are established. Finally, the last section concludes our work.

2 Preliminaries

In this section, we recall some necessary mathematical backgrounds of hyperelliptic curves and hyperelliptic pairings. We mainly focus on high-genus curves, where the definitions of elliptic curves and pairings on elliptic curves can be found in [3,12,34].

2.1 Hyperelliptic Curves

The hyperelliptic curve C of genus g over a field $\mathbb{K}$ with $g \geq 1$ is given by the following equation:

$$C : y^2 + h(x)y = f(x),$$

where $f(x)$ is a monic polynomial of degree $2g+1$, $h(x)$ is a polynomial of degree at most g, and there are no solutions $(x, y) \in \overline{\mathbb{K}} \times \overline{\mathbb{K}}$ simultaneously satisfying the equation $y^2 + h(x)y = f(x)$, the partial derivative equations $2y + h(x) = 0$ and $h'(x)y - f'(x) = 0$. Hence, C is a nonsingular hyperelliptic curve and has only one point ∞ at infinity. The set

$$C(\mathbb{K}) := \{(x, y) \in \mathbb{K} \times \mathbb{K} \mid y^2 + h(x)y = f(x)\} \cup \{\infty\}$$

is called the set of $\mathbb{K}$-rational points on C.

When $g \geq 2$, the set $C(\mathbb{K})$ is not a group, but we can embed C into an abelian variety of dimension g, which is called the Jacobian of C and denoted by J_C. The Jacobian J_C is isomorphic to the divisor class group of degree zero Pic_C^0. Let $\mathcal{O}$ be the identity element of J_C. For $r \in \mathbb{N}$, we define

$$J_C[r] := \{D \in J_C \mid rD = \mathcal{O}\}$$

as the r-torsion subgroup of J_C.

Let $\lambda : C \to J_C$ be an embedding such that $\lambda(\infty) = \mathcal{O}$. The *Theta divisor* Θ on J_C is defined by $\Theta = \lambda(C) + \lambda(C) + \cdots + \lambda(C)$ $(g-1$ times).

2.2 Pairings

Let $\mathbb{F}_q$ be a finite field with q elements and C be a hyperelliptic curve of genus g over $\mathbb{F}_q$. Fixed a subgroup of $J_C(\mathbb{F}_q)$ of order r, the embedding degree is denoted by the smallest integer α satisfying $r \mid q^\alpha - 1$, i.e., the r-th roots of unity μ_r are contained in $\mathbb{F}_{q^\alpha}$.

Let $J_C(\mathbb{F}_{q^\alpha})[r]$ and $J(\mathbb{F}_{q^\alpha})/rJ_C(\mathbb{F}_{q^\alpha})$ be the r-torsion group and the quotient group, respectively. Then the Tate(-Lichtenbaum) pairing is a non-degenerate bilinear map

$$\begin{aligned} J_C(\mathbb{F}_{q^\alpha})[r] \times J_C(\mathbb{F}_{q^\alpha})/rJ_C(\mathbb{F}_{q^\alpha}) &\longrightarrow \mathbb{F}_{q^\alpha}^*/(\mathbb{F}_{q^\alpha}^*)^r \\ (\overline{D}_1, \overline{D}_2) &\longmapsto \langle \overline{D}_1, \overline{D}_2 \rangle_r, \end{aligned}$$

where $\overline{D}_1$ (resp. $\overline{D}_2$) is represented by a divisor D_1 (resp. D_2) with $\mathrm{supp}(D_1) \cap \mathrm{supp}(D_2) = \emptyset$. For implementation, we choose a rational function $f_{r,D_1} \in \mathbb{F}_{q^\alpha}(C)$ such that div $(f_{r,D_1}) = rD_1 - [r]D_1$, then the Tate pairings can be evaluated by

$$\langle \overline{D}_1, \overline{D}_2 \rangle_r = f_{r,D_1}(D_2) = \prod_{P \in C(\overline{\mathbb{F}}_q)} f_{r,D_1}(P)^{ord_P(D_2)}. \tag{2}$$

To achieve cryptographic applications, we also consider the reduced (or modified) Tate pairing

$$t_r(\overline{D}_1, \overline{D}_2) = \langle \overline{D}_1, \overline{D}_2 \rangle_r^{\frac{q^\alpha - 1}{r}}.$$

In practice, Miller's algorithm [28] is still the main approach to evaluating Tate pairings on elliptic curves, which can be generalized to Tate pairings on hyperelliptic curves. Generally, the loop length of Miller's algorithms for Tate pairings is $\lfloor \log_2 r \rfloor$, where there are possible auxiliary additions in every loop.

To reduce the loop length of Miller's algorithms, Hess et al. [20] introduced the Ate pairings on ordinary elliptic curves, where only $\log_2 r/\varphi(\alpha)$ Miller loops are required, then this method was generalized to ordinary hyperelliptic curves [19].

Let C be a hyperelliptic curve over $\mathbb{F}_q$ and $r \mid \#J_C(\mathbb{F}_q)$ be a large prime. Let π be the q-power Frobenius map on C and Frobenius endomorphism on J_C. Consider $\mathbb{G}_1 = J_C[r] \cap \ker(\pi - [1])$ and $\mathbb{G}_2 = J_C[r] \cap \ker(\pi - [q])$, then the Ate pairing is defined by

$$\begin{aligned} a(\cdot,\cdot) : \mathbb{G}_2 \times \mathbb{G}_1 &\longrightarrow \quad \mathbb{F}_{q^\alpha}^* \\ (\overline{D}_2, \overline{D}_1) &\longmapsto f_{q,D_2}(D_1) \end{aligned}$$

such that $\mathrm{supp}(D_1) \cap \mathrm{supp}(D_2) = \emptyset$. Moreover, the relation between the reduced Tate pairings and Ate pairings is

$$t_r(D_2, D_1) = a(D_2, D_1)^{\alpha q^{\alpha-1}},$$

indicating that the additional merit of Ate pairings is unnecessary to evaluate the final exponentiation.

2.3 Hyperelliptic Nets

Now, we recall the definition of hyperelliptic nets [37,39], where the original definition is over $\mathbb{C}$, but can be generalized to arbitrary fields.

To formalize the hyperelliptic nets, the Kleinian sigma function $\sigma \in \mathbb{C}^g$, as the extension of Weierstrass sigma functions for genus one, is the key component. Here, we do not depict its concrete definition, which can be found in [7,8], but recall some properties for hyperelliptic nets.

Let C be a hyperelliptic curve of genus g over $\mathbb{C}$, then its Jacobian is isomorphic to a torus $\mathbb{C}^g/\Lambda$, where Λ is a lattice in $\mathbb{C}^g$. Let $\kappa : \mathbb{C}^g \to J_C$ be the natural projection. We first review two basic properties of Kleinian sigma functions.

Proposition 1 ([30]). *The divisor of* $\sigma(\mathbf{z})$ *is* $\kappa^{-1}(\Theta)$.

Proposition 2 ([29]). *The Kleinian sigma function* σ *is an odd function when* $g \equiv 1, 2 \pmod 4$ *and an even function if* $g \equiv 0, 3 \pmod 4$.

Besides, sigma functions σ satisfy the following property, which is critical to establish the recurrence relations.

Proposition 3 ([38]). *Let* $m > 2^g$ *be an integer and* $\mathbf{z}^{(1)}, \mathbf{z}^{(2)}, \cdots, \mathbf{z}^{(m)} \in \mathbb{C}^g$. *The matrix* A *is defined by*

$$A = \left(\sigma(\mathbf{z}^{(i)} + \mathbf{z}^{(j)})\sigma(\mathbf{z}^{(i)} - \mathbf{z}^{(j)})\right)_{1 \le i,j \le m}.$$

Then it holds that $\det A = 0$. *Furthermore, when* $g \equiv 1, 2 \pmod 4$ *and* m *is even, we have* $Pf\ A = 0$.

Remark 1. The Pfaffian of a skew-symmetric matrix A is defined by

$$\text{Pf } A = \frac{1}{2^n n!} \sum_{\tau \in S_{2n}} sgn(\tau) \prod_{i=1}^{n} a_{\tau_{2i-1}, \tau_{2i}},$$

where S_{2n} is the symmetric group and $sgn(\tau)$ is the symbol of τ. Furthermore, we know that if σ is an odd function, the determinant of matrix A satisfies

$$\det A = (\text{Pf } A)^2.$$

To obtain the hyperelliptic nets, we first construct a new function $\Psi_{\mathbf{v}}$.

Definition 1. *For* $\mathbf{v} = (v_1, \cdots, v_m) \in \mathbb{Z}^m$, *the meromorphic function* $\Psi_{\mathbf{v}}$ *is defined by*

$$\Psi_{\mathbf{v}}(\mathbf{z}^{(1)}, \cdots, \mathbf{z}^{(m)}) = \frac{\sigma(v_1\mathbf{z}^{(1)} + \cdots + v_m\mathbf{z}^{(m)})}{\prod_{i=1}^{m} \sigma(\mathbf{z}^{(i)})^{2v_i^2 - v_i \sum_{j=1}^{v} v_j} \cdot \prod_{1 \le i < j \le m} \sigma(\mathbf{z}^{(i)} + \mathbf{z}^{(j)})^{v_i v_j}}.$$

It is obvious that the function $\Psi_{\mathbf{v}}$ is periodic with respect to Λ in each variable, so $\Psi_{\mathbf{v}}$ is a meromorphic function on J_C^m. Moreover, let $p_i : J_C^m \to J_C$ be the i-th projection and $s_n : J_C^n \to J_C$ be the summation of all components in J_C^n, then the divisor of $\Psi_{\mathbf{v}}$ is formulated as follows:

$$\begin{aligned} \operatorname{div} \Psi_{\mathbf{v}} = ([v_1] \times \cdots \times [v_m])^* s_m^* \Theta - \textstyle\sum_{1 \le i < j \le m} v_i v_j (p_i \times p_j)^* s_2^* \Theta \\ - \textstyle\sum_{i=1}^{m} \left(2v_i^2 - \sum_{j=1}^{m} v_i v_j\right) p_k^* \Theta, \end{aligned} \tag{3}$$

where s_n^* and p_i^* are their pullbacks.

Let $P_i = \kappa(\mathbf{z}^{(i)})$ and $\Psi_{\mathbf{v}}(P_1, \cdots, P_m) = \Psi_{\mathbf{v}}(\mathbf{z}^{(1)}, \cdots, \mathbf{z}^{(m)})$, then one can obtain the recurrence formula of $\Psi_{\mathbf{v}}$ from Theorem 2.

Theorem 2 ([39]). *Let $m > 2^g$ be an integer and $\mathbf{v}_i \in ((1/2)\mathbb{Z})^m$ such that $\mathbf{v}_i + \mathbf{v}_j, \mathbf{v}_i - \mathbf{v}_j \in \mathbb{Z}^m$ for $1 \le i < j \le m$. The matrix*

$$A = \left(\Psi_{\mathbf{v}_i+\mathbf{v}_j} \Psi_{\mathbf{v}_i-\mathbf{v}_j}\right)_{1 \le i,j \le m}$$

satisfies $\det A = 0$. *In particular, if $g \equiv 1, 2 \pmod 4$ and m is even, we have* $Pf\, A = 0$.

The above theorem is true for the functions over $\mathbb{C}$. To broaden its usage, Uchida and Uchiyama [39] have generalized this property to $\Psi_{\mathbf{v}}$ over arbitrary fields $\mathbb{K}$. Thus, the hyperelliptic net associated to a hyperelliptic curve C over arbitrary fields is formalized as follows:

Definition 2. *Let $P_1, \cdots, P_n \in J_C(\mathbb{K})$ such that for all $i \neq j \in \{1, \cdots, n\}$, P_i and $P_i + P_j$ are not contained in Θ. The hyperelliptic net associated to C and $P_1, \cdots, P_n$ is defined by*

$$W_{P_1,\cdots,P_n}(\mathbf{v}) = \Psi_{\mathbf{v}}(P_1, \cdots, P_n),$$

where $\Psi_{\mathbf{v}}$ is a rational function on J_C^n over an arbitrary field $\mathbb{K}$.

Remark 2. A hyperelliptic net associated to an elliptic curve E is called an elliptic net, consistent with the original definition in [35].

Tate pairings and Weil pairings from hyperelliptic nets. The recurrence formula, induced by Theorem 2, can be leveraged for pairings over finite fields. Elliptic nets were first applied to Tate pairings by Stange [35], then the formula of Tate pairings on hyperelliptic curves was formalized via hyperelliptic nets in [37,39]. The unitary formula is depicted as follows.

Theorem 3 ([35,39]). *Let E (resp. C) be an elliptic curve (resp. a hyperelliptic curve) over a field $\mathbb{K}$, and $P, Q \in E(\mathbb{K})$ (resp. $P, Q \in J_C(\mathbb{K})$) such that $[r]P = \mathcal{O}$. In addition, we assume $P, Q, P + Q \notin \Theta$. Let $W_{P,Q}$ be the elliptic net (resp. hyperelliptic net) associated to E (resp. C), P, Q, then we have*

$$\langle P, Q \rangle_r = \frac{W_{P,Q}(r+1,1) W_{P,Q}(1,0)}{W_{P,Q}(r+1,0) W_{P,Q}(1,1)} \in \mathbb{K}^*/(\mathbb{K}^*)^r. \tag{4}$$

To construct Weil pairings $e_r(\cdot,\cdot)$ using hyperelliptic nets, the relation between Weil pairings and Tate pairings is utilized. In particular, with the same notations in Theorem 3 such that $[r]P = [r]Q = \mathcal{O}$, it holds that

$$e_r(P,Q) = \frac{W_{P,Q}(r+1,1)W_{P,Q}(0,r+1)W_{P,Q}(1,0)}{W_{P,Q}(1,r+1)W_{P,Q}(r+1,0)W_{P,Q}(0,1)} \in \mu_r. \tag{5}$$

The generalization of Tate pairings to Weil pairings is almost the same as that in [24], while the main technique of the latter work is the intrinsic property of Theta functions, failing in obtaining other optimized pairings.

3 Pairing Computation via Hyperelliptic Nets

In this section, we establish the unitary formulae of hyperelliptic pairings via hyperelliptic nets.

Apart from Tate pairings and Weil pairings, other optimized pairings on hyperelliptic curves, e.g., Ate pairings, have been introduced for shorter loops. In general, all pairings can be implemented with rational functions $f_{a,P}$ such that

$$\text{div } f_{a,P} = aP - [a]P$$

with $a \in \mathbb{Z}$ and $P \in J_C$. Therefore, we will focus on the pairings generalized from rational functions $f_{a,P}$ and establish their formulae in terms of hyperelliptic nets.

For a divisor P on J_C, we define the translation map

$$\begin{aligned} t_P : J_C &\to J_C \\ Q &\mapsto P + Q. \end{aligned}$$

Moreover, according to the divisors of Kleinian sigma functions $\sigma(\mathbf{z})$, we obtain the following theorem.

Theorem 4. *Let* $\mathbf{v} = (1, v_2, \cdots, v_m) \in \mathbb{Z}^m$, *and* $P_2, \cdots, P_m \in J_C$ *such that* $\pm P_i$ *are all distinct and nonzero. Consider* $\Psi_{\mathbf{v}}(P_1, \cdots, P_m)$ *as a function of* P_1, *then we have*

$$\textit{div } \Psi_{\mathbf{v}} = t^*_{-\sum_{i=2}^m v_i P_i}\Theta - \sum_{i=2}^m v_i t^*_{-P_i}\Theta - (1 - \sum_{i=2}^m v_i)\Theta.$$

Proof. From Eq. (3), we obtain the divisor of $\Psi_{\mathbf{v}}$ immediately. □

Before we give the explicit formula, we introduce some lemmas.

Lemma 1. *Let* a *be an integer and* $S, P, Q \in J_C(\mathbb{K})$ *such that* $S + Q \notin \Theta$. *Then we obtain*

$$\Psi_{1,a,0}(S+Q,P,Q) = \frac{\Psi_{1,a,1}(S,P,Q)}{\Psi_{1,1,1}(S,P,Q)^a}. \tag{6}$$

Proof. Let $\mathbf{v} = (v_1, \cdots, v_m)$ and $\mathbf{P} = (P_1, \cdots, P_n) \in J_C^n$. From [39, Prop. 8], for a $n \times m$ matrix T with entries in $\mathbb{Z}$, we have

$$\Psi_{\mathbf{v}}(\mathbf{P}T) = \frac{\Psi_{T\mathbf{v}}(\mathbf{P})}{\prod_{i=1}^{m} \Psi_{T\mathbf{e}_i}(\mathbf{P})^{2v_i^2 - \sum_{j=1}^{m} v_i v_j} \cdot \prod_{1 \le i < j \le m} \Psi_{T(\mathbf{e}_i + \mathbf{e}_j)}(\mathbf{P})^{v_i v_j}}. \quad (7)$$

Consider

$$T = \begin{pmatrix} 1\,0\,0 \\ 0\,1\,0 \\ 1\,0\,1 \end{pmatrix}, \quad \mathbf{P} = (S, P, Q), \quad \mathbf{v} = (1, a, 0),$$

we obtain

$$\Psi_{1,a,0}(S + Q, P, Q) = \frac{\Psi_{1,a,1}(S, P, Q)}{\Psi_{1,0,1}(S, P, Q)^{1-a} \Psi_{0,1,0}(S, P, Q)^{a^2 - a} \Psi_{1,1,1}(S, P, Q)^a}. \quad (8)$$

Moreover, if $\mathbf{v} = \mathbf{e}_i$ or $\mathbf{v} = \mathbf{e}_i + \mathbf{e}_j$ with $i \neq j$, it holds that $\Psi_{\mathbf{v}} = 1$ [39, Prop. 7]. As a consequence, we get

$$\Psi_{1,a,0}(S + Q, P, Q) = \frac{\Psi_{1,a,1}(S, P, Q)}{\Psi_{1,1,1}(S, P, Q)^a}.$$

□

Lemma 2. *Let $a \in \mathbb{Z}$ and $P_1, P_2, P_3 \in J_C(\mathbb{K})$ such that $P_1, P_2, P_1 + P_2 \notin \Theta$. Then*

(1) $\Psi_{a,0}(P_1, P_2) = \Psi_a(P_1)$;
(2) $\Psi_{1,a,0}(P_1, P_2, P_3) = \Psi_{1,a}(P_1, P_2)$;
(3) $\Psi_{1,a,1}(P_1, P_1, P_2) = \dfrac{\Psi_{1+a,1}(P_1, P_2)}{\Psi_2(P_1)^a}$;
(4) $\Psi_{a,b}(P_1, P_1) = \dfrac{\Psi_{a+b}(P_1)}{\Psi_2(P_1)^{ab}}$.

Proof. (1) and (2) can be easily acquired from the definition of $\Psi_{\mathbf{v}}$.

As for the third equation, we leverage Eq. (7) as well. Namely, let

$$T = \begin{pmatrix} 1\,1\,0 \\ 0\,0\,1 \end{pmatrix}, \quad \mathbf{P} = (P_1, P_2), \quad \mathbf{v} = (1, a, 1),$$

then we have

$$\Psi_{1,a,1}(P_1, P_1, P_2) = \frac{\Psi_{a+1,1}(P_1, P_2)}{\Psi_{1,0}(P_1, P_2)^{a^2 - 3a} \Psi_{0,1}(P_1, P_2)^{-a} \Psi_{1,1}(P_1, P_2)^{a+1} \Psi_{2,0}(P_1, P_2)^a}.$$

Moreover, it holds that $\Psi_{1,0}(P_1, P_2) = \Psi_{0,1}(P_1, P_2) = \Psi_{1,1}(P_1, P_2) = 1$ and $\Psi_{2,0}(P_1, P_2) = \Psi_2(P_1)$, so we obtain

$$\Psi_{1,a,1}(P_1, P_1, P_2) = \frac{\Psi_{1+a,1}(P_1, P_2)}{\Psi_2(P_1)^a}.$$

The last property comes from the following equation.

$$\begin{aligned}\Psi_{a,b}(P_1,P_1) &= \frac{\sigma((a+b)P_1)}{\sigma(P_1)^{a^2-2ab+b^2}\cdot\sigma(2P_1)^{ab}} \\ &= \frac{\sigma((a+b)P_1)}{\sigma(P_1)^{a^2+2ab+b^2}}\cdot\left(\frac{\sigma(P_1)^4}{\sigma(2P_1)}\right)^{ab} \\ &= \frac{\Psi_{a+b}(P_1)}{\Psi_2(P_1)^{ab}}.\end{aligned}$$

□

With the above preparation, it is in a position to present our main theorem.

Theorem 5. *Let $a \in \mathbb{Z}$ and $P, Q \in J_C(\mathbb{K})$. Consider $D_Q = (-P) - (-P-Q)$, then we have*

$$f_{a,P}(D_Q) = \frac{W_{P,Q}(a+1,1)W_{P,Q}(2,0)^a}{W_{P,Q}(a+1,0)W_{P,Q}(2,1)^a}. \tag{9}$$

Furthermore, it is the formula of Ate pairings on hyperelliptic curves.

Proof. From Theorem 4, we consider the function $\Psi_{1,a,0}(-S,P,Q)$ satisfying

$$\operatorname{div}\left(\frac{1}{\Psi_{1,a,0}(-S,P,Q)}\right) = at^*_{-P}\Theta - t^*_{-aP}\Theta - (a-1)\Theta.$$

Thus we fix the function

$$f_{a,P}(S) = \frac{\Psi_{1,0,0}(-S,P,Q)}{\Psi_{1,a,0}(-S,P,Q)}$$

such that

$$\begin{aligned}\operatorname{div} f_{a,P} &= at^*_{-P}\Theta - t^*_{-aP}\Theta - (a-1)\Theta \\ &= a\left(t^*_{-P}\Theta - \Theta\right) - \left(t^*_{-aP}\Theta - \Theta\right),\end{aligned}$$

then $f_{a,P}$ coincides with the rational function for pairings.

Compute the value of $f_{a,P}$ at $D_Q = (-S) - (-S-Q)$, then

$$\frac{f_{a,P}(-S)}{f_{a,P}(-S-Q)} = \frac{\Psi_{1,0,0}(S,P,Q)\Psi_{1,a,0}(S+Q,P,Q)}{\Psi_{1,a,0}(S,P,Q)\Psi_{1,0,0}(S+Q,P,Q)}.$$

From lemma 1, we obtain

$$\frac{f_{a,P}(-S)}{f_{a,P}(-S-Q)} = \frac{\Psi_{1,0,0}(S,P,Q)\Psi_{1,a,1}(S,P,Q)}{\Psi_{1,a,0}(S,P,Q)\Psi_{1,0,1}(S,P,Q)\Psi_{1,1,1}(S,P,Q)^a}.$$

Let $S = P$, then it holds that

$$f_{a,P}(D_Q) = \frac{\Psi_{1,0,0}(P,P,Q)\Psi_{1,a,1}(P,P,Q)}{\Psi_{1,a,0}(P,P,Q)\Psi_{1,0,1}(P,P,Q)\Psi_{1,1,1}(P,P,Q)^a}.$$

From lemma 2, we have

$$f_{a,P}(D_Q) = \frac{\Psi_1(P) \cdot \Psi_{1+a,1}(P,Q)/\Psi_2(P)^a}{\Psi_{1+a}(P)/\Psi_2(P)^a \cdot \Psi_{1,1}(P,Q) \cdot (\Psi_{2,1}(P,Q)/\Psi_2(P))^a}$$
$$= \frac{\Psi_{1+a,1}(P,Q) \cdot \Psi_2(P)^a \cdot \Psi_1(P)}{\Psi_{1+a}(P) \cdot \Psi_{2,1}(P,Q)^a \cdot \Psi_{1,1}(P,Q)}.$$

In addition, it holds that $\Psi_1(P) = \Psi_{1,1}(P,Q) = 1$. Thus,

$$f_{a,P}(D_Q) = \frac{\Psi_{1+a,1}(P,Q) \cdot \Psi_2(P)^a}{\Psi_{1+a}(P) \cdot \Psi_{2,1}(P,Q)^a}.$$

Eventually, we get

$$f_{a,P}(D_Q) = \frac{W_{P,Q}(a+1,1)W_{P,Q}(2,0)^a}{W_{P,Q}(a+1,0)W_{P,Q}(2,1)^a}.$$

According to the definition of Ate pairings, it is evident that the above equation can be leveraged for Ate pairings. □

Remark 3. Since $W_{P,Q}(1,0) = W_{P,Q}(1,1) = 1$, so Eq. (9) is also satisfied for the optimized pairings on elliptic curves using elliptic nets [36].

Now, we reconsider the formula of Tate pairings using hyperelliptic nets. In practice, it is assumed that $W_{P,Q}(1,0) = W_{P,Q}(1,1) = 1$, then we have

$$\langle P,Q\rangle_r = \frac{W_{P,Q}(r+1,1)}{W_{P,Q}(r+1,0)}$$

from Eq. (4). Besides, $W_{P,Q}(2,0)^r$ and $W_{P,Q}(2,1)^r$ are identical in $\mathbb{K}^*/(\mathbb{K}^*)^r$, indicating that

$$\langle P,Q\rangle_r = \frac{W_{P,Q}(r+1,1)W_{P,Q}(2,0)^r}{W_{P,Q}(r+1,0)W_{P,Q}(2,1)^r}.$$

Thus, Theorem 5 provides the unitary formula of Tate pairings and optimized pairings in terms of hyperelliptic nets, i.e., we complete the proof of Theorem 1.

Following the same technique, the formula of Weil pairings is simplified by

$$e_r(P,Q) = \frac{W_{P,Q}(r+1,1)W_{P,Q}(0,r+1)}{W_{P,Q}(1,r+1)W_{P,Q}(r+1,0)},$$

consistent with the formula in Eq. (5).

4 Parallel Algorithms on Hyperelliptic Nets

In this section, we present the parallel algorithm of hyperelliptic nets, where our attention will focus on hyperelliptic nets on genus-2 curves such that the extensively applied pairings on genus-2 curves can be obtained with hyperelliptic nets.

4.1 Optimal Blocks

From Theorem 5, the pairing is evaluated if we obtain $W_{P,Q}(a+1,1)$ and $W_{P,Q}(a+1,0)$ from fixed initial values, where $W_{P,Q}(2,1)$ and $W_{P,Q}(2,0)$ are precomputed. Following the methods in [37,39], we employ the recurrence formulae from hyperelliptic nets for the two values, where the blocks are carefully selected for Double-and-Add algorithms. From now on, we write $W(u,v) = W_{P,Q}(u,v)$ for simplicity.

As $m > 2^g$ is even, we set $m = 2^g + 2$, then the matrix A of Theorem 2 reveals

$$A = (W(u_i + u_j, v_i + v_j)W(u_i + u_j, v_i + v_j))_{1 \leq i,j \leq m},$$

where $u_1, \cdots, u_m, v_1, \cdots, v_m \in \mathbb{Z}$. Moreover, if $g \equiv 1, 2 \pmod 4$, it holds that Pf $A = 0$, indicating that

$$\sum_{i=2}^{m} (-1)^i W(u_1 + u_i, v_1 + v_i) W(u_1 - u_i, v_1 - v_i) \text{Pf } A^{1,i} = 0, \tag{10}$$

where $A^{1,i}$ is the submatrix of A by removing the first and i-th rows and columns. As results, the goal turns to compute $W(a+1,1)$ and $W(a+1,0)$ with Eq. (10).

We now establish the blocks of the hyperelliptic nets of genus 2. We define a block B centered on ℓ as follows:

$$B(\ell) = \begin{bmatrix} W(\ell - k_3, 1), \cdots, W(\ell, 1), \cdots, W(\ell + k_4, 1) \\ W(\ell - k_1, 0), \cdots, W(\ell, 0), \cdots, W(\ell + k_2, 0) \end{bmatrix}.$$

To obtain the required values, we construct two functions on $B(\ell)$.

- Double(ℓ): Returns the block centered on 2ℓ;
- DoubleAdd(ℓ): Returns the block centered on $2\ell + 1$.

To apply the recurrence formulae, we utilize the same strategy in [39] for two functions Double and DoubleAdd. Namely, to compute the terms in the blocks $B(2\ell)$ or $B(2\ell + 1)$, we employ Eq. (10) via the values in Table 1.

Table 1. Chosen values for the recurrence formulae.

Terms	u_1	u_2	u_3	u_4	u_5	u_6	v_1	$v_2, \cdots, v_6$
$W(2\ell, 0)$	$\ell + 1$	$\ell - 1$	3	2	1	0	0	0
$W(2\ell + 1, 0)$	$\ell + 1$	ℓ	3	2	1	0	0	0
$W(2\ell + k, 1)$	ℓ	$\ell + k$	3	2	1	0	1	0

Substituting the values in Table 1, we obtain an instance of Eq. (10), then all the terms in $B(2\ell)$ and $B(2\ell + 1)$ can be represented as a function of elements in the current block and some fixed initial values.

To acquire the complete algorithm, we should first determine four values k_1, k_2, k_3, k_4 such that the block has minimal elements.

Since Double and DoubleAdd return a block centered in 2ℓ and $2\ell+1$, respectively, the term $W(2\ell - k_1, 0)$ must be evaluated from $B(\ell)$ such that k_1 can be fixed. However, the chosen values for $W(2\ell + 2k, 0)$ and $W(2\ell + 2k + 1, 0)$ are different, so we consider two cases:

- If k_1 is odd, i.e., $k_1 = 2k_1' + 1$. To obtain $W(2\ell - 2k_1' - 1, 0)$, the involved term with the minimal index in Eq. (10) is $W(\ell - k_1' - 4, 0)$, which is also the term with the minimal index for $W(2\ell - 2k_1', 0)$. Consequently, we have an inequation
$$\ell - k_1' - 4 \geq \ell - 2k_1' - 1,$$
which indicates $k_1 \geq 7$.
- If k_1 is even, i.e., $k_1 = 2k_1'$. In contrast, to evaluate $W(2\ell - 2k_1', 0)$, the involved term with the minimal index in Eq. (10) is also $W(\ell - k_1' - 4, 0)$, while the term with the minimal index for $W(2\ell - 2k_1' + 1, 0)$ is $W(\ell - k_1' - 3, 0)$. In this situation, we obtain another inequation
$$\ell - k_1' - 4 \geq \ell - 2k_1',$$
which reveals $k_1 \geq 8$.

Overall, the minimal choice of k_1 is 7. With analogous discussion, k_2 can be fixed by 8 such that $W(2\ell + 9, 0)$ can be evaluated from $W(\ell + 1, 0), \cdots, W(\ell + 8, 0)$.

For the formula of $W(2\ell + k, 1)$, the terms involving P and Q are $W(\ell - 3, 1), \cdots, W(\ell + 3, 1)$, while $W(\ell + k - 3, 0), \cdots, W(\ell + k + 3, 0)$, as the remained terms determined by P, have been stored in $B(\ell)$ already. Therefore, we can set $k_3 = k_4 = 3$.

Remark 4. The constructed block is the same as that in [39], where we depict the details. Indeed, there is the analysis of the block choice in [37], but it leads to a wrong conclusion in [37, Theorem 10]. Namely, for $g = 2$ and $m = 6$, the block contains $W(\ell - 8, 0), \cdots, W(\ell + 7, 0)$, but for DoubleAdd, the formula of $W(2\ell + 8, 0)$ demands the term $W(\ell + 8, 0)$, which is not in the current block.

To start this algorithm, one critical thing is to determine the initial values in the block $B(1)$, which has been analyzed in [39].

4.2 Parallel Hyperelliptic Net Algorithms

Now, we establish the parallel algorithm of hyperelliptic nets for hyperelliptic pairings.

As we have illustrated, the minimal block $B(\ell)$ contains 23 elements, namely,

$$B(\ell) = \begin{bmatrix} & W(\ell-3,1), \cdots, & W(\ell,1), \cdots, & W(\ell+3,1) & \\ W(\ell-7,0), & \cdots, & \cdots, W(\ell,0), \cdots, & \cdots, & W(\ell+8,0) \end{bmatrix}.$$

Upon the choice, all terms in block $B(2\ell)$ and $B(2\ell+1)$ can be evaluated on $B(\ell)$ with initial values $W(1,0),\cdots,W(5,0)$ and $W(-4,1),\cdots,W(3,1)$. The generic algorithm [39, Algorithm 1] can be applied to hyperelliptic pairings straightforwardly.

Remark 5. For a j-bit integer a, the minimal loop depth is j. Since even employing $j-1$ times DoubleAdd, the obtained block is $B(2^j-1)$. Thus, apart from some exceptional cases, i.e.,$a=2^\beta$ or $2^\beta+1$, the loop length is optimal.

Note that we need one division by $W(u_1-u_2,v_1-v_2)\text{Pf } A^{1,2}$ through the hyperelliptic nets, where

$$\text{Pf } A^{1,2} = W(5,0) - W(4,0)W(2,0)^3 + W(3,0)^3$$

is only determined by P, so it must hold that $\text{Pf } A^{1,2} \neq 0$. Furthermore, the initial values, i.e., $W(-4,1),\cdots,W(3,1)$ and $W(2,0)$, should be non-zero such that all divisions can proceed successfully.

From now on, we concentrate on the explicit implementations of hyperelliptic net algorithms, i.e., Double and DoubleAdd. We first consider the evaluation of Tate pairings, then this method can be applied to Ate pairings immediately.

For the hyperelliptic net algorithm, the gist is to employ Eq. (10) for the terms in $B(2\ell)$ or $B(2\ell+1)$, where concrete formulae have been established in [37] and the choice of the recurrence formulae can be found in Table 1. With the above consideration, we discuss the explicit cost of hyperelliptic net algorithms, which is measured by the number of basic operations, i.e., multiplications[1]. Let M and M′ be the cost of multiplication over $\mathbb{F}_q$ and $\mathbb{F}_{q^\alpha}$, respectively.

To this aim, the explicit expression of $W(2\ell+2k,0)$, for $k=-3,\cdots,4$, can be derived from Eq. (10), i.e.,

$$W(2\ell+2k,0) = \sum_{3\leq i\neq j\leq 6} \frac{(-1)^{i+j+1}\Delta_{i,j}}{W(2,0)\text{Pf } A^{1,2}}\gamma_{1,i}\gamma_{2,j} \tag{11}$$

with $\gamma_{1,i} = W(\ell+k-i+7,0)W(\ell+k+i-5,0)$ and $\gamma_{2,j} = W(\ell+k-j+5,0)W(\ell+k+j-7,0)$. In addition, $\Delta_{i,j}$ is the Pfaffian of the matrix with the first, second, i-th and j-th rows and columns removed, as a precomputed constant, and it holds that $\Delta_{i,j} = -\Delta_{j,i}$.

In practice, one can select which type of points to evaluate with, where $P \in J_C(\mathbb{F}_q)[r]$ is the acknowledged choice for Tate pairings. Upon this choice, the computation of $W(\ell,0)$ only invokes operations over $\mathbb{F}_q$. More specifically, we first compute $\gamma_{1,j}$ and $\gamma_{2,j}$ for $j=3,\cdots,6$, which needs 8M. Then we need to evaluate $(-1)^{i+j+1}\Delta_{i,j}\gamma_{1,i}\gamma_{2,j}$ for $3\leq i\neq j\leq 6$, demanding 2M for one element and 24M in total. Adding up all the results, the last operation is to multiply the inverse of $W(2,0)\text{Pf } A^{1,2}$, which has been precomputed to avoid inversion. The outline of computation is displayed in Algorithm 1.

[1] Here we do not distinguish the difference between multiplications and squares. In addition, we omit the cost of additions.

Algorithm 1: Computation of $W(2\ell+2k,0)$

Input: Initial values $W(\ell+k+k',0)$ for $-4 \le k' \le 4$ and $\Delta_{i,j}$ for $3 \le i \ne j \le 6$ and $d = (W(2,0)\text{Pf } A^{1,2})^{-1}$.
Output: $W(2\ell+2k,0)$.
1 $\gamma_{1,j} \leftarrow W(\ell+k-j+7,0) \cdot W(\ell+k+j-5,0)$ for $j = 3, \cdots, 6$;
2 $\gamma_{2,j} \leftarrow W(\ell+k-j+5,0) \cdot W(\ell+k+j-7,0)$ for $j = 3, \cdots, 6$;
3 $C_{i,j} \leftarrow \gamma_{1,i} \cdot \gamma_{2,j}$ for $3 \le i, j \le 6$ and $i \ne j$;
4 $D_{i,j} \leftarrow (-1)^{i+j+1}\Delta_{i,j} \cdot C_{i,j}$;
5 $W(2\ell+2k,0) \leftarrow d \cdot (D_{3,4} + D_{3,5} + D_{3,6} + \cdots + D_{6,5})$;
6 **return** $W(2\ell+2k,0)$

It is evident that parallel computation can accelerate evaluations. Namely, the operations in Algorithm 1 can be handled simultaneously once we have enough parallel processors. Indeed, if we obtain more than 12 processors, required for evaluating $C_{i,j}$ and $D_{i,j}$, $\gamma_{1,i}$ and $\gamma_{2,j}$ can be evaluated simultaneously, then 12 processors can be utilized for $C_{i,j}$ and $D_{i,j}$, respectively. At last, one more multiplication is invoked for $W(2\ell+2k,0)$, so the explicit cost of obtaining $W(2\ell+2k,0)$ in parallel is only 4M while the original approach requires 33M. Furthermore, the parallel algorithm with 12 processors is depicted in Table 2, where proc. is the abbreviation of processor and we omit the additions.

Table 2. Parallel algorithm for $W(2\ell+2k,0)$ with 12 processors.

proc. 1	$\gamma_{1,3}$	$C_{3,4}$	$D_{3,4}$	
proc. 2	$\gamma_{1,4}$	$C_{3,5}$	$D_{3,5}$	$W(2\ell+2k,0)$
proc. 3	$\gamma_{1,5}$	$C_{3,6}$	$D_{3,6}$	
proc. 4	$\gamma_{1,6}$	$C_{4,3}$	$D_{4,3}$	
proc. 5	$\gamma_{2,3}$	$C_{4,5}$	$D_{4,5}$	
proc. 6	$\gamma_{2,4}$	$C_{4,6}$	$D_{4,6}$	
proc. 7	$\gamma_{2,5}$	$C_{5,3}$	$D_{5,3}$	
proc. 8	$\gamma_{2,6}$	$C_{5,4}$	$D_{5,4}$	
proc. 9		$C_{5,6}$	$D_{5,6}$	
proc. 10		$C_{6,3}$	$D_{6,3}$	
proc. 11		$C_{6,4}$	$D_{6,4}$	
proc. 12		$C_{6,5}$	$D_{6,5}$	

Similarly, this strategy can be straightforwardly applied to $W(2\ell+2k+1,0)$ for $k = -4, \cdots, 4$, which is a function of $W(\ell+k-3,0), \cdots, W(\ell+k+4,0)$ and fixed initial values. Yet there are some differences, i.e., $d' = (W(1,0)\text{Pf } A^{1,2})^{-1}$ and $\gamma'_{2,j} = W(\ell+k-j+6,0)W(\ell+k+j-6,0)$, it shares the same algorithm with some minor changes. In other words, computing $W(2\ell+2k+1,0)$ only needs 4M with 12 parallel processors.

In contrast, Q, defined over $\mathbb{F}_{q^\alpha}$ in practice, is involved in the formula of $W(2\ell + k, 1)$, so computing $W(2\ell + k, 1)$ invokes the arithmetic over extension fields $\mathbb{F}_{q^\alpha}$. The explicit formula has been generalized from Eq. (10), i.e.,

$$W(2\ell + k, 1) = \sum_{3 \le i \ne j \le 6} \frac{(-1)^{i+j+1} \Delta_{i,j}}{W(-k, 1) \mathrm{Pf}\ A^{1,2}} \gamma_{1,i} \gamma_{2,j}, \tag{12}$$

where $\gamma_{1,i} = W(\ell - i + 6, 1)W(\ell + i - 6, 1)$ and $\gamma_{2,j} = W(\ell + k - j + 6, 0)W(\ell + k + j - 6, 0)$. With an analogous routine, we present Algorithm 2 for $W(2\ell + k, 1)$ with $k = -3, \cdots, 4$.

Algorithm 2: Computation of $W(2\ell + k, 1)$

Input: Initial values $W(\ell + k + k', 0)$ and $W(\ell + k', 1)$ for $-3 \le k' \le 3$ and $\Delta_{i,j}$ for $3 \le i \ne j \le 6$ and $d = (W(-k, 1)\mathrm{Pf}\ A^{1,2})^{-1}$.

Output: $W(2\ell + k, 1)$.

1 $\gamma_{1,j} \leftarrow W(\ell - i + 6, 1) \cdot W(\ell + i - 6, 1)$ for $i = 3, \cdots, 6$;
2 $\gamma_{2,j} \leftarrow W(\ell + k - j + 6, 0) \cdot W(\ell + k + j - 6, 0)$ for $j = 3, \cdots, 6$;
3 $C_{i,j} \leftarrow \gamma_{1,i} \cdot \gamma_{2,j}$ for $3 \le i, j \le 6$ and $i \ne j$;
4 $D_{i,j} \leftarrow (-1)^{i+j+1} \Delta_{i,j} \cdot C_{i,j}$;
5 $W(2\ell + k, 1) \leftarrow d \cdot (D_{3,4} + D_{3,5} + D_{3,6} + \cdots + D_{6,5})$;
6 **return** $W(2\ell + k, 1)$

The concrete cost of Algorithm 2 is $4\mathsf{M} + 29\mathsf{M}'$ since $\gamma_{2,j} \in \mathbb{F}_q$. Once again, the maximal number of processors is achieved for $C_{i,j}$ and $D_{i,j}$ for the ordered pairs (i, j) from $3 \le i, j \le 6$ with $i \ne j$. Therefore, we can evaluate $W(2\ell + k, 1)$ in $4\mathsf{M}'$ within 12 parallel processors, where the parallel algorithm follows the outline in Table 2 as well.

Note that all the terms in $B(2\ell)$ and $B(2\ell+1)$ are irrelevant, i.e., every term can be written as a function of terms in $B(\ell)$ and some initial values, independent of the terms in new blocks, thus parallel computation can be applied immediately. Concretely, the optimized block has 23 elements, which can be evaluated with 23 parallel processors simultaneously.

From the above discussion, we conclude that it only requires $4\mathsf{M}'$ to compute a new block in parallel when there are more than 276 processors. The costs of parallel algorithms under different processors are summarized in Table 3, where the first row refers to the minimal number of processors for different parallelism strategies, i.e., computing terms or blocks in parallel, as the ability in the following two rows, and the remaining rows record the least cost for $W(2\ell + k, 0)$, $W(2\ell + k', 1)$ and one iteration loop under dissimilar parallelism strategies, respectively.

In contrast, general Miller's algorithms [28], as the most applied methods, must invoke Cantor's algorithm for doubling, where the most efficient algorithm

Table 3. Loop costs of parallel algorithms for Tate pairings. Specifically, "term in parallel" means to leverage Algorithms 1 and 2 for terms, while "block in parallel" represents to employ parallel resources to evaluate the terms in blocks simultaneously.

Processors	1	12	23	276
Term in parallel	✗	✓	✗	✓
Block in parallel	✗	✗	✓	✓
$W(2\ell+k,0)$	$33\mathsf{M}$	$4\mathsf{M}$	$33\mathsf{M}$	$4\mathsf{M}$
$W(2\ell+k',1)$	$29\mathsf{M}'+4\mathsf{M}$	$4\mathsf{M}'$	$29\mathsf{M}'+4\mathsf{M}$	$4\mathsf{M}'$
Loop cost	$203\mathsf{M}'+556\mathsf{M}$	$28\mathsf{M}'+64\mathsf{M}$	$29\mathsf{M}'+4\mathsf{M}$	$4\mathsf{M}'$

[21], utilizing the Jacobian coordinates, demands $36\mathsf{M}'$ for non-degenerate divisors, thus the concrete loop cost of Miller's method is more than $36\mathsf{M}'$. Unfortunately, the above approaches have disparate algorithms for the Double-and-Add paradigm so the implementation will be influenced by the Hamming weight of a. Improved from the initial idea in [24], Lubicz and Robert [25] leveraged Theta functions for a modified Miller's algorithm, requiring $11\mathsf{M}'+22\mathsf{M}$, where the cost would reduce if the involved divisor is degenerate. Surprisingly, Double and DoubleAdd for hyperelliptic nets can share an analogous parallel algorithm, resulting in $4\mathsf{M}'$ for one loop in total.

The length of Miller loops for Ate pairings is shortened while $P \in J_C(\mathbb{F}_{q^\alpha})$, so almost all operations are in $\mathbb{F}_{q^\alpha}$ such that the explicit cost of every loop increases. More specifically, based on [19], it requires $\mathbf{I}'+38\mathsf{M}'+7\alpha\mathsf{M}$ for doubling via affine Mumford coordinates with extra cost for possible additions in Miller loops, where $\mathbf{I}'$ is the inversion over $\mathbb{F}_{q^\alpha}$. After that, an improvement was raised with Theta functions [25], where the cost was reduced to $27\mathsf{M}'+9\mathsf{M}$. Under the analogous situation, the hyperelliptic nets can be applied to Ate pairings directly. Furthermore, all the optimization for Tate pairings still works, especially for the aspect in parallel computations. Consequently, yet it requires $759\mathsf{M}'$ in one loop, parallel computations would shrink the evaluation cost. Namely, considering the ability of parallelism, the minimal loop cost is $4\mathsf{M}'$, providing extreme acceleration at the expense of parallel resources.

In a word, compared with other approaches, we discover the irrelevance of elements in the optimal blocks, thus computing pairings in terms of hyperelliptic nets can utilize parallel resources, then the cost reduces dramatically. More specifically, for Tate and Ate pairings, the comparisons of loop cost[2] are depicted in Table 4, where the last column demonstrates the necessity of auxiliary additions in main loops. It is evident that parallel resources can accelerate pairing computation though the current number of parallel processors is excessive, thus it requires more optimization to reduce the number of processors, leaving it for future work.

[2] Here we do not display the original method with Theta functions [24], since it is only applied to Tate pairings with the loop cost in $36\mathsf{M}'$.

Table 4. Comparisons of loop cost of different methods for genus-2 hyperelliptic pairings. We mainly consider the cost of Double in Miller loops, but there are possible extra additions for Miller's algorithm, depicted in the last column, while other approaches can invoke a unified algorithm for Double and DoubleAdd.

Methods	Tate pairings	Ate pairings	Extra additions
Miller's algorithms [19]	$> 36\mathsf{M}'$	$\mathbf{I}' + 38\mathsf{M}' + 7\alpha\mathsf{M}$	✓
Theta functions [25]	$11\mathsf{M}' + 22\mathsf{M}$	$27\mathsf{M}' + 9\mathsf{M}$	✗
Hyperelliptic nets	$556\mathsf{M} + 203\mathsf{M}'$	$759\mathsf{M}'$	✗
Hyperelliptic nets in parallel	$4\mathsf{M}'$	$4\mathsf{M}'$	✗

5 Conclusion

In this work, for hyperelliptic pairings derived from rational functions, we establish the unitary formula in terms of hyperelliptic nets, making it an alternative to traditional Miller's algorithms. Moreover, we construct an optimal block $B(\ell)$, consisting of 23 elements of hyperelliptic nets, such that either $B(2\ell)$ or $B(2\ell + 1)$ can be generated from $B(\ell)$ with the analogous cost. In particular, discovering the intrinsic irrelevance of terms in the optimal block, we leverage parallel resources to reduce the cost of Double-and-Add algorithms, and more optimization can be achieved since new terms can be evaluated in parallel. As a result, the minimal cost for one loop in our parallel algorithm is $4\mathsf{M}'$ with 276 processors.

To further implement this parallel algorithm, more optimizations should be taken into account, especially for the number of required processors. Apart from that, it will be a new topic of implementation on the hardware, where the ultimate goal is to obtain a more efficient pairing computation algorithm.

Acknowledgments. This work is supported by the National Natural Science Foundation of China (No. 62272491) and the Guangdong Major Project of Basic and Applied Basic Research (2019B030302008) and the National R&D Key Program of China under Grant (2022YFB2701500).

References

1. Agrawal, S., Goyal, R., Tomida, J.: Multi-input quadratic functional encryption from pairings. In: Malkin, T., Peikert, C. (eds.) CRYPTO 2021. LNCS, vol. 12828, pp. 208–238. Springer, Cham (2021). https://doi.org/10.1007/978-3-030-84259-8_8
2. Barreto, P.S.L.M., Galbraith, S.D., O'hEigeartaigh, C., Scott, M.: Efficient pairing computation on supersingular abelian varieties. Des. Codes Cryptogr. **42**(3), 239–271 (2007)
3. Blake, I.F., Seroussi, G., Smart, N.P. (eds.): Advances in Elliptic Curve Cryptography. London Mathematical Society Lecture Note Series, vol. 317. Cambridge University Press, Cambridge (2005)

4. Boneh, D., Di Crescenzo, G., Ostrovsky, R., Persiano, G.: Public key encryption with keyword search. In: Cachin, C., Camenisch, J.L. (eds.) EUROCRYPT 2004. LNCS, vol. 3027, pp. 506–522. Springer, Heidelberg (2004). https://doi.org/10.1007/978-3-540-24676-3_30
5. Boneh, D., Franklin, M.K.: Identity-based encryption from the Weil pairing. SIAM J. Comput. **32**(3), 586–615 (2003)
6. Boneh, D., Lynn, B., Shacham, H.: Short signatures from the weil pairing. In: Boyd, C. (ed.) ASIACRYPT 2001. LNCS, vol. 2248, pp. 514–532. Springer, Heidelberg (2001). https://doi.org/10.1007/3-540-45682-1_30
7. Buchstaber, V.M., Enolskiĭ, V.Z., Leĭkin, D.V.: Hyperelliptic Kleinian functions and applications. In: Solitons, geometry, and topology: on the crossroad, American Mathematical Society Translations: Series 2, vol. 179, pp. 1–33. American Mathematical Society, Providence, RI (1997)
8. Bukhshtaber, V.M., Enolskiĭ, V.Z.: Explicit algebraic description of hyperelliptic Jacobians based on Klein's σ-function. Funct. Anal. Appl. **30**(1), 44–47 (1996)
9. Burdges, J., De Feo, L.: Delay encryption. In: Canteaut, A., Standaert, F.-X. (eds.) EUROCRYPT 2021. LNCS, vol. 12696, pp. 302–326. Springer, Cham (2021). https://doi.org/10.1007/978-3-030-77870-5_11
10. Cai, S., Hu, Z., Yao, Z., Zhao, C.: The elliptic net algorithm revisited. J. Cryptogr. Eng. **14**(1), 43–55 (2024). https://doi.org/10.1007/s13389-022-00304-y
11. Chen, B., Zhao, C.: An improvement of the elliptic net algorithm. IEEE Trans. Comput. **65**(9), 2903–2909 (2016)
12. Cohen, H., et al. (eds.): Handbook of Elliptic and Hyperelliptic Curve Cryptography. Chapman and Hall/CRC (2005)
13. De Feo, L., Masson, S., Petit, C., Sanso, A.: Verifiable delay functions from super singular isogenies and pairings. In: Galbraith, S.D., Moriai, S. (eds.) ASIACRYPT 2019. LNCS, vol. 11921, pp. 248–277. Springer, Cham (2019). https://doi.org/10.1007/978-3-030-34578-5_10
14. Duursma, I., Lee, H.-S.: Tate pairing implementation for hyperelliptic curves $y^2 = x^p - x + d$. In: Laih, C.-S. (ed.) ASIACRYPT 2003. LNCS, vol. 2894, pp. 111–123. Springer, Heidelberg (2003). https://doi.org/10.1007/978-3-540-40061-5_7
15. Galbraith, S.D.: Super singular curves in cryptography. In: Boyd, C. (ed.) ASIACRYPT 2001. LNCS, vol. 2248, pp. 495–513. Springer, Heidelberg (2001). https://doi.org/10.1007/3-540-45682-1_29
16. Galbraith, S.D., Hess, F., Vercauteren, F.: Hyperelliptic pairings. In: Takagi, T., Okamoto, T., Okamoto, E., Okamoto, T. (eds.) Pairing-Based Cryptography – Pairing 2007, pp. 108–131. Springer, Berlin, Heidelberg (2007). https://doi.org/10.1007/978-3-540-73489-5_7
17. Galbraith, S.D., Lin, X.: Computing pairings using x -coordinates only. Des. Codes Cryptogr. **50**(3), 305–324 (2009)
18. Galbraith, S.D., Zhao, C.: Self-pairings on hyperelliptic curves. J. Math. Cryptol. **7**(1), 31–42 (2013)
19. Granger, R., Hess, F., Oyono, R., Thériault, N., Vercauteren, F.: Ate pairing on hyperelliptic curves. In: Naor, M. (ed.) EUROCRYPT 2007. LNCS, vol. 4515, pp. 430–447. Springer, Heidelberg (2007). https://doi.org/10.1007/978-3-540-72540-4_25
20. Hess, F., Smart, N.P., Vercauteren, F.: The Eta pairing revisited. IEEE Trans. Inf. Theory **52**(10), 4595–4602 (2006)
21. Hisil, H., Costello, C.: Jacobian coordinates on genus 2 curves. J. Cryptol. **30**(2), 572–600 (2017)

22. Joux, A.: A one round protocol for tripartite Diffie-Hellman. J. Cryptol. **17**(4), 263–276 (2004)
23. Lee, E., Lee, H., Park, C.: Efficient and generalized pairing computation on abelian varieties. IEEE Trans. Inf. Theory **55**(4), 1793–1803 (2009)
24. Lubicz, D., Robert, D.: Efficient pairing computation with theta functions. In: Hanrot, G., Morain, F., Thomé, E. (eds.) ANTS 2010. LNCS, vol. 6197, pp. 251–269. Springer, Heidelberg (2010). https://doi.org/10.1007/978-3-642-14518-6_21
25. Lubicz, D., Robert, D.: A generalisation of Miller's algorithm and applications to pairing computations on abelian varieties. J. Symb. Comput. **67**, 68–92 (2015)
26. Matsuda, S., Kanayama, N., Hess, F., Okamoto, E.: Optimised versions of the ate and twisted ate pairings. In: Galbraith, S.D. (ed.) Cryptography and Coding 2007. LNCS, vol. 4887, pp. 302–312. Springer, Heidelberg (2007). https://doi.org/10.1007/978-3-540-77272-9_18
27. Menezes, A., Vanstone, S.A., Okamoto, T.: Reducing elliptic curve logarithms to logarithms in a finite field. In: Koutsougeras, C., Vitter, J.S. (eds.) STOC 1991, pp. 80–89. ACM (1991)
28. Miller, V.S.: The Weil pairing, and its efficient calculation. J. Cryptol. **17**(4), 235–261 (2004)
29. Mumford, D.: Tata lectures on theta. I. Modern Birkhäuser Classics, Birkhäuser Boston Inc, Boston, MA (2007)
30. Mumford, D.: Tata lectures on theta. II. Modern Birkhäuser Classics, Birkhäuser Boston Inc, Boston, MA (2007)
31. Ogura, N., Kanayama, N., Uchiyama, S., Okamoto, E.: Cryptographic pairings based on elliptic nets. In: Iwata, T., Nishigaki, M. (eds.) IWSEC 2011. LNCS, vol. 7038, pp. 65–78. Springer, Heidelberg (2011). https://doi.org/10.1007/978-3-642-25141-2_5
32. Onuki, H., Teruya, T., Kanayama, N., Uchiyama, S.: Faster explicit formulae for computing pairings via elliptic nets and their parallel computation. In: Ogawa, K., Yoshioka, K. (eds.) IWSEC 2016. LNCS, vol. 9836, pp. 319–334. Springer, Cham (2016). https://doi.org/10.1007/978-3-319-44524-3_19
33. Onuki, H., Teruya, T., Kanayama, N., Uchiyama, S.: The optimal ate pairing over the Barreto-Naehrig curve via parallelizing elliptic nets. JSIAM Lett. **8**, 9–12 (2016)
34. Silverman, J.H.: The Arithmetic of Elliptic Curves. GTM, vol. 106. Springer, New York (2009). https://doi.org/10.1007/978-0-387-09494-6
35. Stange, K.E.: The Tate pairing via elliptic nets. In: Takagi, T., Okamoto, T., Okamoto, E., Okamoto, T. (eds.) Pairing 2007. LNCS, vol. 4575, pp. 329–348. Springer, Heidelberg (2007). https://doi.org/10.1007/978-3-540-73489-5_19
36. Tang, C., Ni, D., Xu, M., Guo, B., Qi, Y.: Implementing optimized pairings with elliptic nets. Sci. China Inf. Sci. **57**(5), 1–10 (2014)
37. Tran, C.: Formulae for computation of Tate pairing on hyperelliptic curve using hyperelliptic nets. In: Pointcheval, D., Vergnaud, D. (eds.) AFRICACRYPT 2014. LNCS, vol. 8469, pp. 199–214. Springer, Cham (2014). https://doi.org/10.1007/978-3-319-06734-6_13
38. Uchida, Y.: Division polynomials and canonical local heights on hyperelliptic Jacobians. Manuscripta Math. **134**(3–4), 273–308 (2011)
39. Uchida, Y., Uchiyama, S.: The Tate-lichtenbaum pairing on a hyperelliptic curve via hyperelliptic nets. In: Abdalla, M., Lange, T. (eds.) Pairing 2012. LNCS, vol. 7708, pp. 218–233. Springer, Heidelberg (2013). https://doi.org/10.1007/978-3-642-36334-4_15
40. Vercauteren, F.: Optimal pairings. IEEE Trans. Inf. Theory **56**(1), 455–461 (2010)

41. Ward, M.: Memoir on elliptic divisibility sequences. Am. J. Math. **70**, 31–74 (1948)
42. Zhang, F.: Twisted Ate pairing on hyperelliptic curves and applications. Sci. China Inf. Sci. **53**(8), 1528–1538 (2010)
43. Zhao, C., Zhang, F., Huang, J.: Efficient Tate pairing computation using double-base chains. Sci. China Ser. F Inf. Sci. **51**(8), 1096–1105 (2008)
44. Zhao, C., Zhang, F., Huang, J.: A note on the Ate pairing. Int. J. Inf. Sec. **7**(6), 379–382 (2008)

AlgSAT—A SAT Method for Verification of Differential Trails from an Algebraic Perspective

Huina Li[1,3], Haochen Zhang[1], Kai Hu[2,3(✉)], Guozhen Liu[3,4(✉)], and Weidong Qiu[1(✉)]

[1] School of Cyber Science and Engineering, Shanghai Jiao Tong University, Shanghai, China
qiuwd@sjtu.edu.cn
[2] School of Cyber Science and Technology, Shandong University, Qingdao, Shandong, China
kai.hu@sdu.edu.sg
[3] School of Physical and Mathematical Sciences, Nanyang Technological University, Singapore, Singapore
liuguozhen@ucas.ac.cn
[4] University of Chinese Academy of Sciences, Beijing, China

Abstract. A good differential is a start for a successful differential attack. However, a differential might be invalid, *i.e.*, there is no right pair following the differential due to some contradictions in the conditions imposed by the differential. In this paper, we present a new automatic model for verifying differential trails from an algebraic perspective. From this algebraic perspective, exact Boolean expressions for differentials over a cryptographic primitive can be conveniently established, allowing the creation of a simpler SAT model. By invoking a SAT solver, the differential trails are verified in full automatically. Compared with the previous MILP models proposed by Liu *et al.* at CRYPTO 2020, our tool is more concise and direct, for there is no need to analyze any differential properties of components of the targets manually. Thus, it is less error-prone and more friendly for programming, significantly improving the efficiency of verifying a given differential trail.

To demonstrate the powerfulness of our new tool, we apply it to Gimli, Keccak-f, and Ascon. For Gimli, our tool takes about one minute to find a semi-free-start collision pair that costs 2^{64} attempts in a random search. Note that Liu *et al.*'s model could not find any results in practical time. For Keccak-f, it is interesting to note that for large-state-size permutations such as Keccak-f[1600], our approach still shows excellent performance. We verify two 4-round differential trails presented at ASIACRYPT 2022 and confirm that both are valid. For Ascon, we check several differential trails reported at FSE 2021. Specifically, we find that a 4-round differential used in the forgery attack on Ascon-128's iteration phase is invalid. Thus, the corresponding forgery attack is also invalid.

T. Zhu and Y. Li (Eds.): ACISP 2024, LNCS 14895, pp. 450–471, 2024.
https://doi.org/10.1007/978-981-97-5025-2_23

Keywords: Cryptographic permutation · SAT · Automatic verification · Semi-free-start collision attacks

1 Introduction

With the rapid development of the Internet of Things (IoT), more and more lightweight mobile devices appear in people's daily life, such as smart cards, wireless sensors, and Radio Frequency IDentification (RFID) tags. As these devices have limited memory and computing resources, it is infeasible to directly apply traditional encryption algorithms such as the Advanced Encryption Standard (AES) in such scenarios. Therefore, lightweight cryptographic algorithms (LWC) have attracted more and more attention. Especially, in August 2018, the National Institute of Standards and Technology (NIST) launched a competition to solicit, evaluate, and standardize LWC algorithms, including authenticated encryption with associated data (AEAD) and lightweight hash functions that are suitable for use in constrained environments. GIMLI [2] and ASCON [7] are both designed based on bit-oriented cryptographic permutation, which is an increasingly popular paradigm for designing LWCs. On February 7th, 2023, ASCON was announced as the final winner.

These permutation-based ciphers usually have high performance in both software and hardware implementations, but simultaneously their novel designs make it difficult for cryptanalysts to fully understand their security properties. As a result, we would always tend to study the security properties of the underlying permutations to deepen our understanding of the whole ciphers. Similar to the classical block ciphers, these permutations are also iterated algorithms consisting of simple round functions. Naturally, some cryptanalytic methods originally developed for block ciphers have been borrowed to evaluate the security of permutations. One of the most important attacks among them is the differential cryptanalysis introduced by Biham and Shamir at CRYPTO 1990 [5]. In a differential attack, the attacker seeks a fixed input difference α_0 that propagates through a r-round primitive (the primitive could be a block cipher or a permutation) to a fixed output difference α_r with a high probability p, the differential is thus represented by (α_0, α_r). To find a proper differential (α_0, α_r) with a high probability for the primitive, we examine the differential property of the i-th $(1 \leq i \leq r)$ round and try to find a local differential for this round denoted by (a_{i-1}, a_i) whose probability is denoted by p_i. With the tacit assumption that differentials of two consecutive rounds are independent, these local differentials for all rounds could be chained into one so-called differential trail (DT) $(\alpha_0, \alpha_1, \ldots, \alpha_r)$, whose probability is computed by $p = \prod_{i=1}^{r} p_i$. For the sake of simplicity, we generally refer to all these underlying assumptions as the Markov assumption in this paper.

With the Markov assumption, many methods [10,11,21,22] such as the so-called automated tools have been invented to search for useful or even optimal DTs. When using automated tools, the differential propagation rules for components of a primitive in each round are modeled by some specific constraints.

All solutions satisfying these constraints are expected to be valid DTs. Based on these constraints, additional constraints, such as describing whether the corresponding S-boxes in the DTs are active or not, would also be added to the constraint pool. The set formed by all these constraints is denoted by $\mathcal{C}$ in this paper. In general, a constraint representing the number of active S-boxes which is the so-called *objective function* denoted by $\mathcal{O}$ is also imposed. Different automated tools handle $(\mathcal{C}, \mathcal{O})$ differently. There are three types of automated tools in the literature that are often used for the search: (a) Boolean satisfiability (SAT) problem [16], where the constraints in $\mathcal{C}$ and the objective function are modeled by the corresponding conjunctive normal forms (CNF). An extension of the SAT called satisfiability modulo theories (SMT) [8] is also available, which generalizes the SAT to more complex formulas involving *e.g.*, the integers and / or bit vectors. (b) Mixed Integer Linear Programming (MILP) [17] where $(\mathcal{C}, \mathcal{O})$ are described by a set of inequalities (including equations). (c) Constraint programming (CP) [9], where users could use more flexible formulas to describe $(\mathcal{C}, \mathcal{O})$.

Although the Markov assumption is generally considered reasonable for block ciphers, unfortunately, sometimes it would not hold for some permutations. At CRYPTO 2020 [13], Liu, Isobe, and Meier pointed out that the 6-round and 2-round DT used for attacking GIMLI-Hash and ASCON-Hash, respectively, found by MILP in [25] is invalid[1]. That means, although these DTs seem legal under the Markov assumption, no conforming right pairs following the predefined DT can be found in practical cryptanalysis. The lack of sophisticated key schedule algorithms or round subkeys are considered as one of the reasons resulting in these incompatibilities.

To guarantee the existence of at least one conforming right pair following the DT. Liu *et al.* [13] developed an improved MILP model that simultaneously takes into account the message pair $(x_0, x_1, \ldots, x_r)$ and the propagations of a DT $(\alpha_0, \alpha_1, \ldots, \alpha_r)$ but without considerations of differential probability. By carefully analyzing the relationship between α_i and x_i, their model traces the hybrid path $((\alpha_0, x_0), (\alpha_1, x_1), \ldots, (\alpha_r, x_r))$. Later, Sadeghi, Rijmen, and Bagheri proposed another MILP model to verify a differential [19]. Unlike the Liu *et al.*'s model that traced the difference and the value, the Sadeghi *et al.* approach directly traced the two encrypted values as $((x_0, x_0'), \ldots, (x_r, x_r'))$ and assigned the input and output differences as $\alpha_0 = x_0 \oplus x_0', \alpha_r = x_r \oplus x_r'$ for differential (α_0, α_r).

Both verification methods mentioned above are based on MILP tools. Till now, few related work on verification DTs with the help of SAT tools. Guo *et al.* [11] at ASIACRYPYT 2022 pointed out that MILP are unlikely to provide an advantage in trail search for cryptographic primitives of large state size like KECCAK-f. They found SAT-based differential trail search of the underlying large-state-size permutation KECCAK-f[1600] of SHA-3 with more flexibility and

[1] A differential can also be considered invalid when its actual probability is (significantly) different from the theoretically estimated one. Cases related to false probability are out of the scope of this paper.

better efficiency. Inspired by Guo *et al.*'s work, we prefer to use SAT tools rather than MILP tools in this paper.

At CRYPTO 2021, Liu *et al.* proposed an algebraic perspective on differential(-linear) cryptanalysis [14]. This novel algebraic perspective pointed out that the output difference of a Boolean function is a special Boolean function of the input difference and input value. For a Boolean function $f : \mathbb{F}_2^n \to \mathbb{F}_2$ representing a certain output bit of a primitive, the output difference of f with respect to the input difference Δ at a point X is

$$\mathcal{D}_\Delta f(X) = f(X) \oplus f(X \oplus \Delta).$$

They defined a new Boolean function f_Δ as

$$f_\Delta(X, x) = f(X \oplus x\Delta),$$

where x is an auxiliary binary variable. Then Liu *et al.* gave the following formula

$$\mathcal{D}_x f_\Delta = D_\Delta f, \tag{1}$$

where $\mathcal{D}_x f_\Delta$ is the partial derivative of f_Δ with respect to x.

In this paper, we present a new and efficient SAT-based search and verification algorithm from an algebraic perspective inspired by Liu *et al.*'s work [14]. Our algorithm is quite generic and easily programmable. To construct the Boolean equations, we only need to simulate the target cipher's update process to obtain the expression of $\mathcal{D}_x f(X \oplus x\Delta)$. These expressions can be conveniently handled through symbolic computation. We take `SageMath` [24] as the symbolic computation tool and the `Bosphorus` [6] as the SAT solver, which supports Boolean equations as its input. Once the input and output DTs (Δ, ∇) are fixed (they are considered as constants), the output difference is completely determined by X. Therefore, to check whether $\mathcal{D}_x f(X \oplus x\Delta) = \nabla \in \mathbb{F}_2$ holds equals to determine whether a solution X exists for this Boolean equation. If the SAT solver returns a right pair $(X, X \oplus \Delta)$, then it confirms the validity of the DT. Also, if the DTs are conditioned (parts of DT set as free variables, *i.e.*, we do not specify their values), any solution (X, Δ, ∇) satisfying $\mathcal{D}_x f(X \oplus x\Delta) = \nabla$ is a valid differential with a right pair $(X, X \oplus \Delta)$. In this way, we can search a valid DT (Δ, ∇) satisfying given conditions and find a right pair $(X, X \oplus \Delta)$ simutenously. Our contributions are presented as below.

Applications to Gimli. We significantly improve the efficiency of search for a valid DT and a message pair simultaneously on GIMLI. Our tool allows us to find a valid 6-round Semi-Free-Start colliding DT and a colliding pair in just 9.74 s, which took Liu *et al.* [13] approximately 4 h. In order to establish a SFS collision attack on the intermediate 8-round GIMLI-Hash, Liu *et al.* designed a conditional 8-round DT,*i.e.*, some round difference bits are known. However, their MILP model cannot search such a DT and colliding pair following this conditional DT in practical time. We apply our tool to search a DT and find a colliding pair simultaneously following Liu *et al.*'s conditional 8-round DT, as a result, it takes us only about one minute. This is a significant improvement compared to Liu *et al.*'s model. With this pair, we successfully mount a practical SFS collision attack on the intermediate 8-round GIMLI-Hash.

Applications to Keccak-f. For large-state-size KECCAK-f, our approach still shows excellent performance. We verify one 4-round DT of KECCAK-f[1600] and one 4-round DT of KECCAK-f[800] in [11], and confirm that both are valid in only 7.86 s and 21.59 s, respectively.

Applications to Ascon. ASCON has a similar non-linear layer as KECCAK $-f$, but with a more lightweight state. We examine some differentials proposed in previous forgery and collision attacks on ASCON-AEAD and ASCON-Hash. One 2-round DT that was used in the improved 2-round collision attack on ASCON-Hash, two 3-round DTs and one 4-round DT used in the forgery attacks on the finalization or iteration phases of ASCON-128 [10] as well as a 5-round truncated DT in [7] are all proved valid. Besides, a 4-round differential leveraged in the forgery attack on ASCON-128 reported in [10] is proven invalid since our verification model proves no right pair exists. Thus, this forgery attack is accordingly invalid.

Table 1. Comparison of our solving times with previous works. Rnd: the round number of differential trails used on attacks. Pre.Time: previous works' Time. Weight: the *weight* of the DT is equal to $log_2(p^{-1})$, where p is the differential probability of the verified DT.

Primitive	Rnd	In Attack	DT from	Validity	$OurTime_{sv}$	Pre. Time
GIMLI-Hash	6	SFS collision	[13]	Valid	9.74 s	4h [13]
	8			Valid	66.71 s	-† [13]
Primitive	**Rnd**	**In Attack**	**DT from**	**Validity**	**Our Time$_v$**	
KECCAK-f[800]	4	Collision	[11]	Valid	7.86 s	
KECCAK-f[1600]				Valid	21.59 s	
ASCON-Hash	2	Collision	[10]	Valid	0.02 s	
ASCON-128A	3	Forgery(final.)		Valid	0.07	
		Forgery(iter.)		Valid	0.31 s	
ASCON-128	3	Forgery(final.)		Valid	0.08 s	
		Forgery(iter.)		Valid	81 s	
	4	Forgery(final.)		Valid	194 s	
		Forgery(iter.)		Invalid	0.05 s	
	5	Distinguisher	[7]	Valid	3894 s	

† The MILP model in [13] could not return any results in practical time.
$Time_{sv}$ is the total time of search and verification.
$Time_v$ is the verification time.

Table 1 summarizes the best verification times achieved for DTs over various cryptographic primitives. All of our times are solved by `Bosphorus` (v3.0). All experiments are conducted on a server with Intel(R) Xeon(R) CPU E5-4650 v3 @ 2.10GHz 12 Core, 65G RAM, and Ubuntu 18.04.5. For the source code please refer to https://github.com/HuinaLi/AlgSAT.

Paper outline. The rest of this paper is organized as follows. In Sect. 2, we give some concepts used in our work. We describe the details of our verification and search approach in Sect. 3. We apply our approach practically to Gimli (Sect. 4.2), Keccak (Section 5.1), and Ascon (Section 6.1), respectively. In Sect. 7, we make a discussion and conclude our work in Sect. 8.

2 Preliminaries

2.1 Differential Cryptanalysis

In a differential attack, the attacker seeks a fixed input difference α_0 that propagates through a r-round primitive (the primitive could be a block cipher or a permutation) to a fixed output difference α_r with a high probability p, the differential is thus represented by (α_0, α_r). If there exists an ordered pair $(x, x \oplus \alpha_0)$ satisfying $f(x) \oplus f(x \oplus \alpha_0) = \alpha_r$, then it is said to follow the differential (α_0, α_r). In this case, we call (α_0, α_r) a valid differential, and $(x, x \oplus \alpha_0)$ is called a right pair. Usually, finding a differential and computing its probability is difficult, so we tend to study the differential properties of every round of the cipher. Let $f = f^{r-1} \circ f^{r-1} \circ \cdots \circ f^0$ be a r-round iterative cipher and α_i, α_{i+1} be the input and output difference of $f^i, 0 \leq i < r$. $(\alpha_0, \alpha_1, \ldots, \alpha_r)$ is called a DT of the cipher f. If there is a vector of variables $(x_0, x_1, \ldots, x_r)$ that satisfies $f^i(x_i) \oplus f^i(x_i \oplus \alpha_i) = \alpha_{i+1}$ for all $0 \leq i < r$, we say that $\{(x_0, x_1, \ldots, x_r), (x_0 \oplus \alpha_0, x_1 \oplus \alpha_1, \ldots, x_r \oplus \alpha_r)\}$ is a right pair following the DT.

Following the Markov cipher assumption [12] where round functions are treated as independent functions, a DT $(\alpha_0, \alpha_1, \ldots, \alpha_r)$, whose probability is computed by $p = \prod_{i=1}^{r} p_i \geq 0$ is valid. However, the independence of round functions might not always hold, especially for permutations without round keys. That means that some differential attacks on certain ciphers might be false since the differentials or DTs used in the attacks might be invalid *i.e.*, there is no right pair following the differential. Therefore, it is necessary to check the validity of DTs of a permutation derived under the Markov assumption.

2.2 SAT-Based Cryptanalysis

Given a Boolean formula $f(x_1, x_2, \ldots, x_n)$, the Boolean satisfiability (SAT) problem is to determine whether there is any assignment of values to these Boolean variables which makes the formula true. The Boolean formula is satisfiable if a valid assignment exists, otherwise it is unsatisfiable. A formula in CNF consists of one or more clauses joined by *conjunctions* ($\wedge$), where each clause is a *disjunction* ($\vee$) of *literals*, each literal represents a positive or negative variable, *e.g.*, x_i or $\neg x_i$. Most of the previously introduced SAT-based cryptanalysis methods [21,22] encode directly the cryptanalysis problem in CNF as a SAT instance under the Markov assumption and then invoke the off-the-shelf SAT solver to solve it. We called such encoding way *direct encoding way* in this paper. There are many off-the-shelf SAT solvers available which have been introduced into cryptanalysis, such as the `CryptoMiniSat` [20] and `CaDiCaL` [4]. These solvers based on *conflict-driven clause learning* (CDCL) [15] support CNF as their input.

However, for cryptanalysts, algebraic normal form (ANF) which consists of $\oplus$ and $\wedge$ is more friendly and preferred to use since the output bits of a cryptographic primitive are naturally written as ANF of its input bits. Unfortunately, compared with CNF solvers, ANF solvers on huge polynomial systems often use more memory that might be infeasible on many computing platforms.

To fill this vacancy, another encoding way—we called it *indirect describing way* presented by Davin *et al.* [6]. They developed an ANF simplification and solving tool, called `Bosphorus`, which bridges between ANF and CNF solving techniques. The `Bosphorus` supports ANF as its input, which could take advantage of the algebra of polynomials naturally. It first uses many optimized mathematical algorithms, including XL, Brickenstein's ANF-to-CNF conversion, Gauss-Jordan elimination, *etc.*, to simplify ANFs and converted these highly optimized ANFs to CNFs. Afterwards, the SAT solver `Cryptominisat` within `Bosphorus` is invoked to solve those CNFs. The `Bosphorus` can be roughly seen as a SAT solver that supports ANFs as input.

3 Verification of a Differential or DT from an Algebraic Perspective

In this section, based on Liu et al.'s differential-linear cryptanalysis from an algebraic perspective [14], we introduce a new approach to efficiently verify a differential or DT.

3.1 SAT Model for Verifying a Differential or Differential Trail

Given a cryptographic primitive $E : \mathbb{F}_2^n \rightarrow \mathbb{F}_2^n$, we denote its n output bits as n ANFs by $(f_0, f_1, \ldots, f_{n-1})$. We introduce our SAT model for verifying a differential and a DT in two cases, respectively.

Simple case. In the simple case, suppose we can derive the ANFs of all output bits of E. According to Eq. 1, to verify a given differential $(\Delta, \nabla) \in \mathbb{F}_2^n \times \mathbb{F}_2^n$, the input value X is set as free variables. We need to compute the ANFs of $(f_0(X \oplus x\Delta), f_1(X \oplus x\Delta), \ldots, f_{n-1}(X \oplus x\Delta))$. The output difference $\nabla = (\nabla_0, \nabla_1, \ldots, \nabla_{n-1}) \in \mathbb{F}_2^n$ is thus $(\mathcal{D}_x f_0(X \oplus x\Delta), \mathcal{D}_x f_1(X \oplus x\Delta), \ldots, \mathcal{D}_x f_{n-1}(X \oplus x\Delta))$. Verifying the differential (Δ, ∇) is equivalent to checking if the following equation set is solvable.

$$\begin{cases} \nabla_0 = \mathcal{D}_x f_0(X \oplus x\Delta) \\ \nabla_1 = \mathcal{D}_x f_1(X \oplus x\Delta) \\ \quad\vdots \\ \nabla_{n-1} = \mathcal{D}_x f_{n-1}(X \oplus x\Delta) \end{cases} \tag{2}$$

Note that $\mathcal{D}_x f_i(X \oplus x\Delta)$, where $0 \le i < n - r$, are the ANFs of X. Equation 2 is naturally a SAT problem that can be solved with a SAT solver.

Example 1. We take the 5-bit S-box of Ascon as an example, and the ANFs of the S-box are shown as follows.

$$\begin{aligned}
f_0 &= x_4x_1 \oplus x_3 \oplus x_2x_1 \oplus x_2 \oplus x_1x_0 \oplus x_1 \oplus x_0\\
f_1 &= x_4 \oplus x_3x_2 \oplus x_3x_1 \oplus x_3 \oplus x_2x_1 \oplus x_2 \oplus x_1 \oplus x_0\\
f_2 &= x_4x_3 \oplus x_4 \oplus x_2 \oplus x_1 \oplus 1\\
f_3 &= x_4x_0 \oplus x_4 \oplus x_3x_0 \oplus x_3 \oplus x_2 \oplus x_1 \oplus x_0\\
f_4 &= x_4x_1 \oplus x_4 \oplus x_3 \oplus x_1x_0 \oplus x_1
\end{aligned} \tag{3}$$

Let the input value be $X = (x_0, x_1, x_2, x_3, x_4)$ and the ANFs of the output bits are denoted by $(f_0, f_1, f_2, f_3, f_4)$. To verify whether (Δ, ∇) is a valid differential where $\Delta = (1, 1, 1, 0, 0)$ and $\nabla = (1, 0, 0, 0, 0)$, $X \oplus x\Delta = (x_0 \oplus x, x_1 \oplus x, x_2 \oplus x, x_3, x_4)$, we compute the expressions of the 5-bit output difference according to Eq. 2 as follows.

$$\begin{cases}
D_x f_0(X \oplus x\Delta) = x_0 \oplus x_2 \oplus x_4 \oplus 1\\
D_x f_1(X \oplus x\Delta) = x_1 \oplus x_2\\
D_x f_2(X \oplus x\Delta) = 0\\
D_x f_3(X \oplus x\Delta) = x_3 \oplus x_4 \oplus 1\\
D_x f_4(X \oplus x\Delta) = x_0 \oplus x_1 \oplus x_4
\end{cases}$$

Since the output difference is $\nabla = (1, 0, 0, 0, 0)$, we obtain the following five equations.

$$\begin{cases}
x_0 \oplus x_2 \oplus x_4 \oplus 1 = 1\\
x_1 \oplus x_2 = 0\\
0 = 0\\
x_3 \oplus x_4 \oplus 1 = 0\\
x_0 \oplus x_1 \oplus x_4 = 0
\end{cases}$$

In Example 1, the five equations are easy to solve even by hand. But most of the time, the equations are much more complicated. We regard them as a SAT problem and use `Bosphorus` to solve these ANFs to decide whether (Δ, ∇) is valid or not by observing whether a solution $(x_0, x_1, x_2, x_3, x_4)$ would be returned.

Complicated case. If the state size of a cryptographic primitive is large, it is computationally infeasible to compute the exact ANFs of the output bits. Inspired by the DATF technique [14], we take advantage of the variable substitutions to simplify the form of the ANFs while retaining the variable x. Suppose E consists of $E = E_{r-1} \circ \cdots \circ E_1 \circ E_0$ where the ANFs of E_i are available, and the output bits of E_i is denoted by $(f_0^{i+1}, f_1^{i+1}, \ldots, f_{n-1}^{i+1})$. To verify a differential (Δ, ∇), we first focus on E_0 and compute the ANFs of

$(f_0^1(X \oplus x\Delta), f_1^1(X \oplus x\Delta), \ldots, f_{n-1}^1(X \oplus x\Delta))$. Subsequently, we introduce $2n$ transitional variables a_j^1, b_j^1, where $0 \leq j < n$ to perform the variable substitutions as follows.

$$\begin{cases} f_j^1(X \oplus x\Delta) = b_j^1 \oplus a_j^1 x \\ a_j^1 = \mathcal{D}_x f_j^1(X \oplus x\Delta) \\ b_j^1 = \mathcal{D}_x f_j^1(X \oplus x\Delta)x \oplus f_j^1(X \oplus x\Delta) \end{cases}, 0 \leq j < \text{n} \quad (4)$$

Based on Eq. 4 (which are the ANFs of transitional variables a^1, b^1 and x), we compute the outputs of E_1, *i.e.*, perform similar variable substitutions by introducing $2n$ new transitional variables $a_j^2, b_j^2, 0 \leq j < n$.

$$\begin{cases} f_j^2(b^1 \oplus a^1 x) = b_j^2 \oplus a_j^2 x \\ a_j^2 = \mathcal{D}_x f_j^2(b^1 \oplus a^1 x) \\ b_j^2 = \mathcal{D}_x f_j^2(b^1 \oplus a^1 x)x \oplus f_j^2(b^1 \oplus a^1 x) \end{cases}, 0 \leq j < \text{n} \quad (5)$$

Repeat this process until the simplified forms of the ANFs of

$$(f_0^r(b^{r-1} \oplus a^{r-1}x), f_1^r(b^{r-1} \oplus a^{r-1}x), \ldots, f_{n-1}^r(b^{r-1} \oplus a^{r-1}x)) \quad (6)$$

is obtained. Likewise, we omit the subscript of f_j^r, and write Eq. 6 as $f^r(b^{r-1} \oplus a^{r-1}x)$. Finally, we add constraints on the overall output difference $\nabla = (\nabla_0, \nabla_1, \ldots, \nabla_{n-1})$ with

$$\mathcal{D}_x f_j^r(b^{r-1} \oplus a^{r-1}x) = \nabla_j, 0 \leq j < n.$$

In this way, we get a set of ANFs that determines whether (Δ, ∇) is a valid differential. Obviously, it is also a SAT problem.

Fast Verification for Differentials or Differential Trails. It is easy to adapt the above verification model for a r-round DT $(\Delta^0, \Delta^1, \ldots, \Delta^r)$ where $\Delta^i, 0 < i \leq r$ is the output difference of the $(i-1)$-th round and Δ^0 is the initial input difference. Similarly, ignoring the ANFs of the intermediate difference, we verify the validity of a differential that uses a differential (Δ^0, Δ^r) rather than a specific DT. The detailed verification process is proposed in Algorithm 1, and we successfully apply it to verify DTs of KECCAK and ASCON in Sect. 5.1 and Sect. 6.1, respectively.

Simultaneously Searching for Differential Trails and Right Pairs. In addition to verifying a given differential (Δ, ∇) or a DT $(\Delta^0, \Delta^1, \ldots, \Delta^r)$, our algorithm can also search simultaneously for DTs and find the right pairs. The only distinction is that we do not add constraints on ∇ or $\Delta^i, i \geq 1$ and let parts of these unknown differences be free variables. On the other hand, if the values of some inner variables are given in advance, *e.g.*, when dealing with a conditional DT (parts of difference are unknown), we can fix those variables accordingly as additional constraints. In this way, every solution to the SAT problem is a valid DT and a right pair. It is especially useful for scenarios where DTs of a specific form, such as collision DTs used in collision attacks, are needed. We apply Algorithm 1 to search for collision DTs of GIMLI in Sect. 4.2 and it shows outstanding performance.

Algorithm 1. Verification of Differential Trails

Require: An unknown message $X = (x_0, \ldots, x_{n-1})$, the primitive $E = E_{r-1} \circ \cdots \circ E_0$, the number r of rounds, a given DT $(\Delta^0, \Delta^1, \ldots, \Delta^r)$, an auxiliary binary variable x.
Ensure: The value of X or "Invalid".
1: Initialize the input variable vector $f^0 = X \oplus x\Delta^0$ and allocate a set $Q = \emptyset$;
2: **for** i from 0 to $r-1$ **do**
3: Compute the output of E_i according to the ANF of E_i , $f^{i+1} \leftarrow E_i(f^i)$.
4: Add $\mathcal{D}_x f^{i+1} = \Delta^{i+1}$ to Q. { For verifying a differential, only when $i = r-1$ we execute this step.}
5: Introduce transitional variables a^{i+1}, b^{i+1}, let $f^{i+1} = a^{i+1}x \oplus b^{i+1}$. {The substitution rule is used after the nonlinear operation by default.}
6: Add $a^{i+1} = \mathcal{D}_x f^{i+1}$ and $b^{i+1} = \mathcal{D}_x f^{i+1} x \oplus f^{i+1}$ to Q.
7: **end for**
8: Convert Q in CNF form by invoking *cnfwrite*() in `Bosphorus`.
9: **if** The SAT problem is feasible **then**
10: **return** X
11: **else**
12: **return** "Invalid"
13: **end if**

3.2 Obtaining and Solving the SAT Model

We exploit `SageMath` to obtain the ANFs of the output bits of a target primitive. `SageMath` is a popular tool in cryptanalysis. For example, in [23], Sun *et al.* took `SageMath` to generate inequalities for a convex hull. `SageMath` also offers good support for calculating Boolean equations (represented by ANFs) over a ring and field. By simulating the round functions of the target cipher with variable substitutions, a set of ANFs linking the input and output differences are established. `Bosphorus` supporting ANFs as its input, and is able to internally transform them into CNFs before solving them. After solving the SAT problem, we examine the returned solution. If no pair exists, the target differential or DT is invalid; otherwise, a confirming right pair will be derived.

4 Gimli

4.1 A Brief Introduction to Gimli

Gimli [1] is one of the second-round candidates of the NIST lightweight cryptography standardization process [18], including an authenticated cipher Gimli-Cipher and a hash function Gimli-Hash. Both of them are built upon the Gimli permutation that applies 24 rounds to a 384-bit state. The state of Gimli permutation is organized as a 3×4 matrix of 32-bit words denoted by $S_{i,j}, 0 \leq i < 3, 0 \leq j < 4$. The j-th column is a sequence of 96 bits such as $S_j = \{S_{0,j}, S_{1,j}, S_{2,j}\}$, the i-th row is a sequence of 128 bits such

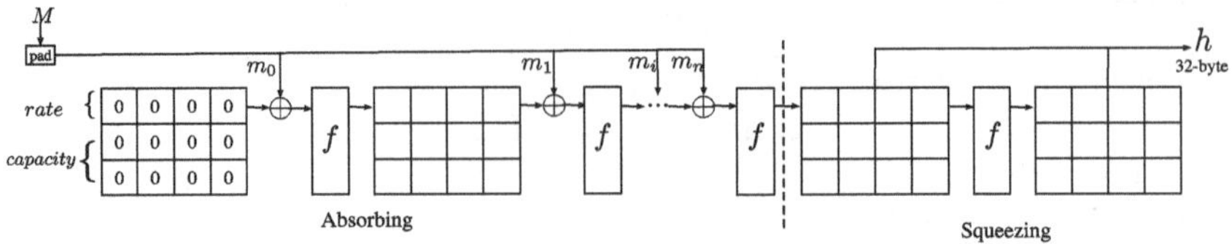

Fig. 1. The illustration of the GIMLI-Hash.

that $S_i = \{S_{i,0}, S_{i,1}, S_{i,2}, S_{i,3}\}$. The bits of each 32-bit word are denoted by $S[32(j+4i)+k], 0 \leq k < 32$, where $S[32(j+4i)]$ indicates the least significant bit (LSB). Each round is a sequence of three operations including a non-linear layer which is a 96-bit SP-box (SP) applied to each column, a linear mixing layer including Small-Swap (S_SW) and Big-Swap (B_SW) in every second round, and a constant addition (AC) in every fourth round. For more details, please refer to [1].

The input state and the intermediate state after i rounds is represented as S^i and $S^{i+1}, 0 \leq i < 24$, respectively. The input difference and the intermediate difference after i rounds be ΔS^i and $\Delta S^{i+1}, 0 \leq i < 24$, respectively.

Gimli-Hash. The GIMLI-Hash scheme is built upon the GIMLI using a sponge construction illustrated in Fig. 1. Firstly, GIMLI-Hash initializes a 48-byte GIMLI state to all-zero, then reads sequentially through a variable-length input as a series of 16-byte input blocks after padding, *i.e.*, $m_0, m_1, \ldots, m_n$. The block size is the so-called absorbing rate, *i.e.*, 128 bits. The remaining c bits of the state are called the capacity which is not directly affected by message bits, nor are they taken as output. After all message blocks are fully processed, a 32-byte hash value h can be obtained. More details of GIMLI-Hash are given in [2].

4.2 Application to Gimli

A Practical SFS Collision Attack on 8-Round Gimli-Hash. The Semi-Free-Start (SFS) collision attack is one of the four types of collision attacks, where the cryptanalyst can choose the initial chain value, *i.e.*, IV as well as a pair of different messages, *i.e.*, M_1, M_2 such that $H(IV, M_1) = H(IV, M_2)$. For the first step of the SFS collision attack on GIMLI-Hash, we need to find a special DT whose input and output differences are both active only in the rate part. In other words, we need to achieve an inner collision in the capacity part of the GIMLI-Hash state. In the second step, by introducing one more pair of message blocks that has the same difference in the rate part, a real SFS collision is successfully converted.

In [13], Liu *et al.* proposed an SFS collision attack on the intermediate 8 rounds of GIMLI-Hash. In this attack, they firstly gave a conditional 8-round DT pattern illustrated in Fig. 2. The input difference is only injected in $\Delta S^1_{0,3}$ and the difference of several internal state words is conditioned. Later, they constructed a MILP model and expect to search for a specific 8-round DT instance

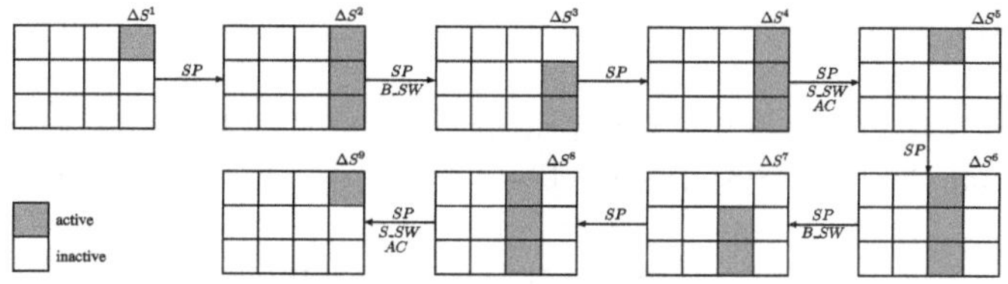

Fig. 2. Semi-free-Start collision attack on the intermediate 8-round GIMLI-Hash.

according to the conditional DT and make an inner collision in the capacity simultaneously. However, it is difficult for their MILP model to find such an 8-round DT instance. The Gurobi solver does not output "INFEASIBLE" or any solution for an acceptable time. Thus, their SFS collision attack is in fact unsuccessful.

In our attack, we aim to search for a valid 8-round DT instance according to the conditional DT in [13]. Once we obtain a right pair, we can launch an SFS collision attack. Our approach, as described in Sect. 3 can easily be adapted to simultaneously search for DT and find the right pair. Based on the 8-round conditioned DT pattern shown in Fig. 2, we present an SFS collision attack.

Firstly, we let active input difference bits be free variables. In [13], the difference $\Delta S^1_{0,3}$ of the conditional DT is active, which means that at least one of the 32 bits of $\Delta S^1_{0,3}$ is nonzero. Therefore, in our attack model, $\Delta S^1_{0,3}$ is represented by 32 unknown binary variables, denoted by $d_0, \ldots, d_{31}$, where the rest of the difference bits of $\Delta S^1_{i,j}$ are zero. For the rest of the round difference, we only add exact ANFs of inactive bits. For example, if $\Delta S^2_{i,j} = 0$, the following 32 ANFs can be obtained

$$\mathcal{D}_x f^1_{32(j+4i)+k} = 0, 0 \leq k < \ 32$$

where $f^1_{32(j+4i)+k}$ is the ANF of $32(j+4i)+k$-th output bit of the first round function. Next, we take `SageMath` to generate all related ANFs that satisfy the condition of this attack model. We use `Bosphorus` to solve these ANFs. Consequently, a valid 8-round DT (Table 4) and an inner collision (Table 5) are successfully found at the same time. Finally, a real collision can be found by introducing one more pair of message blocks $(M, M \oplus \Delta S^9)$ to absorb the difference in the rate part. Compared with the algorithm in [13], our method shows a much higher efficiency. The SAT solver returns a feasible solution in about one minute.

Applications to 6-round Gimli-Hash. Liu *et al.* constructed a MILP model to find a valid 6-round SFS collision DT according to the DT pattern in [25], which cost them about 4 h. Our verification algorithm is more efficient than their MILP model since it took about 9 s for us to find a colliding DT and a right pair that satisfies the same DT pattern (see Table 2 and Table 3).

5 Keccak

KECCAK-$f[b]$ is the underlying permutation of KECCAK hash functions [3], where $b \in \{25, 50, 100, 200, 400, 800, 1600\}$. The b-bit state of KECCAK can be seen as a three-dimensional array, namely A[5][5][w], where $w = \frac{b}{25}$. Each bit of the array is located by (x, y, z) coordinates where $0 \leq x < 5$, $0 \leq y < 5$ and $0 \leq z < w$. At one-dimensional level, A[$*$][y][z] is represented as a row, where A[x][$*$][z] is called a column, A[x][y][$*$] is called a lane. The KECCAK-$f[b]$ permutation consists of the iteration of a round function R denoted by $R = \iota \circ \chi \circ \pi \circ \rho \circ \theta$ with five step mappings *i.e.*, θ, ρ, π, χ, and ι, More details about KECCAK is introduced in [3].

5.1 Application to Keccak

Guo *et al.* at ASIACRYPYT 2022 pointed that MILP are unlikely to provide advantage in trail search for cryptographic primitives of large state size like KECCAK-f and claimed SAT with more flexibility and better efficiency in differential trail search of the underlying permutation KECCAK-f of SHA-3. They first presented one 4-round DT of KECCAK-$f[1600]$ and one 4-round DT of KECCAK-$f[800]$ (see Table 8 and Table 9 in [11]) in 47.76 h and 28.42 h, respectively, Inspired by their work, we apply our verification Algorithm 1 to verify these two DTs for one 4-round DT of KECCAK-$f[1600]$ and one 4-round DT of KECCAK-$f[800]$ with the weight of 133 and 95, respectively, and confirm that both are valid, (see Table 7 and Table 6) which takes us only 7.89 s and 21.59 s, respectively.

6 Ascon

ASCON [7] is the final winner in the lightweight cryptography standardization competition. The ASCON family consists of AEAD and Hash schemes. Both schemes operate on a state of 320 bits which they update with two permutations p^a and p^b whose rounds are respectively $a = 12$ and $b = 6$. For more details about ASCON permutation please refer to [7].

6.1 Application to Ascon

In this section, we show how to leverage our verification algorithm introduced in Sect. 3 to verify some DTs proposed in previous forgery and collision attacks on ASCON-AEAD and ASCON-Hash, respectively.

Verify a DT for 2-round Ascon-Hash. Gerault *et al.* [10] used the CP tool to find a new 2-round DT which could also be used in the collision attack on the 2-round ASCON-Hash. Since the DT proposed in [25] has been proven invalid in [13], a similar case may also occur for this DT. Therefore, it is necessary to check its validity. Using our tool, a right pair following this DT is returned in less than one second. Therefore, we confirm that this DT is valid.

Check the Differentials and DTs in Forgery Attacks on ASCON**.** The authors [10] proposed several forgery attacks against the finalization and iteration phases of ASCON-AEAD. First, they constructed a CP model to search for forgery DTs with different constraints for different phases. Using these DTs, they improved forgery attacks against the finalization phase, as well as the iteration phase of 3- and 4-round ASCON-128. Again, we need to check these DTs to see if they are valid. We apply our approach to verify these forgery DTs. For forgery attacks against the finalization phase of 3-round ASCON-128 with 2^{-32} differential probability (DP), and 4-round ASCON-128 with 2^{-100} DP, we prove that all of these DTs are valid. For forgery attacks against the iteration phase of 3-round ASCON-128 with 2^{-231} DP, we confirm that this DT is also valid. For forgery attacks against the iteration phase of 3-round ASCON-128A (see Table 8), we confirm that this DT is valid too and the conforming right pair is shown in Table 9.

Additionally, we apply our algorithm to check the 4-round forgery DT in the iteration phase, and our program immediately returns "Invalid". This means that this 4-round DT is invalid. We are interested in what results in its invalidity. To find the contradictions hidden among the DT and message value, we separate the whole 4-round DT into two parts (every part contains 2 rounds) and verify them separately. In the first two rounds, we can obtain the right message pair, but in the second two rounds, our program immediately returns "Invalid" in less one second. Therefore, there are some contradictions hidden in the third and fourth rounds.

5-round Results. We verified a 6-round truncated collision-producing DT for ASCON-128 identified in [7]. However, we are not able to find any solution in practical time ($\geq$ 1 month). Therefore, we only checked the first 5 rounds of 6-round truncated DT. The result shows that `Bosphorus` can return a valid 5-round DT and a right pair simultaneously in about one hour.

7 Discussions on Our Verification Algorithm

Similar to previous verification algorithms such as [13,19], our new verification algorithm also traces the propagation of both the values and differences on the target primitive. However, there are some essential differences between ours and [13,19].

Compare to Liu *et al.*'s Algorithm [13]**.** Firstly, the relations between the value transitions and difference transitions are very different. The verification algorithm in [13] derived the relations between the values and differences for the nonlinear functions (the difference transitions and the value transitions are dependent only on the nonlinear operation). For example, to the nonlinear functions of GIMLI, Liu *et al.* derived four types of Boolean relations between the value and difference transitions. More importantly, their manual analysis is not universal, for different cryptographic primitives we need to analyze their nonlinear operations separately, such as S-boxes. Instead, the fundamental theory of our algorithm is the algebraic perspective of differential-linear cryptanalysis proposed recently in [14]. The Boolean expressions of the output difference of a cryptographic primitive can be explicitly presented. As shown in Sect. 3.1, after setting the initial input of the primitive (*i.e.*, the input is $X \oplus x\Delta$), we do not need to worry about the relations between the values and differences over any operation in the process. All we need to do is to simulate the update function by symbolic computations which is friendly to almost all kinds of cryptographic primitives. Secondly, both our algorithm and [13] try to find a solution for a target differential or DT rather than to prove something to be optimal.

Unlike their transformation of the relation into a MILP problem[2], in our algorithm, we use SAT tool to solve this problem from scratch, since the relations derived from our algorithm are an inherited SAT model with ANF forms. By invoking the `Bosphorus`, we can directly simplify and solve this SAT model. As a result, we find the efficiency of our algorithm is significantly higher than [13] as shown in Table 1. For example, an 8-round SFS colliding DT and the conforming colliding pair were obtained in only 66.71 s using our algorithm, while any of such DTs could not be found by [13] in practical time.

Compare to Sadeghi *et al.*'s Algorithm [19]. The authors try to construct two values transitions model at the same time, and add some linear constraints to ensure that the XOR of two values transitions satisfy the given differential trail $(\Delta Y_0, \Delta Y_1, ..., \Delta Y_n)$. It ensures that once the solution is feasible then the given differential trail is valid since it directly retains a right pair. Sadeghi *et al.*'s verification algorithm is natural and simple for most cryptographic primitives.

Since their algorithm, based on the MILP tool, has been applied to ARX structures rather than SPN structures. To better compare the efficiency of our algorithm with [19], we construct a SAT model using their algorithm. The results shows that our algorithm has higher efficiency for large-state-size permutation KECCAK-f, we verify the same 4-round DT in only 7.89 s, but it takes about 79 s using their algorithm. However, when we construct a SAT model on ASCON using their algorithm and apply it to 4-round DT used in the forgery attacks on the finalization, we observed that their algorithm is more efficient than ours. It only takes a couple of seconds to find a pair following the trail, whereas we

[2] Actually, their model can also be transformed into a SAT problem with extra works.

require approximately 3 min. Our efficiency on ASCON is worse than Sadeghi *et al.*'s algorithm.

We think we can explain this phenomenon from the model size aspect with minimal number of clauses. Let's consider an example of modeling KECCAK-f's S-box in CNF. The *direct encoding way* is frequently used in previous works [11,21]. The S-box is described with listing a truth table of 11 variables, including 10 variables that represent input and output value and 1 variable marking compatibility of S-box entries. By putting this truth table into *Logical Friday*[3] , 29 clauses are generated to describe the S-box. Specially, we only take *indirect encoding way* in our algorithm, transforming the S-box's ANF to CNF, 30 clauses are generated. When the quantity of generated clauses from both methods is comparable, our algorithm exhibits a greater array of advantages. And we find that modeling the ASCON's S-box in *direct encoding way* leads to a more compact size compared to the *indirect encoding way.* In detail, the former yields 46 clauses, while the latter generates 81 clauses. We guess this is why our algorithm becomes less efficient than Sadeghi *et al.*'s verification algorithm on ASCON, yet we first provide a quite generic and easily programmable verification tool from an algebraic aspect.

8 Conclusion

In this paper, we propose a new automatic search and verification SAT algorithm for SPN ciphers from an algebraic perspective. Our tool is generic and programmable. We applied it to GIMLI, KECCAK, and ASCON and demonstrate its powerfulness.

As a result, we successfully mounted a SFS collision attack on the intermediate 8-round GIMLI-Hash by searching for a valid DT as well as finding a message pair in about one minute. Besides, our tool could serve as a microscope to conveniently observe the intricate interactions between linear and nonlinear layers in cryptographic primitives. We found that the published forgery attacks of ASCON-128 are invalid because the 4-round forgery DT in the iteration phase is invalid. However, our tool seems infeasible for higher-degree primitives, which will construct a considerable complexity SAT model, SAT solver cannot return any solutions in practical time. We leave these researches as our future work.

Acknowledgments. The authors would like to thank Jian Guo for all valuable feedback. We are grateful to the anonymous reviewers for their valuable comments. This research is supported by the National Natural Science Foundation of China under (Grants No.61972249). Kai Hu is supported by the "ANR-NRF project SELECT".

[3] Logic Friday, a freeware tool for Boolean logic analysis, but only supports Windows. Luckily, the logic module for SymPy allows us to get the same results, please refer to https://docs.sympy.org/latest/modules/logic.html.

A Gimli

A.1 Verified Differential trails and Conforming message pair of Gimli

Table 2. The differential characteristic for SFS 6-round GIMLI-Hash

ΔS^0			
00000000	c803ec98	00000000	c803ec98
00000000	00000000	00000000	00000000
00000000	00000000	00000000	00000000
ΔS^1			
00000000	00000000	00000000	00000000
00000000	a9580034	00000000	a9580034
00000000	98c803ec	00000000	98c803ec
ΔS^2			
00000000	a8c8203e	00000000	a8c8203e
00000000	a0106912	00000000	a0106912
00000000	319026b0	00000000	319026b0
ΔS^3			
00000000	800100f0	00000000	800100f0
00000000	000ae000	00000000	000ae000
00000000	9bc00080	00000000	9bc00080
ΔS^4			
00000000	00000080	00000000	00000080
00000000	00400000	00000000	00400000
00000000	80000000	00000000	80000000
ΔS^5			
00000000	00000000	00000000	00000000
00000000	00000000	00000000	00000000
00000000	80000000	00000000	80000000
ΔS^6			
00000000	80000000	00000000	80000000
00000000	00000000	00000000	00000000
00000000	00000000	00000000	00000000

B Keccak

B.1 Conforming message pair of Keccak

Table 3. Collision message pair for SFS 6-round GIMLI-Hash

X			
e06d07af	445efb01	1a06fc49	47fefb01
46e37879	00d119d0	807e0345	00d11880
33682fd3	03332212	7e8d4676	8334b212
$X \oplus \Delta S^0$			
e06d07af	8c5d1799	1a06fc49	8ffd1799
46e37879	00d119d0	807e0345	00d11880
33682fd3	03332212	7e8d4676	8334b212

Table 4. The differential characteristic for intermediate 8-round GIMLI-Hash: Round 1 to 9

ΔS^1			
00000000	00000000	00000000	81c18ba0
00000000	00000000	00000000	00000000
00000000	00000000	00000000	00000000
ΔS^2			
00000000	00000000	00000000	00000c50
00000000	00000000	00000000	e182408d
00000000	00000000	00000000	a081c18b
ΔS^3			
00000000	00000000	00000000	00000000
00000000	00000000	00000000	b4821adb
00000000	00000000	00000000	13068d32
ΔS^4			
00000000	00000000	00000000	361f001b
00000000	00000000	00000000	0035a72d
00000000	00000000	00000000	3a9b6b80
ΔS^5			
00000000	00000000	99e74180	00000000
00000000	00000000	00000000	00000000
00000000	00000000	00000000	00000000
ΔS^6			
00000000	00000000	004c0800	00000000
00000000	00000000	808967c3	00000000
00000000	00000000	8099e741	00000000
ΔS^7			
00000000	00000000	00000000	00000000
00000000	00000000	13ed158b	00000000
00000000	00000000	03101f8a	00000000
ΔS^8			
00000000	00000000	186bb8bd	00000000
00000000	00000000	dc0b0437	00000000
00000000	00000000	4e8c65a4	00000000
ΔS^9			
00000000	00000000	00000000	0806669c
00000000	00000000	00000000	00000000
00000000	00000000	00000000	00000000

Table 5. Collision message pair for intermediate 8-round GIMLI-Hash: Round 1 to 9

X			
8e57bfca	da90441d	9134941b	3a78650c
cd0c5da0	576ee7cd	7081c41a	df260717
2a98b7a5	02fd11bb	21954066	8e042b58
$X \oplus \Delta S^1$			
8e57bfca	da90441d	9134941b	bbb9eeac
cd0c5da0	576ee7cd	7081c41a	df260717
2a98b7a5	02fd11bb	21954066	8e042b58

Table 6. Conforming pair for 4-round differential trail of KECCAK-f[800] [11]

X				
6e0afffe	fd800d8a	4022d287	b3537a30	d6b65c78
be9ede92	f5464b9a	6cedf67b	38a53a33	0ed777f9
7b0a0f76	680cd690	bba798b6	349ffde5	7e57d84a
f45f264a	41a1e30f	c9101439	a10a3fb2	07e3ba49
a532921a	611a224e	e027aa10	36804867	fc42e2e6
$X \oplus \beta_0$				
2e0afffe	d9800d8a	4022d287	b3537a30	d6b65c78
fe9ede92	f5464b9a	4cedf67b	78a53a33	0ed777f9
3b0a0f76	680cd690	9ba798b6	349ffde5	7e57d84a
f45f264a	45a1e30f	e9101439	a10a3fb2	07e3ba49
e532921a	611a224e	c027aa10	36804867	fc42e2e6

Table 7. Conforming pair for No.1 4-round differential trail of KECCAK-f[1600] [11]

X				
315671d0d0071602	42fd1c41f82b838b	889cdc68f9955a8a	09687aad0335ab19	22c4588ab8ff3e66
3b9e4c0062cdcfc0	5cf91c118a6b9d29	6a02da273a3c06eb	f4a05ed9a49cae3a	aec18c639c6f7a8e
0aec05b00e3d51c0	5e6993b7cab8dc03	98fea95e6a2fb529	05058568b40ea8cf	cd8777e845686eae
9b85ebd1b9ab6716	fd65f99a6f278aa0	3c55b89a8c46af0f	050fad93c11d356c	2baf07920fdec3af
dfe99de2b880f360	55e35833d55f5c2a	f5b1745d702e2291	2db565d2c98140cd	29e6612766e261d6
$X \oplus \beta_0$				
315671d0d0071606	42fd1c41f82b838b	889cdc68f9955a8a	09687aad0335ab1b	22c4588ab8ff3e6e
3b9e4c0062cdcfc4	5cf91c118a6b9d29	6a02da273a3c06eb	f4a05ed9a49cae38	aec18c639c6f7a86
0aec05b00e3d51c4	5e6993b7cab8dc03	98fea95e6a2fb529	05058568b40ea8cd	cd8777e845686ea6
9b85ebd1b9ab6712	fd65f99a6f278aa0	2c55b89a8c46af0f	050fad93c11d356e	2baf07920fdec3af
dfe99de2b880f364	55e35833d55f5c2a	e5b1745d702e2293	2db565d2c98140cd	29e6612766e261de

C Ascon

C.1 Verified Differential trails and Conforming message pair of Ascon

Table 8. Forgery characteristics for round-reduced ASCON-128A with a 3-round permutation in [10]

β_0	α_1	α_2	α_3
0040000400001004	0000000000000000	0240000402001004	2655811c3605b004
0000000000000000	0040000400001004	020080080a002024	2445011424009000
0000000000000000	0000000000000001	0000000000000000	0000000000000000
0000000000000000	0000000000000000	0000000000000000	0000000000000000
0000000000000000	0040000400001004	0a00800800002004	0000000000000000

Table 9. Conforming pair of 3-round differential trail in ASCON-128A permutation phase [10]

DT	X	$X \oplus \beta_0$
Table 8	7c5db1d57b1562b1	7c1db1d17b1572b5
	c4ff06bf3619df87	c4ff06bf3619df87
	40078ce677be3196	40078ce677be3196
	7148974d0af4a995	7148974d0af4a995f
	6f85f592f9f630f0	6f85f592f9f630f0

References

1. Bernstein, D.J., et al.: GIMLI: a cross-platform permutation. In: Fischer, W., Homma, N. (eds.) CHES 2017. LNCS, vol. 10529, pp. 299–320. Springer, Cham (2017). https://doi.org/10.1007/978-3-319-66787-4_15
2. Bernstein, D.J., et al.: Gimli 20190329 (2019). https://csrc.nist.gov/Projects/Lightweight-Cryptography/Round-2-Candidates
3. Bertoni, G., Peeters, M., Van Assche, G., others: The keccak reference (2011). http://keccak.noekeon.org
4. Biere, A.: CADICAL at the SAT Race 2019 (2019). https://github.com/arminbiere/cadical
5. Biham, E., Shamir, A.: Differential cryptanalysis of DES-like cryptosystems. J. Cryptol. **4**(1), 3–72 (1991). https://doi.org/10.1007/BF00630563

6. Choo, D., Soos, M., Chai, K.M.A., Meel, K.S.: BOSPHORUS: bridging ANF and CNF Solvers. In: Design, Automation & Test in Europe Conference & Exhibition, DATE 2019, pp. 468–473 (2019)
7. Dobraunig, C., Eichlseder, M., Mendel, F., Schläffer, M.: Ascon v1.2: lightweight authenticated encryption and hashing. J. Cryptol. **34**(3), 33 (2021). https://doi.org/10.1007/s00145-021-09398-9
8. Ganesh, V., Dill, D.L.: A decision procedure for bit-vectors and arrays. In: Damm, W., Hermanns, H. (eds.) CAV 2007. LNCS, vol. 4590, pp. 519–531. Springer, Heidelberg (2007). https://doi.org/10.1007/978-3-540-73368-3_52
9. Gerault, D., Minier, M., Solnon, C.: Constraint programming models for chosen key differential cryptanalysis. In: Rueher, M. (ed.) CP 2016. LNCS, vol. 9892, pp. 584–601. Springer, Cham (2016). https://doi.org/10.1007/978-3-319-44953-1_37
10. Gérault, D., Peyrin, T., Tan, Q.Q.: Exploring differential-based distinguishers and forgeries for ASCON. IACR Trans. Symmetric Cryptol. **2021**(3), 102–136 (2021). https://doi.org/10.46586/tosc.v2021.i3.102-136
11. Guo, J., Liu, G., Song, L., Tu, Y.: Exploring SAT for cryptanalysis: (quantum) collision attacks against 6-round SHA-3. In: ASIACRYPT (3). Lecture Notes in Computer Science, vol. 13793, pp. 645–674. Springer (2022). https://doi.org/10.1007/978-3-031-22969-5_22
12. Lai, X., Massey, J.L., Murphy, S.: Markov ciphers and differential cryptanalysis. In: Davies, D.W. (ed.) EUROCRYPT 1991. LNCS, vol. 547, pp. 17–38. Springer, Heidelberg (1991). https://doi.org/10.1007/3-540-46416-6_2
13. Liu, F., Isobe, T., Meier, W.: Automatic verification of differential characteristics: application to reduced Gimli. In: Micciancio, D., Ristenpart, T. (eds.) CRYPTO 2020. LNCS, vol. 12172, pp. 219–248. Springer, Cham (2020). https://doi.org/10.1007/978-3-030-56877-1_8
14. Liu, M., Lu, X., Lin, D.: Differential-linear cryptanalysis from an algebraic perspective. In: Malkin, T., Peikert, C. (eds.) CRYPTO 2021. LNCS, vol. 12827, pp. 247–277. Springer, Cham (2021). https://doi.org/10.1007/978-3-030-84252-9_9
15. Marques-Silva, J., Lynce, I., Malik, S.: Conflict-driven clause learning SAT solvers. In: Handbook of Satisfiability - Frontiers in Artificial Intelligence and Applications, vol. 336, 2nd edn., pp. 133–182. IOS Press, Ohmsha (2021)
16. Mouha, N., Preneel, B.: Towards finding optimal differential characteristics for ARX: application to salsa20. IACR Cryptol. **2013**, 328 (2013). https://eprint.iacr.org/2013/328
17. Mouha, N., Wang, Q., Gu, D., Preneel, B.: Differential and linear cryptanalysis using mixed-integer linear programming. In: Wu, C.-K., Yung, M., Lin, D. (eds.) Inscrypt 2011. LNCS, vol. 7537, pp. 57–76. Springer, Heidelberg (2012). https://doi.org/10.1007/978-3-642-34704-7_5
18. NIST: The NIST lightweight cryptography project (2018). https://csrc.nist.gov/Projects/lightweight-cryptography
19. Sadeghi, S., Rijmen, V., Bagheri, N.: Proposing an MILP-based method for the experimental verification of difference-based trails: application to SPECK. SIMECK. Des. Codes Cryptogr. **89**(9), 2113–2155 (2021). https://doi.org/10.1007/s10623-021-00904-5
20. Soos, M., Nohl, K., Castelluccia, C.: Extending SAT solvers to cryptographic problems. In: Kullmann, O. (ed.) SAT 2009. LNCS, vol. 5584, pp. 244–257. Springer, Heidelberg (2009). https://doi.org/10.1007/978-3-642-02777-2_24
21. Sun, L., Wang, W., Wang, M.: More accurate differential properties of LED64 and Midori64. IACR Trans. Symmetric Cryptol. **2018**(3), 93–123 (2018). https://doi.org/10.13154/tosc.v2018.i3.93-123

22. Sun, L., Wang, W., Wang, M.: Accelerating the search of differential and linear characteristics with the SAT Method. IACR Trans. Symmetric Cryptol. **2021**(1), 269–315 (2021). https://doi.org/10.46586/tosc.v2021.i1.269-315
23. Sun, S., Hu, L., Wang, P., Qiao, K., Ma, X., Song, L.: Automatic security evaluation and (related-key) differential characteristic search: application to SIMON, PRESENT, LBlock, DES(L) and other bit-oriented block ciphers. In: Sarkar, P., Iwata, T. (eds.) ASIACRYPT 2014. LNCS, vol. 8873, pp. 158–178. Springer, Heidelberg (2014). https://doi.org/10.1007/978-3-662-45611-8_9
24. The Sage Developers: SageMath, the Sage Mathematics Software System (Version 9.5s) (2022). https://www.sagemath.org
25. Zong, R., Dong, X., Wang, X.: Collision attacks on round-reduced Gimli-Hash/Ascon-Xof/Ascon-Hash. IACR Cryptol. **2019**, 1115 (2019). https://eprint.iacr.org/2019/1115

Hadamard Product Argument from Lagrange-Based Univariate Polynomials

Jie Xie, Yuncong Hu, and Yu Yu(✉)

School of Electronic Information and Electrical Engineering, Shanghai Jiao Tong University, Shanghai, China
{xiejie1006,huyuncong,yyuu}@sjtu.edu.cn

Abstract. Hadamard product is a point-wise product for two vectors. This paper presents a new scheme to prove Hadamard-product relation as a sub-protocol for SNARKs based on univariate polynomials. Prover uses linear cryptographic operations to generate the proof containing logarithmic field elements. The verification takes logarithmic cryptographic operations with constant numbers of pairings in bilinear group. The construction of the scheme is based on the Lagrange-based KZG commitments (Kate, Zaverucha, and Goldberg at Asiacrypt 2010) and the folding technique. We construct an inner-product protocol from folding technique on univariate polynomials in Lagrange form, and by carefully choosing the random polynomials suitable for folding technique, we construct a Hadamard-product protocol from the inner-product protocol, giving an alternative to prove linear algebra relations in linear time, and the protocol has a better concrete proof size than previous works.

Keywords: interactive oracle proofs · SNARKs · Hadamard product

1 Introduction

Succinct non-interactive arguments of knowledge (SNARKs) allow efficient verification for the NP statement. In recent years, there has been a significant increase in interest and research focused on SNARKs. Many researchers have developed sophisticated protocols tailored for industrial applications, each operating under different assumptions. These protocols offer distinct features and efficiencies, making them applicable in various contexts. For example, *Plonk* [13] has a constant proof size and a quasilinear prover running time, while an improved version, *HyperPlonk* [10], sacrifices the constant proof size for a linear prover running time. Since the NP statement to be proven in real-world usage is much more complex in blockchain applications like smart contracts, the proving time is extremely crucial, as generating proofs in a reasonable time is of utmost importance. Thus, in recent years, there has been a significant surge in research focused on linear-proving-time protocols [3–6, 9, 10, 18, 22, 23, 26, 27].

T. Zhu and Y. Li (Eds.): ACISP 2024, LNCS 14895, pp. 472–492, 2024.
https://doi.org/10.1007/978-981-97-5025-2_24

There are numerous SNARKs with quasilinear proving time [11,13,19,21,28], among which proofs for the Hadamard product relation or the inner product relation are necessary, and FFT(Fast Fourier Transform) is required in the proofs. A prevalent strategy to circumvent FFT usage is the adoption of multilinear sum checks [5,10,23], which yield a linear-time prover alongside logarithmic-size proofs and verification. However, due to the inherent discrepancy between low-degree extensions for univariate polynomials and multilinear extensions for multivariate polynomials, certain calculations cannot be seamlessly translated into multilinear polynomial computations. For example, *Spartan* [23], which evolves from *Marlin* [11], incorporates an additional memory-checking sub protocol to authenticate the evaluation of the multilinear extension for a matrix. Consequently, our objective is to construct a Hadamard-product protocol and an inner-product protocol that align with the efficiency of multilinear-setting protocols like *Gemini* [5] and *HyperPlonk* [10], thereby streamlining the construction of linear-prover protocols from their univariate polynomial counterparts.

1.1 Our Results

In this paper, we present a novel Polynomial Interactive Oracle Proof (PIOP) designed for the Hadamard product, leveraging the univariate Lagrange-based polynomial commitment. This innovative approach is contrasted with *Spartan* [23], which utilizes multivariate Lagrange-based polynomial commitment. The protocol facilitates the direct application of the well-established batch KZG commitment scheme [16] for univariate polynomials, resulting in a constant number of pairings in the bilinear group during the verification process. Unlike numerous protocols relying on univariate polynomials [11,19,20], this method eliminates dependencies on FFT, except for the setup and update processes of the KZG commitment scheme, leading to a linear proving time.

As previous researches [1,11], to check Hadamard product $\mathbf{a} \circ \mathbf{b} = \mathbf{c}$, we use a random vector $\mathbf{r}$ and show $\langle \mathbf{a} \circ \mathbf{b}, \mathbf{r} \rangle = \langle \mathbf{c}, \mathbf{r} \rangle$, where $\langle \cdot, \cdot \rangle$ means inner product. The main challenge in the protocol is to deal with the product in sum check. Multivariate protocols [5,26] have a direct generalization from sum check for a single polynomial to sum check for products of multiple polynomials. However, it is not the case in univariate setting. In multivariate, in particular multilinear setting, all variables are naturally separated so that prover can limit the summing space by limit the number of variables and it is easy for prover to compute and prove a linear (or higher degree when it comes to the case of product) function, while in univariate setting we have only one variable. *Gemini* [5] solves the problem by "simulating" multivariate sum check through folding technique. In particular, *Gemini* folds the target polynomial of sum check by *splitting* its even and odd coefficients to two lower degree polynomials and take their random linear combination as the new target polynomial.

Key Observation. In Hadamard product, we need to compute the inner product of two polynomials. Under the restriction of Lagrange-based, we cannot directly compute the product polynomial $\mathbf{a}(x) \cdot \mathbf{b}(x)$ as it has a double degree while the number of evaluation points remains unchanged. We try to apply the idea of *Gemini* to our protocol. Since the splitting is the linear combination of polynomials and the inner product is quadratic, we need to do some modifications. Our observation is that the sum of product has the property $\langle \mathbf{a}, \mathbf{b} \rangle = \sum_{x \in \mathbb{H}} \mathbf{a}(x)\mathbf{b}(x^{-1}) = \sum_{x \in \mathbb{H}} \mathbf{a}_1(x^2)\mathbf{b}_1(x^{-2}) + \mathbf{a}_2(x^2)\mathbf{b}_2(x^{-2})$ for specific groups $\mathbb{H}$ and split polynomial $\mathbf{a}_1, \mathbf{b}_1, \mathbf{a}_2, \mathbf{b}_2$ if we have polynomials $\mathbf{a}(x)\mathbf{b}(x^{-1})$ that agree with $\mathbf{a}, \mathbf{b}$ on $\mathbb{H}$. In particular, $\mathbb{H}$ is a multiplicative subgroup of a finite field. By the observation, the inner product of two polynomials can be turned into the sum of inner product of split polynomials, which allows us to do the similar folding to polynomial $\mathbf{a}, \mathbf{b}$ as *Gemini*.

However, it is still not enough for Hadamard product. The left-hand side of the Hadamard product contains $\mathbf{a} \circ \mathbf{b}$, which is the Hadamard product itself. To avoid the circular argument, we choose a structured vector $\boldsymbol{\rho}$, which we call it *tensor polynomial* instead of a random $\mathbf{r}$. The feature of $\boldsymbol{\rho}$ is that checking $\langle \mathbf{a}, \boldsymbol{\rho} \rangle$ is equivalent to checking folding scheme. When checking $u = \langle \mathbf{a} \circ \boldsymbol{\rho}, \mathbf{b} \rangle = \langle \mathbf{a} \circ \mathbf{b}, \boldsymbol{\rho} \rangle$, we first compute $\mathbf{k} = \mathbf{a} \circ \boldsymbol{\rho}$ and check inner product $u = \langle \mathbf{k}, \mathbf{b} \rangle$ by folding scheme, which is equivalent to checking $u_1 = \langle \mathbf{k}, \boldsymbol{\rho}' \rangle$ and $u_2 = \langle \mathbf{b}, \boldsymbol{\rho}' \rangle$ for some u_1, u_2 computed by u. $u_2 = \langle \mathbf{b}, \boldsymbol{\rho}' \rangle$ can be checked by folding scheme. But when we check $u_1 = \langle \mathbf{k}, \boldsymbol{\rho}' \rangle = \langle \mathbf{a} \circ \boldsymbol{\rho}, \boldsymbol{\rho}' \rangle$ by similar method, it turns out to be slightly different. The problem is that a rational function must be checked in the process. Luckily, only the constant term in the polynomial is involved in a rational function, and we can deal with the constant term separately to avoid checking a rational function. By the adjustment, $u_1 = \langle \mathbf{k}, \boldsymbol{\rho}' \rangle = \langle \mathbf{a} \circ \boldsymbol{\rho}, \boldsymbol{\rho}' \rangle$ can be checked, thus finishing the proof of Hadamard product.

Commitment Scheme. To instantiate polynomial oracles in the protocol, we use KZG commitment [16]. The original KZG commitment is designed for coefficient-based polynomial. To modify it to Lagrange-based, we make some changes to the CRS, in particular, computing Lagrange polynomial instead of x^i terms to construct the CRS. We show that it has exactly the same property as the standard KZG commitment.

Complexity. By applying the modification of the KZG commitment, our Hadamard-product protocol achieves $O(N)$ proving time, $O(\log N)$ proof size, and verification time where N is the length of the vectors in terms of asymptotic complexity. The proof contains $2 \log N + 13$ group elements and $4 \log N + 20$ field elements, and batching k instances of proof only adds $O(k)$ elements to the proof size. The underlying inner-product protocol has proof size of $\log N + 8$ group elements and $4 \log N + 12$ field elements.

Comparison. Due to the fact that most of the univariate Hadamard-product and inner-product protocols [1,11,13–15,19,24,28] are designed in the situation where FFT is available, we mainly make comparison to *Bulletproof* and its improved version. Bulletproof [7] gives an inner-product protocol with $O(N)$ proving time, $O(\log N)$ proof size, and $O(N)$ verification time by a similar folding of our protocol. The improvement of the protocol [9,12,17] gives a $O(\log N)$ verification time, which is the closest to our protocol. The detailed comparison of proof size is listed in Table 1, and our protocol achieves the best proof size for around 30% improvement.

Table 1. Comparison to other Hadamard-product protocols on univariate polynomials. The columns represent asymptotic proving time, proof size and verification time when proving the Hadamard product of two vectors with length $N = 2^n$. For the protocol with logarithm proof size, we list the concrete proof size. By [18], $1\mathbb{G}_T = 4\mathbb{G}_1 = 6\mathbb{F}$ in BLS12-381, thus [9,12,17] have a close proof size.

	Prover	Proof	Verifier
[19] [28]	$O(N \log N)$	$O(1)$	$O(1)$
[4]	$O(N)$	$O(N^\epsilon)$	$O(N^\epsilon)$
[7]	$O(N)$	$2\log N\mathbb{G} + 5\mathbb{F}$	$O(N)$
[12]	$O(N)$	$8\log N\mathbb{G}_1 + 2\log N\mathbb{F}$	$O(\log N)$
[9] [17]	$O(N)$	$2\log N\mathbb{G}_T + 1\mathbb{G}_1 + 2\mathbb{G}_2$	$O(\log N)$
This work	$O(N)$	$(2\log N + 13)\mathbb{G}_1 + (4\log N + 20)\mathbb{F}$	$O(\log N)$

1.2 Related Work

In recent years there have been several inner-product protocols and Hadamard-product protocols applied to distinct SNARK protocols. The inner-product protocols and Hadamard-product protocols can be roughly divided into two groups according to proving time.

The quasilinear-proving-time group contains the directly-computed Hadamard product and inner product with zero polynomial [1,11,13,14,21][1] and Hadamard product and inner product with Laurent polynomials [15,19,20,24,28]. In this group, the protocols always put the entries of vectors in the coefficients of polynomials. The directly-computed inner product and Hadamard product contain all such protocols that are not specifically designed for proof size, as it is easy and direct to compute the polynomial multiplications the quotient polynomial $\mathbf{q}(x)$ in poly$a(x)$poly$b(x)$ − poly$b(x)$ = poly$q(x)$poly$z(x)$ with the help of FFT. The protocols powered by Laurent polynomials use polynomials including x^{-k} in polynomial multiplications to put the results of inner products on a specific coefficient (typically constant) of the product polynomial. The realization of these protocols are slightly different. [15,19] do not directly use the x^{-k} terms.

[1] Most of quasilinear-prover SNARKs contains such type of sub-protocol, thus we only list protocols not containing specifically-designed proofs.

They use the degree of polynomial from x^0 to x^N and place the inner product on the coefficient of x^{N+1}. To maintain soundness, a special SRS for the KZG commitment needs to be provided that does not contain the term $[x^{N+1}]_1$. [24,28] uses an auxiliary polynomial that has zero coefficient in term x^{N+1} to ensure that the coefficient of term x^{N+1} is the alleged value, thus it can use a standard SRS for the KZG commitment. [20] uses bivariate Laurent polynomials poly $f(x, y)$ where y is the variate for the random linear combination and places the alleged value on the coefficient of x^0. The SRS of such a protocol also lacks the term to commit the coefficient of x^0.

The linear-proving-time group contains the Hadamard product and inner product from multivariate sum check and from folding techniques. The protocol based on multivariate sum check includes [5,10,23,26]. *Libra* [26] first gives a dynamic programming method to compute the intermediate results of the sum check for products of multilinear polynomials in linear time and [5,10,23] apply the technique to inner product and Hadamard product. The protocol from folding technique includes [2,7,9,12,17]. The idea is also applying Laurent polynomials, but each time they only compute polynomials with terms x and x^{-1}, halving the length of vector by separating them into two terms. By this technique, they get a linear prover. [2,7] has a linear verifier, since the computations of folding are also required for verifier. [9,12,17] improves the scheme by applying polynomial commitments to [7], using a commit-and-prove technique to delegate verifier's computations to prover. The improvement leads to a logarithm verifier.

2 Preliminaries

2.1 Notation

We use λ to denote the security parameter. Let $\log : \mathbb{N} \rightarrow \mathbb{Z}$ be the base-2 logarithm function with rounding. Let $\mathbb{F}$ be a (finite) field and $\mathbb{F}^*$ be $\mathbb{F}\backslash\{0\}$. We use $|\mathbb{G}|$ to denote the number of entries in group $\mathbb{G}$. We will use $\mathbb{H}$ to represent a multiplicative subgroup of field $\mathbb{F}$. $\mathbb{H}$ is a cyclic group. For a vector $\mathbf{a}$ over $\mathbb{F}$ of length N and a multiplicative subgroup $\mathbb{H} = \{1, \omega, ..., \omega^{N-1}\}$ of size N, let polynomial $\mathbf{a}(x)$ be the unique polynomial of degree at most $N-1$ that defined by $\mathbf{a}(\omega^i) = \mathbf{a}_i$ where $\mathbf{a}_i$ is i-th entry of $\mathbf{a}$. We also call the polynomial Lagrange-based. Sometimes we will also use $\mathbf{a}$ as a polynomial. We use $\deg(\mathbf{a}(x))$ to be the degree of polynomial $\mathbf{a}(x)$. Let $\mathbf{a} \circ \mathbf{b}$ be the Hadamard product of the two vectors and $\mathbf{a} \cdot \mathbf{b} = \langle \mathbf{a}, \mathbf{b} \rangle$ be the inner product of the two vectors. Sometimes we will write inner product in sum, which is $\sum_{x \in \mathbb{H}} \mathbf{a}(x)\mathbf{b}(x)$ for $\mathbf{a}, \mathbf{b} \in \mathbb{F}^N$ and $|\mathbb{H}| = N$. We use PPT as an abbreviation of probabilistic-polynomial-time.

We use $\mathbb{H}^2$ to represent the set of square of all elements in a multiplicative subgroup $\mathbb{H} \subset \mathbb{F}$, and use $\mathbb{G}^k$ for a vector containing k elements of $\mathbb{G}$ otherwise, when $\mathbb{G}$ is not claimed as a multiplicative subgroup of field $\mathbb{F}$. This abuse of notation can be a bit disturbing. For a multiplicative subgroup $\mathbb{H}$ with $|\mathbb{H}| = 2^n$, $\mathbb{H}^{2^n} = \{1\}$.

Definition 1 (even and odd parts of polynomials). *Let* $\mathbf{f}(x) \in \mathbb{F}[x]$. *Let* $\mathbf{f}_e(x)$ *and* $\mathbf{f}_o(x)$ *be the even and odd parts of* $\mathbf{f}(x)$ *i.e. the unique polynomials such that* $\mathbf{f}(x) = \mathbf{f}_e(x^2) + x\mathbf{f}_o(x^2)$.

Remark 1 (compute even and odd parts). Since all polynomials in the protocol are in the Lagrange base on a multiplicative subgroup $\mathbb{H}$, even and odd parts of polynomials cannot be separated directly. Instead, we use following equations to compute the two parts: $\mathbf{f}_e(x^2) = \frac{\mathbf{f}(x)+\mathbf{f}(-x)}{2}$ and $\mathbf{f}_o(x^2) = \frac{\mathbf{f}(x)-\mathbf{f}(-x)}{2x}$. The equations are consistent in the field $\mathbb{F}$. By the equations, we can directly define $\mathbf{f}_e(x)$ and $\mathbf{f}_o(x)$ on $\mathbb{H}^2$, which is also a multiplicative subgroup.

2.2 Polynomial Interactive Oracle Proof

A *polynomial IOP* [5,8,11] over a field family $\mathcal{F}$ for an indexed relation $\mathcal{R}$ is a set of tuples $\mathsf{IOP} = (\mathsf{k}, \mathsf{n}, \mathsf{d}, \mathcal{I}, \mathcal{P}, \mathcal{V})$ where $\mathsf{k}, \mathsf{n}, \mathsf{d} : \{0,1\}^* \to \mathbb{N}$ are polynomial-time computable functions and $\mathcal{I}, \mathcal{P}, \mathcal{V}$ are three algorithms called *indexer, prover, verifier* respectively. The function k,n,d specifics the number of interaction rounds, the number of polynomial oracles in each round and the degree bounds on these polynomials.[2]

In round 0 (offline phase), the indexer $\mathcal{I}$ receives index $\mathbb{i}$ and a field $\mathbb{F} \in \mathcal{F}$ as input, and outputs $\mathsf{n}(0)$ polynomial oracles of degree at most d. In online phase, given an instance $\mathbb{x}$ and a witness $\mathbb{w}$ such that $(\mathbb{i}, \mathbb{x}, \mathbb{w}) \in \mathcal{R}$, prover $\mathcal{P}$ receives $(\mathbb{F}, \mathbb{i}, \mathbb{x}, \mathbb{w})$ and verifier $\mathcal{V}$ receives $(\mathbb{F}, \mathbb{x})$ and oracle access to outputs of $\mathcal{I}(\mathbb{F}, \mathbb{i})$. Prover and verifier interactive for $\mathsf{k} = \mathsf{k}(|\mathbb{i}|)$ rounds.

In each round $j = 1, ..., \mathsf{k}$, verifier sends a message $\boldsymbol{\rho}_j \in \mathbb{F}^*$ to prover, and then prover responds with $\mathsf{n}(j)$ polynomial oracles with degree bound d. Verifier may query any oracle it has received at any time. A query should be an oracle $\mathbf{P}_i^{(j)}$ verifier has already received and an element $z \in \mathbb{F}$ and its answer is $\mathbf{P}_i^{(j)}(z) \in \mathbb{F}$. After all interactions, verifier accepts or rejects by outputting a bit $\{0,1\}$. We say that a prover (probably malicious) $\tilde{\mathcal{P}}$ to be **admissible** that for every oracle prover sends, the degree of the polynomial is bounded by d. A honest prover must be admissible. We also allow prover to send non-oracle message to verifier together with polynomial oracles as in a typical interactive proof.

We say that a polynomial IOP has perfect completeness and soundness error ϵ if the following holds.

- **Completeness.** For every field $\mathbb{F} \in \mathcal{F}$ and tuple $(\mathbb{i}, \mathbb{x}, \mathbb{w}) \in \mathcal{R}$, the probability that prover $\mathcal{P}(\mathbb{F}, \mathbb{i}, \mathbb{x}, \mathbb{w})$ convinces verifier $\mathcal{V}^{\mathcal{I}(\mathbb{F},\mathbb{i})}(\mathbb{F}, \mathbb{x})$ to accept the proof is 1.
- **Soundness.** For every field $\mathbb{F} \in \mathcal{F}$ and tuple $(\mathbb{i}, \mathbb{x}) \notin \mathcal{L}(\mathcal{R})$, the probability for any admissible prover $\tilde{\mathcal{P}}(\mathbb{i}, \mathbb{x})$ to convince verifier $\mathcal{V}^{\mathcal{I}(\mathbb{F},\mathbb{i})}(\mathbb{F}, \mathbb{x})$ to accept is at most ϵ.

[2] The Hadamard-product protocol itself is not necessarily to have indexer. We put it here as it can be a sub-protocol of a SNARK with indexer.

The *proof size* we used in this paper is the sum of numbers of queries, oracles sent by prover and indexer, and non-oracle messages.

The PIOP we construct has a strong property of *knowledge soundness*(against admissible provers). We define the property below.

Knowledge Soundness. We say a PIOP has knowledge error ϵ if there exists a PPT extractor $\mathcal{E}$ for which the following holds. For every field $\mathbb{F} \in \mathcal{F}$, index-instance tuple $(\mathbb{i}, \mathbb{x})$ and admissible prover $\tilde{\mathcal{P}}$, the probability that $\mathcal{E}^{\tilde{\mathcal{P}}}(\mathbb{F}, \mathbb{i}, \mathbb{x}, 1^{\mathsf{l}(\mathbb{i})})$ outputs a valid witness $\mathbb{w}$ for $(\mathbb{i}, \mathbb{x})$ that $(\mathbb{i}, \mathbb{x}, \mathbb{w}) \in \mathcal{R}$ is at least the probability that $\tilde{\mathcal{P}}$ convinces $\mathcal{V}^{\mathcal{I}(\mathbb{F},\mathbb{i})}(\mathbb{F}, \mathbb{x})$ to accept minus ϵ. The notation $\mathcal{E}^{\tilde{\mathcal{P}}}$ means that the extractor $\mathcal{E}$ has a black-box access to a next-message function defined by the algorithm $\tilde{\mathcal{P}}$; in particular, the extractor $\mathcal{E}$ can "rewind" prover $\tilde{\mathcal{P}}$ to any round to get new messages.

The PIOP that we construct has two additional properties.

- **Public-coin.** IOP is *public-coin* if each verifier's message to prover is a uniformly random string of some prescribed length. For public-coin PIOP, all queries can be postponed, w.l.o.g., to a query phase that is after the interactive phase.
- **Non-adaptive Queries.** IOP is *non-adaptive* if all of verifier's query locations are only determined by verifier's inputs and randomness.

2.3 Tensor Polynomials

Definition 2. *Tensor polynomial* $\boldsymbol{\rho}_{\mathbb{H},(r_1,\ldots r_n)}(x)$ *is the unique polynomial defined in a multiplicative subgroup* $\mathbb{H}$ *with* $|\mathbb{H}| = 2^n$ *and* n *random elements in the field* $\mathbb{F}$ *whose degree is less than* 2^n *and* $\boldsymbol{\rho}_{\mathbb{H},(r_1,\ldots r_n)}(x) = \prod_{i=1}^{n} \frac{x^{2^{i-1}}+r_i}{2x^{2^{i-1}}} = \frac{x}{2^n}\prod_{i=1}^{n}(x^{2^{i-1}}+r_i)$ *for every* $x \in \mathbb{H}$. *When* $\mathbb{H}$ *or random elements are clear, we can omit them from* $\boldsymbol{\rho}(x)$.

Since $x\prod_{i=1}^{n}(x^{2^{i-1}}+r_i)$ is a monic polynomial with degree 2^n, $\boldsymbol{\rho}(x)$ can be written as $\boldsymbol{\rho}(x) = \frac{1}{2^n}(x\prod_{i=1}^{n}(x^{2^{i-1}}+r_i) - x^{2^n} + 1)$.

2.4 Lagrange Polynomials

We define Lagrange polynomials in terms of a multiplicative subgroup of a finite field here.

Definition 3 (Lagrange polynomial for multiplicative subgroup). *For a multiplicative subgroup* $\mathbb{H} = \{1, \omega, ..., \omega^{N-1}\}$ *of a finite field* $\mathbb{F}$ *of size* $N = 2^n$, *Lagrange polynomial is the unique polynomial having a degree at most* $N-1$ *that* $\mathrm{poly}L_k(\omega^k) = 1$ *and* $\mathrm{poly}L_k(x) = 0$ *for every* $x \in \mathbb{H}\backslash\{\omega^k\}$.

Proposition 1 ([25], **Proposition 2.5).** *In multiplicative subgroup* $\mathbb{H} = \{1, \omega, ..., \omega^{N-1}\}$ *with* $N = 2^n$, $\mathrm{poly}L_k(x) = \frac{x^N-1}{N}\frac{\omega^k}{x-\omega^k}$.

3 Tensor-Check Protocol

In this section we introduce a sub-protocol involving *tensor polynomial* which is used in our construction. In Sect. 4.4 we show how prover evaluates the tensor polynomial on a multiplicative subgroup $\mathbb{H}$.

Definition 4. *The* ***tensor-check*** *relation $\mathcal{R}_{TC}$ is a set of tuples*

$$(\mathbb{i}, \mathbb{x}, \mathbb{w}) = (\perp, (\mathbb{F}, \mathbb{H}, N, r_1, ..., r_n, u), \mathbf{f})$$

where $\mathbb{H}$ is a multiplicative subgroup of field $\mathbb{F}$ with size $N = 2^n$, $r_1, ..., r_n \in \mathbb{F}$ and $\sum_{x \in \mathbb{H}} \mathbf{f}(x)\boldsymbol{\rho}_{\mathbb{H},(r_1,...,r_n)}(x) = u$ in which $\boldsymbol{\rho}(x)$ is tensor polynomial defined as in Definition 2.

Theorem 1. *For every positive integer $N = 2^n$ and a finite field $\mathbb{F}$ that $N \mid |\mathbb{F}^*|$, there is a PIOP for $\mathcal{R}_{TC}$ with proving time $O(N)$, proof size $O(\log N)$, verification time $O(\log N)$ in terms of field operations and field elements. The PIOP has a perfect completeness and soundness error $\frac{N-1}{|\mathbb{F}^*|}$.*

We prove Theorem 1 by following construction.

Construction 1. *We construct a PIOP with n rounds for indexed $\mathcal{R}_{TC}$. Prover $\mathcal{P}$ takes $(\mathbb{i}, \mathbb{x}, \mathbb{w})$ of the relation as input; verifier $\mathcal{V}$ takes $(\mathbb{i}, \mathbb{x})$ as input.*

- *Let $\mathbf{f}^{(0)}(x) = \mathbf{f}(x)$.*
- *For $i = 1, ..., n$, prover computes $\mathbf{f}^{(i)}(x^2) = \mathbf{f}_e^{(i-1)}(x^2) + r_i \mathbf{f}_o^{(i-1)}(x^2)$ $= \frac{\mathbf{f}^{(i-1)}(x)+\mathbf{f}^{(i-1)}(-x)}{2} + r_i \frac{\mathbf{f}^{(i-1)}(x)-\mathbf{f}^{(i-1)}(-x)}{2x}$ for $x \in \mathbb{H}^{2^{i-1}}$. The degree of $\mathbf{f}^{(i)}(x)$ is less than 2^{n-i}.*
- *Prover sends oracle $\mathbf{f}^{(i)}(x)$ for $i = 0, ..., n$ to verifier.*
- *Verifier sends challenge $\beta \leftarrow \mathbb{F}^*$.*
- *Prover sends $\mathbf{f}^{(i-1)}(\beta)$, $\mathbf{f}^{(i-1)}(-\beta)$ and $\mathbf{f}^{(i)}(\beta^2)$ for $i = 1, ..., n$.*
- *Verifier checks $\mathbf{f}^{(i)}(\beta^2) = \frac{\mathbf{f}^{(i-1)}(\beta)+\mathbf{f}^{(i-1)}(-\beta)}{2} + r_i \frac{\mathbf{f}^{(i-1)}(\beta)-\mathbf{f}^{(i-1)}(-\beta)}{2\beta}$ for $i = 1, ..., n$ and $\mathbf{f}^{(n)}(\beta^2) = u$.*

We call the relation "tensor-check" due to the similarity between Construction 1 and **Construction 1** in [5]. Since our construction has the same computations and verifications as in [5], the soundness and complexity directly follow the proof in [5] and we only need to prove the completeness.

Remark 2. $\mathbf{f}^{(n)}$ in the protocol should be the constant u in the relation, thus it is unnecessary for verifier to query $\mathbf{f}^{(n)}(x)$, i.e. verifier can use u to replace $\mathbf{f}^{(n)}(\beta^2)$ to save one oracle and prover sends n oracles in Construction 1 to verifier.

Lemma 1. *Construction 1 has perfect completeness.*

Proof. Suppose $\sum_{x \in \mathbb{H}} \mathbf{f}(x)\boldsymbol{\rho}_{\mathbb{H},(r_1,...,r_n)}(x) = u$. The verification consists of two parts. The first part is $\mathbf{f}^{(i)}(\beta^2) = \frac{\mathbf{f}^{(i-1)}(\beta)+\mathbf{f}^{(i-1)}(-\beta)}{2} + \frac{\mathbf{f}^{(i-1)}(\beta)-\mathbf{f}^{(i-1)}(-\beta)}{2\beta}$, which is correct if prover is honest, since $\mathbf{f}_e^{(i-1)}(x)$ and $\mathbf{f}_o^{(i-1)}(x)$ are unique and consistent over field $\mathbb{F}$ in terms of $\mathbf{f}^{(i-1)}(x)$.

The second part is $\mathbf{f}^{(n)}(\beta^2) = u$. Since $\mathbf{f}^{(n)}(x)$ is defined on $\mathbb{H}^{2^n} = \{1\}$, it is a constant polynomial, thus it is equivalent to check $\mathbf{f}^{(n)}(1) = u$. We first rewrite $\mathbf{f}^{(i)}(x^2) = \frac{\mathbf{f}^{(i-1)}(x)+\mathbf{f}^{(i-1)}(-x)}{2} + r_i\frac{\mathbf{f}^{(i-1)}(x)-\mathbf{f}^{(i-1)}(-x)}{2x}$ as $\mathbf{f}^{(i)}(x^2) = \frac{x+r_i}{2x}\mathbf{f}^{(i-1)}(x) + \frac{-x+r_i}{2(-x)}\mathbf{f}^{(i-1)}(-x)$. We observe that $\mathbf{f}^{(i)}(x)$ for every $x \in \mathbb{H}^{2^i}$ is determined by 2 unique evaluations of $\mathbf{f}^{(i-1)}(x)$ on $\mathbb{H}^{2^{i-1}}$ for all $i = 1, ..., n$. Therefore, we can construct a full binary tree to describe the evaluations of all $\mathbf{f}^{(i)}$. We put $\mathbf{f}^{(n)}$ at root (level 0), and $\mathbf{f}^{(i)}$ at level $(n-i)$, and $\mathbf{f}^{(i-1)}(x)$ and $\mathbf{f}^{(i-1)}(-x)$ are siblings whose parent is $\mathbf{f}^{(i)}(x^2)$. We additionally put value $\frac{x+r_i}{2x}$ on the edge linking $\mathbf{f}^{(i-1)}(x)$ and $\mathbf{f}^{(i)}(x^2)$. This will automatically place $\frac{-x+r_i}{2(-x)}$ on the edge connecting $\mathbf{f}^{(i-1)}(-x)$ and $\mathbf{f}^{(i)}(x^2)$. Then for every internal node of the tree, the value of the node can be computed by the product of each of its child and the edge between the node and its child respectively and then summing them up. By induction, we can obtain the value of every subtree's root, which equals the sum of the product of each of its leaves and the edge on the path between the root and its leaves, respectively. Observing the path from leave $\mathbf{f}^{(0)}(x)$ to root $\mathbf{f}^{(n)}$, the values on the edges are $\frac{x+r_1}{2x}, \frac{x^2+r_2}{2x^2}, ..., \frac{x^{2^{i-1}}+r_i}{2x^{2^{i-1}}}, ..., \frac{x^{2^{n-1}}+r_n}{2x^{2^{n-1}}}$. Thus, the root of the tree, $\mathbf{f}^{(n)}$, equals $\sum_{x\in\mathbb{H}}(\mathbf{f}^{(0)}(x)\prod_{i=1}^{n}\frac{x^{2^{i-1}}+r_i}{2x^{2^{i-1}}}) = \sum_{x\in\mathbb{H}}\mathbf{f}(x)\boldsymbol{\rho}_{\mathbb{H},(r_1,...,r_n)}(x) = u$. □

Remark 3 (Batching proofs). To batch m proofs for witnesses of $\mathcal{R}_{TC}$ $\mathbf{f}_0, ..., \mathbf{f}_{m-1}$ with the same $\boldsymbol{\rho}$, verifier can simply send a challenge $\xi \leftarrow \mathbb{F}^*$ to prover and run Construction 1 on polynomial poly$f(x) = \sum_{i=0}^{m-1}\xi^i$poly$f_i(x)$. This will add $\frac{m-1}{|\mathbb{F}^*|}$ to the soundness error according to the Schwartz-Zippel lemma. Note that only one challenge[3] is needed in the process.

4 Inner-Product Protocol

In this section, we describe several protocols related to the inner products of polynomials on Lagrange base.

4.1 Inverse Inner Product

Definition 5. *The relation* ***inverse inner product*** $\mathcal{R}_{IIP}$ *is a set of tuple*

$$(\mathbb{i}, \mathbb{x}, \mathbb{w}) = (\perp, (\mathbb{F}, \mathbb{H}, N, u), (\mathbf{f_1}, \mathbf{f_2}))$$

where $\mathbb{H}$ *is a multiplicative subgroup of field* $\mathbb{F}$ *with size* $N = 2^n$, $\mathbf{f_1}, \mathbf{f_2} \in \mathbb{F}^N, u \in \mathbb{F}$ *and* $\sum_{x\in\mathbb{H}}\mathbf{f_1}(x)\mathbf{f_2}(x^{-1}) = u$.

We call it 'inverse inner product' as it uses the inversion of the variable in polynomial polyf_2.

[3] β will be reused, thus it will only be counted in Hadamard product protocol.

Theorem 2. *For every positive integer $N = 2^n$ and a finite field $\mathbb{F}$ that $N \mid |\mathbb{F}^*|$, there is a PIOP for $\mathcal{R}_{IIP}$ with proving time $O(N)$, proof size $O(\log N)$, verification time $O(\log N)$ in terms of field operations and field elements. The PIOP has a perfect completeness and soundness error $\frac{2N}{|\mathbb{F}^*|}$.*

Construction 2. *We construct a PIOP with $n+2$ rounds for indexed $\mathcal{R}_{IIP}$. Prover $\mathcal{P}$ takes $(\mathbb{i}, \mathbb{x}, \mathbb{w})$ of the relation as input; verifier $\mathcal{V}$ takes $(\mathbb{i}, \mathbb{x})$ as input.*

- *Let* $\text{poly}f_i^{(0)}(x) = \text{poly}f_i(x)$, $u^{(0)} = u$, $i = 1, 2$.
- *For each round* $j = 1, ..., n$, *let* $\text{poly}f_{i,e}^{(j-1)}$ *and* $\text{poly}f_{i,o}^{(j-1)}$ *be even and odd parts of polynomial* $\text{poly}f_i^{(j-1)}$, *respectively. Let* $\text{poly}P^{(j)}(r) = \sum_{x\in\mathbb{H}^{2^j}}(\text{poly}f_{1,e}^{(j-1)}(x) + r\text{poly}f_{1,o}^{(j-1)}(x))\cdot$ $(\text{poly}f_{2,e}^{(j-1)}(x^{-1}) + r\text{poly}f_{2,o}^{(j-1)}(x^{-1}))$. *Note that* $\text{poly}P^{(j)}$ *is quadratic. Prover computes* $\text{poly}f_{i,e}^{(j-1)}(x)$ *and* $\text{poly}f_{i,o}^{(j-1)}(x)$ *for* $x \in \mathbb{H}^{2^j}$ *by* $\text{poly}f_i^{(j-1)}(x)$. *Prover computes* $\text{poly}P^{(j)}(0)$, $\text{poly}P^{(j)}(1), \text{poly}P^{(j)}(-1)$ *and send them to verifier.*
- *For each round* $j = 1, ..., n$, *after receiving* $\text{poly}P^{(j)}(0), \text{poly}P^{(j)}(1), \text{poly}P^{(j)}(-1)$, *verifier checks* $\text{poly}P^{(j)}(1) + \text{poly}P^{(j)}(-1) = u^{(j-1)}$, *samples* $r_j \leftarrow \mathbb{F}^*$ *and sends* r_j *to prover. Prover and verifier compute* $u^{(j)} = \text{poly}P^{(j)}(r_j)$.
- *For each round* $j = 1, ..., n$, *after receiving* r_j, *prover computes* $\text{poly}f_i^{(j)}(x) = \text{poly}f_{i,e}^{(j-1)}(x) + r_j\text{poly}f_{i,o}^{(j-1)}(x)$ *for* $x \in \mathbb{H}^{2^j}$ *and* $i = 1, 2$. *Prover sends oracle* $\text{poly}f_i^{(j)}(x)$ *and enters the next round.*
- *In round* $n+1$, *verifier checks relation among 3 constants* $\text{poly}f_1^{(n)} \cdot \text{poly}f_2^{(n)} = u^{(n)}$ *(similar to what is in tensor-check protocol). Verifier samples a challenge* $\beta \leftarrow \mathbb{F}^*$ *and sends it to prover.*
- *In round* $n+2$, *after receiving* β, *prover opens oracles* $\text{poly}f_i^{(j)}(x)$ *where* $j = 0, ..., n-1$ *at* $x = \beta, -\beta$ *and opens* $\text{poly}f_i^{(j)}(x)$ *where* $j = 1, ..., n$ *at* $x = \beta^2$. *Prover sends all openings to verifier.*
- *In round* $n+2$, *after receiving all openings, verifier checks* $\text{poly}f_i^{(j)}(\beta^2) = \frac{\text{poly}f_i^{(j-1)}(\beta)+\text{poly}f_i^{(j-1)}(-\beta)}{2} + r_j\frac{\text{poly}f_i^{(j-1)}(\beta)-\text{poly}f_i^{(j-1)}(-\beta)}{2\beta}$ *where* $i = 1, 2$ *and* $j = 1, ..., n$.

Remark 4. Noticing that the only position we use the oracle $\text{poly}f_i(x)$ is in round $n+2$ and the challenge is sent in round $n+1$, we can postpone all oracle sending to round n, i.e. computing and sending them together. Furthermore, by the observation that the checking equations of $\text{poly}f_i(x)$ have the same form as those of Construction 1, we can use the batch proof of Construction 1 to shrink the proof size with $\frac{1}{|\mathbb{F}^*|}$ extra soundness error. Prover sends $n+1$ oracles, $3n$ field elements for $\text{poly}P(x)$ and 2 field elements for $\text{poly}f_i^{(n)}$ to verifier and verifier sends n challenges for r, 1 challenge for β and 1 challenge for batch proof.

Before proving properties of Construction 2, we need a lemma about $\text{poly}f(x)$.

Lemma 2 ([25], **Lemma 4.5**). *For every finite field* $\mathbb{F}$ *and a multiplicative subgroup* $\mathbb{H}$ *of size* $N = 2^n$ *for a positive integer* n, *and two polynomials* $\text{poly}f_1(x), \text{poly}f_2(x)$ *whose degrees are less than* N *on* $\mathbb{F}$, $\sum_{x\in\mathbb{H}} \text{poly}f_1(x)\text{poly}f_2(x^{-1}) = 2\sum_{x\in\mathbb{H}^2} \text{poly}f_{1,e}(x)\text{poly}f_{2,e}(x^{-1}) + \text{poly}f_{1,o}(x)\text{poly}f_{2,o}(x^{-1})$.

Lemma 3. *Construction 2 has perfect completeness.*

Proof. Suppose $\sum_{x\in\mathbb{H}} \mathbf{f_1}^{(0)}(x)\mathbf{f_2}^{(0)}(x^{-1}) = u^{(0)}$. We first show that for every $j = 1,...,n$, $\text{poly}P^{(j)}(1) + \text{poly}P^{(j)}(-1) = u^{(j-1)} = \sum_{x\in\mathbb{H}^{2^{j-1}}} \text{poly}f_1^{(j-1)}(x)\text{poly}f_2^{(j-1)}(x^{-1})$. It is trivial when $j = 1$. When $j \geq 2$, by the definition of $u^{(j)}$, we have

$$\begin{aligned}
u^{(j-1)} &= \text{poly}P^{(j-1)}(r_{j-1}) \\
&= \sum_{x\in\mathbb{H}^{2^{j-1}}} (\text{poly}f_{1,e}^{(j-2)}(x) + r_{j-1}\text{poly}f_{1,o}^{(j-2)}(x))(\text{poly}f_{2,e}^{(j-2)}(x^{-1}) + r_{j-1}\text{poly}f_{2,o}^{(j-2)}(x^{-1})) \\
&= \sum_{x\in\mathbb{H}^{2^{j-1}}} \text{poly}f_1^{(j-1)}(x)\text{poly}f_2^{(j-1)}(x^{-1}) \\
&\overset{(1)}{=} 2\sum_{x\in\mathbb{H}^{2^j}} \text{poly}f_{1,e}^{(j-1)}(x)\text{poly}f_{2,e}^{(j-1)}(x^{-1}) + \text{poly}f_{1,o}^{(j-1)}(x)\text{poly}f_{2,o}^{(j-1)}(x^{-1}) \\
&\overset{(2)}{=} \text{poly}P^{(j)}(1) + \text{poly}P^{(j)}(-1)
\end{aligned}$$

where (1) is by Lemma 2 and (2) is by definition of $\text{poly}P^{(j)}(r)$. Thus, $\text{poly}P^{(j)}(1) + \text{poly}P^{(j)}(-1) = u^{(j-1)}$.

Then we prove $\text{poly}f_1^{(n)} \cdot \text{poly}f_2^{(n)} = u^{(n)}$. By definition of $u^{(j)}$, $\text{poly}P^{(j)}$ and $\text{poly}f_i^{(j)}$ for $i = 1, 2$, we have $u^{(j)} = \text{poly}P^{(j)}(r_j) = \sum_{x\in\mathbb{H}^{2^j}} (\text{poly}f_{1,e}^{(j-1)}(x) + r_j\text{poly}f_{1,o}^{(j-1)}(x))(\text{poly}f_{2,e}^{(j-1)}(x^{-1}) + r_j\text{poly}f_{2,o}^{(j-1)}(x^{-1})) = \sum_{x\in\mathbb{H}^{2^j}} \text{poly}f_1^{(j)}(x)\text{poly}f_2^{(j)}(x)$ for $j = 1,...,n$. Bringing n into j, we get the proof. The proof of $\text{poly}f_i^{(j)}(\beta^2) = \frac{\text{poly}f_i^{(j-1)}(\beta)+\text{poly}f_i^{(j-1)}(-\beta)}{2} + r_j\frac{\text{poly}f_i^{(j-1)}(\beta)-\text{poly}f_i^{(j-1)}(-\beta)}{2\beta}$ follows the proof of Lemma 1. □

Lemma 4. *Construction 2 has soundness error* $\frac{2N}{|\mathbb{F}^*|}$.

Proof. Suppose $\sum_{x\in\mathbb{H}} \mathbf{f_1}^{(0)}(x)\mathbf{f_2}^{(0)}(x^{-1}) \neq u^{(0)}$. Since $\text{poly}f_1^{(n)} \cdot \text{poly}f_2^{(n)} = u^{(n)}$, there must be some j that $\sum_{x\in\mathbb{H}^{2^{j-1}}} \text{poly}f_1^{(j-1)}(x)\text{poly}f_2^{(j-1)}(x) \neq u^{(j-1)}$ and $\sum_{x\in\mathbb{H}^{2^j}} \text{poly}f_1^{(j)}(x)\text{poly}f_2^{(j)}(x) = u^{(j)}$. Let the largest one be j^*, and the quadratic polynomial in round j^* prover sends is $\text{poly}P'(r)$ (instead of $\text{poly}P^{(j^*)}(r)$). We have $\text{poly}P'(1) + \text{poly}P'(-1) = u^{(j^*-1)} \neq \sum_{x\in\mathbb{H}^{2^{j^*-1}}} \text{poly}f_1^{(j^*-1)}(x)\text{poly}f_2^{(j^*-1)}(x)$. Now we consider two cases.

- $\text{poly}f_i^{(j^*)}(x) = \text{poly}f_{i,e}^{(j^*-1)}(x) + r_{j^*}\text{poly}f_{i,o}^{(j^*-1)}(x)$ for $i = 1, 2$. Let $\text{poly}P^{(j^*)}(r) = \sum_{x\in\mathbb{H}^{2^{j^*}}} (\text{poly}f_{1,e}^{(j^*-1)}(x) + r\text{poly}f_{1,o}^{(j^*-1)}(x))(\text{poly}f_{2,e}^{(j^*-1)}$

$(x^{-1}) + r\text{poly}f_{2,o}^{(j^*-1)}(x^{-1}))$. Since $\text{poly}P'(1) + \text{poly}P'(-1) \neq \sum_{x \in \mathbb{H}^{2^{j^*-1}}} \text{poly}f_1^{(j^*-1)}(x)\text{poly}f_2^{(j^*-1)}(x) = \text{poly}P^{(j^*)}(1) + \text{poly}P^{(j^*)}(-1)$ according to the proof in Lemma 3, $\text{poly}P'(r)$ and $\text{poly}P^{(j^*)}(r)$ are different quadratic polynomials. By $\sum_{x \in \mathbb{H}^{2^{j^*}}} \text{poly}f_1^{(j^*)}(x)\text{poly}f_2^{(j^*)}(x) = u^{(j^*)}$ and $\text{poly}f_i^{(j^*)}(x) = \text{poly}f_{i,e}^{(j^*-1)}(x) + r_{j^*}\text{poly}f_{i,o}^{(j^*-1)}(x)$ for $i = 1, 2$, we have $u^{(j^*)} = \text{poly}P^{(j^*)}(r_{j^*})$. But we also have $u^{(j^*)} = \text{poly}P'(r_{j^*})$ according to verifier's computation. Since r_{j^*} is chosen after sending $\text{poly}P'(r)$, the chance that they have the same value on r is $\frac{2}{|\mathbb{F}^*|}$ as they are both quadratic.
- $\text{poly}f_i^{(j^*)}(x) \neq \text{poly}f_{i,e}^{(j^*-1)}(x) + r_{j^*}\text{poly}f_{i,o}^{(j^*-1)}(x)$ for at least one of $i = 1, 2$. Following Theorem 1, the chance that each one occurs is $\frac{N-1}{|\mathbb{F}^*|}$.

Summing the two cases by union bound, we have the overall soundness error $\frac{2N}{|\mathbb{F}^*|}$. □

Lemma 5. *Prover of Construction 2 can be implemented in $O(N)$ field operations in $\mathbb{F}$ with space complexity $O(N)$ of field elements.*

Proof. In round j, prover needs to compute $\text{poly}f_i^{(j)}, \text{poly}P^{(j)}$ from $\text{poly}f_i^{(j-1)}$ for $i = 1, 2$, which both require $O(2^{n-j})$ field operations. Prover needs to store $\text{poly}f_i^{(j)}$ for $i = 1, 2$ in order to compute $\text{poly}f_i^{(j+1)}$ in the next round, which needs to store $O(2^{n-j})$ field elements. Thus the overall time and space complexity are both $O(2^n) = O(N)$. □

Remark 5 (Batching proofs). For k instances of $\mathcal{R}_{IIP}$ with witness $\text{poly}f_{i,j}(x), i = 0, ..., k-1, j = 1, 2$, verifier first sends a challenge ξ to prover and they prove $\sum_{i=0}^{k-1} \xi^i \text{poly}f_{i,1}(x)\text{poly}f_{i,2}(x^{-1}) = \sum_{i=0}^{k-1} \xi^i u_i$. In round 1 to n, instead of sending $\text{poly}P_i(x)$ for each instance separately, prover sends $\sum_{i=0}^{k-1} \xi^i \text{poly}P_i(x)$. To check the equations for $\text{poly}f_{i,j}(\beta)$, prover and verifier still apply batch scheme for Construction 1. The soundness of such batching scheme can be proved through the same sketch of proof of Lemma 4, and the soundness error will be $\frac{2k(N-1)+2+3(k-1)}{|\mathbb{F}^*|}$. Using the batching scheme, prover sends $n + 2k - 1$ oracles, $3n + 2k$ field elements for $\text{poly}P(x)$ and $\text{poly}f_i^{(n)}$, and 2 challenges are needed.

4.2 Reversing the Vector

The relation $\mathcal{R}_{IIP}$ uses a polynomial $\text{poly}f_2(x^{-1})$, which is not convenient for our full construction. We will introduce a new protocol to solve this problem. We first give a lemma for the new construction.

Lemma 6 ([25], Lemma 4.9). *Suppose $\text{poly}f(x)$ is a polynomial whose degree is less than N defined on a finite field $\mathbb{F}$ with a multiplicative subgroup $\mathbb{H}$ that $|\mathbb{H}| = N$. Let $\text{poly}f'(x^{-1}) = \text{poly}f(x)$ for every $x \in \mathbb{H}$ and $\text{poly}f'(x)$ be a polynomial whose degree is also less than N, then $\text{poly}f(x) = \text{poly}f'(x^{-1}) \cdot x^N - \text{poly}f'(0) \cdot x^N + \text{poly}f'(0)$ for every $x \in \mathbb{F}$.*

We now give the definition of inner-product relation.

Definition 6. *The relation* ***inner product*** $\mathcal{R}_{IP}$ *is a set of tuple*

$$(\mathbb{i}, \mathbb{x}, \mathbb{w}) = (\perp, (\mathbb{F}, \mathbb{H}, N, u), (\mathbf{f_1}, \mathbf{f_2}))$$

where $\mathbb{H}$ *is a multiplicative subgroup of field* $\mathbb{F}$ *with size* $N = 2^n$, $\mathbf{f_1}, \mathbf{f_2} \in \mathbb{F}^N, u \in \mathbb{F}$ *and* $\sum_{x \in \mathbb{H}} \mathbf{f_1}(x)\mathbf{f_2}(x) = u$.

Theorem 3. *For every positive integer* $N = 2^n$ *and a finite field* $\mathbb{F}$ *that* $N \mid |\mathbb{F}^*|$, *there is a PIOP for the* $\mathcal{R}_{IP}$ *with proving time* $O(N)$, *proof size* $O(\log N)$, *verification time* $O(\log N)$ *in terms of field operations and field elements. The PIOP has a perfect completeness and soundness error* $\frac{3N}{|\mathbb{F}^*|}$.

Construction 3. *Let* $\text{poly}f_2'(x^{-1}) = \text{poly}f_2(x)$ *on* $\mathbb{H}$. *Prover first sends oracle* $\text{poly}f_2'(x)$ *and prover and verifier follow Construction 2 on relation* $\mathcal{R}_{IIP}$: $(\mathbb{i}, \mathbb{x}, \mathbb{w}) = (\perp, (\mathbb{F}, \mathbb{H}, N, u), (\mathbf{f}_1, \text{poly}f_2'))$. *After the sub-protocol, verifier sends a challenge* $\xi \leftarrow \mathbb{F}^*$ *to prover and prover opens* $\text{poly}f_2'(x)$ *at point* ξ *and* $\text{poly}f_2(x)$ *at point* ξ^{-1} *along with point* 0. *Verifier checks* $\text{poly}f_2'(\xi) = \xi^N \text{poly}f_2(\xi^{-1}) - \xi^N \text{poly}f_2(0) + \text{poly}f_2(0)$.

By Lemma 6 and Theorem 2, the completeness and proving complexity of Construction 3 are inferred directly. The additional soundness error is due to checking $\text{poly}f_2'(\xi) = \xi^N \text{poly}f_2(\xi^{-1}) - \xi^N \text{poly}f_2(0) + \text{poly}f_2(0)$, which is $\frac{N}{|\mathbb{F}^*|}$ by Schwartz-Zippel lemma. 1 more oracle is needed in the proof than in Construction 2.

Remark 6 (Batching proofs). For k instances of $\mathcal{R}_{IP}$ with witness $\text{poly}f_{i,j}(x), i = 0, ..., k-1, j = 1, 2$, prover only needs to call batching proof of $\mathcal{R}_{IIP}$, and checks $\text{poly}f_{i,2}'(\xi) = \xi^N \text{poly}f_{i,2}(\xi^{-1}) - \xi^N \text{poly}f_{i,2}(0) + \text{poly}f_{i,2}(0)$ using random linear combination as for Construction 1. This will add $\frac{k}{|\mathbb{F}^*|}$ to soundness error. $n + 3k - 1$ oracles, $3n + 2k$ field elements, and 3 challenges are needed to batch k proofs.

Remark 7. If we are proving Hadamard product in vector form, we can directly construct an inverse polynomial to use Construction 2 instead of Construction 3 to avoid to check $\text{poly}f_2'(\xi) = \xi^N \text{poly}f_2(\xi^{-1}) - \xi^N \text{poly}f_2(0) + \text{poly}f_2(0)$.

4.3 Inner Product with Tensor Polynomial

To construct a Hadamard product, we need to do an "inner product" of 3 polynomials. We use the inner-product protocol along with the tensor polynomial to construct the "inner product".

Definition 7. *The relation* ***triple inner product*** $\mathcal{R}_{TIP}$ *is a set of tuple:*

$$(\mathbb{i}, \mathbb{x}, \mathbb{w}) = (\perp, (\mathbb{F}, \mathbb{H}, N, r_1, ..., r_n, u), (\mathbf{f}_1, \text{poly}f_2))$$

,where $\mathbb{H}$ *is a multiplicative subgroup of size* $N = 2^n$, $\mathbf{f}_1, \text{poly}f_2 \in \mathbb{F}^N, u \in \mathbb{F}$, *and* $\sum_{x \in \mathbb{H}} \mathbf{f}_1(x) \text{poly}f_2(x) \boldsymbol{\rho}_{\mathbb{H},(r_1,...,r_n)}(x) = u$.

Theorem 4. *For every positive integer $N = 2^n$ and a finite field $\mathbb{F}$ that $N \mid |\mathbb{F}^*|$, there is a PIOP for the $\mathcal{R}_{TIP}$ with proving time $O(N)$, proof size $O(\log N)$, verification time $O(\log N)$ in terms of field operations and field elements. The PIOP has perfect completeness and soundness error $\frac{5N+n-2}{|\mathbb{F}^*|}$.*

Construction 4. *We construct a PIOP with $2n+3$ rounds for indexed $\mathcal{R}_{TIP}$. Prover $\mathcal{P}$ takes $(\mathbb{i}, \mathbb{x}, \mathbb{w})$ of the relation as input; verifier $\mathcal{V}$ takes $(\mathbb{i}, \mathbb{x})$ as input.*

- *Prover computes* $\text{poly}k = \text{poly}f_1 \circ \boldsymbol{\rho}_{\mathbb{H},(r_1,...,r_n)}$ *and sends oracle* $\text{poly}k(x)$ *to verifier.*
- *Prover and verifier invoke protocol in Construction 3 to verify relation* $\mathcal{R}_{IP} : (\mathbb{i}, \mathbb{w}, \mathbb{x}) = (\perp, (\mathbb{F}, \mathbb{H}, N, u), (\text{poly}k, \text{poly}f_2))$. *Let randomness in round* $1, 2, ..., n$ *be* $r'_1, ..., r'_n$.
- *Let* $\text{poly}f_1^{(0)}(x) = \text{poly}f_1(x)$. *Prover computes oracle* $\text{poly}f_1^{(j)}(x^2) = \frac{x+r_j}{2x}\frac{x+r'_j}{2x}\text{poly}f_1^{(j-1)}(x) + \frac{-x+r_j}{2x}\frac{-x+r'_j}{2x}\text{poly}f_1^{(j-1)}(-x)$ *for* $j = 1, ..., n-1$ *and* $x \in \mathbb{H}^{2^{j-1}}$. *Prover sends oracle* $\text{poly}f_1^{(1)}, ..., \text{poly}f_1^{(n-1)}$ *to verifier.*
- *Verifier samples* $\gamma \leftarrow \mathbb{F}^*$ *and sends it to prover.*
- *Prover opens* $\text{poly}f_1^{(j)}$ *at* $\gamma, -\gamma, \gamma^2$ *and* 0 *for* $j = 0, ..., n-1$ *and sends it to verifier.*
- *For* $j = 1, ..., n$, *verifier checks* $\text{poly}f_1^{(j)}(\gamma^2) = \frac{r_j+r'_j}{4\gamma}(\text{poly}f_1^{(j-1)}(\gamma) - \text{poly}f_1^{(j-1)}(-\gamma)) + \frac{\text{poly}f_1^{(j-1)}(\gamma)+\text{poly}f_1^{(j-1)}(-\gamma)}{4} + \frac{r_jr'_j}{4\gamma^2}(\text{poly}f_1^{(j-1)}(\gamma)+\text{poly}f_1^{(j-1)}(-\gamma)-2\text{poly}f_1^{(j-1)}(0))+\frac{r_jr'_j}{2}\text{poly}f_1^{(j-1)}(0)\gamma^{\frac{N}{2^{j-1}}-2}$ *and* $\text{poly}f_1^{(n)} = \text{poly}k^{(n)}$.

Before proving the completeness, we need the following lemma related to oracle checking.

Lemma 7 ([25], Lemma 4.16). *For every field $\mathbb{F}$, a multiplicative subgroup $\mathbb{H}$ with size $N = 2^n$ and a polynomial $\text{poly}f(x)$ whose degree is less than N, defining a polynomial $\text{poly}f'(x)$ with degree less than $\frac{N}{2}$ that $\text{poly}f'(x^2) = \frac{x+r}{2x}\frac{x+r'}{2x}\text{poly}f(x) + \frac{-x+r}{2(-x)}\frac{-x+r'}{2(-x)}\text{poly}f(-x)$ for every $x \in \mathbb{H}$, then for every point $x \in \mathbb{F}$, $\text{poly}f'(x^2) = \frac{r+r'}{4x}(\text{poly}f(x) - \text{poly}f(-x)) + \frac{\text{poly}f(x)+\text{poly}f(-x)}{4} + \frac{rr'}{4x^2}(\text{poly}f(x) + \text{poly}f(-x) - 2\text{poly}f(0)) + \frac{rr'}{2}\text{poly}f(0)x^{N-2}$.*

Lemma 8. *Construction 4 has perfect completeness.*

Proof. Suppose $\sum_{x\in\mathbb{H}} \mathbf{f}_1(x)\mathbf{f}_2(x)\boldsymbol{\rho}_{\mathbb{H},(r_1,...,r_n)}(x) = u$. Then $\text{poly}k = \text{poly}f_1 \circ \boldsymbol{\rho}_{\mathbb{H},(r_1,...,r_n)}$, $\sum_{x\in\mathbb{H}} \text{poly}k(x)\text{poly}f_2(x) = u$. Thus $(\text{poly}k, \text{poly}f_2)$ is witness for relation $\mathcal{R}_{IP}$, and the sub-protocol can be correctly implemented. By the proof of Lemma 1, $\text{poly}k^{(n)} = \sum_{x\in\mathbb{H}} \text{poly}k(x)\boldsymbol{\rho}_{\mathbb{H},(r'_1,...,r'_n)}(x) = \sum_{x\in\mathbb{H}} \text{poly}f_1(x)\boldsymbol{\rho}_{\mathbb{H},(r_1,...,r_n)}(x)\boldsymbol{\rho}_{\mathbb{H},(r'_1,...,r'_n)}(x)$.

Recall the tree we construct in the proof of Lemma 1. When we replace the value on the edge $\frac{x+r_i}{2x}$ with $\frac{x+r_i}{2x}\frac{x+r'_i}{2x}$, we can get

$$\begin{aligned}\text{poly}f_1^{(n)}(x) &= \sum_{x\in\mathbb{H}}(\mathbf{f}^{(0)}(x)\prod_{i=1}^{n}\frac{x^{2^{i-1}}+r_i}{2x^{2^{i-1}}}\prod_{i=1}^{n}\frac{x^{2^{i-1}}+r_i'}{2x^{2^{i-1}}})\\ &= \sum_{x\in\mathbb{H}}\mathbf{f}(x)\boldsymbol{\rho}_{\mathbb{H},(r_1,\dots,r_n)}(x)\boldsymbol{\rho}_{\mathbb{H},(r_1',\dots,r_n')}(x) = \text{poly}k^{(n)}\end{aligned}$$

. The oracle checking follows Lemma 7. □

To prove soundness error, we introduce a lemma to show $\boldsymbol{\rho}_{\mathbb{H},(r_1',\dots,r_n')}$ is "random enough" to mask a polynomial.

Lemma 9 ([25], **Lemma 4.18**). *For every field $\mathbb{F}$, a multiplicative subgroup $\mathbb{H}$ with size $N = 2^n$ and two polynomials $\text{poly}f_1(x)$ and $\text{poly}f_2(x)$ whose degrees are less than N, if $(r_1,\dots,r_n)$ are all randomly chosen from $\mathbb{F}$ and $\sum_{x\in\mathbb{H}}\text{poly}f_1(x)\boldsymbol{\rho}_{\mathbb{H},(r_1,\dots,r_n)}(x) = \sum_{x\in\mathbb{H}}\text{poly}f_2(x)\boldsymbol{\rho}_{\mathbb{H},(r_1,\dots,r_n)}(x)$, then $\text{poly}f_1(x) = \text{poly}f_2(x)$ except for probability $\frac{n}{|\mathbb{F}^*|}$.*

Lemma 10. *Construction 4 has soundness error $\frac{5N+n-2}{|\mathbb{F}^*|}$.*

Proof. Suppose $\sum_{x\in\mathbb{H}}\mathbf{f}_1(x)\mathbf{f}_2(x)\boldsymbol{\rho}_{\mathbb{H},(r_1,\dots,r_n)}(x) \neq u$. Since the verification of sub-protocol is passed, $\sum_{x\in\mathbb{H}}\text{poly}k(x)\text{poly}f_2(x) = u$ except for $\frac{3N}{|\mathbb{F}^*|}$ probability. Along with $\sum_{x\in\mathbb{H}}\mathbf{f}_1(x)\mathbf{f}_2(x)\boldsymbol{\rho}_{\mathbb{H},(r_1,\dots,r_n)}(x) \neq u$, we have $\text{poly}f_1(x)\boldsymbol{\rho}_{\mathbb{H},(r_1,\dots,r_n)} \neq \text{poly}k(x)$ at some point $x\in\mathbb{H}$ except for $\frac{3N}{|\mathbb{F}^*|}$ probability. Also, by proof of soundness of Construction 1 in [5], $\sum_{x\in\mathbb{H}}\text{poly}k(x)\boldsymbol{\rho}_{\mathbb{H},(r_1',\dots,r_n')} = \text{poly}k^{(n)}$, except for the probability $\frac{N-1}{|\mathbb{F}^*|}$.

Now we consider the oracle that checks $\text{poly}f_1^{(j)}$ on $j = 1,\dots,n$. By Lemma 7, following the proof of soundness of Construction 1, $\sum_{x\in\mathbb{H}}\text{poly}f_1(x)\boldsymbol{\rho}_{\mathbb{H},(r_1,\dots,r_n)}\boldsymbol{\rho}_{\mathbb{H},(r_1',\dots,r_n')} = \text{poly}f_1^{(n)}$ except for probability $\frac{N-1}{|\mathbb{F}^*|}$. Then $\sum_{x\in\mathbb{H}}\text{poly}f_1(x)\boldsymbol{\rho}_{\mathbb{H},(r_1,\dots,r_n)}\boldsymbol{\rho}_{\mathbb{H},(r_1',\dots,r_n')}$ $= \text{poly}f_1^{(n)} = \text{poly}k^{(n)} = \sum_{x\in\mathbb{H}}\text{poly}k(x)\boldsymbol{\rho}_{\mathbb{H},(r_1',\dots,r_n')}$ except for probability $\frac{2N-2}{|\mathbb{F}^*|}$.

We then suppose that $\text{poly}f_1(x)\boldsymbol{\rho}_{\mathbb{H},(r_1,\dots,r_n)} \neq \text{poly}k(x)$ and $\sum_{x\in\mathbb{H}}\text{poly}f_1(x)\boldsymbol{\rho}_{\mathbb{H},(r_1,\dots,r_n)}\boldsymbol{\rho}_{\mathbb{H},(r_1',\dots,r_n')} = \text{poly}k^{(n)} = \sum_{x\in\mathbb{H}}\text{poly}k(x)\boldsymbol{\rho}_{\mathbb{H},(r_1',\dots,r_n')}$. Since $\text{poly}k(x)$ is sent before $(r_1',\dots,r_n')$ are chosen, by Lemma 9, $\text{poly}f_1(x)\boldsymbol{\rho}_{\mathbb{H},(r_1,\dots,r_n)} = \text{poly}k(x)$ except for probability $\frac{n}{|\mathbb{F}^*|}$. Thus, the overall soundness error is $\frac{3N}{|\mathbb{F}^*|} + \frac{2N-2}{|\mathbb{F}^*|} + \frac{n}{|\mathbb{F}^*|} = \frac{5N+n-2}{|\mathbb{F}^*|}$. □

Lemma 11. *Prover of Construction 4 can be implemented in $O(N)$ field operations in $\mathbb{F}$ with space complexity $O(N)$ of field elements.*

Proof. The sub-protocol of $\mathcal{R}_{IP}$ and computing and storing $\mathbf{k}(x)$ cost $O(N)$ field operations $O(N)$ space in terms of field elements. Computing and storing $\text{poly}f_1^{(j)}(x)$ cost $O(2^{n-j})$ field operations and $O(2^{n-j})$ space in terms of field elements. Thus the overall time and space complexity for Construction 4 are both $O(N)$. □

Remark 8 (Batching proofs). It is easy to batch the proofs on sub-protocol level. The sub-protocol for $\mathcal{R}_{IP}$ can be batched by the scheme in Remark 6. The rest part $\sum_{x\in H}\text{poly}f_1(x)\text{poly}\rho_{\mathbb{H},r_1,\dots,r_n}(x)\text{poly}\rho_{\mathbb{H},r'_1,\dots,r'_n}(x) = k^{(n)}$ can be batched by a random linear combination as in Remark 3. Furthermore, Construction 4 can be 'partially batched', which means the sub-protocol for $\mathcal{R}_{IP}$ in the protocol can be batched with an independent instance of Construction 3.

4.4 Computing Tensor Polynomial

In Construction 4, prover needs to compute $k = f_1(x) \circ \boldsymbol{\rho}_{\mathbb{H},(r_1,\dots,r_n)}$. To maintain the proving time of $O(N)$, prover needs to compute $\boldsymbol{\rho}_{\mathbb{H},(r_1,\dots,r_n)} = \frac{1}{2^n}(x\prod_{i=1}^{n}(x^{2^{i-1}}+r_i))$ on $\mathbb{H}$ in $O(N)$ time. Directly computing requires $O(N\log N)$ time. Observing that all elements $x\in\mathbb{H}$ only has 2^j different evaluations on $x^{2^{n-j}}$, we can compute $\boldsymbol{\rho}_{r_1,\dots,r_n}(x)$ in a reverse direction.

We set a full binary tree to save the intermediate results. The tree has $n+1$ levels from 0 to n, and each node saves 2 values. Level 0 (root) has value (1,0). Level i has 2^i nodes indexed from 0 to 2^i-1. We use $u_{i,j}$ to represent node j at level i. Let $\mathbb{H} = (1,\omega,\dots,\omega^{N-1})$. Let $u_{i,j} = (u_{i-1,\lfloor j/2\rfloor,0}\cdot(\omega^{u_{i,j,1}}+r_{n-i+1}), u_{i-1,\lfloor j/2\rfloor,1}/2 + lsb(j)\cdot 2^{n-1})$ where $lsb(j)$ is the least significant bit of j. Since the tree has $O(N)$ nodes and the time to compute each node (i,j) from its parent $(i-1,\lfloor j/2\rfloor)$ is $O(1)$, we can compute the tree in $O(N)$ time.

Lemma 12 ([25], Lemma 4.22). $u_{i,j,0} = \prod_{k=n-i+1}^{n}(\omega^{u_{i,j,1}\cdot 2^{k-(n-i+1)}}+r_k)$ *for every node* (i,j) *in the tree, and* $(u_{i,j,1})$ *for all nodes in level* i *is a permutation of* $(0\cdot\frac{N}{2^i},\dots,(2^i-1)\cdot\frac{N}{2^i})$.

The lemma can be easily proved by induction. Considering level n and definition of $\boldsymbol{\rho}_{\mathbb{H},(r_1,\dots,r_n)}$, we have the following corollary.

Corollary 1. $\boldsymbol{\rho}_{\mathbb{H},(r_1,\dots,r_n)}(\omega^{u_{n,j,1}}) = \frac{\omega^{u_{n,j,1}}}{2^n}u_{n,j,0}$.

5 Hadamard Product

In this section, we will introduce a PIOP for Hadamard product based on PIOP for $\mathcal{R}_{TIP}$, $\mathcal{R}_{IP}$ and $\mathcal{R}_{TC}$.

Definition 8. *The* ***Hadamard-product*** *relation* $\mathcal{R}_{HP}$*is a set of tuples*

$$(\mathbb{i},\mathbb{x},\mathbb{w}) = (\bot,(\mathbb{F},\mathbb{H},N),(\mathbf{a},\mathbf{b},\mathbf{c}))$$

where $\mathbb{H}$ *is a multiplicative subgroup with size* $N=2^n$, $\mathbf{a},\mathbf{b},\mathbf{c}\in\mathbb{F}^N$, $u\in\mathbb{F}$, *and for every* $x\in\mathbb{H}$, *we have* $\mathbf{a}(x)\mathbf{b}(x)=\mathbf{c}(x)$.

Remark 9. The indexed relation defined above does not contain index. It is reasonable when it is used as a sub-protocol of R1CS proving as [5]. But when it is considered as a standalone protocol, there should be oracles polya,b,c as induces or being sent by prover depending on whether they are viewed as induces or witnesses.

Theorem 5. *For every positive integer $N = 2^n$ and a finite field $\mathbb{F}$ that $N \mid |\mathbb{F}^*|$, there is a PIOP for the $\mathcal{R}_{HP}$ with proving time $O(N)$, proof size $O(\log N)$, verification time $O(\log N)$ in terms of field operations and field elements. The PIOP has perfect completeness and soundness error $\frac{9N+2n-4}{|\mathbb{F}^*|}$.*

Construction 5. *We construct a PIOP with $3n+4$ rounds for indexed $\mathcal{R}_{HP}$. Prover $\mathcal{P}$ takes $(\mathbb{i}, \mathbb{x}, \mathbb{w})$ of the relation as input; verifier $\mathcal{V}$ takes $(\mathbb{i}, \mathbb{x})$ as input.*

- *Verifier sends $(r_1, ..., r_n) \leftarrow (\mathbb{F}^*)^n$ to prover.*
- *Prover computes $u = \sum_{x \in \mathbb{H}} \mathrm{polyc}(x)\boldsymbol{\rho}_{\mathbb{H},(r_1,...,r_n)}(x)$ by applying prover side of Construction 1.*
- *Prover sends u and oracle $\boldsymbol{\rho}_{\mathbb{H},(r_1,...,r_n)}(x)$ to verifier.*
- *Prover and verifier invoke protocol in Construction 3 to verify relation $\mathcal{R}_{IP} : (\mathbb{i}, \mathbb{x}, \mathbb{w}) = (\perp, (\mathbb{F}, \mathbb{H}, N, u), (\mathbf{c}, \boldsymbol{\rho}_{\mathbb{H},(r_1,...,r_n)}(x)))$ and Construction 4 to verify relation $\mathcal{R}_{TIP} : (\mathbb{i}, \mathbb{x}, \mathbb{w}) = (\perp, (\mathbb{F}, \mathbb{H}, N, r_1, ..., r_n, u), (\mathbf{a}, \mathbf{b}))$. Verifier checks the consistency of oracle $\boldsymbol{\rho}_{\mathbb{H},(r_1,...,r_n)}(x)$ by querying at point β and locally computing $\boldsymbol{\rho}_{\mathbb{H},(r_1,...,r_n)}(\beta)$. Note that the two sub-protocol can be batched by Remark 8.*

Remark 10 (Public-coin). Since the protocol is public-coin, we can postpone all queries to oracles in each sub-protocol to the end of the protocol and choose the same challenge $\beta \leftarrow \mathbb{F}^*$ for all equations, which will be better for batch queries.

The completeness of Construction 5 can be directly inferred from the perfect completeness of Construction 1,Construction 3 and Construction 4. We only need to prove soundness and complexity.

Lemma 13. *Construction 5 has soundness error $\frac{9N+2n-4}{|\mathbb{F}^*|}$.*

Proof. Suppose $\mathbf{a}(x)\mathbf{b}(x) \neq \mathbf{c}(x)$ for some $x \in \mathbb{H}$. Then by Lemma 9, $\sum_{x \in \mathbb{H}} \mathbf{a}(x)\mathbf{b}(x)\boldsymbol{\rho}_{\mathbb{H},(r_1,...,r_n)}(x) \neq \sum_{x \in \mathbb{H}} \mathrm{polyc}(x)\boldsymbol{\rho}_{\mathbb{H},(r_1,...,r_n)}(x)$ except for $\frac{n}{|\mathbb{F}^*|}$ probability. Suppose prover sends u' in step 2, then $\sum_{x \in \mathbb{H}} \mathbf{a}(x)\mathbf{b}(x)\boldsymbol{\rho}_{\mathbb{H},(r_1,...,r_n)}(x) \neq u'$ or $\sum_{x \in \mathbb{H}} \mathrm{polyc}(x)\boldsymbol{\rho}_{\mathbb{H},(r_1,...,r_n)}(x) \neq u'$ except for $\frac{n}{|\mathbb{F}^*|}$ probability. By soundness of Construction 3 and Construction 4, and the soundness error by checking oracle $\boldsymbol{\rho}_{\mathbb{H},(r_1,...,r_n)}(x)$, we get the overall soundness error $\frac{3N}{|\mathbb{F}^*|} + \frac{5N+n-2}{|\mathbb{F}^*|} + \frac{N-1}{|\mathbb{F}^*|} + \frac{n}{|\mathbb{F}^*|} = \frac{9N+2n-4}{|\mathbb{F}^*|}$. □

Remark 11 (Knowledge soundness). The knowledge soundness for standalone Hadamard-product protocol is trivial, since verifier can rewind and query $\mathbf{a}, \mathbf{b}, \mathbf{c}$ for N time at distinct points if verifier accepts over the probability greater than soundness error above.

Remark 12. In order to reduce the randomness sampled from verifier, we may use $(r^{2^0}, r^{2^1}, ..., r^{2^{n-1}})$ to replace $(r_1, ..., r_n)$. The change will cause the soundness error of Lemma 9 increase from $\frac{n}{|F*|}$ to $\frac{N-1}{|F*|}$, and the soundness error of Construction 5 will increase from $\frac{9N+2n-4}{|\mathbb{F}^*|}$ to $\frac{11N-6}{|\mathbb{F}^*|}$. Besides, by using the same r, we can easily batch instances of Hadamard products.

Lemma 14. *Prover of Construction 5 can be implemented in $O(N)$ field operations in $\mathbb{F}$ with space complexity $O(N)$ of field elements.*

Proof. Prover needs to run prover side of Construction 1, Construction 3 and Construction 4, which will all cost $O(N)$ field operations and $O(N)$ space for field elements. Thus, prover's time and space complexity in terms of field operations and field elements, respectively, are all $O(N)$. □

Remark 13 (Batching equation checkings). There are $3n+2$ equations to check in Construction 5. After choosing the challenge β, verifier can also choose a challenge $\alpha \leftarrow \mathbb{F}^*$. Since β is chosen, all the equations can be written in form $c_{i,1}\text{poly}f_{i,1}(\beta^2) = c_{i,2}\text{poly}f_{i,2}(\beta) + c_{i,3}\text{poly}f_{i,3}(-\beta) + c_{i,4}\text{poly}f_{i,4}(0)$ for $i = 0, ..., 3n-1$ where $\text{poly}f_{i,j}$ is one of the oracles sent by prover and $c_{i,j}$ is a fixed field element that can be computed in $O(n)$ by both prover and verifier, except the oracle checking in Construction 3 and oracle checking in Construction 5. By choosing α, prover and verifier can only check the equation $\sum_{i=0}^{3n-1} \alpha^i c_{i,1}\text{poly}f_{i,1}(\beta^2) = \sum_{i=0}^{3n-1} \alpha^i c_{i,2}\text{poly}f_{i,2}(\beta) + \sum_{i=0}^{3n-1} \alpha^i c_{i,3}\text{poly}f_{i,3}(-\beta) + \sum_{i=0}^{3n-1} \alpha^i c_{i,4}\text{poly}f_{i,4}(0)$. The linear combination will lead to $\frac{3n}{|\mathbb{F}^*|}$ extra soundness error, but it will be useful if the instantiation of oracle is additive.

6 Polynomial Commitment Scheme

To instantiate the oracle in PIOP above, we will use the polynomial commitment scheme. In this section, we will describe a polynomial commitment scheme, which is a slight modification of the KZG commitment, to commit polynomials in Lagrange base. We first define polynomial commitment scheme following [5,16]. The polynomial commitment scheme over a field family $\mathcal{F}$ contains a tuple of algorithm $\mathsf{PC} = (\mathsf{Setup}, \mathsf{Com}, \mathsf{Open}, \mathsf{Verify})$ defined by following syntax.

- **Setup** $\mathsf{PC.Setup}(1^\lambda, D) \rightarrow (ck, vk)$. On input a security parameter λ in unary and a degree D, $\mathsf{PC.Setup}$ outputs a commitment key ck and a verification key vk containing a description of $\mathbb{F} \in \mathcal{F}$.
- **Commit** $\mathsf{PC.Com}(ck, \text{poly}P) \rightarrow C$. On input the commitment key ck and a polynomial $\text{poly}P$ of degree at most D, $\mathsf{PC.Com}$ outputs a commitment C for $\text{poly}P$.
- **Open** $\mathsf{PC.Open}(ck, \text{poly}P, z) \rightarrow \pi$. On input the commitment key ck, a polynomial $\text{poly}P$ of degree at most D and an evaluation point z, $\mathsf{PC.Open}$ outputs a proof π for the evaluation.
- **Verify** $\mathsf{PC.Verify}(rk, C, \pi, z, v) \rightarrow \{0, 1\}$. On input the verification key rk, the commitment C, the proof π, an evaluation point z and an alleged value v, $\mathsf{PC.Verify}$ outputs 1 if π is a valid proof that $\text{poly}P$ is a polynomial having degree at most D, committed in C and $\text{poly}P(z) = v$, and 0 otherwise.

To prove knowledge soundness, a polynomial commitment scheme must satisfy completeness and extractability (as a stronger notion of evaluation binding in [5,11]).

Definition 9 (Completeness). *For every degree bound $D \in \mathbb{N}$ and every PPT adversary $\mathcal{A}$,*

$$\Pr\left[\begin{array}{c|r} & (ck, rk) \leftarrow \mathsf{PC.Setup}(1^\lambda, D) \\ \deg(\text{poly}P) \leq D & (\mathbf{P}, z) \leftarrow \mathcal{A}(ck, rk) \\ \downarrow & C \leftarrow \mathsf{PC.Com}(ck, \text{poly}P) \\ \mathsf{PC.Verify}(rk, C, \pi, z, v) = 1 & v = \text{poly}P(z) \\ & \pi \leftarrow \mathsf{PC.Open}(ck, \text{poly}P, z) \end{array}\right] \geq 1 - \text{negl}(\lambda)$$

Definition 10 (Extractability). *For any degree bound $D \in \mathbb{N}$ and any PPT adversary $\mathcal{A}$, there exists an extractor $\mathcal{E}$ such that for every round bound $r \in \mathbb{N}$, PPT query sampler $\mathcal{Q}$ and PPT adversary $\mathcal{B}$,*

$$\Pr\left[\begin{array}{c|r} & (ck, rk) \leftarrow \mathsf{PC.Setup}(1^\lambda, D) \\ \mathsf{PC.Verify}(rk, C, \pi, z, v) = 1 & C \leftarrow \mathcal{A}(ck, rk) \\ \downarrow & \text{poly}P \leftarrow \mathcal{E}(ck, rk) \\ \deg(\mathbf{P}) \leq D \wedge v = \text{poly}P(z) & z \leftarrow \mathcal{Q}(ck, rk) \\ & (\pi, v, st) \leftarrow \mathcal{B}(ck, rk, z) \end{array}\right] \geq 1 - \text{negl}(\lambda)$$

Standard KZG commitment [5,11,16] has both completeness and extractability, but it works on coefficient-based polynomials. We do a slightly modification on KZG commitment to make it work on Lagrange-based polynomials. In short, the SRS of KZG has $(G, \tau G, \tau^2 G, ..., \tau^D G)$ in group G_1 of a bilinear group. We use FFT to turn the SRS into $(\text{poly}L_0(\tau)G, ..., \text{poly}L_{D-1}(\tau)G)$, where $\text{poly}L_k(x)$ is the k-th Lagrange polynomial of a specific multiplicative subgroup $\mathbb{H}$. Then we can commit and open a Lagrange-based polynomial as what we do in standard KZG commitment, getting the exactly same commitment and proof as if it is in the coefficient base. They have the same security propositions due to their SRS can be transformed mutually. The only quasilinear overhead of such scheme is the FFT, but it can be done during setup, keeping the linear overhead of commit and open.[4]

Remark 14 (Batching proofs). In Remark 13, we combine all equations except for the two in Construction 3 and Construction 5. Thanks to the additive property of KZG commitment, prover and verifier can sum up the commitments in the protocol to get an equation for only 4 polynomials with their commitments. Along with the equation in Construction 3 and Construction 5, which needs 4 polynomial openings, prover only needs to open 8 polynomials at 8 points, respectively.

[4] Details are displayed in [25].

References

1. Ben-Sasson, E., Chiesa, A., Riabzev, M., Spooner, N., Virza, M., Ward, N.P.: Aurora: transparent succinct arguments for R1CS. In: Ishai, Y., Rijmen, V. (eds.) EUROCRYPT 2019. LNCS, vol. 11476, pp. 103–128. Springer, Heidelberg (2019). https://doi.org/10.1007/978-3-030-17653-2_4
2. Bootle, J., Cerulli, A., Chaidos, P., Groth, J., Petit, C.: Efficient zero-knowledge arguments for arithmetic circuits in the discrete log setting. In: Fischlin, M., Coron, J. (eds.) EUROCRYPT 2016. LNCS, vol. 9666, pp. 327–357. Springer, Heidelberg (2016). https://doi.org/10.1007/978-3-662-49896-5_12
3. Bootle, J., Cerulli, A., Ghadafi, E., Groth, J., Hajiabadi, M., Jakobsen, S.K.: Linear- time zero-knowledge proofs for arithmetic circuit satisfiability. In: Takagi, T., Peyrin, T. (eds.) ASIACRYPT 2017. LNCS, vol. 10626, pp. 336–365. Springer, Heidelberg (2017). https://doi.org/10.1007/978-3-319-70700-6_12
4. Bootle, J., Chiesa, A., Groth, J.: Linear-time arguments with sublinear verification from tensor codes. In: Pass, R., Pietrzak, K. (eds.) TCC 2020. LNCS, vol. 12551, pp. 19–46. Springer, Heidelberg (2020). https://doi.org/10.1007/978-3-030-64378-2_2
5. Bootle, J., Chiesa, A., Hu, Y., Orrù, M.: Gemini: elastic SNARKs for diverse environments. In: Dunkelman, O., Dziembowski, S. (eds.) EUROCRYPT 2022. LNCS, vol. 13276, pp. 427–457. Springer, Heidelberg (2022). https://doi.org/10.1007/978-3-031-07085-3_15
6. Bootle, J., Chiesa, A., Liu, S.: Zero-Knowledge IOPs with Linear-Time Prover and Polylogarithmic-Time Verifier, Cryptology ePrint Archive, Paper 2020/1527 (2020)
7. Bünz, B., Bootle, J., Boneh, D., Poelstra, A., Wuille, P., Maxwell, G.: Bulletproofs: short proofs for confidential transactions and more. In: SP 2018, pp. 315–334. IEEE Computer Society (2018)
8. Bünz, B., Fisch, B., Szepieniec, A.: Transparent SNARKs from DARK compilers. In: Canteaut, A., Ishai, Y. (eds.) EUROCRYPT 2020. LNCS, vol. 12105, pp. 677–706. Springer, Cham (2020). https://doi.org/10.1007/978-3-030-45721-1_24
9. Bünz, B., Maller, M., Mishra, P., Tyagi, N., Vesely, P.: Proofs for inner pairing products and applications. In: Tibouchi, M., Wang, H. (eds.) ASIACRYPT 2021. LNCS, vol. 13092, pp. 65–97. Springer, Cham (2021). https://doi.org/10.1007/978-3-030-92078-4_3
10. Chen, B., Bünz, B., Boneh, D., Zhang, Z.: HyperPlonk: plonk with linear-time prover and high-degree custom gates. In: Hazay, C., Stam, M., EUROCRYPT 2023. LNCS, vol. 14005, pp. 499–530. Springer, Heidelberg (2023). https://doi.org/10.1007/978-3-031-30617-4_17
11. Chiesa, A., Hu, Y., Maller, M., Mishra, P., Vesely, N., Ward, N.: Marlin: preprocessing zkSNARKs with universal and updatable SRS. In: Canteaut, A., Ishai, Y. (eds.) EUROCRYPT 2020. LNCS, vol. 12105, pp. 738–768. Springer, Cham (2020). https://doi.org/10.1007/978-3-030-45721-1_26
12. Daza, V., Ràfols, C., Zacharakis, A.: Updateable inner product argument with logarithmic verifier and applications. In: Kiayias, A., Kohlweiss, M., Wallden, P., Zikas, V. (eds.) PKC 2020. LNCS, vol. 12110, pp. 527–557. Springer, Cham (2020). https://doi.org/10.1007/978-3-030-45374-9_18
13. Gabizon, A., Williamson, Z.J., Ciobotaru, O.: PLONK: permutations over lagrangebases for oecumenical noninteractive arguments of knowledge. Cryptology ePrint Archive, Report 2019/953 (2019)

14. Gennaro, R., Gentry, C., Parno, B., Raykova, M.: Quadratic span programs and succinct NIZKs without PCPs. In: Johansson, T., Nguyen, P.Q. (eds.) EUROCRYPT 2013. LNCS, vol. 7881, pp. 626–645. Springer, Heidelberg (2013). https://doi.org/10.1007/978-3-642-38348-9_37
15. Izabachène, M., Libert, B., Vergnaud, D.: Block-wise P-signatures and non-interactive anonymous credentials with efficient attributes. In: Chen, L. (ed.) IMACC 2011. LNCS, vol. 7089, pp. 431–450. Springer, Heidelberg (2011). https://doi.org/10.1007/978-3-642-25516-8_26
16. Kate, A., Zaverucha, G.M., Goldberg, I.: Constant-size commitments to polynomials and their applications. In: Abe, M. (ed.) ASIACRYPT 2010. LNCS, vol. 6477, pp. 177–194. Springer, Heidelberg (2010). https://doi.org/10.1007/978-3-642-17373-8_11
17. Kothapalli, A., Masserova, E., Parno, B.: A direct construction for asymptotically optimal zkSNARKs. Cryptology ePrint Archive, Report 2020/1318 (2020)
18. Lee, J.: Dory: efficient, transparent arguments for generalised inner products and polynomial commitments. Cryptology ePrint Archive, Report 2020/1274 (2020)
19. Lipmaa, H., Siim, J., Zajac, M.: Counting Vampires: from univariate sum check to updatable ZK-SNARK. In: S. Agrawal and D. Lin. ASIACRYPT 2022. LNCS, vol. 13792, pp. 249–278. Springer, Heidelberg (2022). https://doi.org/10.1007/978-3-031-22966-4_9
20. Maller, M., Bowe, S., Kohlweiss, M., Meiklejohn, S.: Sonic: zero-knowledge SNARKs from linear-size universal and updateable structured reference strings. Cryptology ePrint Archive, Paper 2019/099 (2019)
21. Ràfols, C., Zapico, A.: An algebraic framework for universal and updatable SNARKs. In: Malkin, T., Peikert, C. (eds.) CRYPTO 2021. LNCS, vol. 12825, pp. 774–804. Springer, Cham (2021). https://doi.org/10.1007/978-3-030-84242-0_27
22. Ron-Zewi, N., Rothblum, R.D.: Proving as fast as computing: succinct arguments with constant prover overhead. In: Leonardi, S., Gupta, A., STOC '22, pp. 1353–1363. ACM (2022)
23. Setty, S.: Spartan: efficient and general-purpose zkSNARKs without trusted setup. In: Micciancio, D., Ristenpart, T. (eds.) CRYPTO 2020. LNCS, vol. 12172, pp. 704–737. Springer, Cham (2020). https://doi.org/10.1007/978-3-030-56877-1_25
24. Szepieniec, A., Zhang, Y.: Polynomial IOPs for linear algebra relations. In: Hanaoka, G., Shikata, J., Watanabe, Y., PKC 2022. LNCS, vol. 13177, pp. 523–552. Springer, Heidelberg (2022).https://doi.org/10.1007/978-3-030-97121-2_19
25. Xie, J., Hu, Y., Yu, Y.: Hadamard product argument from lagrange-based univariate polynomials. Cryptology ePrint Archive, Paper 2024/613 (2024)
26. Xie, T., Zhang, J., Zhang, Y., Papamanthou, C., Song, D.: Libra: succinct zero-knowledge proofs with optimal prover computation. In: Boldyreva, A., Micciancio, D. (eds.) CRYPTO 2019. LNCS, vol. 11694, pp. 733–764. Springer, Cham (2019). https://doi.org/10.1007/978-3-030-26954-8_24
27. Zhang, J., et al.: Doubly efficient interactive proofs for general arithmetic circuits with linear prover time. In: Kim, Y., Kim, J., Vigna, G., Shi, E., CCS '21, pp. 159–177. ACM (2021)
28. Zhang, Y., Szepieniec, A., Zhang, R., Sun, S., Wang, G., Gu, D.: VOProof: efficient zkSNARKs from vector oracle compilers. In: Yin, H., Stavrou, A., Cremers, C., Shi, E., CCS 2022, pp. 3195–3208. ACM (2022)

Author Index

T. Zhu and Y. Li (Eds.): ACISP 2024, LNCS 14895, pp. 493–496, 2024.
https://doi.org/10.1007/978-981-97-5025-2

GPSR Compliance

The European Union's (EU) General Product Safety Regulation (GPSR) is a set of rules that requires consumer products to be safe and our obligations to ensure this.

If you have any concerns about our products, you can contact us on ProductSafety@springernature.com

In case Publisher is established outside the EU, the EU authorized representative is:

Springer Nature Customer Service Center GmbH
Europaplatz 3
69115 Heidelberg, Germany

Zeitfracht Medien GmbH
Ferdinand-Jühlke-Straße 7
99095 Erfurt, Deutschland
produktsicherheit@kolibri360.de